Lecture Notes on

Electron Correlation and Magnetism

SERIES IN MODERN CONDENSED MATTER PHYSICS

Editors-in-charge: I. Dzyaloshinski and Yu Lu

Published

Vol. 1: Exotic Properties of Superfluid Helium 3
by G. E. Volovik

Vol. 2: Quantum Dissipative Systems
by U. Weiss

Vol. 4: Simple Views on Condensed Matter
by P.-G. de Gennes

Vol. 5: Lectures on Electron Correlation & Magnetism
by P. Fazekas

Vol. 6: Low-Dimensional Quantum Field Theories for Condensed Matter Physicists
eds. L. Yu, S. Lundqvist and G. Morandi

Vol. 7: The Physics of Quantum Well Infrared Photodetectors
by K. K. Choi

Vol. 8: Simple Views on Condensed Matter (2nd Edition)
by P.-G. de Gennes

Vol. 9: Theory of Critical Phenomena in Finite-Size System: Scaling and Quantum Effects
by J. G. Brankov, D. M. Danchev and N. T. Tonchev

Vol. 10: Quantum Dissipative Systems (2nd Edition)
by U. Weiss

Vol. 11: Magnetism in Heavy Fermions
by H. B. Radousky

Vol. 12: Simple Views on Condensed Matter (3rd Edition)
by P.-G. de Gennes

Vol. 13: Quantum Dissipative Systems (3rd Edition)
by U. Weiss

Series in Modern Condensed Matter Physics – Vol. 5

Lecture Notes on Electron Correlation and Magnetism

Patrik Fazekas
Research Institute for Solid State Physics & Optics, Budapest

World Scientific
Singapore • New Jersey • London • Hong Kong

Published by

World Scientific Publishing Co. Pte. Ltd.

5 Toh Tuck Link, Singapore 596224

USA office: 27 Warren Street, Suite 401-402, Hackensack, NJ 07601

UK office: 57 Shelton Street, Covent Garden, London WC2H 9HE

Library of Congress Cataloging-in-Publication Data
Fazekas, Patrik,
Lecture notes on electron correlation and magnetism / P. Fazekas.
p. cm. -- (Series on modern condensed matter physics : vol. 5)
Includes bibliographical references and index.
ISBN-13 978-981-02-2474-5
ISBN-10 981-02-2474-5
1. Magnetism, Band theory of. 2. Electron configuration.
3. Electron-electron interactions. 4. Metal-insulator transitions.
I. Title. II. Series: Series in modern condensed matter physics ;
vol. 5
QC754.2.B35F39 1999 98-51617
CIP

British Library Cataloguing-in-Publication Data
A catalogue record for this book is available from the British Library.

Contents

Preface

In the academic years 1991–1992 and 1992–1993, the author had the opportunity to give a series of lectures entitled *Electron Correlation and Magnetism* at the Diploma Course organized by the International Center for Theoretical Physics (ICTP, Trieste). The Diploma Course is a one-year teaching program organized for students and young physicists from developing countries, with the intention of standardizing and upgrading their education to a level that is equivalent to that expected in the developed countries. The staff sought to achieve this by elaborating one-term courses to give a solid background in one of the standard subjects, but which also open ways into areas of current interest thereby enabling the students to begin research work.

The one-term, 4-hours-a-week course on magnetism presented a challenge known to all physicists in the field: research interests in the past half a century have been dominated by the effects of strong electron–electron interaction, while standard solid state physics textbooks remain within the bounds of band theory which is a suitable language for weakly correlated systems, and then add a chapter on Heisenberg magnets whose very existence is in contradiction with the rest of the material, and gets never properly justified. The usual way of clarifying these matters is to go through a formal education in many-body theory, and to learn about strong correlation effects piecemeal from its applications (and breakdowns). This, however, is usually the beginning of the professional career of a theoretician, and it may not be the most recommendable approach for others. One takes a long time to discover that there is a unified, non-formal way of thinking about strong correlation phenomena that has long been shared by experimentalists and theoreticians in the field; it can be called elementary and should be accessible to all – but it cannot be found in the well-known textbooks.

The author has attempted to fill this gap in his own way in the lectures mentioned above. The provision of notes for the courses was compulsory, and the 300-page handwritten version of the manuscript became an internal ICTP publication (DICTP CMP–MN(1,...,8), 1992). Yu Lu (who organized the course) kindly suggested that it should be rechecked, typed, whereupon it might be published by World Scientific, Singapore as *Lecture Notes on Electron Correlation and Magnetism*.

Though eager to accept the offer, I missed a series of deadlines, and was unable to avoid a considerable increase in the number of pages. However, this was made inevitable by (among other things) the nature of the undertaking: gaps in the reasoning had to be filled in, illustrative material collected, and the delay turned out not to be a bad thing bearing in mind the developments of the intervening years. I was astonished to discover that the subject of ferromagnetism which (in teaching) one tends to skip over as the easiest and dullest part of the subject magnetism, had in fact been the least understandable in the standard approach which uses the Hubbard model as its starting point. The developments of 1995–1998 brought significant advances in our understanding of ferromagnetism. Another subject which did not appear (though it should have done) in the older version of my notes, is the role of band degeneracy and orbital ordering, which enjoys great renewed interest.

I had the chance to experiment with new versions of presenting my material in the following years while giving courses for students specializing in solid state physics at the R. Eötvös University (Budapest) in 1993–1995, and at the Technical University of Budapest from 1996 to the present day. Usually, I give a two-semester course with two hours a week. Obviously the material assembled here is too much for that, so it should rather be understood that different versions of the course can be put together around the same central themes. The essential chapters are 4 and 5 (chapters 2 and 3 give background material with somewhat updated emphasis, and may be skipped). Basically, the most important correlation effect is the Mott transition, and spin models of magnetic ordering should be understood as strong-coupling effective Hamiltonians derived from an underlying fermion model. What follows afterwards is a matter of taste. One can continue with Ch. 6, or go straight to either Ch. 7 or Ch. 9. The mean field (Stoner) theories described in Ch. 7 should be supplemented with their criticism, especially with regard to ferromagnetism (see the discussion on p. 458, and Sec. 8.5). Any survey of magnetic ordering phenomena benefits from including a discussion of orbital ordering (sections 5.4 and 8.5). As for chapters 9 and 10, the author found that the Gutzwiller–Brinkman–Rice approach, though it requires a fair amount of preparation, offers the easiest way to discussing how the Mott transition is approached from the metallic side, and thus has some pedagogical advantages. Similarly, the most elementary way to discuss the fractional Quantum Hall Effect is by reference to Jastrow-type correlated many-electron states (Ch. 12).

Occasionally, when the general derivation of a result would have meant an undue burden on the main text, a simplified demonstration is posed as one of the *Problems*, and outlined in the *Solutions*. Searching for these is facilitated by listing the problems as separate *Contents* items.

I am aware of the many omissions, some of which are hard to justify. First, essentially the entire solid state physics could be approached by means of density functional theory (DFT). Some of the problems we consider to be particularly difficult (for instance, the ferromagnetism of iron) have been non-problems for DFT experts for thirty years. However, DFT is an art which one either practices, or one does not; and in the latter case, it is difficult to do anything more than to acknowledge what comes out of the calculations. At least, the author found it difficult to interpret the results by direct insight, or to establish links with the arguments which are straightforward in terms of the Hubbard model framework. Thus DFT is not described in any detail, but one has to keep it in mind that it is a very successful approach within the domain of its applicability. This domain is certainly getting wider, but we prefer to take the view that the Mott insulator and other strong-coupling states had better be described in the strong correlation framework. Second, even within the scope of lattice fermion models, a different author might well have written a completely different book with the same title, in which exactly solvable models (the one-dimensional Hubbard and Heisenberg model, and the Anderson/Kondo impurity problems) would have been dominant themes. In the present text, the emphasis is on motivating the introduction of models, and characterizing their overall behaviour (whenever known), rather than on their exact solution in particular cases. Third, we generally consider lattice problems, and discuss the (admittedly very important) impurity problems only very briefly.

A word should be added about the references which are many but even so, the list should not be regarded as complete, representative, or fair. This is meant to be an introductory text, not a professional review on any of the subjects. To illustrate a point, I usually chose the first source I came across, and did not attempt to survey the literature. I tried to cite reviews, or alternatively some of the latest works on a subject (in the hope that their reference lists lead on to the earlier publications). Nevertheless, I would like to apologize to all whose contributions are not properly represented here.

A book like this cannot be written without a great deal of help from others. Discussions with, and criticism from, colleagues and students have played a decisive role in shaping the text. I am particularly indebted to Attila Virosztek who has carefully read almost the entire manuscript, not seldom in several versions, and kindly pointed out an alarmingly large number of mistakes, inaccuracies, and cases of nebulous formulation. Tamás Kemény and Kazumasa Itai have also given great help by reading and checking most of the text. Karlo Penc has given valuable advice about certain subjects, and also helped me with some of the figures. G. Aeppli, K. Held, M. Jarrell, Y. Tokura and M. Ulmke sent me postscript files of (in some cases, still unpublished) illustrative material, as well as valuable comments. It would be impossible to recall all the occasions when I received assistance from colleagues who commented upon a part of the manuscript, or discussed with me one of the numerous issues, but let me mention at least C. Andreani, A. Asamitsu, D. Baeriswyl, S. Daul, P. Erdős, F. Gebhard, B. Győrffy, J. Hajdu, A. Jánossy, H.-Y. Kee, A. Klümper, J. Kollár, G. Kriza, P. Monachesi, E. Müller–Hartmann, T. Pruschke, T. Sakakibara, P. Santini, H. Shiba, J. Sólyom, A. Sütő, H. Tasaki, I. Tüttő, O. Újsághy and D. Vollhardt. I also thank all the authors who permitted me to display figures from their publications.

The bulk of the work was done in the years 1994–1997 at my home institute, the Research Institute for Solid State Physics and Optics of the Hungarian Academy of Sciences. I gratefully acknowledge that during this period, I received financial support from the Hungarian National Research Foundation, under Grant No. OTKA T-014201. Since the beginning of this year, I have been supported by Grant No. OTKA T-025505.

I am indebted to World Scientific for their ongoing encouragement and support.

Special thanks do I owe to my wife for her constant support during all these years.

Budapest, September 1998 *Patrik Fazekas*

Chapter 1

Introduction

1.1 Magnetism and Other Effects of Electron–Electron Interaction

Magnetism is a vast subject and only a small part of it can be dealt with in these notes. Our first aim is to learn how to describe the effect of an external magnetic field H on the behaviour of a solid substance. In the limit of weak fields, the strength of the response is measured by the susceptibility

$$\chi = \frac{M}{H},$$

where M is the magnetization density. Generally, we can probe the system by a field which is space-, and time-dependent. Decomposing both H and M into Fourier-components, the relevant response function is the wavevector-, and frequency-dependent generalized susceptibility $\chi(\mathbf{q}, \omega)$ which should tell us all we want to know about the weak-field behaviour of the system. But what lies beyond? Need we worry about what happens in stronger fields?

The calculation of the magnetization requires solving a quantum-mechanical eigenvalue problem where the interaction of the probe with the external field is added to the Hamiltonian. The microscopic energy scale of this interaction is $\mu_B H$, where

$$\mu_B = \frac{e\hbar}{2mc} \approx \begin{cases} 9.27 \times 10^{-21} \text{ergG}^{-1} & \text{(CGS)} \\ 9.27 \times 10^{-24} \text{JT}^{-1} & \text{(SI)} \end{cases}$$

is the Bohr magneton. (G stands for 'gauss', while T for 'Tesla'). μ_B is very nearly equal to the spin magnetic moment of an electron in vacuum.

Let us try to get a feeling for the energy scale $\mu_B H$. In condensed matter physics, one often uses the electron volt 1eV $\equiv 1.6\times10^{-19}$J as the energy unit. Furthermore, it is useful to relate energies to equivalent temperatures, remembering that Boltzmann's constant is $k_B = 1.38\times10^{-23}\mathrm{JK}^{-1} = 8.6\times10^{-5}\mathrm{eVK}^{-1}$; thus 1eV corresponds to the temperature 11604K. 5T $\equiv$ 5×10^4G is a standard laboratory field; the corresponding interaction energy is of the order $\mu_B H \sim 3$K which is, by solid state standards, rather small. Let us recall that typical bandwidths are 1–10eV$\sim$ 10^4–10^5K; the larger Coulomb matrix elements are in the same range; phonons are characterized by Debye temperatures $\theta_D \approx$100–1000K; even the spin–orbit coupling tends to be quite a bit stronger than $\mu_B H$. Common values of $\mu_B H$ are in the same range as (conventional "low-temperature") superconducting T_cs; indeed, the suppression of superconductivity by a sufficiently strong magnetic field is a long-known phenomenon. However, apart from that, it might seem that practicable magnetic fields cannot cause a drastic change in the behaviour of solids.

That would be a rash conclusion because it may happen that all the other terms in the Hamiltonian conspire to produce a set of more-or-less degenerate levels and, for (nearly) degenerate levels, any perturbation is a large perturbation. For instance, all the Bloch states on the Fermi surface are degenerate, and they are not even approximate eigenstates once an external field is turned on. The study of the resulting plethora of galvanomagnetic phenomena is a science in itself. – It is even possible to create situations where the coupling to the magnetic field is the dominating term of the Hamiltonian: such will be the case with the Quantum Hall Effect.

Another case when laboratory fields can have drastic effects arises when the Fermi temperature T_F turns out be very small, say, in the range 1–10K. There can be diverse reasons for this; marginal band overlap in semimetals like Bi can be one such reason. We will be more closely interested in a different situation, namely when strong electron correlations cause the system to develop narrow effective bands, as in *heavy fermion materials.*

Conventionally, one speaks of weak magnetism when the magnetization of the substance is induced by the external field and vanishes in its absence. There are, however, also cases of strong magnetism[1], when the specimen shows a spontaneous magnetization even in the absence of an external field. Formally, the onset of magnetic ordering is signalled by a divergent static susceptibility $\chi(\mathbf{q}, \omega = 0)$.

In contrast to the field-induced magnetization, the arising of spontaneous magnetic order can *not* be considered as the effect of some small perturbation. We have already indicated that realizable magnetic fields are relatively weak; the dipole fields of atomic moments at atomic distances are even weaker. In ferromagnetic iron, the magnetic moment of the atoms is $\mu_{\text{at}} \sim 2.1\mu_{\text{B}}$, and the nearest-neighbour distance is $a \approx 0.255\text{nm} \sim 5a_{\text{H}}$ where $a_{\text{H}} = \hbar^2/me^2$ is the Bohr radius. Recalling that $\alpha \equiv e^2/\hbar c \sim 1/137$ is the fine structure constant, and that the Rydberg is $1\text{Ry} \equiv e^4m/2\hbar^2 \sim 13.6\text{eV}$, the energy of the classical dipole–dipole interaction can be estimated as $\mu_{\text{at}}^2/a^3 \sim (\mu_{\text{at}}/\mu_{\text{B}})^2(a/a_{\text{H}})^{-3}\alpha^2\,\text{Ry} \sim 10^{-5}\text{eV} \sim 0.1\text{K}$ which certainly cannot account for a Curie temperature $T_{\text{C}} \sim 1043\text{K}$! The magnetism of iron is *not* due to tiny classical moments kept aligned by sitting in each other's magnetic fields. (Though the idea may sound quite plausible, and was actually put forward in the middle of the last century. It fails on two accounts: the aligning fields would be far too weak, and magnetism is, in any case, essentially non-classical.)

In the overwhelming majority of cases, *spontaneous magnetic order is a manifestation of strong electron–electron interactions.* It is not a secondary effect due to a weak perturbation which you could superimpose on some underlying zeroth-order state. The magnetism of strongly magnetic substances is due to the large terms in the Hamiltonian. It arises in a natural manner whenever (from band theory) you would expect to have a metal with a relatively narrow conduction band. You often find that the substance under consideration is not a metal at all, but some kind of a magnetic insulator. *The arising of magnetic order is intimately related to the question of metal–insulator phase transitions.*

It should be immediately remarked that although it is a general rule that by becoming magnetic, a substance gets "less metallic", it does not follow that it has to be an insulator: that depends also on the band

[1]This classification of substances as weakly or strongly magnetic, corresponds to an older usage of these words. A different terminology will be discussed later.

filling and other circumstances. The best-known ferromagnets Fe, Co, and Ni are metallic. So are the magnetic rare earths Gd, and Dy, or the ferromagnetic oxide CrO_2.

Thus the proper setting for learning about magnetism is the study of electron–electron interactions in general. Once electron–electron interactions start to play an important role, the development of magnetic order becomes a natural option. Frequently, it is not the only interesting thing the system is doing: it is likely to be accompanied by a metal–insulator phase transition, a structural distortion, or even by (occasionally high-temperature) superconductivity. For a number of systems, the interesting question is how they become magnetic; for some others, it might be even more interesting how they avoid becoming magnetic. It would be unnatural to separate these questions from each other. Our interest must lie in electron correlation phenomena[2] in general: whether the outcome is magnetism, or something else, depends on the circumstances.

This book is intended to be a rather elementary introduction to the subject of electron correlation phenomena. There are, of course, many excellent treatises[3] on correlation effects but these usually require substantial pre-education in many-body theory. The intention here is to start at the level (and sometimes, inevitably, with the subject matter) of standard solid state physics textbooks [32, 209] and, with no more than a nodding acquaintance with second quantization and response functions, reach a qualitative understanding of the questions of current research on Mott transitions, heavy fermion systems, and some exotic magnets. We will, however, speak very little about high-temperature superconductivity.

It must be clear from the foregoing that our primary interest is in the microscopic mechanisms of magnetism, i.e., in arriving at the models which are capable of magnetic ordering. The subsequent description of the magnetic ordering transitions, especially for insulators, can often be done without any reference to the underlying quantum-mechanical

---

[2]At this stage, "correlation" has to remain a catchword. Though we sense that *strongly interacting* systems are likely to be *strongly correlated*, these are not synonyms. The strength of the interaction can be read off from the Hamiltonian while "correlation" has a technical meaning which should become clearer gradually.

[3]We cite just a few: [33], [117], [119], [123], and [304].

many-body problem, and it is a highly developed, largely closed subject outside the scope of this volume. As a rule, we are not going to discuss the classical statistical physics of magnetic phase transitions; for instance, we will not speak about the celebrated exact solution of the two-dimensional Ising model. We will see, however, that motivating the introduction of a spin model with Ising-like anisotropy from an underlying electronic model is quite a tricky problem.

Though we almost always mean the "magnetism of electrons" when we speak about magnets, it should not be forgotten that solid He^3 provides a spectacular case of nuclear magnetism caused by quantum-mechanical exchange! In other materials for which, by supreme feats of experimental effort, the ordering of nuclear moments has been demonstrated, the otherwise usually negligible dipole–dipole interaction becomes important.

1.2 Sources of Magnetic Fields

The range of intensity of the magnetic fields at our disposal largely determines which physical phenomena we are able to see. There are certain threshold values for various effects. Every time a more powerful magnet becomes available, the discovery of new field-induced effects follows (from the reasonably expected metamagnetic transitions to the quite unexpected Quantum Hall Effect). It is thus useful to have an idea of the field strengths that can be produced by various means [49].

As for units, we will be usually following the habit of the particular subject under discussion. Much of the time, this means using c.g.s. units, since their use is widespread in the scientific literature. The c.g.s. unit of the magnetic field H is the oersted; to convert to SI, use $1\text{Oe} = (10^3/4\pi)\text{Am}^{-1}$. The unit of the field $B = H + 4\pi M$ is numerically and dimensionally the same, but then it is called gauss. Since here we are essentially speaking about the field in vacuum, we will rather freely use gauss (G) as the unit of the magnetic field. The related SI unit is $1\text{Tesla} \equiv 10^4\text{G}$. Weak fields are conveniently expressed in G, while standard laboratory fields are usually given in Tesla (T).

The Earth's magnetic field is about half a gauss. It was extraordinarily important in the history of physics [429] to have a source of much more intense fields to experiment with: the naturally magnetic magnetite, and magnetized iron. Ordinary iron-based permanent magnets like those used to stick up papers on bulletin boards, provide a few hundred gauss. The most powerful permanent magnets (like the samarium-cobalt or neodymium-iron-boron magnets) have fields of 3000-4000G.

The density of polarizable electrons in solids puts a theoretical limit of $\sim$ 3T to fields we can expect from permanent magnets. This is still far away, and the on-going effort to find better magnets is important for applications. However, this is not the way to produce large fields for research. In our laboratories, we now routinely have fields of $5 - 30$T. These are produced by electromagnets, i.e., they arise from a current flowing around a loop or coil of wire.

Though it might seem that there is no limit to the fields you can produce by currents, it does not work that way. The field exerts a force on the current-carrying coils which eventually exceeds the tensile strength of the material, however strong it may be. At the same time, the current heats the wire fiercely. If you try to drive too much current, your electromagnet breaks or melts. The current record for a DC resistive magnet is $\sim$30T [49].

We need not worry about the heating effect if we use superconducting coils. Here, however, the critical field of the superconductor limits the performance (to below $\sim$20T). A possibility to circumvent this limitation is to build a *hybrid magnet*, putting a resistive electromagnet inside a larger superconducting one. Hybrid magnets with fields in excess of 20T are now commercially available, and the current record field from such a source is $\sim$38.5T.

So much for the constant magnetic fields from DC magnets. Higher fields can be produced in pulses of duration from μseconds to $\sim$100 milliseconds. These may look short for a human observer but many solid state relaxation phenomena take much less time, thus for research, a field pulse is often as good as a DC field. With a *non-destructive pulsed magnet* you have a pulse, and then wait for the magnet to cool off. Fields exceeding 70T can be achieved. A quite radical approach is to build a self-destructing magnet which disrupts upon producing the pulse. Apparently, $\sim$150T can be produced with relative ease and regularity. In extreme circumstances, fields approaching $\sim$1000T have been detected.

Whether it is with DC or pulsed magnets, producing near-record fields takes clever design, and a lot of money and electric power. There are renowned high-field laboratories (Grenoble, Tallahassee, Tokyo, Amsterdam, etc.) which are visited by researchers wishing to do experiments in intense magnetic fields. However, these human achievements are dwarfed by what Nature can do: the magnetic field at the surface of a pulsar is estimated to be $\sim 10^8$T.

1.3 Getting Acquainted: Magnetite

By the time we finish our studies of classical electromagnetism we appreciate that magnetism, and magnetic fields, are not among the most straightforward phenomena we meet in this world. It is a piece of good luck that nature offered mankind an immediate sensuous experience[4] of what magnetism is about in the form of a mineral that you can pick up from the ground. A strongly magnetic material, magnetite, has been known since millenia[5]; its old-fashioned name "lodestone" (or "loadstone") should be understood as "leading stone", referring to the magnetic compass. Most of the ancient civilizations knew about lodestone; it never ceased to be a source of wonder for people who cared to wonder about such things. Do we, by now, know all about magnetite that there is to know?

Not at all, you will discover. Certainly we know a lot, but the more carefully you look, the more complicated the substance seems to get. This is a not uncommon experience, but magnetite appears to offer a larger collection of curious features than many of the related compounds. In view of this, and of the role magnetite played in the history of the subject, it is perhaps not a bad idea to begin a course on magnetism with a first attempt to understand the gross features of magnetite[6].

A word of warning should be given before we start. Magnetite is complicated. Some of the conceptual difficulties posed by its description

[4]To really appreciate this, let us recall that it is commonly held that the world of tangible objects, which we directly feel and see, is classical and non-relativistic. Well, magnetism turns out to be essentially quantum-mechanical and relativistic!

[5]For a fascinating account of the historical background, cf. Ch. 1. of [264]. See also [429], and [50].

[6]Here we assume the reader's familiarity with the crystalline, and magnetic, structure of Fe_3O_4 from the basic course on solid state physics (see Ch. 16 of [209]).

are the difficulties of contemporary many-body physics. Magnetite is certainly not the hottest subject at present but it is a subject which people have kept on returning to, in all epochs of solid state physics. At the moment, we just want to get an impression of the kind of the questions asked in this field, not of the answers. Do not worry if you do not understand completely the arguments yet; this is what the rest of the book is supposed to be about. As we progress, our understanding of the problems posed here should improve.

1.3.1 Charge States

The chemical formula of magnetite is Fe_3O_4. It is an iron oxide but not the only one. There are also FeO, and Fe_2O_3 (hematite). The picture of ionic bonding being quite sufficient for our present purposes, we can say that FeO $\rightsquigarrow$ $Fe^{++}O^{--}$ is the oxide of divalent iron, while $Fe_2O_3 \rightsquigarrow Fe_2^{+++}O_3^{--}$ is that of trivalent iron. In the case of magnetite, $Fe_3O_4 \rightsquigarrow Fe^{++}Fe_2^{+++}\,O_4^{--}$, both Fe^{++} and Fe^{+++} appear, therefore magnetite is called a *mixed valent* oxide.

As a chemical formula, this looks clear enough but if you think of the band theory of solids, it appears meaningless. In terms of d-shell filling, Fe^{3+} is $3d^5$, and Fe^{2+} is $3d^6$. These are configurations of a partially filled d-shell, so surely they must give rise to a partially filled d-band? Let us assume that we can think of the resulting band as a tight-binding band with nearest neighbour hopping. Starting with a pair of $3d^5$ and $3d^6$ sites, we might envisage the following hopping event between sites i and j

$$|3d^5\rangle_i|3d^6\rangle_j \longrightarrow |3d^4\rangle_i|3d^7\rangle_j \tag{1.1}$$

which can be realized by a minority-spin electron jumping away from a $3d^5$ site as shown in the following picture:

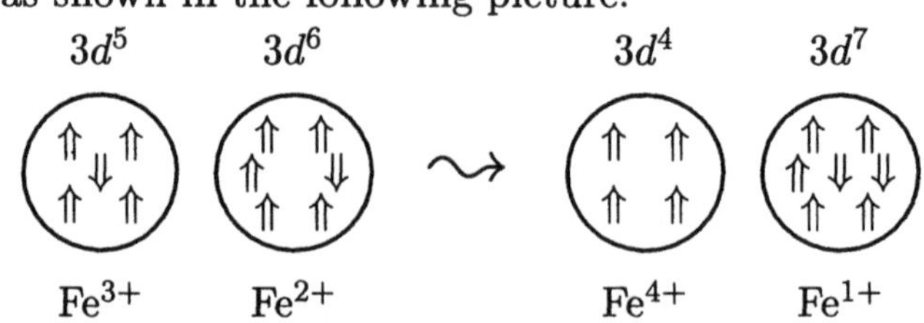

In chemical language, monovalent Fe^{+}, and tetravalent Fe^{4+}, ions are also expected to appear! This is, however, not what we wish to describe.

Our picture of magnetite is that it contains only divalent and trivalent iron[7]. Mixed valence is a widespread phenomenon in transition metal, and rare earth compounds, and it means that the state of the substance is dominated by the coexistence of *two* valence states, with the exclusion of the others. In most cases, you should think of two neighbouring valence states, as in our present example. There are, however, atoms which prefer to "skip" a valence state, so some mixed valent compounds are based on the coexistence of, say, Bi^{3+} with Bi^{5+}, or Pt^{2+} with Pt^{4+}.

To return to magnetite, the restriction to two valence states means that we allow only such hopping processes which amount to an interchange of Fe^{2+} and Fe^{3+}:

$$|3d^5\rangle_i |3d^6\rangle_j \longleftrightarrow |3d^6\rangle_i |3d^5\rangle_j \,. \tag{1.2}$$

1.3.2 Spin States

It is immediately clear that restricting hopping in the above sense is not what we mean by the band motion of electrons. Rather, it is an example of *correlated motion*. However, Eq. (1.2) is still "too permissive". Fe^{2+} and Fe^{3+} are magnetic ions in the sense that they retain the total spin, and the concomitant magnetic moment, that they would have in free space. In our studies of magnetism, we first have to understand why the free ions Fe^{2+} and Fe^{3+} have total spins $S = 2$, and $S = 5/2$, respectively. Then we have to understand whether, and how, this state of affairs can survive in a solid state environment.

Now consider the hopping process

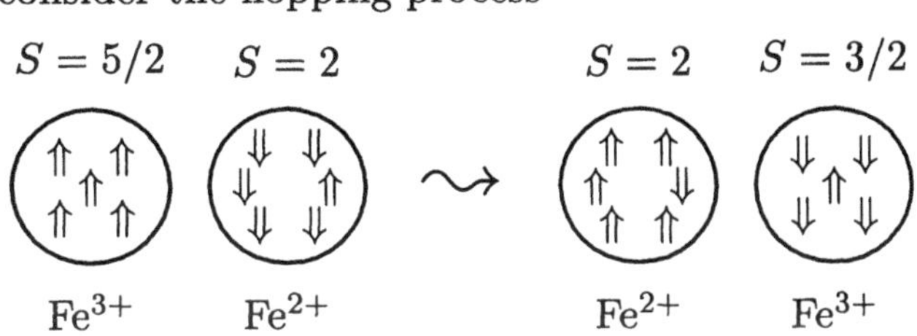

Though this is compatible with the restriction (1.2), we consider it

[7]Such statements cannot be strict. What we really mean is that the fraction of mono-, and tetravalent sites is by far not so high as independent-electron theory would lead us to think. It is still not clear what is the relative abundance of various valence states in magnetite.

inadmissible because it leads to a $S = 3/2$ low-spin state[8] of the Fe^{3+} ion. The physical reason for the exclusion of such processes is of the same nature as the valence restriction formulated in (1.2): it would lead from a low-energy subspace to a high-energy subspace. We will learn later both the reason for the energetical separation of subspaces, and the formal way of enforcing such constraints.

In our particular case, there is an additional factor which helps to give a simple form to both constraints, meaning Eq. (1.2), and the restriction to high-spin states. Magnetite is ferrimagnetic, with the ordering temperature $T_N \sim 858$K. If we restrict ourselves to temperatures, say, $T < 200$ K, it should be an acceptable approximation to think of the electrons as moving on the background of a frozen pattern of spin order.

We recall the basic features of ferrimagnetic order in the inverse spinel structure of magnetite [209]. There are two sublattices of crystallographically inequivalent lattice sites. Below T_N, the spins on the two sublattices are oppositely polarized. From the chemical unit of three iron ions, the sublattice A of tetrahedrally coordinated sites has one Fe^{3+} ion, and the sublattice B of sites with octahedral local coordination, has one Fe^{3+} ion, and one Fe^{2+} ion. The spins of Fe^{3+} ions thus compensate each other; the residual ferrimagnetic moment of magnetite can be ascribed to the Fe^{2+} ions which are distributed over the octahedral sublattice B.

Thus at sufficiently low temperatures, where magnetic order is nearly perfect, the only allowed hopping is that of the minority-spin electrons, transmuting Fe^{2+} and Fe^{3+} sites:

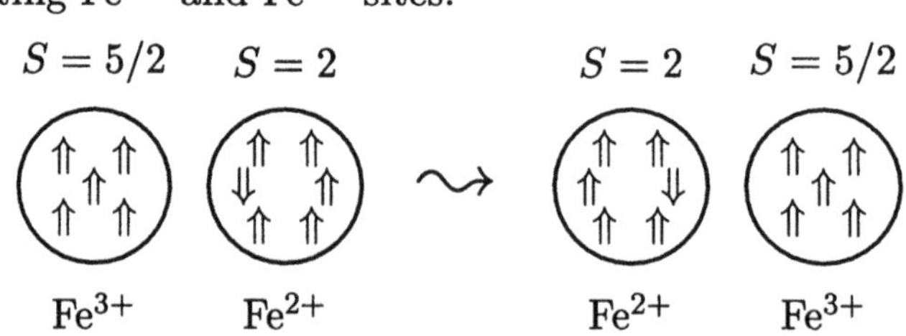

Having fixed the spin order means that, concentrating our attention to the mobile electrons, the spin degree of freedom can be forgotten,

---

[8]It is called low-spin in comparison to $S = 5/2$ which is the largest possible value of the spin for this ion.

and the problem becomes reduced to the nearest-neighbour hopping of spinless fermions:

$$\mathcal{H}_{\mathrm{hop}} = -t \sum_{<\mathrm{i,j}>} (c_{\mathrm{i}}^{\dagger} c_{\mathrm{j}} + c_{\mathrm{j}}^{\dagger} c_{\mathrm{i}}) \ . \tag{1.3}$$

1.3.3 Charge Ordering

By the end of the foregoing discussion, we have reduced the problem of electron motion in magnetite to that of a half-filled band of spinless fermions moving on the B-sublattice. We have come a long way from the original problem of 16 d-electrons per chemical unit cell, but still not long enough. A half-filled band is supposed to go with a metallic ground state. Look, however, at the temperature dependence of the electrical conductivity of magnetite, shown in Fig. 1.1.

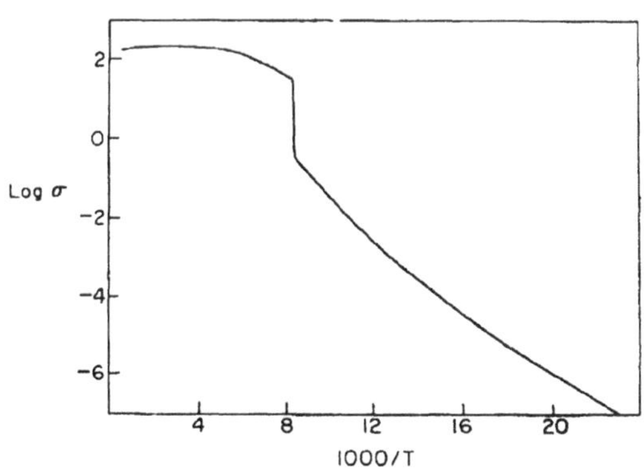

Figure 1.1: The dependence of the conductivity of Fe_3O_4 on the inverse temperature $1/T$, after Miles *et al.* [275]. The discontinuity is the Verwey transition.

Apparently, the material is not a good conductor at any temperature, and it is definitely a semiconductor at low temperatures. There is a two-orders-of-magnitude jump in the resistivity as the sample is cooled below the so-called Verwey temperature $T_{\mathrm{V}} \sim 122\mathrm{K}$ [293]. This first-order transition is a notable example of *metal–insulator transitions*[9].

[9] Actually, here it looks more like a semiconductor–semiconductor transition. How-

Why is magnetite not a metal? Because of the electron–electron repulsion. As we are going to see, we have taken substantial parts of the electron–electron interaction into account by restricting the allowed valence states to Fe^{2+} and Fe^{3+}; by ascribing definite (maximum) spins to these valence states; and by imposing magnetic order. Still, there must be a relevant part missing.

In a band ground state, one electron sees only the effective (averaged) field of the others, but there are no dynamical correlations: the instantaneous state of one electron does not depend on what another electron is doing "just then". In our residual band model (1.3), the spinless fermions are prohibited (by the Pauli principle) to share the same site but nothing prevents them from sitting at nearest-neighbour sites. This is, however, energetically undesirable because electrons repel most strongly at nearest-neighbour distance a. Let the associated Coulomb energy be $e^2/a = V$. We have to include this Coulomb term in the Hamiltonian

$$\mathcal{H} = \mathcal{H}_{\text{hop}} + \mathcal{H}_{\text{el-el}} = -t \sum_{<\mathbf{i},\mathbf{j}>} (c_\mathbf{i}^\dagger c_\mathbf{j} + c_\mathbf{j}^\dagger c_\mathbf{i}) + V \sum_{<\mathbf{i},\mathbf{j}>} \hat{n}_\mathbf{i} \hat{n}_\mathbf{j} \tag{1.4}$$

where $\hat{n}_\mathbf{i} = c_\mathbf{i}^\dagger c_\mathbf{i}$ is the site occupation number.

The Hamiltonian $\mathcal{H}$ defined above is far from incorporating all the electronic states whose energies lie reasonably close to (say, not more than a few eV from) the Fermi energy. In fact, it describes only one fermion per Fe_3O_4 unit, and even of that only the charge degree of freedom. It is a very stripped-down, single-band model of the electronic structure of magnetite. Still, we think it should suffice for the simplest description of the Verwey transition. The long way leading from the characterization of the real, complicated material to (1.4) is an example how *model Hamiltonians* are devised. Such model Hamiltonians are hoped to embody the essential aspects of the physics on the one hand, and to be simple enough to handle on the other.

The two terms in (1.4) stand for *competing tendencies*. Let us try to visualize what each of the terms "wants" the ground state to be, first on the example of a linear chain. If V is small, the kinetic energy dominates,

---

ever, the similarity in the underlying mechanism leads us to include the Verwey transition among the correlation-driven metal–insulator transitions.

and we get essentially a half-filled band, with a little suppression of the simultaneous occupancy of nearest-neighbour sites. On the other hand, if we set $t = 0$, for half as many particles as sites the term

$$\mathcal{H}_{\mathrm{el-el}} = V \sum_{<\mathrm{i,j}>} \hat{n}_{\mathrm{i}} \hat{n}_{\mathrm{j}} \tag{1.5}$$

will want to set up a superstructure in which only every other site is occupied:

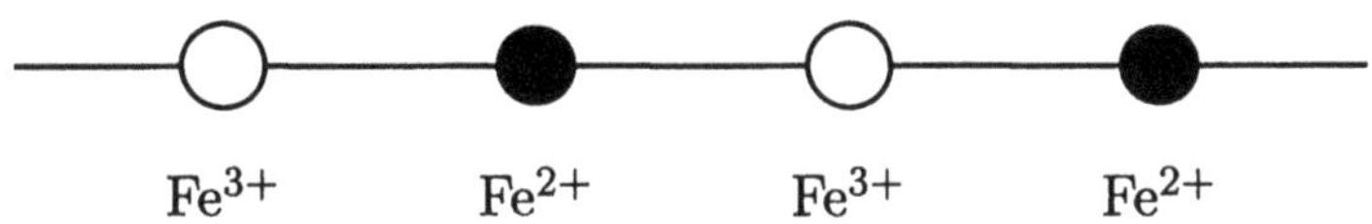

Such an *electron crystal* can be considered as a large-amplitude *charge density wave.* The Verwey transition in magnetite has been considered as a candidate for the long-sought phenomenon of Wigner crystallization, the formation of an electron crystal. In our particular case, it also means that the two different valence states tend to form a superstructure: this can be referred to as *heterogeneous mixed valence.* The phenomenon is widespread in transition metal oxides.

The Wigner crystal is a prime example for *strongly correlated states.* Here it is clear what strong correlation means: the electrons do their utmost to avoid each other in direct space. It is easy to accept that the electron crystal is insulating (in contrast, the ground state of $\mathcal{H}_{\mathrm{hop}}$ would be a conducting electron fluid). Another interesting point is that the charge-ordered ground state shown above does not have the full translational invariance of the underlying lattice: it is *symmetry-breaking.* Quite generally, one of the most important effects of electron–electron interaction is the appearance of symmetry-breaking ground states, i.e., of ground states whose symmetry is lower than that of the Hamiltonian.

Returning now to the real, three-dimensional magnetite, let us accept that the first-order transition manifested in the spectacular jump in conductivity is a charge ordering transition of the kind we have been discussing. Below the Verwey temperature $T_{\mathrm{V}} \sim 122\mathrm{K}$, a superstructure of different valence states should be seen. Once it was thought that it can be envisaged as a simple alternation of Fe^{2+} and Fe^{3+} sites when looking down along certain rows of atoms [18]. However, we would now doubt this. The exact nature of the (probably quite complicated) ordering pattern is still in dispute.

1.4 Variety of Correlated Systems: An Outline of the Course

Looking back: the consideration of such a long-studied material as magnetite confronted us with a wealth of the effects of electron–electron interaction. We have met magnetism, mixed valence, charge ordering transition, Wigner crystals, and metal–nonmetal transitions. We learned that if electron–electron interactions are strong enough to make a system magnetic, then the system is likely to do other remarkable things as well (or if it won't, you can prod it by some means: doping antiferromagnetic La_2CuO_4 with Sr turned it into a high-temperature superconductor!) So the main thrust of our studies of magnetism should be the study of *strongly correlated electron systems.*

Large groups of materials are thought to manifest interesting correlation effects:

- transition metal compounds, mainly oxides;
- the transition metals themselves (we should be able to say something about the ferromagnetism of iron, or the strange spin density wave state of chromium);
- rare earth elements and actinides, their compounds and alloys;
- arguably also a number of organic compounds.

We will discuss a number of correlation effects, trying to emphasize their qualitative aspects. Specific materials will be often mentioned to illustrate the phenomena under discussion, but no attempt is made to describe any class of substances exhaustively. At this point, a bias should be admitted. Most of the illustrative examples are either Mott insulators, or strongly correlated metals which are in some sense near to a Mott-insulating state. Conventional metallic ferromagnets like iron are, of course, also treated but they receive a bit less attention. This corresponds to a bias on the theoretical side: our arguments will be usually based on *lattice models of interacting fermions*, such as the Hubbard model. In these studies, it is assumed that valuable insight into the behaviour of complicated systems can be gained by using models with a few parameters, and a small number of states per unit cell. Whenever applicable, this approach offers us a detailed understanding of the mechanism of collective phenomena. The approach works quite wonderfully for antiferromagnetic Mott insulators, or heavy fermion metals, but at the time of writing, it is debated which lattice model would be most suitable to explain itinerant ferromagnetism. Some of the options are described in Ch. 8.

Though only macroscopic systems undergo phase transitions and show spontaneous magnetic order, correlation effects arise as soon as we have two interacting electrons. Both the relative simplicity of the problem, and the importance for the interpretation of experiments, justify spending some time on the study of few-electron systems: atoms, ions, and molecules (Ch. 2). Since correlation tends to localize the electrons onto ions, what we learn about the ionic properties, will be very helpful in understanding the paramagnetic phase of strongly correlated d- and f-electron systems. The free-ion behaviour is, however, modified by crystal field effects: these will be described in Ch. 3. Much of this is, of course, fairly standard stuff but even in these introductory chapters, we are trying to bring examples which correspond to our current interest in Mott insulators and heavy fermion systems.

Solving the eigenvalue problem of an electron moving on a homogeneous background yields Landau quantization and provides the setting for an introduction into the Quantum Hall Effect, both integer and fractional (Ch. 12).

The main thrust of the course is to develop a qualitative understanding of strongly correlated lattice systems (Chs. 4 to 11). We are going to see that the major effect of a sufficiently strong electron–electron interaction is to cause a *Mott transition*, a correlation-driven metal–insulator transition (Chs. 4 and 9). We will understand that once a system has become a Mott insulator, it will also want to be an ordered magnet (usually an antiferromagnet, see Ch. 5). However, this is not the only regime where magnetism occurs (Ch. 7). First, special features of the band structure may allow even a weakly interacting system to become a magnetic insulator. More important is that, in order to have a correlated insulator, the band filling has to be at (or at least near to) special discrete values. Non-integral filling arising from doping, or the overlapping of bands, stabilizes the metallic state even at very high interaction strengths. It might be quite natural to expect that such strongly correlated metals are magnetic. Well, sometimes they are, and sometimes they are not. It is interesting to observe that doping the antiferromagnetic Mott insulator $LaMnO_3$ with Sr yields a metallic ferromagnet with fascinating properties. However, doping the similar Mott insulator $LaTiO_3$ produces a non-magnetic metal. Nevertheless, the strength of correlations is still evident in features like the large mass enhancement of the Fermi liquid (Ch. 10). Extreme examples of this phenomenon are found in the rare-earth-based heavy fermion systems (Ch. 11).

As we proceed with our studies, it becomes gradually clear that the various spin Hamiltonians which are habitually used to describe magnetic ordering, are but low-energy effective Hamiltonians for Mott insulators. We will learn how the consideration of interaction effects in an underlying fermionic model gives the microscopic justification for introducing the Heisenberg model, spin anisotropy, biquadratic coupling, etc. Once this is given, further investigations can be pursued by the special techniques which were developed for spin systems. Spin wave theory (Ch. 6) is built on the assumption of a ground state which breaks spin-rotational invariance. We will also discuss whether the ground state of the antiferromagnetic Heisenberg model can be some different kind of collective spin state.

Finally, a disclaimer is in order. The discovery of high-T_c superconductors has given a huge impetus to the theory of correlation effects. This fact notwithstanding, we renounce discussing superconductivity. Our aim is rather to prepare the ground for such studies, by describing various materials as correlated systems capable of a variety of collective phenomena. In any case, we will frequently speak about systems which can become superconducting. Though copper is not a transition metal, we can treat the layered copper oxide systems along with the transition metal compounds, because it is a common feature that their behaviour is largely determined by the unfilled *d*-shells. We repeatedly cite the CuO_2 plane as an illustrative example. However, we do not address the question whether high-T_c superconductivity itself is an electron correlation effect, and what its particular mechanism may be.

Chapter 2

Atoms, Ions, and Molecules

The study of small systems – atoms, ions, and molecules – is an important part of the theory of magnetism. Many of the mechanisms which are thought to give rise to magnetic ordering in macroscopic systems, begin to work as soon as we have two interacting electrons. Here we will be mainly interested in finding out whether the electron spins wish to line up, or they rather prefer to be antiparallel. A system with a high-spin ground state exemplifies the tendency to ferromagnetism, while discussing the singlet ground state of the hydrogen molecule will lead us to recognizing the same kind of interaction which causes the antiferromagnetism of Mott insulators.

A further reason to examine small systems is that in many magnets, the electrons turn out to be essentially localized to ions, molecules, or small clusters of atoms. Thus what we learned about the magnetic properties of an isolated ion, may be directly applicable also to the case when the ion is embedded in a solid. This is often the case with rare earth ions.

The character of the ground state becomes manifest in the manner in which the system reacts to an external magnetic field. Therefore we will have to study the quantum-mechanical eigenvalue problem of electrons subject to a magnetic field. Since the field causes only a weak perturbation, the zero-field problem has to be solved with great care. Relativistic effects (in particular, the spin–orbit coupling) have to be taken into account.

2.1 Hydrogen Atom in a Magnetic Field

Much can be learned about the effect of a magnetic field on the quantum-mechanical motion of localized electrons[1] by studying the simplest realistic problem: that of the hydrogen atom. We give just a brief summary of those results which are covered in standard courses on quantum mechanics; for more details, you may consult [232], [355], [399], and [465].

2.1.1 Non-Relativistic Treatment

We consider a single electron (with charge $-e$) moving in the field of a nucleus with charge Ze. ($Z=1$ for a H atom, $Z=2$ for He^+, etc.) In the absence of external fields, the non-relativistic Hamilton operator is

$$\mathcal{H}_0 = \frac{\hat{\mathbf{p}}^2}{2m} - \frac{Ze^2}{r}. \tag{2.1}$$

The eigenstates are labelled with the (integer) quantum numbers n, l, and m

$$\psi_{nlm}(r,\vartheta,\varphi) = R_{nl}(r)Y_l^m(\vartheta,\varphi). \tag{2.2}$$

The possible values are $n=1,2,\ldots$; for a given n: $l=0,1,\ldots,n-1$, and for a given l: $m=-l,-l+1,\ldots,l-1,l$. However, the energy depends only on n

$$E_n^0 = -\frac{mZ^2e^4}{2\hbar^2n^2} = -\frac{(Ze)^2}{2a_0n^2} \tag{2.3}$$

where $a_0 = \hbar^2/(me^2) \approx 0.529 \cdot 10^{-8}$cm is the Bohr radius. The characteristic energy scale is given by the ionization energy of the H atom $e^2/2a_0 \equiv 1$Ry (1 Rydberg≈13.6eV). Disregarding spin degeneracy, the level E_n^0 is n^2-fold degenerate.

We will not need the general expression of $R_{nl}(r)$ but it is worth noting that it contains the exponential factor $e^{-r/\xi}$, where $\xi = na_0/Z$. Radial dependences for the first few shells are listed in App. A.

It follows from the spherical symmetry of the problem that the ψ_{nlm} are eigenstates of the (orbital) angular momentum, and its z-component:

$$\mathbf{L}^2\psi_{nlm} = l(l+1)\hbar^2\psi_{nlm} \quad \text{and} \quad L^z\psi_{nlm} = m\hbar\psi_{nlm}. \tag{2.4}$$

[1]The motion of an otherwise free electron in a magnetic field poses a somewhat different problem. It will be described in Ch. 12.

($\mathbf{L}$ is often defined to be dimensionless; then $\mathbf{L}^2\psi_{nlm} = l(l+1)\psi_{nlm}$ and $L^z\psi_{nlm} = m\psi_{nlm}$). The corresponding angular dependence is described by the spherical harmonics $Y_l^m(\vartheta, \varphi)$. Again, we omit the general discussion of these functions but since we are going to need Y_l^m up to $l = 4$, we give a list of them in App. A.

Motion in a Magnetic Field

The classical Lagrangian governing the motion of a particle with charge Q and mass M in arbitrary electric and magnetic fields is [209]

$$\mathcal{L} = \frac{1}{2}M\dot{\mathbf{r}}^2 - Q\phi(\mathbf{r}) + \frac{Q}{c}\dot{\mathbf{r}} \cdot \mathbf{A}(\mathbf{r}) \tag{2.5}$$

where ϕ is the scalar potential, and $\mathbf{A}$ is the vector potential. The canonical momentum is

$$\mathbf{p} \equiv \frac{\partial \mathcal{L}}{\partial \dot{\mathbf{r}}} = M\dot{\mathbf{r}} + \frac{Q}{c}\mathbf{A} \tag{2.6}$$

wherefrom the Hamiltonian is found to be

$$\mathcal{H}(\mathbf{r}, \mathbf{p}) \equiv \mathbf{p} \cdot \dot{\mathbf{r}} - \mathcal{L} = \frac{1}{2M}\left(\mathbf{p} - \frac{Q}{c}\mathbf{A}\right)^2 + Q\phi. \tag{2.7}$$

We use the correspondence principle to arrive at the quantum-mechanical Hamilton operator: we consider (2.7) as a function of the canonically conjugate operators $\hat{\mathbf{r}} \equiv \mathbf{r}$ and $\hat{\mathbf{p}} = (\hbar/i)\nabla$

$$\mathcal{H} = \frac{1}{2M}\left(\frac{\hbar}{i}\nabla - \frac{Q}{c}\mathbf{A}\right)^2 + Q\phi. \tag{2.8}$$

When using (2.8), care must be taken in expanding the square because $\mathbf{A}$ and $\hat{\mathbf{p}}$ do not commute. In fact,

$$\hat{\mathbf{p}} \cdot \mathbf{A} + \mathbf{A} \cdot \hat{\mathbf{p}} = 2\mathbf{A} \cdot \hat{\mathbf{p}} - i\hbar\, \nabla \cdot \mathbf{A}. \tag{2.9}$$

We will be following custom and use the *transverse gauge*, for which $\nabla \cdot \mathbf{A} = 0$.

Let us consider the motion of an electron in a uniform magnetic field[2] **H**. It is usual[3] to choose

$$\mathbf{A} = \frac{1}{2}\,\mathbf{H} \times \mathbf{r} \tag{2.10}$$

(verify that this is a transverse gauge!) Replacing $Q = -e$, $M = m$, and using (2.9)

$$\frac{1}{2m}\left(\hat{\mathbf{p}} + \frac{e}{c}\mathbf{A}\right)^2 = \frac{\hat{\mathbf{p}}^2}{2m} + \frac{e}{2mc}(\mathbf{H}\times\mathbf{r})\cdot\hat{\mathbf{p}} + \frac{e^2}{8mc^2}(\mathbf{H}\times\mathbf{r})^2. \tag{2.11}$$

In the second term, we can use the cyclic property $(\mathbf{H}\times\mathbf{r})\cdot\hat{\mathbf{p}} = (\mathbf{r}\times\hat{\mathbf{p}})\cdot\mathbf{H}$, and recognize that $\mathbf{r}\times\hat{\mathbf{p}} = \hbar\mathbf{L}$, where $\mathbf{L}$ is the (dimensionless) operator of the orbital angular momentum. Furthermore, the coefficient

$$\frac{e\hbar}{2mc} = \mu_{\mathrm{B}} \tag{2.12}$$

is just the Bohr magneton, the atomic unit of the magnetic moment for electrons. Now the second term can be written as $\mu_{\mathrm{B}}\mathbf{L}\cdot\mathbf{H}$.

Let the field point in the z-direction, then we find $(\mathbf{H}\times\mathbf{r})^2 = H^2(x^2+y^2)$. All in all, the coupling of the orbital motion of the electron to a uniform magnetic field is described by the field-dependent terms of the kinetic energy operator

$$\frac{1}{2m}\left(\hat{\mathbf{p}} + \frac{e}{c}\mathbf{A}\right)^2 = \frac{\hat{\mathbf{p}}^2}{2m} + \mu_{\mathrm{B}}HL^z + \frac{e^2H^2}{8mc^2}\,(x^2+y^2). \tag{2.13}$$

Note that we have not yet made any assumption as to how the electron moves; the form of the coupling $\mu_{\mathrm{B}}\mathbf{L}\cdot\mathbf{H}$ is valid for any conceivable orbital motion.

We may introduce the operator of the orbital magnetic moment as

$$\hat{\mu}^z_{\mathrm{orb}} = -\frac{\partial\mathcal{H}}{\partial H} = -\mu_{\mathrm{B}}L^z - \frac{e^2}{4mc^2}\,(x^2+y^2)H. \tag{2.14}$$

[2]In a magnetically polarizable medium, we should distinguish between **B** and **H**, and remember that it is **B** which is related to **A** via $\mathbf{B} = \nabla\times\mathbf{A}$. We consider non-magnetic substances with $\mathbf{B} = \mathbf{H}$ and denote the field by **H** as it is often done in the literature.

[3]But we may choose otherwise. In Ch. 12 we will see that a different choice (the Landau gauge) offers certain advantages.

The first term is the paramagnetic term and it describes a *permanent moment* $-\mu_{\rm B}\mathbf{L}$. The moment is called permanent because it exists irrespective of whether an external field is applied or not. In contrast, the second (diamagnetic) term corresponds to an *induced moment* $\sim H$ which is proportional, and antiparallel, to the field. We will soon argue that if there is a non-zero permanent moment, the diamagnetic contribution can be neglected.

Zeeman Effect (I)

Let us return to the hydrogen atom problem (2.1). Henceforth, we assume that the $\mathbf{r}$ appearing in the vector potential (2.10) is measured from the nucleus[4]. For a field pointing in the z-direction $\mathbf{H} = (0, 0, H)$

$$\begin{aligned} \mathcal{H} &= \mathcal{H}_0 + \mathcal{H}_{\rm field} \\ &= \frac{\hat{\mathbf{p}}^2}{2m} - \frac{Ze^2}{r} + \mu_{\rm B} H L^z + \frac{e^2H^2}{8mc^2}\,(x^2+y^2). \end{aligned} \tag{2.15}$$

Let us estimate the relative importance of the two field-dependent terms! For a typical atomic orbital, the ratio of the second-order and the first-order term is

$$\frac{e^2H^2}{8mc^2}\cdot\frac{1}{\mu_{\rm B}H}\cdot\frac{\langle x^2+y^2\rangle}{\langle L^z\rangle} \approx \frac{\hbar^3}{4m^2e^3}\cdot H \approx 10^{-6} H[{\rm T}]. \tag{2.16}$$

For this order-of-magnitude estimate, we used $\langle x^2+y^2\rangle \sim a_0^2$, and $\langle L^z\rangle = 1$. For a laboratory field of 1T (Tesla), the ratio is $\sim 10^{-6}$.

Similarly, we can estimate the relative strength of the first-order term compared to typical electronic energies. The size of the latter may be estimated as the potential energy of an electron in the $1s$ state

$$\frac{|\mu_{\rm B}\langle L^z\rangle| H}{e^2/a_0} \approx 2\cdot 10^{-6} H[{\rm T}]. \tag{2.17}$$

---

[4]Naturally, gauge freedom would permit us to choose any other point as the origin. The physical significance of the individual terms in (2.13) would then change but physical results (e.g., the susceptibility) would not. Our choice is the most convenient to work with, and it allows to ascribe a transparent physical meaning to the terms of (2.11), and (2.14).

It follows that in magnetic fields attainable in the laboratory, the typical energy shifts are very small and thus it is usually correct to treat the field-dependent terms as small perturbations. Furthermore, whenever the first-order term gives a non-zero contribution, the second-order term becomes negligible.

Returning to ordinary field strengths: for $l \neq 0$ shells it is permissible to approximate $\mathcal{H}_{\text{field}}$ by its first-order term (the so-called Zeeman term). Furthermore, the estimate (2.16) allows to neglect the mixing between different hydrogenic levels. Within a given (n, l) subspace, the Hamiltonian is diagonal in the basis (2.2)

$$(\mathcal{H}_0 + \mu_\mathrm{B} H L^z)\psi_{nlm} = \left(E_n^0 + mH\mu_\mathrm{B}\right)\psi_{nlm} \,. \tag{2.18}$$

The magnetic field lifts the $(2l + 1)$-fold degeneracy of a given (nl) shell. For instance, a p-level will be split into three nondegenerate levels. This phenomenon is the *Zeeman splitting*, or (first-order) Zeeman effect. Note that the energy difference between two consecutive levels $(\mu_\mathrm{B} H)$ is independent of n and l.

The form (2.15) does not give the complete single-electron Hamiltonian. It follows from a fully relativistic treatment [439, 465] which we do not review here, that the electron must have also an intrinsic angular momentum with $\hbar/2$-quantization, the electron spin. In the limit of low-energy solutions, a correction to (2.15) is found, namely the additional Zeeman term $2\mu_\mathrm{B}\mathbf{S}\cdot\mathbf{H}$ which describes the coupling of the electron spin to the magnetic field. (The very fact that the factor of 2 appears here, indicates that the spin of the electron cannot be modelled by the orbital motion of some internal parts of the electron.) The full Zeeman term reads

$$\mathcal{H}_\mathrm{Z} = \mu_\mathrm{B}(\mathbf{L} + 2\mathbf{S})\cdot\mathbf{H} \,. \tag{2.19}$$

To evaluate the consequences of the appearance of the spin degree of freedom, we have to take into account relativistic effects, foremost the spin–orbit coupling.

2.1.2 Relativistic Effects

At moderate velocities, relativistic effects should appear as corrections in higher orders of v/c (in fact, the leading terms of the energy shifts are of the order $(v/c)^2$). To have an idea of the order of magnitude of v/c, we make a simple estimate. Let us assume that the angular momentum quantum $\hbar$ arises from a motion with orbital velocity v at the radius a_0/Z. Then $v \sim \hbar Z/a_0 m = e^2 Z/\hbar$, and

$$\frac{v}{c} \sim \frac{e^2 Z}{\hbar c} = Z\alpha \quad \text{where} \quad \alpha \approx \frac{1}{137.036} . \tag{2.20}$$

For reasons which we shall soon understand, α is called the *fine structure constant.*

Expanding the relativistic energy–momentum relationship

$$\mathcal{E} = \sqrt{\mathbf{p}^2 c^2 + m^2 c^4} \approx mc^2 + \frac{\mathbf{p}^2}{2m} - \frac{1}{8}\frac{(\mathbf{p}^2)^2}{m^3 c^2} , \tag{2.21}$$

we are led to add the perturbation

$$\mathcal{H}^{(1)} = -\frac{1}{8}\frac{(\mathbf{p}^2)^2}{m^3 c^2} , \tag{2.22}$$

to the zero-field Hamiltonian (2.1). $\mathcal{H}^{(1)}$ is often called the mass–velocity term. Realizing that

$$\mathcal{H}^{(1)} = -\frac{1}{2mc^2}\left(\mathcal{H}_0 + \frac{Ze^2}{r}\right)^2 \tag{2.23}$$

and that $\mathcal{H}^{(1)}$ is isotropic and thus diagonal in the basis (2.2), we find the first-order energy shifts

$$\begin{aligned} \Delta E_{nlm} &= \int d\mathbf{r} \psi^*_{nlm}(\mathbf{r}) \mathcal{H}^{(1)}(r) \psi_{nlm}(\mathbf{r}) && (2.24) \\ &= -\frac{1}{2mc^2}\left((E_n^0)^2 + 2E_n^0 Ze^2 \left\langle \frac{1}{r} \right\rangle_{nl} + Z^2 e^4 \left\langle \frac{1}{r^2} \right\rangle_{nl}\right) . \end{aligned}$$

$\langle 1/r \rangle = Z/(a_0 n^2)$ is l-independent but it can be shown [355] that

$$\left\langle \frac{1}{r^2} \right\rangle_{nl} = \frac{Z^2}{a_0^2 n^3 (l + 1/2)} . \tag{2.25}$$

One of the important relativistic effects is the lifting of the degeneracy of the hydrogenic levels (2.3): now the energies do depend on l:

$$\Delta E_{nlm} = \frac{(Z\alpha)^2}{n^2}\left(\frac{n}{l + 1/2} - \frac{3}{4}\right) \cdot E_n^0 . \tag{2.26}$$

One might say that a fine structure appeared in the level scheme, the scale of which is controlled by the fine structure constant α. The level splittings (2.26) are of the order $\sim (Z\alpha)^2\text{Ry}$. Note that increasing the ionic charge Z leads to a rapid increase of relativistic corrections.

(2.26) is in itself not relevant because it is merely one of the contributions to the fine structure. Before discussing the result, we must consider the contribution arising from the spin–orbit coupling.

Spin–Orbit Coupling

The nature of the spin–orbit coupling can be understood by the following argument: the electric field of the nucleus is $\mathbf{E} = -\nabla\phi = -(\mathbf{r}/r)d\phi/dr$. As seen by the electron, the nucleus is orbiting around it with velocity $\mathbf{v}$, creating a magnetic field $\mathbf{H}_{\text{nucl}} = -(1/c)\mathbf{v} \times \mathbf{E}$. The electron spin couples to this magnetic field in the form given in (2.19)

$$2\mu_{\text{B}}\mathbf{S} \cdot \mathbf{H}_{\text{nucl}} = -\frac{e\hbar}{mc^2}\mathbf{S} \cdot (\mathbf{v} \times \mathbf{r})\frac{1}{r}\frac{d\phi}{dr} = -\frac{e\hbar^2}{m^2c^2}\mathbf{S} \cdot \mathbf{L}\frac{1}{r}\frac{d\phi}{dr} . \tag{2.27}$$

The simple reasoning led to almost the right result. A careful argument shows that the correct expression for the spin–orbit coupling differs by a factor of 2. Inserting the Coulomb potential appearing in (2.1), for the hydrogen problem we find

$$\mathcal{H}_{\text{SO}} = \frac{\hbar^2}{2m^2c^2}\frac{Ze^2}{r^3}\mathbf{S} \cdot \mathbf{L} . \tag{2.28}$$

(Remember that in the above equation, $\mathbf{S}$ and $\mathbf{L}$ are dimensionless operators).

Once $\mathcal{H}_{\text{SO}}$ is introduced, $\mathbf{S}$ and $\mathbf{L}$ are no longer separately conserved quantities. Rotational invariance guarantees merely the conservation of the total angular momentum $\mathbf{J} = \mathbf{L} + \mathbf{S}$. It is often useful to rewrite $\mathcal{H}_{\text{SO}}$ as

$$\mathcal{H}_{\text{SO}} = \frac{\lambda}{2}(\mathbf{J}^2 - \mathbf{L}^2 - \mathbf{S}^2) \tag{2.29}$$

where λ can be identified from (2.28). Note that $\lambda > 0$.

Low-order perturbation theory shows that combining the result (2.26) with that coming from spin–orbit coupling, the shifts of the hydrogenic levels are given by

$$\Delta E_{nlm} = \frac{(Z\alpha)^2}{n^2}\left(\frac{n}{j+1/2} - \frac{3}{4}\right) \cdot E_n^0 \tag{2.30}$$

where $j = l \pm 1/2$ is the total angular momentum quantum number. We just mention that another relativistic correction, the Darwin term, gives a further contribution for s-states. This completes the description of the fine structure. All in all, the eigenstates are classified according to n, $\mathbf{J}^2$, J^z, and $\mathbf{L}^2$ (formally, we could have added $\mathbf{S}^2$ to the list but in our particular case, it is just a constant).

To quote an example, the $n = 2$, $l = 1$ level is split into the higher-lying fourfold degenerate $j = 3/2$ level, and the lower-lying twofold degenerate $j = 1/2$ level (their spectroscopic notations[5] are $^2P_{3/2}$ and $^2P_{1/2}$, respectively). Interestingly, the $^2P_{1/2}$ level remains degenerate (to all orders of α) with the $^2S_{1/2}$ level arising from $n = 2$, $l = 0$. However, it turns out that this degeneracy is lifted by the Lamb shift, a quantum electrodynamical effect which is about a factor ~ 10 smaller than the $^2P_{3/2}-{}^2P_{1/2}$ fine structure splitting.

This is still not the end of the fine details of the level structure. The magnetic field created by the magnetic moment of the nucleus interacts with both the spin and the orbital magnetic moment of the electron, giving rise to the *hyperfine structure*. In the simplest case, the coupling can be written as

$$\mathcal{H}_{\mathrm{hf}} = A\,\mathbf{I} \cdot \mathbf{S} \tag{2.31}$$

where $\mathbf{I}$ is the spin of the nucleus. We do not discuss this, apart from pointing it out that the characteristic scale of the hyperfine splittings is $m/M \sim 10^{-3} - 10^{-5}$ times that of the fine structure.

Zeeman Effect (II)

Considering the hydrogen atom in a magnetic field, we take the Zeeman term from (2.19) but neglect the diamagnetic term. Assuming that the field $\mathbf{H}$ is pointing in the z-direction, the Hamiltonian is

$$\mathcal{H} = \mathcal{H}_0 + \mathcal{H}_{\mathrm{rel}} + \mu_{\mathrm{B}}(J^z + S^z)H. \tag{2.32}$$

[5]The notation is $^{2S+1}L_J$, where $L = S, P, D, F, G, H, \ldots$ for $l = 0, 1, 2, 3, 4, 5, \ldots$.

Though $\mathcal{H}_{\rm rel}$ should be really the sum of all the relativistic corrections, it is often represented by $\mathcal{H}_{\rm SO} = \lambda \mathbf{S} \cdot \mathbf{L}$. We wish to diagonalize $\mathcal{H}$.

The difficulty arises from the fact that switching on the magnetic field destroys the spherical symmetry of the $H = 0$ problem, replacing it with a merely axial symmetry (about the field direction). It follows that though j^z is still a good quantum number, j is not: S^z mixes states belonging to different js. Besides, there is no reason to declare that this mixing is weak. Relativistic energy shifts being (for $Z = 1$) of the order α^2Ry $\sim$ 1T, Zeeman energies in laboratory fields can easily be stronger than the fine structure splittings. Thus it is of practical importance to determine the change of the spectrum as one gradually passes from the weak-field case to the strong-field limit. We will often meet similar problems. The beauty of the present case is that it is quite easy to get an exact closed-form solution. However, we prefer to relegate the full solution to Problem 2.1, and treat here the limiting cases.

First, let us consider the *weak-field case* when the Zeeman term is a small perturbation in comparison to the spin–orbit coupling. The zero-field Hamiltonian is diagonal in the $|JJ^zLS\rangle$ basis[6]. Adopting the simplified notation $|J = l{\pm}1/2, J^z = j^z, L = l, S = 1/2\rangle = |l{\pm}1/2, j^z, l\rangle$, we quote from App. A the general expressions

$$|l \pm 1/2, j^z, l\rangle = \alpha_\pm |l, j^z - 1/2\rangle|\uparrow\rangle + \beta_\pm |l, j^z + 1/2\rangle|\downarrow\rangle \qquad (2.33)$$

where

$$\alpha_\pm = \pm\sqrt{\frac{l \pm j^z + 1/2}{2l+1}} \quad \text{and} \quad \beta_\pm = \sqrt{\frac{l \mp j^z + 1/2}{2l+1}}\,. \qquad (2.34)$$

For a weak field, the mixing between states with different js can be neglected, thus we have only the diagonal matrix elements

$$\langle l \pm 1/2, j^z, l|S^z|l \pm 1/2, j^z, l\rangle = \frac{\alpha_\pm^2 - \beta_\pm^2}{2} = \pm\frac{j^z}{2l+1}. \qquad (2.35)$$

[6]In the single-electron case, the spin–orbit coupling $\mathbf{L} \cdot \mathbf{S} = (\mathbf{J}^2 - \mathbf{L}^2 - \mathbf{S}^2)/2$ commutes with both $\mathbf{L}^2$ and $\mathbf{S}^2$. Remember that the analogous statement is not true for many-electron systems.

The Zeeman splittings are given by

$$\langle l \pm 1/2, j^z, l|\mathcal{H}_Z|l \pm 1/2, j^z, l\rangle = \begin{cases} \dfrac{2(l+1)}{2l+1}\mu_B H j^z & \text{for } j = l + 1/2 \\ \dfrac{2l}{2l+1}\mu_B H j^z & \text{for } j = l - 1/2. \end{cases} \tag{2.36}$$

It is interesting to observe that now (in contrast to (2.18)) the energy spacing depends on l. Applied to the $n = 2$, $l = 1$ states, we find a higher-lying quartet with the energy spacing $4\mu_B H/3$, and a lower-lying doublet split by $2\mu_B H/3$.

The result (2.36) can be rewritten as

$$\langle l \pm 1/2, j^z, l|\mathcal{H}_Z|l \pm 1/2, j^z, l\rangle = g\mu_B H j^z \tag{2.37}$$

where g is the so-called *Landé factor*

$$g = \frac{3}{2} + \frac{s(s+1) - l(l+1)}{2j(j+1)} . \tag{2.38}$$

We are going to see that this result can be generalized to many-electron ions.

Replacing $l = 0$ for s-electrons we recover that the spin-only value of the Landé factor is $g_S = 2$. This is what follows from the Dirac equation. However, turning to quantum electrodynamics one finds that there is a small radiative correction

$$g_S = 2\left(1 + \frac{\alpha}{2\pi} + O(\alpha^2)\right) \sim 2.00232 . \tag{2.39}$$

We will usually neglect this correction and calculate with $g_S = 2$.

Next, we consider the case of *strong fields*, which is also known as the *Paschen–Back limit*. If the Zeeman term is much stronger than the relativistic terms, we can use the basis of decoupled orbital and spin momenta $|L, L^z, S, S^z\rangle \rightsquigarrow |l, m\rangle|s^z\rangle$. $\mathcal{H}_0 + \mathcal{H}_Z$ is diagonal, and the zeroth-order Zeeman splittings are given by

$$\langle l, m|\langle s^z|\mathcal{H}_Z|l, m\rangle|s^z\rangle = (m + 2s^z)\mu_B H. \tag{2.40}$$

Taking again the $n = 2$, $l = 1$ states as an example, we find five equidistant levels, with the energy spacing $\mu_B H$. The middle level is twofold degenerate.

Since this levels scheme is quite different from the weak-coupling behaviour (2.36), it is clear that increasing the field from weak to strong values causes a major reshuffling of the energy levels. It is interesting to see how this happens in detail, and the reader may find out about it by solving the following

Problem 2.1. Solve the eigenvalue problem of (2.32) for arbitrary strengths of the magnetic field, taking the example of the $n = 2$, $l = 1$ states. Use the basis states (2.33). The only relevant data about the fine structure is Δ, the zero-field value of the ${}^2P_{3/2}$– ${}^2P_{1/2}$ splitting. Discuss the transition from the weak-field limit to the Paschen–Back limit!

2.2 Direct Exchange

Going over from the hydrogen atom to atoms with several electrons, the essential new difficulty is the appearance of the electron–electron interaction. The Coulomb force between the electrons can act in many ways. Here we treat one of its important effects: it can lift the degeneracy of states with different total spin. This manifestation of the electron–electron interaction is called the *exchange interaction*.

Let us consider two electrons in the same shell, say, in the configuration $3d^2$. The state of each electron is described by a spin–orbital $\varphi(\mathbf{r})\chi(\mathbf{s})$, where φ is one of the $3d$ states, and $\chi = \alpha$ for $\uparrow$-spin electrons, or $\chi = \beta$ for $\downarrow$-spin electrons. It is quite obvious that the Coulomb energy depends on which orbital states are occupied. Less apparently, even if we fix the orbitals, the interaction energy still depends on the spin arrangement. It is this problem that we want to examine closely now. To formulate it generally, let us assume that the two-electron Hamiltonian

$$\mathcal{H} = h_0(\mathbf{r}_1) + h_0(\mathbf{r}_2) + \frac{e^2}{|\mathbf{r}_1 - \mathbf{r}_2|} \tag{2.41}$$

consists of the sum of the (spin-independent) one-electron Hamiltonians $h_0(\mathbf{r})$, and the Coulomb interaction. We assume that the one-electron problem has been solved, yielding an orthonormal basis of eigenstates $\varphi(\mathbf{r})$. An orthonormal basis of two-electron states is formed by the 2×2 Slater determinants composed of one-electron spin–orbitals. These are, however, not necessarily two-electron eigenstates because we have not considered the effect of the Coulomb interaction yet. The eigenstates of (2.41) have to be sought as linear combinations of Slater determinants.

Let φ_a and φ_b be orthogonal eigenstates of h_0, corresponding to the one-electron eigenenergies ϵ_a and ϵ_b: $h_0(\mathbf{r})\varphi_a(\mathbf{r}) = \epsilon_a\varphi_a(\mathbf{r})$, and $h_0(\mathbf{r})\varphi_b(\mathbf{r}) = \epsilon_b\varphi_b(\mathbf{r})$, and $\int \varphi_a^*(\mathbf{r})\varphi_b(\mathbf{r})d\mathbf{r} = 0$.

Let us assume that some further reasoning tells us that it is meaningful to consider the action of $\mathcal{H}$ within the subspace where the orbital character of the two-electron state is fixed: there is one a electron and one b electron, and it is only the spins that can be varied. Later we are going to see that this "spin degeneracy problem" (and its generalization for more than two electrons) arises frequently, and its solution is important for the theory of magnetism. In our present case, only four two-electron states are permitted: two with parallel spins, and two with antiparallel spins. Taking only the one-electron energies, all the four states have the same energy contribution $\epsilon_a + \epsilon_b$. The interaction energy will, however, lift the degeneracy. We want to find out whether the electron–electron interaction "wants" the spins to be parallel, or antiparallel. More precisely, the question is whether the total spin $S = 0$ (singlet), or $S = 1$ (triplet) state has lower energy.

Let us enumerate the four basis states. The parallel spins can be both up

$$\begin{aligned} \psi_1 &= \frac{1}{\sqrt{2}} \begin{vmatrix} \varphi_a(\mathbf{r}_1)\alpha(s_1) & \varphi_a(\mathbf{r}_2)\alpha(s_2) \\ \varphi_b(\mathbf{r}_1)\alpha(s_1) & \varphi_b(\mathbf{r}_2)\alpha(s_2) \end{vmatrix} \\ &= \frac{1}{\sqrt{2}}\alpha(s_1)\alpha(s_2)[\varphi_a(\mathbf{r}_1)\varphi_b(\mathbf{r}_2) - \varphi_a(\mathbf{r}_2)\varphi_b(\mathbf{r}_1)] \end{aligned} \tag{2.42}$$

or both down

$$\psi_4 = \frac{1}{\sqrt{2}}\beta(s_1)\beta(s_2)[\varphi_a(\mathbf{r}_1)\varphi_b(\mathbf{r}_2) - \varphi_a(\mathbf{r}_2)\varphi_b(\mathbf{r}_1)]. \tag{2.43}$$

These have a simple appearance. In contrast, the states with antiparallel spins do not possess the property that the spin and orbital dependence factorizes. We find

$$\begin{aligned} \psi_2 &= \frac{1}{\sqrt{2}} \begin{vmatrix} \varphi_a(\mathbf{r}_1)\beta(s_1) & \varphi_a(\mathbf{r}_2)\beta(s_2) \\ \varphi_b(\mathbf{r}_1)\alpha(s_1) & \varphi_b(\mathbf{r}_2)\alpha(s_2) \end{vmatrix} \qquad (2.44) \\ &= \frac{1}{\sqrt{2}}[\varphi_a(\mathbf{r}_1)\varphi_b(\mathbf{r}_2)\beta(s_1)\alpha(s_2) - \varphi_b(\mathbf{r}_1)\varphi_a(\mathbf{r}_2)\alpha(s_1)\beta(s_2)] \end{aligned}$$

and

$$\psi_3 = \frac{1}{\sqrt{2}}[\varphi_a(\mathbf{r}_1)\varphi_b(\mathbf{r}_2)\alpha(s_1)\beta(s_2) - \varphi_b(\mathbf{r}_1)\varphi_a(\mathbf{r}_2)\beta(s_1)\alpha(s_2)]. \quad (2.45)$$

$\mathcal{H}$ has to be diagonalized in the subspace $\psi_1, \ldots, \psi_4$. Let us evaluate the matrix elements

$$\begin{aligned}\langle\psi_1|\frac{e^2}{|\mathbf{r}_1-\mathbf{r}_2|}|\psi_1\rangle &= \frac{1}{2}\int d\mathbf{r}_1 \int d\mathbf{r}_2[\varphi_a^*(\mathbf{r}_1)\varphi_b^*(\mathbf{r}_2) - \varphi_a^*(\mathbf{r}_2)\varphi_b^*(\mathbf{r}_1)] \\ &\quad \frac{e^2}{|\mathbf{r}_1-\mathbf{r}_2|}[\varphi_a(\mathbf{r}_1)\varphi_b(\mathbf{r}_2) - \varphi_a(\mathbf{r}_2)\varphi_b(\mathbf{r}_1)] \\ &= C_{ab} - J_{ab} \qquad (2.46)\end{aligned}$$

where we have introduced the Coulomb integral

$$C_{ab} = e^2 \int d\mathbf{r}_1 \int d\mathbf{r}_2 \frac{|\varphi_a(\mathbf{r}_1)|^2|\varphi_b(\mathbf{r}_2)|^2}{|\mathbf{r}_1-\mathbf{r}_2|} \quad (2.47)$$

and the exchange integral

$$J_{ab} = e^2 \int d\mathbf{r}_1 \int d\mathbf{r}_2 \frac{\varphi_a^*(\mathbf{r}_1)\varphi_b(\mathbf{r}_1)\varphi_b^*(\mathbf{r}_2)\varphi_a(\mathbf{r}_2)}{|\mathbf{r}_1-\mathbf{r}_2|} . \quad (2.48)$$

C_{ab} is obviously positive; less obviously, so is J_{ab} because it can be regarded as the self-interaction energy of the exchange charge density $\varphi_a^*(\mathbf{r})\varphi_b(\mathbf{r})$ (a formal proof will be given at the end of the Section).

Note that J_{ab} arose as a "cross-term": e.g., we can take the second term from the left-hand factor and the first term from the right-hand factor in (2.46). There is no such cross-term in the corresponding matrix element with ψ_2, because the spin functions are orthogonal

$$\langle\psi_2|\frac{e^2}{|\mathbf{r}_1-\mathbf{r}_2|}|\psi_2\rangle = C_{ab} . \quad (2.49)$$

There is, however, an off-diagonal matrix element connecting ψ_2 and ψ_3

$$\langle\psi_2|\frac{e^2}{|\mathbf{r}_1-\mathbf{r}_2|}|\psi_3\rangle = -J_{ab} . \quad (2.50)$$

It is interesting to realize that the Coulomb interaction can induce transitions between different localized configurations[7]. There are many more off-diagonal Coulomb processes which we will meet in the solution to Problem 2.2.

The above results can be combined into the following expression for the matrix of $\mathcal{H}$

$$(\epsilon_a + \epsilon_b)\begin{pmatrix} 1 & 0 & 0 & 0 \\ 0 & 1 & 0 & 0 \\ 0 & 0 & 1 & 0 \\ 0 & 0 & 0 & 1 \end{pmatrix} + \begin{pmatrix} C_{ab} - J_{ab} & 0 & 0 & 0 \\ 0 & C_{ab} & -J_{ab} & 0 \\ 0 & -J_{ab} & C_{ab} & 0 \\ 0 & 0 & 0 & C_{ab} - J_{ab} \end{pmatrix} \tag{2.51}$$

ψ_1 and ψ_4 are immediately eigenstates, being the $S^z = \pm 1$ components of the triplet. The triplet eigenenergy is

$$\epsilon_{\mathrm{tr}} = \epsilon_a + \epsilon_b + C_{ab} - J_{ab} \,. \tag{2.52}$$

There remains a 2×2 problem to solve. One of the eigenvalues is ϵ_{tr}. The other eigenvalue is

$$\epsilon_{\mathrm{s}} = \epsilon_a + \epsilon_b + C_{ab} + J_{ab} \tag{2.53}$$

which corresponds to the non-degenerate singlet level.

Note that because of $J_{ab} > 0$, the triplet is lower in energy than the singlet. Our essential finding is that *direct exchange in orthogonal orbitals favours ferromagnetism.*

There is a very simple argument to support this result. Two parallel-spin electrons cannot be at the same point $\mathbf{r}$ in space because then they would occupy the same quantum-mechanical state, and this is forbidden by the Pauli principle. Since the wave function has a continuous dependence on the relative coordinate $\mathbf{r}_1 - \mathbf{r}_2$, this means that the amplitude is gradually suppressed if we decrease $|\mathbf{r}_1 - \mathbf{r}_2|$, vanishing when $|\mathbf{r}_1 - \mathbf{r}_2| = 0$. This is called the *exchange hole effect*, and it says that parallel-spin electrons tend to keep away from each other in real space. There is no such effect for antiparallel-spin electrons whose meeting at

[7] If we use the spin-orbitals as the one-electron basis. Note that this property is associated with the fact that the one-electron orbitals are extended in space and therefore can overlap. The Coulomb interaction is strictly diagonal in the $\mathbf{r}$-representation which uses fully localized δ-functions as a basis.

the same point is not forbidden by the Pauli principle. Therefore electrons in the singlet state feel more of the mutual Coulomb repulsion: this pushes the singlet energy up.

Since the orbital part of the wave function is rigid, the states can be labelled by their spin component. We use the basis $|S, S^z\rangle$, where S is the total spin, and S^z its z-component. Three eigenstates form a triplet

$$|1,1\rangle = \psi_1 \qquad |1,0\rangle = \frac{1}{\sqrt{2}}(\psi_2 + \psi_3) \qquad |1,-1\rangle = \psi_4 \,. \tag{2.54}$$

The fourth eigenstate is a singlet

$$|0,0\rangle = \frac{1}{\sqrt{2}}(\psi_2 - \psi_3) \,. \tag{2.55}$$

The action of $\mathcal{H}$ can be expressed in terms of spin operators. Let us note that $(\mathbf{S}_1 + \mathbf{S}_2)^2$ gives 0 for a singlet and 2 for a triplet. Hence

$$2\mathbf{S}_1\mathbf{S}_2 + \frac{1}{2} = (\mathbf{S}_1 + \mathbf{S}_2)^2 - 1 = \begin{cases} 1 & \text{for a triplet} \\ -1 & \text{for a singlet.} \end{cases} \tag{2.56}$$

Using this identity, (2.52) and (2.53) can be conveniently combined into a single expression

$$\begin{aligned} \mathcal{H} &\to \frac{\epsilon_{\rm s} + \epsilon_{\rm tr}}{2} - \frac{\epsilon_{\rm s} - \epsilon_{\rm tr}}{2}\left(2\mathbf{S}_1\mathbf{S}_2 + \frac{1}{2}\right) \\ &= \text{constant} - 2J_{ab}\mathbf{S}_1 \cdot \mathbf{S}_2 \,. \end{aligned} \tag{2.57}$$

Within the 4-element subspace, the electron–electron interaction can be described as an *exchange Hamiltonian*. The possibility of this appealing reformulation was first noticed by Dirac.

As we are going to see, a similar result is found in a variety of cases: Let us assume that some further reasoning[8] tells us that the orbital component of a many-electron wave function can be considered rigid. Build a subspace of the Hilbert space by keeping the orbital component fixed, and choosing the spins in all possible ways. Within this subspace,

[8]Usually stating that the electrons are Mott-localized.

the effect of the electron–electron interaction can be expressed by spin operators and it is often enough to keep the pair interactions

$$\mathcal{H}_{\text{exch}} = \sum_{i,j} J_{ij} \mathbf{S}_i \mathbf{S}_j \, . \tag{2.58}$$

This is the most familiar form of *exchange Hamiltonians.*

The form (2.58) may mislead us into thinking that there is the Coulomb force and in addition some mysterious spin–spin interaction. This would be quite wrong. As the above derivation made it clear, exchange is not a primary interaction. In fact, exchange is a manifestation of the Coulomb interaction *and* quantum-mechanical indistinguishability: the matrix elements of the Coulomb interaction have to be taken between antisymmetrized many-electron states. It is essential that the electrons inhabit the same region of space[9]. If we separated the electrons by classical walls, then the exchange charge density $\varphi_a^*(\mathbf{r})\varphi_b(\mathbf{r})$ would be identically zero, and the exchange interaction would vanish.

Let us emphasize that there is nothing sacred about the (bilinear) pair form of the exchange Hamiltonian (2.58), however convenient and useful it may be. If we were treating a many-electron problem like we did with the two-electron problem, we may be wondering whether expressing the Coulomb interaction would not necessitate the inclusion of more complicated terms; generally, it would. However, regarding that exchange Hamiltonians are difficult to derive from "first principles", in solid state physics one often takes a pragmatic approach, keeping the form of the Hamiltonian simple, and fitting the coupling constants to experiments.

Before closing this subsection, let us discuss some simple features of the two-electron eigenstates. The $S^z = 0$ component of the triplet is

$$|1,0\rangle = \frac{1}{2}[\varphi_a(\mathbf{r}_1)\varphi_b(\mathbf{r}_2) - \varphi_a(\mathbf{r}_2)\varphi_b(\mathbf{r}_1)][\alpha(s_1)\beta(s_2) + \beta(s_1)\alpha(s_2)] \tag{2.59}$$

[9] In a small Hilbert space spanned by two orbitals, it is not possible to distinguish between "the two electrons should be able to visit the same points of space" and "with a finite probability density, the two electrons should be at the same point of space". The first requirement is essential, while the second is not. Two strongly interacting fermions can exhange places even if they strictly avoid each other. The description of such a correlated motion requires a larger Hilbert space. We return to this problem in our study of ring exchange processes (Sec. 5.1.6).

while the singlet eigenstate is

$$|0,0\rangle = \frac{1}{2}[\varphi_a(\mathbf{r}_1)\varphi_b(\mathbf{r}_2)+\varphi_a(\mathbf{r}_2)\varphi_b(\mathbf{r}_1)][\alpha(s_1)\beta(s_2)-\beta(s_1)\alpha(s_2)]. \quad (2.60)$$

These illustrate the statement that, generally speaking, the many-electron eigenstates cannot be expressed as a single Slater determinant composed of the solutions of the one-electron problem: they have to be sought as the linear combination of several (usually many) determinants.

The antisymmetrical nature of the two-electron eigenstates appears in an appealingly simple manner. Looking at (2.42), (2.43), and (2.59), we can say that in the triplet, the space dependence is odd, and the spin function is even. Conversely, (2.60) shows that in the singlet the spin wave function is odd, and the space dependence is even. It is well to remember that these useful properties of the two-electron states have no straightforward generalization to three or more electrons. To be precise: it is still possible to find a state with a totally symmetric spin dependence, and an antisymmetric orbital dependence; this is the maximum-spin state. However, it is not possible to construct a totally antisymmetric state of three or more $S = 1/2$ spins. Consequently for general, less-than-maximum values of the total spin, the antisymmetry of the many-electron state is not manifested in a factorization into simple spin- and orbital dependences.

Finding a reason for the exchange interaction to be ferromagnetic, we may interpret it as a hopeful sign that eventually we can understand ferromagnets[10]. However, we are aware that in nature, antiferromagnetism is much more widespread than ferromagnetism so we may well be wondering how antiferromagnetism can arise. As we are going to see, antiparallel spin alignment becomes favoured if we drop either of the starting assumptions for the above result. Direct exchange for *non-orthogonal orbitals* tends to be antiferromagnetic (as in the hydrogen molecule problem, Sec. 2.5. See also the *Digression* at the end of this Section). Relaxing the constraint that only one electron can occupy either orbital, we find the *kinetic exchange* which is basically antiferromagnetic (Sec. 2.5.2). Finally, we mention that exchange is often *indirect*, involving a third orbital in between the two we were consid-

[10] Actually, the connection is anything but immediate: it is quite difficult to understand ferromagnetism (see Ch. 8).

ering. In this context, the mechanism we have been discussing here is called direct exchange.

Proof of $J_{ab} \geq 0$. We rephrase the proof given in [99]. Introducing $w(\mathbf{r}-\mathbf{r}') = e^2/|\mathbf{r}-\mathbf{r}'|$ and $f(\mathbf{r}) = \varphi_a(\mathbf{r})\varphi_b^*(\mathbf{r})$, let us rewrite J_{ab} as

$$J_{ab} = \int d\mathbf{r} \int d\mathbf{r}' f^*(\mathbf{r}) w(\mathbf{r}-\mathbf{r}') f(\mathbf{r}') . \qquad (2.61)$$

The integral $\int d\mathbf{r}'$ is a convolution, thus in terms of the Fourier transforms it can be rewritten as

$$\int w(\mathbf{r}-\mathbf{r}') f(\mathbf{r}') d\mathbf{r}' = \int d\mathbf{k} e^{i\mathbf{k}\mathbf{r}} w(\mathbf{k}) f(\mathbf{k}),$$

and J_{ab} can be expressed as the Fourier integral

$$J_{ab} = \int f^*(-\mathbf{k}) w(\mathbf{k}) f(\mathbf{k}) d\mathbf{k} = \int |f(\mathbf{k})|^2 w(\mathbf{k}) d\mathbf{k}. \qquad (2.62)$$

Recalling that $w(\mathbf{k}) = 4\pi e^2/k^2$, the proof of $J_{ab} \geq 0$ is completed.

Digression: Relationship to Wigner's Theorem about the Two-Electron Problem. The apparent generality of the previous result should not mislead us into thinking that the ground state of an interacting two-electron problem is always a triplet. Beware: it is not this what we proved! In fact, the exact opposite is true: it can be shown that the ground state of the general two-electron Hamiltonian

$$\mathcal{H} = \frac{\mathbf{p}_1^2}{2m} + \frac{\mathbf{p}_2^2}{2m} + \mathcal{V}(\mathbf{r}_1, \mathbf{r}_2) \qquad (2.63)$$

(where the real, symmetrical $\mathcal{V}(\mathbf{r}_1, \mathbf{r}_2)$ contains both the external potential and the interaction) is *nodeless*, and therefore belongs to $S = 0$ [245, 264]. The proof (ascribed to Wigner) is indirect: let us assume that the ground state wave function $\psi(\mathbf{r}_1, \mathbf{r}_2)$ does have a node, and the corresponding energy is E_0. One can show that taking $|\psi(\mathbf{r}_1, \mathbf{r}_2)|$ as a variational trial state, we get exactly the same energy. However, $|\psi(\mathbf{r}_1, \mathbf{r}_2)|$ cannot be the ground state because it has discontinuous derivatives at the nodal surface. Therefore, the true ground state energy must be lower than E_0, which contradicts the starting assumption $\leadsto$ The true ground state must be nodeless. The nodeless singlet ground states of the

He atom, and the H_2 molecule (for arbitrary proton–proton distance) illustrate the theorem. Note that in both cases, the electrons occupy non-orthogonal (or possibly even identical) orbitals.

There is no real contradiction between this theorem, and our previous result about the ferromagnetic nature of direct exchange. The point is that our starting example of a shell with the $3d^2$ configuration is not really a two-electron problem: there must be quite a few electrons in the shells underneath, otherwise it would not be meaningful to pose the question of the "ground state" of the two d-electrons. The $3d$ functions must be orthogonal to the inner-shell states, and therefore have nodes to begin with. The proof of Wigner's theorem is just not meant for such a situation [264]. We can also say that for two electrons, $3d^2$ would be an excited-state configuration, and therefore the theorem is not applicable.

Problem 2.2. A possible realization of the two-electron problem considered in this Section is that it describes a diatomic molecule, and φ_a and φ_b are (Wannier-type) orthogonalized atomic orbitals belonging to sites a and b, resp. The triplet (2.52) and singlet (2.53) eigen-energies describe the effect of the Coulomb interaction restricted to the subspace of neutral (Heitler–London) configurations in which each orbital is occupied by one electron. We will find several more Coulomb matrix elements if we lift this restriction and allow the two electrons to occupy the orbitals in an arbitrary manner, subject only to the Pauli principle. This enlarges the Hilbert space by the inclusion of the ionized configurations.

Use the formalism of second quantization to classify all Coulomb contributions! The Hilbert space is spanned by the two orbitals, i.e., the field operator can be written as

$$\Psi^\dagger(\mathbf{r},\sigma) = \varphi_a^*(\mathbf{r})c_{a\sigma}^\dagger + \varphi_b^*(\mathbf{r})c_{b\sigma}^\dagger \,. \tag{2.64}$$

Assume that a and b are energetically equivalent, so that you do not have to introduce different couplings to distinguish situations where the role of a and b is reversed.

2.3 Many-Electron Ions

For an atom (ion) with several electrons, we partition the Hamilton operator into a non-relativistic part, relativistic corrections, and the coupling to the external magnetic field

$$\mathcal{H} = \mathcal{H}_{\text{nr}} + \mathcal{H}_{\text{rel}} + \mathcal{H}_{\text{field}} \,. \tag{2.65}$$

The non-relativistic part contains the sum of the one-electron terms, and the all-important Coulomb repulsion between the electrons

$$\mathcal{H}_{\rm nr} = \mathcal{H}_0 + \mathcal{H}_{\rm el-el} = \sum_i \frac{\hat{\mathbf{p}}_i^2}{2m} - \sum_i e\phi(\mathbf{r}_i) + \sum_{i<j} \frac{e^2}{|\mathbf{r}_i - \mathbf{r}_j|} \,. \tag{2.66}$$

Extensive work on the two-electron case (the He atom, the H^- ion, etc.) has uncovered a complicated structure of the wave function. A full solution for more than two electrons is hopeless or impracticable. However, we will take it for granted that a mean-field solution (obtained in the Hartree–Fock approximation) gives at least a good starting point. In this approach one assumes that, in addition to the field $\phi(\mathbf{r})$ of the nucleus, each electron feels also the spherically symmetrical average field of the other electrons, so that it will move in some hydrogen-like orbit. Using the Pauli principle, we construct the ground state by filling the lowermost available states. Thus we arrive at the concept of the ground state *configuration* of an atom: it is a list of the occupied states, labelling them with the names of hydrogenic orbitals. For instance, the sodium atom has the configuration $1s^2 2s^2 2p^6 3s^1$. However, such full lists are seldom given. Conveying the idea that inner shells with a noble gas configuration form an imperturbable part of the ion core, we may also say that Na has the configuration [Ne]$3s^1$, or simply $3s^1$.

What is left out in this scheme? It is just the possibility that configurations may get mixed. The interaction term $\mathcal{H}_{\rm el-el}$ has non-vanishing matrix elements between any two two-electron states if they have the same values of the relevant quantum numbers. Thus, for example's sake, the ^1D state ($S = 0$, $L = 2$) may be realized either in a $n_1 s^1 n_2 d^1$ configuration, or in a $n_1 p^1 n_2 p^1$ configuration. It follows that the configuration $n_1 s^1 n_2 d^1$ does not appear pure, but has an admixture of $n_1 p^1 n_2 p^1$ configurations. We trust that such effects are usually weak and will refer to them only occasionally[11].

In what follows, we discuss the level structure belonging to a given configuration. First, let us consider two electrons in the same shell,

[11] At least in our present discussion. Since atoms have discrete spectra, configuration mixing has to act across large energy differences, and therefore it is a weak effect. In solids, a vast number of states lie infinitesimally close, and the story may be different. Ch. 9 will be devoted to what is essentially a configuration interaction calculation for the Hubbard model.

say, the configuration $3d^2$. Since there are $C_2^{10} = 45$ ways to choose two states out of ten, the Hilbert space is 45-dimensional. At the level of $\mathcal{H}_0$, all the 45 states are degenerate, with energy $2\epsilon_d$. However, the Coulomb interaction between two electrons depends on the precise nature of the one-electron states (spin–orbitals) which the electrons occupy, thus "switching on" the electron–electron interaction lifts the degeneracy. The 45 states will be arranged in several levels, with the energy splittings between them given by certain matrix elements of the Coulomb interaction. The remaining degeneracies are those required by symmetries.

Since both the total spin $\mathbf{S} = \sum_i \mathbf{S}_i$, and the total orbital angular momentum $\mathbf{L} = \sum_i \mathbf{L}_i$ commute with $\mathcal{H}_{\text{nr}}$, the levels can be labelled by the quantum numbers L and S, and the eigenstates can be chosen as $|L, S, L^z, S^z\rangle$. This scheme is called the L–S or Russell–Saunders coupling. The possible values of L and S are determined by the Pauli principle and the antisymmetrical nature of the wave function (see Problem 2.3). The corresponding $(2L+1)(2S+1)$-fold degenerate levels are also called the terms of the configuration. These degeneracies will, however, be partially lifted by the relativistic part of $\mathcal{H}$ which is usually taken as the sum of spin–orbit coupling terms

$$\mathcal{H}_{\text{rel}} \rightsquigarrow \mathcal{H}_{\text{SO}} = \sum_i \lambda_i \, \mathbf{L}_i \cdot \mathbf{S}_i \tag{2.67}$$

where the λ_i are of the form seen in (2.28)[12], i.e., $\lambda_i > 0$. Generally, switching on the spin–orbit interaction leads to the breakdown of the L–S coupling because $\mathcal{H}_{\text{SO}}$ does not commute with either $\mathbf{L}^2$ or $\mathbf{S}^2$. However, for light and moderately heavy atoms (say, up to $Z \sim 50$) it is an acceptable approximation to consider $\mathcal{H}_{\text{SO}}$ as a weak perturbation compared to the splittings between the terms. Then we can neglect the non-diagonal terms of $\mathcal{H}_{\text{SO}}$, and keep only the diagonal ones. It follows from the Wigner–Eckart theorem that, *acting within a given (L, S) subspace*, $\mathcal{H}_{\text{SO}}$ can be replaced by a much simpler operator:

$$\mathcal{H}_{\text{SO}} = \sum_i \lambda_i \, \mathbf{L}_i \cdot \mathbf{S}_i \rightsquigarrow \lambda(L, S)\mathbf{L} \cdot \mathbf{S} \, . \tag{2.68}$$

Furthermore, it can be shown that $\lambda > 0$ for a less-than-half-filled shell, while $\lambda < 0$ for a more-than-half-filled shell.

[12] Or, more fundamentally, in (2.27), where ϕ is now the self-consistent potential.

Remark. It may sound perplexing that λ can be either positive or negative, while the individual λ_is are all positive. We may try to convince ourselves by the following simple argument which works for $S = S_{\max}$ (later we will see that this choice corresponds to the ground state). For less-than-half-filling, we can choose all spins parallel, and then $\mathbf{S}_i = \mathbf{S}/N$ where N is the number of electrons. Let $L^z = L$, then

$$\langle LSLS|\sum_i \lambda_i\, \mathbf{L}_i \cdot \mathbf{S}_i|LSLS\rangle = \frac{\lambda_0}{N}\langle LSLS|\mathbf{S}\sum_i \mathbf{L}_i|LSLS\rangle = \lambda(L,S)\cdot LS$$

where $\lambda_0 > 0$ is the common value of λ_i. It follows that

$$\lambda(L,S) = \frac{\lambda_0}{N} = \frac{\lambda_0}{2S} > 0\,. \tag{2.69}$$

On the other hand, for a completely filled shell, the spin–orbit interaction energy must be zero. Since for more-than-half-filling, the spins of the missing electrons (holes) can be taken fully aligned, repeating the previous argument for

$$\langle LSLS|\sum_i^{\text{electrons}} \lambda_i\, \mathbf{L}_i \cdot \mathbf{S}_i|LSLS\rangle = -\langle LSLS|\sum_i^{\text{holes}} \lambda_i\, \mathbf{L}_i \cdot \mathbf{S}_i|LSLS\rangle$$

we find

$$\lambda(L,S) = -\frac{\lambda_0}{2S} < 0 \tag{2.70}$$

what we wanted to show.

Once (2.68) is accepted, the spin–orbit interaction can be rewritten in a form analogous to the single-electron result (2.29)

$$\mathcal{H}_{\text{SO}} = \frac{\lambda}{2}(\mathbf{J}^2 - L(L+1) - S(S+1))\,. \tag{2.71}$$

The (L,S)-term undergoes a multiplet splitting, with the individual levels belonging to definite values of J, satisfying $|L-S| \le J \le L+S$. The ground state belongs to $J = |L-S|$ for less-than-half-filled shells, and to $J = L+S$ for more-than-half-filled shells. The energy splittings satisfy the Landé interval rule

$$E(L,S,J) - E(L,S,J-1) = \lambda J\,. \tag{2.72}$$

As we see from (2.28), the strength of the spin–orbit coupling is a quickly increasing function of the atomic number Z, thus we expect that the L–S coupling scheme breaks down for heavy elements. This is indeed the case: for large Z, J and J^z are the only relevant quantum numbers, and the mixing of states with different (L,S) values becomes noticeable. However, one really never gets to the extreme case of the so-called J–J coupling where the spin–orbit interaction would be much stronger than the energy differences between the terms. Heavy atoms can rather be characterized as intermediate-coupling cases. This manifests itself in, say, 10%-like deviations from the Landé interval rule. For our purposes, it will be good enough to use the results of the L–S coupling even for heavy atoms, and point out numerical inaccuracies when the need arises.

We should have a rough idea of the numerical values. The λ_0 appearing in (2.69) or in (2.70) is: $\sim 3.7 \cdot 10^{-3}$eV for C ($Z = 6$); $\sim 6 \cdot 10^{-2}$eV for Co ($Z = 27$); ~ 0.125eV for Rh ($Z = 45$); and ~ 1.25eV for Bi ($Z = 83$). An extensive list can be found in [465]. (Remember that the λ figuring in (2.71) may be only 1/5th of the corresponding λ_0).

Problem 2.3. Determine the possible (L,S) value pairs a) for a p^2 configuration, and b) for a d^2 configuration!

2.3.1 Coupling to the Magnetic Field

We repeat the argument which led from (2.7) to (2.13) in the single-electron case. The classical Hamiltonian is now

$$\mathcal{H}_{\mathrm{cl}}(\mathbf{r}_1,\ldots,\mathbf{r}_N;\mathbf{p}_1,\ldots,\mathbf{p}_N) = \frac{1}{2M}\sum_{j=1}^{N}\left(\mathbf{p}_j - \frac{Q}{c}\mathbf{A}(\mathbf{r}_j)\right)^2 + \mathcal{V}(\mathbf{r}_1,\ldots,\mathbf{r}_N) \tag{2.73}$$

where $\mathcal{V}$ is the sum of the potential and electron–electron interaction energies. (2.73) is quantized following the rules $[\hat{p}_m, r_n] = (\hbar/i)\delta_{mn}$. Assuming a constant field and using the gauge (2.10), we find that, in first order, the magnetic field couples to

$$\sum_i (\mathbf{l}_i + 2\mathbf{s}_i) = \mathbf{L} + 2\mathbf{S}$$

where $\mathbf{L}$ is the total orbital angular momentum, and $\mathbf{S}$ is the total spin. The full expression of $\mathcal{H}_{\mathrm{field}}$ is given by the sum of the Zeeman term and the diamagnetic term

$$\mathcal{H}_{\text{field}} = \mathcal{H}_{\text{Z}} + \mathcal{H}_{\text{dia}} = \mu_{\text{B}}(\mathbf{L} + 2\mathbf{S}) \cdot \mathbf{H} + \frac{e^2 H^2}{8mc^2} \sum_i (x_i^2 + y_i^2) \,. \tag{2.74}$$

First let us consider the weak-field case when $H\mu_{\text{B}} \ll \lambda$. In the Russell–Saunders coupling scheme we can confine our interest to a given (L, S, J) subspace. Assuming $J \neq 0$, there is a contribution $\sim H$, and then we can neglect the diamagnetic term. The difficulty in treating the paramagnetic term arises from the fact that the conserved vector in the absence of the field is $\mathbf{J} = \mathbf{L} + \mathbf{S}$ but the field couples to $\mathbf{L} + 2\mathbf{S}$. We invoke the Wigner–Eckart theorem [399] which tells us that the matrix elements of a vector operator can be calculated as if the vector were parallel to the conserved vector $\mathbf{J}$, but its length has to be replaced with its component along $\mathbf{J}$. Applying this to $\mathbf{L} + 2\mathbf{S}$, we find that the first-order Zeeman effect is described by

$$\langle JJ^z LS | H\mu_{\text{B}}(\mathbf{L} + 2\mathbf{S}) | JJ^z LS \rangle = gH\mu_{\text{B}} J^z \tag{2.75}$$

where the Landé factor g expresses to which extent is $\mathbf{L} + 2\mathbf{S}$ parallel to $\mathbf{J}$:

$$g = \frac{(\mathbf{L} + 2\mathbf{S})\mathbf{J}}{\mathbf{J}^2} = \frac{(\mathbf{L} + 2\mathbf{S})(\mathbf{L} + \mathbf{S})}{\mathbf{J}^2} = \frac{3}{2} + \frac{S(S+1) - L(L+1)}{2J(J+1)} \,. \tag{2.76}$$

(We used $2\mathbf{LS} = (\mathbf{L}+\mathbf{S})^2 - \mathbf{L}^2 - \mathbf{S}^2 = \mathbf{J}^2 - L(L+1) - S(S+1)$.) We can envisage the situation in this way: since the only conserved vector is $\mathbf{J}$, the magnetic field causes $\mathbf{L} + 2\mathbf{S}$ to precess about the fixed $\mathbf{J}$. Averaged over a precession period, the remaining contribution is the projection of $\mathbf{L} + 2\mathbf{S}$ onto the direction of $\mathbf{J}$. – Notice that (2.76) is analogous to the single-electron result (2.38).

If $H\mu_{\text{B}}$ is not negligibly small compared to the fine structure splittings, it becomes important that $\mathbf{L} + 2\mathbf{S}$ has also non-diagonal matrix elements $\langle J' J^z LS | (\mathbf{L} + 2\mathbf{S}) | JJ^z LS \rangle$, where the dipole selection rules require $J' = J \pm 1$. In the limit of very strong fields, one should see a Paschen–Back effect analogous to what we found for the H atom. For moderately strong fields, second-order perturbation theory is sufficient to describe the level displacements

$$E(J, J^z, L, S) = E^{(0)}(J) + gH\mu_{\text{B}} J^z + \frac{|\langle JJ^z LS | \mathcal{H}_{\text{Z}} | J+1, J^z LS \rangle|^2}{E^{(0)}(J) - E^{(0)}(J+1)}$$

$$+\frac{|\langle JJ^zLS|\mathcal{H}_Z|J-1,J^zLS\rangle|^2}{E^{(0)}(J)-E^{(0)}(J-1)}\,. \tag{2.77}$$

For the sake of future applications, we quote the matrix element [465]

$$\frac{1}{\mu_B H}\langle JJ^zLS|\mathcal{H}_Z|J+1,J^zLS\rangle = f_1(J,L,S)f_2(J,J^z) \tag{2.78}$$

where

$$f_1(J,L,S)=\sqrt{\frac{[(J+1)^2-(S-L)^2](J+2+S+L)(-J+S+L)}{4(J+1)^2(2J+1)(2J+3)}} \tag{2.79}$$

and

$$f_2(J,J^z)=\sqrt{(J+1-J^z)(J+1+J^z)}\,. \tag{2.80}$$

The quadratic terms in (2.77) describe the *second-order Zeeman effect.*

Digression: The Bohr–Van Leeuwen Theorem

Let us return to the classical Hamiltonian (2.73). The corresponding partition function is

$$\mathcal{Z}_{\rm cl}=\int d\mathbf{p}_1\ldots d\mathbf{p}_N\int d\mathbf{r}_1\ldots d\mathbf{r}_N e^{-\beta\mathcal{H}_{\rm cl}}\,. \tag{2.81}$$

It is clear that for each of the $\mathbf{p}$-integrations, the replacement $\mathbf{p}\to\mathbf{p}-(Q/c)\mathbf{A}$ can be made and then the magnetic field no longer appears in the expression of the classical free energy

$$\mathcal{F}_{\rm cl}=-k_BT\ln\mathcal{Z}_{\rm cl}\,.$$

It follows that the magnetization

$$M_{\rm cl}=-\frac{\partial\mathcal{F}_{\rm cl}}{\partial H}\equiv 0 \tag{2.82}$$

vanishes identically. Classical physics cannot explain either the spontaneous magnetization of materials like iron, or the field-induced magnetization of non-magnetic substances. This remarkable result is known as the Bohr–Van Leeuwen theorem[13].

[13] Apparently, it was first derived by N. Bohr in a Thesis work (Copenhagen, 1911). J.H. Van Leeuwen rediscovered and thoroughly discussed it in her Thesis work (Leiden, 1919). The theorem became widely known only when Van Vleck re-examined and carefully explained it in his book ([423] was first published in 1932).

The theorem may strike us as implausible because it is long known[14] that a collection of classical dipoles $\vec{\mu}$ (with the definite magnitude $\mu = |\vec{\mu}|$) possesses the paramagnetic susceptibility (2.91). The point is, however, that the existence of such permanent magnetic dipoles cannot be justified within classical physics. According to classical statistics, the definition of $\mathcal{Z}_{\text{cl}}$ implies integration over $-\infty < \mu_x < \infty$, etc. If we proceed in this manner, the Bohr–Van Leeuwen theorem follows.

It is interesting to realize that essentially everything that we find in our studies of magnetism, is a pure quantum effect. We may be wondering where is the point where $\hbar \neq 0$ makes itself felt; after all, the classical and quantum Hamiltonians look exactly the same! It can be shown (though it takes a lengthy argument [465]) that the appearance of a finite equilibrium value of M can be traced back to the fact that $\hat{\mathbf{p}}$ and $\mathbf{A}$ do not commute. Another essential ingredient is the electron spin which is a purely quantum phenomenon.

2.3.2 Hund's Rules

Let us assume that we want to find out about the magnetic moment of a Mn^{3+} ion which has the configuration $3d^4$. What is the value of S, L, and J in this case? And what about Ho^{3+} with its configuration $4f^{10}$? We certainly would not like to do a ten-electron calculation to get the answer! And we do not need to, either, because the answer to such questions is given by *Hund's rules.* They assert that, within a given configuration of one-electron orbitals, the ground state of an isolated atom or ion

- Has the largest value of the total spin S (Hund's first rule)
- Has the largest value of the total orbital angular momentum L which is permitted by the first rule (Hund's second rule)
- The total angular momentum is $J = |L - S|$ for less-than-half-filled shells, and $J = L + S$ for more-than-half-filled shells (Hund's third rule)

For Mn^{3+} we would predict $S = 2$, $L = 2$, and $J = 0$. However, we are seldom interested in the properties of an isolated ion swimming all

[14] The derivation was given by P. Langevin in 1905.

alone in free space. What about Mn^{3+} in, say, $LaMnO_3$? We are going to see that the prediction $S = 2$ is absolutely relevant, $L = 2$ might be true but it is irrelevant, and Hund's third rule no longer applies. In contrast, the predictions $S = 2$, $L = 6$, and $J = 8$ remain valid also for Ho^{3+} in, say, HoF_3. The reason for these complications is that an ion embedded in a solid is subject to the *crystal field* of the surrounding atoms (see Ch. 3). The outcome is decided by the competition between crystal fields, and the interactions which are responsible for Hund's rules.

This brings us to the question: what is the reasoning which leads to Hund's rules? For the third rule, the answer is simple: it is a relativistic effect due to the spin–orbit interaction. Within the context set by the first and second rule, we have (2.71) to diagonalize, and the third rule immediately follows. However, the situation is much more complicated with the first two rules.

As far as the first rule is concerned, we can be largely satisfied with the insight gained in Sec. 2.2: for orthogonal orbitals, the exchange is ferromagnetic. It might seem that this is all that we need. All the $3d$ (or $4f$, etc.) electrons have the same kinetic energy, and the same potential energy in the central field of the ion, thus there is only the energy of their mutual repulsion to minimize. Parallel-spin electrons avoid each other most effectively because of the exchange hole effect; Hund's first rule follows. Is there any reason to doubt the validity of this argument for real (but isolated) atoms or ions? We can, in fact, think of such reasons. We have to remember that, in general, it is merely an approximation to assign to the ground state a fixed configuration in terms of a pre-determined set of one-electron orbitals. The electron–electron interaction can give rise to configuration mixing; symmetry requires only that J (and parity) should be preserved. And for a given state, orbitals may relax so as to minimize the total energy. These effects can put even the first rule in jeopardy. Things get much worse with the second rule for which we would not have even a plausibility argument to begin with. We cannot go into the details here (see, however, [229] for a thorough up-to-date discussion).

We have to appreciate that though Hund's first and second rule are not theorems, they are almost always true. It has been argued that the ground state of the neutral Ce atom violates the first rule [287] but the author is not aware of any other example. As far as Ce is concerned, we might pinpoint the anomalous feature associated with the violation: the configuration[15] is $(4f)^2(5d)^1(6s)^1$, i.e., there are *two* incomplete shells with $l \neq 0$. Apparently, Hund [176] devised his rules for systems with only one incomplete $l \neq 0$ shell.

The order of the three rules reflects also a hierarchy of energy scales. The first rule is guaranteed by the exchange matrix elements (2.48), the largest of which are easily of the order $\sim$ 1eV. The second rule is coming from similar matrix elements which are, however, numerically smaller (say, $\sim 0.1 - 0.3$eV). The third rule arises from the significantly weaker spin–orbit coupling. Thus, when a further energy scale (e.g., crystal field) begins to compete with Hund's rule energies, the third rule is invalidated first, the second may be put in doubt, while Hund's first rule is a tough survivor.

It is tempting to generalize Hund's first rule to molecules. After all, there are many molecules wich (in the molecular orbital scheme) have an incomplete degenerate shell and then the exchange hole effect prefers a high-spin ground state. We may be heartened to find that proceeding in this spirit, we predict the O_2 molecule to be magnetic as it actually is (Problem 2.4). However, relaxation phenomena play a much bigger role for molecules than for atoms, and we often find low-spin ground states. For instance, C_{60}^{2-} has a singlet ground state, not a triplet.

Problem 2.4. Work out the molecular orbital scheme for the oxygen molecule. Assume that the exchange hole effect lifts the degeneracy in an open-shell situation. What is the total spin in the ground state?

2.4 Paramagnetism and Diamagnetism

Let us consider N independent atoms (or ions, or molecules) in a volume V. We wish to calculate the temperature dependence of the magnetic susceptibility $\chi(T)$, defined as

$$\chi(T) = -\frac{1}{V}\frac{\partial^2 \mathcal{F}}{\partial H^2} \tag{2.83}$$

[15] This must be a delicate point since several standard texts [32, 209] give the configuration as $(4f)^2(6s)^2$.

where $\mathcal{F}$ is the free energy.

Though this may sound as if the problem were posed for gases or perhaps liquids, we mean to apply the results also for solids. It will turn out that, at sufficiently high temperatures (which, for many rare earth compunds, may well mean T merely in excess of $\sim 10 - 100\text{K}$), it is a good approximation to treat the moment-carrying ions as independent. Besides, the free-ion formulas are important as a starting point for the mean field treatment of ion–ion interactions.

Our characterization of the ionic spectra relied on a hierarchy of energy scales. We usually assume

$$\underbrace{\text{Coulomb energies}}_{\text{Hund's 1 and 2}} \quad > \quad \underbrace{\text{spin–orbit } \lambda}_{\text{Hund's 3}} \quad > \quad \text{Zeeman splittings.} \tag{2.84}$$

However, we have seen that the separation of the scales is not always so neat. For large Z, λ is in the range of the smaller Coulomb matrix elements which leads to the breakdown of the L–S coupling. And for $Z = 1$ (the H atom), it is easy to reach the Paschen–Back limit where the Zeeman energy is rather larger than λ.

Now we have to find the place of the thermal energy scale $k_{\text{B}}T$. There are a few important cases where $k_{\text{B}}T$ is of the order of multiplet splittings, so that states with different J are thermally excited; these will be mentioned later. Now we assume $k_{\text{B}}T \ll \lambda$ so that as far as thermal averaging is concerned, we can confine our interest to the ground state values of L, S, and J (observe, however, that we do not neglect the field-induced mixing of states with differing J; this is evident in the quadratic terms of (2.77)). Since a field of $\sim 30\text{T}$ (Tesla) can be taken to correspond to $T = 30\mu_{\text{B}}\text{T}/k_{\text{B}} \sim 20\text{K}$ which is not a particularly low temperature, it is meaningful to assume that H/T can be arbitrary. The susceptibility (2.83) is defined in the weak-field limit. Plotting the H-dependence of the magnetization, we get the *magnetization curve.*

2.4.1 Paramagnetic Susceptibility

In the limit of weak fields, it is sufficient to keep terms up to order H^2. The energy expression (2.77) can be written as

$$E(J, J^z, L, S) = E_0 + aJ^z H + [b_0 - b_2(J^z)^2]H^2 \tag{2.85}$$

where S, L, and J have their ground state values. The free energy is

$$\mathcal{F} = -k_\mathrm{B}T \ln \mathcal{Z} = -k_\mathrm{B}TN \ln \mathcal{Z}_1 \tag{2.86}$$

where $\mathcal{Z}_1$ is the single-ion partition function. Expanding to order H^2 yields

$$\mathcal{Z}_1 = \sum_{J^z=-J}^{J} \exp\left\{-\frac{E(J, J^z, L, S)}{k_\mathrm{B}T}\right\} \tag{2.87}$$

$$\approx (2J+1)e^{-\frac{E_0}{k_\mathrm{B}T}} \left\{1 - \frac{b_0 H^2}{k_\mathrm{B}T} + \frac{J(J+1)H^2}{3}\left(\frac{a^2}{2(k_\mathrm{B}T)^2} + \frac{b_2}{k_\mathrm{B}T}\right)\right\}.$$

Using the definition (2.83), the susceptibility can be obtained from the free energy

$$\frac{1}{N}\mathcal{F} \approx E_0 - k_\mathrm{B}T \ln(2J+1) - \left(\frac{a^2 J(J+1)}{6k_\mathrm{B}T} - \frac{b_2 J(J+1)}{3} + b_0\right) H^2 \tag{2.88}$$

by replacing $a = g\mu_\mathrm{B}$, and b_0 and b_2 according to (2.78)–(2.80), as

$$\chi = \chi_\mathrm{C} + \chi_\mathrm{VV}\,. \tag{2.89}$$

The first term is the strongly temperature-dependent *Curie susceptibility*

$$\chi_\mathrm{C} = \frac{N}{V}\frac{g^2\mu_\mathrm{B}^2 J(J+1)}{3k_\mathrm{B}T} = \frac{C}{T} \tag{2.90}$$

where C is the Curie constant. Remembering that the susceptibility of a collection of classical dipoles $\vec{\mu}$ would be [209]

$$\chi = \frac{N}{V}\frac{\mu^2}{3k_\mathrm{B}T} \tag{2.91}$$

it is usual to rewrite χ_C as

$$\chi_\mathrm{C} = \frac{N}{V}\frac{p^2\mu_\mathrm{B}^2}{3k_\mathrm{B}T} \tag{2.92}$$

where

$$p = g\sqrt{J(J+1)} \tag{2.93}$$

is the *effective number of Bohr magnetons.*

Recalling our discussion of the Bohr–van Leeuwen theorem, it is clear that the Langevin formula (2.91) does not really belong to classical physics. A consequent application of classical statistics would exclude the possibility of postulating the existence of classical dipoles with a given magnitude $|\vec{\mu}|$. In retrospect, one might say that the success of Langevin's theory (1905) is due to the fact that it implicitly used quantization.

The second term in (2.89) is the T-independent *Van Vleck susceptibility.* It results from the field-induced admixture of the higher-lying levels of the multiplet. Its full expression can be derived from (2.77)–(2.80) (for an application, see Problem 2.6). Here we merely note that it has the form

$$\chi_{\mathrm{VV}} \sim \frac{N}{V} \frac{(g\mu_{\mathrm{B}})^2}{\lambda} \tag{2.94}$$

since the spin–orbit coupling constant λ gives the order of magnitude of the energy denominators in (2.77).

We may ask whether (2.89) is applicable for the high-temperature (paramagnetic) phase of solids. Plotting the temperature dependence of the susceptibility, we often find that it can be fitted with the sum of a constant and a *Curie–Weiss term*

$$\chi(T) \sim \chi_0 + \frac{C'}{T+\theta} \, . \tag{2.95}$$

Simple mean field theory predicts $\theta = -T_{\mathrm{C}}$ for ferromagnets and $\theta = T_{\mathrm{N}}$ for antiferromagnets [32, 209]. However, C' is not necessarily close to the C defined in (2.90), and for many antiferromagnets, θ is rather different from T_{N}. A realistic description of crystal field effects and intersite interactions is necessary for resolving the discrepancies. Let us, however, note that there are examples of the Curie law (2.90) being followed with extraordinary accuracy down to quite low temperatures, for instance for $Gd(C_2H_5SO_4)_3{\cdot}9H_2O$ [209]. The reason is twofold: first, the large non-magnetic molecules keep the Gd ions widely separated which greatly reduces the ion–ion interaction. Second, according to Hund's rules, the $Gd^{3+} \rightsquigarrow 4f^7$ ion is a spin-only ion with $S = 7/2$ and $L = 0$, thus spin–orbit corrections are very weak[16].

[16]In view of the purported $L = 0$, we may well ask why do we have spin–orbit

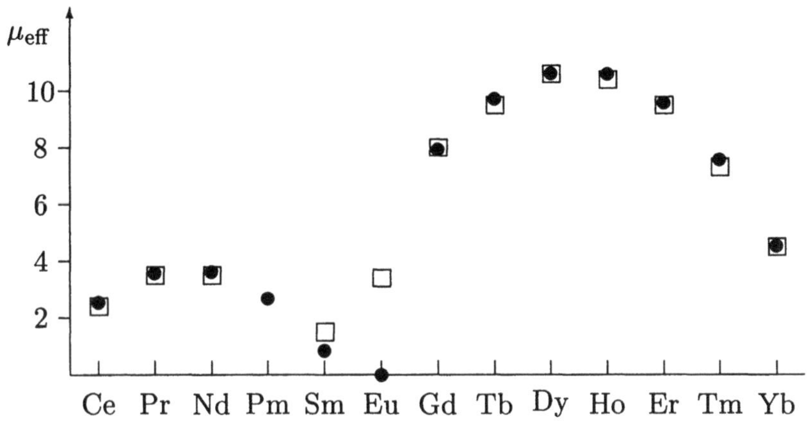

Figure 2.1: Effective magnetic moments (in units of μ_B) of the trivalent rare earth ions. Full circles: theoretical values for Hund's rule ground states, squares: measured moments at room temperature. (Based on data taken from [209]).

Fig. 2.1 shows the effective Bohr magneton numbers p for trivalent rare earth ions. We see that the theoretical predictions obtained from inserting Hund's rules values for S, L, and J into (2.76) and (2.93), are in excellent overall agreement with the experimental data deduced from room temperature measurements of the susceptibility. We will learn later that the compactness of the $4f$ shell explains why the $4f$ electrons are well shielded from both the crystal fields and ion–ion interactions, and thus rare earth ions tend to show essentially free-ion behaviour at moderately high temperatures.

Even the two cases where substantial disagreement is found, namely Sm and Eu, can be accounted for within the independent-ion picture. We have to consider the effect of thermal excitation across fine-structure splittings. The discrepancy is most noticeable for the configuration $4f^6$ which has a $J = 0$ ground state, thus its Curie susceptibility vanishes. However, the $J = 1$ excited state of Eu^{3+} is a mere $\Delta \sim 500$K away, and it makes a large contribution to the room-temperature susceptibility

corrections at all. The reason is configuration interaction: only $J = 7/2$ is a good quantum number but $L = 0$ is not strictly true because and ${}^6P_{7/2}$, etc. get admixed to the Hund's rule ground state ${}^8S_{7/2}$. This explains why one often finds $g \approx 1.99$ in experiments, though the free-spin value would be 2.0023 [2].

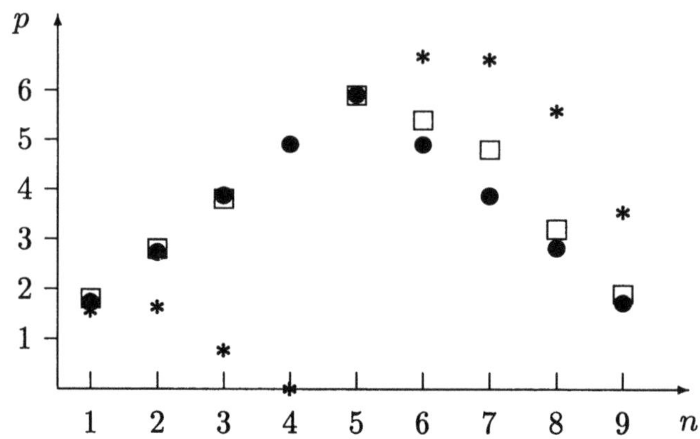

Figure 2.2: Effective Bohr magneton numbers for the $3d^n$ configurations. Stars: theoretical values with (2.93), full circles: spin-only values, empty boxes: representative experimental values taken from [209]. The overall agreement between spin-only values and experiments indicates that the orbital moment is quenched.

[237]. The susceptibility of the isoelectronic ion Sm^{2+} is discussed in Problem 2.6. – We should not be surprised that it is just for these ions that the thermal excitation of states with different Js becomes important. Combining the Landé interval rule (2.72) with (2.69), we estimate $\Delta \sim (J+1)\lambda_0/2S$, thus Δ is smallest in the neighbourhood of the filling $4f^6$ where J is small and S is large.

In sharp contrast to its success with rare earth systems, the free-ion approach fails quite badly for the $3d$ transition metal ions (Fig. 2.2). The prediction (2.93) based on applying all three of Hund's rules, is in striking disagreement with representative experimental values[17]. Instead, the latter are often in good agreement with the *spin-only value* of the effective Bohr magneton number

$$p_S = 2\sqrt{S(S+1)} \tag{2.96}$$

where S is predicted by Hund's first rule. We will learn in Ch. 3 the reason why the crystalline environment tends to *quench* the orbital moment of d-electrons.

[17] Since the valence of transition metals is less stable than that of rare earths, each $3d^n$ stands for a number of ions. E.g., $3d^4$ can be either Mn^{3+} or Cr^{2+}.

Magnetization Curve

In small fields, the magnetization is $M = \chi H$ but this linear H-dependence cannot go on for ever. Eventually, all moments are lined up along the field, and the magnetization saturates. The magnetization curve $M(H)$ shows how M passes from the linear to the saturated regime.

We make standard simplifying assumptions to calculate the simplest example of a full magnetization curve. We assume that either λ is very large, or $L = 0$, so that the quadratic terms in (2.77) can be neglected. However, we allow arbitrarily large values of the ratio $x = g\mu_B HJ/k_B T$.

The free energy (per ion) is now

$$\begin{aligned} \frac{1}{N}\mathcal{F} &= E_0 - k_B T \ln \left\{ \sum_{J^z=-J}^{J} e^{-xJ^z/J} \right\} \\ &= E_0 - k_B T \ln \frac{\sinh[x(2J+1)/2J]}{\sinh[x/2J]} . \end{aligned} \tag{2.97}$$

The magnetization is then

$$M = -\frac{\partial \mathcal{F}}{\partial H} = N g \mu_B J B_J(x) \tag{2.98}$$

where

$$B_J(x) = \frac{2J+1}{2J} \coth \frac{2J+1}{2J} x - \frac{1}{2J} \coth \frac{x}{2J} \tag{2.99}$$

is the Brillouin function. The law (2.98) is well-obeyed in substances with widely separated magnetic ions which have spin-only moments; see [209] for illustrative examples.

Expanding $B_J(x)$ in a power series, we can identify the higher-order susceptibilities (in this context we rename the ordinary χ as χ_1)

$$M = \chi_1 H + \frac{1}{3!}\chi_3 H^3 + \ldots = \sum_{n=1} \frac{1}{(2n-1)!}\chi_{2n-1} H^{2n-1} . \tag{2.100}$$

χ_1 is given in (2.92). The third-order free-ion susceptibility

$$\chi_3 = -\frac{N}{V} \frac{(g\mu_B)^4}{15(k_B T)^3} J(J+1)(2J^2+2J+1) = -\frac{C^{(3)}}{T^3} \tag{2.101}$$

is negative, corresponding to the fact that the Brillouin function approaches saturation in a smooth manner. However, the experimentally measured χ_3 can be either positive or negative, or it may change sign at some temperature. This is the consequence of a combination of crystal field and intersite interaction effects [288]. It has been argued that the behaviour of χ_3 may provide valuable clues in certain cases where the nature of the ordering transition is debated [331]. We begin to familiarize ourselves with this subject in Problem 2.8.

Problem 2.5. In many interesting compounds, Ce ions are trivalent. At sufficiently high temperatures[18], the dominant contribution to the susceptibility of the solid is the Curie susceptibility of independent Ce ions. What is the effective number of Bohr magnetons?

Problem 2.6. SmB_6 is a *mixed valent* material, containing about 60% trivalent, and 40% divalent samarium. Let us assume that only the Sm $4f$-electrons contribute to the magnetic susceptibility, and that the Sm-ions can be treated as independent (it is a good approximation from $T \sim 100$K upwards).

a) Evaluate the expression (2.89) for the paramagnetic susceptibility! The gap $\Delta = E^{(0)}(J+1) - E^{(0)}(J)$ separating the lowest-lying excited level from the ground state in the multiplet structure is $\Delta \approx 410$K for the Sm^{2+} ion, and $\Delta \approx 1450$K for the Sm^{3+} ion. What is the relative importance of the Curie and Van Vleck terms at room temperature?

b) Because at least one of the Δs is anomalously small, it is no longer a good approximation to neglect thermal excitation to the next level in the multiplet structure. Write down the corresponding correction to the paramagnetic susceptibility.

Problem 2.7. Many uranium compounds are of great current interest as heavy fermion or mixed valent systems. The first thing you want to know is the valence state of uranium and, surprisingly, it is quite difficult to decide whether it is U^{3+} or U^{4+}. You might think that measuring the high-temperature susceptibility which shows a Curie-type behaviour, should clinch the point. Let us assume that a measurement tells us that the effective moment is $3.6\mu_B$, with 1% accuracy. Can you decide whether the valence state is U^{3+} or U^{4+}?

Problem 2.8. Consider a lattice of $S = 1$ ions which have both bilinear and biquadratic nearest-neighbour interactions ($J > 0$, $J_Q > 0$)

$$\mathcal{H} = -J \sum_{i,j} S_i^z S_j^z - J_Q \sum_{i,j} \left(3(S_i^z)^2 - 2\right) \left(3(S_j^z)^2 - 2\right) \qquad (2.102)$$

(the form of the biquadratic term is chosen to correspond to the quadrupole–

[18]For certain heavy fermion compounds, $T > 10$K may be reckoned as a high temperature. See Ch. 11.

quadrupole interaction discussed in Sec. 3.10.1). Derive the mean field expression of χ_3 for the disordered phase.

2.4.2 Diamagnetism

For a closed-shell atom or ion with $S = L = J = 0$ (e.g., for a Ne atom or for the ion Na^+), the coupling to the magnetic field is described by the diamagnetic term in (2.74):

$$\mathcal{H}_{\text{dia}} = \frac{e^2 H^2}{8mc^2} \sum_i (x_i^2 + y_i^2) \,. \tag{2.103}$$

Because of the spherical symmetry of a closed shell, it is easy to accept that

$$\langle \sum_i (x_i^2 + y_i^2) \rangle = \frac{2}{3} \langle \sum_i (x_i^2 + y_i^2 + z_i^2) \rangle = \frac{2}{3} \langle \sum_i r_i^2 \rangle \,. \tag{2.104}$$

Since it takes a large energy to excite an electron out of a closed shell, it is usually an excellent approximation to replace $T = 0$ in the definition (2.83). The resulting diamagnetic susceptibility is

$$\chi_{\text{dia}} = -\frac{N}{V} \frac{e^2}{3mc^2} \sum_{\text{fs}} (2l_{\text{fs}} + 1) \langle n_{\text{fs}} l_{\text{fs}} | r^2 | n_{\text{fs}} l_{\text{fs}} \rangle \tag{2.105}$$

where the summation is over filled shells. ($\langle r^2 \rangle$ depends only on the quantum numbers n and l).

Even if there are electrons outside the closed shells, the *core diamagnetism* given in (2.105) gives a correction term which should not be forgotten if we want to interpret high-precision measurements of the magnetic susceptibility.

To get a feeling for the order of magnitude of χ_{dia}: for Na, the core contribution is $\sim -0.18 \cdot 10^{-6}$c.g.s., which constitutes a sizeable contribution to the total $\chi \sim 0.48 \cdot 10^{-6}$c.g.s. of metallic sodium [465]. For He, $\chi_{\text{dia}} \sim -1.8 \cdot 10^{-6}$c.g.s., while for Xe $\chi_{\text{dia}} \sim -43 \cdot 10^{-6}$c.g.s., reflecting the increase in the size, and number, of closed shells.

The obvious increase of $|\chi_{\text{dia}}|$ with $\langle r_\perp^2 \rangle$ ($r_\perp$ being the radius in the plane perpendicular to the field) suggests the idea: why do not we take larger objects[19] to get stronger diamagnetism? To some extent, the idea works. For instance, the planar benzene molecule is known to have a large diamagnetic susceptibility if the field is applied perpendicular to its plane. Going further, we may be wondering whether small metallic particles (with radii $\sim 1-10$nm) are strongly diamagnetic. This possibility was suggested [177] long before the investigation of such mesoscopic (or nanoscale) systems became commonplace. As it turns out, there is an effect of the sort we envisage but the overall magnetic behaviour is quite complicated [146].

Digression: Superstrong Fields

In all of the previous discussion we assumed that $\mathcal{H}_{\text{dia}}$ is a small perturbation when compared to the zero-field term $\mathcal{H}_0$. Since laboratory fields seldom exceed 30 Tesla, using (2.16) and (2.17), we can convince ourselves that $\langle\mathcal{H}_{\text{dia}}\rangle \sim 10^{-7} - 10^{-9}$Ry which is indeed minute. However, the magnetic field of neutron stars can reach 10^8T. For an atom on the surface of a pulsar, $\mathcal{H}_{\text{dia}}$ is the large term and $\mathcal{H}_0$ the small perturbation. The full Hamiltonian has merely axial symmetry and thus though m is still a good quantum number, l is not. The hydrogenic wave functions are strongly mixed and the ground state charge distribution is cigar-shaped rather than spherical. The binding of solids (if there are any) becomes very anisotropic, resulting in a bizarre world [341].

But we do not have to go as far as the pulsars: a similar situation may arise in solid state physics. Impurity states in semiconductors can be described by hydrogenic wave functions [209], only the Bohr radius changes to $a_0^* = \kappa\hbar^2/(m^*e^2)$ where m^* is the effective mass, and κ the dielectric constant. Correspondingly, the effective Rydberg becomes $E_0^* = e^4m^*/2\kappa^2\hbar^2$. For semiconductors like InSb, or $Hg_xCd_{1-x}Te$, $\kappa \sim 10$ and $m^*/m \sim 0.01$, thus $\langle\mathcal{H}_{\text{dia}}\rangle$ can become bigger than E_0^*.

Let us take a lightly doped semiconductor like InSb, and increase the magnetic field from small to very large values. What is it that we are supposed to observe? We use the Heisenberg uncertainty relations $\Delta x \cdot \Delta p_x \geq \hbar$, etc. to get a naive estimate for the field-dependence of the ground state. Since the ground state evolves continuously from the

---

[19] Assuming that there exist quantum states with orbits along the circumference of the system.

hydrogenic $1s$ state, we can assume $\langle L^z \rangle = 0$ when writing down the expectation value of $\mathcal{H}$ shown in (2.15)

$$E \sim \frac{\hbar^2}{2m^*}\left(\frac{2}{a_\perp^2}+\frac{1}{a_z^2}\right)-\frac{e^2}{\kappa r}+\frac{e^2H^2a_\perp^2}{4mc^2}\,. \tag{2.106}$$

We assumed that the radius of the ground state orbit is $a_\perp$ perpendicularly to the field, and a_z parallel to the field. We may assume $r = \min(a_\perp, a_z)$. Minimizing with respect to a_z, we find that whatever the ratio $a_\perp/a_z$, $a_z \sim \kappa\hbar^2/(m^*e^2) = a_0^*$, the effective Bohr radius. However, $a_\perp$ is strongly field-dependent. At $H = 0$, we get the Bohr radius. At large H, we may drop the Coulomb term, and arrive at

$$a_\perp \sim \sqrt{\frac{2\hbar c}{eH}} = \sqrt{2}l_{\mathrm{c}} \tag{2.107}$$

where l_{c} is the so-called cyclotron radius. In the plane perpendicular to the field, the orbit is shrinking with increasing H. Replacing $a_\perp$ back into (2.106), we find $E = \hbar eH/m^*c$. However, this is just a new energy zero for a free electron. If the electron's orbit is centered on the impurity charge, then it gains the Coulomb energy $e^2/\kappa a_\perp$. Thus, in the limit of large fields, the impurity ionization energy behaves like

$$\Delta E \sim \frac{e^2}{\kappa a_\perp} \sim \sqrt{H}\,. \tag{2.108}$$

The square root dependence is not unlike the predictions of sophisticated calculations [233]. In any case, ΔE is an increasing function of the field (we understand, though, that the small-H behaviour need not be accurately given by (2.108)).

Let us return to n-type InSb. In zero field, the donors are ionized at a temperature of the order of $\Delta E(H = 0)$, and the sample is reasonably well-conducting. If a large field is switched on, various measurements indicate that the conduction electrons disappear: this is the *magnetic freeze-out.* We can understand the phenomenon as the result of the field-dependence of the ionization energy: with increasing $\Delta E(H)$, the electrons become bound to the impurities. Large magnetic fields have a drastic effect in $Hg_xCd_{1-x}Te$ samples as well [340]. It is disputed whether this is due to magnetic freeze-out, or perhaps Wigner crystallization [121].

2.5 Hydrogen Molecule

Up to now, we have seen cases where the parallel alignment of spins is favoured energetically. The case of exchange in orthogonal orbitals is illustrated by electrons in the same (say, $3d$-) shell. All the electrons sit in the same potential well, so in order to minimize the total energy, they just have to avoid each other as much as possible. This is achieved by taking advantage of the exchange hole effect which works for parallel spin alignment. Hence Hund's first rule, and the precedence of the high-spin states.

However, in nature antiferromagnetism is more frequently seen than ferromagnetism so we must find similarly cogent reasons for electrons preferring the antiparallel alignment of spins. Such a reason is found if we consider a two-center problem, for instance, that of the hydrogen molecule. Each of the nuclei gives rise to a potential well. Midway between the protons, the overlapping of the attractive atomic potentials creates a region of low potential energy. The electrons prefer to accumulate in this central region. There is, however, only one orbital state which has a large amplitude in the interatomic space, so the electron spins must be antiparallel. Generally, low-spin states are expected whenever the gain in one-electron energy outweighs the energy cost of electron–electron repulsion.

Let us study the hydrogen molecule in some detail. The $1s$ state of an electron in an isolated hydrogen atom satisfies

$$\mathcal{H}_{\rm at}(\mathbf{r}-\mathbf{R})\varphi_{1s}(\mathbf{r}-\mathbf{R}) = \epsilon_{1s}\varphi_{1s}(\mathbf{r}-\mathbf{R}) \tag{2.109}$$

where $\mathbf{R}$ is the position of the proton.

Let us consider a system consisting of two hydrogen atoms, one centered at $\mathbf{R}_a$, and the other at $\mathbf{R}_b$. The Hamiltonian consists of several terms:

$$\begin{aligned}\mathcal{H} &= H_{\rm at}(\mathbf{r}_1-\mathbf{R}_a) + H_{\rm at}(\mathbf{r}_2-\mathbf{R}_b) && \text{isolated atoms}\\ &\quad -\frac{e^2}{|\mathbf{r}_1-\mathbf{R}_b|} - \frac{e^2}{|\mathbf{r}_2-\mathbf{R}_a|} && \text{the electrons feel the field of both protons}\\ &\quad +\frac{e^2}{|\mathbf{r}_1-\mathbf{r}_2|} && \text{electron–electron interaction}\end{aligned}$$

$$+\frac{e^2}{|\mathbf{R}_a - \mathbf{R}_b|} \qquad \text{proton–proton interaction} \tag{2.110}$$

Since exciting an electron out of the $1s$ shell requires a large energy, it looks safe to assume that the low-lying states of the molecule can be sought in the six-dimensional Hilbert space spanned by the atomic $1s$-orbitals. The basis states are graphically shown below. There are four neutral configurations

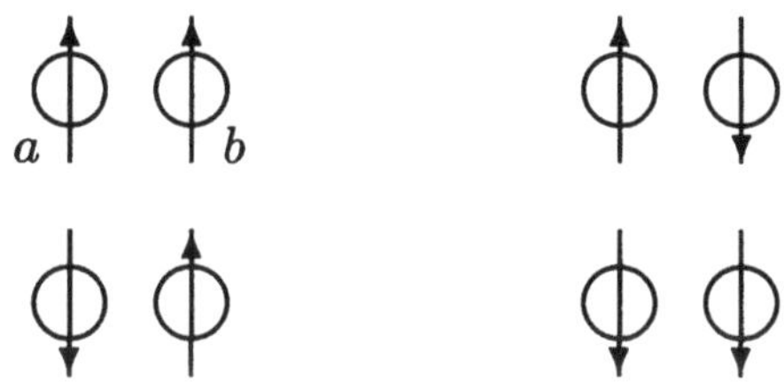

(2.111)

and two ionized configurations

(2.112)

In what follows, we use the simplified notations $\varphi_{1s}(\mathbf{r}-\mathbf{R}_a) = \varphi_a(\mathbf{r})$, and $\varphi_{1s}(\mathbf{r} - \mathbf{R}_b) = \varphi_b(\mathbf{r})$. An essential quantity is the overlap integral

$$\int \varphi_a^*(\mathbf{r})\varphi_b(\mathbf{r})d\mathbf{r} = l. \tag{2.113}$$

2.5.1 Direct Exchange in Non-Orthogonal Orbitals

We expect that for sufficiently large internuclear separations, the low-energy states belong to local neutrality, with one electron residing in each orbital. This is the so-called *Heitler–London scheme.* Our first task is to diagonalize the Hamiltonian (2.110) in the subspace of the four neutral configurations. Let us observe that $\mathcal{H}$ is manifestly isotropic in spin-space, thus the total spin S, and its z-component S^z are good quantum numbers. We are looking for linear combinations of the neutral configurations (2.111), which are S-, and S^z-eigenstates.

The first neutral configuration shown in (2.111) is described by the Slater determinant

$$\begin{aligned}
|\psi_1\rangle &= \frac{1}{\sqrt{2(1-l^2)}}\begin{vmatrix} \varphi_a(\mathbf{r}_1)\alpha(s_1) & \varphi_a(\mathbf{r}_2)\alpha(s_2) \\ \varphi_b(\mathbf{r}_1)\alpha(s_1) & \varphi_b(\mathbf{r}_2)\alpha(s_2) \end{vmatrix} \qquad (2.114) \\
&= \frac{1}{\sqrt{2(1-l^2)}}\alpha(s_1)\alpha(s_2)[\varphi_a(\mathbf{r}_1)\varphi_b(\mathbf{r}_2) - \varphi_a(\mathbf{r}_2)\varphi_b(\mathbf{r}_1)].
\end{aligned}$$

This is the $S^z = 1$ component of a triplet. The $S^z = 0$ and $S^z = -1$ components are

$$|\psi_2\rangle = \frac{1}{2\sqrt{1-l^2}}[\varphi_a(\mathbf{r}_1)\varphi_b(\mathbf{r}_2) - \varphi_a(\mathbf{r}_2)\varphi_b(\mathbf{r}_1)][\alpha(s_1)\beta(s_2) + \beta(s_1)\alpha(s_2)] \qquad (2.115)$$

and

$$|\psi_3\rangle = \frac{1}{\sqrt{2(1-l^2)}}\beta(s_1)\beta(s_2)[\varphi_a(\mathbf{r}_1)\varphi_b(\mathbf{r}_2) - \varphi_a(\mathbf{r}_2)\varphi_b(\mathbf{r}_1)]. \qquad (2.116)$$

The fourth neutral state is the singlet

$$|\psi_4\rangle = \frac{1}{2\sqrt{1+l^2}}[\varphi_a(\mathbf{r}_1)\varphi_b(\mathbf{r}_2) + \varphi_a(\mathbf{r}_2)\varphi_b(\mathbf{r}_1)][\alpha(s_1)\beta(s_2) - \beta(s_1)\alpha(s_2)]. \qquad (2.117)$$

It is obvious that within the restricted Hilbert space of neutral configurations $|\psi_1\rangle$, $|\psi_2\rangle$, $|\psi_3\rangle$, and $|\psi_4\rangle$ are eigenstates of the Hamiltonian because they carry different symmetry labels. The triplet energy is

$$\epsilon_t = \langle\psi_1|H|\psi_1\rangle = 2\epsilon_{\text{at}} + \frac{C_{ab} - I_{ab}}{1-l^2} \qquad (2.118)$$

where the Coulombic term contains, in addition to the familiar Coulomb integral, also contributions arising from an electron in one of the atomic orbitals feeling the attraction of the other nucleus

$$\begin{aligned}
C_{ab} &= \int d\mathbf{r}_1 \int d\mathbf{r}_2 |\varphi_a(\mathbf{r}_1)|^2 \frac{e^2}{|\mathbf{r}_1 - \mathbf{r}_2|}|\varphi_b(\mathbf{r}_2)|^2 \qquad (2.119) \\
&\quad - \int d\mathbf{r}_1 \frac{e^2}{|\mathbf{r}_1 - \mathbf{R}_b|}|\varphi_a(\mathbf{r}_1)|^2 - \int d\mathbf{r}_2 \frac{e^2}{|\mathbf{r}_2 - \mathbf{R}_a|}|\varphi_b(\mathbf{r}_2)|^2.
\end{aligned}$$

The exchange term is augmented by similar contributions

$$I_{ab} = \int d\mathbf{r}_1 \int d\mathbf{r}_2 \frac{e^2}{|\mathbf{r}_1 - \mathbf{r}_2|} \varphi_a^*(\mathbf{r}_1)\varphi_b(\mathbf{r}_1)\varphi_b^*(\mathbf{r}_2)\varphi_a(\mathbf{r}_2) \tag{2.120}$$

$$-l \int d\mathbf{r}_1 \frac{e^2}{|\mathbf{r}_1 - \mathbf{R}_b|} \varphi_a^*(\mathbf{r}_1)\varphi_b(\mathbf{r}_1) - l \int d\mathbf{r}_2 \frac{e^2}{|\mathbf{r}_2 - \mathbf{R}_a|} \varphi_b^*(\mathbf{r}_2)\varphi_a(\mathbf{r}_2).$$

The singlet eigenenergy is

$$\epsilon_{\mathrm{s}} = \langle\psi_4|H|\psi_4\rangle = 2\epsilon_{\mathrm{at}} + \frac{C_{ab} + I_{ab}}{1 + l^2} \tag{2.121}$$

and the singlet–triplet splitting is given by

$$\epsilon_{\mathrm{tr}} - \epsilon_{\mathrm{s}} = 2\frac{l^2 C_{ab} - I_{ab}}{1 - l^4}. \tag{2.122}$$

Using the same argument that we used to write down (2.57), we can assert that in the subspace of neutral configurations, the full interaction Hamiltonian can be replaced by the effective Hamiltonian

$$\mathcal{H}_{\mathrm{eff}} = \frac{\epsilon_{\mathrm{s}} + \epsilon_{\mathrm{tr}}}{2}\mathcal{E} + J_{12}\, \mathbf{S}_1 \cdot \mathbf{S}_2 \tag{2.123}$$

where $\mathcal{E}$ is the 4×4 unit matrix, and the exchange coupling is

$$J_{12} = 2\frac{l^2 C_{ab} - I_{ab}}{1 - l^4}. \tag{2.124}$$

The essential point is that, depending on the values of C_{ab}, I_{ab}, and l, J_{12} can be either positive or negative. At $l = 0$ $J_{12} = -I_{ab} \rightsquigarrow -J_{ab} < 0$, in agreement with (2.57). Increasing l, J_{12} changes sign and becomes positive for $l^2 C_{ab} > I_{ab}$. *For sufficiently large overlap, the exchange coupling is antiferromagnetic.* The ground state of the hydrogen molecule is a spin-singlet.

If we are not interested in the absolute position of the ground state level, we can drop the constant term from (2.123), arriving at the standard form of the Heisenberg–Dirac exchange Hamiltonian

$$\mathcal{H}_{\mathrm{eff}} = J_{12}\, \mathbf{S}_1 \cdot \mathbf{S}_2. \tag{2.125}$$

(2.125) is the two-site version of the antiferromagnetic Heisenberg model.

2.5.2 Kinetic Exchange

Having described the Heitler–London treatment of the hydrogen molecule, let us discuss its limitations. Problems arise both at large, and at small, interatomic separations.

It is interesting to note that (2.124) predicts that J_{12} changes sign from antiferromagnetic to ferromagnetic coupling as the interproton distance is increased, because l vanishes as the atoms are pulled apart. Such a behaviour certainly follows from the Heitler–London scheme but it is not the correct conclusion for the hydrogen molecule which has a singlet ground state at all values of $|\mathbf{R}_a - \mathbf{R}_b|$. We can invoke Wigner's theorem (p. 35) to argue that this must be so but the detailed argument for $J_{12} > 0$ even at large $|\mathbf{R}_a - \mathbf{R}_b|$ is far from being simple [133]. In the region where the atomic states overlap, the electrons are relatively far from both nuclei but near to each other so that the electron–electron interaction becomes relatively strong. The proper description of the correlated motion of the electrons amounts to taking the admixture of higher-lying atomic states into account and thus requires the consideration of configuration interaction effects. — This also teaches us that it is often difficult, even for the simplest systems, to get accurate results for the exchange coupling. The safest course is to assume a plausible form of the exchange Hamiltonian and to get the coupling constants from a mixture of theoretical reasoning and experimental fit.

The Heitler–London picture meets another kind of trouble at relatively short interatomic distances $|\mathbf{R}_a - \mathbf{R}_b|$ (including the equilibrium separation). Here, we have to worry about having neglected the ionized configurations shown in (2.112). In one of them, both electrons are on the atom 'a'

$$|\psi_a\rangle = \frac{1}{\sqrt{2}}\begin{vmatrix} \varphi_a(\mathbf{r}_1)\alpha(s_1) & \varphi_a(\mathbf{r}_2)\alpha(s_2) \\ \varphi_a(\mathbf{r}_1)\beta(s_1) & \varphi_a(\mathbf{r}_2)\beta(s_2) \end{vmatrix}$$

$$= \frac{1}{\sqrt{2}}[\alpha(s_1)\beta(s_2) - \beta(s_1)\alpha(s_2)]\varphi_a(\mathbf{r}_1)\varphi_a(\mathbf{r}_2) \quad (2.126)$$

and similarly, when both electrons are on atom 'b'

$$|\psi_b\rangle = \frac{1}{\sqrt{2}}[\alpha(s_1)\beta(s_2) - \beta(s_1)\alpha(s_2)]\varphi_b(\mathbf{r}_1)\varphi_b(\mathbf{r}_2). \quad (2.127)$$

Putting two electrons on the same atom surely costs a lot of energy: it involves the on–site Coulomb interaction (also called "intrasite interaction")

$$U_{1s} = \langle \psi_a | \frac{e^2}{|\mathbf{r}_1 - \mathbf{r}_2|} | \psi_a \rangle = \int d\mathbf{r}_1 \int d\mathbf{r}_2 |\varphi_a(\mathbf{r}_1)|^2 \frac{e^2}{|\mathbf{r}_1 - \mathbf{r}_2|} |\varphi_a(\mathbf{r}_2)|^2. \tag{2.128}$$

This interaction parameter is usually called the Hubbard U. Since the mutually repelling electrons now share the same atomic orbital, U_{1s} is rather larger than the interatomic Coulomb energy C_{ab} which is, in turn, larger than the (interatomic) exchange term I_{ab}. But in spite of the energy cost, the ionized conigurations must get mixed into the ground state because there is a matrix element

$$-t = -\int d\mathbf{r}\ \varphi^*_{\text{at}}(\mathbf{r} - \mathbf{R}_a) \frac{e^2}{|\mathbf{r} - \mathbf{R}_b|} \varphi_{\text{at}}(\mathbf{r} - \mathbf{R}_b) \tag{2.129}$$

which connects the ionized and neutral configurations. It is the usual tight-binding hopping integral [32].

$-t$ is the amplitude for the single-electron[20] hopping process

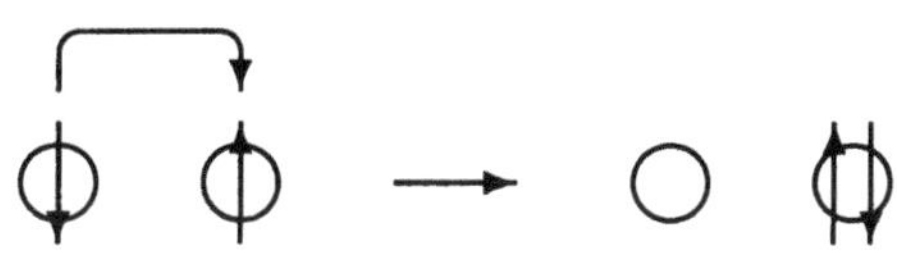

(2.130)

The relevant matrix element is

$$\langle \psi_4 | \mathcal{H} | \psi_5 \rangle = -2t \tag{2.131}$$

where $|\psi_5\rangle$ is the even combination of the on-site singlets

$$|\psi_5\rangle = \frac{1}{\sqrt{2}}(|\psi_a\rangle + |\psi_b\rangle). \tag{2.132}$$

This is the first time that we made explicit use of a geometrical symmetry of the molecule: reflection in the plane bisecting the internuclear

[20]The electron–electron interaction gives a further contribution to the one-electron hopping, as well as the possibility of a two-electron hopping process, which we neglect here. See, however, Problem 2.2.

separation. Let us remember that $\mathcal{H}$ has non-vanishing matrix elements only between states of the same symmetry. The Heitler–London ground state $|\psi_4\rangle$ is connected to the only other even-parity spin-singlet $|\psi_5\rangle$. The odd-parity ionized state

$$|\psi_6\rangle = \frac{1}{\sqrt{2}}(|\psi_a\rangle - |\psi_b\rangle) \tag{2.133}$$

has no partner to mix with and therefore its energy remains the diagonal U_{1s}. The triplet level (2.118) remains unaltered because parallel-spin electrons cannot share the same site, and thus U_{1s} can have no effect on triplet states.

There remains the problem of diagonalizing the full Hamiltonian in the even-parity singlet subspace spanned by $|\psi_4\rangle$ and $|\psi_5\rangle$. In order to see the effect of "switching on" t and U_{1s} more clearly, let us consider a simplified problem in which we neglect the intersite Coulomb matrix elements C_{ab} and I_{ab}, and also the non-orthogonality effects arising from $l \neq 0$. With these assumptions, the remaining two singlet eigenenergies are

$$\epsilon_{\mathrm{s}}^{\pm} \approx 2\epsilon_{\mathrm{at}} + \frac{1}{2}(U_{1s} \pm \sqrt{U_{1s}^2 + 16t^2})\,. \tag{2.134}$$

$\epsilon_{\mathrm{s}}^{-}$ is the ground state energy. It is interesting to write down its approximate form for the physically relevant limit $U_{1s} \gg t$

$$\epsilon_{\mathrm{s}}^{-} \approx 2\epsilon_{\mathrm{at}} - \frac{4t^2}{U_{1s}}\,. \tag{2.135}$$

Since in the same approximation $\epsilon_{\mathrm{tr}} \approx 2\epsilon_{\mathrm{at}}$, the singlet–triplet splitting is estimated as

$$\epsilon_{\mathrm{tr}} - \epsilon_{\mathrm{s}}^{-} \approx \frac{4t^2}{U_{1s}}\,. \tag{2.136}$$

In other words, the admixture of the ionic configurations leads to a further lowering of the singlet ground state energy: it helps to stabilize the (in the hydrogen molecule, still embryonic) antiferromagnetic state. This mechanism of creating a singlet–triplet splitting is called the *kinetic exchange*.

In a full treatment, we should consider both effects: those arising from the intersite Coulomb matrix elements, and those described by

(2.136). Though the full expression is more complicated, we may approximate the exchange coupling as a sum of direct exchange and kinetic exchange contributions:

$$J_{12} \sim J_{\rm dir} + J_{\rm kin} = 2\frac{l^2 C_{ab} - I_{ab}}{1 - l^4} + \frac{4t^2}{U_{1s}} . \qquad (2.137)$$

Fig. 2.3 shows the development of the level scheme as first the intersite Coulomb interaction, and then t is switched on.

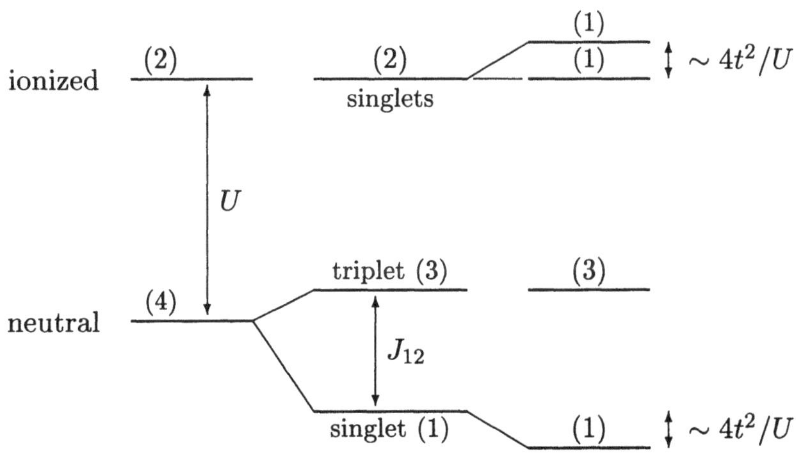

Figure 2.3: The schematic development of the level scheme of the hydrogen molecule as first the exchange J_{12}, and then the hopping t is switched on. Numbers in parentheses give the level degeneracies.

Though (2.125) is expressed in terms of localized spins, the kinetic term implies that we have gone beyond the Heitler–London scheme. By first-order perturbation theory, the ground state is

$$|\psi_4'\rangle \approx |\psi_4\rangle - \frac{2t}{U_{1s}} |\psi_5\rangle . \qquad (2.138)$$

The kinetic exchange contribution can be explained by saying that the electron gains some kinetic energy by hopping out of its localized atomic orbital. The energy shift (2.136) is synonymous with the presence of the ionized configurations.

2.5.3 Molecular Orbitals versus Heitler–London

The role of the ionized configurations is increasing if the interatomic distance $|\mathbf{R}_a - \mathbf{R}_b|$ is reduced, because t becomes larger. At sufficiently small $|\mathbf{R}_a - \mathbf{R}_b|$ we may argue that the single-electron hopping is the dominating effect and electron–electron interaction can be treated as a perturbation. This scenario also leads to a singlet ground state.

In this *molecular orbital treatment*, we start with the single-electron eigenstates of the molecule: these are the even, and odd, molecular orbitals:

$$\begin{aligned}\phi_e(\mathbf{r}) &= \frac{1}{\sqrt{2}}[\varphi_a(\mathbf{r}) + \varphi_b(\mathbf{r})] \\ \phi_o(\mathbf{r}) &= \frac{1}{\sqrt{2}}[\varphi_a(\mathbf{r}) - \varphi_b(\mathbf{r})]\end{aligned} \tag{2.139}$$

The corresponding single-electron energies are

$$\begin{aligned}\epsilon_e &= \epsilon_{1s} - t \\ \epsilon_o &= \epsilon_{1s} + t\end{aligned} \tag{2.140}$$

Obviously, we are doing best by putting both electrons into the even (bonding) state, and leaving the odd (antibonding) state empty

$$|\psi_{\mathrm{MO}}\rangle = \frac{1}{\sqrt{2}}\begin{vmatrix} \phi_e(\mathbf{r}_1)\alpha(s_1) & \phi_e(\mathbf{r}_2)\alpha(s_2) \\ \phi_e(\mathbf{r}_1)\beta(s_1) & \phi_e(\mathbf{r}_2)\beta(s_2) \end{vmatrix} . \tag{2.141}$$

This is the prescription of the molecular orbital method for the ground state. We recognize that this is band theory applied to the H_2 molecule. We solved the one-electron problem, and filled the energy levels from bottom upwards.

Where remain all the problems with the electron–electron interaction? We will spot them immediately if we expand the molecular orbital (MO) ground state

$$|\psi_{\mathrm{MO}}\rangle = \frac{1}{\sqrt{2}}(|\psi_4\rangle + |\psi_5\rangle) . \tag{2.142}$$

Thus the independent electron treatment predicts a ground state which is an equal admixture of ionized and neutral configurations! The corresponding ground state energy[21] must be roughly

$$E_{\rm MO} \approx 2\epsilon_{1s} - 2t + \frac{U_{1s}}{2} \,. \tag{2.143}$$

If $U_{1s} > 4t$ (which, we imagine, may easily be the case) then the above estimate is lying higher than the energy of two independent atoms! It means that for large U_{1s} the Heitler–London scheme which (in the same rough approximation) yields

$$E_{\rm HL} \approx 2\epsilon_{1s} \tag{2.144}$$

provides a more acceptable description of the ground state.

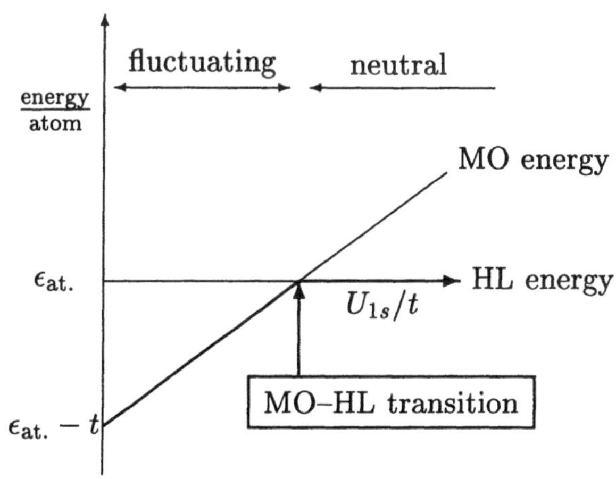

Figure 2.4: Ground state transition from the molecular orbital to the Heitler–London regime, based on the naivest energy estimates (in an improved description, we would find that the ground state evolves continuously from the MO limit to the HL limit).

It should be emphasized that the above estimates (which neglect even direct exchange) are very naive. In writing down (2.135), we were

[21]Without the proton–proton interaction term which is unrelated to the electronic structure.

doing much better for the Heitler–London energy; and surely, the MO energy could also be improved. However, the largest interaction term is U_{1s} and the essential difference between the HL and MO schemes is that U_{1s} gives a large contribution to the MO energy and a (perturbatively) small contribution to the HL energy. If we stick to the naive estimates (2.143) and (2.144), we have to conclude that there is a ground state transition from the molecular orbital regime to the Heitler–London regime as U_{1s} is increased (Fig. 2.4.).

We know, of course, that in a small system like the hydrogen molecule there can be no sharp transition, and the true ground state evolves continuously from the MO limit to the HL limit as U_{1s} is increased from 0 to ∞ (the $\epsilon_{\rm S}^{-}$ given in (2.134) gives a pretty good idea what this continuous U/t-dependence looks like). However, molecular orbital theory is like band theory, and the Heitler–London picture is like what we use to describe magnetic insulators, so the interest of Fig. 2.4 is that it gives a foretaste of the Mott-transition in certain solids. In the sense of (2.142), the MO ground state shows strong charge fluctuations and is thus similar to the metallic state. The Heitler–London regime is governed by the exchange Hamiltonian (2.125) which is like the Heisenberg model of antiferromagnetic insulators.

Solutions to the Problems

Problem 2.1. Since $\mathcal{H}_0+\mathcal{H}_{\rm rel}$ is diagonal in the basis $|j=l\pm 1/2, j^z, l\rangle$, only the Zeeman term $\mathcal{H}_{\rm Z} = (J^z+S^z)\mu_{\rm B}H$ (or more precisely, its S^z-term) can have non-diagonal matrix elements. Even this is severely restricted. Taking the matrix elements of $[\mathbf{L}^2, J^z+S^z]=0$ between states derived from different l, we get (very schematically written)

$$\langle l|[\mathbf{L}^2, J^z+S^z]|l'\rangle = (l(l+1)-l'(l'+1))\langle l|J^z+S^z|l'\rangle = 0,$$

wherefrom the matrix element vanishes for $l\neq l'$. Similarly, from $[\mathcal{H}_0, J^z+S^z]=0$ it follows that there are no matrix elements of J^z+S^z which would connect states with different n. Finally, axial symmetry guarantees that J^z must be conserved. For our present case, it means that the six states derived from $2p$ states are not coupled either to the $2s$-like states, or to states with $n\neq 2$. The only non-vanishing matrix elements are between states with $j^z = 1/2$, and between states with $j^z=-1/2$. Using (2.33) and (2.34)

$$\langle 3/2, \pm 1/2, 1|S^z|1/2, \pm 1/2, 1\rangle = -\sqrt{2}/3 \qquad (2.145)$$

The diagonal elements were given in (2.36).

The details of $\mathcal{H}_{\rm rel}$ are not important now; it is enough to know that they give rise to a $^2P_{3/2}$– $^2P_{1/2}$ fine structure splitting whose size Δ is of the order of λ, the spin–orbit coupling constant. We set the energy zero at the $H = 0$ value of the $^2P_{1/2}$ energy.

The states $|3/2, \pm 3/2, 1\rangle$ are eigenstates of the full Hamiltonian, with eigenvalues $\epsilon_{\pm 3/2} = \Delta \pm 2h$, where $h = H\mu_{\rm B}$. In the $j^z = 1/2$ subspace we have the 2×2 eigenvalue problem

$$\begin{pmatrix} \Delta + 2h/3 & -\sqrt{2}h/3 \\ -\sqrt{2}h/3 & h/3 \end{pmatrix} \tag{2.146}$$

to solve. The roots are

$$\epsilon^{\pm}_{1/2} = \frac{1}{2}[\Delta + h \pm \sqrt{\Delta^2 + h^2 + (2\Delta h/3)}]. \tag{2.147}$$

An $h \ll \Delta$ expansion recovers two levels of the weak-field Zeeman effect (2.36). For $h \gg \Delta$, $\epsilon^{+}_{1/2} \sim h$, while $\epsilon^{-}_{1/2} \sim \Delta/3$ is vanishingly small on the scale of the Zeeman splitting. The remaining two roots

$$\epsilon^{\pm}_{-1/2} = \frac{1}{2}[\Delta - h \pm \sqrt{\Delta^2 + h^2 - (2\Delta h/3)}] \tag{2.148}$$

can be similarly discussed. The six eigenvalues continuously interpolate between the weak-field limit where the levels are grouped into a doublet and quartet, and the high-field limit of equidistant levels at $0, \pm h, \pm 2h$, the level at 0 being twofold degenerate.

A straightforward plot of the h-dependence of the energies would fail to show the Paschen–Back limit convincingly, because it lies at $h \to \infty$, where the splittings diverge. Therefore we decided to use the reduced field strength $y = h/(h + \Delta)$ which compresses the entire $0 \le h \le \infty$ semi-axis into the interval $y \in [0, 1]$. At the same time, we use the reduced energies $\epsilon \to \epsilon/(h+\Delta)$. At $y = 0$, the levels $^2P_{1/2}$ and $^2P_{3/2}$ are lying at 0 and 1, respectively, and at small y we see the essentially undistorted weak-field behaviour. In the high-field limit, the Zeeman energies are found at ± 2, ± 1, and 0. The plot is shown in Fig. 2.5. We discover that getting from one limit to the other involves level crossings at intermediate fields, and the arising of the twofold degeneracy of the zero-energy level at $y \to 1$ is also clearly seen.

Finally, it should be emphasized that though we were formally taking the $h \to \infty$ limit, this means merely a field sufficiently stronger than the spin–orbit coupling but still weak enough, so that it was right to neglect the diamagnetic term. For a truly strong field, $\mathcal{H}_{\rm dia}$ would become important.

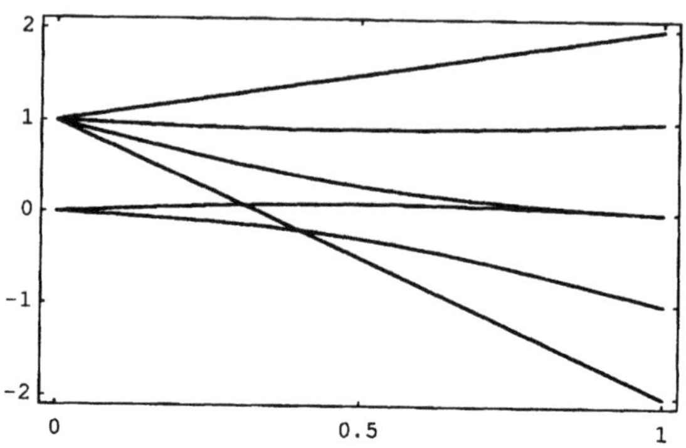

Figure 2.5: The reduced energies of the ^2P levels of the H atom in a magnetic field, plotted as a function of the reduced field y ($y = 0$ is zero field, $y = 1$ corresponds to infinite field. The energies are scaled to get a finite splitting at all y. See text.) With increasing field, the level scheme evolves from a six-level picture to the Paschen–Back limit with five equidistant levels.

Problem 2.2. Inserting the expansion

$$\Psi^\dagger(\mathbf{r},\sigma) = \varphi_a^*(\mathbf{r})c_{a\sigma}^\dagger + \varphi_b^*(\mathbf{r})c_{b\sigma}^\dagger \tag{2.149}$$

into the general expression of the Coulomb interaction

$$\mathcal{H}_{\text{Coulomb}} = \frac{1}{2}\sum_\sigma\sum_{\sigma'}\int d\mathbf{r}\int d\mathbf{r}'\,\Psi^\dagger(\mathbf{r},\sigma)\Psi^\dagger(\mathbf{r}',\sigma')\frac{e^2}{|\mathbf{r}-\mathbf{r}'|}\Psi(\mathbf{r}',\sigma')\Psi(\mathbf{r},\sigma) \tag{2.150}$$

we could expect 64 terms. However, the choice $\sigma = \sigma'$ yields only 8 non-vanishing terms, the remaining 24 being prohibited by the Pauli principle. The 40 non-vanishing contributions can be classified as follows:

$\sigma = \sigma' = \uparrow$ restricts us to the Heitler–London configurations. First we have from the choice *abba* (i.e., choosing the a-term from $\Psi^\dagger(\mathbf{r})$, the b-term from $\Psi^\dagger(\mathbf{r}')$ and $\Psi(\mathbf{r}')$, and again the a-term from $\Psi(\mathbf{r})$)

$$\frac{1}{2}\int d\mathbf{r}\int d\mathbf{r}'\,\varphi_a^*(\mathbf{r})\varphi_b^*(\mathbf{r}')\frac{e^2}{|\mathbf{r}-\mathbf{r}'|}\varphi_b(\mathbf{r}')\varphi_a(\mathbf{r})\,c_{a\uparrow}^\dagger c_{b\uparrow}^\dagger c_{b\uparrow}c_{a\uparrow} \rightsquigarrow \frac{C_{ab}}{2}\hat{n}_{a\uparrow}\hat{n}_{b\uparrow}\,. \tag{2.151}$$

baab gives the same, cancelling the factor 1/2. A similar duplication of contributions is found in all the remaining cases, too.

Direct exchange contributions arise from *abab* and *baba*, e.g.,

$$\frac{1}{2}\int d\mathbf{r}\int d\mathbf{r}'\,\varphi_a^*(\mathbf{r})\varphi_b^*(\mathbf{r}')\frac{e^2}{|\mathbf{r}-\mathbf{r}'|}\varphi_a(\mathbf{r}')\varphi_b(\mathbf{r})\,c_{a\uparrow}^\dagger c_{b\uparrow}^\dagger c_{a\uparrow}c_{b\uparrow} \rightsquigarrow -\frac{J_{ab}}{2}\hat{n}_{a\uparrow}\hat{n}_{b\uparrow}\,. \tag{2.152}$$

The discussion is similar for the case $\sigma = \sigma' =\downarrow$.

We find new Coulomb matrix elements if we turn to the case $\sigma \neq \sigma'$. Taking first $\sigma =\uparrow$, $\sigma' =\downarrow$, let us notice that now we can choose all orbitals the same, as in *aaaa*:

$$\frac{1}{2}\int d\mathbf{r}\int d\mathbf{r}'\,\varphi_a^*(\mathbf{r})\varphi_a^*(\mathbf{r}')\frac{e^2}{|\mathbf{r}-\mathbf{r}'|}\varphi_a(\mathbf{r}')\varphi_a(\mathbf{r})\,c_{a\uparrow}^\dagger c_{a\downarrow}^\dagger c_{a\downarrow}c_{a\uparrow} \rightsquigarrow \frac{U}{2}\hat{n}_{a\uparrow}\hat{n}_{a\downarrow}\,, \tag{2.153}$$

where the on-site Coulomb interaction

$$U = \int d\mathbf{r}\int d\mathbf{r}'\,|\varphi_a(\mathbf{r})|^2\frac{e^2}{|\mathbf{r}-\mathbf{r}'|}|\varphi_a(\mathbf{r}')|^2 \tag{2.154}$$

is usually called the Hubbard U. This time the 1/2 will be cancelled by an identical contribution from $\sigma =\downarrow$, $\sigma' =\uparrow$. We also find $U\hat{n}_{b\uparrow}\hat{n}_{b\downarrow}$.

The remaining diagonal contributions are the intersite Coulomb interaction terms $C_{ab}(\hat{n}_{a\uparrow}\hat{n}_{b\downarrow} + \hat{n}_{a\downarrow}\hat{n}_{b\uparrow})$.

Off-diagonal exchange terms are coming from *abab* and *baba*. Taking the first

$$\frac{1}{2}\int d\mathbf{r}\int d\mathbf{r}'\,\varphi_a^*(\mathbf{r})\varphi_b^*(\mathbf{r}')\frac{e^2}{|\mathbf{r}-\mathbf{r}'|}\varphi_a(\mathbf{r}')\varphi_b(\mathbf{r})\,c_{a\uparrow}^\dagger c_{b\downarrow}^\dagger c_{a\downarrow}c_{b\uparrow} \rightsquigarrow -\frac{J_{ab}}{2}S_a^+S_b^-\,, \tag{2.155}$$

where we identified $c_{a\uparrow}^\dagger c_{a\downarrow} = S_a^+$, and $c_{b\downarrow}^\dagger c_{b\uparrow} = S_b^-$ as the spin-flip operators. Collecting such contributions, we find $-J_{ab}(S_a^+S_b^- + S_b^+S_a^-)$.

There remain 20 terms to account for. These describe the Coulomb-assisted hopping of either one electron, or two electrons. Consider the *aaab* term

$$\frac{1}{2}\int d\mathbf{r}\int d\mathbf{r}'\,\varphi_a^*(\mathbf{r})\varphi_a^*(\mathbf{r}')\frac{e^2}{|\mathbf{r}-\mathbf{r}'|}\varphi_a(\mathbf{r}')\varphi_b(\mathbf{r})\,c_{a\uparrow}^\dagger c_{a\downarrow}^\dagger c_{a\downarrow}c_{b\uparrow} \rightsquigarrow \frac{X_{ab}}{2}c_{a\uparrow}^\dagger c_{b\uparrow}\hat{n}_{a\downarrow}\,, \tag{2.156}$$

where

$$X_{ab} = \int d\mathbf{r}\int d\mathbf{r}'\,\varphi_a^*(\mathbf{r})\varphi_b(\mathbf{r})\frac{e^2}{|\mathbf{r}-\mathbf{r}'|}|\varphi_a(\mathbf{r}')|^2 \tag{2.157}$$

is the amplitude of the interaction-assisted single-electron hopping. Further 15 terms give similar contributions.

Finally, consider *aabb*

$$\frac{1}{2}\int d\mathbf{r}\int d\mathbf{r}'\,\varphi_a^*(\mathbf{r})\varphi_a^*(\mathbf{r}')\frac{e^2}{|\mathbf{r}-\mathbf{r}'|}\varphi_b(\mathbf{r}')\varphi_b(\mathbf{r})\,c_{a\uparrow}^\dagger c_{a\downarrow}^\dagger c_{b\downarrow}c_{b\uparrow} \rightsquigarrow \frac{J_{ab}}{2}c_{a\uparrow}^\dagger c_{a\downarrow}^\dagger c_{b\downarrow}c_{b\uparrow}\,. \tag{2.158}$$

A singlet electron pair is removed from b and placed on a: it is a double hopping process. It is interesting that the amplitude is given by the familiar exchange integral.

Before summarizing our findings, it is useful to collect the exchange contributions into a standard expression. We found the diagonal contribution $-J_{ab}(\hat{n}_{a\uparrow}\hat{n}_{b\uparrow}+\hat{n}_{a\downarrow}\hat{n}_{b\downarrow})$ which can be understood as $-2J_{ab}(S_a^z S_b^z+1/4)$. However, the "1/4" is really meant to express that there is one electron in a, and one in b, with whatever spin, so we should replace the 1/4 by $(\hat{n}_{a\uparrow}+\hat{n}_{a\downarrow})(\hat{n}_{b\uparrow}+\hat{n}_{b\downarrow})/4$.

Collecting first the contributions from the Heitler–London (neutral) configurations, we find

$$\begin{aligned}\mathcal{H}^{(1)}_{\text{Coulomb}} &= C_{ab}(\hat{n}_{a\uparrow}+\hat{n}_{a\downarrow})(\hat{n}_{b\uparrow}+\hat{n}_{b\downarrow}) \\ &\quad -2J_{ab}\left(\mathbf{S}_a\cdot\mathbf{S}_b+\frac{1}{4}(\hat{n}_{a\uparrow}+\hat{n}_{a\downarrow})(\hat{n}_{b\uparrow}+\hat{n}_{b\downarrow})\right) .\end{aligned} \tag{2.159}$$

Up to this point, the second-quantized formalism provided us with an elegant rederivation of the results of Sec. 2.2.

The other contributions arise from including the ionized configurations

$$\begin{aligned}\mathcal{H}^{(2)}_{\text{Coulomb}} &= U(\hat{n}_{a\uparrow}\hat{n}_{a\downarrow}+\hat{n}_{b\uparrow}\hat{n}_{b\downarrow}) \\ &\quad +X_{ab}\sum_\sigma(c^\dagger_{a\sigma}c_{b\sigma}+c^\dagger_{b\sigma}c_{a\sigma})(\hat{n}_{a,-\sigma}+\hat{n}_{b,-\sigma}) \\ &\quad +J_{ab}(c^\dagger_{a\uparrow}c^\dagger_{a\downarrow}c_{b\downarrow}c_{b\uparrow}+c^\dagger_{b\uparrow}c^\dagger_{b\downarrow}c_{a\downarrow}c_{a\uparrow}) .\end{aligned} \tag{2.160}$$

The results derived here will be of great help in constructing the extended Hubbard model (Sec. 5.3).

Problem 2.3. a) p-electrons have $|\mathbf{l}|=1$, so in a p^2 configuration the total angular momentum $|\mathbf{L}|=|\mathbf{l}_1+\mathbf{l}_2|$ can take on the values $L=2$, 1, and 0. The two 1/2-spins can be combined into a triplet ($S=1$), or a singlet ($S=0$). However, we have to be careful because the Pauli principle restricts which L value can be paired with a given S.

If the spins are parallel ($S=1$), the orbital states must be different, thus the largest L^z corresponds to the simultaneous occupation of the $m=1$ and $m=0$ states $\leadsto L=1$. Further states can be generated by acting on this state either with L^-, or with S^-; this gives the $(2L+1)(2S+1)=9$-fold degenerate ^3P term. Taking the spin–orbit coupling into account, we find the multiplet splitting $^3P\to{}^3P_0+{}^3P_1+{}^3P_2$. The Hund's rule ground state set is $^3P_0$.

Antiparallel-spin electrons ($S=0$) can simultaneously occupy the $m=1$ orbital state: this gives the $L^z=2$ component of the 5-fold degenerate $^1D_2$.

Up to now, we have accounted for 14 states. The total degeneracy of the p^2 configuration is $C_2^6 = 15$, thus the remaining state must have $L = 0$ and $S = 0$. All in all, we found $p^2 \to {}^3P + {}^1D + {}^1S$.

b) The classification of the $C_2^{10} = 45$ states of the d^2 configuration requires more care. We use the $|LL^zSS^z\rangle$ basis, and try to find the terms by identifying the state $|LLSS\rangle$ which has the largest possible values of L^z and S^z.

Since the quantum numbers n and l are fixed, we can denote the single-electron states with ψ_m where $m = -2, -1, \ldots, 2$. There is one $L^z = 4$ state:

$$|4400\rangle = \psi_2(\mathbf{r}_1)\psi_2(\mathbf{r}_2)[\alpha(s_1)\beta(s_2) - \beta(s_1)\alpha(s_2)]. \tag{2.161}$$

This led to finding the ${}^1G$ term (9 states).

There are two $L^z = 3$ states: the symmetrized and the antisymmetrized product of ψ_2 and ψ_1. The symmetrized product must go with a spin singlet thus it is $|4300\rangle$. The antisymmetrized orbital function must be paired with a spin triplet:

$$|3311\rangle = \frac{1}{\sqrt{2}}\alpha(s_1)\alpha(s_2)[\psi_2(\mathbf{r}_1)\psi_1(\mathbf{r}_2) - \psi_2(\mathbf{r}_2)\psi_1(\mathbf{r}_1)]. \tag{2.162}$$

From this state, the $(2L+1)(2S+1) = 21$-fold degenerate ${}^3F$ term can be generated.

There are three ways to get $L^z = 2$: with the symmetrized (antisymmetrized) product of ψ_2 and ψ_0, and with the symmetrical $\psi_1\psi_1$. Turning down $L = 4$ and $L = 3$ states accounts for a symmetrical and an antisymmetrical orbital function, thus there must exist a ${}^1D$ term (5 states).

The reader can easily convince herself that the same pattern continues and a new triplet appears with $L = 1$ (9 states), and a new singlet with $L = 0$ (1 state). Altogether we found $d^2 \to {}^1G + {}^3F + {}^1D + {}^3P + {}^1S$. The ground state level is ${}^3F_2$.

Problem 2.4. The oxygen atom has 8 electrons: $1s^2 2s^2 2p^4$. Forgetting the essentially unperturbed $1s$ electrons, we form bonding and antibonding combinations of the $2s$ and $2p$ orbitals, as we did in writing down (2.139) for the hydrogen molecule. The level scheme [68] is shown in Fig. 2.6. Antibonding levels are denoted by a $*$. Choosing the molecule axis as the z-axis, it is clear that the atomic $2p_z$ orbitals overlap strongly, resulting in a wide bonding-antibonding splitting between the $\sigma 2p$ and $\sigma^* 2p$ levels. The $2p_x$ and $2p_y$ orbitals have smaller overlaps, giving rise to the weaker π-bonds.

Filling up the MO levels with the available 12 electrons, we are left with two electrons to be put into the antibonding $\pi_y^*\pi_z^* 2p$ level which, however, could accomodate four electrons. Accepting that the spin degeneracy is lifted by the exchange hole effect, we predict a triplet ground state: the oxygen molecule has a permanent magnetic moment corresponding to $S = 1$. At low

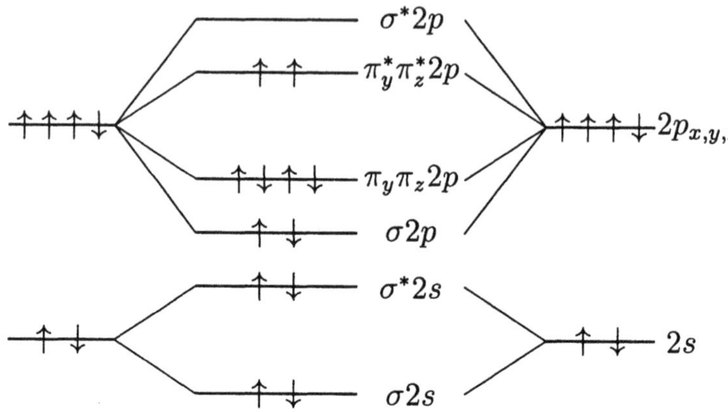

Figure 2.6: Schematic representation of the electronic structure of the oxygen molecule in the MO scheme, following [68]. The atomic states (indicated on left and right) combine into bonding and antibonding molecular orbitals (middle). The ground state is a triplet.

temperatures, solid β-O_2 is found to be an antiferromagnet with complicated structure [186].

Problem 2.5. Ce^{3+} has the configuration $4f^1$, thus $S = 1/2$, $L = 3$, and $J = 5/2$. From (2.76), the Landé factor is $g = 6/7$. The effective number of Bohr magnetons is $p \approx 2.535$.

Problem 2.6. Sm^{3+} has the configuration $4f^5$; the ground state term is ${}^6H_{5/2}$, and the first excited level has $J = 7/2$. The ionic susceptibility has both Curie and Van Vleck contributions. The Sm^{2+} ion with its configuration $4f^6$ represents the curious case of an $L \neq 0$, $S \neq 0$ ion without a Curie susceptibility: namely here the Hund's rule values are $S = 3$, $L = 3$ and $J = 0$. However, we find a substantial Van Vleck susceptibility.

a) We give a general discussion of the ratio of Curie and Van Vleck susceptibilities for light rare earth ions. For the configurations $4f^1$ to $4f^6$, we can use $J = L - S$. Replacing (2.78)–(2.80) into (2.88), we find that the Van Vleck susceptibility is given by

$$\chi_{\mathrm{VV}} = \frac{2(L+1)S\mu_{\mathrm{B}}^2}{3(J+1)\Delta} \,. \tag{2.163}$$

The Curie susceptibility being (2.90), the ratio is

$$\frac{\chi_{\mathrm{VV}}}{\chi_{\mathrm{C}}} = \frac{2(L+1)S}{g^2 J(J+1)^2} \frac{k_{\mathrm{B}}T}{\Delta} \tag{2.164}$$

where g is given by (2.76). Using the values appropriate for Sm^{3+}, we find that at room temperature, the Van Vleck susceptibility is about 2.5 times as large as the Curie susceptibility! Furthermore, it follows from (2.163) that χ_{VV} for Sm^{2+} is about 10 times as large as for Sm^{3+}. Since SmB_6 contains 60% Sm^{3+} and 40% Sm^{2+}, we conclude that the Van Vleck contribution to the susceptibility is (at $T = 300$K) an order of magnitude larger than the Curie term.

b) Since the Boltzmann factor $e^{-\beta\Delta}$ is not negligibly small for the low-lying levels of the multiplet, the definition (2.87) of the single-ion partition function $\mathcal{Z}_1$ has to be extended to include at least the lowest excited level with $J' = |L - S| + 1$ (for less-than-half filling)

$$\mathcal{Z}_1 = \sum_{J^z=-J}^{J} \exp\left\{-\frac{E(J, J^z, L, S)}{k_B T}\right\} + \sum_{J^z=-J'}^{J'} \exp\left\{-\frac{E(J', J^z, L, S)}{k_B T}\right\} . \tag{2.165}$$

Since we merely want to get an impression of the character of the result, we neglect the second-order Zeeman effect. Then we need only the diagonal matrix elements of $\mathcal{H}_Z$ which are of the form (2.75), with the g-factor depending on J. Choosing the energy zero at the zero-field value of $E(J, ...)$, the energies can be written as

$$\begin{aligned} E(J, J^z, L, S) &= gH\mu_B J^z \\ E(J', J^z, L, S) &= \Delta + g'H\mu_B J^z . \end{aligned} \tag{2.166}$$

The free energy can be expanded like we did in (2.88). Using the definition (2.83), we readily obtain

$$\chi = \frac{\mu_B^2}{3k_B T} \frac{g^2 J(J+1)(2J+1) + (g')^2 J'(J'+1)(2J'+1)e^{-\Delta/k_B T}}{2J+1+(2J'+1)e^{-\Delta/k_B T}} \tag{2.167}$$

For $\Delta/T \to \infty$, (2.90) is recovered. Here we found an additional, activated Curie-like contribution. The full expression (containing also activated Van Vleck type terms) was given by Van Vleck (see Ch. IX of [423]).

Thinking of the Sm^{2+} ion which does not have a proper Curie susceptiblity in the literal sense, (2.167) can be interpreted as an activated Curie-type susceptibility which (at room temperature) corresponds to the effective Bohr magneton number $p = (3/2)(6\exp(-\Delta/k_B T)/(1+3\exp(-\Delta/k_B T)))^{1/2} \sim 1.4$. It is a sizeable contribution.

Problem 2.7. According to Hund's rules, for U^{4+} ($5f^2$) the ground state has $S = 1$, $L = 5$, and $J = 4$, while for U^{3+} ($5f^3$) we find $S = 3/2$, $L = 6$, and $J = 9/2$. The Landé factor (2.76) is 0.8 for $5f^2$, and 0.727 for $5f^3$. The effective number of Bohr magnetons $p = g\sqrt{J(J+1)}$ is 3.58 for U^{4+}, and 3.62 for U^{3+}. Since both values lie well within the 1% range about 3.6, such a measurement could not decide between the two valence states.

Problem 2.8. We discuss in Sec. 3.10.1 that the biquadratic coupling appearing in (2.102) models the interaction of the quadrupolar moments $Q_j = 3(S_j^z)^2 - 2$. Q has the eigenvalues 1 (for $S^z = \pm 1$), and -2 (for $S^z = 0$). The single-site mean field Hamiltonian is

$$\mathcal{H}_{\rm MF} = -(h + \lambda\langle S^z\rangle)S^z - \eta\langle Q\rangle Q \tag{2.168}$$

where $\lambda = zJ$, $\eta = zJ_Q$ (z being the coordination number), $h = Hg\mu_{\rm B}$, and $\langle\,\rangle$ denotes thermal expectation value. The self-consistency conditions are

$$\langle S^z\rangle = \frac{1}{\mathcal{Z}} e^{\beta\eta\langle Q\rangle}\left(e^{\beta(h+\lambda\langle S^z\rangle)} - e^{-\beta(h+\lambda\langle S^z\rangle)}\right) \tag{2.169}$$

and

$$\langle Q\rangle = \frac{1}{\mathcal{Z}}\left\{e^{\beta\eta\langle Q\rangle}\left(e^{\beta(h+\lambda\langle S^z\rangle)} + e^{-\beta(h+\lambda\langle S^z\rangle)}\right) - 2e^{-2\beta\eta\langle Q\rangle}\right\} \tag{2.170}$$

where $\beta = 1/k_{\rm B}T$ and

$$\mathcal{Z} = e^{\beta\eta\langle Q\rangle}\left(e^{\beta(h+\lambda\langle S^z\rangle)} + e^{-\beta(h+\lambda\langle S^z\rangle)}\right) + e^{-2\beta\eta\langle Q\rangle}\,.$$

At sufficiently low T, one may find either magnetic order ($\langle S^z\rangle \neq 0$), or quadrupolar order ($\langle Q\rangle \neq 0$), or simultaneously both. Ferromagnetism sets in at $T_{\rm C} = 2\lambda/3$ if $J > 3J_Q$, or quadrupolar ordering occurs first at $T_Q = 2\eta$ if $J < 3J_Q$. In the disordered phase, the external magnetic field induces a magnetization $M \approx \chi_1 H + (\chi_3/3!)H^3$, and also a quadrupolar moment $\langle Q\rangle \approx \chi_Q H^2$. Replacing these into (2.169) and (2.170), and taking C and $C^{(3)}$ from (2.90) and (2.101), we find that $\chi_1 = C/(T - T_{\rm C})$ is unaffected by J_Q, while

$$\chi_3 = -\frac{C^{(3)}T}{(T-T_{\rm C})^4} + \frac{C^{(3)}T}{(T-T_{\rm C})^4}\cdot\frac{T_Q}{T-T_Q} \tag{2.171}$$

contains an important quadrupolar contribution. This is a special case of the result derived in [288].

In the absence of quadrupolar interactions $T_Q = 0$, and $\chi_3 < 0$, diverging at $T \to T_{\rm C} + 0$ ($\chi_3 \to -\infty$ and at the same time $\chi_1 \to \infty$). However, for $J_Q > 0$ we find also a positive term which may change the overall sign of χ_3. Furthermore, if $T_Q > T_{\rm C}$, the onset of quadrupolar order is signalled by $\chi_3 \to +\infty$, while χ_1 remains finite[22].

[22] The appealing simplicity of this result is usually lost if we include crystal field effects (see Ch. 3), but χ_3 remains a sensitive probe of multipolar interactions [288].

Chapter 3

Crystal Field Theory

3.1 Incomplete Shells in an Anisotropic Environment: Crystal Fields

As shown in Fig. 2.1, the formulae (2.76) and (2.89)–(2.93) which describe the combined response of orbital and spin momenta to external fields, are spectacularly successful in describing the paramagnetic susceptibilities of rare earth systems. The high-temperature susceptibilities of insulating transition metal compounds are described by similar formulae; however, as we can see from Fig. 2.2, the magnetic moment seems to arise almost solely from the spins of the $3d$ ions. The orbital contribution is completely (or in any case, largely) missing.

The reason for the discrepancy is that the above-cited formulae were derived using free ion eigenstates, while the d-shells in question belong to ions embedded in a solid. It is intuitively clear that even in an insulator, the motion of the electrons must be influenced by the crystalline environment. Lattice effects are certainly even larger in metals, so much, that we may be wondering whether the concept of ionic-like d- or f-electrons has any use at all. There are, however, important classes of systems where this is the case[1].

[1]Electrons occupying the compact $4f$-shell can remain localized even in metals. Furthermore, the high-temperature susceptibility of transition metals or their alloys often contains a Curie-like (or Curie–Weiss) component. We will understand in later chapters why d- and f-electrons can show localized behaviour in solids; here we assume that they can do so, and explore the consequences.

In this Chapter, we will be speaking about systems where the d- or f-electrons in partially filled shells occupy well-localized, ionic-like states. In other words, we assume that the d-electron systems we have in mind are Mott-insulators[2]. f-electrons are often localized even if the surrounding solid is a metal; this phenomenon, like the Mott-localization of d-electrons, is an electron correlation effect. In either case, the localized electrons must be affected by the electrostatic field of the surrounding ions: *the crystal fields.*

A typical situation is shown in Fig. 3.1. Many interesting compounds with the composition ABO_3 (where A and B stand for different metals) have the well-known perovskite structure (see, e.g., Ch. 14 of [209]). The ideal perovskite structure is cubic; the actual structure can often be understood as a slightly distorted perovskite. For a first orientation, it is perfectly alright to pick out a small detail of the ideal structure: the left panel of Fig. 3.1 shows the typical local environment "seen" by the metal ion B. If we are paying attention to the B atoms only (say, because only they carry magnetic moments), we can envisage the structure as a cubic array of corner-sharing BO_6 octahedra. The horizontal planes can be drawn as the BO_2 planes shown in the right panel of the figure. These planes are separated by AO planes.

The Mott insulator $LaTiO_3$ has attracted considerable interest [401]. Though the material is not really cubic, the gross features can be understood by assuming the ideal perovskite structure, identifying the filled circles in Fig. 3.1 with Ti ions. When discussing the electronic structure of a material like this, it is useful to start with the assumption that it is an ionic compound, and try to guess the ionic charges. La is very stable as a trivalent ion, and oxygen usually likes to be divalent, so the nominal valence states can be summarized in the formula $La^{3+}Ti^{3+}O_3^{2-}$. La^{3+} and O^{2-} have closed shells. A closed-shell ion is non-magnetic[3] and more-or-less tightly bound. It means that a finite, often considerable energy is needed to excite an electron out of the shell. As far as the low-temperature, low-energy electrical and magnetic properties are concerned, all the action is taking place in the incomplete shells (or in the bands derived from them). In our particular case, the magnetic moment is carried by the remaining single d-electron of the Ti^{3+} ion. Now,

[2]For a detailed treatment, cf. Ch. 4

[3]In the sense of having no permanent moment.

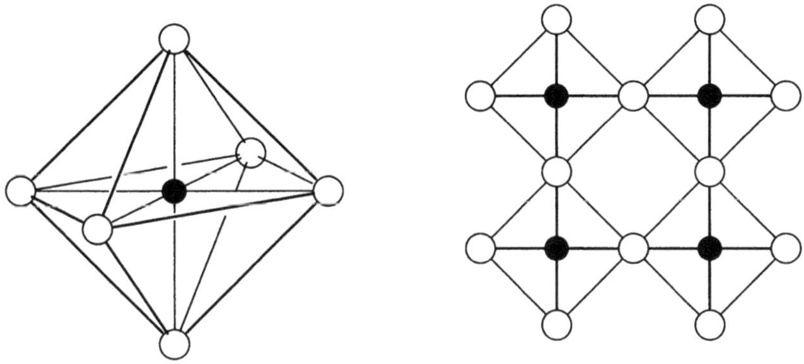

Figure 3.1: *Left*: Schematic representation of the local environment of transition metal ions in the undistorted perovskite structure, and in related structures such as the K_2NiF_4 structure. In $LaTiO_3$, a Ti ion (filled circle) is sitting in the octahedral environment of surrounding oxygen ions (empty circles); in La_2CuO_4, the black circle is a Cu ion. *Right*: the horizontal plane through the octahedron centers is the characteristic CuO_2 plane of the cuprates.

what is the ground state of the $3d^1$ configuration? If the Ti^{3+} ion were alone in vacuum, and if we neglected the spin–orbit coupling (which is indeed small for Ti), the electron could be in any of the five $3d$ orbital states, and its ground state would be tenfold degenerate (counting also the spin degeneracy). If we included the spin–orbit interaction, Hund's third rule would tell us that the ground state is the fourfold degenerate $J = 3/2$ level.

The essential point, though, is that the motion of the d-electron is governed not only by the potential of the Ti^{3+} ion core, which we include in the ionic Hamiltonian $\mathcal{H}^{(Ti)}$, but also by the electrostatic potential of the surrounding six oxygen ions

$$\mathcal{H}_{3d}(\mathbf{r}) = \mathcal{H}^{(\mathrm{Ti})}(\mathbf{r}) + \sum_{j=1}^{6} V^{(\mathrm{oxy})}(\mathbf{r} - \mathbf{R}_j). \tag{3.1}$$

The very notion of a $3d$-state derives from the eigenvalue problem of the spherically symmetrical $\mathcal{H}^{(\mathrm{Ti})}$. If the anisotropic field produced by the oxygen ions were very strong, we should start the solution of the Schrödinger equation from scratch. Fortunately, it is an acceptable approximation to say that the *crystal field* due to the surrounding ions is not particularly strong, and thus it does not mix different shells. It will, however, strongly mix states within the same shell because these are degenerate at the level of $\mathcal{H}^{(\mathrm{Ti})}$. Our task is to diagonalize (3.1) in the subspace of the ionic $3d$-states: hence the subscript "$3d$" given to $\mathcal{H}$. Essentially, we are doing atomic physics in an anisotropic external potential: this is the subject of crystal field theory.

We might start with adopting some concrete expression for $V^{(\mathrm{oxy})}$ (it would be a standard approximation to replace the oxygen ions by point charges) but we do not really need it. It turns out that a lot can be learnt without specifying the details of the Hamiltonian. The most essential feature of $\mathcal{H}_{3d}$ is its *symmetry*. The ionic Hamiltonian $\mathcal{H}^{(\mathrm{Ti})}$ has full rotational symmetry; in contrast, $\mathcal{H}_{3d}$ has only the symmetries of the octahedron shown in Fig. 3.1. As we are going to see, this is enough to deduce some essential features of the level scheme of the d-electrons.

Putting emphasis on the symmetries immediately rids us of a possible worry: what about the farther neighbourhood of our Ti ion? Beyond the octahedron of the first-neighbour oxygen ions lies a cube of eight La^{3+} ions: perhaps their effect is not entirely negligible? The point is that as long as we rely on symmetry considerations, such complications do not matter. The field of the La-ions has the same symmetry as that of the oxygen ions, and the same is true about the more distant environment[4]. The octahedron in Fig. 3.1 has the symmetries of the entire lattice, as seen from the central Ti-ion. The symmetry operations (rotations, reflexions, and inversion) which leave the central Ti-ion in place, form a group, namely, the so-called *point group*. The point group is the symmetry group of the crystal field Hamiltonian $\mathcal{H}_{3d}$. – The next section reviews as much of the theory of group representations as is needed to deal with the problem at hand.

Before going on, let us mention another example of crystal-field effects. The Mott insulator La_2CuO_4 is the parent compound of the first-discovered high-temperature superconductor [47]. La_2CuO_4 has an orthorombic distortion for reasons which we will understand later. The ideal, undistorted structure of many A_2BO_4 compounds is the so-

[4]For complicated structures, it may not be obvious that successive coordination shells have the same symmetry, and this has to be checked carefully.

called K_2NiF_4 structure [84] which differs from the perovskite structure by having *two* AO planes (rather than one AO plane) in between two consecutive BO_2 planes. The CuO_2 plane of La_2CuO_4 (shown in the right panel of Fig. 3.1) is a famous object because it is thought to be the relevant structural unit of the high-T_c cuprates. Thus the local environment of the Cu ions is approximately octahedral as in Fig. 3.1. Identifying the valence states as $La_2^{3+}Cu^{2+}O_4^{2-}$, we immediately see that the only "interesting" ion (i.e., one with a partially filled shell) is Cu^{2+} which has the configuration $3d^9$. Since $3d^{10}$ would be a completely filled shell, it is justified to consider Cu^{2+} as having one d-hole. The problem of the electronic structure of Cu^{2+} is thus the mirror image of that of the structure of Ti^{3+}: there we were interested in the only d-state which is occupied; here in the only state which is unoccupied.

Crystal field effects will also give an explanation of the missing orbital contribution to the susceptibility of transition metal ions. This raises, however, another question: why have crystal fields seemingly no effect on $4f$-electrons? In fact they do have an effect, only a much weaker one than on $3d$-electrons. The reason is that the $4f$ orbitals are lying so deep within the ion core that other occupied shells of the same ion largely screen out the potential of the surrounding ions. $4f$ electrons indeed hardly notice whether they are in an ion embedded in a crystal, or in a free ion. The $4f$ electrons in rare-earth-based solids give an example of a *weak crystal field* satisfying

$$\text{exchange splittings} \quad > \quad \text{spin–orbit coupling} \quad > \quad \text{crystal fields}$$

Because of this order of the relevant coupling strengths, even Hund's third rule takes precedence over lattice effects. Crystal field effects (and quite subtle ones at that!) act only within a given J-manifold.

Transition metal ions (particularly $3d$-ions) belong to the category of *intermediate crystal field* strengths

$$\text{exchange splittings} \quad > \quad \text{crystal fields} \quad > \quad \text{spin–orbit coupling}$$

Since crystal fields now dominate over the spin–orbit coupling, Hund's third rule ceases to apply. This means that though L and S remain valid quantum numbers, and their values are still given by Hund's first and second rule, J is no longer a good quantum number. An intermediate crystal field mixes states within a given (L, S) term.

Strong crystal fields are comparable to (or larger than) the exchange splittings giving rise to Hund's first and second rule, thus they mix even states belonging to different terms. It is quite usual to find strong crystal fields in $4d$ and $5d$ transition metal compounds. On the other hand, there are only rare instances of insulating solids where $3d$ ions are subject to such strong crystal fields that even Hund's first rule is put out of action. In any case, the description of strong effects of surrounding anions usually requires going beyond the simple ionic model, and taking into account the mixing of the d-orbitals of the central cation with the p-orbitals of the neighbouring anions. The binding of the substance becomes partially covalent, and the details of the electronic structure are rather different from the simple ionic picture we are using. However, the symmetry analysis of the orbitals remains essentially the same which we introduce in this Chapter. As a matter of terminology, let us remark that considering the effects of the mixing between cation- and anion-orbitals is the subject of *ligand field theory* [379], while the term "crystal field theory" is reserved for the purely ionic description.

3.2 The Role of Symmetry Arguments in Quantum Mechanics

A detailed account of the theory of representations of symmetry groups, and their application to solid state physics, would require an entire course[5]. Here we assume that the reader is already familiar with the elements of group theory and recall just the arguments why group theory is useful in solving the quantum mechanical eigenvalue problem.

Let us assume that there exists a set of geometric transformations $\mathcal{P} : \mathbf{r} \to \mathbf{r}'$ which leave the Hamiltonian $\mathcal{H}$ invariant: $\mathcal{H}(\{\mathbf{r}'\}) = \mathcal{H}(\{\mathbf{r}\})$. (In general, $\mathcal{H}$ depends on a number of position vectors $\mathbf{r}$, which are all transformed by $\mathcal{P}$. $\{\mathbf{r}\}$ stands for the set of all the $\mathbf{r}$ vectors. However,

[5] In the author's experience, the typical student will have taken courses in group theory, which were usually rather more advanced than what we need here, but put less emphasis on applications to finite groups. The aim of the present text is to refresh the reader's memory about group theory, going into details about what we are going to use, and omitting (as far as the continuity of the argument allows) what we do not need. For a systematic treatment, see [399] or [195]. Other texts are quoted elsewhere.

in what follows, we use the simple notation $\mathbf{r}$ for a collection of position vectors). We have in mind translations, rotations and reflections as well as their suitable combinations (screw axes and glide planes). These are volume- and angle-preserving transformations which always have an inverse $\mathcal{P}^{-1}$. These transformations form the symmetry group of the Hamilton operator (also called the group of the Schrödinger equation). In addition, the Hamiltonian can have non-geometrical symmetries like time reversal invariance; this will be discussed separately.

The transformation of $\mathbf{r}$ in coordinate space induces a transformation in the space of functions of $\mathbf{r}$

$$\hat{\mathcal{P}}\phi(\mathbf{r}) = \phi(\mathcal{P}^{-1}\mathbf{r}). \tag{3.2}$$

(A momentary reflection convinces us that putting $\mathcal{P}^{-1}\mathbf{r}$ into the argument changes the contours of ϕ according to $\mathcal{P}$). The group of the operators $\hat{\mathcal{P}}$ is isomorphic to the group of the symmetry operations $\mathcal{P}$; we denote both by G.

By definition

$$\begin{aligned}(\hat{\mathcal{P}}\mathcal{H})\phi &= \hat{\mathcal{P}}(\mathcal{H}(\mathbf{r})\phi(\mathbf{r})) = \mathcal{H}(\mathcal{P}^{-1}\mathbf{r})\phi(\mathcal{P}^{-1}\mathbf{r}) = \mathcal{H}(\mathbf{r})(\hat{\mathcal{P}}\phi) \\ &= (\mathcal{H}\hat{\mathcal{P}})\phi \end{aligned} \tag{3.3}$$

i.e., $\mathcal{H}$ commutes with all symmetry operators $\hat{\mathcal{P}}$

$$[\hat{\mathcal{P}}, \mathcal{H}] = 0. \tag{3.4}$$

The solution of the eigenvalue problem is often greatly simplified by introducing the symmetry classification of the eigenstates. Let us single out one symmetry element $\hat{\mathcal{P}}$, and work in a basis which consists of the eigenstates of $\hat{\mathcal{P}}$. (3.4) becomes

$$\begin{aligned}\langle i|[\hat{\mathcal{P}}, \mathcal{H}]|k\rangle &= \sum_j \langle i|\hat{\mathcal{P}}|j\rangle \cdot \langle j|\mathcal{H}|k\rangle - \sum_j \langle i|\mathcal{H}|j\rangle \cdot \langle j|\hat{\mathcal{P}}|k\rangle \\ &= \langle i|\mathcal{H}|k\rangle \cdot (\langle i|\hat{\mathcal{P}}|i\rangle - \langle k|\hat{\mathcal{P}}|k\rangle) = 0. \end{aligned} \tag{3.5}$$

It follows that $\mathcal{H}$ can have non-vanishing matrix elements only between states of the same symmetry, i.e., between states belonging to the same eigenvalue of $\hat{\mathcal{P}}$. To cite a familiar example: if $\hat{\mathcal{P}}$ is the inversion, then $\mathcal{H}$ does not connect even states with odd states.

If all symmetry operators commute with each other then according to (3.4) we can seek a complete set of common eigenstates of $\mathcal{H}$, and of all the $\hat{\mathcal{P}}$ operators. Within a common eigensubspace of the $\hat{\mathcal{P}}$'s, the matrix elements of $\mathcal{H}$ will be, in general, non-vanishing, and all the eigenenergies are different. In other words: if the symmetry group is Abelian (commutative) then the energy levels are expected to be non-degenerate. If we still find a degeneracy then it is not symmetry-related and is therefore called accidental.

It is often found that not all symmetry operators commute: such is the case with rotations about different axes, or rotations and translations. Now symmetry operations have an immediate usefulness because they permit us to generate new eigenstates. Namely, we must be able to find an energy eigenstate ψ_n for which $\hat{\mathcal{P}}\psi_n$ is linearly independent of ψ_n. Then what about $\hat{\mathcal{P}}\psi_n$? We apply (3.3) to find

$$\mathcal{H} \cdot \hat{\mathcal{P}}\psi_n = \hat{\mathcal{P}} \cdot \mathcal{H}\psi_n = E_n \cdot \hat{\mathcal{P}}\psi_n. \tag{3.6}$$

If ψ_n is an eigenstate with eigenvalue E_n, then $\hat{\mathcal{P}}\psi_n$ is also an eigenstate with the same eigenvalue. Once we have found an eigenstate ψ_n, we can apply the symmetry operations to generate further eigenstates. The degeneracy of the eigenstates $\hat{\mathcal{P}}\psi_n$ is symmetry-related. However, it does not follow that all the states $\hat{\mathcal{P}}\psi_n$ should be independent; the dimensionality of the eigensubspace $\{\hat{\mathcal{P}}\psi_n | \ \hat{\mathcal{P}} \in G\}$ is usually much smaller than the number $\mathcal{N}$ of the group elements ($\mathcal{N}$ is called the order of the group). The question arises whether group theory allows us to make general statements about the degree of degeneracy. Such is indeed the case: The dimensionalities of the irreducible representations of G give the possible degeneracies.

3.2.1 Irreducible Representations

It should be stressed again that it is not our purpose here to give a systematic development of group representations. We just want to recall those results which permit us to proceed straight to the treatment of the crystal field splittings of ionic levels.

Starting from an eigenstate ψ_n of $\mathcal{H}$, let us apply to it all the symmetry operators $\hat{\mathcal{P}}$. The resulting set of functions forms a d-dimensional subspace of degenerate eigenstates with the property that applying any

$\hat{\mathcal{P}}$ to any of the states leads to a state which is lying in the same subspace. Choosing a d-element basis $\{\psi_j\}$, the action of the $\hat{\mathcal{P}}$ can be described by the $d \times d$ matrices $\Gamma(\mathcal{P})$

$$\hat{\mathcal{P}} \cdot \psi_j = \sum_k \Gamma_{kj}(\mathcal{P})\psi_k. \tag{3.7}$$

The matrices form a group which is homomorphic to G since

$$\begin{aligned} \hat{\mathcal{P}}_1\hat{\mathcal{P}}_2 \cdot \psi_j &= \hat{\mathcal{P}}_1 \cdot \sum_k \Gamma_{kj}(\mathcal{P}_2)\psi_k = \sum_k \sum_l \Gamma_{lk}(\mathcal{P}_1)\Gamma_{kj}(\mathcal{P}_2)\psi_l \\ &= \sum_l [\Gamma(\mathcal{P}_1) \cdot \Gamma(\mathcal{P}_2)]_{lj}\psi_l. \end{aligned} \tag{3.8}$$

We say that the Γ-matrices give a d-dimensional representation of G. The representation characterizes the symmetry properties of the subspace in question. These properties are of course independent of the choice of the basis $\{\psi_j\}$. Since a change of basis is effected by a unitary transformation U, we are led to consider the representations Γ and Γ' equivalent if they are connected by a unitary transformation:

$$\Gamma'(\mathcal{P}) = U^{-1}\Gamma(\mathcal{P})U. \tag{3.9}$$

Two representations are inequivalent, if such a connecting transformation cannot be found. Two states have different symmetries if they transform according to unitarily inequivalent representations.

Since the trace of a matrix is left unchanged by the transformation (3.9), it is useful to introduce the *character of a representation* defined by

$$\chi(\mathcal{P}) = \sum_i \Gamma_{ii}(\mathcal{P}) = \mathrm{Tr}(\Gamma(\mathcal{P})). \tag{3.10}$$

For the unit element $\mathcal{E}$ of the group $\chi(\mathcal{E}) = d$, the dimensionality of the representation.

It may happen that we can find a transformation U which brings all matrices to the same block form

$$U^{-1}\Gamma(\mathcal{P})U = \begin{pmatrix} \Gamma_1(\mathcal{P}) & 0_1 \\ 0_2 & \Gamma_2(\mathcal{P}) \end{pmatrix} \tag{3.11}$$

where $\Gamma_1(\mathcal{P})$ is a $d_1 \times d_1$ matrix, and $\Gamma_2(\mathcal{P})$ a $d_2 \times d_2$ matrix, d_1 and d_2 being the same for all $\mathcal{P}$. 0_1 is the $d_1 \times d_2$ null matrix, and 0_2 the $d_2 \times d_1$

null matrix. – In this case, we call the representation Γ reducible: it has been decomposed into the representations Γ_1 and Γ_2.

What has happened is that instead of the original basis $\psi_1, \ldots, \psi_d$ we managed to find new basis states $\phi_1^{(1)}, \ldots, \phi_{d_1}^{(1)}$, and $\phi_1^{(2)}, \ldots, \phi_{d_2}^{(2)}$, so that the states $\phi_j^{(1)}$ are mixed only among themselves (and similarly for $\phi_j^{(2)}$). Clearly, the symmetry does not require a d-fold degeneracy; it may require a d_1-fold (and a d_2-fold) degeneracy – unless the representations Γ_1 and Γ_2 are themselves reducible.

If it is impossible to find a transformation U with the property (3.11), we call the representation irreducible. The degeneracy of an eigenvalue of $\mathcal{H}$ should thus be the dimensionality of one of the irreducible representations of the symmetry group[6].

We are frequently faced with the following problem: we have identified a (perhaps rather small) number of basis states which span the physically relevant subspace of the Hilbert space. We find, however, that these form the basis of a reducible representation. Quite often, it is not so much the detailed solution of the actual eigenvalue problem that we are interested in but rather in predicting how many different energy levels are expected, with which degeneracies. We want a machinery for the decomposition of reducible representations.

Returning to (3.11): for the characters,

$$\chi(\mathcal{P}) = \chi_1(\mathcal{P}) + \chi_2(\mathcal{P}) \tag{3.12}$$

holds. This suggests that the decomposition of a representation can be found by looking for a corresponding decomposition of its character. In this, we are helped by the following orthogonality theorem

$$\sum_{\mathcal{P}} \chi_i^*(\mathcal{P})\chi_j(\mathcal{P}) = \mathcal{N}\delta_{ij} \tag{3.13}$$

where $\mathcal{N}$ is the order of the group.

[6]These are the degeneracies which we can explain as following from the symmetry. It is possible that additional degeneracies are found; these should be considered (and are called) accidental, resulting from a special choice of the parameters of the Hamiltonian. It may happen, however, that a seemingly inexplicable degeneracy can be explained as the consequence of a previously unrecognized greater symmetry of the problem. The degeneracy of the different l-levels in the non-relativistic hydrogen atom problem is a case in point.

The character of a general (reducible) representation can be expressed as

$$\chi(\mathcal{P}) = \sum_j a_j \chi_j(\mathcal{P}). \tag{3.14}$$

The sum runs over the indices of the irreducible representations; the jth irreducible representation occurs a_j times. From (3.13) it follows that

$$a_j = \frac{1}{\mathcal{N}} \cdot \sum_{\mathcal{P}} \chi_j^*(\mathcal{P})\chi(\mathcal{P}). \tag{3.15}$$

Thus what we need is the character table of the irreducible representations. The form of the character table is quite remarkably constrained by two theorems:

Theorem 1. The sum of the squares of the dimensionalities of the irreducible representations is equal to the order of the group

$$\sum_{j=1}^{r} d_j^2 = \mathcal{N} . \tag{3.16}$$

To formulate the second theorem, we have to recall the notion of conjugacy classes. Two elements $\mathcal{P}$ and $\mathcal{Q}$ of a group are conjugate if there exists a group element $\mathcal{R}$ such that $\mathcal{P} = \mathcal{R}\mathcal{Q}\mathcal{R}^{-1}$. The set of all elements conjugate to $\mathcal{P}$ forms a conjugacy class. Since conjugacy is a symmetric relation (in fact, an equivalence relation), the element $\mathcal{P}$ does not play any distinguished role, the conjugacy class can be associated with any of its elements. It also follows that the group can be unambiguously decomposed into conjugacy classes. Now we can formulate

Theorem 2: The number of inequivalent irreducible representations of a finite group is equal to the number of its conjugacy classes.

In the context of the point groups which are of immediate interest to us, conjugacy classes have a simple meaning: they are comprised of symmetry operations whose nature is the same. For instance, rotations by the same angle about equivalent axes belong to the same conjugacy class.

It should be obvious that if $\mathcal{P}$ and $\mathcal{Q}$ are conjugate then they have the same character: $\chi(\mathcal{P}) = \chi(\mathcal{Q})$. This permits us to rewrite (3.15) in the useful form

$$a_j = \frac{1}{\mathcal{N}} \cdot \sum_{k=1}^{r} \mathcal{N}_k \chi_j^*(\mathcal{C}_k)\chi(\mathcal{C}_k) \tag{3.17}$$

where the summation is over the classes $\mathcal{C}$ and $\mathcal{N}_k$ is the number of elements in the kth conjugacy class $\mathcal{C}_k$ ($\sum \mathcal{N}_k = \mathcal{N}$, the order of the group).

3.3 The Octahedral Group

One of the most important symmetry groups in solid state physics is the octahedral group $\mathcal{O}$ which consists of all the rotations which take an octahedron into itself. The group characterizes, as far as rotations are concerned, the symmetry of the crystal field produced by six oxygen ions surrounding a central metal ion (Fig 3.1). The very same group would describe the symmetry of a crystal field produced by eight atoms situated at the corners of a cube. The cube and the octahedron are, of course, different geometric objects but their symmetry (and thus symmetry group) is the same (a glance at Fig. 3.2 should convince us). For this reason we may speak of the octahedron of atoms as a "cubic environment" which gives rise to a "cubic field".

Naturally, the octahedron (or cube) has further symmetries, including foremost the inversion $\mathcal{I}$, and reflections. Since $\mathcal{I}$ commutes with any rotation, a larger symmetry group including $\mathcal{I}$ can be built as the direct product[7] $\mathcal{O}_h = \mathcal{O} \otimes \{\mathcal{E}, \mathcal{I}\}$. The resulting full octahedral group $\mathcal{O}_h$ is the complete symmetry group of the octahedron (or cube). This may not be immediately obvious because we did not explicitely include such symmetry elements as reflection planes. We can, however, convince ourselves that the reflections, and the relevant combinations of reflections and rotations[8], can be identified as the product of $\mathcal{I}$ and a rotation, and are thus contained in $\mathcal{O}_h$.

Fig. 3.2 helps us to enumerate the elements of $\mathcal{O}$, and divide them into conjugacy classes. We have to remember that rotations by the same angle about equivalent axes form a conjugacy class. Two axes are equivalent if one is taken into the other by a symmetry operation.

---

[7]The general definition: if all elements a of the group A commute with all the elements b of the group B, and the only common element of A and B is the unit element, then the set of all the products ab is the direct product group $A \otimes B$.

[8]These are called "improper rotations", in contrast to the simple rotations which are then called "proper".

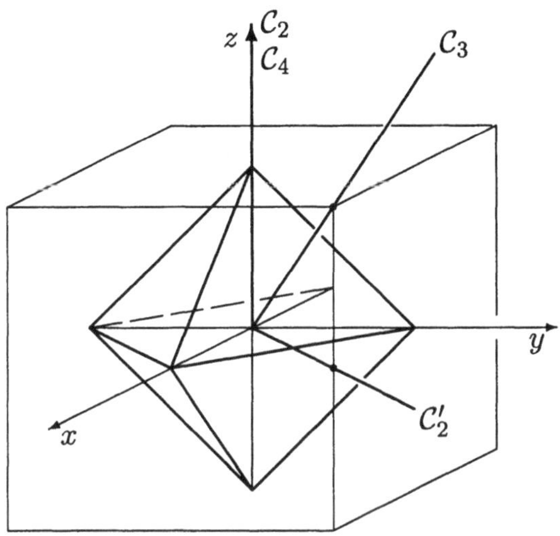

Figure 3.2: Rotational symmetries of the octahedron and the cube. To keep the figure simple, only one C_3 axis and one C_2' axis is shown.

Consider first rotations about the z-axis. It is clear that a $\pi/2$-rotation C_4 is a symmetry element; so are the repeated transformations C_4^2, C_4^3, and $C_4^4 = \mathcal{E}$, the unit element. The z-axis is a fourfold axis; and so are the x-, and y-axes. These axes are equivalent because, for instance, a $\pi/2$ rotation about the z-axis takes the x-axis into the y-axis. Hence $\pi/2$-rotations about the three axes belong to the same conjugacy class. Furthermore, a π-rotation about the x-axis takes the z-axis into the $(-z)$-axis, and a $\pi/2$-rotation about the $-z$-axis is the same as a $3\pi/2$-rotation about the z-axis, thus the $3\pi/2$-rotations are in the same conjugacy class as the $\pi/2$-rotations. There are altogether six elements in the conjugacy class $6C_4$. The π-rotations form the three-element class $3C_2$.

There is a different kind of π-rotation: it is about axes which are passing through the centre of the cube, parallel to the face diagonals (one of these axes, denoted by C_2', is shown in Fig. 3.2). Since there is no element of $\mathcal{O}$ which would take a coordinate axis into C_2', the six π-rotations discussed here form a conjugacy class of their own, which we denote by $6C_2'$.

Table 3.1: Character table of the octahedral group $\mathcal{O}$. (Two of the several standard notations of the representations are shown.)

		basis	$\mathcal{E}$	$8C_3$	$3C_2$	$6C_2'$	$6C_4$
A_1	Γ_1	$\{x^2+y^2+z^2\}$	1	1	1	1	1
A_2	Γ_2	$\{xyz\}$	1	1	1	-1	-1
E	Γ_3	$\{x^2-y^2, 3z^2-r^2\}$	2	-1	2	0	0
T_1	Γ_4	$\{x, y, z\}$	3	0	-1	-1	1
T_2	Γ_5	$\{xy, yz, zx\}$	3	0	-1	1	-1

Body diagonals of the cube are threefold axes C_3. Since threefold axes can be turned upside down by C_2'-rotations, $\pi/3$- and $2\pi/3$-rotations are in the same class, building the eight-element class $8C_3$.

The unit element forms a conjugacy class in itself.

There being five conjugacy classes, the Theorem 2. tells us that there are five irreducible representations. Since there is only one way to satisfy (3.16), their dimensionalities must be $d_1 = 1$, $d_2 = 1$, $d_3 = 2$, $d_4 = 3$, and $d_5 = 3$. The first one-dimensional representation which we denote by A_1, must be the identity representation where $\Gamma(\mathcal{P}) = 1$ for each element of the group.

The other representations are easily constructed once suitable basis functions have been found. In our case, there is a simple standard choice for these, which we included in Table 3.1. For the simplest applications, however, we do not need the detailed form of the representations; it is enough to know the characters.

To construct the character table, let us remember that conjugated elements have the same character, thus it is enough to consider one element from each class. For example's sake, let us consider the three-dimensional irreducible representation T_2. $\mathcal{E}$ is represented by the 3×3 unit matrix, so $\chi(\mathcal{E}) = 3$. A $2\pi/3$-rotation about the C_3-axis shown in Fig. 3.2 has the effect $x \to y \to z \to x$, so its matrix in the basis $\{xy, yz, zx\}$ is

$$R_1 = \begin{pmatrix} 0 & 1 & 0 \\ 0 & 0 & 1 \\ 1 & 0 & 0 \end{pmatrix}$$

and Trace$(R_1) = \chi(C_3) = 0$. Similarly, a π-rotation about the C_2'-axis shown in the figure effects $x \to y \to x,\ \ z \to -z$, so its matrix is

$$R_2 = \begin{pmatrix} 1 & 0 & 0 \\ 0 & 0 & -1 \\ 0 & -1 & 0 \end{pmatrix}$$

with Trace$(R_2) = \chi(C_2') = 1$. Other entries of the character table can be determined similarly.

For subsequent applications, we have to be familiar with some of the lower-symmetry environments (tetragonal, etc.). The discussion of these is relegated to the Problems:

Problem 3.1: *Tetragonal symmetry.* A tetragonal distortion of an originally cubic environment can be envisaged as the elongation of the octahedron along the z-axis (Fig. 3.6). Alternatively, it is the symmetry of a rectangular prism whose base is a square of side a, and its height is $c \neq a$. As before, we are interested only in the subgroup of proper rotations.

a) Enumerate the group elements, and the conjugate classes! Determine the number, and the dimensionalities, of the irreducible representations.

b) Construct the character table, and find convenient basis states!

c) What happens to the irreducible representations of the octahedral group when the symmetry is lowered to tetragonal?

Problem 3.2: The same as Problem 3.1, for *orthorombic* symmetry. An orthorombic distortion lowers the symmetry from tetragonal by keeping the prism rectangular, but making all three sides a, b, and c unequal.

3.4 Symmetry Properties of Atomic States

The angular dependence of the atomic wave functions is described by the familiar spherical harmonics $Y_l^m(\vartheta, \varphi) \sim P_l^m(\cos\vartheta)e^{im\varphi}$ (see Appendix A). This results from the fact that the atomic potential is spherically symmetrical: the symmetry group of the Hamiltonian is[9] the continuous group $SO(3)$, the group of all rotations in three-dimensional space. It follows from the full rotational symmetry that angular momentum is a good quantum number, and that the atomic eigenstates can be sought as angular momentum eigenstates. In fact, l and m are just the eigenvalues

[9]There is also inversion symmetry which does not concern us at the moment.

(in units of $\hbar$) of the orbital angular momentum, and its z-component

$$\begin{aligned} \mathbf{L}^2 Y_l^m(\vartheta,\varphi) &= \hbar^2 l(l+1) Y_l^m(\vartheta,\varphi) \\ L^z Y_l^m(\vartheta,\varphi) &= \hbar m Y_l^m(\vartheta,\varphi). \end{aligned} \tag{3.18}$$

We need to know how wave functions transform under the action of rotations. We consider first an infinitesimal rotation by the angle $\delta\varphi$

$$\mathcal{R}_{\delta\varphi} \cdot \mathbf{r} = \mathbf{r} + \delta\mathbf{r} = \mathbf{r} + \vec{\delta\varphi} \times \mathbf{r}\,. \tag{3.19}$$

As usual, the vector $\vec{\delta\varphi}$ is pointing along the axis of rotation, and its length is $\delta\varphi$. The origin of the vector space $\mathbf{r}$ is a point lying on the rotation axis. Performing the corresponding transformation on the wave function

$$\begin{aligned} \hat{\mathcal{R}}_{\delta\varphi}\psi(\mathbf{r}) &= \psi(\mathbf{r}-\delta\mathbf{r}) = \psi(\mathbf{r}) - (\vec{\delta\varphi}\times\mathbf{r})\nabla\psi \\ &= \psi - \vec{\delta\varphi}\cdot(\mathbf{r}\times\nabla)\psi = \left[1 - \frac{i}{\hbar}\vec{\delta\varphi}\cdot(\mathbf{r}\times\mathbf{p})\right]\psi(\mathbf{r}) \end{aligned} \tag{3.20}$$

where $\mathbf{p}$ is the momentum operator, allows us to identify

$$\hat{\mathcal{R}}_{\delta\varphi} = 1 - \frac{i}{\hbar}\vec{\delta\varphi}\cdot\mathbf{L}. \tag{3.21}$$

The dimensionless angular momentum $\mathbf{L}/\hbar$ is the infinitesimal generator for rotations.

A finite rotation by angle φ is made up of a large number of infinitesimal rotations

$$\hat{\mathcal{R}}_{\varphi} = \lim_{n\to\infty}\left(1 - \frac{i}{\hbar}\frac{\vec{\varphi}}{n}\cdot\mathbf{L}\right)^n = \exp\left\{-\frac{i}{\hbar}\vec{\varphi}\cdot\mathbf{L}\right\}. \tag{3.22}$$

Since the transformation operators $\hat{\mathcal{R}}_{\varphi}$ are solely functions of the angular momentum operator $\mathbf{L}$ (in fact, they can be expressed as Taylor series in $\mathbf{L}$), eigensubspaces of $\mathbf{L}$ are automatically eigensubspaces of all rotations. It follows that the spherical harmonics Y_l^m, for $m = l, l-1, \ldots, -l$ form the basis of a $(2l+1)$-dimensional irreducible representation of $SO(3)$. For instance, the d-states provide a five-dimensional representation. – It can be shown that thereby we have found all the (infinitely many) odd-dimensional representations.

$SO(3)$ has infinitely many conjugacy classes. Since for an isotropic system, all rotation axes are equivalent, a conjugacy class consists of all rotations by a given angle α. To determine the character of a representation, it is enough to consider one element from each conjugacy class. It is very convenient to perform rotations about the z-axis, because this influences only φ, leaving ϑ unchanged. Then the factor P^l_m plays no role and we have a representation over the functions $e^{im\varphi}$. Replacing $\mathbf{L}$ by $L^z = \frac{\hbar}{i}\frac{\partial}{\partial\varphi}$ in (3.22), the effect of an α-rotation is

$$\begin{aligned} \exp\left(-\alpha\frac{i}{\hbar}\frac{\hbar}{i}\frac{\partial}{\partial\varphi}\right)e^{im\varphi} &= \left[1-\alpha\frac{\partial}{\partial\varphi}+\frac{\alpha^2}{2!}\frac{\partial^2}{\partial\varphi^2}+\ldots\right]e^{im\varphi} \\ &= \left[1-i\alpha m+\frac{(i\alpha m)^2}{2!}+\ldots\right]e^{im\varphi} = e^{im(\varphi-\alpha)}. \end{aligned} \tag{3.23}$$

The matrix representing the α-rotation is

$$\begin{pmatrix} e^{-il\alpha} & 0 & \ldots & 0 \\ 0 & e^{-i(l-1)\alpha} & \ldots & 0 \\ & & \ddots & \\ 0 & 0 & \ldots & e^{il\alpha} \end{pmatrix} \tag{3.24}$$

with the trace

$$\begin{aligned} \chi^{(l)}(\alpha) &= \mathrm{Tr}(\Gamma^{(l)}(\alpha)) = e^{-il\alpha}+e^{-i(l-1)\alpha}+\ldots+e^{il\alpha} \\ &= \frac{e^{i(l+1/2)\alpha}-e^{-i(l+1/2)\alpha}}{e^{i\alpha/2}-e^{-i\alpha/2}} = \frac{\sin(l+1/2)\alpha}{\sin(\alpha/2)}. \end{aligned} \tag{3.25}$$

3.5 Splitting of a *d*-Level in Cubic Field

The octahedral group is a subgroup of the rotation group $\mathcal{O} \subset SO(3)$, so any representation of $SO(3)$ provides a representation of $\mathcal{O}$. However, irreducible representations of $SO(3)$ will be in general reducible representations of $\mathcal{O}$. The question of reducibility is easily decided by checking the characters. In Table 3.2 we list the characters of the representations provided by the s-, p-, d-, and f-functions. The table was constructed using the formula (3.25) which gives the character of any

Table 3.2: Character table of the representations of the octahedral group $\mathcal{O}$, using s-, p-, d-, and f-like atomic orbitals as basis.

	l	E	$8C_3$	$3C_2$	$6C_2'$	$6C_4$
Γ_s	0	1	1	1	1	1
Γ_p	1	3	0	-1	-1	1
Γ_d	2	5	-1	1	1	-1
Γ_f	3	7	1	-1	-1	-1

rotation, including those which make up the octahedral group $\mathcal{O}$. For instance, for C_4

$$\chi(C_4) = \chi\left(\frac{\pi}{2}\right) = \frac{\sin\left((2l+1)\pi/4\right)}{\sin\left(\pi/4\right)} = \begin{cases} 1 & l = 0, 1, 4, 5, \ldots \\ -1 & l = 2, 3, 6, 7, \ldots \end{cases}$$

The notations $\Gamma_s, \Gamma_p, \ldots$ are self-explanatory but not standard.

Let us now return to the question posed in Section 3.1: what happens to the d-states in a cubic environment? The representation Γ_d is five-dimensional, and the irreducible representations of the octahedral group are at most three-dimensional, so Γ_d must be reducible. The way it gets decomposed can be determined using the formula (3.17), and the characters given in Tables 3.1 and 3.2. We find

$$\Gamma_d = E \oplus T_2 \tag{3.26}$$

Thus the fivefold degeneracy of the d-states is lifted by the crystal field and the d-level is split into a twofold degenerate E and a threefold degenerate T_2 level (also commonly called e_g and t_{2g} levels where the subscript g reminds us[10] that here we are speaking about even functions).

Group theory can only tell us into how many levels, with which degeneracies, a given set of states gets split. It cannot tell us the size of the splitting, neither the order of the levels. To learn about these, the details of the Hamiltonian have to be specified and an actual calculation has to be done, which is simplified, but not supplanted, by symmetry considerations. The splitting of the d-level is, however, such a simple

[10]from the German word *gerade*.

affair that with a little more effort, it is possible to guess the order of the levels.

At the level of the atomic (ionic) Hamiltonian, we are free to choose any five linearly independent states as a basis. (A.9)–(A.11) (or (A.12)–(A.14)) give the standard complex form of the $l = 2$ spherical harmonics. Here, we find it instructive to go over to a basis of *real wave functions.* When writing these down, we include the common factor r^2 which describes a part of the hitherto neglected radial dependence of the actual $3d$-states (but we still omit the factor $e^{-r/\xi}$ which is uninteresting from our present point of view). The real basis functions are the so-called *tesseral harmonics* [178] of the general form $Y_l^m \pm Y_l^{-m}$

$$\left.\begin{array}{lcl} r^2[Y_2^2(\vartheta,\varphi) - Y_2^{-2}(\vartheta,\varphi)] & \sim & xy \\ r^2[Y_2^1(\vartheta,\varphi) - Y_2^{-1}(\vartheta,\varphi)] & \sim & xz \\ r^2[Y_2^1(\vartheta,\varphi) + Y_2^{-1}(\vartheta,\varphi)] & \sim & yz \end{array}\right\} \quad T_2$$

$$\left.\begin{array}{lcl} r^2[Y_2^2(\vartheta,\varphi) + Y_2^{-2}(\vartheta,\varphi)] & \sim & x^2 - y^2 \\ r^2 Y_2^0(\vartheta,\varphi) & \sim & 3z^2 - r^2 \end{array}\right\} \quad E\,. \tag{3.27}$$

Introducing real basis states has immediately led to finding the standard basis of the T_2 and E representations (see Table 3.1).

The charge distribution of one representative of the E-type, and the T_2-type, functions is shown in Fig. 3.3. We see that the E functions have lobes pointing along the coordinate axes, while the lobes of the T_2 basis states point along the (110) directions bisecting the angle between two coordinate axes. Recalling Fig. 3.1, we readily understand the reason for the energy difference between the E and T_2 levels. The E (or e_g) states have electron charge maxima near to the negatively charged oxygen ions, while the T_2 (or t_{2g}) states manage to avoid these repelling charges, as much as it is possible for d-states. We conclude that in an octahedral environment of anions, the T_2 level is lying lower than the E level. On the other hand, if we place anions at the corners of the cube shown in Fig. 3.2, we find the T_2 orbitals getting nearer to the repulsive charge of the ions, hence the order of the levels must be reversed (Fig. 3.4).

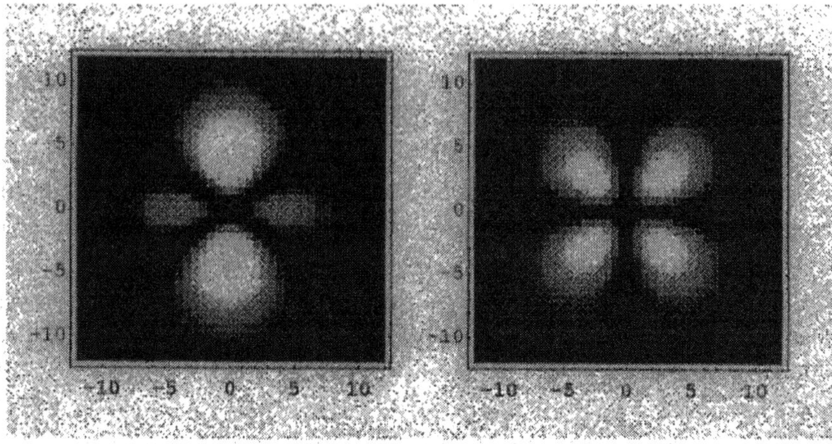

Figure 3.3: d-electron clouds in a cubic crystal field. The lighter areas correspond to higher electron density. Left: the charge distribution of a $(3z^2 - r^2)$-orbital in the x-z plane. The charge is strongly peaked along the z-axis. Right: the charge distribution of a xy-orbital in the x-y plane. Charge maxima lie in between the principal axes.

3.5.1 Quenching the Orbital Angular Momentum

Now we are in the position to reconsider the question posed at the beginning of this Chapter: why is it that incomplete d-shells often do not have an orbital contribution to their static susceptibility? We might be tempted to argue like this: for each of the states (3.27), $\langle L^z \rangle = 0$. The cubic symmetry permits us to conclude that the same must be true for the other two components of **L**. The orbital angular momentum (and the corresponding magnetic moment) is quenched.

This would be, however, not really conclusive because within a degenerate set of levels, one is free to choose any complex linear combinations as a new basis. One often uses the real basis functions (3.27) but this is not compulsory. Once this is realized, it makes a big difference whether the partially occupied level is the E level, or the T_2 level. A momentary reflection convinces us that no matter how we combine the two E states, the expectation value of all the three components of **L** vanishes. On the other hand, the basis of the T_2 representation can be redefined to contain the complex angular momentum eigenstates $Y_2^{\pm 1}$. If we have just one d-electron, nothing prevents it from occupying the

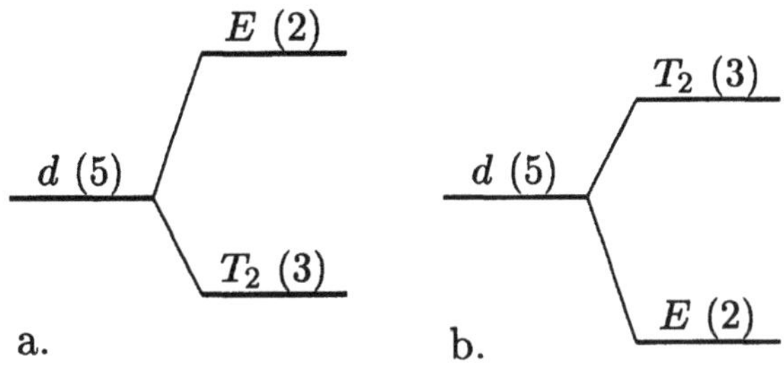

Figure 3.4: The splitting of a d-level in the field of $a.$: six anions forming an octahedron, and $b.$: eight anions forming a cube.

T_2-type state Y_2^{+1}, and thus having $L^z = 1$. Thus it cannot surprise us that a single d-electron in the T_2 level responds to a magnetic field as if it had the pseudo angular momentum $l = 1$. Why the angular momentum is only "pseudo" and not a true one, is explained in more detail in the solution to Problem 3.3 (see p. 135).

The quenching of the orbital angular momentum is a curious affair. By construction, each of the basis states is still an $l = 2$ state. However, the matrix elements which could give rise to a Zeeman effect, vanish. The essential observation is that quenching certainly occurs whenever crystal field effects force us to use real wave functions. We have just learned that as long as the relevant representation has a dimensionality higher than 1, the use of real wave functions is not compelling, and only a detailed argument can decide whether the orbital angular momentum is quenched or not. However, if the crystal field has a sufficiently low symmetry, all the irreducible representations become one-dimensional, and the wave functions are essentially real[11]. Given that the wave function $|\psi\rangle$ is real, it is easy to check that $\langle\psi|L^z|\psi\rangle$ must be vanishing. $\langle\psi|L^z|\psi\rangle = -i\hbar\langle\psi|\partial/\partial\phi|\psi\rangle$ is pure imaginary. On the other hand, a measurable quantity (an observable) should be real. Both statements can only be true if $\langle\psi|L^z|\psi\rangle = 0$.

It is at least clear that the orbital moment is fully quenched in a crystal field of orthorombic symmetry because all the irreducible representations are one-dimensional (see Problem 3.2).

[11]We could give the wave functions an overall complex phase factor $e^{i\alpha}$ but that does not change the physics.

Let us return to the case of a cubic field, assuming that the sequence of levels is as shown in Fig. 3.4.a and that Hund's first rule applies, and survey what happens to the orbital momentum for each of the configurations d^n. For d^1 and d^2, the T_2 level is partially filled and for the reasons described above, we expect that the quenching of the orbital momentum is only partial. For d^3, all the three T_2 states are occupied, and the total L^z must be zero. At d^4, the configuration must be $t_{2g}^3 e_g^1$, the angular momentum is quenched. d^5 with five parallel spins has no orbital angular momentum at all. For d^6 and d^7, the filling of the T_2 level begins again, and the quenching is expected to be partial. On the other hand, the quenching is complete for d^8 and d^9.

3.5.2 Partial Restoration of Orbital Momentum by Spin–Orbit Coupling

One of our standard examples, the Cu^{++} ion has the configuration $3d^9$. From the above discussion it would follow that even in the high-symmetry cubic environment, this ion has a spin-only magnetic moment corresponding to $S = 1/2$. It should be stressed that this is true only if we neglect the spin–orbit coupling. The spin–orbit coupling has two important effects. First, it acts to restore some of the orbital angular momentum. Second, by tying the spin to the orbital motion which is influenced by the crystal fields, it provides a mechanism whereby the spin can "feel" the orientation of the crystal axes. This is the way how the phenomenon of *magnetic anisotropy* arises.

A well-studied case is that of the ionic insulator $CuSO_4{\cdot}5H_2O$, where the Cu^{++} ions are sitting in a tetragonal crystal field [465]. As one can learn by solving Problem 3.1, superimposing a small tetragonal component on a large cubic field splits the twofold degenerate E level into two non-degenerate levels (page 135). The single d-hole occupies the $d_{x^2-y^2}$ level. The wave function is real and there could be no doubt about the complete quenching of the orbital momentum if it were not for the spin–orbit coupling. Relegating the details of the derivation to the solution of Problem 3.4 (see page 137), we quote the final result: switching on a magnetic field H, the Zeeman energies are

$$\mathcal{E} = \pm \frac{\mu_B H}{2} \sqrt{g_\perp^2 \sin\theta^2 + g_\parallel^2 \cos\theta^2} \tag{3.28}$$

where θ is the angle between the magnetic field and the tetragonal four-fold axis. The isotropic g-factor seen in (2.38) (and, more particularly, in (2.39)) became replaced by an anisotropic quantity

$$g_{\perp} = 2\left(1 + \frac{\lambda}{\mathcal{E}_{E_t} - \mathcal{E}_{B_{1t}}}\right) \tag{3.29}$$

$$g_{\parallel} = 2\left(1 + \frac{4\lambda}{\mathcal{E}_{B_{2t}} - \mathcal{E}_{B_{1t}}}\right) \tag{3.30}$$

where λ is the spin–orbit coupling constant, and the crystal field energy levels carry the indices of the tetragonal point group (see p. 135).

For a spin-only moment, we would expect $g = 2$. If the level in question is non-degenerate, the deviation from 2 can be estimated by first-order perturbation theory, and it is small if λ is sufficiently smaller than the relevant crystal field splittings. In this case, the magnetic moment is still mainly a spin moment, and the restoration of the orbital magnetic moment is a weak effect. (As a matter of fact, for Cu^{++} in $CuSO_4{\cdot}5H_2O$ the corrections are quite substantial: $g_{\parallel} = 2.4$ and $g_{\perp} = 2.06$. In any case, the deviation of g from 2 due to solid-state effects is expected to be rather larger than the radiative correction shown in (2.39)).

The above results can be reformulated in a more elegant manner by introducing the tensor

$$\mathbf{g} = \begin{pmatrix} g_{\perp} & 0 & 0 \\ 0 & g_{\perp} & 0 \\ 0 & 0 & g_{\parallel} \end{pmatrix} \tag{3.31}$$

which allows us to write the coupling to the field as

$$\mathcal{H} = \mu_{\rm B}\mathbf{S} \cdot \mathbf{g} \cdot \mathbf{H}\,. \tag{3.32}$$

This is a simple example of single-ion *effective spin Hamiltonians*. Spin Hamiltonians are useful for the description of systems like $CuSO_4{\cdot}5H_2O$: the angular momentum is almost quenched by the crystal field, thus the magnetic field acts on the spin, but the weak spin–orbit coupling manages to transmit the effect of the anisotropic environment. Doing calculations like that in Problem 3.4, but for ions with spin $S > 1/2$, one finds that the general form of the spin Hamiltonian is [465]

$$\mathcal{H} = \mathcal{E}^{(0)} + \mu_{\rm B}\mathbf{S} \cdot \mathbf{g} \cdot \mathbf{H} - \lambda^2 \mathbf{S} \cdot \Lambda \cdot \mathbf{S} - \mu_{\rm B}^2 \mathbf{H} \cdot \Lambda \cdot \mathbf{H} \tag{3.33}$$

where the tensor Λ is simply related to $\mathbf{g}$: $g_{ij} = 2(\delta_{ij} - \lambda\Lambda_{ij})$. Like in the free-ion energy (2.74), one finds terms of order H and H^2, only now they acquire tensorial character. In general, the tensors have non-vanishing off-diagonal elements. The third (H-independent!) term of $\mathcal{H}$ deserves particular attention: it is called the *crystal field anisotropy term*. For ions with $S > 1/2$, it can give rise to *zero field splittings*: the spin may have certain preferred orientations even in the absence of a magnetic field. To see this more clearly, let us imagine that all $\Lambda_{ij} = 0$ with the exception of Λ_{zz}. Then the anisotropy term is just

$$\mathcal{H}_{\text{anis}} = -\Lambda_{zz}\,(S^z)^2 \ . \tag{3.34}$$

For $\Lambda_{zz} > 0$, the $S^z = \pm S$ states are preferred; for $\Lambda_{zz} < 0$ (for integer S) $S^z = 0$ has the lowest energy. This is an example of the so-called *single-ion anisotropy*[12].

Naturally, we cannot expect that the (electrostatic) crystal field makes a distinction between the $S^z = S$ and $S^z = -S$ states! Time reversal invariance guarantees that these remain degenerate if $\mathbf{H} = 0$. It also follows that this single-ion term cannot lead to anisotropy for $S = 1/2$ ions like Cu^{++}.

Problem 3.3. Write down the matrix of the orbital angular momentum for d-electrons in a cubic field, and discuss the quenching of the orbital momentum for t_{2g}, and e_g, electrons!

Problem 3.4. The $3d$ wave functions corresponding to the tetragonal-field-split levels are shown in Fig. 3.7 (p. 102). Consider the case of a Cu^{2+} ion (a single d-hole). Let us switch on a weak spin–orbit coupling.

a) Determine the lowest-order corrections to the d-hole ground state wave function and ground state energy! Is the ground state degenerate?

b) Switch on a magnetic field pointing in an arbitrary direction. Determine the anisotropy of the Zeeman splitting!

3.5.3 High-Spin versus Low-Spin States

Before closing our discussion of d-states in an octahedral environment, we should mention that the $3d^6$ configuration (realized by Fe^{2+} and Co^{3+} ions) is occasionally found to violate Hund's first rule. The rule

[12] The spin–orbit coupling can introduce anisotropy in the effective spin–spin interaction as well. We do not discuss this so-called exchange anisotropy.

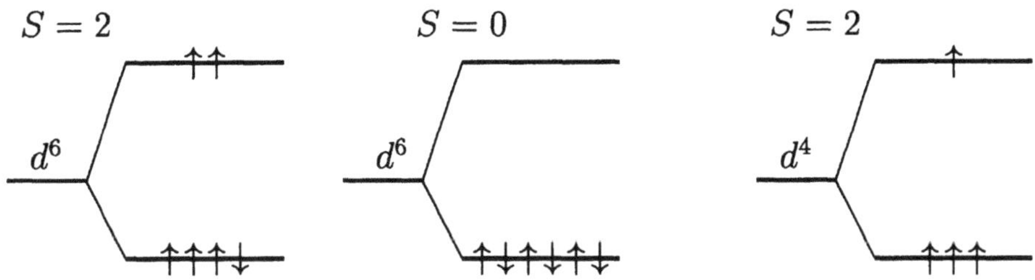

Figure 3.5: *Left and middle*: High-spin and low-spin configurations of a Co^{3+} (or Fe^{2+}) ion in cubic environment. *Right*: In $LaMnO_3$, the Mn^{3+} ion preserves its high-spin state but the large crystal field splitting leads to a clear disctinction between "core" (t_{2g}) and "lightly bound" (e_g) electrons.

would predict the configuration $t_{2g}^4 e_g^2$, with $S = 2$; we may call this the high-spin state (Fig. 3.5). On the other hand, with six d-electrons one can just fill the t_{2g} level, thereby avoiding the occupation of e_g states. The configuration t_{2g}^6 goes with $S = 0$; this is the low-spin state. The t_{2g}-e_g gap is the largest of crystal-field splittings, and it is apparently large enough to compete with the intra-atomic exchange energy which favours $S = 2$. In nature, sometimes the high-spin state, and sometimes the low-spin state is found to be stable. A careful investigation of $LiCoO_2$ provided a clear-cut case of the low-spin state of Co^{3+} [421].

We may ask whether the competition between high-spin and low-spin states does not extend to the $3d^7$ configuration where $S = 3/2$ is the high spin and $S = 1/2$ would be the low spin. Indeed it has been suggested that Ni^{3+} is in its $S = 1/2$ low-spin state in the curious antiferromagnet $LiNiO_2$.

The perovskite $LaCoO_3$ is an interesting system where the low-spin and high-spin configurations have comparable stability, and therefore both are seen in the range of accessible temperatures. Given that the low-spin configuration is slightly lower in energy (which seems to be the case), we can think of at least two reasons why the high-spin configuration should appear in large concentration at elevated temperatures. One is the large spin entropy associated with $S = 2$. Another is that as soon as the ions acquire spin, they can gain energy from spin–spin interactions, or may even order magnetically. We may be wondering whether these mechanisms are capable to drive a low-spin–high-spin phase transition. Indeed $LaCoO_3$ shows complicated phase transitions but their description requires a lot more sophistication [1] than the simple picture we have been contemplating.

$LaMnO_3$ is another highly interesting perovskite. The Mn^{3+} ion clearly preserves its $S = 2$ high-spin state (Fig. 3.5, right), but the large t_{2g}–e_g gap is not without consequences: it enables us to distinguish between the electrons which form essentially stable t_{2g}^3 "cores", and the lightly bound e_g electrons. This distinction plays an important role in theories of the magnetism of doped $LaMnO_3$ (see Sec. 8.5.3).

3.6 Jahn–Teller Effect

Returning now to the case of a Cu^{2+} ion in an octahedral environment: the d-hole is in an e_g-state. As long as the environment is truly octahedral, we cannot decide whether the orbital character is $(3z^2 - r^2)$-like or $(x^2 - y^2)$-like: the ionic ground state is predicted to be orbitally degenerate. Now, it is just this what is not allowed to happen, as stated in the celebrated

Jahn–Teller theorem (1936): if the symmetry of the crystal field is so high that the ground state of an ion is predicted to be orbitally degenerate then it will be energetically preferable for the crystal to distort in such a way that the orbital degeneracy is lifted [182]. The appearance of the distortion is the *Jahn–Teller effect*[13].

For the sake of simplicity, let us assume that the octahedron gets elongated in the z-direction (Fig. 3.6). Since now the two oxygen ions on the z-axis are lying farther from the Cu-ion than the oxygen ions in the x-y-plane, an electron in the $(3z^2 - r^2)$-orbital is less repelled by other charges, and thus lying at a lower energy level, than an electron in the $(x^2 - y^2)$-orbital. The twofold degenerate E-level has been split by the tetragonal distortion into (orbitally) non-degenerate levels (see Fig. 3.7).

It should be stressed that we were ascribing the Jahn–Teller splitting to a tetragonal distortion mainly because it is the easiest to envisage. Any other distortion which splits the e_g level, would do equally well.

[13]See [309] for a recent pedagogical review.

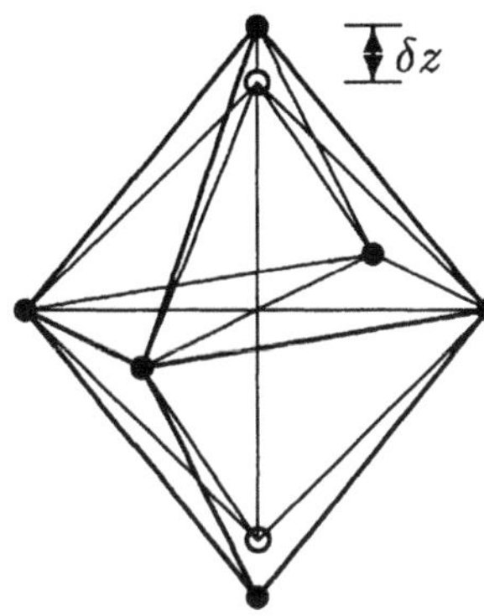

Figure 3.6: Tetragonal distortion from an ideal (thin lines) to an elongated octahedron (thick lines).

As a matter of fact, the low-temperature phase of undoped La_2CuO_4 is orthorombic [384]. Since the orthorombic point group is a subgroup of the tetragonal one, our arguments are valid for this case, too.

The energetical reason for the Jahn–Teller distortion is easy to grasp. The distortion of the octahedron costs an elastic energy $\sim \alpha(\delta z)^2$. The splitting of the E level is proportional to δz. Two electrons gain energy by going to the lower-lying $(3z^2 - r^2)$-level, and only one electron loses by having to go up to the $(x^2 - y^2)$-level, thus there is an electronic energy gain $\sim -\beta\delta z$. The total energy $\alpha(\delta z)^2 - \beta\delta z$ is minimized by the finite distortion $\delta z = \beta/2\alpha$.

We have been speaking about solids because that is our immediate interest. The theorem was originally formulated for molecules, stating that if for a high-symmetry structure the electronic ground state were orbitally degenerate then the molecule spontaneously distorts to lift the degeneracy. Actually, the simple energetical argument given above had been proposed by Landau prior to the work of Jahn and Teller. However, it rests on the naturally-sounding *assumption* that the distortion-induced splitting of the electronic levels is $\sim \delta z$. At the same time (the early 1930s), a perplexing counter-example was also known: for the CO_2 molecule which has a *linear* geometry, the splitting in question is $\sim (\delta z)^2$, and therefore the molecule does *not* show a Jahn–Teller distortion even though it does have an orbitally degenerate electronic

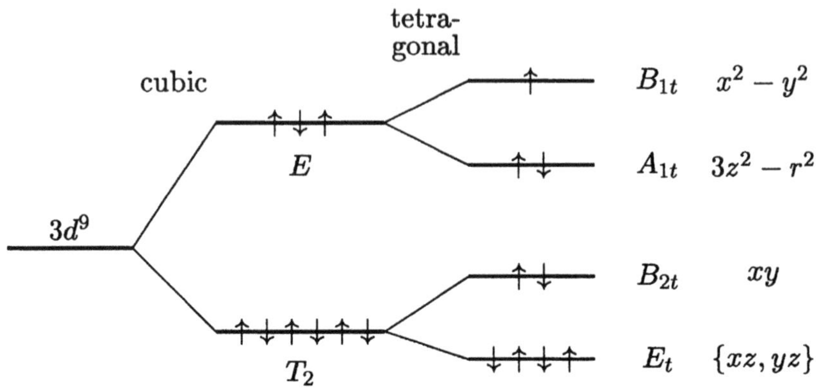

Figure 3.7: Schematic representation of the level structure of a Cu^{2+} ion in a nearly octahedral environment. The large cubic component of the crystal field gives rise to the T_2 and E levels, which would leave the ground state orbitally degenerate. The small tetragonal distortion leads to a further splitting of both levels, making it unambiguous that the hole is in the $(x^2 - y^2)$-type level.

ground state. This made the status of the energetical argument look rather uncertain. It has taken the careful group-theoretical work by Jahn and Teller [182] to prove that molecules with a linear geometry constitute the only exception from the theorem. The story of these developments is told by Teller in [394].

Let us stress that in the previous discussion, we were neglecting spin–orbit coupling, which permitted us to separate the orbital and spin components of the ground state. The Jahn–Teller theorem can be generalized to the case of non-vanishing spin–orbit coupling, as it will be discussed later.

Returning to our example from solid state physics: we find it natural that, because of the original equivalence of the coordinate axes, the distortion may equally well be taking place along the x-, or y-axis. If the octahedra were independent, we might expect a random mixture of distortions along the x-, y-, and z-axes. It seems to be, however, intuitively clear that in a periodic array of closely spaced (or even corner-sharing) octahedra, the total energy of the crystal will be minimized if the distortions form a well-defined pattern. If they happen all in the same direction, we speak about a *ferrodistortive* state. La_2CuO_4 gives an example for this. It is interesting that the (in many respects) very similar[14] K_2CuF_4 displays an alternating sequence of distortions in two different directions, providing an example for the *antiferrodistortive* state [181].

It is usual to find that the Jahn–Teller distortion appears at a structural phase transition[15]. We can even envisage the following scenario: at high temperatures, randomly oriented Jahn–Teller complexes[16] are favourable, because randomness brings a large entropy. At low temperatures, the distortions have to get ordered, to minimize their interaction energy. The ordering changes the global symmetry, thus it has to set in at a well-defined transition temperature. This scenario for a structural phase transition is known as the *cooperative Jahn–Teller effect*. It is thought that the cubic to tetragonal transition of TmCd at $T_{cr} = 3.16K$ is a clear-cut example of this phenomenon [251]. The extensive literature on phase transitions of similar nature is reviewed in [129].

A glance at Fig. 3.5 shows that $LaMnO_3$ should be a Jahn–Teller system, too! And so it is: it was found to undergo a cubic–tetragonal transition at $T_{tr} \approx 800K$ [200]. The 'fourth' electron of Mn^{3+} can occupy either a $d_{x^2-y^2}$ or a $d_{3z^2-r^2}$ orbital. Similar is the case of the Cu-based perovskite $KCuF_3$ in which the d-hole can occupy either of the two e_g states. Both compounds are antiferrodistortive: the (simple cubic) sublattice of $3d$ ions gets divided into two alternating sublattices. However, our original choice of the e_g basis is no longer relevant: the distortion lowered the symmetry from cubic to something else, and the corresponding basis functions have to be used. In $KCuF_3$, the order can be envisaged as the formation of chains with z^2-x^2 and z^2-y^2 character [165] (Fig. 3.8). There is an interesting interplay between orbital and magnetic ordering; we return to this question in Sec. 5.4. At the same time, we will learn that not only the Jahn–Teller mechanism (electron–phonon coupling) can lead to orbital ordering but the purely electronic

---

[14]The ideal (undistorted) structure of La_2CuO_4 and K_2CuF_4 is the same K_2NiF_4 structure (see p. 79), thus we start with Cu^{2+} ions in geometrically identical environments. Nevertheless, the physics of the two substances turns out to be startlingly different (see Sec. 5.4).

[15]The alternative being that the distortion is present at all accessible temperatures. In this case, the high-temperature symmetrical phase is a purely theoretical concept.

[16]A complex is a characteristic, usually highly symmetrical atom group carved out from the crystal. For instance, the octahedron shown in Fig. 3.1 can be called the CuO_6^{10-} complex.

Figure 3.8: The alternation of $d_{z^2-x^2}$ and $d_{z^2-y^2}$ orbitals in the ordered state of $KCuF_3$.

mechanism of kinetic exchange as well.

In the whole of the previous discussion, we have tacitly assumed that the atomic positions are parameters of the problem, not dynamical variables: the atoms get displaced to some new positions and stay there. This is meant by the static Jahn–Teller effect. However, allowing for the coupling between electrons and the ionic motions one finds that a small system (a molecule) can tunnel between energetically equivalent distortions (say between distortions along the x-, y-, and z-axis if the ideal structure of the molecule is cubic). This phenomenon is the *dynamic Jahn–Teller effect.*

3.7 Time Reversal Invariance

Before continuing the investigation of the consequences of spatial symmetries, it is useful to consider briefly a non-geometrical symmetry: time reversal invariance.

It is well-known what time reversal invariance means in classical physics. Newton's law

$$m\ddot{\mathbf{r}} = \mathbf{F} \tag{3.35}$$

is invariant under $t \to -t$ if the force $\mathbf{F}$ is independent of the velocity. Note that the velocity $\mathbf{v} = (d\mathbf{r}/dt)$ changes sign upon time reversal, and so does the angular momentum $\mathbf{L} = \mathbf{r} \times \mathbf{p}$. So we are back at the popular formulation that if we make a movie of a classical mechanical system, and run it backwards, then it will also look like physical reality.

This does not hold in the presence of a magnetic field $\mathbf{H}$ because the Lorentz force $(e/c)(\mathbf{v} \times \mathbf{H})$ changes sign if $t \to -t$. However, if we simultaneously reverse the field $\mathbf{H} \to -\mathbf{H}$, invariance is restored (it is consistent with reversing the current in the coil which produces the field).

In quantum mechanics, taking the complex conjugate of both sides of the time-dependent Schrödinger equation

$$\mathcal{H}\psi = i\hbar\frac{\partial\psi}{\partial t} \tag{3.36}$$

and assuming that $\mathcal{H}$ is real[17], we get

$$\mathcal{H}\psi^* = i\hbar\frac{\partial\psi^*}{\partial(-t)} \tag{3.37}$$

which tells us that ψ^* develops in the negative time direction the same way as ψ develops in the positive time direction. This does not influence the probability density $|\psi|^2 = |\psi^*|^2$ but reverses currents which are proportional to $\Im(\psi^*\nabla\psi)$. Orbital angular momenta are similarly reversed.

Taking the complex conjugate of the eigenvalue equation $\mathcal{H}\psi_n = E_n\psi_n$, we get

$$\mathcal{H}\psi_n^* = E_n\psi_n^* \tag{3.38}$$

i.e., ψ^* is also an eigenstate, and with the same eigenvalue as ψ_n. It may happen that we find $\psi^* \parallel \psi$ in which case we learned nothing new. However, if ψ^* is linearly independent of ψ, we can draw the conclusion that the energy level in question must be *degenerate.*

Actually, it is perfectly possible that this does not add anything new to what we already know from studying the spatial symmetries. For instance, the degeneracy of the p-states with φ-dependences $(e^{i\varphi}, 1, e^{-i\varphi})$

[17]This amounts to stating that the external magnetic field is zero. At the end of the Section, we discuss complex Hamiltonians.

follows from the spherical symmetry of the atom. Complex conjugation merely rearranges the states to $(e^{-i\varphi}, 1, e^{i\varphi})$.

Up to now, we have not considered the spin of the electron. With this restriction, the time reversal operator $\mathcal{T}_t$ acts as the complex conjugation $\mathcal{K}$

$$\mathcal{T}_t\psi(\mathbf{r}) = \mathcal{K}\psi(\mathbf{r}) = \psi^*(\mathbf{r}). \tag{3.39}$$

This is a very simple operator but not a linear operator. Note that

$$\mathcal{K}(c\psi) = c^*\psi^* \neq c\psi^* = c(\mathcal{K}\psi). \tag{3.40}$$

An operator with this property is called antilinear. Since it preserves the absolute value of the scalar product: $|(\mathcal{K}\psi, \mathcal{K}\phi)| = |(\psi^*, \phi^*)| = |(\psi, \phi)|$, the operator is also antiunitary.

If we include the spin degree of freedom, $\mathcal{T}_t$ can no longer be identified simply with $\mathcal{K}$ because it must also reverse the spin: $\mathbf{S} \to -\mathbf{S}$. We express the spin as $\mathbf{S} = (1/2)\vec{\sigma}$, $\vec{\sigma}$ being the vector of Pauli matrices

$$\sigma_x = \begin{pmatrix} 0 & 1 \\ 1 & 0 \end{pmatrix} \quad \sigma_y = \begin{pmatrix} 0 & -i \\ i & 0 \end{pmatrix} \quad \sigma_z = \begin{pmatrix} 1 & 0 \\ 0 & -1 \end{pmatrix}. \tag{3.41}$$

It must hold that $\mathcal{T}_t\vec{\sigma} \rightsquigarrow -\vec{\sigma}$.

Complex conjugation is enough to achieve $\sigma_y \rightsquigarrow -\sigma_y$, but nothing happened yet to σ_x and σ_z. It is clear that σ_x and σ_z can be reversed by performing a π-rotation about the y-axis. We have learned the form (3.22) of the operator for a rotation by some arbitrary angle φ. Making use of the fact that

$$\sigma_y^2 = \begin{pmatrix} 1 & 0 \\ 0 & 1 \end{pmatrix} = E,$$

the 2×2 unit matrix, we find

$$\begin{aligned} e^{i\varphi S_y} &= \exp(i\frac{\varphi}{2}\sigma_y) = E + i\frac{\varphi}{2}\sigma_y + \frac{1}{2!}(i\frac{\varphi}{2})^2\sigma_y^2 + \frac{1}{3!}(i\frac{\varphi}{2})^3\sigma_y^3 + \ldots \\ &= E\left(1 + \frac{1}{2!}(i\frac{\varphi}{2})^2 + \frac{1}{4!}(i\frac{\varphi}{2})^4 + \ldots\right) + \sigma_y\left(i\frac{\varphi}{2} + \frac{1}{3!}(i\frac{\varphi}{2})^3 + \ldots\right) \\ &= E \cdot \cos\frac{\varphi}{2} + i\sigma_y \cdot \sin\frac{\varphi}{2}. \end{aligned} \tag{3.42}$$

For $\varphi = \pi$, we get

$$\exp\left(i\frac{\pi}{2}\sigma_y\right) = i\sigma_y. \tag{3.43}$$

The time reversal operator should complex conjugate the wave function and reverse all spin components. For one-electron states, this is accomplished by[18]

$$\mathcal{T}_t = \exp\left(i\frac{\pi}{2}\sigma_y\right) \cdot \mathcal{K}. \tag{3.44}$$

Being the product of a unitary operator, and the antiunitary operator $\mathcal{K}$, $\mathcal{T}_t$ itself is antiunitary.

The generalization to a number of particles with arbitrary spins $\mathbf{S}_1$, ... $\mathbf{S}_n$ is straightforward

$$\mathcal{T}_t = e^{i\pi S_1^y} e^{i\pi S_2^y} \ldots e^{i\pi S_n^y} \cdot \mathcal{K}. \tag{3.45}$$

Let us consider the effect of $\mathcal{T}_t^2$. One might naively expect that this could have no effect at all. However, we have to check how

$$\mathcal{T}_t^2 = e^{i2\pi S_1^y} e^{i2\pi S_2^y} \ldots e^{i2\pi S_n^y} \tag{3.46}$$

acts on an n-electron state. We have to recall that the 2π-rotation of an electron spin (or, generally, a half-integral spin) is not identity but instead brings a factor -1

$$\left(e^{i\pi S^y}\right)^2 = \left(e^{i\frac{\pi}{2}\sigma_y}\right)^2 = (i\sigma_y)^2 = (-1) \cdot E. \tag{3.47}$$

It is worrying only at the first sight that the state of an electron does not return into itself after a 2π-rotation. We have to remember that measurable quantities are matrix elements, i.e., bilinear combinations of state vectors, and these are left unchanged by 2π-rotations.

For a system of n electrons, combining (3.47) with (3.46) gives[19]

$$\mathcal{T}_t^2 = (-1)^n = \begin{cases} 1 & \text{if } n \text{ is even} \\ -1 & \text{if } n \text{ is odd.} \end{cases} \tag{3.48}$$

This has a profound consequence if n is odd. We have already remarked that $\mathcal{T}_t\psi$ is degenerate with ψ – or rather, that it has the same

[18] For an easy check, take the action on the $\downarrow$-spin state: $i\sigma_y|\downarrow\rangle = |\uparrow\rangle$.

[19] In (3.48) we omit the unit matrix acting in the n-spin space.

eigenvalue as ψ. The question is whether $\mathcal{T}_t\psi$ is really different from (linearly independent of) ψ. The answer depends on the parity of n.

Denoting the unitary factor in (3.45) as

$$\mathcal{U} = e^{i\frac{\pi}{2}\sigma_{1,y}} e^{i\frac{\pi}{2}\sigma_{2,y}} \dots e^{i\frac{\pi}{2}\sigma_{n,y}} \tag{3.49}$$

we can write $\mathcal{T}_t = \mathcal{U}\mathcal{K}$. For general ψ and ϕ

$$(\mathcal{T}_t\psi, \mathcal{T}_t\phi) = (\mathcal{U}\mathcal{K}\psi, \mathcal{U}\mathcal{K}\phi) = (\mathcal{K}\psi, \mathcal{K}\phi) = (\psi^*, \phi^*) = (\phi, \psi). \tag{3.50}$$

Using now (3.48)

$$(\mathcal{T}_t\psi, \psi) = (\mathcal{T}_t\psi, \mathcal{T}_t^2\psi) = (-1)^n(\mathcal{T}_t\psi, \psi). \tag{3.51}$$

It follows that for odd n

$$(\mathcal{T}_t\psi, \psi) = 0. \tag{3.52}$$

$\mathcal{T}_t\psi$ and ψ are orthogonal. Combining with (3.38), it means that we have found an actual degeneracy: if ψ_n is an energy eigenstate with eigenvalue E_n, then $\mathcal{T}_t\psi_n$ is another eigenstate with the same eigenvalue. Thus we have proven

Kramers' Theorem: if the Hamiltonian is time-reversal invariant then the eigenstates of a system with an odd number of electrons are at least twofold degenerate.

The theorem comes into its own in cases of low spatial symmetry. The degeneracies we treated in the previous section resulted from the rather high (for instance, cubic) symmetry of the crystal field. It stands to reason that these degeneracies would be lifted by lowering the symmetry by further distortions. However, the electric fields described by the crystal field Hamiltonian are time reversal invariant, thus Kramers degeneracies cannot be removed by crystal fields, no matter how low the symmetry is.

In the absence of spin–orbit coupling the statement can be regarded as rather obvious: an odd number of electrons will necessarily have a non-zero total spin "which can point either up or down". It is important to realize that Kramers' Theorem holds also in the presence of spin–orbit coupling because $\mathcal{H}_{SO} = \xi\mathbf{S}\cdot\mathbf{L}$ is invariant under time reversal (since both $\mathbf{S}$ and $\mathbf{L}$ get reversed). Taking one-electron states,

$$\psi_1 = \varphi_1(\mathbf{r})\alpha + \varphi_2(\mathbf{r})\beta$$

and

$$\psi_2 = \varphi_2^*(\mathbf{r})\alpha + \varphi_1^*(\mathbf{r})\beta$$

form a Kramers doublet. (In the solution to Problem 3.4 (p. 137) we work out the structure of a Kramers doublet, discussing the effect of the spin–orbit coupling on a B_{1t} level in tetragonal field.)

With the spin–orbit coupling included, the "orbital degeneracy" discussed previously looses its simple meaning. However, we can check whether the degeneracy of a level is the minimum value required by Kramers' Theorem, or higher. If the latter is the case, it must be the consequence of a sufficiently high spatial symmetry. – Now we are in the position to restate the Jahn–Teller Theorem in its precise form: if the degeneracy of a level is higher than the Kramers minimum then it will be energetically favourable for the system to distort in such way as to reduce the degeneracies.

We can also add a remark to our previous discussion of the quenching of the orbital angular momentum. Disregarding spin–orbit coupling, Kramers' minimum degeneracy is just spin degeneracy, i.e., in sufficiently low-symmetry crystal fields the orbital states are nondegenerate. Then it follows that the orbital functions must be real. We have shown (on page 95) that $\langle L^z \rangle$ vanishes for real wave functions.

Kramers' Theorem has some subtle consequences also if n is even. We have been arguing that for rare earth ions, all three of Hund's Rules follow from energy terms rather larger than the crystal field splittings; it would then seem that rare earth substances must be strong magnets, with the momenta corresponding to a specific J. In this context, it is interesting to consider substances like the cubic semimetal PrSb [286], or the orthorombic insulator HoF_3. Hund's rules give $J = 4$ for Pr^{3+} which has two f-electrons, and $J = 8$ for the ionic configuration $4f^{10}$ which corresponds to Ho^{3+}. The crystal field has a low symmetry in the case of HoF_3 and all the levels are singlets; but even in a cubic field (pertinent to PrSb), Kramers' Theorem permits that at least one of the levels be a singlet. HoF_3 just barely manages to develop magnetic order at $T_{cr} = 0.53$K, due to very complex reasons [190]; while PrSb is (as far as one can say) non-magnetic. It should be stressed that these spectacular cases of the suppression of magnetism occur in spite of the fact that Hund's rules remain valid! We are going to discuss the f^2 configuration in detail in the next Section.

Conventional wisdom would have it that for large J, the vectors $\mathbf{J}$ begin to behave in a unified manner as almost classical vectors. However, since the large-J limit is approached through a sequence integer→half-integer→integer→ ..., this limit cannot be smooth; in a low-symmetry crystal field, every integer-J ion has a finite Van Vleck susceptibility, and every half-integer-J ion a diverging Curie susceptibility. The comparison of the behaviour of Dy^{3+} ($J = 15/2$) and Ho^{3+} ($J = 8$) ions in similar environments offers a nice illustration [199]. Nevertheless, it is possible to work out a unified quasi-classical treatment by ensuring that 2π-rotations give a change of sign for half-integer J, but not for integer J [199].

Finally, a word should be added about the case when time reversal invariance is absent. Just as in classical mechanics, switching on a magnetic field destroys the invariance. Kramers' Theorem ceases to hold, and – in the absence of other, geometrical symmetries – the eigenstates are all expected to be non-degenerate. The familiar picture of Zeeman-split levels illustrates the statement.

A more subtle consequence is that the statement about the real nature of the eigenstates no longer holds. In fact, in our discussion of Landau states with symmetrical gauge (see Sec. 12.2.3), we are arguing that the eigenstates are essentially complex, and that this is a sign of broken time reversal invariance. It is educative to recall that the Hamiltonian (12.55) is not a real operator ($\mathcal{H}^* \neq \mathcal{H}$). Thereby the basic assumption leading to (3.38) has fallen and the statement about the degeneracy of ψ and ψ^* becomes invalid. In fact, we bring an example in which both ψ and ψ^* are eigenstates, but they belong to different Landau levels.

On the other hand, (12.55) shows the invariance property $\mathcal{H}^*(-H) = \mathcal{H}(H)$. It is a particular case of the well-known statement that turning over the magnetic field simultaneously with time reversal constitutes a symmetry transformation.

3.8 The f^2 Configuration

Crystal fields can have a profound effect on the magnetic properties of rare earth elements and compounds (especially for the so-called light rare earths, i.e., elements in the first half of the series). Our starting point is that for these systems the crystal field is weak, and thus the ground state level belongs to the total angular momentum J predicted by Hund's third rule. For a free ion, the degenerate ground states would span a $(2J+1)$-dimensional representation of the symmetry group. However, we usually find that $2J+1$ is far too large to be the dimension of any irreducible representation of the point group! It follows that the crystal field, however weak, has an essential effect: it splits the $(2J+1)$-fold degenerate free ion level into a number of sublevels. For a crystal field of sufficiently low symmetry, the remaining degeneracies are given by the Kramers minimum: for an even number n of electrons, all the crystal field levels are non-magnetic singlets! But even if the point group symmetry is cubic, the ground state is usually far less magnetic than a free ion would be.

As an example, let us consider the f^2 configuration. It is realized as $4f^2$ for the trivalent praseodymium ion Pr^{3+}. Elemental Pr is a most astonishing antiferromagnet with a Néel temperature of about 50 millikelvin [191, 303]. Magnetic ordering in Pr involves the nuclear moments in an essential way, and we do not cover that subject, but we discuss a number of Pr compounds whose behaviour is conceptually simpler. – The configuration $5f^2$ is found for U^{4+} which is the essential ionic species in a number of systems of great physical interest.

3.8.1 Cubic Crystal Field

For the configuration f^2 Hund's rules give $L=5$, $S=1$, and $J=4$, thus the free ion ground state is the ninefold degenerate $^3H_4$. First let us see what happens if we place the ion in a cubic crystal field. The example is less relevant for elemental praseodymium (where a fraction of the lattice sites has approximately cubic symmetry) than for the compound PrSb or cubic uranium intermetallic compounds like UBe_{13}. (The examples are rather arbitrarily chosen: there are many rare earth, and actinide, systems to illustrate the points we wish to make).

We have to realize that we are not really well-equipped to solve this problem. We have seen how to treat the splitting of single-electron f-states but here we are speaking about two-electron states. The systematic approach would be to construct direct-product representations based on single-electron states. Besides, since the two electrons are in the same shell, we should worry about the effects of antisymmetrization. For a careful treatment, we should follow Wigner [441] and include the permutations of indistinguishable particles among the symmetry operations of the system. Finally, we should remember that J is a composite angular momentum, formed by the addition of spin and orbital angular momenta.

Doing all this properly would be a lot of work but we can spare the effort by a convenient shortcut. Deriving $J = 4$ from Hund's rules respected the rules of the addition of angular momenta *and* the Pauli principle. The statement that $J = 4$ means that the components of $\mathbf{J}$ transform like the components of a single-electron orbital state with $L = 4$. The expression (3.25) was based on this transformation property. Thus we can simply take (3.25) to derive the characters

	$\mathcal{E}$	$8\mathcal{C}_3$	$3\mathcal{C}_2$	$6\mathcal{C}_2'$	$6\mathcal{C}_4$
$\Gamma_{J=4}$	9	0	1	1	1

Using Table 3.1 and the rule (3.17), we obtain the splitting scheme

$$\Gamma_{J=4} = \Gamma_1 \oplus \Gamma_3 \oplus \Gamma_4 \oplus \Gamma_5. \tag{3.53}$$

The ninefold degenerate ${}^3H_4$ term is split into the singlet[20] Γ_1, the doublet Γ_3, and the triplets Γ_4 and Γ_5.

Group theory alone cannot tell us the sequence of levels. A detailed examination [286] predicts that in PrSb, the ground state is the singlet Γ_1 and the first excited state the triplet Γ_4. Other systems behave differently. In the high-temperature phase of UO_2 , the Γ_5 triplet is the lowest-lying crystal field level [13]. It is argued that the ground state of UBe_{13} is built upon the Γ_3 doublet, which has far-reaching consequences for the physics of that system [73]. Let us emphasize that Γ_3 (or E) is a *non-Kramers doublet*; a glance at (3.27) will convince us that the two E states are not related to each other by time reversal (in fact, both states are time-reversal invariant, thus the doublet can be split into two singlets by lowering the crystal field symmetry). In a Kramers doublet, the two states carry opposite magnetic moments; in contrast,

[20]Here the terms singlet, doublet, triplet, etc. simply mean that a level is, respectively, non-degenerate, or twofold, threefold, etc. degenerate. The same terms have a narrower meaning when applied to pure spin states.

the Γ_3 doublet is non-magnetic. It can be, however, verified that it is quadrupolar doublet. We return to this question in Sec. 3.10.

Let us first focus on the case of PrSb. Alone the fact that the ground state is a singlet tells us that its magnetic moment must be zero (if it were not, the time-reversed state would have the same energy and the ground state should be degenerate). Previously we were making comments about the quenching of the orbital momentum of d-electrons, which, however, did not affect the spin momentum. Here we have a more drastic effect: *the total angular momentum can be quenched*, even in a cubic field! Clearly, whenever such an effect occurs, it can be ascribed to the spin–orbit coupling which transmits the effect of the crystal field from the directly affected orbital momentum to the spin momentum.

The quenching of **J** (as observed in the magnetic behaviour) is not in contradiction with the ground state having the total angular momentum J required by Hund's rules: all the crystal-field-split states are eigenstates of J^2 with the eigenvalue 4(4+1)=20. To see this more clearly, let us construct the singlet ground state which, according to (3.53), must be a basis state for the identity representation Γ_1.

The transformation properties of the $^3H_4$ multiplet are the same as those of the $l = 4$ spherical harmonics Y_4^m. We quote from App. A

$$\begin{aligned} Y_4^{\pm 4} &= \sqrt{\frac{9}{4\pi}}\sqrt{\frac{35}{128}}\,\frac{(x \pm iy)^4}{r^4} \\ Y_4^0 &= \sqrt{\frac{9}{4\pi}}\sqrt{\frac{1}{64}}\,\frac{35z^4 - 30z^2r^2 + 3r^4}{r^4}. \end{aligned} \tag{3.54}$$

The z-axis is a C_4-axis.

The above states are invariant under the C_2-rotations. A C_4-rotation about the z-axis has the effect $x \to y$, $y \to -x$, $z \to z$. This transformation also leaves (3.54) invariant. Since no other of the functions Y_4^m has this property, we can seek the basis state in the form

$$|\Gamma_1\rangle_{l=4} = \alpha Y_4^0 + \frac{\beta}{\sqrt{2}}\,(Y_4^4 + Y_4^{-4}). \tag{3.55}$$

Demanding that $|\Gamma_1\rangle$ is invariant under the C_3-rotation which effects $x \to y \to z \to x$, we find $\alpha/\beta = \sqrt{7/5}$. We could now write down the Γ_1-type state in terms of the spherical harmonics Y_4^m. What we need, however, is the crystal field ground state of the f^2 configuration which must be the same expression in terms of the $J = 4$, $J^z = m$ angular momentum eigenstates which we denote by $|m\rangle$

$$|\Gamma_1\rangle_{J=4} = \sqrt{\frac{7}{12}}\,|0\rangle + \sqrt{\frac{5}{24}}\,[\,|4\rangle + |-4\rangle\,]. \tag{3.56}$$

(Henceforth, understanding that symmetry arguments permit us to deduce $J = 4$ states from the corresponding $l = 4$ states, we drop the cumbersome subscripts). It is easy to check that $\langle\Gamma_1|J^z|\Gamma_1\rangle = \langle\Gamma_1|J^x|\Gamma_1\rangle = \langle\Gamma_1|J^y|\Gamma_1\rangle = 0$. There is no Curie-like term in the susceptibility of PrSb.

On the other hand, the ground state is connected to higher-lying states by components of $\mathbf{J}$. For instance, acting with J^z

$$J^z|\Gamma_1\rangle = \sqrt{\frac{10}{3}}\,[\,|4\rangle - |-4\rangle\,] = \sqrt{\frac{20}{3}}\,|\Gamma_4^0\rangle \tag{3.57}$$

we obtain one of the states corresponding to the Γ_4 triplet level. The two other Γ_4 states can be reached if we let J^x act on $|\Gamma_1\rangle$

$$\begin{aligned} J^x|\Gamma_1\rangle &= \sqrt{\frac{10}{3}}\left[\left(\sqrt{\frac{7}{8}}\,|1\rangle + \sqrt{\frac{1}{8}}\,|-3\rangle\right) + \left(\sqrt{\frac{7}{8}}\,|-1\rangle + \sqrt{\frac{1}{8}}\,|3\rangle\right)\right] \\ &= \sqrt{\frac{10}{3}}\,(|\Gamma_4^+\rangle + |\Gamma_4^-\rangle). \end{aligned} \tag{3.58}$$

The equations (3.57) and (3.58) tell us that there is a Van Vleck type susceptibility which, because of the overall cubic symmetry of the problem, must be isotropic. At $T = 0$ (using the Lande factor $g = 0.8$ which belongs to the $^3H_4$ multiplet)

$$\chi_{\mathrm{VV}} \equiv \chi^z = \frac{2g^2\mu_{\mathrm{B}}^2|\langle\Gamma_4^0|J^z|\Gamma_1\rangle|^2}{\Delta}. \tag{3.59}$$

Since the Γ_1–Γ_4 splitting is very small: $\Delta \sim 7$meV [140], we predict that *the Van Vleck susceptibility is huge.* This is indeed the case: the measured value is $\chi_{\mathrm{VV}} \sim 0.09\mu_{\mathrm{B}}/\mathrm{T}$ (T: Tesla). It follows that for this system, laboratory fields can produce drastic effects. Extrapolating from the low-field susceptibility, we expect that a field of ~ 10T should induce a very respectable moment $\sim 1\mu_{\mathrm{B}}$, which is indeed spectacular for a non-magnetic substance. The magnetization curve can be calculated [286] by including all crystal-field levels, and turns out to be non-linear but the naive estimate is not that far off. Experimentally [351], PrSb acquires a magnetic moment $0.89\mu_B$ per Pr ion in a field of 10T at the temperature $T = 4.2$K. The quenching of the angular momentum by crystal fields is a weak effect and it can be lifted by a strong magnetic field which mixes the crystal-field levels and thereby liberates the magnetic moment.

Here we can make a preliminary comment about the nature of rare earth magnetism. Let us formally split the lattice Hamiltonian into single-site and intersite terms

$$\mathcal{H} = \sum_{j}^{\text{sites}} \mathcal{H}_j + \sum_{i,j} \mathcal{H}_{ij}. \tag{3.60}$$

Crystal fields at site i are lumped into $\mathcal{H}_i$. Our previous discussion shows that the ground state for an even number n of f-electrons is often a non-magnetic singlet. Even if it turns out to be a doublet or a triplet, the angular momentum J is largely quenched. Similarly, if n is odd, the Kramers minimum of the degeneracy belongs to a doublet. Therefore, in most of the cases, we conclude that the single-ion ground state is either non-magnetic, or carries a much smaller magnetic moment than what would correspond to the free-ion J. On the other hand, it is well-known that a large number of rare earth substances are strongly magnetic, and that neither the high-temperature paramagnetic moments (Fig. 2.1), nor the low-temperature ordered moments, are very different from the free ion values. The resolution of the apparent paradox comes from realizing that intersite (exchange) terms can act like a field and, as we are going to discuss in Sec. 3.8.4, they can induce sizeable ordered moments. Also, $k_B T$ at room temperature easily exceeds some of the crystal field splittings and thus thermal averaging over the crystal field levels can account for the almost full-sized paramagnetic moments (see also Problem 3.5).

3.8.2 Tetragonal Crystal Field

More complicated level schemes arise in environments of lower symmetry. We do not treat these in any detail, except to point out cases where

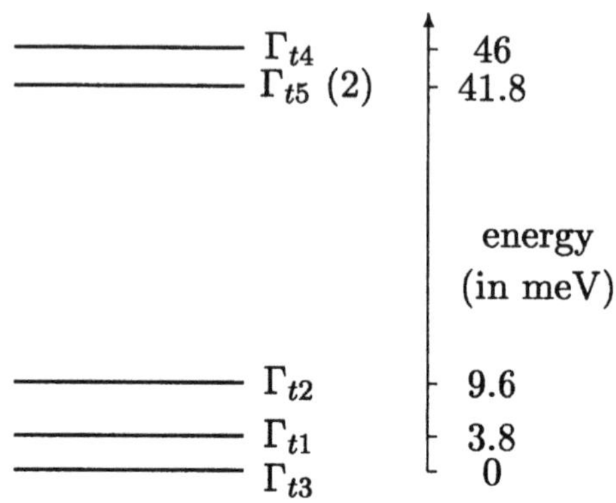

Figure 3.9: The five lowest levels of a uranium ion in URu_2Si_2, after [345]. The order of the levels depends on assumptions about the crystal potential.

new effects appear.

Using the character table 3.5 (page 135), we can easily check that in a tetragonal crystal field, the f^2 configuration splits into five singlets and two doublets

$$\Gamma_{J=4} = 2\Gamma_{t1} \oplus \Gamma_{t2} \oplus \Gamma_{t3} \oplus \Gamma_{t4} \oplus 2\Gamma_{t5}. \tag{3.61}$$

For an application, we may consider the interesting and much-studied substance URu_2Si_2, in which U^{4+} ions are sitting in a tetragonal environment. Remarkably, in spite of long and painstaking work, the precise sequence of the crystal field levels is still not quite certain [345]. There seems to be consensus that the lowest (three or four) levels are singlets, and these are followed by a higher-lying doublet, but it is not clear which singlet is the ground state. For the sake of the present argument, let us assume [345] that the ground state is Γ_{t3}, and that this is followed by the low-lying singlets Γ_{t1}, and Γ_{t2}, the somewhat higher-lying doublet Γ_{t5}, and the singlet Γ_{t4} (Fig. 3.9). The remaining two levels are irrelevant for our argument. – We emphasize that the physics which we will be describing, could similarly arise from other level schemes.

Recalling (A.22), and using the same notations as in (3.56), we identify the Γ_{t3} ground state as

$$|\Gamma_{t3}\rangle = \frac{1}{\sqrt{2}}[\, |2\rangle + |-2\rangle \,] \tag{3.62}$$

while the Γ_{t4}-type state is

$$|\Gamma_{t4}\rangle = \frac{1}{\sqrt{2}}[\ |2\rangle - |-2\rangle\ ]. \tag{3.63}$$

The Γ_{t1}-type state, belonging to the identical representation, has the structure familiar from (3.56)

$$|\Gamma_{t1}\rangle = \epsilon|0\rangle + \sqrt{\frac{1-\epsilon^2}{2}}\ [\ |4\rangle + |-4\rangle\ ]. \tag{3.64}$$

We need not specify ϵ which depends on the form of the crystal potential.

Finally, the Γ_{t2}-type state can be shown to be

$$|\Gamma_{t2}\rangle = \frac{1}{\sqrt{2}}[\ |4\rangle - |-4\rangle\ ]. \tag{3.65}$$

Let us now consider the problem of the Van Vleck susceptibility at $T = 0$. As we are going to see, the inequivalence of the axes manifests itself in a (possibly strong) *anisotropy of the susceptibility*.

If the field is applied in the z-direction, the relevant matrix element is

$$J^z|\Gamma_{t3}\rangle = 2|\Gamma_{t4}\rangle. \tag{3.66}$$

The ground state is connected to the excited singlet $|\Gamma_{t4}\rangle$, and the z-component of the susceptibility is

$$\chi^z = \frac{2g^2\mu_{\mathrm{B}}^2|\langle\Gamma_{t4}|J^z|\Gamma_{t3}\rangle|^2}{\Delta_4} \tag{3.67}$$

where Δ_4 is the energy of the Γ_{t4} level (we have chosen the Γ_{t3} level as the energy zero).

Applying the field in the x-direction, we have to consider

$$J^x|\Gamma_{t3}\rangle = \frac{1}{2}[\sqrt{7}|3\rangle + 3|-1\rangle\ ] + \frac{1}{2}[\sqrt{7}|-3\rangle + 3|1\rangle\ ]. \tag{3.68}$$

The function on the right-hand side belongs to the higher-lying doublets, thus the dominant contribution to χ^x is

$$\chi^x \approx \frac{2g^2\mu_{\mathrm{B}}^2|\langle\Gamma_{t5}|J^x|\Gamma_{t3}\rangle|^2}{\Delta_5}. \tag{3.69}$$

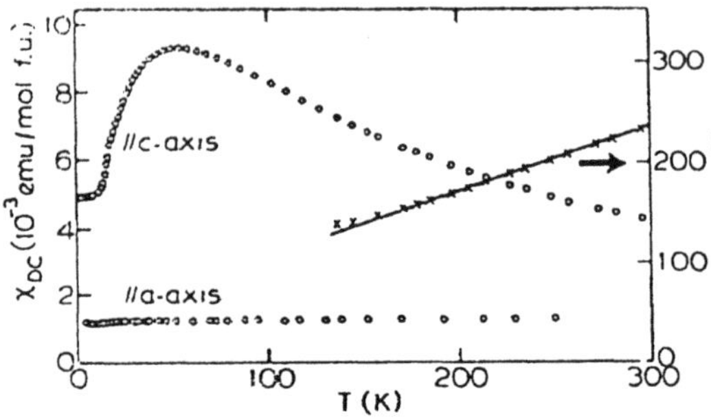

Figure 3.10: Temperature dependence of the susceptibility of URu_2Si_2, along the tetragonal c-axis (z-axis in the text), and in a perpendicular direction, after [318]. Note that χ_z is large and strongly temperature dependent, while χ_x is relatively small and almost temperature-independent.

Comparing (3.67) with (3.69), we see that both the matrix elements and the energy denominators differ, thus it becomes (in principle) understandable that the susceptibility of URu_2Si_2 is strongly anisotropic, in the sense that χ^z is considerably larger than χ^x. Experimental data are shown in Fig. 3.10. In addition to the anisotropy seen at low temperatures, another remarkable feature is the strong temperature dependence of χ^z. Let us note that the susceptibility passes through a maximum as the temperature is raised. A similar behaviour is often seen, in widely varying systems, and it can have a number of quite different explanations. Among other things, it can be a crystal field effect. This is demonstrated in the solution to Problem 3.6 (see p. 142).

3.8.3 Metamagnetic Transition

In the systems under consideration, laboratory magnetic fields are often comparable to the smaller crystal field splittings, and therefore they can have a drastic effect on the level structure, including the possibility of a field-induced transition due to level crossing.

Let us start from (3.66) which tells us that in a magnetic field pointing in the z-direction, the states $|\Gamma_{t3}\rangle$ and $|\Gamma_{t4}\rangle$ are coupled. The matrix of the 2×2 eigenvalue problem is

$$\begin{pmatrix} 0 & -2g\mu_B H^z \\ -2g\mu_B H^z & \Delta_4 \end{pmatrix}$$

and the eigenvalues are

$$\lambda_{3,4} = \frac{\Delta_4 \mp \sqrt{\Delta_4^2 + 16(g\mu_B H^z)^2}}{2}. \tag{3.70}$$

If the field becomes very strong, the Zeeman splitting dominates over the crystal field splitting, and the eigenvalues approach $\pm 2g\mu_B H^z$. The corresponding eigenstates evolve from $|\Gamma_{t3}\rangle$, $|\Gamma_{t4}\rangle$ towards the pure J^z eigenstates $|2\rangle$, $|-2\rangle$.

Similarly, $|\Gamma_{t1}\rangle$ is coupled to $|\Gamma_{t2}\rangle$. However, $|\Gamma_{t2}\rangle$ couples also to the second, higher-lying $|\Gamma_{t1}\rangle$ state whose existence we deduce from (3.61). We do not wish to treat this 3×3 eigenvalue problem formally because the outcome is clear: in high fields, the lowest of the three levels belongs to $|4\rangle$, and its energy is $\sim -4g\mu_B H^z$. At some threshold field, this becomes lower than λ_3. Thus there must be a *level crossing as a function of the magnetic field* at some critical value H_{cr}, where the nature of the ground state changes from $|\Gamma_{t3}\rangle$-like to $|\Gamma_{t1}\rangle$-like, and it is accompanied by a *sudden jump in the magnetization*. This is an example of the so-called *metamagnetic transition*[21].

As a matter of fact, URu_2Si_2 shows several closely spaced metamagnetic transitions around ~ 36T (Fig. 3.11) [81, 380]. The full story is obviously quite complicated but the kind of level-crossing we have been discussing, is likely to be part of an eventual explanation.

As we are going to see in the solution to Problem 3.6 in pp. 142–144, even a simplified four-level "toy model" suffices to predict a static magnetic behaviour which looks remarkably similar to the experimental findings on URu_2Si_2. In particular, the result for the temperature dependence of the z-component of the susceptibility (Fig. 3.16) is in nice

[21]We will speak about a metamagnetic transition whenever we find that the applied field causes a discontinuous (or at least an anomalously fast) increase of magnetization. Such behaviour can result from a great variety of causes; in addition to crystal field effects, we will discuss transitions caused by the nearness to a ferromagnetic phase (p. 396 and p. 484), or to the Mott-localized state (p. 592). See [160] for a concise treatise on metamagnetism.

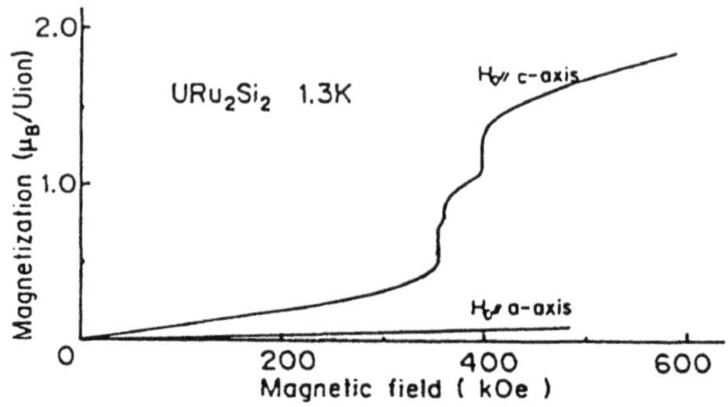

Figure 3.11: The magnetization curve of URu_2Si_2, with the field applied in the z direction (after [380].). Three closely spaced metamagnetic transitions are seen.

qualitative agreement with the experiment (Fig. 3.10). With the same set of parameters, the predicted metamagnetic transition (Fig. 3.15) is roughly in the right place but we could not explain why there are actually three transitions (Fig. 3.11). It goes without saying that improved agreement, with a larger set of measured data, could be obtained by refining the crystal field picture but that is not the point we wish to make. Looking at the figures 3.13–3.16 proves that interesting, and varied, physical behaviour can result from crystal field effects. The occasional close similarity to the results of measurements should encourage us to investigate, to which extent is the observed behaviour the consequence of crystal field effects, and what is left for other reasons to explain. Naturally, we cannot expect that our simple arguments could have provided us with *the* eventual explanation of any feature of such a complicated substance as URu_2Si_2.

This may be the right place to mention a few salient facts about URu_2Si_2. Up to now, we have been speaking about this system as if it were a collection of independent $5f^2$ ions, but this cannot be right at low temperatures. URu_2Si_2 shows a rather mysterious-looking transition at $T_{tr} \approx 17K$, and it becomes superconducting at $T_c \approx 0.8K$ [318, 350]. These phase transitions are cooperative phenomena involving the f-electrons, and require the consideration of inter-f-shell interactions. It had been thought that the 17K transition is basically magnetic but later, it was realized that the puny $0.03\mu_B$ antiferromagnetic ordered moment cannot drive the transition, and the primary order parameter is possibly quadrupolar. The fact that the third-order magnetic susceptibility χ_3 shows a positive divergency at $T_{\rm tr}$ [331], may (but again, may not) support the quadrupolar scenario (the relationship between a χ_3 anomaly and quadrupolar ordering is demonstrated in the solution to Problem 2.8, p. 74). We return to the question of quadrupolar ordering in Secs. 3.9 and 3.10.1. We are going to see that there are well-established cases of quadrupole order, and also of coexisting quadrupole *and* magnetic order[22]. However, the case of URu_2Si_2 is less clear-cut than some of the others, and the nature of its ordered phases is still debatable.

3.8.4 Exchange Induced Magnetism

$PrRu_2Si_2$ is isostructural with URu_2Si_2 and it contains $Pr^{3+} \leadsto 4f^2$ ions, thus we might think that the two systems ought to show similar behaviour. In fact, they could hardly be more different. $PrRu_2Si_2$ is a ferromagnet below $T_C = 14$K, and the $T = 0$ ionic moment[23] is fairly large: $2.76\mu_B$ [300].

It is thought that the crystal field ground state of a Pr^{3+} ion is a singlet; however, this time it is a Γ_{t1} state, separated by a gap Δ from the Γ_{t2} singlet [300]. Apparently, it is a good approximation to take $|\Gamma_{t1}\rangle = (1/\sqrt{2})(|4\rangle + |-4\rangle)$, and $|\Gamma_{t2}\rangle = (1/\sqrt{2})(|4\rangle - |-4\rangle)$, as before. Thus the large moments are there, only they are mixed into non-magnetic combinations. *If* we can de-mix across the crystal field gap, we should expect that the resulting moments interact with a strongly anisotropic exchange coupling like $-IJ_i^zJ_j^z$. Using the mean field approximation, the single-ion energy expression is

$$\mathcal{H} = \Delta|\Gamma_{t2}\rangle\langle\Gamma_{t2}| - \alpha J^z \tag{3.71}$$

where $\alpha = zI\langle J^z\rangle$, z being the coordination number. Doing the calculation, one finds that the system has a magnetically ordered ground

[22] As a rule, these two kinds of order should set in at different temperatures since they are of completely different nature: magnetism breaks time reversal invariance, while quadrupole order does not. It is possible that there are two closely spaced transitions in URu_2Si_2, which cannot be resolved experimentally.

[23] This comes from Mössbauer measurements. Bulk magnetization would give the somewhat smaller $2.69\mu_B$. Though it is clear that Pr makes the substance magnetic, the Ru $4d$ bands become polarized, and this accounts for the discrepancy [300].

state if the interaction I exceeds a threshold value $I_{\text{tr}} \propto \Delta$. This is not a particularly stringent condition, and exchange induced magnetism is a widespread phenomenon. The saturation magnetization $M(T = 0)$ and the Curie temperature T_{C} increase continuously from 0 as I/Δ is increased beyond its threshold value, obeying the relationship[24]

$$M(T = 0) = 4g\mu_{\text{B}}\sqrt{1 - \tanh^2 \frac{\Delta}{2k_{\text{B}}T_{\text{C}}}}. \tag{3.72}$$

Though the agreement with experimental data is not better than 30%, we may feel confident that the picture of exchange induced magnetic moments (and their consequent ordering) gives a good qualitative understanding of the behaviour of $PrRu_2Si_2$, and many similar systems.

Problem 3.5. Let us consider a simplified treatment of an f^2 ion in a cubic field, in which only the lowest two levels are kept. Let us assume that the ground state is the singlet Γ_1, and the first excited state the triplet Γ_4. The Γ_4 states can be taken from (3.57) and (3.58).

a) Determine the temperature dependence of the susceptibility! What is the effective paramagnetic moment in the high-temperature regime?

b) Consider the effect of a strong magnetic field at $T = 0$! What is the saturation moment at very high fields? Is it the same as the effective moment at high temperatures?

Taking the crystal field splitting Δ=7meV found for PrSb, what is the moment induced by an external field of 10T?

c) *Schottky anomaly*. Plot the specific heat as the function of temperature.

Problem 3.6. *A toy model of the metamagnetic transition*. Of the nine states of the f^2 configuration, we keep only the four singlets shown in Fig. 3.9. For the sake of simplicity, let us set $\epsilon = 0$ in (3.64) (actually, this may not be a bad approximation).

We switch on a magnetic field in the z-direction.

a) Determine the magnetization curve at $T = 0$.

b) Determine the temperature dependence of χ^z.

3.9 Double Groups

Many cerium compounds are interesting as magnetic or heavy fermion systems, thus it is worth getting acquianted with the electronic structure

[24]The expression would get slightly modified if we used the unrestrained form (3.64) for $|\Gamma_{t1}\rangle$.

of the Ce^{3+} ion which has the configuration $4f^1$. According to Hund's third rule, the total angular momentum in the ground state is $J = L - S = 3 - \frac{1}{2} = 5/2$, a half-integer value. The ground state multiplet of an isolated Ce^{3+} ion is sixfold degenerate.

In Section 3.5 we learned how to treat the splitting of the odd-dimensional representations of the rotation group in a crystal field. Now let us consider the splitting of the 6-fold degenerate ground state level in a cubic crystal field. A thorough discussion would be lengthy but we will be taking convenient shortcuts.

To start with, let us accept the plausible result that the formula (3.25) for the character of a rotation, which we derived for integer angular momenta, is valid for half-integer values of J as well

$$\chi_J(\alpha) = \frac{\sin{(J+1/2)\alpha}}{\sin{(\alpha/2)}}. \tag{3.73}$$

It is easy to verify that

$$\chi_J(\alpha + 2\pi) = (-1)^{2J}\chi_J(\alpha). \tag{3.74}$$

We might have expected that the character is the same for an $(\alpha + 2\pi)$-rotation as for an α-rotation, since a 2π-rotation should leave things unchanged. This is not always so: *for half-integral* J, we find that *the character is double-valued*: $\chi_J(\alpha + 2\pi) = -\chi_J(\alpha)$. In particular, for $\alpha = 0$, $\chi_J(2\pi) = -(2J+1)$. – The double-valuedness of the character has the same origin as the minus sign found for the 2π-rotation of the electron spin in (3.47).

Actually, we should note that the character is double-valued only for $\alpha \neq \pi$. From (3.73) it follows that

$$\chi_J(\pi) = \chi_J(\pi + 2\pi) = 0. \tag{3.75}$$

Bethe showed how to deal with this strange situation by enlarging the group to a *double group* so that every character becomes single-valued. Let us introduce the fiction that the 2π-rotation $\mathcal{R}$ is a new group element, different from the identity. On the other hand, the 4π-rotation behaves like the identity since

$$\chi_J(\alpha \pm 4\pi) = \chi_J(\alpha). \tag{3.76}$$

We double the group by introducing along with the α-rotations $\mathcal{R}_\alpha$ $(0 \leq \alpha < 2\pi)$ the $(\alpha + 2\pi)$-rotations $\mathcal{R}\mathcal{R}_\alpha = \mathcal{R}_\alpha\mathcal{R}$. Furthermore, $\mathcal{R}^2 = \mathcal{E}$, the identity.

Let us consider the example of the octahedral (or cubic) double group $\mathcal{O}'$. It has $2 \times 24 = 48$ elements but it does not follow that it should have twice as many classes as the octahedral group $\mathcal{O}$. (3.75) permits that $\mathcal{R}_\pi$ be in the same class as $\mathcal{R}\mathcal{R}_\pi$; we have to find out whether this is indeed the case.

Let us observe that $\mathcal{R}\mathcal{R}_\pi$ is the inverse of $\mathcal{R}_\pi$, hence it should belong to a π-rotation about the reversed axis. On the other hand, for each C_2-axis in the cubic group there is a rotation which reverses that axis. Then, according to our usual criterion, $\mathcal{R}\mathcal{R}_\pi$ and $\mathcal{R}_\pi$ are in the same conjugacy class. This argument is, of course, a bit flimsy but it is a valid conclusion that in a double group, $\mathcal{R}\mathcal{R}_\pi$ and $\mathcal{R}_\pi$ are in the same class if there is another C_2-axis perpendicular to the rotation axis in question.

For $\alpha \neq \pi$, $\mathcal{R}_\alpha$ and $\mathcal{R}\mathcal{R}_\alpha$ have different characters so they certainly belong to different classes. In the group $\mathcal{O}'$ we can identify eight conjugacy classes which are enumerated in Table 3.3. (3.16) is satisfied by

$$1^2 + 1^2 + 2^2 + 3^2 + 3^2 + 2^2 + 2^2 + 4^2 = 48. \tag{3.77}$$

Five irreducible representations are essentially identical to those we found for $\mathcal{O}$ (Table 3.1), with the additional feature that $\chi(\mathcal{R}) = \chi(\mathcal{E})$. There are three new double-valued representations: two two-dimensional, and one four-dimensional representation, whose characters are different in sign for $\mathcal{R}_\alpha$ and $\mathcal{R}\mathcal{R}_\alpha$. The characters are given in Table 3.3, while in Table 3.4, we give the characters of the representations of $\mathcal{O}'$ by the $J = 1/2$, $3/2$, $5/2$, and $7/2$, angular momentum states; these can be obtained from (3.73). The level splittings in a cubic field can be derived using (3.15)

$$\begin{aligned}
D_{1/2} &\longrightarrow \Gamma_6 \\
D_{3/2} &\longrightarrow \Gamma_8 \\
D_{5/2} &\longrightarrow \Gamma_7 \oplus \Gamma_8 \\
D_{7/2} &\longrightarrow \Gamma_6 \oplus \Gamma_7 \oplus \Gamma_8
\end{aligned} \tag{3.78}$$

A cubic field does not split the $J = 1/2$, and $J = 3/2$, levels. The $J = 5/2$ level which corresponds to the ground state of a $4f^1$ ion,

Table 3.3: Character table of the octahedral double group O'.

	$\mathcal{E}$	$\mathcal{R}$	$8\mathcal{C}_3$	$8\mathcal{RC}_3$	$3\mathcal{C}_2 + 3\mathcal{RC}_2$	$6\mathcal{C}_2' + 6\mathcal{RC}_2'$	$6\mathcal{C}_4$	$6\mathcal{RC}_4$
Γ_1	1	1	1	1	1	1	1	1
Γ_2	1	1	1	1	1	-1	-1	-1
Γ_3	2	2	-1	-1	2	0	0	0
Γ_4	3	3	0	0	-1	-1	1	1
Γ_5	3	3	0	0	-1	1	-1	-1
Γ_6	2	-2	1	-1	0	0	$\sqrt{2}$	$-\sqrt{2}$
Γ_7	2	-2	1	-1	0	0	$-\sqrt{2}$	$\sqrt{2}$
Γ_8	4	-4	-1	1	0	0	0	0

Table 3.4: Character table for representations by the $J = 1/2$, $3/2$, $5/2$, and $7/2$ states.

	$\mathcal{E}$	$\mathcal{R}$	$8\mathcal{C}_3$	$8\mathcal{RC}_3$	$3\mathcal{C}_2 + 3\mathcal{RC}_2$	$6\mathcal{C}_2' + 6\mathcal{RC}_2'$	$6\mathcal{C}_4$	$6\mathcal{RC}_4$
$D_{1/2}$	2	-2	1	-1	0	0	$\sqrt{2}$	$-\sqrt{2}$
$D_{3/2}$	4	-4	-1	1	0	0	0	0
$D_{5/2}$	6	-6	0	0	0	0	$-\sqrt{2}$	$\sqrt{2}$
$D_{7/2}$	8	-8	1	-1	0	0	0	0

gets split into the fourfold degenerate Γ_8, and the twofold degenerate Γ_7 states. It should be emphasized that all resulting degeneracies are at least twofold. This is a consequence of Kramers' theorem: a half-integral J can only result from an odd number of electrons. In contrast to the non-Kramers doublet Γ_3 which we met in the discussion of the f^2 configuration, Γ_7 is a Kramers doublet.

As shown in Fig. 3.12, Γ_7 and Γ_8 states have markedly different charge clouds. The figure also gives the basis states which are expressed in the $|J^z\rangle$ basis, so $|1/2\rangle$ means $|L = 3, S = 1/2, J = 5/2, J^z = 1/2\rangle$, etc. Γ_7 lobes point towards repulsive negative charges situated at the corners of a cube, while Γ_8 lobes manage to avoid these directions. We are reminded of Fig. 3.3 which shows the charge density contours of the

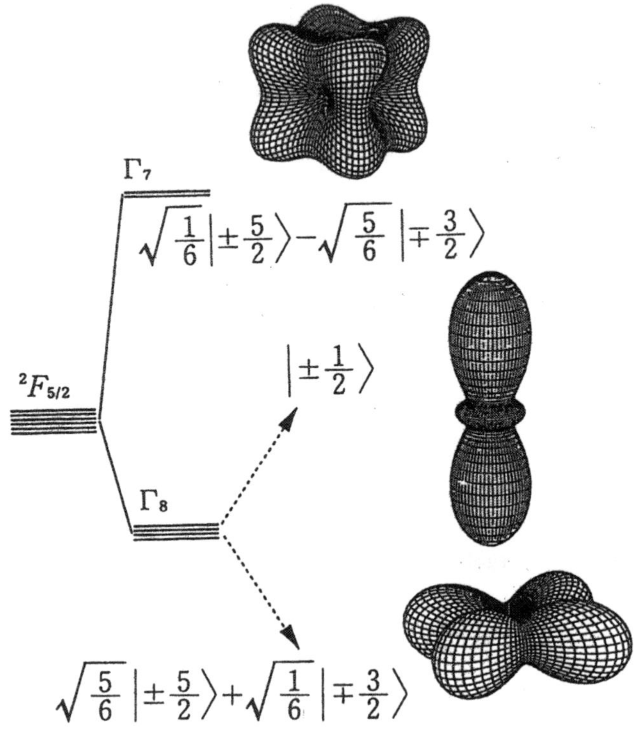

Figure 3.12: The characteristic shapes of electron clouds with Γ_7 and Γ_8 symmetries (by courtesy of T. Sakakibara). The level scheme corresponds to the crystal field of eight anions situated in the [111] directions as in CeB_6.

x^2-y^2 and $3z^2-r^2$ d-states. The resemblance is not merely superficial: we will shortly see (p. 132) that the two kinds of Γ_8 charge clouds are conveniently characterized by their $(3z^2-r^2)$-like (or, if we prefer, (x^2-y^2)-like) *quadrupole moments*[25].

The ground state of a Ce^{3+} ion in a cubic environment may be either Γ_7 or Γ_8; group theory alone cannot give the answer. However, looking at the shape of the charge clouds, we may adopt the simple rule which

[25]The quadrupole moment measures what we may loosely call the orbital character of these states. However, while for d-states the orbital state is well-defined (Fig. 3.8 illustrates the point), the basis states discussed here are linear combinations of several $|L^z, L=3\rangle$ orbital states (see App. A).

we used for Fig. 3.4: if the crystal field arises from an octahedron (a cube) of six (eight) surrounding negative charges, then the low-lying level is Γ_7 (Γ_8).

$CeAl_2$ is an ordered magnet with a modulated structure below $T_N = 3.2K$ [116]. The average ordered moment is $0.63\mu_B$, indicating a Γ_7 ground state (see Problem 3.7).

The crystal structure of CeB_6 can be envisaged as a bcc lattice in which Ce^{3+} ions sit at the central site of the cubic unit cell, and eight B_6 groups occupying the corner sites generate the crystal field. The ground state level is the Γ_8 quartet [471]. It is apparent from Fig. 3.12 that Γ_8 electrons have both magnetic and electric (or charge) degrees of freedom. There are two pairs of time-reversed states, thus we may ascribe a "spin" $\sigma^z = 1/2$ ($\sigma^z = -1/2$) to the $\langle J^z \rangle > 0$ ($\langle J^z \rangle < 0$) states (note that this σ has nothing to do with spin-1/2 states; it is just a reminder that there is a magnetic moment which can point either up or down). The two charge clouds can be distinguished with a pseudospin $\tau^z = \pm 1/2$. We follow [415] in relabelling the Γ_8 states as $|\tau^z, \sigma^z\rangle$ states

$$|1/2, \pm 1/2\rangle = \sqrt{5/6}|\pm 5/2\rangle + \sqrt{1/6}|\mp 3/2\rangle \qquad |-1/2, \pm 1/2\rangle = |\pm 1/2\rangle. \tag{3.79}$$

In the absence of an external magnetic field, CeB_6 has two phase transitions. At $T_Q = 3.3K$ it enters a state with staggered ("antiferro") quadrupolar order which can be envisaged as an alternation of the two kinds of Γ_8 charge clouds, but without magnetic moments ($\langle \tau^z \rangle \neq 0$, $\langle \sigma^z \rangle = 0$). Below the Néel temperature $T_N = 2.4K$, there is also a complicated magnetic order ($\langle \tau^z \rangle \neq 0$, $\langle \sigma^z \rangle \neq 0$). We should not be surprised that the two transitions do not coincide since quadrupolar ordering does not break time reversal invariance, while magnetic ordering does. Since the Γ_8–Γ_7 separation is fairly large (~47meV), it is not a bad approximation to describe the phase transitions as an interplay of electric quadrupolar[26], and magnetic dipolar, ordering within the basis (3.79) [415]. This is in contrast to the induced-moment phenomenon we met in Sec. 3.8.4, where coupling to higher-lying levels was essential.

What should one see at $2.4K < T < 3.3K$ where there is only

[26] A more accurate treatment of the T–H phase diagram reveals that octupoles are also important [363]. The fact that interactions do not get substantially weaker with increasing multipole rank, is a source of great complexity.

quadrupolar, but no magnetic, order? We may fear that nothing too much[27], since each Ce site has the same number of f-electrons, only their spatial distribution is cigar-shaped on one sublattice, and flattened on the other. To be sure, there should be a corresponding shift of the positions of the B_6 groups[28], but the stiffness of the boron network and the compactness of f orbitals make this effect essentially unobservable. However, we are rescued by a curious phenomenon: a *uniform* magnetic field produces (in addition to a uniform polarization) a *staggered* magnetic order which has the same pattern as the quadrupolar order, and this is easily seen by neutrons. The reader can understand the basic mechanism of this effect by solving Problem 3.8.

Problem 3.7. Determine the magnetic moment in the Γ_7 ground state. Compare the result with the ordered moment $0.63\mu_B$ observed in $CeAl_2$.

Problem 3.8. Derive the Zeeman coupling for Γ_8 electrons, working in the basis (3.79). Discuss the response to a uniform external field in the presence of staggered quadrupolar ordering.

3.10 Crystal Field Potentials

Symmetry arguments enable us to learn much about crystal field effects without having to specify the crystal field Hamiltonian. However, the details of the crystal field become important if we want to determine the sequence of the levels, or the size of the splittings. Finding realistic crystal field potentials is a difficult task which we cannot deal with here. Our present aim is to get an impression of the reasoning which leads to the form of the crystal field Hamiltonians found in the literature. For more details, the reader is referred to [178].

Let us assume that the crystal field arises from point charges lying outside the region that we want to describe. The crystal field potential $\mathcal{V}$ satisfies Laplace's equation, the solution of which can be expanded in

---

[27] We will meet the same difficulty when discussing the analogous orbital ordering in d-electron compounds (see sections 5.4 and 8.5).

[28] In fact, a lattice distortion is required by the Jahn–Teller theorem, since the degeneracy exceeds the Kramers minimum.

terms of spherical harmonics

$$\mathcal{V}(\mathbf{r}) = \sum_l \sum_m A_l^m Y_l^m(\vartheta, \varphi) r^l. \tag{3.80}$$

$\mathcal{V}$ must transform according to the identical representation of the point group. It would seem that we have to know the basis function of the identical representation for each l but for practical purposes we usually need just a few of them.

Usually, we are interested in how $\mathcal{V}$ acts within a subspace spanned by the d-states (or f-states) of an unfilled shell. To find the crystal-field-split states of a single d-electron, we have to evaluate matrix elements of the form $\int \psi^* \phi \mathcal{V} d\mathbf{r}$, where ψ and ϕ are d-states. Because of the addition properties of the angular momentum, the expansion of $\psi^* \phi$ cannot contain higher than $l = 4$ components. Because of the orthogonality properties of the spherical harmonics, the matrix element vanishes for the $l > 4$ components of $\mathcal{V}$; hence for this particular problem, it is sufficient to expand $\mathcal{V}$ to fourth order in l. By similar arguments, for f-states it is sufficient to consider the terms with $l \leq 6$.

Whether we are speaking about d-, or f-states, ψ and ϕ have the same parity, hence $\psi^* \phi$ has even parity. Therefore we can drop the odd-parity (l odd) terms in the expansion of $\mathcal{V}$.

In what follows, we construct the Hamiltonian of a cubic crystal field.

The largest term is the spherically symmetrical $l = 0$ term $\mathcal{V}_0$. It is an important contribution to the total energy of the crystal but it does not influence the spectrum. Neither does the $l = 2$ term which has the same symmetry (cf. Table 3.1). The first non-trivial contribution comes from $l = 4$, and we do not have to go beyond sixth-order, thus

$$\mathcal{V} = \mathcal{V}_0 + \mathcal{V}_4 + \mathcal{V}_6. \tag{3.81}$$

Let us determine $\mathcal{V}_4$. We know the basis function (3.55) of the identical (Γ_1) representation. Replacing (3.54), the fourth-order term is found to be

$$\mathcal{V}_4 = A_4 \left(x^4 + y^4 + z^4 - \frac{3}{5} r^4 \right). \tag{3.82}$$

A similar reasoning gives the more complicated sixth-order term

$$\mathcal{V}_6 \quad = \quad A_6 \left[x^6 + y^6 + z^6 + \frac{15}{4} \left(x^2 y^4 + x^2 z^4 \right. \right.$$

$$+ \quad y^2x^4 + y^2z^4 + z^2x^4 + z^2y^4\Big) - \frac{15}{14}r^6\Big]. \tag{3.83}$$

The prefactors depend on the geometry of the charges: for instance, for six point charges q at the positions $(\pm a, 0, 0)$, $(0, \pm a, 0)$, $(0, 0, \pm a)$, one finds

$$\mathcal{V}_0 = \frac{6q}{a} \qquad A_4 = \frac{35q}{4a^5} \qquad A_6 = \frac{-21q}{2a^7}. \tag{3.84}$$

If the cubic field arises from eight charges situated at the corners of a cube (Fig. 1.2) then

$$\mathcal{V}_0 = \frac{8q}{a} \qquad A_4 = -\frac{70q}{9a^5} \qquad A_6 = \frac{-224q}{9a^7}. \tag{3.85}$$

It should be noted that the sign of A_4 differs for the two configurations.

The formulae (3.82), (3.83) look pleasingly symmetrical but they are cumbersome to use: it is not quite simple to evaluate the matrix element of, say, x^2z^4 between f-states. The situation gets worse if we have several electrons in the shell and the matrix elements have to be taken between Slater determinants. Noticing that all the relevant matrix elements have to be taken between states of a given L- (or for f-electrons, J-) manifold, Stevens suggested that $\mathcal{V}_4$, etc., should be replaced by their *operator equivalents* which are expressed in terms of *total angular momentum operators.*

If we have in mind d-electrons, we may neglect the spin–orbit coupling and express the operator equivalent in terms of the components of the (total) orbital angular momentum $\mathbf{L}$. The simple recipe is to perform the replacements $x \leadsto L_x$, $y \leadsto L_y$, $z \leadsto L_z$, and $(x \pm iy) \leadsto L^{\pm}$. Furthermore, if we meet an expression like xy, we have to remember that though x and y commute, L_x and L_y do not, thus we have to use the symmetrical replacement $xy \leadsto (L_xL_y + L_yL_x)/2$. – The justification for introducing operator equivalents comes from the Wigner–Eckart theorem.

To see an example, let us consider the operator equivalent of $\mathcal{V}_4$ in terms of L_x, L_y, and L_z. It is useful to go back to (3.55) and (3.54). Omitting the prefactor, the relevant replacement is

$$35z^4 - 30z^2r^2 + 3r^4 + \frac{5}{2}[(x+iy)^4 + (x-iy)^4]$$

$$\rightsquigarrow \quad 35L_z^4 - 30L(L+1)L_z^2 + 3L^2(L+1)^2 + 25L_z^2 - 6L(L+1)$$
$$+\frac{5}{2}[(L^+)^4 + (L^-)^4]. \tag{3.86}$$

It may not be obvious where the last two terms in the second line are coming from. The reason is that, for instance, z^2r^2 cannot be simply replaced by $L_z^2L^2$. Actually, $z^2r^2 = z^2(x^2+y^2+z^2) = z^2x^2+z^2y^2+z^4$, so we need first the equivalent of $z^2x^2 = zzxx$. Remembering the rule of symmetrizing the expression of L-operators, we have to write

$$z^2x^2 \rightsquigarrow \frac{1}{6}(L_z^2L_x^2 + L_x^2L_z^2 + L_xL_zL_xL_z +$$
$$L_xL_zL_zL_x + L_zL_xL_xL_z + L_zL_xL_zL_x)\,. \tag{3.87}$$

Using the standard commutation relations, each term can be expressed in terms of squared operators and $L_xL_yL_z$, for example

$$L_xL_zL_xL_z = L_x(iL_y + L_xL_z)L_z = iL_xL_yL_z + L_x^2L_z^2. \tag{3.88}$$

Proceeding like this, one finds [178]

$$r^2z^2 \rightsquigarrow \frac{1}{6}[6L(L+1)L_z^2 + L(L+1) - 5L_z^2]. \tag{3.89}$$

Decomposing r^4 in a similar fashion, (3.86) can be confirmed.

The operator in the second line of (3.86) is the Stevens operator O_4^0, and $(1/2)[(L^+)^4 + (L^-)^4] = O_4^4$. Thus in standard notations

$$\mathcal{V}_4 = B_4^0\;[O_4^0 + 5O_4^4]. \tag{3.90}$$

To take the simplest example, let us reconsider the splitting of a d-level in a cubic field. The functions $Y_2^m = |2,m\rangle$ are eigenstates of O_4^0

$$O_4^0|2,0\rangle = 72|2,0\rangle \qquad O_4^0|2,\pm1\rangle = -48|2,\pm1\rangle \qquad O_4^0|2,\pm2\rangle = 12|2,\pm2\rangle. \tag{3.91}$$

The only nonvanishing matrix elements of O_4^4 are

$$\langle 2,\mp2|O_4^4|2,\pm2\rangle = 12. \tag{3.92}$$

$|2,\pm1\rangle$ and $|2,0\rangle$ are eigenstates of (3.90), with eigenvalues $-48B_4^0$, and $72B_4^0$, respectively. $|2,\pm2\rangle$ are combined by the non-diagonal matrix elements into $(1/\sqrt{2})(|2,2\rangle + |2,-2\rangle)$, with eigenvalue $72B_4^0$, and $(1/\sqrt{2})(|2,2\rangle - |2,-2\rangle)$, with the eigenvalue $-48B_4^0$.

To introduce another standard notation: the splitting is often set equal to $10Dq$. Assuming that the cubic field originated from six point charges at the corners of an octahedron (3.84), $A_4 \sim B_4^0 > 0$, and our previous result can be restated in the following manner: the t_{2g} states $(1/\sqrt{2})(|2,1\rangle \pm |2,-1\rangle)$ and $(1/\sqrt{2})(|2,2\rangle - |2,-2\rangle)$ are lying at the energy $-4Dq$, while the e_g states$(1/\sqrt{2})(|2,2\rangle + |2,-2\rangle)$ and $|2,0\rangle$ have the energy $6Dq$. This was shown in Fig. 1.3.a. If the crystal field arises from eight point charges forming a cube, A_4 changes sign (3.85), and the sequence of the levels is reversed (Fig. 1.3.b).

3.10.1 Quadrupole Moments

We have seen on the example of CeB_6 that the electric quadrupole moment can play the role of the order parameter in certain phase transitions. The matrix elements of the components of the quadrupole moment tensor are defined as

$$\langle\psi'|Q_{ij}|\psi\rangle = e\int d\mathbf{r}\psi'(\mathbf{r})(3x_ix_j - r^2\delta_{ij})\psi(\mathbf{r}) \qquad (3.93)$$

(in what follows, we use $x_1 \to x$, $x_2 \to y$, $x_3 \to z$). The Stevens operator equivalent is

$$\langle\psi'|Q_{ij}|\psi\rangle \propto \langle\psi'|(1/2)(J_iJ_j + J_jJ_i) - \mathbf{J}^2\delta_{ij}|\psi\rangle\,. \qquad (3.94)$$

We will not be interested in the constant of proportionality, and henceforth call the right hand side of (3.94) the quadrupole moment (QM). Some of the QM components are often-used Stevens operators: $Q_{zz} = 3J_z^2 - J(J+1) = O_0^2$, and $(Q_{xx} - Q_{yy})/\sqrt{3} = \sqrt{3}(J_x^2 - J_y^2) = O_2^2$.

Assuming cubic symmetry, we have the chance to stumble upon the non-Kramers doublet Γ_3 which is conveniently understood as a QM doublet. For $J = 4$, the basis functions are

$$|\phi_{31}^{J=4}\rangle = \sqrt{\frac{7}{24}}|4\rangle - \sqrt{\frac{5}{12}}|0\rangle + \sqrt{\frac{7}{24}}|-4\rangle \quad \text{and} \quad |\phi_{32}^{J=4}\rangle = \frac{|2\rangle + |-2\rangle}{\sqrt{2}},$$

thus $\langle\phi_{31}^{J=4}|Q_{zz}|\phi_{31}^{J=4}\rangle = 8$ and $\langle\phi_{32}^{J=4}|Q_{zz}|\phi_{32}^{J=4}\rangle = -8$. For $J = 6$, the Γ_3 basis can be chosen as [239]

$$|\phi_{31}^{J=6}\rangle = \sqrt{\frac{11}{32}}|6\rangle + \sqrt{\frac{5}{32}}|2\rangle + \sqrt{\frac{5}{32}}|-2\rangle + \sqrt{\frac{11}{32}}|-6\rangle$$

$$|\phi_{32}^{J=6}\rangle = \frac{1}{4}(|4\rangle + \sqrt{2}|0\rangle + |-4\rangle),$$

yielding $\langle\phi_{31}^{J=6}|Q_{zz}|\phi_{31}^{J=6}\rangle = 36$ and $\langle\phi_{32}^{J=6}|Q_{zz}|\phi_{32}^{J=6}\rangle = -36$. Let us note that the intermetallic compound TmCd is supposed to contain Tm^{3+} ions for which $S = 1$, $L = 5$, and $J = 6$. At high temperatures the system has the CsCl structure but at $T_Q = 3.16$K it undergoes a cubic to tetragonal structural phase transition [251]. In this system the quadrupole–quadrupole interaction seems to be the driving force of the cooperative Jahn–Teller effect [12]. It had been thought [251] that the crystal field ground state in the cubic phase is the Γ_3 doublet[29]. At $T > T_Q$, $\langle Q_{zz}\rangle = 0$, but $\langle Q_{zz}\rangle \neq 0$ appears at $T < T_Q$. Writing down only the terms which contain $Q_{zz} \propto O_2^0$, the simplest approach would be to describe the transition to a uniformly distorted phase by the mean field Hamiltonian[30]

$$\mathcal{H} = \lambda O_2^0 \langle O_2^0\rangle + \mu\epsilon O_2^0\,, \tag{3.95}$$

where the second term describes the coupling of the quadrupolar moments to the appropriate component ϵ of the lattice strain.

Now let us return to URu_2Si_2, with its supposed $|\Gamma_{t3}\rangle = 1/\sqrt{2}(|2\rangle + |-2\rangle)$ crystal field ground state (Sec. 3.8.2). For a tetragonal system, a $\sim O_2^0$ term is part of the ground state energy, and the ordering QM component may be $O_2^2 = \sqrt{3}(J_x^2 - J_y^2) = (\sqrt{3}/4)[(J^+)^2 + (J^-)^2]$ [345]. $\langle\Gamma_{t3}|O_2^2|\Gamma_{t3}\rangle = 0$ but there is a $\langle\Gamma_{t1}|O_2^2|\Gamma_{t3}\rangle \neq 0$ matrix element to the first excited state. Thus here we have to envisage the ordering of interaction-induced quadrupolar moments, in complete analogy with induced-moment magnetism which we learned about in Sec. 3.8.4.

Finally, what about CeB_6 ? The pseudospin τ we used in (3.79) is related to the quadrupolar moment: e.g., we find $\langle\pm 1/2,\sigma|Q_{zz}|\pm 1/2,\sigma\rangle = \pm 8$, i.e., we may write $Q_{zz} = 16\tau^z$. However, it is not unambiguous that the order parameter should be chosen as $\langle Q_{zz}\rangle$; one may also think of a suitable combination of $\langle Q_{xy}\rangle$-like quantities.

---

[29] Later, it was argued [12] that it is more likely to be one of the triplets, which has both magnetic and quadrupolar moments. It would be beside our present point to enter an argument about the actual nature of TmCd. It is, however, a general lesson that completely different crystal field schemes can give equally satisfying fits to the same set of measured data. We met similar difficulties with URu_2Si_2.

[30] Quadrupolar ordering may be accompanied by anomalous behaviour (possibly divergency) of the third-order magnetic susceptibility [288]. A simple mean field calculation is given in Problem 2.8 (see p. 74).

To see the behaviour of non-diagonal matrix elements, consider

$$\langle 1/2, \pm 1/2 | Q_{xy} | -1/2, \pm 1/2 \rangle = \mp i\sqrt{3}.$$

This is reproduced by replacing $Q_{xy} = 4\sqrt{3}\sigma^z \tau^y$, thus the magnetic degree of freedom σ appears in the quadrupole moment [415]. It is interesting to observe that the time reversal invariance of Q_{xy} is sustained by having $\mathcal{T}_t \sigma^z = -\sigma^z$ and $\mathcal{T}_t \tau^y = -\tau^y$ (however, τ^x and τ^z are time reversal invariant).

Assuming an intersite quadrupole–quadrupole interaction term like $\lambda Q_{xy}(\mathbf{R}_1) Q_{xy}(\mathbf{R}_2)$, and decoupling σ from τ, we are left with

$$\lambda \langle \tau^y(\mathbf{R}_1) \tau^y(\mathbf{R}_2) \rangle \sigma^z(\mathbf{R}_1) \sigma^z(\mathbf{R}_2).$$

A piece of the quadrupole–quadrupole interaction may appear as a "spin Hamiltonian". Naturally, there are also many other contributions to consider. See [415] and [363] for details about the behaviour of CeB_6.

Solutions to the Problems

Problem 3.1. Let us use Fig. 3.2 to decide which proper rotations belong to the tetragonal point group. The z-axis remains a fourfold axis; it is also a C_2-axis which is now inequivalent to the other axes. The x-, and y-axes are equivalent C_2-axes which we denote as C_2', while the axes bisecting the angle between the x- and y-axes are the C_2''-axes. Thus the group of proper rotations (in Tinkham's book [399], it is called $D_4(422)$) contains (including also $\mathcal{E}$) 8 elements, and 5 classes. It follows that it has five irreducible representations, four one-dimensional, and one two-dimensional.

It is easy to find simple basis functions by trying those we met in the discussion of the octahedral group, and simplifying them further, whenever possible. For instance, take the E states. z^2 is now a basis of the identical representation, while $x^2 - y^2$ turns out to be a basis function of a one-dimensional representation. Of the T_2 states, xy can serve as the basis of another one-dimensional representation, while $\{xz, yz\}$ can be chosen as the basis of the two-dimensional representation. A momentary reflection will, however, convince us that the still simpler set $\{x, y\}$ serves the same purpose. These considerations also tell us how the E and T_2 representations get split but that can also be derived using (3.17), and the character table 3.5, which is now straightforward to construct. (Since some of the standard notations for representations, such as A_1, E, Γ_3, etc. are also used for the octahedral group where they have a different meaning, the subscript t has been added to avoid confusion.)

Table 3.5: Character table of the subgroup of proper rotations of the tetragonal point group.

		basis	$\mathcal{E}$	C_2	$2C_4$	$2C_2'$	$2C_2''$
A_{1t}	Γ_{t1}	$\{z^2\}$	1	1	1	1	1
A_{2t}	Γ_{t2}	$\{z\}$	1	1	1	-1	-1
B_{1t}	Γ_{t3}	$\{x^2-y^2\}$	1	1	-1	1	-1
B_{2t}	Γ_{t4}	$\{xy\}$	1	1	-1	-1	1
E_t	Γ_{t5}	$\{x,y\}$	2	-2	0	0	0

Table 3.6: Character table of the subgroup of proper rotations of the orthogonal point group.

	basis	$\mathcal{E}$	C_{2x}	$2C_{2y}$	$2C_{2z}$
A_{1o}	$\{z^2\}$	1	1	1	1
A_{2o}	$\{x\}$	1	1	-1	-1
B_{1o}	$\{y\}$	1	-1	1	-1
B_{2o}	$\{z\}$	1	-1	-1	1

It should be mentioned that the character table is the same as that of the C_{4v} group which is wellknown from the literature. C_{4v} is isomorphic, but not identical, to our $D_4(422)$. C_{4v} characterizes tetragonal symmetry by including, in addition to the (vertical) fourfold axis, two sets of vertical mirror planes.

The irreducible representations of the octahedral group are split by a tetragonal distortion according to

$$\begin{aligned} E &\rightarrow A_{1t} \oplus B_{1t} \\ T_1 &\rightarrow A_{2t} \oplus E_t \\ T_2 &\rightarrow B_{2t} \oplus E_t. \end{aligned} \tag{3.96}$$

Problem 3.2. The group consists of three inequivalent C_2 axes, pointing respectively along the x-, y-, and z-directions, and the unit element. The group is denoted by $D_2(2,2,2)$ and it is traditionally called the "Vierergruppe" [399].

There are four one-dimensional irreducible representations, whose basis functions can be chosen either as z^2, xy, yz, xz, or as z^2, x, y, z. The characters are given in Table 3.6.

Problem 3.3. We will find it useful to introduce new basis states for the

t_{2g} subspace (omitting the r-dependent factors)

$$\phi_1 = Y_2^{-1} \qquad \phi_0 = \frac{1}{\sqrt{2}}(Y_2^2 - Y_2^{-2}) \qquad \phi_{-1} = -Y_2^1. \tag{3.97}$$

For the e_g states we keep the original

$$\psi_1 = Y_2^0 \qquad \psi_2 = \frac{1}{\sqrt{2}}(Y_2^2 + Y_2^{-2}). \tag{3.98}$$

In the basis $\{\phi_1, \phi_0, \phi_{-1}, \psi_1, \psi_2\}$ (strictly in this order), the angular momentum operators are represented by the matrices

$$L_x = -\frac{1}{\sqrt{2}}\left(\begin{array}{ccc|cc} 0 & 1 & 0 & -\sqrt{3} & -1 \\ 1 & 0 & 1 & 0 & 0 \\ 0 & 1 & 0 & \sqrt{3} & 1 \\ \hline -\sqrt{3} & 0 & \sqrt{3} & 0 & 0 \\ -1 & 0 & 1 & 0 & 0 \end{array}\right)$$

$$L_y = -\frac{i}{\sqrt{2}}\left(\begin{array}{ccc|cc} 0 & 1 & 0 & \sqrt{3} & -1 \\ -1 & 0 & 1 & 0 & 0 \\ 0 & -1 & 0 & \sqrt{3} & -1 \\ \hline -\sqrt{3} & 0 & -\sqrt{3} & 0 & 0 \\ 1 & 0 & 1 & 0 & 0 \end{array}\right)$$

$$L_z = -\left(\begin{array}{ccc|cc} 1 & 0 & 0 & 0 & 0 \\ 0 & 0 & 0 & 0 & -2 \\ 0 & 0 & -1 & 0 & 0 \\ \hline 0 & 0 & 0 & 0 & 0 \\ 0 & -2 & 0 & 0 & 0 \end{array}\right) \tag{3.99}$$

First, let us consider the action of $\mathbf{L}$ within the t_{2g} subspace. The 3×3 diagonal blocks remind us of the $l = 1$ angular momentum operators. In fact, in the standard p-basis $\{Y_1^1, Y_1^0, Y_1^{-1}\}$, the components of $\mathbf{L}$ look like

$$L_x^{(p)} = \frac{1}{\sqrt{2}}\begin{pmatrix} 0 & 1 & 0 \\ 1 & 0 & 1 \\ 0 & 1 & 0 \end{pmatrix} \qquad L_y^{(p)} = \frac{i}{\sqrt{2}}\begin{pmatrix} 0 & 1 & 0 \\ -1 & 0 & 1 \\ 0 & -1 & 0 \end{pmatrix}$$

$$L_z^{(p)} = \begin{pmatrix} 1 & 0 & 0 \\ 0 & 0 & 0 \\ 0 & 0 & -1 \end{pmatrix} \tag{3.100}$$

Thus we can assert that, *within the subspace of t_{2g} functions,*

$$\mathbf{L}(t_{2g}) = -\mathbf{L}(p). \tag{3.101}$$

Furthermore

$$(\mathbf{L}(t_{2g}))^2 = (\mathbf{L}(p))^2 = 2. \tag{3.102}$$

In a sense, d-electrons in t_{2g} states behave as if they were p-electrons carrying $l = 1$: their angular momentum is not fully quenched.

Caution has to be excercised in using the t_{2g}–p equivalence: naturally, the d-electrons cannot have become p-electrons! After all, all the states are (by construction) pure d-shell states so that for them $\mathbf{L}^2 = 6$ remains strictly true. The resolution of the apparent contradiction is that the true angular momentum matrices are the 5×5 matrices listed in (3.99). By restricting the Hilbert space to the three-dimensional subspace of the t_{2g}-states, we have discarded off-diagonal elements of $\mathbf{L}$ which connect the t_{2g} and e_g subspaces. $\mathbf{L}(t_{2g})$ is only a "pseudo angular momentum"; it is easy to convince ourselves that its components do not satisfy the commutation relations expected of a "true" angular momentum. However, if the t_{2g}–e_g splitting is large, the off-diagonal elements cannot matter much, and the equivalence has practical usefulness.

On the other hand, within the e_g subspace all matrix elements are zero; this means that the angular momentum of the e_g-electrons is fully quenched.

Problem 3.4. Regarding the $3d^9$ configuration as a single-hole system, we have a one-particle problem to solve. For the hole, the spin–orbit coupling is (with $\lambda > 0$, the negative sign is according to (2.70))

$$\mathcal{H}_{SO} = -\lambda \mathbf{S} \cdot \mathbf{L} = -\lambda S^z L^z - \frac{\lambda}{2}(S^+ L^- + S^- L^+) \tag{3.103}$$

where we have to take $L = 2$, $S = 1/2$.

a) In the absence of the spin–orbit coupling, the ground state of the hole is twofold degenerate, the two states being $\psi(B_{1t})\alpha$ and $\psi(B_{1t})\beta$, where α and β are the $\uparrow$ and $\downarrow$ spin eigenstates. Recalling that $\psi(B_{1t}) \sim (Y_2^2 + Y_2^{-2})/\sqrt{2}$ (and similarly $\psi(B_{2t}) \sim (Y_2^2 - Y_2^{-2})/\sqrt{2}$), we find that

$$\mathcal{H}_{SO} \cdot \psi(B_{1t})\alpha = -\frac{\lambda}{\sqrt{2}}(Y_2^2 - Y_2^{-2})\alpha - \frac{\lambda}{\sqrt{2}} Y_2^{-1}\beta. \tag{3.104}$$

The spin–orbit term couples the crystal field levels with the matrix elements

$$\langle \psi(B_{2t})\alpha | \mathcal{H}_{SO} | \psi(B_{1t})\alpha \rangle = -\lambda \tag{3.105}$$

and

$$\langle \psi(E_t^-)\beta | \mathcal{H}_{SO} | \psi(B_{1t})\alpha \rangle = -\frac{\lambda}{\sqrt{2}} \tag{3.106}$$

where $\psi(E_t^{\pm}) \sim Y_2^{\pm 1}$. – Similarly

$$\langle \psi(B_{2t})\beta | \mathcal{H}_{SO} | \psi(B_{1t})\beta \rangle = \lambda \tag{3.107}$$

and

$$\langle\psi(E_t^+)\alpha|\mathcal{H}_{SO}|\psi(B_{1t})\beta\rangle = -\frac{\lambda}{\sqrt{2}}. \tag{3.108}$$

Note that $\mathcal{H}_{SO}$ mixes different spin directions, the states can no longer be indexed by α and β. Starting with $\psi(B_{1t})\alpha$ and $\psi(B_{1t})\beta$, respectively, first-order perturbation theory gives the following new states

$$\begin{aligned}\psi_A &= \psi(B_{1t})\alpha + \frac{\lambda}{\mathcal{E}(B_{1t}) - \mathcal{E}(B_{2t})}\,\psi(B_{2t})\alpha \\ &\quad + \frac{1}{\sqrt{2}} \cdot \frac{\lambda}{\mathcal{E}(B_{1t}) - \mathcal{E}(E_t)}\,\psi(E_t^-)\beta \end{aligned}\tag{3.109}$$

and

$$\begin{aligned}\psi_B &= \psi(B_{1t})\beta - \frac{\lambda}{\mathcal{E}(B_{1t}) - \mathcal{E}(B_{2t})}\,\psi(B_{2t})\beta \\ &\quad + \frac{1}{\sqrt{2}} \cdot \frac{\lambda}{\mathcal{E}(B_{1t}) - \mathcal{E}(E_t)}\,\psi(E_t^+)\alpha. \end{aligned}\tag{3.110}$$

It should be pointed out that though we are solving the single-hole problem, we did not find it necessary to complicate matters by introducing the energy levels of the d-hole in the crystal field formally. The hole spectrum could be obtained by turning the spectrum shown in Fig. 3.7 upside down. Here we are expressing everything with the electron energies which can be identified from Fig. 3.7.

To second order in λ, the energies are found to be

$$\mathcal{E}_A = \mathcal{E}_B = \mathcal{E}(B_{1t}) - \frac{\lambda^2}{\mathcal{E}(B_{1t}) - \mathcal{E}(B_{2t})} - \frac{\lambda^2}{2(\mathcal{E}(B_{1t}) - \mathcal{E}(E_t))}. \tag{3.111}$$

Observe that though the spin–orbit coupling mixes the spin directions, it does not lift the twofold degeneracy which, in the absence of the coupling, we associated with the arbitrariness of the spin direction. The deeper reason is that the entire Hamiltonian, including $\mathcal{H}_{SO}$, preserves time reversal invariance. Kramers' Theorem (cf. Sec. 3.7) asserts that for a time reversal invariant system consisting of an odd number of electrons, each level is at least twofold degenerate.

b) Let us introduce the Zeeman term

$$\begin{aligned}\mathcal{H}_Z &= -\mu_B \mathbf{H}\cdot(\mathbf{L} + 2\mathbf{S}) \qquad (3.112)\\ &= -\mu_B[H^z(L^z + 2S^z) + \frac{1}{2}(H^+(L^- + 2S^-) + H^-(L^+ + 2S^+))]\end{aligned}$$

where $H^\pm = H^x \pm iH^y$.

The susceptibility of our tetragonal system turns out to be axially symmetrical thus without loss of generality we can take a field in the x–z plane: $H^+ = H^- = H\sin\theta$, $H^z = H\cos\theta$. Considering the ground state set, the Zeeman term is found to have both diagonal and off-diagonal matrix elements

$$\langle\psi_{A,B}|\mathcal{H}_Z|\psi_{A,B}\rangle = \mp\mu_B H\cos\theta\left(1+\frac{4\lambda}{\mathcal{E}(B_{2t})-\mathcal{E}(B_{1t})}\right) \quad (3.113)$$

$$\langle\psi_B|\mathcal{H}_Z|\psi_A\rangle = -\mu_B H\sin\theta\left(1+\frac{\lambda}{\mathcal{E}(E_t)-\mathcal{E}(B_{1t})}\right). \quad (3.114)$$

Diagonalizing the 2×2 matrix gives a Zeeman splitting whose magnitude depends on the direction of the field

$$\mathcal{E} = \pm\frac{\mu_B H}{2}\sqrt{g_\perp^2\sin\theta^2 + g_\parallel^2\cos\theta^2}. \quad (3.115)$$

The perpendicular and parallel components of the g-tensor can be identified as

$$g_\perp = 2\left(1+\frac{\lambda}{\mathcal{E}_{E_t}-\mathcal{E}_{B_{1t}}}\right) \quad (3.116)$$

$$g_\parallel = 2\left(1+\frac{4\lambda}{\mathcal{E}_{B_{2t}}-\mathcal{E}_{B_{1t}}}\right). \quad (3.117)$$

Problem 3.5. a) Let the magnetic field point in the z-direction, so the Zeeman term is

$$\mathcal{H}_Z = -g\mu_B H J^z \quad (3.118)$$

where $g = 0.8$ is the Lande factor for the $^3H_4$ multiplet.

In the basis $|\Gamma_1\rangle$, $|\Gamma_4^0\rangle$, $|\Gamma_4^+\rangle$, $|\Gamma_4^-\rangle$, the Hamiltonian matrix is

$$\begin{pmatrix} 0 & -2\sqrt{5/3}h & 0 & 0 \\ -2\sqrt{5/3}h & \Delta & 0 & 0 \\ 0 & 0 & \Delta+5h/3 & 0 \\ 0 & 0 & 0 & \Delta-5h/3 \end{pmatrix} \quad (3.119)$$

where Δ is the crystal field splitting, and we introduced the notation $h = g\mu_B H$.

Within the four-dimensional subspace, $|\Gamma_4^\pm\rangle$ remain eigenstates, only their energies are shifted by $\langle\Gamma_4^\pm|\mathcal{H}_Z|\Gamma_4^\pm\rangle$.

$|\Gamma_1\rangle$ and $|\Gamma_4^0\rangle$ are mixed by the magnetic field. Let us denote the new eigenstates by $|\tilde{\Gamma}_1\rangle$ and $|\tilde{\Gamma}_4^0\rangle$. We find that the new energy eigenvalues are $(\Delta \pm \sqrt{\Delta^2 + (80/3)h^2})/2$.

The magnetic polarization along the z-direction is

$$M^z = \frac{g\mu_B}{\mathcal{Z}}\left\{\langle\Gamma_4^+|J^z|\Gamma_4^+\rangle\cdot\exp\left[-\frac{\Delta}{k_BT}-\frac{5h}{3k_BT}\right]\right. \tag{3.120}$$
$$+\langle\Gamma_4^-|J^z|\Gamma_4^-\rangle\cdot\exp\left[-\frac{\Delta}{k_BT}+\frac{5h}{3k_BT}\right]$$
$$+\langle\tilde{\Gamma}_1|J^z|\tilde{\Gamma}_1\rangle\cdot\exp\left[-\frac{1}{2k_BT}(\Delta-\sqrt{\Delta^2+(80/3)h^2})\right]$$
$$\left.+\langle\tilde{\Gamma}_4^0|J^z|\tilde{\Gamma}_4^0\rangle\cdot\exp\left[-\frac{1}{2k_BT}(\Delta+\sqrt{\Delta^2+(80/3)h^2})\right]\right\}$$

where $\mathcal{Z}$ is the partition function of the four-state system

$$\mathcal{Z} = 2\exp\left[-\frac{\Delta}{k_BT}\right]\cdot\cosh\frac{5h}{3k_BT} \tag{3.121}$$
$$+2\exp\left[-\frac{\Delta}{2k_BT}\right]\cdot\cosh\frac{1}{2k_BT}(\Delta-\sqrt{\Delta^2+(80/3)h^2}).$$

If we want to calculate the susceptibility, it is sufficient to keep the terms which are linear in h. We find

$$\chi = \frac{(g\mu_B)^2}{1+3\exp(-\Delta/k_BT)}\cdot\left\{2\left(\frac{5}{3}\right)^2\frac{1}{k_BT}\exp\left[-\frac{\Delta}{k_BT}\right]\right.$$
$$\left.+\frac{40}{3\Delta}\cdot\left(1-\exp\left[-\frac{\Delta}{k_BT}\right]\right)\right\}. \tag{3.122}$$

The result is plotted in Fig. 3.13. χ has a finite limiting value at low temperatures: it approaches the Van Vleck susceptibility (3.59) as $T\to 0$. At the same time, it contains terms which can be loosely described as an activated Curie-like susceptibility. Note, however, that the contribution of the Γ_4 level is not a purely thermal effect: the field-induced mixing of the levels is also playing an essential role.

Let us assume that Δ is much smaller than the other crystal field splittings (though this is not very realistic for PrSb). It is then possible to reach an "intermediate high-temperature range" where Δ is sufficiently small compared to k_BT but the occupation of the higher-lying levels is still negligible. The Boltzmann factor can be expanded to order $1/T$ in the second term of the numerator of χ, and replaced by 1 in elsewhere. The high-temperature susceptibility will be of the Curie form

$$\chi \approx \frac{(g\mu_B)^2}{4k_BT}\left[2\left(\frac{5}{3}\right)^2+2\left(2\sqrt{\frac{5}{3}}\right)^2\right]. \tag{3.123}$$

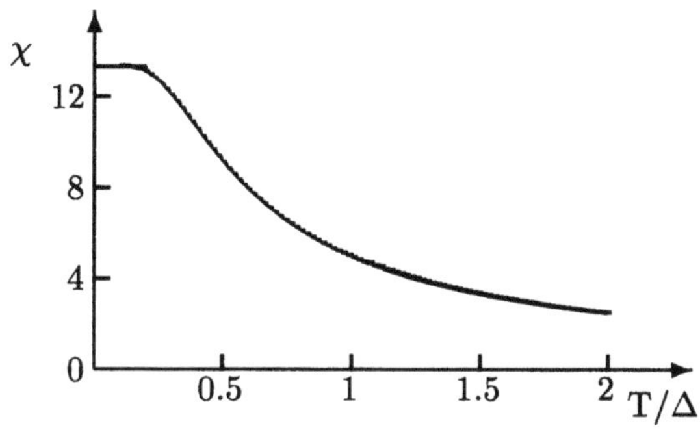

Figure 3.13: The temperature dependence of the susceptibility (in arbitrary units) given in (3.122).

We can identify the effective moment $3.01\mu_B$, which is about 84% of the free-ion value $3.58\mu_B$. (3.123) is a special case of *Van Vleck's formula for the paramagnetic moment* belonging to a bunch of low-lying crystal field levels [423, 62]. The general expression can be formulated as follows: let us assume that the crystal field states $|\Gamma\rangle$ are at approximately zero energy. Then the effective moment μ_P is given by

$$\mu_P^2 = \frac{3}{N}\sum_{\Gamma}\sum_{\Gamma'}|\langle\Gamma'|M^z|\Gamma\rangle|^2 \tag{3.124}$$

where N is the number of the low-lying states, M^z is the component of the magnetic moment along the applied field, and the independent summations are over the N states.

b) Generally, the magnetic moment of the maximally polarized state is different from the paramagnetic moment which can be deduced from the high-temperature susceptibility. Even for a free ion, the moment's maximum z-component is $g\mu_B J$, while the paramagnetic moment is $g\mu_B\sqrt{J(J+1)}$. Crystal field effects can make this difference larger.

In our particular case, the $|\tilde{\Gamma}_1\rangle$ state (which becomes $|\Gamma_1\rangle$ when $h \to 0$) is the ground state at all values of h. The polarization caused by the field is

$$\langle\tilde{\Gamma}_1|M^z|\tilde{\Gamma}_1\rangle = \frac{40g\mu_B}{\sqrt{3}} \cdot \frac{h}{\sqrt{3\Delta^2 + 80h^2}} . \tag{3.125}$$

In the limit of high fields, this approaches

$$\lim_{h\to\infty}\langle\tilde{\Gamma}_1|M^z|\tilde{\Gamma}_1\rangle = \sqrt{\frac{20}{3}}g\mu_B \approx 2.06\mu_B \tag{3.126}$$

which is about 69% of the paramagnetic moment.

Replacing Δ=7meV, and $g = 0.8$, the moment induced by a field of 10T is predicted to be about $0.58\mu_B$. This falls short of the measured value of $0.89\mu_B$. We cite two reasons to account for the discrepancy. First, we neglected the higher-lying levels. Second, as we are going to discuss in a later chapter, intersite exchange interactions enhance the susceptibility, and also the high-field magnetization, beyond the single-ion value [286].

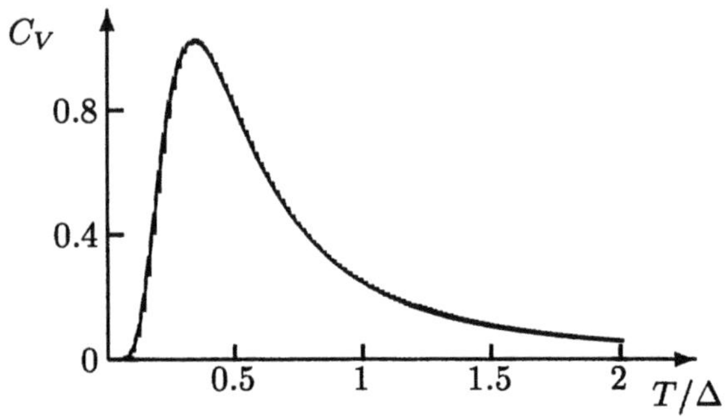

Figure 3.14: The specific heat (in arbitrary units) plotted as a function of temperature shows the characteristic Schottky anomaly.

c) For a single ion with a set of discrete energies ϵ_i, the specific heat is

$$C_V = \frac{1}{k_B T^2}\left\{ \frac{\sum \epsilon_i^2 e^{-\epsilon_i/k_B T}}{\sum e^{-\epsilon_i/k_B T}} - \left(\frac{\sum \epsilon_i e^{-\epsilon_i/k_B T}}{\sum e^{-\epsilon_i/k_B T}} \right)^2 \right\}. \tag{3.127}$$

In zero magnetic field, we have $\epsilon_1 = 0$, $\epsilon_2 = \epsilon_3 = \epsilon_4 = \Delta$. The plot of C_V against the temperature (Fig. 3.14) shows a rather sharp maximum which is called the *Schottky anomaly.* If such a feature is seen in the specific heat of a material, it is often argued to be an experimental proof that crystal field effects are important. It is easy to convince ourselves that the peak broadens in a magnetic field.

Problem 3.6. The Zeeman term $-g\mu_B J^z H^z$ couples $|\Gamma_{t3}\rangle$ with $|\Gamma_{t4}\rangle$, and $|\Gamma_{t1}\rangle$ with $|\Gamma_{t2}\rangle$, so we have two 2×2 eigenvalue problems to solve. The $|\Gamma_{t3}\rangle$–$|\Gamma_{t4}\rangle$ problem was diagonalized in (3.70); the $|\Gamma_{t1}\rangle$–$|\Gamma_{t2}\rangle$ problem can be solved similarly to yield

$$\lambda_{1,2} = \frac{\Delta_2 - \Delta_1 \mp \sqrt{(\Delta_2 - \Delta_1)^2 + 64(g\mu_B H^z)^2}}{2}. \tag{3.128}$$

a) As H^z is increased from zero, the ground state evolves smoothly from $|\Gamma_{t3}\rangle$, with the weight of the $|2\rangle$ component increasing gradually. In large fields, λ_3 is approximately $-2g\mu_B H^z$. However, well before this asymptotic expression becomes valid, the state derived from $|\Gamma_{t3}\rangle$ ceases to be the ground state because λ_1 gets lower than λ_3. The reason is that the corresponding state will be, in high fields, dominated by its $|4\rangle$ component, and its Zeeman energy lowering rate $\sim -4g\mu_B H^z$ is faster by a factor of 2. The critical field where the metamagnetic transition is taking place, can be determined from $\lambda_3 = \lambda_1$. Using the crystal field levels shown in Fig. 3.9, $g = 0.8$, and $\mu_B \approx 5.8 \cdot 10^{-2}$meV/T, we predict a metamagnetic transition at $H_{cr.} = 34.2$ Tesla. (In URu_2Si_2, there are several metamagnetic transitions around 36 Tesla; see Fig. 3.11).

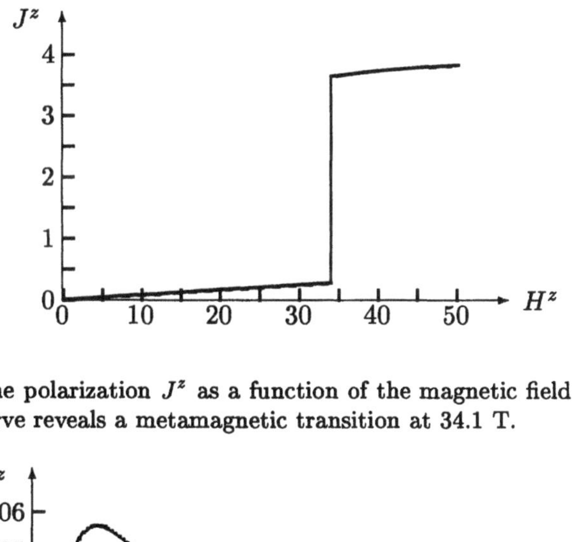

Figure 3.15: The polarization J^z as a function of the magnetic field H^z (in units of Tesla). The curve reveals a metamagnetic transition at 34.1 T.

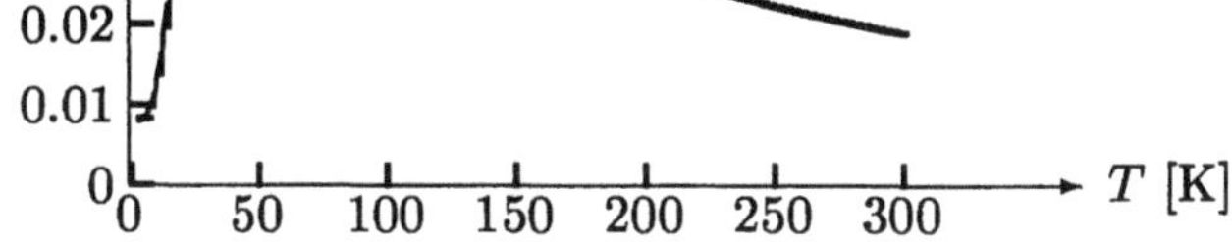

Figure 3.16: The z-direction susceptibility (in arbitrary units) of an f^2 ion in a tetragonal crystal field, based on the simplified model of Problem 3.6. Note the non-monotonic temperature dependence.

The metamagnetic transition is visible in the magnetization curve as a sudden jump of the magnetization at $H_{cr.}$. For $H < H_{cr.}$, the polarization follows

$$J^z_< = \frac{8g\mu_B H^z}{\sqrt{\Delta_4^2 + 16(g\mu_B H^z)^2}} \tag{3.129}$$

while for $H > H_{cr.}$ we find

$$J^z_> = \frac{32g\mu_B H^z}{\sqrt{(\Delta_2 - \Delta_1)^2 + 64(g\mu_B H^z)^2}}. \tag{3.130}$$

The complete curve is shown in Fig. 3.15.

b) We omit the straightforward details of the derivation of χ^z, the temperature dependence of which is shown in Fig. 3.16. The behaviour should be contrasted with what we have seen in Fig. 3.13. Noteworthy is the strong peak around ~ 50K (which is rather similar to the one seen in experiments on URu_2Si_2: see Fig. 3.10). The presently discussed *non-monotonic temperature dependence* of the susceptibility is a single-ion effect[31], resulting from thermal excitation to the relatively low-lying, highly polarizable Γ_{t1} and Γ_{t2} levels.

Sticking to the four-level model, we would predict $\chi^x = 0$. In reality, the in-plane susceptibility χ^x is small, and essentially independent of temperature (Fig. 3.10). We can understand this by recalling (3.69): the most important contribution to χ^x comes from the coupling of the ground state to the higher-lying Γ_{t5} doublet, and thus thermal excitation plays a minor role in the temperature range discussed here.

Problem 3.7. If the ground state Γ_7 is well separated from the Γ_8 excited state, then the magnetic moment is determined by symmetry alone, and the crystal potential parameters do not appear. Referring to the character table 3.4, it can be checked that the states

$$\psi_\pm = \sqrt{1/6}|\pm 5/2\rangle - \sqrt{5/6}|\mp 3/2\rangle \tag{3.131}$$

can be used as a basis for the representation Γ_7. In the above expression, $|5/2\rangle$, etc. should be understood as the $|L, S, J, J^z\rangle \rightsquigarrow |J^z\rangle$ states. The z-component of the magnetic moment is $g\mu_B\langle\psi_\pm|J^z|\psi_\pm\rangle = \mp 5/7\mu_B$, thus we deduce the effective magnetic moment $0.71\mu_B$. It is compatible with the site-averaged ordered moment $0.63\mu_B$ quoted for $CeAl_2$ [116].

It has been argued that the agreement may be fortuitously good [27]. Taking the mixing with the higher-lying Γ_8 level into account gives a larger single-ion value ($\sim 0.9\mu_B$). However, in metallic $CeAl_2$ the Kondo effect reduces this ionic moment by about 30% to its observed size.

[31] Later we are going to see that an apparently similar behaviour follows from antiferromagnetic interactions between localized spins.

Problem 3.8. Within the $L = 3$, $S = 1/2$, $J = 5/2$ set of levels, the (first order) response to an external magnetic field is given by the Zeeman term (2.75)

$$\mathcal{H}_{\rm Z} = \mu_{\rm B}\mathbf{H}(\mathbf{L}+2\mathbf{S}) \rightsquigarrow g\mu_{\rm B}\mathbf{H}\cdot\mathbf{J} \tag{3.132}$$

where $g = 6/7$ (see Problem 2). We need the matrix elements of J^x, J^y, and J^z in the basis (3.79). Our solution follows [415].

As for J^z, it is easy to verify that

$$\langle 1/2, \pm 1/2|J^z|1/2, \pm 1/2\rangle = \pm 11/6 \qquad \langle -1/2, \pm 1/2|J^z|-1/2, \pm 1/2\rangle = \pm 1/2. \tag{3.133}$$

Wishing to express the same in terms of our σ and τ, we notice that this is achieved by the replacement

$$J^z \rightsquigarrow \frac{7}{3}\sigma^z\left(1+\frac{8}{7}\tau^z\right). \tag{3.134}$$

Now take the matrix elements of J^+ and J^-. For instance,

$$J^+|1/2,-1/2\rangle = (5/\sqrt{6})|-3/2\rangle + \sqrt{5/6}|5/2\rangle$$

thus $\langle 1/2, 1/2|J^+|1/2,-1/2\rangle = 5/3$. We might be tempted to claim that a piece of J^+ acts like $(5/3)\sigma^+$, but this would not be correct because

$$J^+|-1/2,-1/2\rangle = 3|-1/2, 1/2\rangle.$$

Both results can be summarized in the replacement

$$J^+ \rightsquigarrow \frac{5}{3}\left(\frac{1}{2}+\tau^z\right)\sigma^+ + 3\left(\frac{1}{2}-\tau^z\right)\sigma^+ = \left(\frac{7}{3}-\frac{4}{3}\tau^z\right)\sigma^+.$$

But this is only one of the contributions to J^+, because we still have

$$J^+|1/2,1/2\rangle = (2/\sqrt{3})|-1/2\rangle \quad\rightsquigarrow\quad \langle -1/2,-1/2|J^+|1/2,1/2\rangle = 2/\sqrt{3}$$

and

$$J^+|-1/2,1/2\rangle = 2\sqrt{2}|3/2\rangle \quad\rightsquigarrow\quad \langle 1/2,-1/2|J^+|-1/2,1/2\rangle = 2/\sqrt{3}$$

to account for, which is achieved by $J^+ \rightsquigarrow (2/\sqrt{3})\sigma^+(\tau^+ + \tau^-)$. – Repeating the same for J^-, we arrive at

$$J^x \rightsquigarrow \sigma^x\left(\frac{7}{3}-\frac{4}{3}\tau^z+\frac{4}{\sqrt{3}}\tau^x\right) \quad \text{and} \quad J^y \rightsquigarrow \sigma^y\left(\frac{7}{3}-\frac{4}{3}\tau^z-\frac{4}{\sqrt{3}}\tau^x\right). \tag{3.135}$$

Following [415], we rewrite the Zeeman term as

$$\mathcal{H}_{\rm Z} = 2\mu_{\rm B}\left\{\sigma^x\left(1+\frac{8}{7}T^x\right)H_x + \sigma^y\left(1+\frac{8}{7}T^y\right)H_y + \sigma^z\left(1+\frac{8}{7}T^z\right)H_z\right\} \tag{3.136}$$

where

$$T^x = \frac{\sqrt{3}}{2}\tau^x - \frac{1}{2}\tau^z \qquad T^y = -\frac{\sqrt{3}}{2}\tau^x - \frac{1}{2}\tau^z \qquad T^z = \tau^z\,. \tag{3.137}$$

In spite of the apparent simplicity of (3.132), the action of the Zeeman term is quite complicated within the Γ_8 quartet.

Now let us consider a field in the z direction. Above $T_{\rm Q} = 3.3$K, $\langle\tau^z\rangle = 0$, and the magnetization is uniform. However, in the temperature range $2.4\text{K} < T < 3.3\text{K}$, $\langle\tau^z\rangle = Q(T)$ on one sublattice, while $\langle\tau^z\rangle = -Q(T)$ on the other sublattice. According to (3.136) and (3.137), the uniform external field produces a staggered magnetization whose spatial pattern is the same as that of the quadrupolar order.

Chapter 4

Mott Transition and Hubbard Model

Our principal subject is the study of electron correlation effects. But what are correlation effects? The definition works by exclusion: any purely electronic phenomenon which cannot be explained within the independent-electron (Hartree–Fock) approximation, will be regarded as a correlation effect. Since we are primarily interested in crystalline materials, for us, independent-electron theory means just the band theory of solids.

Arguably, the greatest triumph of band theory was that, in the years following its inception, it successfully classified a huge number of crystalline solids as metals or insulators. Not much later, it was also realized that the most spectacular failure of band theory is in exactly the same field: it predicts that certain insulators should be metals [294].

4.1 Metals and Insulators: Breakdown of the Independent-Electron Description

We call a material insulating if its static electrical conductivity approaches zero as the temperature is lowered

$$\lim_{T \to 0} \sigma(T) = 0. \tag{4.1}$$

If the conductivity remains finite (or even diverges, in the case of superconductors) then we regard the system metallic.

Band theory, however sophisticated in its method, will lead to an independent-particle description of the ground state: for N electrons, we have to fill the $N/2$ lowest-lying band states with two electrons each, and form the corresponding Slater determinant. Leaving magnetism aside for a moment, we find the ground state possessing the full translational invariance of the lattice as well as spin-rotational and time-reversal invariance.

In the independent-electron description of crystals, the material is predicted to be metallic if the density of states at the Fermi level is non-vanishing: $\rho(\epsilon_F) \neq 0$. If $\rho(\epsilon_F) = 0$, the material must be an insulator.

Translational invariance (lattice periodicity) is important for this criterion to hold: for non-periodic (random) potentials it may happen that $\rho(\epsilon_F) \neq 0$ and still $\lim_{T\to 0} \sigma(T) = 0$. Disorder-induced localization of the one-electron states (the so-called Anderson localization[1]) is one way to turn a prospective metal into an insulator. We will occasionally invoke this concept to interpret experimental findings. However, here we confine our attention to systems with a periodic potential[2].

In band theory, $\rho(\epsilon_F) = 0$ arises if ϵ_F lies between the top of the uppermost filled band and the bottom of the lowermost empty band. Usually, there is a gap separating these two bands. However, there are a few examples (such as grey tin) of zero-gap semiconductors where the valence and conduction bands touch.

In any case, band theory asserts that for $\rho(\epsilon_F) = 0$ it is necessary to have an even number of electrons per unit cell since a band with orbital degeneracy n can hold $2n$ electrons per unit cell.

Note that the reverse statement does not hold: even if we have just the right (even) number of electrons to fill the valence band, $\rho(\epsilon_F) \neq 0$ may still occur because of an overlap between the valence and conduction bands. Such is the case with the alkali earths (Ca, Sr, Ba). A tiny overlap of bands makes Bi a very interesting semimetal.

---

[1]Disorder-induced insulating behaviour is reviewed in [293] and [330].

[2]To be more precise, we allow for the presence of weak disorder in three-dimensional systems, so as to avoid non-physical features which would arise from a truly free acceleration of the conduction electrons in an external electric field. We consider it natural that a metal has a finite residual resistivity.

On the other hand, *with an odd number of electrons per unit cell we necessarily have a partially filled band, so $\rho(\epsilon_F) \neq 0$, and consequently the material must be a metal.*

This basic prediction of band theory spectacularly fails in a number of cases [5, 293, 330]. One of the oldest-known, and clearest, examples is provided by the transition metal oxide CoO. The material has a (slightly distorted) rocksalt structure so the unit cell contains one Co atom, and one O atom. Remembering that the outer shells of Co have the configuration $3d^7 4s^2$, and for oxygen we can take $2s^2 2p^4$, the number of electrons[3] per unit cell is 9+6=15, an odd number. According to band theory, CoO should be a metal. In fact, it is one of the toughest insulators known. Band theory has failed.

Unfortunately, the argument cannot be kept as simple as this, neither so transparent. Independent electron theory in principle allows that the symmetry of the mean field be lower than the symmetry of the crystal lattice. In our particular case, this is realized by the "different orbitals for different spins" scheme[4]. We take the sublattice of Co ions and subdivide it further into magnetic sublattices so that $\uparrow$-spins settle predominantly on one sublattice, and $\downarrow$-spins on another[5]. Having done this, we find an even number of electrons per magnetic supercell and thus the independent electron description has, in principle at least, an escape route to avoid contradiction with experiment. It is a quite different question whether the band structure calculation based on the assumption of a magnetic superstructure is then able to come up with the right size of the gap to show CoO to be as insulating as it actually is. Without wishing to enter the argument, we note that it is at least doubtful that it can [306].

So much for the ground state. At sufficiently high temperatures, the antiferromagnetic order sustaining the magnetic supercell vanishes, so we are back at the chemical unit cell with an odd number of elec-

---

[3]Electrons in deeper-lying closed shells could be included in the count without affecting the argument.

[4]Also called the unrestricted Hartree–Fock method. A particular version of it will be described in detail in Sec. 7.6.

[5]From our present point of view, it is just a cumbersome detail that this subdivision is not easy. The Co sublattice is the close-packed fcc lattice so it cannot be divided into two alternate sublattices; at the very least, it must be divided into four. It is sufficient to know that the number of magnetic sublattices is even.

trons. Just counting the number of electrons in the unit cell would lead us to expect that the antiferromagnetic–paramagnetic transition is accompanied by an insulator-to-metal transition. However, CoO remains a staunch insulator also in its paramagnetic phase. You might expect that this must mean the eventual demise of the independent-electron approach. However, it is just imaginable that the finding of an insulating paramagnetic state may still be described in the independent-electron framework as an effect of strong disorder[6]. Arguments of this kind may never end. We prefer to follow a different path: to show that strong electron–electron interaction can cause a metal–insulator transition in a half-filled band, and that correlation effects are important both on the metallic, and on the insulating, side of the transition. The correlation-driven metal–insulator transition is the Mott transition.

4.2 Mott Transition

Let us invoke a well-known *Gedankenexperiment* to convince ourselves that

- correlation effects necessarily arise, and
- the consideration of electron correlation effects provides the clearest and most economical way to understand phenomena like the insulating nature of CoO.

Let us consider the tight binding picture of the band structure of Na, a solid with one atom per unit cell (Fig. 4.1).

All atomic levels give rise to bands, of which only the $3s$ band achieves a sizeable width. The deeper-lying bands are very narrow and completely filled. The $3s$ band is half-filled with one electron/unit cell and thus Na is expected to be a metal. This statement should, according to band theory, continue to hold as we (in thought) increase the lattice constant from its equilibrium value a_0 to larger interatomic separations. The $3s$ band becomes quickly narrower but it is always half-filled, so Na should remain a metal.

[6]The exchange fields of the randomly oriented spins add up to a wildly fluctuating potential. A current-carrying electron which tries to pass through the crystal will be strongly scattered by this and may become Anderson-localized.

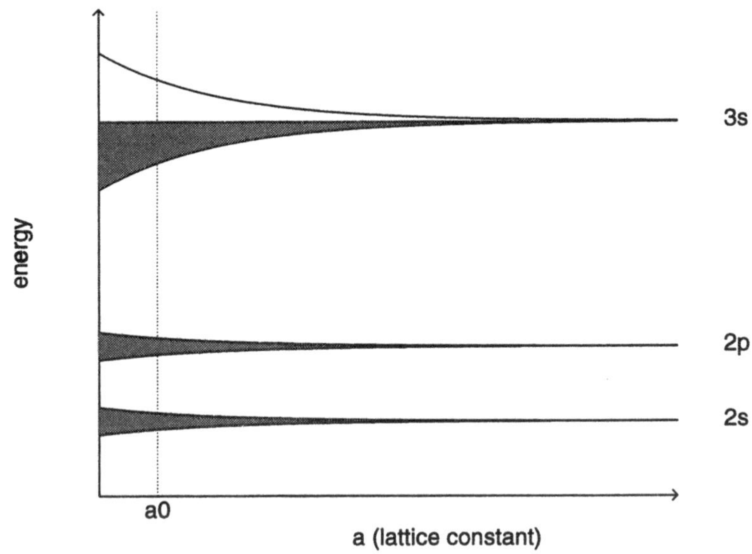

Figure 4.1: The schematic representation of the lattice constant dependence of the tight binding band structure of Na. The bands derived from the atomic $2s$, $2p$, and $3s$ levels are shown. Shading indicates filled states. At the equilibrium lattice constant a_0 the $3s$-band is fairly wide. Band theory would lead us to expect that the half-filled $3s$ band gives rise to metallic conduction, no matter how narrow the $3s$ band becomes.

It is clear that eventually this leads to an absurdity. At large enough atom–atom distances, we no longer have a metal but an array of neutral Na atoms, i.e., a non-conductor.

Why is a hypothetical periodic array of widely separated Na atoms with a lattice constant of, say, 10nm not a metal? For electrical conduction, electrons should propagate through the lattice, leading to *charge fluctuations*

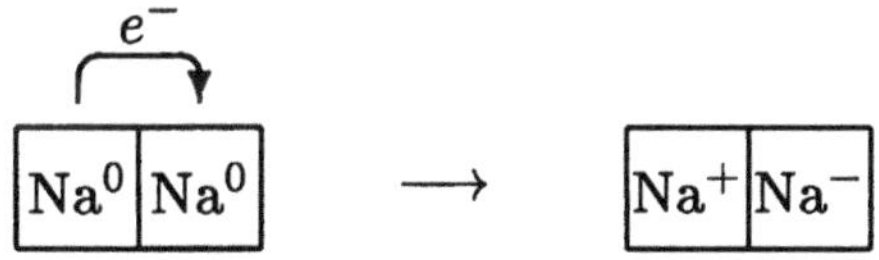

$$\epsilon_{3s} \quad \epsilon_{3s} \qquad\qquad\qquad 0 \quad 2\epsilon_{3s} + U_{3s} \tag{4.2}$$

In the first line, the process of electron hopping is schematically shown. In the second line, we give the corresponding energies of the neutral, and ionized configurations, respectively. An electron sitting in a $3s$ state possesses the atomic energy ϵ_{3s}. If two electrons share the same atomic shell, as at the Na^- site shown on the right-hand side, their strong Coulomb repulsion goes with the intraatomic Coulomb energy[7]

$$U_{3s} = \int d\mathbf{r}_1 \int d\mathbf{r}_2 |\phi_{3s}(\mathbf{r}_1)|^2 \frac{e^2}{|\mathbf{r}_1 - \mathbf{r}_2|} |\phi_{3s}(\mathbf{r}_2)|^2 \,. \tag{4.3}$$

This allows us to deduce that electron hopping requires the energy investment U_{3s}. Since sharing the same atomic orbital confines the two electrons to within a short distance (~ 1Å) from each other, U_{3s} must be large.

The intraatomic Coulomb energy, usually called the Hubbard U, is the most important energy term for a system of interacting electrons moving in a narrow band. For various atomic orbitals in a solid state environment, its typical range is 1-10eV. The charge transfer process shown in (4.2) costs a lot of energy so it just does not want to happen. The charge fluctuations necessary for metallic conduction are suppressed by the intrasite Coulomb interaction – at least above some critical value[8] of the lattice constant (i.e., for sufficiently narrow bands). This transition from a metallic to an insulating state would be an example of the *Mott transition.*

The energy terms needed to discuss the *Gedankenexperiment* in Fig. 4.1 can be collected in the following simple Hamiltonian

$$\begin{aligned} \mathcal{H}_{3s} &= \epsilon_{3s} \sum_{\mathrm{j}} \sum_{\sigma} \hat{n}_{\mathrm{j}\sigma} - t_{3s} \sum_{\langle \mathrm{j,l} \rangle} \sum_{\sigma} (c^{\dagger}_{\mathrm{j}\sigma} c_{\mathrm{l}\sigma} + \mathrm{H.c.}) + U_{3s} \sum_{\mathrm{j}} \hat{n}_{\mathrm{j}\uparrow} \hat{n}_{\mathrm{j}\downarrow} \\ &= \mathcal{H}_{\mathrm{band}} + \mathcal{H}_{\mathrm{Coulomb}} \end{aligned} \tag{4.4}$$

[7] The analogous quantity for the hydrogen molecule was introduced in (2.128).

[8] For Na, this would be larger than the equilibrium lattice constant (so the system remains, in fact, a metal) but not absurdly larger. We have in mind interatomic distances $\sim$ several Ås. – Essentially the same argument was put forward in 1937 by R. Peierls to explain that NiO is an insulator. He argued that it is due to the large Coulomb repulsion between the *d*-electrons, and that each Ni^{2+} ion keeps exactly 8 *d*-electrons, and only the spin degree of freedom survives. We have chosen to speak about the hypothetical dilated Na because this leads on straight to the introduction of the non-degenerate Hubbard model.

where $\mathcal{H}_{\rm band}$ contains the first two sums. $\mathbf{j}$ is a lattice site index, $\sigma =\uparrow,\downarrow$ denotes the spin, and the sum over $\langle \mathbf{j},\mathbf{l}\rangle$ runs over nearest neighbours. $c^\dagger_{\mathbf{j}\sigma}$ creates an electron in the $3s$ state at site $\mathbf{j}$ with spin σ, and $\hat{n}_{\mathbf{j}\sigma} = c^\dagger_{\mathbf{j}\sigma}c_{\mathbf{j}\sigma}$ is the corresponding occupation number. The tight binding matrix element for the hopping process shown in (4.2) is $-t_{3s}$. With reference to the band motion of the electrons, $\mathcal{H}_{\rm band}$ is also called the "kinetic energy term" (even though, strictly speaking, the band energy has contributions from both the kinetic energy, and the atomic potentials). $\mathcal{H}_{\rm band}$ can be rewritten in terms of the Bloch operators which correspond to the $3s$ band centered about ϵ_{3s}

$$\mathcal{H}_{\rm band} = \sum_{\mathbf{k}}\sum_{\sigma} \epsilon_{\mathbf{k}} c^\dagger_{\mathbf{k}\sigma} c_{\mathbf{k}\sigma} \,. \tag{4.5}$$

The ground state of $\mathcal{H}_{\rm band}$ is the metallic Fermi sea

$$|{\rm FS}\rangle = \prod_{\mathbf{k}}^{\epsilon_{\mathbf{k}}<\epsilon_{3s}} c^\dagger_{\mathbf{k}\uparrow} c^\dagger_{\mathbf{k}\downarrow}|0\rangle. \tag{4.6}$$

$|{\rm FS}\rangle$ is, by definition, *uncorrelated*, meaning that the occupation of an orbital by an $\uparrow$-spin electron is independent of its occupation by a $\downarrow$-spin electron. It is easy to see that this implies that $|{\rm FS}\rangle$ has a lot of charge fluctuations. Since the half-filled band ground state $|{\rm FS}\rangle$ is homogeneous, the probability that a site is occupied by an $\uparrow$-spin electron must be 1/2; the same is the probability of its occupation by a $\downarrow$-spin electron. The probability of finding a site doubly occupied (i.e., in the configuration Na^-) is thus 1/4. On the average 25% of the sites are Na^-, 25% are Na^+, and 50% are neutral. The Na^0 sites have a singly occupied $3s$-shell, carrying either an $\uparrow$-spin, or a $\downarrow$-spin electron. In other words, if we were making a "snapshot" of $|{\rm FS}\rangle$, to "see where the electrons are", typically we would find something like the situation shown on the left of Fig. 4.2.

The expectation value of the Coulomb energy is

$$\langle {\rm FS}|\mathcal{H}_{\rm Coulomb}|{\rm FS}\rangle = \frac{LU_{3s}}{4} \tag{4.7}$$

where L is the number of lattice sites. The band energy is given by the average energy of the occupied part of the band,

$$\langle {\rm FS}|\mathcal{H}_{\rm band}|{\rm FS}\rangle = L(\epsilon_{3s} - \alpha t_{3s}) \,. \tag{4.8}$$

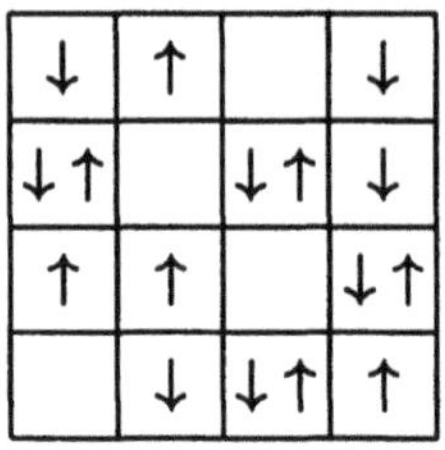
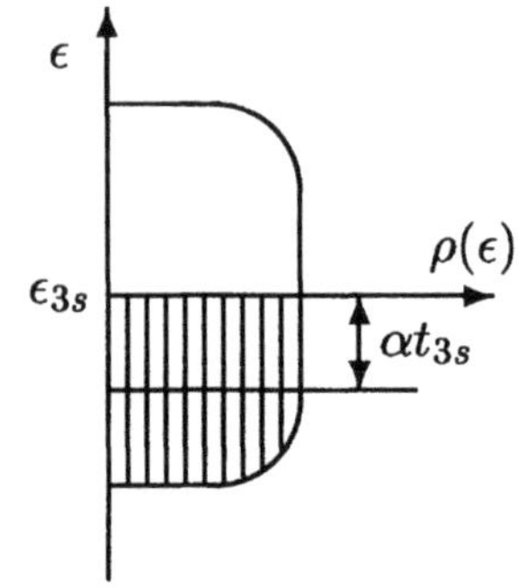

Figure 4.2: *Left*: a snapshot of the Fermi sea ground state. *Right*: schematic representation of the half-filled $3s$ band. The average energy of the filled part of the band lies a distance αt_{3s} below the centre of the band.

where α is a number of order unity, and it is determined by the shape of the band. (4.8) expresses that the average kinetic energy gain due to delocalization into Bloch states is of order t_{3s} (see the right-hand side of Fig. 4.2).

The total energy density corresponding to the uncorrelated band ground state is thus

$$\frac{1}{L}\langle \mathrm{FS}|\mathcal{H}_{3s}|\mathrm{FS}\rangle = \epsilon_{3s} - \alpha t_{3s} + \frac{U_{3s}}{4} \,. \tag{4.9}$$

By increasing the lattice constant a, the ratio U_{3s}/t_{3s} increases steeply[9]. In doing so, it passes through the critical value $(U_{3s}/t_{3s})_{\mathrm{cr}} = 4\alpha$, where the Coulomb energy cost of charge fluctuations begins to exceed the kinetic energy gain arising from occupying the lower half of a band. In other words, the Fermi sea ground state energy is predicted to exceed the energy of an assembly of L neutral Na atoms, which is just $L\epsilon_{3s}$. Consequently, the formation of the band, i.e., of an itinerant ground state, is no longer favourable: it is energetically better to localize exactly one electron at each lattice site. If we allow only these two extremes for the ground state: either the uncorrelated $|\mathrm{FS}\rangle$, or a Heitler–London-type state of fully localized electrons, then we are led to

[9] U_{3s} remains relatively unaffected but t_{3s} which behaves like the overlap integral between atomic orbitals centered at nearest-neighbour sites, falls off exponentially.

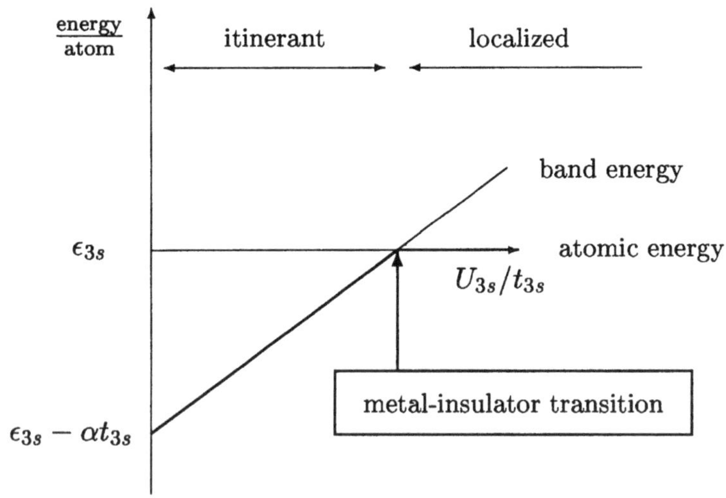

Figure 4.3: The most naive scenario for the Mott transition: The energy of the uncorrelated Fermi sea exceeds the energy of independent Na atoms if $U_{3s}/t_{3s} > 4\alpha$. At the critical value, the nature of the many-electron ground state changes from itinerant to localized (the ground state follows the bold line).

predict a phase transition: for $(U_{3s}/t_{3s}) < (U_{3s}/t_{3s})_{\rm cr}$, $|\text{FS}\rangle$ is accepted as ground state, while for $(U_{3s}/t_{3s}) > (U_{3s}/t_{3s})_{\rm cr}$, the ground state is an array of neutral atoms. Thus we have a transition from a delocalized metallic state to a fully localized insulating state (illustrated in Fig. 4.3). It goes without saying that this utterly naive scenario of a Mott transition will have to be replaced by a more sophisticated argument. However, the picture remains basically the same: *the Mott transition is an interaction-induced, collective localization of all the electrons.*

Remark: Remember that we had a very similar story with the transition from the Heitler–London picture to the molecular orbital picture for the hydrogen molecule (Sec. 2.5.3). We are, of course, aware that the nature of the ground state of H_2 changes continuously between the two limits, and the sharp transition shown in Fig. 2.4 is spurious. In contrast, we assert that the sharp transition seen in Fig. 4.3 is physically genuine; it is only its characterization which is unacceptably simple-minded. The reason for this difference is that we need a macroscopically large system (the so-called thermodynamic limit; see Sec. 7.2.1) to have a sharp phase transition. Mean-field-like arguments like our simple estimate are not sensitive to the difference between finite and infinite systems.

In what follows, we systematically refine the description of both the metallic and the insulating state. This will result in a shift of $(U/t)_{\rm cr}$, and bears also on the question whether the transition is first- or second-order.

The insulating state is highly *correlated*: the electrons minimize the Coulomb interaction energy by avoiding each other as much as possible. For one electron/site, this is best done by letting one electron to stay at each lattice site.

Note that while $|{\rm FS}\rangle$ is unique, the envisaged insulating ("neutral") ground state is vastly (to be precise, 2^L-fold) degenerate because at each site, the electron spin can be either $\uparrow$, or $\downarrow$. This degeneracy of the $U_{3s}/t_{3s} \to \infty$ ground state should be lifted for any finite U_{3s}/t_{3s} by small quantum-mechanical excursions of the electrons. The process of mixing in the virtual charged states picks out an optimal spin arrangement which will be understood as the ground state of an effective spin Hamiltonian acting in the low-energy subspace of the large-U Hubbard model. *The Mott insulator is a magnet*[10]. However, its insulating nature is *not* due to magnetism but due to the intrasite Coulomb repulsion which operates in the paramagnetic state as well.

Our interpretation of the findings about CoO is that this material has a sufficiently large lattice constant to start out (at atmospheric pressure) on the insulating side of the Mott transition: it is a Mott insulator to begin with. This implies that it should be possible to drive CoO through an insulator-to-metal transition under pressure which diminishes the lattice constant, and thus decreases U/t. Just in the case of CoO, this may require very high pressures. There are, however, substances which start out near enough to the metal–insulator phase boundary so that an easily manageable change of pressure or temperature (or composition, magnetic field, etc.) takes them across a metal–insulator transition.

---

[10]In the sense of being describable in terms of an effective spin Hamiltonian. It need not follow that it is magnetically ordered. However, most Mott insulators turn out to be antiferromagnets.

4.3 The Hubbard Model

The celebrated Hubbard model is the simplest[11] many-body Hamiltonian which allows a meaningful description of two opposing tendencies:

- the kinetic energy (electron hopping) wishes to delocalize the electrons into itinerant states (Bloch states), leading to metallic behaviour
- the electron–electron interaction (approximated here by the on-site Coulomb interaction) wants to localize the electrons onto sites, driving the transition to a Mott insulator which is usually magnetic.

The Hubbard model was introduced essentially simultaneously by Gutzwiller [142], Hubbard [174], and Kanamori [201] at a time when great effort was made to understand the magnetism of transition metals. This sudden burst of activity may lead us to suspect that the idea had been taking shape for some time before and indeed, there had been forerunners. Anderson's earlier discussion of superexchange [20, 25] also relied on introducing a Hubbard-like Hamiltonian which he considered in the large-U limit at half-filling. At even earlier times, essentially the same model was known as the Pariser–Pople–Parr model of quantum chemistry, where it was used to describe aromatic molecules. However, we may date from 1963 the ambitious program that mapping out the phase diagram of the Hubbard model in its entire parameter space should give us a basic understanding of Mott insulators, antiferromagnetism, and itinerant ferromagnetism.

The usual form of the one-band Hubbard model is essentially the same (4.4) that we introduced in our discussion of the *Gedankenexperiment* illustrated in Fig. 4.1

$$\mathcal{H} = -t \sum_{\langle \mathrm{j},\mathrm{l} \rangle} \sum_{\sigma} (c^{\dagger}_{\mathrm{j}\sigma} c_{\mathrm{l}\sigma} + c^{\dagger}_{\mathrm{l}\sigma} c_{\mathrm{j}\sigma}) + U \sum_{\mathrm{j}} \hat{n}_{\mathrm{j}\uparrow} \hat{n}_{\mathrm{j}\downarrow} \,. \tag{4.10}$$

[11] There is, actually, an even simpler model to demonstrate these opposing tendencies: the spinless fermion model (1.4). However, we find that including the spin degree of freedom brings so much important physics that we do not wish to omit it.

Here $c^\dagger_{\mathbf{j}\sigma}$ creates an electron in the Wannier state $\phi(\mathbf{r}-\mathbf{R_j})$ with spin σ, $\mathbf{j}$ being a lattice site index. $\hat{n}_{\mathbf{j}\sigma} = c^\dagger_{\mathbf{j}\sigma}c_{\mathbf{j}\sigma}$ is the corresponding occupation number operator (the "hat" serves to distinguish it from its expectation value $n_{\mathbf{j}\sigma} = \langle \hat{n}_{\mathbf{j}\sigma}\rangle$. For other operators, where the danger of mixing them up with expectation values is less immediate, we do not bother to write hats.)

The operators satisfy the anticommutation relations

$$\{c_{\mathbf{j}\sigma}, c^\dagger_{\mathbf{l}\sigma'}\} = \delta_{\mathbf{jl}}\delta_{\sigma\sigma'} \tag{4.11}$$

$$\{c_{\mathbf{j}\sigma}, c_{\mathbf{l}\sigma'}\} = 0\,. \tag{4.12}$$

The sum over the nearest neighbour pairs $\langle\mathbf{j},\mathbf{l}\rangle$ describes the kinetic energy (hopping) of electrons

$$\mathcal{H}_{\text{band}} = -t\sum_{\langle\mathbf{j},\mathbf{l}\rangle}\sum_{\sigma}(c^\dagger_{\mathbf{j}\sigma}c_{\mathbf{l}\sigma} + c^\dagger_{\mathbf{l}\sigma}c_{\mathbf{j}\sigma})\,. \tag{4.13}$$

The atomic level about which the tight-binding band is centered is usually chosen as the energy zero.

The interaction term

$$\mathcal{H}_U = U\sum_{\mathbf{j}}\hat{n}_{\mathbf{j}\uparrow}\hat{n}_{\mathbf{j}\downarrow} \tag{4.14}$$

describes the Coulomb repulsion between electrons sharing the same orbital, with the Hubbard U defined analogously to (4.3)

$$U = \int d\mathbf{r}_1\int d\mathbf{r}_2|\phi(\mathbf{r}_1-\mathbf{R_j})|^2\frac{e^2}{|\mathbf{r}_1-\mathbf{r}_2|}|\phi(\mathbf{r}_2-\mathbf{R_j})|^2\,. \tag{4.15}$$

This definition makes it obvious that we are speaking about the positive-U Hubbard model. We mention that in the literature, one discusses also the negative-U Hubbard model. Though at the first sight it looks physically unmotivated, the case $U < 0$ arises in two different contexts. The first may be regarded as purely formal: there is a canonical transformation which connects $U > 0$ and $U < 0$ Hamiltonians (see [244] and Problem 4.6). The second possibility is that you regard the negative-U Hubbard model as an effective Hamiltonian (similar to the

BCS model Hamiltonian) which describes a net electron–electron attraction due to processes which are not characterized in detail. Following this interpretation, the model is used for the description of certain superconductors.

In contrast to the Hubbard on-site interaction, the true Coulomb interaction is long-ranged. A realistic Hamiltonian should contain lots of intersite terms. The pair terms[12] include, in addition to the direct Coulomb and (ferromagnetic) exchange interactions, also Coulomb-assisted hopping processes. We should be aware that the inclusion of these terms can lead to new physics. However, this is not systematically explored yet and therefore we will usually stick to the simple Hubbard model (4.10). A trivial point: we may be worried that if the long-range part of the Coulomb force is "switched on", the system should explode, because it seems to carry a macroscopic total charge. To avoid such a catastrophy, it is understood that whichever electron number we are considering, charge neutrality is guaranteed by a compensating background charge which comes from ions which are not explicitely described in the model.

The band term may be rewritten in its (diagonal) Bloch representation

$$\mathcal{H}_{\text{band}} = \sum_{\mathbf{k}\sigma} \epsilon_{\mathbf{k}} \hat{n}_{\mathbf{k}\sigma} \tag{4.16}$$

without making the form of $\epsilon(\mathbf{k})$ explicit. $\epsilon(\mathbf{k})$ is easy to express for the nearest-neighbour tight binding model (4.13) but generally, we may admit any band structure $\epsilon(\mathbf{k})$ in the Hubbard model, requiring only that $\mathcal{H}_U$ is expressed in terms of the corresponding Wannier operators. We take it for granted that a tight binding parametrization for the band can be found, even if it may necessitate the inclusion of farther-neighbour hoppings. On the other hand, it was somewhat flimsy to assume that only one band appears in the Bloch representation. One customarily speaks of the *one-band (or single-band) Hubbard model* and writes $\mathcal{H}_{\text{band}}$ either as in (4.13) or as in (4.16). However, recent work on Hubbard models on non-Bravais lattices made us conscious of the fact that what we really mean is the *single-orbital Hubbard model* in which there is only one Wannier state for each site. For Bravais lattices

---

[12]They will be discussed in Sec. 5.3.

(such as the D-dimensional cubic lattice which we will study most of the time) the single-orbital model has only one band; for non-Bravais lattices, it has several bands. The latter possibility will be illustrated in Problems 4.3–4.

The zero-temperature behaviour of the Hubbard model is governed by two parameters: the relative interaction strength U/t, and the electron density $n = N/L$. Here N denotes the total number of electrons, and L is the number of lattice sites (for non-Bravais lattices, the number of unit cells). For the single-orbital model, $0 \leq n \leq 2$; $n = 1$ belongs to half-filling. One often uses U/W instead of U/t, W being the bandwidth.

In a loose manner of speaking, the expression for the grand-canonical potential

$$\mathcal{H} - \mu\hat{N} = \sum_{\mathbf{k}\sigma}(\epsilon_{\mathbf{k}} - \mu)\hat{n}_{\mathbf{k}\sigma} + U\sum_{\mathbf{j}}\hat{n}_{\mathbf{j}\uparrow}\hat{n}_{\mathbf{j}\downarrow} \tag{4.17}$$

is also referred to as the Hubbard model. Here μ is the chemical potential, and $\hat{N} = \sum_{\mathbf{k}\sigma}\hat{n}_{\mathbf{k}\sigma}$ is the operator of the total number of particles. This form is suitable if we wish to work in the 4^L-dimensional Hilbert space comprising the states with all possible $0 \leq N \leq 2L$ values of the total number of electrons. The Hilbert space for a fixed number N of the electrons has the much lower dimensionality C_N^{2L}.

The Hubbard model is the most-studied *lattice fermion model.* There are other lattice models of considerable interest which we will briefly discuss in later chapters: the periodic Anderson model allows us to study homogeneous valence mixing, and the arising of a heavy Fermi liquid in certain f-electron systems (Ch. 11); two-band and many-band Hubbard models might be relevant for itinerant ferromagnetism (Ch. 7) and maybe for high-temperature superconductivity; and we have already introduced (Ch. 1) a spinless fermion model for Wigner crystallization. A common feature of lattice models is the existence of certain magic values of the band filling where new correlated phases may appear. We expect a Mott insulator only if n is an integer; Wigner crystallization was associated with $n = 1/2$, etc. Loosely speaking, at a magic filling a density wave of the electron system is in registry with the underlying lattice; such commensurability effects can arise only in lattice models[13]. A quite different class of models is that of *continuum models* in which an interacting electron fluid is moving on a uniform background of positive charge. Interaction effects in continuum models have been extensively studied [327, 90, 257], culminating in the development of Fermi liquid theory. It is reasonable to expect that at low band fillings, the lattice models behave like continuum models. Even the typical strong correlation features (heavy mass, etc.) found in the immediate vicinity of $n = 1$ for the Hubbard model can be reformulated in the Fermi liquid terminology [431]. We will, however, confine our interest to the lattice models.

4.3.1 Local Basis

The Hubbard model is a four-state model, in the sense that each lattice site can be found in any of the following four local basis states:

$$|0\rangle_{\mathrm{j}} \qquad \text{site j is empty} \tag{4.18}$$

$$|\uparrow\rangle_{\mathrm{j}} = c^{\dagger}_{\mathrm{j}\uparrow}|0\rangle_{\mathrm{j}} \qquad \text{site j occupied by an } \uparrow\text{-electron} \tag{4.19}$$

$$|\uparrow\rangle_{\mathrm{j}} = c^{\dagger}_{\mathrm{j}\downarrow}|0\rangle_{\mathrm{j}} \qquad \text{site j occupied by an } \downarrow\text{-electron} \tag{4.20}$$

$$|d\rangle_{\mathrm{j}} = c^{\dagger}_{\mathrm{j}\uparrow}c^{\dagger}_{\mathrm{j}\downarrow}|0\rangle_{\mathrm{j}} \qquad \text{site j is doubly occupied} \tag{4.21}$$

The definition of $|d\rangle$ fixes a convention about the order of the two creation operators. Merely saying that the site is doubly occupied would be ambiguous: it could also be understood to mean $c^{\dagger}_{\mathrm{j}\downarrow}c^{\dagger}_{\mathrm{j}\uparrow}|0\rangle = -c^{\dagger}_{\mathrm{j}\uparrow}c^{\dagger}_{\mathrm{j}\downarrow}|0\rangle = -|d\rangle$.

We will find it useful to introduce the corresponding local projectors

$$\hat{P}_{\mathrm{j}0} = |0\rangle_{\mathrm{j}}\,{}_{\mathrm{j}}\langle 0| = (1-\hat{n}_{\mathrm{j}\uparrow})(1-\hat{n}_{\mathrm{j}\downarrow}) \tag{4.22}$$

$$\hat{P}_{\mathrm{j}\uparrow} = |\uparrow\rangle_{\mathrm{j}}\,{}_{\mathrm{j}}\langle\uparrow| = \hat{n}_{\mathrm{j}\uparrow}(1-\hat{n}_{\mathrm{j}\downarrow}) \tag{4.23}$$

$$\hat{P}_{\mathrm{j}\downarrow} = |\downarrow\rangle_{\mathrm{j}}\,{}_{\mathrm{j}}\langle\downarrow| = \hat{n}_{\mathrm{j}\downarrow}(1-\hat{n}_{\mathrm{j}\uparrow}) \tag{4.24}$$

[13] This statement seems to be contradicted with the finding of an incompressible Laughlin state at $\nu = 1/3$ in a continuum model (Sec. 12.4.1). We can, however, argue that there the external magnetic field gave rise to a length scale, and thus to the possibility of magic values of filling.

$$\hat{P}_{\mathrm{j}d} = |d\rangle_{\mathrm{j}}\,{}_{\mathrm{j}}\langle d| = \hat{n}_{\mathrm{j}\uparrow}\hat{n}_{\mathrm{j}\downarrow}\,. \tag{4.25}$$

The identity

$$\hat{P}_{\mathrm{j}0} + \hat{P}_{\mathrm{j}\uparrow} + \hat{P}_{\mathrm{j}\downarrow} + \hat{P}_{\mathrm{j}d} = \hat{1} \tag{4.26}$$

where $\hat{1}$ is the unit operator, expresses the completeness of the local basis.

The statement that correlations are strong, essentially means that an $\uparrow$-spin electron observes very carefully whether it is sharing a lattice site with a $\downarrow$-spin electron. The local projectors (4.22)–(4.25) provide a tool to keep track of these local events.

4.3.2 Which Electrons Do We Mean?

It is often claimed that the Hubbard model gives an acceptable description of complicated systems like La_2CuO_4 or V_2O_3. What is really meant thereby? Certainly, there must be a lot of bands: oxygen $2p$ bands, $3d$ and $4s$ bands derived from the transition metal and we still did not mention the electronic states introduced by the La atoms. Where is the supposedly non-degenerate narrow band which is *the* band figuring in the Hubbard model?

We wish to understand the electrical and magnetic properties of the materials; these are low-energy (low-frequency) characteristics measured at relatively low temperatures. Consequently, we will be interested only in the electronic states lying in the immediate vicinity of the Fermi level. Let us take the example of La_2CuO_4. In this system, as in most $3d$ transition metal oxides, it is usual to start with the assumption that the (relevant part of the) bands derived from the Cu $3d$ states are lying in between oxygen $2p$ bands which are completely filled, and the higher-lying Cu $4s$ band which is empty. As for the $3d$-type bands, we assume that they are narrow enough so that they are split by crystal fields (Fig. 3.7). The T_2 subband is completely filled and so is the A_{1t} subband produced by the tetragonal component of the crystal field. We are left with one x^2-y^2 orbital, and one electron, per Cu atom. The broadening of the uppermost, crystal-field-split level produces a half-filled, narrow (x^2-y^2)-type band. It is perfectly reasonable to use the Hubbard model for the description of the electrons in this band. In view of the narrowness of the band we expect a largish $U_{x^2-y^2}/t_{x^2-y^2}$, and we are not surprised that the system turns out to be a Mott insulator. Thus though La_2CuO_4 has a lot of different kinds of electrons, many of them in relatively wide bands, the question whether the substance is metallic or insulating is decided by a subsystem of strongly correlated electrons which happen to occupy the states near the Fermi level.

To take another example, let us consider cerium compounds where the strongly correlated subsystem is that of $4f$-electrons. For $4f$ electrons, the spin–orbit coupling is much stronger than the crystal fields, so the first step towards identifying the low-energy subspace is to restrict the state of each Ce^{3+} ion to the $J = 5/2$ Hund's rule ground state. According to (3.78), a cubic crystal field splits the sixfold degenerate ionic level into a twofold degenerate Γ_7, and a fourfold degenerate Γ_8 state. If Γ_7 happens to lie lower, we may say that a single f-electron at a Ce site has two states available, as if it had a (pseudo)spin 1/2. In this sense, we may once again use the single-orbital Hubbard model as the low-energy Hamiltonian.

We do not necessarily manage to narrow the discussion to a last subband which has only Kramers degeneracy. For instance, in $LaTiO_3$ the cubic crystal field may be assumed to split the Ti d-band into T_2- and E-type subbands but we do not think that the lower-symmetry components of the crystal field would be strong enough to cause further splittings. Thus we are left with a 1/6-filled T_2-band which may be identified as a threefold degenerate Hubbard band with $n = 1$. Since now the full range of band filling is $0 \leq n \leq 6$, we seem to be facing a new problem. However, we can convince ourselves that our qualitative argument for a Mott transition works for any intermediate integer value of n thus we find no difficulty in understanding that $LaTiO_3$ can be a Mott insulator. Similar is the case of V_2O_3 in which, however, there are $n = 2$ d-electrons per V-site.

4.4 Limiting Cases

It is worthwhile to discuss the two limiting cases: $U = 0$ when we have just non-interacting band electrons, and $t = 0$ when the system decomposes into a set of isolated atoms. The latter can be also understood as the limit of extremely strong interaction with $U/t = \infty$.

4.4.1 The Band Limit

Most of the literature about the Hubbard model uses the simple form (4.13) of the band term, on a D-dimensional hypercubic lattice. There is a single band with the dispersion relation

$$\epsilon(\mathbf{k}) = -2t \sum_{j=1}^{D} \cos k_j a \tag{4.27}$$

where a is the lattice constant. We assume familiarity with the essential features of the one-, and three-dimensional cases but we will pay particular attention to $D = 2$ (the square lattice) and $D \to \infty$.

It is conventional to assume that in (4.13), $t > 0$. Does the sign of t really matter? We can convince ourselves that it does not if the lattice is *bipartite*, i.e., it can be divided into sublattices A and B in such a manner that the hopping connects an A-site only to B-sites, and vice versa. Namely, we can redefine the Wannier functions on sublattice B by incorporating a minus sign: $\phi(\mathbf{r} - \mathbf{R_j}) \to -\phi(\mathbf{r} - \mathbf{R_j})$ if $\mathbf{j} \in B$. This change of definition[14] cannot change the physical properties of the system but it changes the sign of t. A glance at (4.27) shows that the spectrum of the tight binding models with t and $-t$ is the same. Then for any number of electrons, we find the same ground state energy and excitation energies, and thus the same free energy. We should remember, though, that the argument does not work for non-bipartite lattices like the triangular lattice, and for those the $t > 0$ and $t < 0$ models have different physics.

In the grand-canonical version

$$\mathcal{H}_{\text{band}} = \sum_{\mathbf{k}\sigma} [\epsilon(\mathbf{k}) - \mu] \hat{n}_{\mathbf{k}\sigma} \,. \tag{4.28}$$

Changing n from 0 to 2, μ is increasing in a smooth fashion from $-2tD$ to $2tD$. For a non-interacting metal, the *charge compressibility* (or

---

[14] We could have argued more formally by performing the canonical transformation $c^\dagger_{\mathbf{j}\sigma} \to -c^\dagger_{\mathbf{j}\sigma}$ for $\mathbf{j} \in B$, but no change ($c^\dagger_{\mathbf{l}\sigma} \to c^\dagger_{\mathbf{l}\sigma}$) for $\mathbf{l} \in A$. This leads to $\mathcal{H}_{\text{band}} \to -\mathcal{H}_{\text{band}}$.

charge susceptibility), defined at $T = 0$ as the inverse second derivative of the ground state energy with respect to the particle density[15]

$$\kappa = \left(\frac{\partial^2 E}{\partial n^2}\right)^{-1} = 2\rho(\mu) \tag{4.29}$$

is directly expressible with the density of states, and it is finite for $\rho(\mu) \neq 0$. The system is *compressible.* For an interacting system, this simple relationship between κ and ρ no longer holds but it is generally true that a finite charge compressibility signifies a metal.

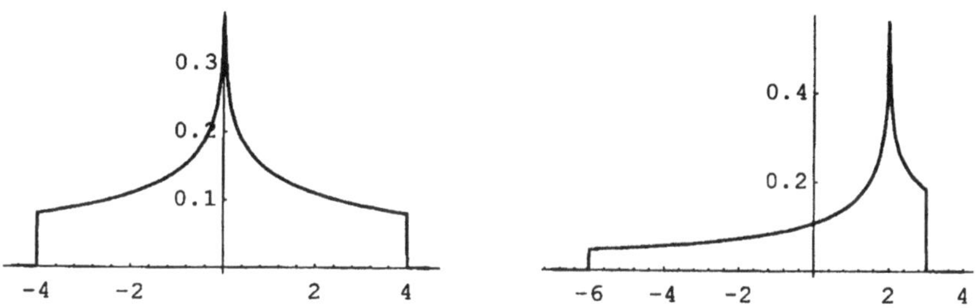

Figure 4.4: The density of states of the square lattice (left), and the triangular lattice (right) tight binding band, for $t = 1$.

The *square lattice Hubbard model* has become the subject of intense research effort when it was brought in connection with the high-temperature superconductivity of $La_{2-x}Sr_xCuO_4$ [23]. Increased attention was paid also to the features which are related to the peculiarities of the band structure. The density of states (DOS) of the square lattice tight binding band can be given in closed form

$$\rho(\epsilon) = \frac{1}{2\pi^2 t}\Theta(4t - |\epsilon|)\, K\left(1 - \frac{\epsilon^2}{16t^2}\right) \tag{4.30}$$

where K is the complete elliptic integral of the first kind[16]. The shape of the DOS is shown in Fig. 4.4 (left). We observe two striking features:

[15] E is the energy density per lattice site (or unit cell). The factor of 2 appears in (4.29) because of the spin.

[16] See, e.g., Ch. 17 of M. Abromowitz and I.A. Stegun: *Handbook of Mathematical Functions*, Dover Publ., New York (1968).

the step-like discontinuities at the band edges, and the logarithmic singularity in the centre. This latter can be obtained from the small-ϵ expansion of (4.30):

$$\rho(\epsilon) \sim \frac{1}{2\pi^2 t} \ln \frac{t}{|\epsilon|} . \tag{4.31}$$

Both of these features are van Hove *singularities*: the steps belong to the minimum and maximum of ϵ, while the logarithmic divergence turns out to be a saddle point singularity (see Problem 4.1). The existence of these singularities follows from the two-dimensional nature of the electronic structure, and is thus not unique to the square lattice: we find similar behaviour, for example's sake, for the triangular lattice[17] (Fig. 4.4, right). However, for the non-bipartite triangular lattice the saddle point singularity is no longer found in the centre of the band.

Every finite dimension D has its own characteristic set of van Hove singularities; those for $D = 3$ are discussed in introductory textbooks [32, 209]. The singularities get milder with increasing D, and disappear completely when $D \to \infty$ (Sec. 4.6).

Problem 4.1. *Square lattice.* Use an approximate form of the equi-ϵ lines in **k**-space to derive the van Hove singularities!

Funny band structures. For future reference, we need a few not-so-well-known band structures. The following problems are about tight binding models on non-Bravais lattices.

Problem 4.2. Characterize the density of states of the nearest-neighbour hopping tight binding band on the fcc lattice. The hopping amplitude is t_1. Does the sign of t_1 influence the DOS? Try to find out whether the DOS diverges somewhere and in case yes, which law does it follow.

Now switch on (in addition to t_1) a second-neighbour hopping $t_2 = t_1/2$ along the principal axes. Can you make any simple statement about the DOS? What about the singularity?

Problem 4.3. The linear chain with first- and second-neighbour hoppings can be redrawn as the "railroad trestle" lattice in Fig. 4.5 (the nearest-neighbour hopping is v, while the second-neighbour hopping is t for the even sublattice, and t' for the odd sublattice). Determine the band structure.

---

[17] Again, the DOS can be expressed with a complete elliptic integral but the formula [153] is much more complicated than (4.30). The author is grateful to E. Müller–Hartmann for calling his attention to this result.

a) *Tasaki's parametrization* [392]: Is it possible to chose such a $v \neq 0$ that the dispersion relation follows a pure cosine law? – What happens at $t' = 0$?

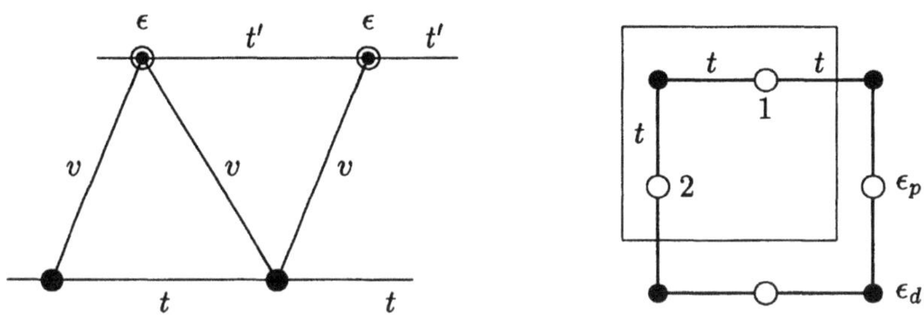

Figure 4.5: *Left*: the railroad trestle lattice. Hopping matrix elements are shown on the connecting lines. The top sites have an on-site energy ϵ. *Right*: the CuO_2 lattice. The empty circles are O atoms with site energy ϵ_p, the filled circles Cu atoms with site energy ϵ_d. All hopping amplitudes are t. The unit cell (thin lines) contains a base of three atoms.

b) *Degenerate minima*: Discuss the case $\epsilon = 0$, $t' = t$. What is the number of minima? Give the DOS for $v = 1/2$, $t = -1/2$.

Problem 4.4. The CuO_2 lattice (right-hand side of Fig. 4.5) is a simplified picture of the CuO_2 planes in La_2CuO_4. Here we replace the actual oxygen *p*- and Cu *d*-orbitals with *s*-orbitals, and neglect the direct hopping between oxygen atoms. We notice that this is an example for an *asymmetric bipartite lattice*: there are twice as many O-sites as Cu-sites.

Characterize the band structure!

Set $\epsilon_p = \epsilon_d = 0$. What is the total spin of the ground state at half-filling?

4.4.2 The Atomic Limit

In discussing the atomic limit $t = 0$, we find it useful to explicitely include the atomic level ϵ_{at} associated with the Wannier states

$$\mathcal{H}_{\text{at}} = \sum_{\mathbf{j}} \left[\epsilon_{\text{at}} \sum_{\sigma} \hat{n}_{\mathbf{j}\sigma} + U \hat{n}_{\mathbf{j}\uparrow} \hat{n}_{\mathbf{j}\downarrow} \right] . \tag{4.32}$$

(Unless it is our intention to draw attention to ϵ_{at}, we may choose it as the energy zero. In most of this text, $\epsilon_{\text{at}} = 0$).

$\mathcal{H}_{at} = \sum_j \mathcal{H}_j$ describes a set of L independent sites. The spectrum of a single site

$$\mathcal{H}_j = \epsilon_{at} \sum_\sigma \hat{n}_{j\sigma} + U \hat{n}_{j\uparrow} \hat{n}_{j\downarrow} \tag{4.33}$$

looks deceptively simple (Fig. 4.6, left). The first electron going to site j has the energy ϵ_{at}. With the addition of a second electron, the site has the energy $2\epsilon_{at} + U$; this can be interpreted as the result of having put the second electron to the energy level $\epsilon_{at} + U$. It is important to notice that this level scheme is fundamentally different from the usual single-electron spectra. In a single-electron level scheme, we are free to put "the first electron" to any level we wish. In contrast, the spectrum shown in Fig. 4.6 is *occupation number dependent*: the upper level at $\epsilon_{at} + U$ is not there at all, unless there is already one electron occupying the Wannier state!

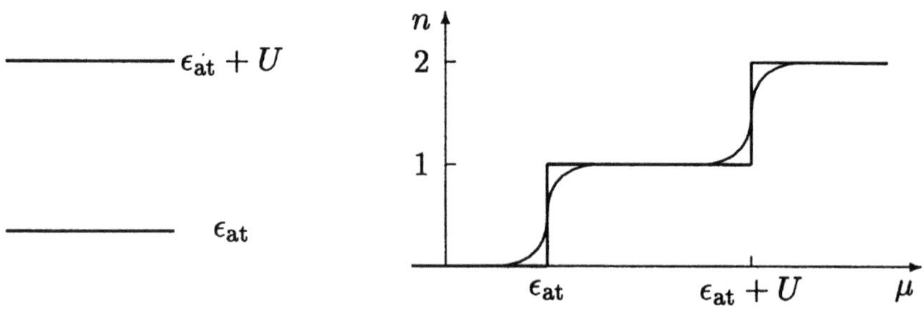

Figure 4.6: ***Left***: **Single-site spectrum in the Hubbard model. The existence of the upper level depends on the occupation of the lower level.** ***Right***: **In the atomic limit, the electron density n plotted as a function of the chemical potential μ shows jumps at ϵ_{at} and $\epsilon_{at} + U$ (thick line). We envisage that for a finite hopping t the plateaus are still recognisable (thin line).**

The ground state of $\mathcal{H}_{at}$ is tremendously degenerate. For $N < L$ electrons the singly occupied sites can be picked out in C_N^L ways, and each of these sites can harbour either an $\uparrow$- or a $\downarrow$-spin, so the degeneracy is $C_N^L 2^N$-fold. At half-filling ($N = L$) we find only the spin degeneracy 2^L. The exponentially large ground state degeneracy of the $U = \infty$ limit makes the description of the large-U behaviour by means of a t/U-expansion a very difficult problem.

The atomic limit of the grand-canonical potential

$$\mathcal{H}_{\rm at} - \mu \hat{N} = \sum_{\rm j} \left[(\epsilon_{\rm at} - \mu) \sum_{\sigma} \hat{n}_{\rm j\sigma} + U \hat{n}_{\rm j\uparrow} \hat{n}_{\rm j\downarrow} \right] \tag{4.34}$$

is minimized by

$$n = \begin{cases} 0 & \text{for } \mu < \epsilon_{\rm at} \\ 1 & \text{for } \epsilon_{\rm at} < \mu < \epsilon_{\rm at} + U \\ 2 & \text{for } \epsilon_{\rm at} + U < \mu . \end{cases} \tag{4.35}$$

The strictly step-function-like character of the dependence of n on μ (Fig. 4.6, right, thick line) is naturally an artefact of the atomic limit. We may, however, envisage that switching on a not too strong hopping t, the plateau regions remain recognizable (thin line). The strong Hubbard U causes the electron density to stay near to (stoichiometric) integer values in wide ranges of the chemical potential. In this sense, taking $n = 1$ (which is necessary for a Mott transition) is not such an arbitrary assumption at all: we may assume that μ is in one of the plateau regions[18].

What Causes the Ordering?

It belongs to folklore that "the non-interacting system does not order, interaction drives the system towards ordering". Just with the Hubbard model with its purely on-site interaction, one needs special care in formulating such a statement because the $t = 0$ system does not order, either: there is interaction but no communication from site to site. We conclude that ordering can arise only from an interplay of the on-site interaction, and the band motion of electrons.

We emphasize that the story would be quite different for models with intersite interactions, such as (1.4): there the interaction term taken in itself has an ordered ground state.

[18]For instance, μ may be controlled by the filling of other (wide) bands which overlap the Hubbard band and which we left out of the model. Such multi-band situations will be discussed in Ch. 11.

4.5 Symmetries

We learned in Ch. 3 that symmetry considerations can be very helpful in solving the eigenvalue problem. Assuming that U is the same at every lattice site, $\mathcal{H}_U$ is invariant under all the symmetry operations of the lattice and the Hubbard model as a whole will have the same geometrical symmetries as the band term $\mathcal{H}_{\rm band}$. In addition to the purely geometrical symmetries, there are further symmetries for which the physical character of the fermion operators has to be considered. The continuous spin-rotational invariance is extremely important both for classifying the eigenstates, and because the appearance of magnetic order in the Hubbard model can be understood as the spontaneous breaking of this symmetry. Not all but at least a large class of Hubbard models possesses an important discrete symmetry: the electron–hole symmetry. Finally, we mention that in the absence of a magnetic field, the model shows time-reversal invariance.

4.5.1 Spin-Rotational Invariance

The two terms of the Hubbard model describe the kinetic energy of the electrons, and the most essential part of the Coulomb force which depends only on the distance between the electrons. Neither of these constituents have anything to do with the spin of the electrons, thus we may take it for granted that the Hubbard Hamiltonian is invariant under the simultaneous rotation of all the spins.

While this may be considered as evident for the band term, we may be worried about the interaction term because it contains $\hat{n}_\uparrow$ and $\hat{n}_\downarrow$, and thus a seemingly explicit dependence on the choice of the quantization axis. To dispel this fear, it is useful to rewrite $\hat{n}_{\rm j\uparrow}\hat{n}_{\rm j\downarrow}$ as follows:

$$\begin{aligned}\hat{n}_{\rm j\uparrow}\hat{n}_{\rm j\downarrow} &= c^\dagger_{\rm j\uparrow}c_{\rm j\uparrow}c^\dagger_{\rm j\downarrow}c_{\rm j\downarrow} = c^\dagger_{\rm j\uparrow}c_{\rm j\uparrow}(1-c_{\rm j\downarrow}c^\dagger_{\rm j\downarrow}) \\ &= \hat{n}_{\rm j\uparrow} - c^\dagger_{\rm j\uparrow}c_{\rm j\downarrow}c^\dagger_{\rm j\downarrow}c_{\rm j\uparrow} = \hat{n}_{\rm j\uparrow} - S^+_{\rm j}S^-_{\rm j}\,. \end{aligned} \tag{4.36}$$

Similarly

$$\hat{n}_{\rm j\uparrow}\hat{n}_{\rm j\downarrow} = \hat{n}_{\rm j\downarrow} - S^-_{\rm j}S^+_{\rm j}\,. \tag{4.37}$$

Finally, by taking the square of the equation $S^z_{\rm j} = \frac{1}{2}(\hat{n}_{\rm j\uparrow} - \hat{n}_{\rm j\downarrow})$, and noting that $\hat{n}_{\mathrm{j}\sigma}^2 = \hat{n}_{\mathrm{j}\sigma}$, we can derive

$$\hat{n}_{\mathrm{j}\uparrow}\hat{n}_{\mathrm{j}\downarrow} = \frac{\hat{n}_{\mathrm{j}}}{2} - 2(S_{\mathrm{j}}^z)^2 \tag{4.38}$$

where $\hat{n}_{\mathrm{j}} = \hat{n}_{\mathrm{j}\uparrow} + \hat{n}_{\mathrm{j}\downarrow}$. Combining the above three expressions, we arrive at

$$\hat{n}_{\mathrm{j}\uparrow}\hat{n}_{\mathrm{j}\downarrow} = \frac{\hat{n}_{\mathrm{j}}}{2} - \frac{2}{3}\mathbf{S}_{\mathrm{j}}^2 \tag{4.39}$$

so $\mathcal{H}_U$ (4.14) can be expressed as

$$\mathcal{H}_U = \frac{UN}{2} - \frac{2U}{3}\sum_{\mathrm{j}} \mathbf{S}_{\mathrm{j}}^2 . \tag{4.40}$$

This form makes the spin-rotational invariance manifest.

Recalling that

$$\mathbf{S}_{\mathrm{j}}^2 = \begin{cases} 3/4 & \text{for } |\uparrow\rangle,\ |\downarrow\rangle \\ 0 & \text{for } |0\rangle,\ |d\rangle \end{cases} \tag{4.41}$$

in (4.40) we find another indication that the interaction term "wants as many uncompensated spins as possible".

4.5.2 Electron–Hole Symmetry

For the purposes of this discussion, we rewrite the interaction term as

$$\mathcal{H}_U = U\sum_{\mathrm{j}} \left(\hat{n}_{\mathrm{j}\uparrow} - \frac{1}{2}\right)\left(\hat{n}_{\mathrm{j}\downarrow} - \frac{1}{2}\right) + \frac{UN}{2} - \frac{UL}{4} \tag{4.42}$$

while the band term is taken in its most general form

$$\mathcal{H}_{\mathrm{band}} = -\frac{1}{2}\sum_{\mathrm{j,l}}\sum_{\sigma}(t_{\mathrm{jl}}c_{\mathrm{j}\sigma}^{\dagger}c_{\mathrm{l}\sigma} + t_{\mathrm{jl}}^{*}c_{\mathrm{l}\sigma}^{\dagger}c_{\mathrm{j}\sigma}). \tag{4.43}$$

An $\uparrow$-spin electron is turned into an $\uparrow$-spin hole by the canonical transformation $c_{\mathrm{j}\uparrow}^{\dagger} \to c_{\mathrm{j}\uparrow}$, $c_{\mathrm{j}\uparrow} \to c_{\mathrm{j}\uparrow}^{\dagger}$; similarly for $\downarrow$-spins. For the occupation numbers $\hat{n}_{\mathrm{j}\sigma} \to c_{\mathrm{j}\sigma}c_{\mathrm{j}\sigma}^{\dagger} = 1 - \hat{n}_{\mathrm{j}\sigma}$, and $N \to 2L - N$. Under this transformation, the operator term in (4.42) remains unchanged, we get merely an additive constant $\mathcal{H}_U \to \mathcal{H}_U - U(N-L)$. In contrast, applying the transformation to (4.43) we find that the hopping term gets transformed into something which looks completely different:

$$t_{\mathbf{jl}} c^{\dagger}_{\mathbf{j}\sigma} c_{\mathbf{l}\sigma} + t^{*}_{\mathbf{jl}} c^{\dagger}_{\mathbf{l}\sigma} c_{\mathbf{j}\sigma} \longrightarrow -t^{*}_{\mathbf{jl}} c^{\dagger}_{\mathbf{j}\sigma} c_{\mathbf{l}\sigma} - t_{\mathbf{jl}} c^{\dagger}_{\mathbf{l}\sigma} c_{\mathbf{j}\sigma}. \tag{4.44}$$

Generally speaking, the Hubbard model does not show electron–hole symmetry.

There is, however, an important class of Hubbard models which do show particle–hole symmetry: bipartite lattices in the absence of an external magnetic field. The latter condition permits to choose $t_{\mathbf{jl}}$ as real, while a bipartite lattice can be decomposed into sublattices A and B in such a manner that $t_{\mathbf{jl}} \neq 0$ only if $\mathbf{j}$ and $\mathbf{l}$ belong to different sublattices. For the sake of simplicity, we will assume nearest-neighbour hopping, with only one kind of t (i.e., we are back at the form (4.13)). We have seen on p. 164 (in footnote 14) that the minus sign appearing in (4.44) can be compensated by a further canonical transformation. Henceforth, when we speak about the electron–hole transformation for bipartite lattices, we will mean the combination of these two transformations, defined as

$$c^{\dagger}_{\mathbf{j}\sigma} \to \eta_{\mathbf{j}} c_{\mathbf{j}\sigma} \qquad c_{\mathbf{j}\sigma} \to \eta_{\mathbf{j}} c^{\dagger}_{\mathbf{j}\sigma} \tag{4.45}$$

where

$$\eta_{\mathbf{j}} = \begin{cases} 1 & \text{if } \mathbf{j} \in A \\ -1 & \text{if } \mathbf{j} \in B. \end{cases} \tag{4.46}$$

Since for nearest neighbours $\eta_{\mathbf{j}}\eta_{\mathbf{l}} = -1$, under (4.45) $\mathcal{H}_{\text{band}} \to \mathcal{H}_{\text{band}}$, and for the entire Hubbard Hamiltonian with total electron number N

$$\mathcal{H}(N) \longrightarrow \mathcal{H}(2L-N) - U(N-L) \tag{4.47}$$

Since $\mathcal{H}(N)$ and $\mathcal{H}(2L-N)$ differ only by an additive constant, the excitation energies are the same for N and $2L-N$ electrons. There is, however, a difference of the ground state energy densities (per lattice site)

$$E(2-n) = E(n) + U(n-1) \tag{4.48}$$

It follows that the temperature dependence of the free energies is the same, and that equivalent ordering transitions happen at the same temperatures. Granting the assumptions stated before, *the phase diagram of the Hubbard model is symmetrical about half-filling*. This is what we observe in Fig. 4.11.

The same conditions which lead to electron–hole symmetry (bipartite structure, no magnetic field), give also a new kind of continuous symmetry: a global $SU(2)$ pseudospin symmetry. This will be briefly described in the solution to Problem 4.5 (see p. 193).

The symmetry properties discussed before were quite general, being (under stated conditions) valid for arbitrary dimension, or band filling. It should be, however, mentioned that the one-dimensional Hubbard model has further U-dependent, non-Abelian symmetry groups [244].

Problem 4.5. Consider a bipartite lattice. Examine the consequences of applying the particle–hole transformation (4.45)–(4.46) only to the $\downarrow$-spin electrons!

4.6 Infinite-Dimensional Hubbard Model

One of the basic questions which can be addressed starting from the Hubbard model, is that of magnetic ordering. In statistical physics, magnetic ordering is often treated using classical lattice models of interacting spins, e.g., the Ising model. The limit of infinite spatial dimensionality plays an important role for the Ising model: for $D \to \infty$ the mean field theory becomes exact. The question of magnetism is posed at a more fundamental level by introducing the Hubbard model which is a quantum lattice model of interacting fermions. It is natural to ask whether going to $D \to \infty$ brings some simplification in the treatment of the Hubbard Hamiltonian as well.

It turns out that one has to be circumspect in defining the limit $D \to \infty$ [272]. First, the Hubbard model is a lattice model and the structure of the lattice matters very much. There are different versions of the $D \to \infty$ Hubbard model depending on which lattice structure is suitably generalized (see the Problems at the end of this section). Here we consider the D-dimensional simple cubic lattice (which we also call the hypercubic lattice): taking $a = 1$, the basis vectors are $\mathbf{e}_1 = (1,0,...,0)$, $\mathbf{e}_2 = (0,1,...,0)$, ..., $\mathbf{e}_D = (0,0,...,1)$. According to (4.27), the band edges are lying at $\pm 2tD$. Thus, if we were assuming a D-independent t, in the limit of infinite D the average band energy $\langle \mathcal{H}_{\mathrm{band}} \rangle$ would certainly become infinite. Since the value of D does not influence the (finite) on-site interaction energy $\langle \mathcal{H}_U \rangle$, we would find that $\langle \mathcal{H}_U \rangle / \langle \mathcal{H}_{\mathrm{band}} \rangle \to 0$ at $D \to \infty$, i.e., the model would be driven to its non-interacting limit. This is not what we want: the physics of the Hubbard model is non-trivial only if $\langle \mathcal{H}_U \rangle / \langle \mathcal{H}_{\mathrm{band}} \rangle$ remains finite. We have to find a way to scale t so that the kinetic energy remains finite when $D \to \infty$.

There is a simple argument to tell us how the scaling has to be done. Let us measure the kinetic energy of the half-filled Fermi sea by $\sqrt{\langle \mathcal{H}^2_{\mathrm{band}} \rangle}$. We find

$$\langle \mathcal{H}^2_{\mathrm{band}} \rangle = 4t^2 \left(\sum_{j=1}^{D} \langle \cos^2 k_j \rangle + \sum_{j \neq l}^{D} \langle \cos k_j \cos k_l \rangle \right) = 2Dt^2. \tag{4.49}$$

We can convince ourselves that the cross-terms should drop out (it is easiest for $D = 2$). For a half-filled band, $\cos^2 k = (1+\cos 2k)/2$ samples the entire range $-\pi \leq k < \pi$, thus we may set $\langle \cos^2 k_j \rangle = 1/2$. It is remarkable that the kinetic energy $\sqrt{\langle \mathcal{H}^2_{\mathrm{band}} \rangle}$ diverges like $\sqrt{D}$ and not as D as we might have naively assumed from the behaviour of the band edges. We will understand that this arises from the gradual change of the shape of the band density of states $\rho_D(\epsilon)$: as D increases, less and less weight is going into the far-away tails.

It follows from (4.49) that the kinetic energy remains finite at $D \to \infty$ if we use the scaled hopping parameter

$$t^* = \frac{t}{\sqrt{2D}}. \tag{4.50}$$

In calculating $\rho_D(\epsilon)$, we follow the analytical treatment described in [295] which, in addition to solving the $D = \infty$ problem, also permits to calculate finite-D corrections by expanding around $D = \infty$ in powers of

$1/D$. The density of states (DOS) is, by definition,

$$\rho_D(\epsilon) = \prod_{n=1}^{D} \int_{-\pi}^{\pi} \frac{dk_n}{2\pi} \delta(\epsilon - \epsilon(\mathbf{k})) \tag{4.51}$$

where

$$\epsilon(\mathbf{k}) = -2t^* \sum_{n=1}^{D} \cos k_n. \tag{4.52}$$

We exploit the additivity of the kinetic energy by taking the Fourier transform

$$\begin{aligned} \Phi_D(s) &= \int_{-\infty}^{\infty} \rho_D(\epsilon) e^{is\epsilon} d\epsilon = \prod_{n=1}^{D} \left(\int_{-\pi}^{\pi} \frac{dk_n}{2\pi} e^{-2it^* s \cos k_n} \right) \\ &= \left(\int_{-\pi}^{\pi} \frac{dk}{2\pi} e^{-2it^* s \cos k} \right)^D = (J_0(2st^*))^D \end{aligned} \tag{4.53}$$

where J_0 is the zeroth-order Bessel function. $J_0(x)$ has its highest peak at $x = 0$. This is hugely magnified by raising to a large power D thus it is justified to confine our attention to the region of small s. Expanding the exponential inside the integral in the second line of (4.51) in a Taylor series, performing the k-integrals, and replacing (4.50), we get[19]

$$\Phi_D(s) = \exp\left\{ -\frac{s^2 t^2}{2} - \frac{s^4 t^4}{16D} + O(D^{-2}) \right\} \tag{4.54}$$

which can be Fourier transformed back to yield

$$\rho_D(\epsilon) = \frac{1}{\sqrt{2\pi t^2}} \exp\left\{ -\frac{\epsilon^2}{2t^2} - \frac{1}{16D} \left(\frac{\epsilon^4}{t^4} - \frac{6\epsilon^2}{t^2} + 3 \right) \right\}. \tag{4.55}$$

The infinite-dimensional tight binding band has a Gaussian DOS[20]

$$\rho_\infty(\epsilon) = \frac{1}{\sqrt{2\pi t^2}} \exp\left(-\frac{\epsilon^2}{2t^2} \right). \tag{4.56}$$

[19] The k-integration gives $(1 - (st^*)^2 + (st^*)^4/4)^D$. It can be checked that this agrees with the power expansion of (4.54).

[20] The Gaussian tails may introduce some artificial features at exponentially small densities, because the first few electrons in the band can be put "infinitely deep in energy". It is understood that to see reasonable behaviour, we need a finite concentration of electrons.

$\rho_\infty(\epsilon)$ is represented by a perfectly smooth curve (Fig. 4.7, left). In contrast, $\rho_D(\epsilon)$ has some non-analyticities for any finite D: the van Hove singularities. The price we have to pay for the simplifications brought by the $D = \infty$ limit is the loss of all physical features which would be associated with van Hove singularities.

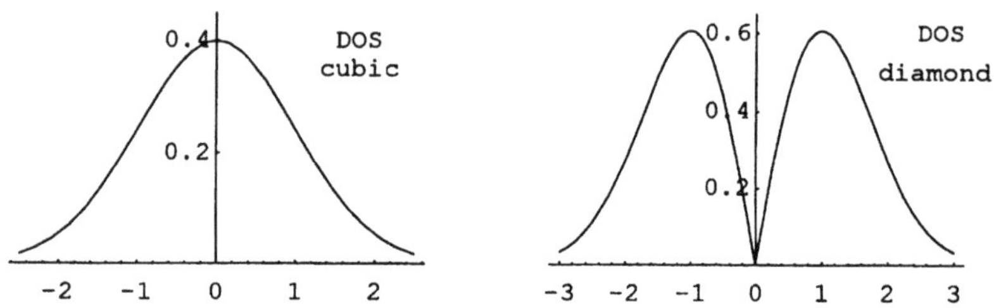

Figure 4.7: Density of states for infinite dimensional tight binding models ($t = 1$). *Left*: $\rho(\epsilon)$ (given by (4.56)), shown in units of $1/\sqrt{2\pi}$) for the $D = \infty$ cubic lattice. *Right*: $\rho(\epsilon)$ for the $D = \infty$ diamond lattice (see (4.103)).

It is worth pointing out that the $1/D$-expansion seen in (4.55) arises from the small-s behaviour of the Bessel function in (4.53) and therefore it can not recover the van Hove singularities of the large-D limit. These would appear in the contributions from the lesser maxima of the Bessel function and are thus represented by terms of order $e^{-\alpha D}$ [295], i.e., they are weaker than any power of $1/D$.

Up to this point, introducing the limit $D \to \infty$ may appear harmless enough, but searching further, we soon discover a few alarming features. We are accustomed to the idea that the kinetic energy $\epsilon(\mathbf{k})$ is a smooth function of $\mathbf{k}$: nearby points in $\mathbf{k}$-space have nearby energies. To decide whether this remains true at $D \to \infty$, we introduce the energy–energy correlation function [123]

$$C_{\mathbf{q}}(\epsilon_1, \epsilon_2) = \frac{1}{L} \sum_{\mathbf{k}} \delta(\epsilon_1 - \epsilon(\mathbf{k}+\mathbf{q}))\delta(\epsilon_2 - \epsilon(\mathbf{k})) \,. \qquad (4.57)$$

At the spanning vector $\mathbf{Q} = (\pi, \pi, \ldots)$, we find $C_{\mathbf{Q}}(\epsilon_1, \epsilon_2) \sim \delta(\epsilon_1 + \epsilon_2)$, expressing that the perfect nesting property $\epsilon(\mathbf{k}+\mathbf{Q}) = -\epsilon(\mathbf{k})$, which holds for the cubic lattice in any dimension, continues to be valid at $D \to \infty$. Furthermore, some correlation is found along the line $\mathbf{q} = \lambda \mathbf{Q}$

($|\lambda| \leq 1$). However, ϵ_1 and ϵ_2 are found to be completely uncorrelated for any other $\mathbf{q}$ [295]. The complete lack of smoothness in the $\mathbf{k}$-dependence of $\epsilon(\mathbf{k})$ means that though the Fermi energy remains a useful concept, the Fermi surface (and particularly, the "inside" and "outside" of the Fermi surface) does not [123]. The great formal appeal of the limit $D \to \infty$ notwithstanding, it is well to remember that in certain respects, an infinite dimensional model is unlike any finite-dimensional model.

It has to be stressed that the limit $D \to \infty$ can be taken only for lattice models. In $D = 1$, 2, and 3, the continuum model (an electron liquid moving on a uniform neutralizing background) is a popular choice for studying the effects of electron–electron interaction. The corresponding band DOS is $\rho \sim \epsilon^{(D-2)/2}$, and there is no way to make it behave meaningfully at $D = \infty$.

Problem 4.6. The band term of the $D \to \infty$ hypercubic Hubbard model can be extended by farther-neighbour hoppings *along the coordinate axis*. Thus let t_2 be the amplitude of hopping by $2a$ along any of the main axes:

$$\mathcal{H}_{\text{hop}}^{(2)} = -t_2 \sum_{\mathbf{R}} \sum_{j} (c^{\dagger}_{\mathbf{R}\pm 2a\mathbf{e}_j,\sigma} c_{\mathbf{R},\sigma} + \text{H.c.}) \tag{4.58}$$

etc. How has t_2 to be scaled to give a finite contribution as $D \to \infty$? What is the form of the DOS?

Problem 4.7. The fcc lattice can be generalized to arbitrary dimension D in the following manner: consider a cubic array of lattice constant 1, and pick the sites for which the sum of the coordinates is even[21] [152, 296, 413, 416]. How should the hopping be scaled to get a meaningful model at $D \to \infty$? Consider also the t_1–t_2 generalization of the fcc lattice which we introduced in Problem 4.2. Determine the density of states of the infinite-dimensional fcc lattice. Which expression do you find for $t_2 = t_1/2$?

Problem 4.8. *The infinite-dimensional diamond lattice* can be generated in the $D \to \infty$ limit of a construction which gives the honeycomb lattice for $D = 2$, and the diamond lattice for $D = 3$ [346]. For general D, each lattice site has $D+1$ neighbours to which the unit vectors $\mathbf{e}_1,\ldots,\mathbf{e}_{D+1}$ are pointing. Any two vectors subtend the same angle: $\mathbf{e}_i \cdot \mathbf{e}_j = \alpha_D$.

What is the value of α_D? Work out the details of the lattice structure, and of the band structure with nearest-neighbour hopping! What is the $D = \infty$ form of the density of states?

[21] This lattice is often called the half-hypercubic lattice. A different D-dimensional generalization of the fcc lattice is suggested in [413].

4.7 Hubbard Subbands

The Hubbard model becomes interesting when both U and t are finite. Now we try to understand the finite-U behaviour starting from the atomic limit and switching on a small t. Strictly at $t = 0$, the spectrum consists of L copies of the two-level spectrum shown in Fig. 4.6 (left). How is this picture modified when the electrons start to hop around?

To have a simple case, let us assume that the system contains $L+1$ electrons. At $t = 0$, the ground states have L electrons in the lower level ϵ_{at}, and the $(L+1)$th electron can be in any of the upper levels at $\epsilon_{\mathrm{at}} + U$. For further simplification, assume that the L lower-level electrons have $\downarrow$-spins, and the $(L+1)$th electron has $\uparrow$-spin; then the $t = 0$ ground state configurations are

$$|\varphi_{\mathbf{j}}\rangle = c_{\mathbf{j}\uparrow}^{\dagger} \prod_{\mathbf{i}}^{L} c_{\mathbf{i}\downarrow}^{\dagger}|0\rangle. \tag{4.59}$$

It is easy to see that

$$\mathcal{H}_{\mathrm{band}}|\varphi_{\mathbf{j}}\rangle = -t\sum_{\mathbf{m}}|\varphi_{\mathbf{m}}\rangle$$

where the $\mathbf{m}$ are nearest neighbours of $\mathbf{j}$. In the $(L+1)$-electron basis $|\varphi_{\mathbf{j}}\rangle$, the hopping term acts as it does in the single-electron Wannier basis, thus the hopping of the $(L+1)$th electron broadens the upper atomic level into a tight binding band of width $\sim 2zt$. Assuming $L-1$ spin-polarized electrons, and considering the motion of a single hole, a similar argument can be given for the arising of a band centered on the lower atomic level. We end up with the picture of two bands of states, each with a width $\sim 2zt$, centered at the atomic levels ϵ_{at} and $\epsilon_{\mathrm{at}} + U$ (Fig. 4.8). We will refer to these as the Hubbard subbands.

It must be emphasized that our handwaving argument does not give more than the merest indication of the arising of Hubbard subbands. If we had not made our artificial assumption about (almost) complete spin polarization, we would have met considerable technical difficulties. For a proper derivation, the reader is referred to the third of Hubbard's classic series of papers [174]. But all in all, the emerging picture is rather similar to what we show in Fig. 4.8. For t sufficiently smaller

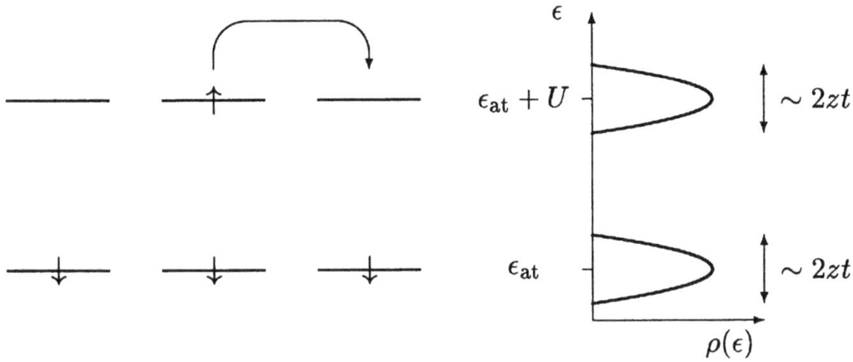

Figure 4.8: Schematic representation of the origin of the Hubbard subbands. The hopping of an ↑-spin electron on the background of ↓-spin electrons gives rise to a band of width $\sim 2zt$ centered on $\epsilon_{at} + U$. Similarly, the motion of a ↓-hole leads to the appearance of the lower subband.

than U, if we add an electron (or a hole) to the system, the energy levels of the quasiparticle belong to one of the well-separated Hubbard subbands, whose widths are of the order of zt. The essential change with respect to our naive result is that the two-subband-picture can be used for electrons with arbitrary spins, including a paramagnetic mixture of ↑-spin and ↓-spin electrons.

Though at the first sight, the two subbands may look like the valence and conduction band of a semiconductor, we must repeat the warning given about the level scheme in Fig. 4.6: this is an essentially occupation-number dependent band structure. The very existence of the upper Hubbard subband depends on the presence of electrons in the lower Hubbard subband. The difference from the usual semiconductor band structure becomes clearer if we recall that an "ordinary" band can hold up to $2L$ electrons, while the lower Hubbard subband cannot contain more than L electrons (for more than L electrons, some of the sites must be doubly occupied which is associated with an energy U, so the occupation of the upper Hubbard subband has in fact begun). Similarly, the upper Hubbard subband has place only for not more than L electrons. This does not arise from a reduction of the **k**-space domain: the subbands have the same set of **k**-vectors as the original one-electron band in the $U = 0$ limit. The splitting into two Hubbard subbands is a correlation effect, beyond the capabilities of independent-electron theories to describe.

4.7.1 The Mott–Hubbard Transition

The concept of Hubbard subbands offers a direct way to describe a correlation-induced metal–insulator transition (Fig. 4.9). For a large enough U, the subbands are well-separated. There is a critical value $U_{\rm cr} \sim zt$ where the subbands touch and we have a situation like that in a zero-gap semiconductor. For $U < U_{\rm cr}$ the subbands overlap, there is a finite density of states $\rho(\epsilon_{\rm F})$ at the Fermi level, and the system is metallic. To justify this scenario, it has to be ascertained that the Hubbard subbands remain recognizable even when U becomes less than $U_{\rm cr}$. The first derivation was given by Hubbard, using a Green's function decoupling method which has become known as the Hubbard III approximation [174]. Later work, especially on the infinite-dimensional Hubbard model [130], essentially confirmed[22] the validity of the picture shown in Fig. 4.9.

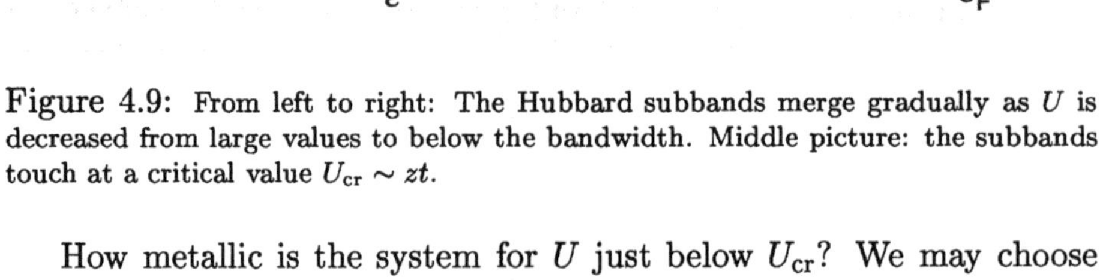

Figure 4.9: From left to right: The Hubbard subbands merge gradually as U is decreased from large values to below the bandwidth. Middle picture: the subbands touch at a critical value $U_{\rm cr} \sim zt$.

How metallic is the system for U just below $U_{\rm cr}$? We may choose $\rho(\epsilon_F)$ to measure this. With a three-dimensional form of the DOS at the subband edges, $\rho(\epsilon_{\rm F})$ *increases continuously from 0* as U decreases below

[22]There is one important addition: in the metallic regime immediately below $U_{\rm cr}$, there is a sharp peak sitting at ϵ_F, indicating that the almost localized metal is a heavy Fermi liquid. This will be discussed in detail in Ch. 10.

U_{cr} (Fig. 4.10, left). We have found a *second-order metal–insulator transition* at $U = U_{cr}$.

The idea of a correlation-driven metal–insulator transition was introduced in 1949 in a seminal paper by Mott [292]. However, he envisaged a *first-order transition.* Using our picture, the argument can be rephrased as follows. For a slight overlap between the upper and lower subbands, we would have a small concentration δn of "electrons" in the upper subband and "holes" in the lower subband[23]. The Thomas–Fermi screening length $\lambda_{TF} \sim 1/\sqrt{\delta n}$ being much larger than the radius ξ_{exc} of the electron–hole bound state (the exciton radius), the few carriers combine into bound pairs, and the supposedly weakly metallic system is, in fact, not a metal at all but an (excitonic) insulator. The transition to the metallic state occurs abruptly when the overlap of the subbands is large enough so that $\lambda_{TF} \sim \xi_{exc}$. In the original sense, the *Mott transition* is a discontinuous transition from an insulator to a fully developed metallic state with a largish density of charge carriers (Fig. 4.10, right).

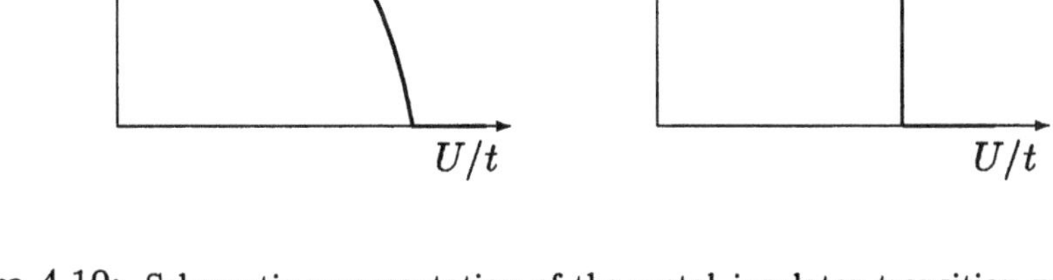

Figure 4.10: Schematic representation of the metal–insulator transition without, and with, the inclusion of screening. *Left*: For the Hubbard model, $\rho(\epsilon_F)$ approaches 0 gradually as the Hubbard subbands separate. *Right*: the inclusion of long-range Coulomb forces gives rise to a first-order Mott-transition (δn is the density of charge carriers).

[23]Remember that it is far from being obvious that this "semiconductor language" can be used for the Hubbard subbands (Mott's original argument was phrased differently). We are using this formulation in order to connect to the previous discussion.

Let us notice that starting with the Hubbard model, we should not expect to recover the above result. Screening comes from the long-range part of the Coulomb force, the part which is left out from the Hubbard model. We feel certain that with the on-site term (4.14), we included the most essential part of the Coulomb interaction: $\mathcal{H}_U$ is the reason why we eventually must have an insulator. But the details of getting there depend on what else is there in the Hamiltonian, and we will have to keep in mind that the omission of the long-range part of $\mathcal{H}_{\rm Coulomb}$ changed the character of the transition. This is too bad but we will find that the Hubbard model is difficult enough as it is.

On the experimental side, metal–insulator transitions in crystalline systems[24] are strongly discontinuous [5, 293, 330]. The first-order transitions are often accompanied by substantial changes of the atomic distances and this should remind us of another feature missing from the Hubbard model: the coupling of the correlated electrons to the lattice. The experimentally observable Mott-transitions of real systems are complex phenomena and only a part of the rich behaviour is captured by the Hubbard scenario (which we may call the Mott–Hubbard transition). We contend, however, that the local correlations caused by the large U constitute the most essential aspect of Mott phenomena.

4.8 Ground State Phase Diagram

Much of this text is devoted to mapping out the behaviour of the Hubbard model in the n–U/t plane. This gives us the ground state phase diagram. The full phase diagram would be three-dimensional, with the temperature T as the third axis. However, a systematic study of the finite-temperature behaviour would require more sophisticated methods than what we develop here. We will be satisfied if we manage to find, and describe, the various ground states and then make educated guesses about what happens at finite T (see Sec. 5.5.1).

The first question is whether the ground state is metallic or insulat-

[24] There is a quite different class of metal–insulator transitions in strongly disordered systems, e.g. in doped semiconductors like Si:P, where disorder has to be included in the description from the outset. Experimentally, most systems show aspects of correlation *and* disorder. Nevertheless, we treat only correlation phenomena.

ing. Previously, we have convinced ourselves that at half-filling ($n = 1$), a Mott-transition should set in at some critical value $(U/t)_{\rm cr}$, so the system is metallic for $(U/t) < (U/t)_{\rm cr}$, and insulating for $(U/t) > (U/t)_{\rm cr}$.

What happens at $n \neq 1$? Let us make the plausible assumption that if the system is metallic at $U/t = \infty$ then it is metallic at any finite U/t. It is easy to argue that the ground state should be conducting for any $0 < n < 1$ and $1 < n < 2$.

For less electrons than sites $N < L$ (or $n < 1$), a large U forbids double occupation, so some sites are empty (positively charged), while others have one electron (neutral sites). It looks plausible[25] that the strong Hubbard U does not inhibit the jumping of an electron to a neighbouring empty site:

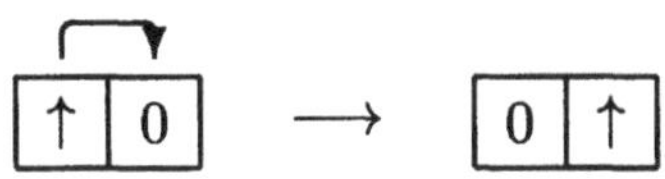

(4.60)

The process can be looked upon as the hopping of the empty site which is a charge carrier. This costs no energy thus the system is a metal.

This finding may raise doubts about the relevance of arguments for the Mott transition as an observable phenomenon. Since n can vary continuously between 0 and 2, it would seem that the point $n = 1$ is a set of zero measure, utterly without interest. This would be a false conclusion. On the one hand, we have seen (Fig. 4.6) that the filling tends to be pinned near integer values of n, thus $n = 1$ is not so unexpected. On the other hand, for n sufficiently near to 1, disorder effects (Anderson localization) come into play and make the system a non-conductor by localizing the few ($\sim |n - 1|$) carriers implied by our arguments. As a matter of experimental fact, Mott insulators (or if you like, Mott–Anderson insulators) are robust, and their nature can be largely understood by studying the case $n = 1$.

Next: which kinds of ordered ground states are possible? Expressly avoiding the thorny question whether the positive-U Hubbard model

[25] Actually, there is a subtle difficulty here. While it is true that we cannot identify an activation energy of the order U, the hopping of an empty site on the background of spins may leave a trail of disturbed spin arrangement, so it does not look like the free propagation of a hole, either. The argument works simply if we assume that all spins are $\uparrow$.

can have a superconducting ground state, we are looking for *magnetic ground states*. We assume that the system is at least two-dimensional ($D \geq 2$) so that there are no general arguments against the existence of ground-state ordering[26]. As for the Mott-insulating state, we will show in Sec. 5.1 that it is characterized by antiferromagnetic nearest-neighbour interactions and therefore it has (almost always) antiferromagnetic long-range order. At $n \neq 1$, for a first orientation let us look at the mean-field results. Most Hartree–Fock calculations were done for the nearest-neighbour hopping model on a D-dimensional cubic lattice. The general appearance[27] of the resulting phase diagram is shown in Fig. 4.11. In addition to the fully symmetrical (non-ordered) paramagnetic state (P), we find a two-sublattice antiferromagnetic (AF) phase straddling the $n = 1$ line, and two extended regions of ferromagnetism (F). The $n = 1$ states are insulating for $U > 0$; away from half-filling, all the phases are metallic.

This phase diagram has a feature which we should consider unexpected. Namely, we have found good reasons to think that the transition to a Mott insulator occurs at a finite $U \sim t$, while in Fig. 4.11 the metal–insulator transition sets in at infinitesimally small U. We will see in Ch. 7 that the reason is the so-called perfect nesting property of the band structure which is common to D-dimensional cubic lattices, and which makes the P state unstable against the formation of a spin density wave (SDW). The "true Mott transition" would be an effect of strong correlations and we marked by a black dot where it should have occured, had it not been pre-empted by the SDW instability. (Under the circumstances discussed here, nothing special happens at the black dot). The striking mirror symmetry of the phase diagram about half-filling was explained in Sec. 4.5.2 by the bipartite nature of cubic lattices.

Fig. 4.12 gives an "artist's impression" of the corresponding Hartree–Fock phase diagram for a non-nesting, non-bipartite situation. This picture embodies our expectations about the "generic" features of the

[26]In $D = 1$, quantum fluctuations destroy long-range order even at $T = 0$. We return to this question in later chapters.

[27]This is just a sketchy drawing and not the result of a calculation. Nevertheless, Fig. 4.11 closely resembles the Hartree–Fock phase diagram for the square lattice [166], and the $D = 3$ cubic lattice Hubbard model [324].

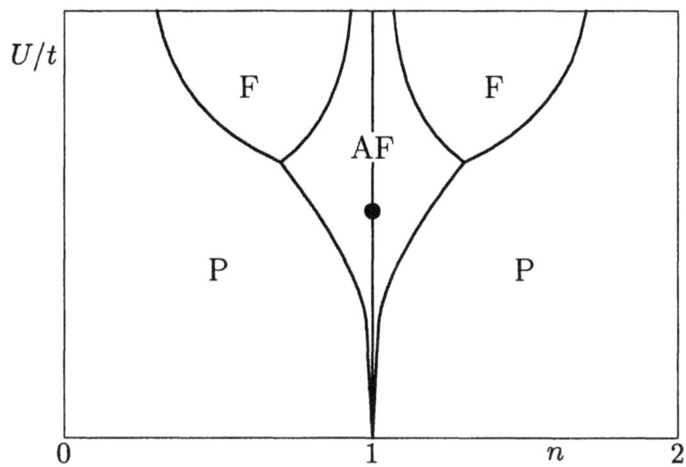

Figure 4.11: Schematic representation of the $T = 0$ mean-field phase diagram of the D-dimensional cubic lattice Hubbard model in the n–U/t plane. The phases are F: ferromagnetic, AF: two-sublattice antiferromagnetic, P: paramagnetic. The states on the $n = 1$ line are insulating, away from half-filling all phases are metallic. The black dot shows where the "true" Mott transition should have set in.

Hubbard model (i.e., those which do not arise from a special choice of the lattice and the band structure). Its most important feature is that the appearance of magnetism at any n requires that U/t exceeds a finite threshold value. On the $n = 1$ line, this threshold value is not necessarily the same at which the ground state becomes insulating: magnetic ordering may precede the Mott transition. Even on the $n = 1$ line, AF will no longer mean the simplest two-sublattice antiferromagnetic structure but some (in general non-collinear) order which fits the lattice structure. Not wishing to enter this subject now, we merely hinted at the problem by using the notation AF–SDW for the region between the two pockets of F ordering. Fig. 4.12 was designed to imitate certain features[28] of the mean field phase diagram of the triangular lattice

[28] Primarily the strong asymmetry about half-filling which arises from the non-bipartite nature. However, the triangular lattice with nearest-neighbour hopping does have the perfect nesting property, though not at $n = 1$. In the true Hartree–Fock phase diagram the boundary of the AF–SDW phase swoops down to $U = 0$ at

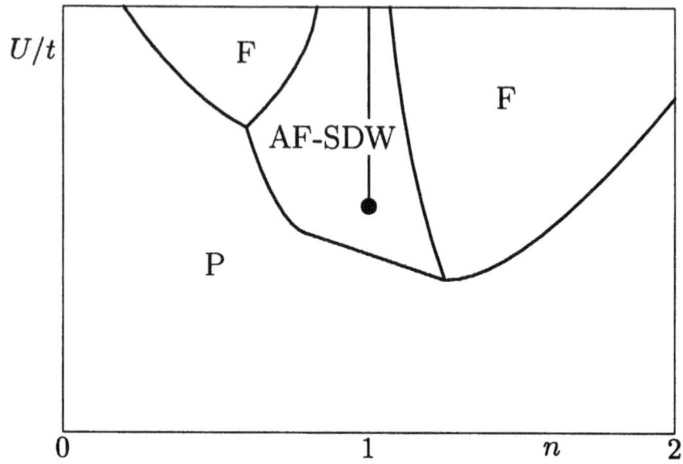

Figure 4.12: Schematic Hartree–Fock ground state phase diagram of the Hubbard model on a non-bipartite lattice. The region between two asymmetric ferromagnetic (F) regions is occupied by commensurate antiferromagnetic or incommensurate spin density wave states (AF-SDW). In general, the Mott-insulating state (thick line starting at the dot marking the Mott transition) is lying inside the AF phase.

Hubbard model [153].

Returning to Fig. 4.11, why only these two kinds of magnetism? Could the system not choose some more complicated ordering patterns? It certainly could and up-to-date mean-field phase diagrams show a wealth of curious ordered phases. In the early calculations, the restriction to the F and AF solutions was done just for calculational convenience. While AF is the correct solution for $n = 1$, away from half-filling, our phases may get replaced by more complicated ones. In any case, the overall extent of the region with some kind of ordering can only become larger[29].

However, even with the restriction to simple kinds of ordering, Figs. 4.11 and 4.12 look really hopeful: they promise us the understanding of antiferromagnetic Mott insulators and itinerant ferromagnets! What

---

$n = 1.5$.

[29]In the mean field approximation. It is difficult to judge how much of this order is really there. Hartree–Fock is notorious for overestimating ordering tendencies!

is more, we did not have to make any special efforts to "fine-tune" the Hubbard model to get these predictions: seemingly, the two major types of magnetism are to be found in the phase diagram of most Hubbard models. Sure enough, we know that a mean-field theory has to be corrected with the inclusion of fluctuation effects but we may have surmised that for $D = 3$, fluctuations play a minor role and would merely shift the phase boundaries without changing the qualitative appearance of the phase diagram.

Nowadays, we are more cautious. After decades of intense research effort, the emerging picture might be summarized like this: The Hubbard model does indeed give us an understanding of the Mott transition, and the nature of (most) Mott insulators. This is comforting because we consider the existence of Mott insulators as the major correlation effect: it is associated with the large energy scale U ($\sim \Delta$, the Mott–Hubbard gap). Magnetic ordering is a secondary effect, associated with smaller energies. Of the many kinds of magnetism, antiferromagnetism emerges very naturally from the Hubbard model. We can be happy about this, too, because antiferromagnetism is the most commonly found kind of order. It would have been nice if, in addition, even the one-band Hubbard model would have enabled us to reach a trouble-free qualitative understanding of the ferromagnetism of the iron group elements[30]. However, we are slowly becoming convinced that the two large F regions in Fig. 4.11 are mere artefacts of the Hartree–Fock approximation: either they are not there at all, or get shrunk to a physically irrelevant part of the phase diagram. It is the subject of on-going research which modification or extension of the simple Hubbard model would allow us to grasp the essential features of itinerant ferromagnetism. As discussed later (see especially Ch. 8), asymmetry of the band (non-bipartite lattices), band degeneracy, and the hybridization of correlated bands can be favourable for the appearance of ferromagnetism.

Solutions to the Problems

Problem 4.1 The DOS is the line integral in **k**-space over an equipotential

[30]The periodic table contains two more ferromagnetic metals: Gd and Dy. The physics of those is completely different and gives no reason to worry.

line

$$\rho(\epsilon) = \frac{1}{4\pi^2} \oint \frac{d\mathbf{l}}{|\nabla_{\mathbf{k}}\epsilon|} \,. \tag{4.61}$$

Near the bottom of the band $\epsilon(\mathbf{k})/2t \approx (-4+\mathbf{k}^2)/2$, so the equipotential lines are nearly circular and

$$\rho(\epsilon) \approx \frac{1}{4\pi^2} \int_0^{2\pi} \frac{k \cdot d\varphi}{2t \cdot k} = \frac{1}{4\pi t} \Theta(\epsilon + 4t) \tag{4.62}$$

gives the characteristic step-function behaviour.

For $\epsilon \ll t$ either $\pi - k_x \ll \pi$, and then

$$\frac{1}{2t}\epsilon(k_x, 0) = -(\cos k_x + 1) \approx -\frac{1}{2}(\pi - k_x)^2 \,, \tag{4.63}$$

or $k_y \ll \pi$, and then

$$\frac{1}{2t}\epsilon(\pi, k_y) \approx \frac{k_y^2}{2} \,. \tag{4.64}$$

Introducing the new variable $k_y' = \pi - k_x$, in a small environment of $(\pi, 0)$

$$\epsilon' = \frac{\epsilon}{t} \approx (k_y')^2 - k_x^2 \,, \tag{4.65}$$

so the equipotential lines are hyperbolae. Since $k_y' = \sqrt{\epsilon' + k_x^2}$, we find

$$dl = dk_x \sqrt{1 + \left(\frac{dk_y}{dk_x}\right)^2} = dk_x \sqrt{\frac{\epsilon' + 2k_x^2}{\epsilon' + k_x^2}} \,.$$

Furthermore,

$$|\nabla_{\mathbf{k}}\epsilon'| = 2\sqrt{\epsilon' + 2k_x^2} \,.$$

Integrating for both branches

$$\int \frac{dl}{|\nabla_{\mathbf{k}}\epsilon'|} \approx \int_0^{O(1)} \frac{dk_x}{\sqrt{\epsilon' + k_x^2}} = [\ln(k_x + \sqrt{k_x^2 + \epsilon'})]_0^{O(1)} \rightsquigarrow \ln\frac{t}{\epsilon} \,. \tag{4.66}$$

All in all, we found that for $|\epsilon| \ll t$, the DOS is dominated by a *logarithmic singularity* at the centre of the band

$$\rho(\epsilon) \sim \frac{1}{2\pi^2 t} \ln\frac{t}{|\epsilon|} \,. \tag{4.67}$$

Problem 4.2. The nearest-neighbour vectors being (in units of the edge of the cubic cell) (1,1,0), (1,-1,0),...,(0,-1,-1), the familiar form of the dispersion

$$\epsilon_1(\mathbf{k}) = 4t_1 \left(\cos\frac{k_x}{2}\cos\frac{k_y}{2} + \cos\frac{k_y}{2}\cos\frac{k_z}{2} + \cos\frac{k_z}{2}\cos\frac{k_x}{2} \right) \tag{4.68}$$

follows. Choosing $t_1 > 0$, the band is lying in the interval $-4t_1 \leq \epsilon(\mathbf{k}) \leq 12t_1$, thus it is asymmetric about the atomic level. The fcc lattice is not bipartite, thus we have no reason to expect a symmetrical DOS. Changing the sign of t_1 turns the band upside down.

The canonical form of the DOS of the fcc lattice[31] shows square root behaviour at the band extrema and three kinks (corresponding to saddle points) in between [196]. One does not expect a diverging DOS anywhere.

The fcc tight binding DOS which we can derive from (4.68), is markedly different [168]. One finds, in addition to the square root behaviour at the top of the band, and a kink at $\epsilon = 0$, a divergency at the lower band edge. The basic reason is that the minimum energy $\epsilon = -4t_1$ is reached not merely at points, but at pairs of perpendicular lines crossing at $(0,2\pi,2\pi)$, and at other X points. It can be checked that $\epsilon = -4t_1$ for $(0,2\pi,k_z)$ and for $(0,k_y,2\pi)$. Expanding

$$\frac{1}{4t_1}\epsilon(2\delta k_x,\ 2(\pi - \delta k_y),\ 2(\pi - \delta k_z)) \approx -1 + \frac{\delta k_x{}^2}{2} + \frac{\delta k_y{}^2 \delta k_z{}^2}{4}\,, \tag{4.69}$$

we conclude that the equi-ϵ lines in the $\delta k_x = 0$ plane are hyperbolae

$$\delta k_z \sim \frac{\sqrt{1 + (\epsilon/4t_1)}}{\delta k_y}\,. \tag{4.70}$$

We want to determine the number of states $\Delta N(\epsilon)$ with energies lying between $-4t_1$ and ϵ, where ϵ is very near to $-4t_1$. We approximate the corresponding $\mathbf{k}$-space volume as a body whose base area is the area under the hyperbole (4.70), and whose height is $\delta k_x \sim \sqrt{1 + (\epsilon/4t_1)}$

$$\Delta N(\epsilon) \sim \sqrt{1 + (\epsilon/4t_1)} \cdot \ln\left(\sqrt[4]{1 + (\epsilon/4t_1)} \right)\,. \tag{4.71}$$

(4.71) is consistent with a density of states which diverges logarithmically[32] at

[31] Meaning features which follow from the symmetry of the fcc lattice, without relying on the particular dispersion relation (4.68).

[32] While a logarithmic divergency of the DOS is typical for $D = 2$, it requires a "fine-tuning" of the band structure in $D = 3$. The hcp lattice (with nearest-neighbour hopping) has the same DOS as the fcc lattice, and thus the same logarithmic divergency. The DOS of the bcc tight binding band diverges at $\epsilon = 0$.

the bottom of the band

$$\rho(\epsilon) \sim \frac{1}{t_1} \ln \left(1 + \frac{\epsilon}{4t_1}\right) . \tag{4.72}$$

Naturally, the above approximate form can be used only in the immediate neighbourhood of $\epsilon = -4t_1$.

Switching on a second-neighbour hopping t_2 along the principal axes, the band is modified to

$$\begin{aligned} \epsilon_2(\mathbf{k}) &= \epsilon_1(\mathbf{k}) + 2t_2(\cos k_x + \cos k_y + \cos k_z) \\ &= \frac{1}{2t_1} \cdot \epsilon_{sc}^2(\mathbf{k}) + (2t_2 - t_1)\epsilon_{sc}(2\mathbf{k}) - 3t_1 , \end{aligned} \tag{4.73}$$

where we introduced the dispersion relation of the simple cubic lattice with nearest neighbour hopping t_1

$$\epsilon_{sc}(\mathbf{k}) = 2t_1[\cos(k_x/2) + \cos(k_y/2) + \cos(k_z/2)] . \tag{4.74}$$

The second term in the second line of (4.73) drops out if $t_2 = t_1/2$. It follows that the DOS of our t_1–t_2 fcc lattice can be expressed with the DOS of the sc lattice

$$\rho(\epsilon) = t_1 \cdot \frac{\rho_{sc}\left(\sqrt{2t_1\epsilon + 6t_1^2}\right)}{\sqrt{2t_1\epsilon + 6t_1^2}} . \tag{4.75}$$

The DOS of the $t_2 = t_1/2$ model has an algebraic ($1/\sqrt{\epsilon}$-like) singularity at the lower band edge $\epsilon = -3t_1$.

Though the DOS diverges at the minimum energy both for $t_2 = 0$ and $t_2 = t_1/2$, one finds the usual $\sqrt{\epsilon - \epsilon_{min}}$-type behaviour for $0 < t_2 < t_1/2$. However, the DOS is strongly peaked near the lower band edge. This kind of DOS turns out to be favourable for the appearance of ferromagnetism [416].

Problem 4.3. The unit cell contains one bottom and one top site. We denote the corresponding Wannier operators by c and d. The Hamiltonian reads

$$\begin{aligned} \mathcal{H} &= \sum_j [t(c_j^\dagger c_{j+1} + c_{j+1}^\dagger c_j) + t'(d_j^\dagger d_{j+1} + d_{j+1}^\dagger d_j) \\ &\quad + v(c_j^\dagger d_j + d_j^\dagger c_j + d_j^\dagger c_{j+1} + c_{j+1}^\dagger d_j) + \epsilon d_j^\dagger d_j] \end{aligned} \tag{4.76}$$

(spin indices are irrelevant in a one-electron problem. Alternatively, we can think of spinless fermions.)

a) In k-space, the model corresponds to the hybridization of the c and d tight binding bands. For each k, we have to solve the eigenvalue problem of a 2×2 matrix:

$$\begin{pmatrix} 2t\cos k & v(e^{ik}+1) \\ v(e^{-ik}+1) & 2t'\cos k + \epsilon \end{pmatrix} \tag{4.77}$$

We do not write down the cumbersome expression for the dispersion relation. However, the reader will find it easy to check that the parametrization given by Tasaki [392]

$$v = \lambda(t - t') \qquad \epsilon = (\lambda^2 - 2)(t - t') \tag{4.78}$$

for arbitrary λ simplifies the square root and leads to two cosine bands. Chosing $t' < 0$, the lower band has the dispersion $2t' \cos k - 2(t - t')$, while the upper band follows $2t \cos k + \lambda^2 (t - t')$. It is curious that in spite of a non-zero hybridization, the shape of the two bands can be preserved.

At $t' = 0$ the lower band becomes flat. In terms of Fig. 4.5, this is the case of a "sawtooth lattice".

b) The case $\epsilon = 0$, $t' = t$ can be conveniently reformulated (with $t = t' = -t_1 < 0$ and $v = -t_2$) as the tight binding chain with first- and second-neighbour hopping

$$\mathcal{H} = -t_1 \sum_j \sum_\sigma (c^\dagger_{j\sigma} c_{j+1\sigma} + \text{H.c.}) - t_2 \sum_j \sum_\sigma (c^\dagger_{j\sigma} c_{j+2\sigma} + \text{H.c.}) \,. \tag{4.79}$$

Note that now the unit cell contains only one site. The dispersion relation is given by

$$\epsilon(k) = -2t_1 \cos k - 2t_2 \cos 2k \,. \tag{4.80}$$

$\cos k$ has one minimum for $k \in [-\pi, \pi]$, but $\cos 2k$ has two. The same should be true of $\epsilon(k)$ if the second term is sufficiently strong. $\epsilon(k)$ has extrema at $k = 0$ and $k = \pm \arccos(-t_1/4t_2)$. The two minima merge at $k = 0$ when $t_2 = -t_1/4$. $\epsilon(k)$ has two minima if $t_2 < -t_1/4$ (Fig. 4.13, left).

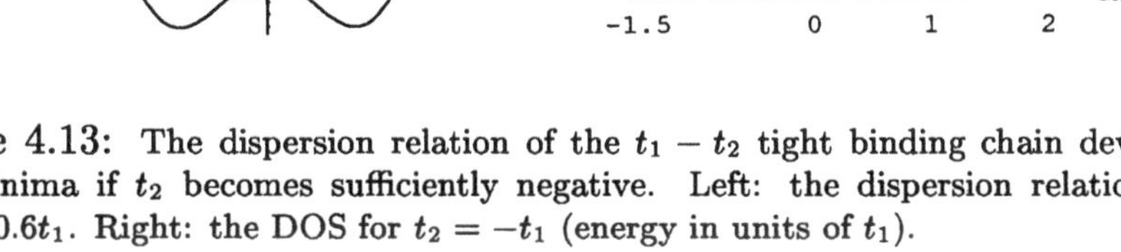

Figure 4.13: The dispersion relation of the $t_1 - t_2$ tight binding chain develops two minima if t_2 becomes sufficiently negative. Left: the dispersion relation for $t_2 = -0.6t_1$. Right: the DOS for $t_2 = -t_1$ (energy in units of t_1).

Though at the first sight, this may look merely curious, at the second, it should look alarming. There is a range of energies where the eigenvalues are fourfold degenerate! If we think of $\epsilon(k)$ as arising from solving the Schrödinger equation for a strictly one-dimensional system, this is not allowed to happen:

an ordinary differential equation of second order has only two independent solutions for any eigenvalue. Consequently, using $t_2 < -t_1/4$ in (4.79) cannot be looked upon as the parametrization of the band structure of a strictly one-dimensional system. However, the objection does not apply to the zig-zag geometry in Fig. 4.5. There is no obvious reason why the dispersion with two minima could not be realized in a quasi-one-dimensional arrangement of three-dimensional atoms.

The $t_2 = t_1$ case is shown in Fig. 4.13 (right). Setting $t_1 = 1/2$, the dispersion is $\epsilon(k) = -\cos k + \cos 2k$. The band has degenerate minima at $\pm \arccos(1/4)$, where $\epsilon = -9/8$, and a local minimum at $k = 0$ where $\epsilon = 0$. The DOS is readily expressed as

$$\rho(\epsilon) = \frac{2\sqrt{2}}{\pi} \frac{\Theta(\epsilon + 9/8)}{\sqrt{9+8\epsilon}} \left\{ \frac{\Theta(-\epsilon)}{\sqrt{3-4\epsilon-\sqrt{9+8\epsilon}}} + \frac{\Theta(2-\epsilon)}{\sqrt{3-4\epsilon+\sqrt{9+8\epsilon}}} \right\} . \tag{4.81}$$

We observe that, in addition to the customary band edge singularities, there is also a $\Theta(-\epsilon)/\sqrt{-\epsilon}$ singularity inside the band (Fig. 4.13, right).

Finally, it should be mentioned that the corresponding t_1–t_2–U Hubbard chain is of considerable interest as a simple model showing itinerant ferromagnetism (see p. 439).

Problem 4.4. The unit cell $\mathbf{j}$ contains three atoms (Fig. 4.5, right). Let the corresponding Wannier states are created by $d^\dagger_{\mathbf{j}\sigma}$, $p^\dagger_{\mathbf{j}1\sigma}$, and $p^\dagger_{\mathbf{j}2\sigma}$. The elementary lattice translation vectors are $\hat{\mathbf{x}}$ and $\hat{\mathbf{y}}$. The tight binding Hamiltonian can be written as

$$\begin{aligned} \mathcal{H} &= -t\sum_{\mathbf{j}\sigma} \left(d^\dagger_{\mathbf{j}\sigma} p_{\mathbf{j}1\sigma} + d^\dagger_{\mathbf{j}\sigma} p_{\mathbf{j}2\sigma} + d^\dagger_{\mathbf{j}+\hat{\mathbf{x}}\sigma} p_{\mathbf{j}1\sigma} + d^\dagger_{\mathbf{j}-\hat{\mathbf{y}}\sigma} p_{\mathbf{j}2\sigma} + \text{H.c.} \right) \\ &\quad + \sum_{\mathbf{j}\sigma} \left[\epsilon_d d^\dagger_{\mathbf{j}\sigma} d_{\mathbf{j}\sigma} + \epsilon_p (p^\dagger_{\mathbf{j}1\sigma} p_{\mathbf{j}1\sigma} + p^\dagger_{\mathbf{j}2\sigma} p_{\mathbf{j}2\sigma}) \right] \end{aligned} \tag{4.82}$$

Going over to Bloch representation, the matrix of the 3×3 eigenvalue problem for subspace $\mathbf{k} = (k_x, k_y)$ is

$$\begin{pmatrix} \epsilon_d & -t(e^{ik_x}+1) & -t(e^{ik_y}+1) \\ -t(e^{-ik_x}+1) & \epsilon_p & 0 \\ -t(e^{-ik_y}+1) & 0 & \epsilon_p \end{pmatrix} \tag{4.83}$$

One of the three bands is dispersionless

$$\lambda^0(\mathbf{k}) = \epsilon_p \tag{4.84}$$

while the other two bands can be thought to result from a $\mathbf{k}$-dependent hybridization of the d- and p-levels

$$\lambda^{\pm}(\mathbf{k}) = \frac{1}{2}\left\{\epsilon_p + \epsilon_d \pm \sqrt{(\epsilon_p - \epsilon_d)^2 + 16t^2\left(\cos^2\frac{k_x}{2} + \cos^2\frac{k_y}{2}\right)}\right\} . \tag{4.85}$$

Setting $\epsilon_p = \epsilon_d = 0$, the spectrum becomes symmetrical about the dispersionless band $\lambda^0(\mathbf{k}) = 0$. Half-filling the system with three electrons per unit cell, the ground state is obtained by filling the lower band $\lambda^-(\mathbf{k})$ completely, and half-filling the central band $\lambda^0(\mathbf{k})$. Because of the flatness of $\lambda^0(\mathbf{k})$, all spin configurations are degenerate. If the number of unit cells is L, filling half of the $2L/3$ spin-orbitals of the central band can be done in $C_{L/3}^{2L/3}$ ways: this is the ground state degeneracy. The total spin of the ground states ranges from 0 to the macroscopic $L/6$. We will exploit this feature in our discussion of Lieb's ferrimagnetism (see p. 426).

Problem 4.5. For a bipartite lattice, the transformation

$$c_{\mathbf{j}\downarrow}^{\dagger} \to \eta_{\mathbf{j}} c_{\mathbf{j}\downarrow} \qquad c_{\mathbf{j}\downarrow} \to \eta_{\mathbf{j}} c_{\mathbf{j}\downarrow}^{\dagger} \tag{4.86}$$

with the familiar η (4.46) maps the kinetic energy of the $\downarrow$-spin electrons into itself. Since the $\uparrow$-spin operators are not transformed, the total kinetic energy remains unchanged. On the other hand, for the interaction

$$\mathcal{H}_U = U\sum_{\mathbf{j}}\left(\hat{n}_{\mathbf{j}\uparrow} - \frac{1}{2}\right)\left(\hat{n}_{\mathbf{j}\downarrow} - \frac{1}{2}\right) \to -\mathcal{H}_U \tag{4.87}$$

i.e., the $U > 0$ Hubbard model is mapped onto a Hubbard model with attractive interaction $U < 0$. Studying the repulsive Hubbard model, we can make statements about the attractive Hubbard model, and vice versa. However, we have to make a little dictionary to translate the results. For instance, the particle of one model is, via $N_\uparrow + N_\downarrow \to N_\uparrow - N_\downarrow + L$, related to the magnetization of the other.

Combining (4.86) with the $SU(2)$ symmetry, we generate a further $SU(2)$ symmetry. Namely, our original $U > 0$ Hubbard model is related to a $U < 0$ Hubbard model which, however, possesses an $SU(2)$ (spin-rotational) symmetry of its own. Applying the transformation backwards, the $SU(2)$ generators of the $U < 0$ model get translated into the following isospin-rotation generators of the original model:

$$\mathcal{I}^z = \frac{1}{2}(N_\uparrow + N_\downarrow - L) \qquad \mathcal{I}^+ = \sum_{\mathbf{j}} \eta_{\mathbf{j}}\, c_{\mathbf{j}\uparrow}^{\dagger} c_{\mathbf{j}\downarrow}^{\dagger} \qquad \mathcal{I}^- = (\mathcal{I}^+)^{\dagger} . \tag{4.88}$$

These are also $SU(2)$ generators because they obey the standard commutation relations $[\mathcal{I}^z, \mathcal{I}^\pm] = \pm\mathcal{I}^\pm$, $[\mathcal{I}^+, \mathcal{I}^-] = 2\mathcal{I}^z$. Before drawing any conclusions, we have to remember that the electron–hole symmetrical form of the interaction which we introduced in (4.42) is not the full $\mathcal{H}_U$: there is also $UN/2$ which introduces splittings between states with different N. However, if we define our model with the grand-canonical potential $\mathcal{H} - \mu N$, and set $\mu = -U/2$, then the newly found isospin $SU(2)$ symmetry predicts the degeneracy of certain states whose particle numbers differ by $\pm$ 2 [469]. The physical significance of this finding is not entirely clear, but we mention that it enables us to construct eigenstates possessing off-diagonal long range order, and this may be looked upon as a hint of superconductivity [452].

We have learned that Hubbard models on bipartite lattices, and in the absence of a magnetic field, have two global $SU(2)$ symmetries, not merely one. However, the enlarged symmetry group is not $SU(2) \otimes SU(2)$ but rather $SO(4) = SU(2) \otimes SU(2)/Z_2$ [453], the essential reason being that the two nominally different 2π-rotations amount to the same element **-1** when acting on the Hilbert space.

Problem 4.6. Setting $a = 1$, the band structure is

$$\epsilon(\mathbf{k}) = -2\tilde{t}_1 \sum_{n=1}^{D} \cos k_n - 2\tilde{t}_2 \sum_{n=1}^{D} \cos 2k_n \tag{4.89}$$

where $\tilde{t}_1$ and $\tilde{t}_2$ are the unscaled hopping amplitudes. Using the definition (4.51), the Fourier transform of the DOS is

$$\Phi_D(s) = \left(\int_{-\pi}^{\pi} \frac{dk}{2\pi} e^{-2i\tilde{t}_1 s \cos k - 2i\tilde{t}_2 s \cos 2k} \right)^D . \tag{4.90}$$

We use the small-s expansion, like we did in the derivation of (4.54). Going to second order, we get

$$\Phi_D(s) = \exp\left\{-2Ds^2(\tilde{t}_1^2 + \tilde{t}_2^2)\right\}. \tag{4.91}$$

Thus $\tilde{t}_1$ and $\tilde{t}_2$ appear in equivalent ways, and the proper scaling is

$$t_1 = \frac{\tilde{t}_1}{\sqrt{2D}} \qquad t_2 = \frac{\tilde{t}_2}{\sqrt{2D}}. \tag{4.92}$$

The $D = \infty$ density of states is Gaussian like (4.56) [295]

$$\rho_\infty(\epsilon) = \frac{1}{\sqrt{2\pi(\tilde{t}_1^2 + \tilde{t}_2^2)}} \exp\left\{ -\frac{\epsilon^2}{2(\tilde{t}_1^2 + \tilde{t}_2^2)} \right\}. \tag{4.93}$$

An analogous result is obtained with farther-neighbour hoppings along the main axes. This generalization is interesting because it offers an example of an analytically simple DOS for a non-bipartite lattice: (4.89) does not show either electron–hole symmetry, or the perfect nesting property (7.49).

Problem 4.7. Denoting the unit vectors of the D-dimensional simple cubic lattice by $\mathbf{e}_i$ ($i = 1, \ldots, D$), the D-dimensional fcc lattice is defined in the following manner: the nearest neighbours of a site can be reached by the vectors $\pm\mathbf{e}_i \pm \mathbf{e}_j$, where $i \neq j$. The coordination number is $z = 2D(D-1)$.

Our solution follows [296] and [152]. Analogously to (4.73), we immediately find

$$\epsilon(\mathbf{k}) = \frac{1}{2t_1} \cdot \epsilon_{\rm sc}^2(\mathbf{k}) + (2t_2 - t_1)\epsilon_{\rm sc}(2\mathbf{k}) - Dt_1 \ , \tag{4.94}$$

where $\epsilon_{\rm sc}(\mathbf{k})$ is the DOS of the D-dimensional simple cubic lattice. The second term drops out if $t_2 = t_1/2$, and we find the generalization of (4.75)

$$\rho(\epsilon) = t_1 \cdot \frac{\rho_{\rm sc}\left(\sqrt{2t_1(\epsilon + Dt_1)}\right)}{\sqrt{2t_1(\epsilon + Dt_1)}} \ . \tag{4.95}$$

In the present case, the consistent scaling is $t_1 = \tilde{t}_1/\sqrt{z} \to \tilde{t}_1/D$. Recalling (4.56), we find that in the limit $D \to \infty$

$$\rho(\epsilon) = \frac{1}{\sqrt{\pi\tilde{t}_1(\sqrt{2}\epsilon + \tilde{t}_1)}} \cdot \exp\left(-\frac{\sqrt{2}\epsilon + \tilde{t}_1}{2\tilde{t}_1}\right) . \tag{4.96}$$

The DOS falls off in the characteristic Gaussian manner at high energies, but (like in the $D = 3$ case) it has an inverse square root divergency at the lower band edge $\epsilon = -\tilde{t}_1/\sqrt{2}$. The finite lower cut-off makes the physics of the $D = \infty$ fcc lattice substantially different from that of the infinite dimensional simple cubic lattice (see Sec. 8.4).

Though we used the assumption $t_2 = t_1/2$ in our reasoning, it is easy to see that the $D = \infty$ result (4.96) must be true for any t_1–t_2 model, as long as t_2/t_1 remains finite. Namely, the second-neighbour hopping involves only $2D$ neighbours, and thus its contribution vanishes if $t_2 \sim 1/D$ (remember that we would need $t_2 \sim 1/\sqrt{D}$ to keep this contribution finite).

Problem 4.8. Our derivation of the DOS of the infinite-dimensional diamond lattice closely follows [346]. Let us assume that for $D-1$, we found a set of D vectors $\mathbf{e}_1^{(D-1)}, \ldots, \mathbf{e}_D^{(D-1)}$ which point from a site to its nearest neighbours, satisfying $\mathbf{e}_i^{(D-1)} \cdot \mathbf{e}_j^{(D-1)} = \alpha_{D-1}$ for $1 \le i < j \le D-1$. In the next step, this allows us to construct the $D+1$ D-dimensional unit vectors

$$\mathbf{e}_j^{(D)} \quad = \quad \left(\sqrt{1-\alpha_D^2}\, \mathbf{e}_j^{(D-1)}, \alpha_D\right) \qquad \text{for } j\text{=1,...,}D$$

$$\mathbf{e}_{D+1}^{(D)} = (\underbrace{0,\ldots,0}_{D-1},1). \tag{4.97}$$

The property $\mathbf{e}_i^{(D)} \cdot \mathbf{e}_j^{(D)} = \alpha_D$ requires

$$\alpha_D = \frac{\alpha_{D-1}}{1-\alpha_{D-1}}. \tag{4.98}$$

It is easy to check that this equation is solved by

$$\alpha_D = -\frac{1}{D} \tag{4.99}$$

$\alpha_2 = -1/2$ corresponds to the honeycomb lattice and $\alpha_3 = -1/3$ to the familiar diamond lattice. Actually, it is permitted to start the series with the linear chain with its $\alpha_1 = -1$; then the construction (4.97) gives the honeycomb lattice for $D = 2$.

In what follows, we can let ourselves be led by the familiar results for the honeycomb lattice. For any D, the lattice is bipartite with a two-atom base. The basis vectors can be chosen as $\mathbf{a}_j = \mathbf{e}_{D+1} - \mathbf{e}_j$, for $j = 1,\ldots,D$. The reciprocal basis vectors $\mathbf{b}_j$ satisfy $\mathbf{a}_i \cdot \mathbf{b}_j = 2\pi\delta_{ij}$. The electronic structure can be understood as the result of the hybridization of two degenerate, dispersionless bands (one for each sublattice), with the $\mathbf{k}$-dependent hybridization amplitude $v_{\mathbf{k}} = -\tilde{t}\sum_{j=1}^{D+1} \exp(i\mathbf{k}\mathbf{e}_j) = -\tilde{t}(1 + \sum_{j=1}^{D} \exp(i2\pi k_j))$, where $\mathbf{k} = \sum_j k_j \mathbf{b}_j$. Since $v_{\mathbf{k}}$ vanishes at certain points in the Brillouin zone (e.g., at $\mathbf{k} = \mathbf{b}_1/2$), there is no proper gap between the two bands, only a V-shaped pseudogap. The dispersion relation is

$$\epsilon^{(\pm)}(\mathbf{k}) = \pm|v_{\mathbf{k}}| = \pm\tilde{t}\sqrt{X^2+Y^2} \tag{4.100}$$

where

$$X = 1 + \sum_{j=1}^{D} \cos(2\pi k_j) \quad \text{and} \quad Y = \sum_{j=1}^{D} \sin(2\pi k_j). \tag{4.101}$$

Both X and Y are sums like the sum of cosines we considered in Sec. 4.6, and thus in the limit $D \to \infty$ they follow the Gaussian distribution[33]. Introducing

$$t = \sqrt{\frac{2}{D}}\,\tilde{t} \tag{4.102}$$

[33]This looks fairly obvious about Y but superficially it may appear as if the center of distribution of X were 1, not 0. We have to remember that $1 + X = \Re\left(\sum_{j=1}^{D+1} \exp(i\mathbf{k}\mathbf{e}_j)\right)$ and that it is the vectors $\mathbf{e}_j$ which are symmetrically distributed around the origin. The basis $\mathbf{b}_j$ is kind of lopsided.

the DOS belonging to (4.100) can be written as

$$
\begin{aligned}
\rho(\epsilon) &= \frac{1}{2\pi t^2}\int dX \int dY \exp\left(-\frac{X^2+Y^2}{2t^2}\right) \\
&\cdot(\delta(\epsilon - t\sqrt{X^2+Y^2}) + \delta(\epsilon + t\sqrt{X^2+Y^2})) \\
&= \frac{1}{t^2}\int_0^\infty dR\, R\exp\left(-\frac{R^2}{2t^2}\right)(\delta(R-\epsilon)+\delta(R+\epsilon)) \\
&= \frac{|\epsilon|}{t^2}\exp\left(-\frac{\epsilon^2}{2t^2}\right). \qquad (4.103)
\end{aligned}
$$

The result is shown in Fig. 4.7 (right). At half filling (one electron per site, two electrons per unit cell) the non-interacting system is a zero-gap semiconductor.

Chapter 5

Mott Insulators

5.1 The Large-U Limit

We gain valuable insight into the nature of the correlated motion of electrons by studying the regime $U/t \gg 1$. It is in this limit that the broadening of the atomic levels into Hubbard subbands is taking place. Away from half-filling, the electrons manage to sustain a propagating motion even at $U \to \infty$. At $n = 1$, a Mott transition occurs at a critical $U_{\mathrm{cr}} \sim t$, so the large-U regime is well inside the insulating phase. Still, $\mathcal{H}_{\mathrm{band}}$ has an important effect: it gives rise to antiparallel nearest-neighbour spin correlations.

Actually, we learned on the example of the hydrogen molecule (see Sec. 2.5) that this situation can be dealt with by introducing an effective spin Hamiltonian; the approximate solution of the underlying electron problem gives the exchange coupling constant. As far as the case $n = 1$ is concerned, we will do something quite similar here: we derive an antiferromagnetic Heisenberg model to describe the low-energy physics of Mott insulators. Though we could try to guess the form of the spin Hamiltonian by simple reasoning (as it will be done in Problem 5.1), it is better to use this opportunity to learn a systematic treatment. We seek the canonical transformation which rotates to the Hilbert space basis in which the separation into low-energy and high-energy subspaces becomes clear-cut.

As it turns out, the canonical transformation works for arbitrary filling. For $n < 1$, it leads to the $t-J$ model, describing the propagation of holes on the background of interacting spins.

The effective Hamiltonian method has a long history [157, 64, 137]. Our treatment is partly based on the short review [88].

5.1.1 Classification of Hopping Events

We have discussed that, strictly in the atomic limit ($t = 0$), the energy levels have a huge degeneracy. For instance, for $N < L$, there are $C_N^L 2^N$ ground states. When a small hopping term $t \ll U$ is switched on, these states are mixed and the sharp atomic level broadens into the lower Hubbard subband. The motion of the electrons is constrained by having to avoid the creation of doubly occupied sites. Now we wish to describe this in detail.

Here we are faced with the basic difficulty of the theory of strongly correlated systems. At the first sight, the solution may seem straightforward since there is a small parameter $t/U \ll 1$, and one could attempt to treat the effect of $\mathcal{H}_{\text{band}}$ by perturbation theory. However, the task at hand is very different from that of standard perturbation theory which is set up to handle problems with a large single-electron term and a weak interaction. In our case, the zeroth order term is the interaction term $\mathcal{H}_U$, and the one-electron term $\mathcal{H}_{\text{band}}$ is the perturbation. In the standard case, the zeroth-order ground state is nondegenerate; here, the ground state of $\mathcal{H}_U$ has a large degeneracy. Within a degenerate set of levels, even a weak perturbation has a drastic effect: it lifts the degeneracy.

There is, however, a well-known way to treat consecutive orders of t/U systematically: it can be accomplished by a suitable *canonical transformation.* The idea is something like this: our zeroth-order eigenstates are mixed by the perturbation $\mathcal{H}_{\text{band}}$. If we knew the true eigenstates, we could do a Hilbert-space rotation to that basis, and should not worry about the mixing of states any more. This full solution is usually not available, and therefore we proceed iteratively. In the first step, we rotate to a basis whose states are not mixed in order t. In the second step, we want to avoid mixing in order t^2, etc. While this is the general principle, we will be, in fact, content with the lowest-order result.

Let us consider how a hopping process changes the local configuration. Starting from a neutral state, the process we have seen in (4.2)

(5.1)

creates a doubly occupied site, and thus leads from a low-energy situation to a high-energy state. We can think of it as a transition from the lower Hubbard subband to the upper one. If we want to single out this process from all the possible hopping processes from **i** to **j**, we have to consider (see (4.22)–(4.25))

$$\hat{P}_{\mathbf{i}0}\hat{P}_{\mathbf{j}d}c^{\dagger}_{\mathbf{j}\uparrow}c_{\mathbf{i}\uparrow}\hat{P}_{\mathbf{i}\uparrow}\hat{P}_{\mathbf{j}\downarrow}\,. \tag{5.2}$$

This is an example of a *projected hopping term.* The central theme of the treatment of strong correlations is that we want to be able to tell whether a change in the state of the system has led to an increase, a decrease, or no change, in the interaction energy. For this, it is not enough to say: "an ↑-spin electron jumped from **i** to **j**". What is required is to specify the local configuration of the sites whose state can have changed, both before and after the hopping. Using projected hopping operators like the one above, is the most straightforward way to achieve this.

Actually, we were overcautious in fully specifying the state of both sites both before, and after, the hopping. The same hopping event is obtained if we require only that site **i** is in the ↑-state before, and site **j** is in the d-state after, the hopping:

$$\hat{P}_{\mathbf{j}d}c^{\dagger}_{\mathbf{j}\uparrow}c_{\mathbf{i}\uparrow}\hat{P}_{\mathbf{i}\uparrow} = \hat{n}_{\mathbf{j}\uparrow}\hat{n}_{\mathbf{j}\downarrow}c^{\dagger}_{\mathbf{j}\uparrow}c_{\mathbf{i}\uparrow}\hat{n}_{\mathbf{i}\uparrow}(1-\hat{n}_{\mathbf{i}\downarrow}) = \hat{n}_{\mathbf{j}\downarrow}c^{\dagger}_{\mathbf{j}\uparrow}c_{\mathbf{i}\uparrow}(1-\hat{n}_{\mathbf{i}\downarrow}). \tag{5.3}$$

Collecting all the similar contributions, the Hamilton operator of hopping processes which *increase* the number of doubly occupied sites by one, can be written as

$$\mathcal{H}_t^{+} = -t\sum_{\langle\mathbf{i},\mathbf{j}\rangle}\sum_{\sigma}[\hat{n}_{\mathbf{i}-\sigma}c^{\dagger}_{\mathbf{i}\sigma}c_{\mathbf{j}\sigma}(1-\hat{n}_{\mathbf{j}-\sigma}) + \hat{n}_{\mathbf{j}-\sigma}c^{\dagger}_{\mathbf{j}\sigma}c_{\mathbf{i}\sigma}(1-\hat{n}_{\mathbf{i}-\sigma})]. \tag{5.4}$$

Similarly,

$$\mathcal{H}_t^- = -t \sum_{\langle \mathbf{i},\mathbf{j} \rangle} \sum_{\sigma} [(1 - \hat{n}_{\mathbf{i}-\sigma}) c_{\mathbf{i}\sigma}^{\dagger} c_{\mathbf{j}\sigma} \hat{n}_{\mathbf{j}\sigma} + (1 - \hat{n}_{\mathbf{j}-\sigma}) c_{\mathbf{j}\sigma}^{\dagger} c_{\mathbf{i}\sigma} \hat{n}_{\mathbf{i}\sigma}] \tag{5.5}$$

describes hopping events in which the number of doubly occupied sites is *decreased* by one. An example is shown below

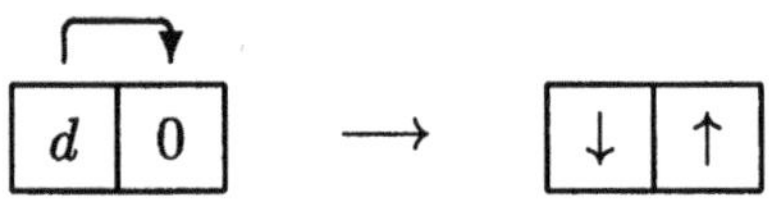

(5.6)

Finally, there are processes which do not change the number of doubly occupied sites. First, it is possible that there is no double occupation to begin with, and it is also not created in the hopping process, as in (4.60). In this process, the electron remains in the lower Hubbard subband. Another possibility is that it starts out, and remains, in the upper subband

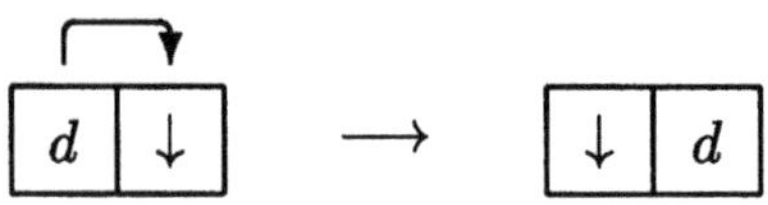

(5.7)

The processes (4.60) and (5.7) are summarized in

$$\mathcal{H}_t^0 = -t \sum_{\langle \mathbf{i},\mathbf{j} \rangle} \sum_{\sigma} [(1 - \hat{n}_{\mathbf{i}-\sigma}) c_{\mathbf{i}\sigma}^{\dagger} c_{\mathbf{j}\sigma} (1 - \hat{n}_{\mathbf{j}-\sigma}) + \hat{n}_{\mathbf{i}-\sigma} c_{\mathbf{i}\sigma}^{\dagger} c_{\mathbf{j}\sigma} \hat{n}_{\mathbf{j}-\sigma} + \text{H.c.}]. \tag{5.8}$$

We have enumerated all possible types of hopping events and thereby decomposed the kinetic energy into three terms

$$\mathcal{H}_{\text{band}} = \mathcal{H}_t^+ + \mathcal{H}_t^- + \mathcal{H}_t^0 . \tag{5.9}$$

A comparison of (4.13) with (5.9) shows the kind of difficulties arising from strong Coulomb interaction. If we do not have to pay attention to local correlations, we can simply say things like "an ↑-electron jumps from $\mathbf{i}$ to $\mathbf{j}$ ", meaning that this is taking place in an average environment and we do not have to follow what the other electrons are doing in the meantime. However, with a strong Hubbard U, this is no longer permissible: depending on the configuration of the sites, the hopping may be accompanied with a substantial change of energy, or it may be Coulomb energy preserving. Some of these processes will be energetically preferable, others will be suppressed. To keep track of this, we have to distinguish between hopping processes according to the local environment in which they are taking place. Thereby $\mathcal{H}_{\rm band}$, the seemingly uncomplicated one-electron part of the Hamiltonian, acquires a complicated many-body character.

5.1.2 The Canonical Transformation

We have the intuitive notion that low-energy excitations are propagating in the lower Hubbard subband. The partitioning of the hopping Hamiltonian (5.9) shows that the pure band motion of the electrons does not correspond to this picture: it mixes states from the two subbands via the terms $\mathcal{H}_t^+$ and $\mathcal{H}_t^-$. The "unmixing" of the Hubbard subbands can be achieved by introducing suitable linear combinations of the uncorrelated basis states. Rotating (in the many-electron Hilbert space) to such a new basis can be performed by a suitable canonical transformation. In the new basis, the Hubbard Hamiltonian

$$\mathcal{H} = \mathcal{H}_t^+ + \mathcal{H}_t^- + \mathcal{H}_t^0 + \mathcal{H}_U \tag{5.10}$$

acts like the *effective Hamiltonian*

$$\begin{aligned}
\mathcal{H}_{\rm eff} &= e^{iS}\mathcal{H}e^{-iS} = \mathcal{H} + i[S,\mathcal{H}] + \frac{i^2}{2}[S,[S,\mathcal{H}]] + \ldots \\
&= \mathcal{H}_U + \mathcal{H}_t^+ + \mathcal{H}_t^- + \mathcal{H}_t^0 + i[S,\mathcal{H}_U] \\
&\quad + i[S,\mathcal{H}_t^+ + \mathcal{H}_t^- + \mathcal{H}_t^0] + \frac{i^2}{2}[S,[S,\mathcal{H}]] + \ldots
\end{aligned} \tag{5.11}$$

We want to choose the generator S in such a way that $\mathcal{H}_{\rm eff}$ does not connect different subbands. Actually, this cannot be achieved fully in a finite number of steps.

First, we will be content with eliminating the largest cross-terms $\mathcal{H}_t^+$ and $\mathcal{H}_t^-$ which are of the order t. Since the largest term of $i[S,\mathcal{H}]$ is obviously $i[S,\mathcal{H}_U]$, we must require that this cancels $\mathcal{H}_t^+ + \mathcal{H}_t^-$. It follows that S must be of the order of t/U. To learn its form, try commuting a piece of $\mathcal{H}_t^+$

$$[\hat{n}_{\mathrm{i}\downarrow}c_{\mathrm{i}\uparrow}^\dagger c_{\mathrm{j}\uparrow}(1-\hat{n}_{\mathrm{j}\downarrow}),\, \hat{n}_{\mathrm{i}\uparrow}\hat{n}_{\mathrm{i}\downarrow}] = -\hat{n}_{\mathrm{i}\downarrow}c_{\mathrm{i}\uparrow}^\dagger c_{\mathrm{j}\uparrow}(1-\hat{n}_{\mathrm{j}\downarrow}) \tag{5.12}$$

from which you correctly guess that

$$[\mathcal{H}_t^+, \mathcal{H}_U] = -U\mathcal{H}_t^+ \,. \tag{5.13}$$

Similarly

$$[\mathcal{H}_t^-, \mathcal{H}_U] = U\mathcal{H}_t^- \,. \tag{5.14}$$

It follows that we should use

$$S = -\frac{i}{U}(\mathcal{H}_t^+ - \mathcal{H}_t^-) \tag{5.15}$$

because

$$i[S, \mathcal{H}_U] = \frac{1}{U}[\mathcal{H}_t^+ - \mathcal{H}_t^-, \mathcal{H}_U] = -(\mathcal{H}_t^+ + \mathcal{H}_t^-) \tag{5.16}$$

cancels the unwanted pieces of $\mathcal{H}_{\mathrm{band}}$. However, we also generated new terms from

$$i[S, \mathcal{H}_t^+ + \mathcal{H}_t^-] = \frac{1}{U}[\mathcal{H}_t^+ - \mathcal{H}_t^-, \mathcal{H}_t^+ + \mathcal{H}_t^-] = \frac{2}{U}[\mathcal{H}_t^+, \mathcal{H}_t^-] \tag{5.17}$$

which is of the order of t^2/U. There is also $i[S, \mathcal{H}_t^0]$ to consider, which we defer till the end of the discussion. And finally, we should not forget the third term on the right-hand side of (5.11) which provides a contribution of the same order of magnitude

$$\frac{i^2}{2}[S,[S,\mathcal{H}_U]] = -\frac{i}{2}[S, \mathcal{H}_t^+ + \mathcal{H}_t^-] = -\frac{1}{U}[\mathcal{H}_t^+, \mathcal{H}_t^-] \,. \tag{5.18}$$

Combining the above two equations, and omitting, with later justification, any contributions that may be coming from $i[S, \mathcal{H}_t^0]$, the effective Hamiltonian (to order t^2/U) is found to be

$$\mathcal{H}_{\mathrm{eff}} = \mathcal{H}_t^0 + \mathcal{H}_U + \frac{1}{U}[\mathcal{H}_t^+, \mathcal{H}_t^-] \,. \tag{5.19}$$

We could, of course, evaluate $[\mathcal{H}_t^+, \mathcal{H}_t^-]$ as it stands, by replacing the expressions (5.4) and (5.5). However, we find it more educative to do this calculation using the so-called Hubbard operators.

5.1.3 Hubbard Operators

To treat correlations properly, one has to be able to say whether at the beginning or at the end of a process, a particular site is doubly occupied, or not. The introduction of the so-called Hubbard operators provides us with the tools necessary for such a full description of the local environment. Any process affecting the state of the site $\mathbf{j}$ can be described as a transition between the local basis states listed in equations (4.18)–(4.21)

$$X_{\mathbf{j}}^{b\leftarrow a} = |b\rangle_{\mathbf{j}}\ {}_{\mathbf{j}}\langle a| \,. \tag{5.20}$$

The projection operators introduced in (4.22)–(4.25) are, if you like, "diagonal" Hubbard operators: $\hat{P}_a = |a\rangle\langle a| = X^{a\leftarrow a}$. The off-diagonal X-operators are projected creation/annihilation operators. For instance,

$$X_{\mathbf{j}}^{\sigma\leftarrow 0} = c_{\mathbf{j}\sigma}^{\dagger}(1 - \hat{n}_{\mathbf{j}-\sigma})\,. \tag{5.21}$$

Products of Hubbard operators behave like

$$X_{\mathbf{j}}^{c\leftarrow f} X_{\mathbf{j}}^{b\leftarrow a} = \delta_{bf} X_{\mathbf{j}}^{c\leftarrow b} X_{\mathbf{j}}^{b\leftarrow a} = \delta_{bf} X_{\mathbf{j}}^{c\leftarrow a}\,. \tag{5.22}$$

The Hubbard operators provide an alternative to the standard description which works with the ordinary creation and annihilation operators. Naturally, the same ends can be reached by either means; the choice is a matter of convenience. To understand the relationship between the two descriptions, let us consider $c_{\mathbf{j}\uparrow}^{\dagger}$. $c_{\mathbf{j}\uparrow}^{\dagger}$ either turns an empty site into $|\uparrow\rangle$, or a $\downarrow$-site into $|d\rangle$. Thus we can write

$$c_{\mathbf{j}\uparrow}^{\dagger} = X_{\mathbf{j}}^{\uparrow\leftarrow 0} + X_{\mathbf{j}}^{d\leftarrow\downarrow}\,. \tag{5.23}$$

One might think that $c_{\mathbf{j}\downarrow}^{\dagger}$ is given by the same expression if we replace $\uparrow$ with $\downarrow$ but in fact we find a sign difference in the second term

$$c_{\mathbf{j}\downarrow}^{\dagger} = X_{\mathbf{j}}^{\downarrow\leftarrow 0} - X_{\mathbf{j}}^{d\leftarrow\uparrow}\,. \tag{5.24}$$

The difference arises because $|d\rangle$ was defined as $|d\rangle = c_{\mathbf{j}\uparrow}^{\dagger} c_{\mathbf{j}\downarrow}^{\dagger}|0\rangle$. Acting with $c_{\mathbf{j}\downarrow}^{\dagger}$ on $|\uparrow\rangle_{\mathbf{j}}$ implies that $c_{\mathbf{j}\downarrow}^{\dagger}$ acts after $c_{\mathbf{j}\uparrow}^{\dagger}$, which would give $-|d\rangle$ if it were not compensated by the extra negative sign.

(5.23) and (5.24) can be summarized in

$$c^\dagger_{\mathrm{j}\sigma} = X_{\mathrm{j}}^{\sigma\leftarrow 0} + \eta(\sigma) X_{\mathrm{j}}^{d\leftarrow -\sigma} . \tag{5.25}$$

where we introduced the sign factor

$$\eta(\sigma) = \begin{cases} +1 & \text{if } \sigma = \uparrow \\ -1 & \text{if } \sigma = \downarrow . \end{cases} \tag{5.26}$$

Constraints imposed by strong local correlations can be expressed in a compact form if we use the Hubbard operators. The price we have to pay for this convenience is that the Hubbard operators do not obey the canonical commutation rules but instead follow

$$[X_{\mathrm{i}}^{b\leftarrow a}, X_{\mathrm{j}}^{d\leftarrow c}]_- = \delta_{\mathrm{i,j}}(\delta_{ad} X_{\mathrm{j}}^{b\leftarrow c} - \delta_{bc} X_{\mathrm{j}}^{d\leftarrow a}). \tag{5.27}$$

A hopping process changes the local state at two sites, thus it can be represented by a pair of Hubbard operators. For instance,

$$\mathcal{H}_t^0 = -t \sum_{\langle \mathrm{i,j} \rangle} \sum_{\sigma} [X_{\mathrm{i}}^{\sigma\leftarrow 0} X_{\mathrm{j}}^{0\leftarrow\sigma} + X_{\mathrm{i}}^{d\leftarrow -\sigma} X_{\mathrm{j}}^{-\sigma\leftarrow d} + \text{H.c.}] \tag{5.28}$$

where the first term corresponds to processes like (4.60), and the second to processes like (5.7).

To express $\mathcal{H}_t^+$ in terms of Hubbard operators, let us use (5.25) and $(X_{\mathrm{i}}^{b\leftarrow a})^\dagger = X_{\mathrm{i}}^{a\leftarrow b}$ to get

$$\begin{aligned} &\hat{n}_{\mathrm{i}-\sigma} c^\dagger_{\mathrm{i}\sigma} c_{\mathrm{j}\sigma} (1 - \hat{n}_{\mathrm{j}-\sigma}) \\ \rightsquigarrow\ & (X_{\mathrm{i}}^{\sigma\leftarrow 0} + \eta(\sigma) X_{\mathrm{i}}^{d\leftarrow -\sigma})(X_{\mathrm{j}}^{0\leftarrow\sigma} + \eta(\sigma) X_{\mathrm{j}}^{-\sigma\leftarrow d}) X_{\mathrm{j}}^{\sigma\leftarrow\sigma} X_{\mathrm{i}}^{-\sigma\leftarrow -\sigma} \\ =\ & \eta(\sigma) X_{\mathrm{i}}^{d\leftarrow -\sigma} X_{\mathrm{j}}^{0\leftarrow\sigma} . \end{aligned} \tag{5.29}$$

Hence

$$\mathcal{H}_t^+ = -t \sum_{\langle \mathrm{i,j} \rangle} \sum_{\sigma} \eta(\sigma) [X_{\mathrm{i}}^{d\leftarrow -\sigma} X_{\mathrm{j}}^{0\leftarrow\sigma} + X_{\mathrm{j}}^{d\leftarrow -\sigma} X_{\mathrm{i}}^{0\leftarrow\sigma}], \tag{5.30}$$

and similarly

$$\mathcal{H}_t^- = -t \sum_{\langle \mathrm{i,j} \rangle} \sum_{\sigma} \eta(\sigma) [X_{\mathrm{i}}^{\sigma\leftarrow 0} X_{\mathrm{j}}^{-\sigma\leftarrow d} + X_{\mathrm{j}}^{\sigma\leftarrow 0} X_{\mathrm{i}}^{-\sigma\leftarrow d}]. \tag{5.31}$$

5.1.4 The $t - J$ Model

Let us consider the action of $\mathcal{H}_{\text{eff}}$ on the low-energy subspace of neutral configurations. The full $\mathcal{H}_{\text{eff}}$ is given in (5.19). In the lower Hubbard subband, there is no contribution from $\mathcal{H}_U$, and from $\mathcal{H}_t^0$, only the part

$$\mathcal{H}_t^0 \rightsquigarrow -t \sum_{\langle \mathbf{i},\mathbf{j} \rangle} \sum_{\sigma} [X_{\mathbf{i}}^{\sigma \leftarrow 0} X_{\mathbf{j}}^{0 \leftarrow \sigma} + X_{\mathbf{j}}^{\sigma \leftarrow 0} X_{\mathbf{i}}^{0 \leftarrow \sigma}]$$

will be active.

Evaluating $\mathcal{H}_t^+ \mathcal{H}_t^- - \mathcal{H}_t^- \mathcal{H}_t^+$, let us first recall that $\mathcal{H}_t^+$ and $\mathcal{H}_t^-$ are sums of pair operators. Since it is clear that operations at disjoint pairs commute, non-vanishing contributions arise only if the pairs have one or two sites in common:

$$[\mathcal{H}_t^+, \mathcal{H}_t^-] = \sum_{\langle \mathbf{i},\mathbf{j} \rangle} \sum_{\langle \mathbf{k},\mathbf{l} \rangle} [\mathcal{H}_{t,\mathbf{ij}}^+, \mathcal{H}_{t,\mathbf{kl}}^-] = \sum_{\langle \mathbf{i},\mathbf{j} \rangle} [\mathcal{H}_{t,\mathbf{ij}}^+, \mathcal{H}_{t,\mathbf{ij}}^-] + \sum_{\langle \mathbf{i},\mathbf{j},\mathbf{k} \rangle} [\mathcal{H}_{t,\mathbf{ij}}^+, \mathcal{H}_{t,\,\mathbf{jk}}^-] \tag{5.32}$$

where the sum over $\langle \mathbf{i}, \mathbf{j}, \mathbf{k} \rangle$ is over trimers so that $\mathbf{j}$ is a nearest neighbour of both $\mathbf{i}$ and $\mathbf{k}$.

Let us first deal with the pair contribution. From the commutator $\mathcal{H}_{t,\mathbf{ij}}^+ \mathcal{H}_{t,\mathbf{ij}}^- - \mathcal{H}_{t,\mathbf{ij}}^- \mathcal{H}_{t,\mathbf{ij}}^+$, the first term drops out because the number of d sites is already zero, so it cannot be decreased any further. As for the second term, we consider first the contribution arising from an electron with spin σ jumping from site $\mathbf{j}$ to $\mathbf{i}$, creating a d-site there; then $\mathcal{H}_t^-$ makes d disappear again by letting an electron with spin σ' jump from $\mathbf{i}$ to $\mathbf{j}$:

$$\begin{aligned}
\frac{1}{U}[\mathcal{H}_{t,\mathbf{ij}}^+, \mathcal{H}_{t,\mathbf{ij}}^-] &\rightsquigarrow -\frac{1}{U} \mathcal{H}_{t,\mathbf{ij}}^- \mathcal{H}_{t,\mathbf{ij}}^+ \\
&\rightsquigarrow -\frac{t^2}{U} \sum_{\sigma} \sum_{\sigma'} \eta(\sigma)\eta(\sigma') X_{\mathbf{j}}^{\sigma' \leftarrow 0} \underbrace{X_{\mathbf{i}}^{-\sigma' \leftarrow d} X_{\mathbf{i}}^{d \leftarrow -\sigma}}_{X_{\mathbf{i}}^{-\sigma' \leftarrow -\sigma}} X_{\mathbf{j}}^{0 \leftarrow \sigma} \\
&= -\frac{t^2}{U} \sum_{\sigma} \sum_{\sigma'} \eta(\sigma)\eta(\sigma') X_{\mathbf{i}}^{-\sigma' \leftarrow -\sigma} X_{\mathbf{j}}^{\sigma' \leftarrow \sigma}.
\end{aligned} \tag{5.33}$$

Let us first observe that the above operator acts only if in the initial state, both $\mathbf{i}$ and $\mathbf{j}$ are occupied by a single electron, and the two spins are antiparallel; in any other case, the contribution is zero.

Let us do the sum over σ'. If $\sigma' = \sigma$, the original spin arrangement will be reproduced; if $\sigma' = -\sigma$, the spins will be interchanged. (5.33) can be continued as

$$\rightsquigarrow -\frac{t^2}{U}\sum_{\sigma} X_{\mathbf{i}}^{-\sigma\leftarrow-\sigma} X_{\mathbf{j}}^{\sigma\leftarrow\sigma} + \frac{t^2}{U}\sum_{\sigma} X_{\mathbf{i}}^{\sigma\leftarrow-\sigma} X_{\mathbf{j}}^{-\sigma\leftarrow\sigma}. \tag{5.34}$$

The second term above is obviously nothing but $(t^2/U)(S_{\mathbf{i}}^+ S_{\mathbf{j}}^- + S_{\mathbf{j}}^+ S_{\mathbf{i}}^-)$: it flips nearest neighbour spins if they are antiparallel, and gives no contribution if they are parallel. The first term is diagonal so it has the character of $S_{\mathbf{i}}^z S_{\mathbf{j}}^z$. Since the contribution has to be $-t^2/U$ for antiparallel spins, and zero for parallel spins (or if at least one of the sites is empty), we can identify it as

$$\frac{2t^2}{U}\cdot\left(S_{\mathbf{i}}^z S_{\mathbf{j}}^z - \frac{\hat{n}_{\mathbf{i}}\hat{n}_{\mathbf{j}}}{4}\right).$$

All in all, we can rewrite (5.33) as

$$\frac{1}{U}[\mathcal{H}_{t,\mathbf{ij}}^+, \mathcal{H}_{t,\mathbf{ij}}^-] \rightsquigarrow \frac{2t^2}{U}\left(\mathbf{S}_{\mathbf{i}}\cdot\mathbf{S}_{\mathbf{j}} - \frac{\hat{n}_{\mathbf{i}}\hat{n}_{\mathbf{j}}}{4}\right). \tag{5.35}$$

We have to recall that in (5.33), we started to collect the contributions from processes in which the first jump is from $\mathbf{j}$ to $\mathbf{i}$. The same contribution arises if the first jump is from $\mathbf{i}$ to $\mathbf{j}$. Hence in the low-energy subspace $\mathcal{H}_{\text{eff}}$ can be expressed as

$$\begin{aligned}\mathcal{H}_{\text{eff}} \rightsquigarrow \mathcal{H}_{tJ} &= -t\sum_{\langle\mathbf{i},\mathbf{j}\rangle}\sum_{\sigma}[(1-\hat{n}_{\mathbf{i}-\sigma})c_{\mathbf{i}\sigma}^\dagger c_{\mathbf{j}\sigma}(1-\hat{n}_{\mathbf{j}-\sigma}) + \text{H.c.}] \\ &\quad + \frac{4t^2}{U}\sum_{\langle\mathbf{i},\mathbf{j}\rangle}\left[\mathbf{S}_{\mathbf{i}}\cdot\mathbf{S}_{\mathbf{j}} - \frac{\hat{n}_{\mathbf{i}}\hat{n}_{\mathbf{j}}}{4}\right] \\ &\quad + \text{three-site terms.}\end{aligned} \tag{5.36}$$

This is the so-called $t-J$ *model.* (Actually, it is a matter of taste whether one understands the three-site terms as part of the model, or keeps only the pair terms). It is a three-state model: a lattice site can be in the states $|\uparrow\rangle$, $|\downarrow\rangle$, or $|0\rangle$. The first term describes hopping constrained to the subspace with no doubly occupied sites: it is essentially the empty sites (holes) that can hop. Nearest-neighbour spins interact antiferromagnetically, with the coupling constant $J = 4t^2/U$. At exact half-filling, the $t-J$ model reduces to the antiferromagnetic Heisenberg model. (See p. 595 for an alternative derivation of the latter result).

Let us return to the case of general filling, and discuss the structure of the three-site terms. The **ijk** trimer of sites can be taken either in a collinear, or a non-collinear shape, with **j** as the common endpoint of the edges **ij** and **jk**. If the starting configuration is $0 \downarrow\uparrow$, acting with $[\mathcal{H}^{+}_{t,\mathbf{jk}}, \mathcal{H}^{-}_{t,\mathbf{ij}}] \rightsquigarrow -\mathcal{H}^{-}_{t,\mathbf{ij}}\mathcal{H}^{+}_{t,\mathbf{jk}}$ can lead to two different final states:

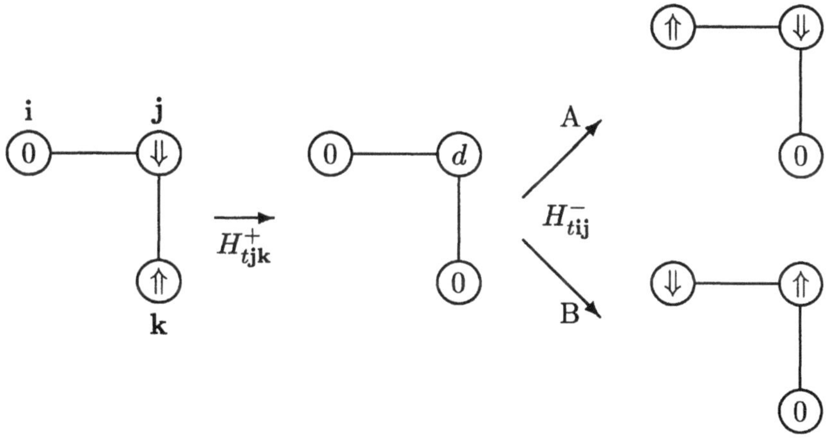

Process A amounts to a second-neighbour hopping via an intermediate site which is singly occupied by an electron whose spin is antiparallel to that of the hopping electron. The corresponding term in the Hamiltonian is

$$\mathcal{H}^{\mathrm{A}}_{3-\mathrm{site}} = -\frac{t^2}{U}\sum_{\mathbf{ijk}}\sum_{\sigma}(\tilde{c}^{\dagger}_{\mathbf{i}\sigma}\hat{n}_{\mathbf{j},-\sigma}\tilde{c}_{\mathbf{k}\sigma} + \mathrm{H.c.}) \qquad (5.37)$$

where we introduced the customary notation for the projected fermion operators

$$\tilde{c}_{\mathbf{k}\sigma} = c_{\mathbf{k}\sigma}(1 - \hat{n}_{\mathbf{k},-\sigma}) \qquad (5.38)$$

etc.

Process B can be described as a second-neighbour spin-flip hopping process, accompanied by flipping the spin at the intermediate site:

$$\mathcal{H}^{\mathrm{B}}_{3-\mathrm{site}} = \frac{t^2}{U}\sum_{\mathbf{ijk}}\sum_{\sigma}(\tilde{c}^\dagger_{\mathbf{i},-\sigma}\tilde{c}^\dagger_{\mathbf{j},-\sigma}\tilde{c}_{\mathbf{j}\sigma}\tilde{c}_{\mathbf{k}\sigma} + \mathrm{H.c.})\,. \tag{5.39}$$

Finally, let us discuss the contributions arising from $i[S,\mathcal{H}_t^0]$, which are of order t^2/U. It is immediately clear that the pair-term vanishes and that $[\mathcal{H}^-_{t,\mathbf{ij}},\mathcal{H}^0_{t,\mathbf{jk}}]$ cannot contribute at all. As for the remaining three-site contributions, let us take the example of the initial configuration $\Downarrow\Downarrow\Uparrow$, on which $[\mathcal{H}^+_{t,\mathbf{ij}},\mathcal{H}^0_{t,\mathbf{jk}}] \rightsquigarrow \mathcal{H}^0_{t,\mathbf{ij}}\mathcal{H}^+_{t,\mathbf{jk}}$ can act:

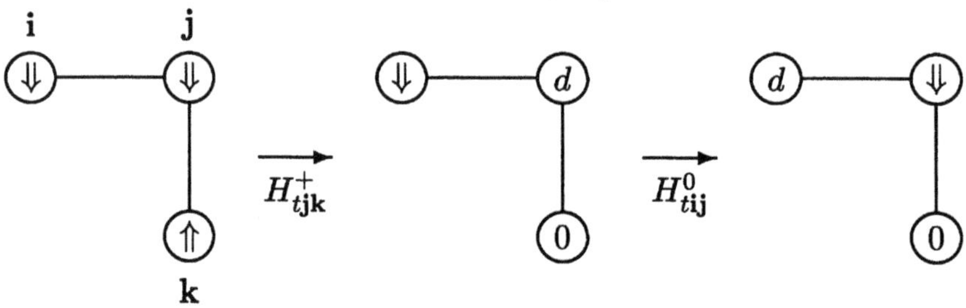

This three-site process which is active even at $n = 1$, produces an $0d$ pair at next-to-nearest neighbour distance. Other three-site processes need an intermediate empty site to achieve the same. Thus we learned that the canonical transformation eliminated $\mathcal{H}_t^+ + \mathcal{H}_t^-$ from the original Hamiltonian at the price of generating new mixing terms which still connect the low-energy subspace (lower Hubbard subband) to the high-energy states. However, these mixing terms are of order t^2/U, i.e., in the sense of a t/U-expansion one order smaller than the terms eliminated. Even these higher-order mixing terms can be got rid of, if we are taking the trouble to go one step further in the expansion of S:

$$S = -\frac{i}{U}(\mathcal{H}_t^+ - \mathcal{H}_t^-) + S^{(2)} \tag{5.40}$$

where $S^{(2)}$ is of the order $(t/U)^2$. We need not specify the form of $S^{(2)}$; it is enough to know that the operator equation

$$i[S^{(2)},\mathcal{H}_U] = -\frac{1}{U}[\mathcal{H}_t^+ - \mathcal{H}_t^-,\mathcal{H}_t^0] \tag{5.41}$$

does have a solution [157]. Then, to order t^2/U, the effective Hamiltonian is the $t - J$ model given in (5.36), including the three-site terms (5.37) and (5.39).

5.1.5 Half-Filled Band: The Heisenberg Model

At exact half-filling, the large-U effective Hamiltonian is the well-known antiferromagnetic Heisenberg model

$$\mathcal{H} = J \sum_{\langle \mathrm{i},\mathrm{j}\rangle} \mathbf{S}_\mathrm{i} \cdot \mathbf{S}_\mathrm{j} \tag{5.42}$$

with $J = 4t^2/U$. It is the $n = 1$ case of the $t - J$ model[1]. J arises from the same kind of *kinetic exchange* process that we learned about in connection with the H_2 problem (see (2.136) in Sec. 2.5.2). The reason for the antiferromagnetic nature of the coupling is the following[2]: if neighbouring spins are antiparallel, a virtual hopping process can create an intermediate $0d$ pair state with energy U; the associated energy gain is (from second-order perturbation theory) $\sim -t^2/U$. If the spins are parallel, the hopping is prohibited by the Pauli principle.

The effective Hamiltonian (5.42) acts on pure spin states. The original four-state Hubbard model of interacting fermions has been replaced by a two-state model of interacting spins. This has been achieved by the canonical transformation (5.11), i.e., by adopting a new basis in the Hilbert space. We have to remember that in terms of our original fermions, this does not mean that charge fluctuations are completely suppressed. On the contrary, it constitutes a description of the effects of the virtual admixture of doubly occupied sites.

To see how this works, let us determine the density of the doubly occupied sites in the large-U limit! We do this by performing the canonical transformation from the ground state of the Hubbard model $|\Psi\rangle$ to the ground state of the effective Hamiltonian $|\tilde{\Psi}\rangle = e^{iS}|\Psi\rangle$:

$$\begin{aligned}
\langle\Psi|\hat{n}_{\mathrm{j}\uparrow}\hat{n}_{\mathrm{j}\downarrow}|\Psi\rangle &= \langle\Psi|X_\mathrm{j}^{d\leftarrow d}|\Psi\rangle = \langle\Psi|e^{-iS}e^{iS}X_\mathrm{j}^{d\leftarrow d}e^{-iS}e^{iS}|\Psi\rangle \\
&= \langle\tilde{\Psi}|(1 + iS - \frac{1}{2}S^2 + \ldots)X_\mathrm{j}^{d\leftarrow d}(1 - iS - \frac{1}{2}S^2 + \ldots)|\tilde{\Psi}\rangle \\
&= \langle\tilde{\Psi}|X_\mathrm{j}^{d\leftarrow d}|\tilde{\Psi}\rangle + i\langle\tilde{\Psi}|(SX_\mathrm{j}^{d\leftarrow d} - X_\mathrm{j}^{d\leftarrow d}S)|\tilde{\Psi}\rangle
\end{aligned}$$

---

[1] In the absence of holes, pair hopping and the three-site processes are blocked. The density–density interaction term $\hat{n}_\mathrm{i}\hat{n}_\mathrm{j}$ seen in (5.36) would give only an uninteresting additive constant which we omit.

[2] See also the solution to Problem 5.1 (p. 250).

$$+\langle\tilde{\Psi}|SX_{\mathbf{j}}^{d\leftarrow d}S|\tilde{\Psi}\rangle + \ldots$$

Since $|\tilde{\Psi}\rangle$ contains no doubly occupied sites, only the last term gives a finite contribution up to order t^2. We rewrite it using (5.15), and remembering that $\mathcal{H}_t^-|\tilde{\Psi}\rangle = 0$, the previous equation is continued as

$$\begin{aligned}
&\rightsquigarrow -\frac{1}{U^2}\langle\tilde{\Psi}|(\mathcal{H}_t^+ - \mathcal{H}_t^-)X_{\mathbf{j}}^{d\leftarrow d}(\mathcal{H}_t^+ - \mathcal{H}_t^-)|\tilde{\Psi}\rangle \\
&= \frac{1}{U^2}\langle\tilde{\Psi}|\mathcal{H}_t^- X_{\mathbf{j}}^{d\leftarrow d}\mathcal{H}_t^+|\tilde{\Psi}\rangle \rightsquigarrow \frac{t^2}{U^2}\sum_{\mathbf{l}}\langle\tilde{\Psi}|(X_{\mathbf{j}}^{\downarrow\leftarrow d}X_{\mathbf{l}}^{\uparrow\leftarrow 0} - X_{\mathbf{j}}^{\uparrow\leftarrow d}X_{\mathbf{l}}^{\downarrow\leftarrow 0}) \\
&\cdot X_{\mathbf{j}}^{d\leftarrow d}(X_{\mathbf{j}}^{d\leftarrow\downarrow}X_{\mathbf{l}}^{0\leftarrow\uparrow} - X_{\mathbf{j}}^{d\leftarrow\uparrow}X_{\mathbf{l}}^{0\leftarrow\downarrow})|\tilde{\Psi}\rangle
\end{aligned}$$

where $\mathbf{l}$ is a nearest neighbour of $\mathbf{j}$. Evaluating term-by-term, we find

$$\begin{aligned}
&\rightsquigarrow \frac{t^2}{U^2}\sum_{\mathbf{l}}\langle\tilde{\Psi}|(X_{\mathbf{j}}^{\downarrow\leftarrow\downarrow}X_{\mathbf{l}}^{\uparrow\leftarrow\uparrow} - X_{\mathbf{j}}^{\downarrow\leftarrow\uparrow}X_{\mathbf{l}}^{\uparrow\leftarrow\downarrow} \\
&- X_{\mathbf{j}}^{\uparrow\leftarrow\downarrow}X_{\mathbf{l}}^{\downarrow\leftarrow\uparrow} + X_{\mathbf{j}}^{\uparrow\leftarrow\uparrow}X_{\mathbf{l}}^{\downarrow\leftarrow\downarrow})|\tilde{\Psi}\rangle = \frac{t^2}{U^2}\sum_{\mathbf{l}}\langle\tilde{\Psi}|\left(\frac{1}{2} - S_{\mathbf{j}}^z\right)\left(\frac{1}{2} + S_{\mathbf{l}}^z\right) \\
&- S_{\mathbf{j}}^- S_{\mathbf{l}}^+ - S_{\mathbf{j}}^+ S_{\mathbf{l}}^- + \left(\frac{1}{2} - S_{\mathbf{j}}^z\right)\left(\frac{1}{2} + S_{\mathbf{l}}^z\right)|\tilde{\Psi}\rangle. \qquad (5.43)
\end{aligned}$$

The leading contribution to the density of doubly occupied sites is

$$n_d = \langle\Psi|\hat{n}_{\mathbf{j}\uparrow}\hat{n}_{\mathbf{j}\downarrow}|\Psi\rangle = 2z\frac{t^2}{U^2}\langle\tilde{\Psi}|\left(\frac{1}{4} - \mathbf{S}_1\cdot\mathbf{S}_2\right)|\tilde{\Psi}\rangle \qquad (5.44)$$

where z is the coordination number. It is very reasonable that n_d should be found proportional to the spin correlation function shown above. If nearest-neighbour spins are parallel, no hopping can take place, and there is no concomitant admixture of double occupation. But then there can be no energy gain from virtual hopping processes, either.

The key quantity to calculating the band energy is the band state occupation number. It was found [385] that, to order t/U

$$n_{\mathbf{k}} = \langle\Psi|\hat{n}_{\mathbf{k}\uparrow} + \hat{n}_{\mathbf{k}\downarrow}|\Psi\rangle \approx 1 - \frac{\epsilon(\mathbf{k})}{U}\left(1 - 4\langle\tilde{\Psi}|\mathbf{S}_1\cdot\mathbf{S}_2|\tilde{\Psi}\rangle\right) \qquad (5.45)$$

where $\epsilon(\mathbf{k})$ is the tight binding band energy. The proof can be done analogously to that of (5.44), and it is deferred to Problem 5.2.

Now we gain some insight into how the effective Hamiltonian – in the present case, the Heisenberg model – works. (5.42) is a model Hamiltonian for interacting spins. On the other hand, in terms of the original Hubbard model, we should be speaking about band energy, and Coulomb energy, contributions! Exactly how are these related to the apparent spin–spin interaction energy which is the only thing that the Heisenberg model can tell us about?

The answer (to order t^2/U) is given in our previous results. For the sake of simplicity, let us consider the one-dimensional Hubbard model, for which $z = 2$, and $\epsilon(k) = -2t\cos k$. From (5.44), the electron–electron interaction energy density is

$$\frac{\langle\Psi|\mathcal{H}_U|\Psi\rangle}{L} = Un_d = \frac{t^2}{U}\left(1 - 4\langle\tilde{\Psi}|\mathbf{S}_1\cdot\mathbf{S}_2|\tilde{\Psi}\rangle\right). \tag{5.46}$$

The band energy density, remembering that $\sum\cos k = 0$, and $\sum(\cos k)^2 = 1/2$, is found to be

$$\frac{\langle\Psi|\mathcal{H}_{\text{band}}|\Psi\rangle}{L} = -\frac{2t^2}{U}\left(1 - 4\langle\tilde{\Psi}|\mathbf{S}_1\cdot\mathbf{S}_2|\tilde{\Psi}\rangle\right). \tag{5.47}$$

Thus both the band energy and the Coulomb interaction energy are proportional to the Heisenberg interaction term. (5.46) and (5.47) add up to

$$\frac{\langle\Psi|\mathcal{H}|\Psi\rangle}{L} = \frac{4t^2}{U}\cdot\langle\tilde{\Psi}|\left(\mathbf{S}_1\cdot\mathbf{S}_2 - \frac{1}{4}\right)|\tilde{\Psi}\rangle \tag{5.48}$$

which reiterates the statement about the effective Hamiltonian. The mechanism of the effective exchange interaction is now explicitely seen to be that the kinetic energy gain due to the virtual admixture of doubly occupied sites outweighs the energy cost of the on-site interaction.

Problem 5.1. The basic idea of the long derivation leading to the Heisenberg Hamiltonian is that in the low-energy subspace of a rotated basis, the full Hubbard Hamiltonian acts like (5.42). To make this result more tangible, use low-order perturbation theory to construct the new basis states for a nearest-neighbour pair of sites, and verify that the matrix elements of the Hubbard

Hamiltonian can be identified as those of a Heisenberg coupling.

Problem 5.2. Consider the half-filled Hubbard model in the limit $U/t \gg 1$. Calculate the ground state expectation value of the band state occupation number $n_{\mathbf{k}} = n_{\mathbf{k}\uparrow} + n_{\mathbf{k}\downarrow}$ to order t/U.

5.1.6 Higher-Order Exchange

The simple form (5.42) of the Heisenberg model is the result of second-order perturbation theory in t. We may be wondering what do we find if we go to higher orders.

Let us review the fourth-order terms. The spin-flip terms arise from processes in which four consecutive hoppings lead from one spin configuration to another. For the sake of simplicity let us consider a four-site ring with nearest-neighbour hopping. Denoting the eigenstates as $|S_1^z S_2^z S_3^z S_4^z\rangle$, etc., we find, for example's sake

$$|\uparrow\downarrow\uparrow\downarrow\rangle \rightsquigarrow |0d\uparrow\downarrow\rangle \rightsquigarrow |\downarrow\uparrow\uparrow\downarrow\rangle \rightsquigarrow |\downarrow\uparrow d0\rangle \rightsquigarrow |\downarrow\uparrow\downarrow\uparrow\rangle \tag{5.49}$$

Such a change is effected by $S_1^- S_2^+ S_3^- S_4^+$. Since the total Hamiltonian is spin-rotationally invariant, this must be part of the isotropic term $(\mathbf{S}_1 \cdot \mathbf{S}_2)(\mathbf{S}_3 \cdot \mathbf{S}_4)$. By symmetry, the term $(\mathbf{S}_2 \cdot \mathbf{S}_3)(\mathbf{S}_4 \cdot \mathbf{S}_1)$ must have the same coefficient. Finally, we can convince ourselves that there must exist a term like $(\mathbf{S}_1 \cdot \mathbf{S}_3)(\mathbf{S}_2 \cdot \mathbf{S}_4)$ as well (you may check that there are processes which connect $|\uparrow\uparrow\downarrow\uparrow\rangle$ to $|\downarrow\uparrow\uparrow\uparrow\rangle$). Enumerating all the processes, one finds the four-site term (or plaquette exchange) [385]

$$\mathcal{H}_{4-\mathrm{site}} = \frac{5t^4}{U^3}[(\mathbf{S}_1 \cdot \mathbf{S}_2)(\mathbf{S}_3 \cdot \mathbf{S}_4) + (\mathbf{S}_2 \cdot \mathbf{S}_3)(\mathbf{S}_4 \cdot \mathbf{S}_1) - (\mathbf{S}_1 \cdot \mathbf{S}_3)(\mathbf{S}_2 \cdot \mathbf{S}_4)] \,. \tag{5.50}$$

This is the most interesting fourth-order term. It is easy to see that we find also second-neighbour pair exchange

$$\mathcal{H}_{\mathrm{nnn}} = \frac{4t^4}{U^3}[(\mathbf{S}_1 \cdot \mathbf{S}_3) + (\mathbf{S}_2 \cdot \mathbf{S}_4)] \tag{5.51}$$

as well as a fourth-order correction to the nearest-neighbour coupling.

All this was good enough for the non-degenerate Hubbard model, i.e., for Mott insulators with localized $S = 1/2$ spins. For $S > 1/2$

magnets (which may be described by degenerate Hubbard models), more possibilities have to be considered. For instance, we are going to see in the solution to Problem 5.8 (see, in particular, the discussion on pp. 261–262) that fourth-order processes give rise to a nearest-neighbour biquadratic coupling $\sim Q(\mathbf{S}_1 \cdot \mathbf{S}_2)^2$. There will be also three-site terms of the type $(\mathbf{S}_1 \cdot \mathbf{S}_2)(\mathbf{S}_2 \cdot \mathbf{S}_3)$.

Are the higher-order terms important? Well, we have just convinced ourselves that their coupling constants must be quite a bit smaller than J (5% may be a generous estimate), thus we should say that usually they are not important. However, there is always a chance that they play a role in frustrated magnets, where the competition between leading-order terms leaves the question of order for the weaker terms to decide [460].

There is, however, a system for which higher-order terms like (5.50) are known to be important: solid helium-3 [74]. Because of the lightness of the ^{3}He atoms, zero-point motion prevents the system from solidifying under atmospheric pressure; it takes at least 30 atmospheres to make a solid. The zero-point oscillations are extremely vigorous, so much that roughly once in 40000 vibrations, neighbouring atoms are actually exchanging places. Since the atomic wave functions have a substantial overlap, quantum-mechanical indistinguishability and the Fermi statistics of ^{3}He atoms become important. As the atoms move, they take their $S = 1/2$ nuclear spins with them; the exchange of two atoms with antiparallel spins is manifested as the simultaneous spin-flip at the two lattice sites. This can be described by the spin-flip part of the Heisenberg model: in the present problem

$$\mathcal{P}_{ij} = 2\left(\mathbf{S}_i \cdot \mathbf{S}_j + \frac{1}{2}\right) \tag{5.52}$$

is the permutation operator for atoms i and j.

If it were only for this term, there could be no question about the nature of the order: ^{3}He has the bipartite bcc structure which is ideally suited for the formation of a two-sublattice antiferromagnet. However, experiments show that ^{3}He has a more complicated order which can be explained by invoking the presence of higher-order exchange terms. We can argue as follows: Though helium-3 is an extremely dilated solid, two neighbouring atoms still find it rather hard to exchange places by squeezing past each other, thus the pair coupling constant J_2 is relatively small. However, three or four atoms forming a loop will find it quite easy to move around, each of them chasing the tail of the previous fellow. The four-particle *ring exchange* process corresponds to the cyclic permutation [397]

$$\begin{aligned} \mathcal{H}_4 &= \sum_{ijkl} \mathcal{P}_{ij}\mathcal{P}_{jk}\mathcal{P}_{kl}\mathcal{P}_{li} \qquad (5.53) \\ &\rightsquigarrow J_4 \sum_{ijkl} [(\mathbf{S}_i\cdot\mathbf{S}_j)(\mathbf{S}_k\cdot\mathbf{S}_l) + (\mathbf{S}_j\cdot\mathbf{S}_k)(\mathbf{S}_l\cdot\mathbf{S}_i) - (\mathbf{S}_i\cdot\mathbf{S}_k)(\mathbf{S}_j\cdot\mathbf{S}_l)] \,. \end{aligned}$$

There are similar three-atom exchange terms. The phase diagram of solid helium-3 is reasonably well described by suitably chosen J_2, J_3, and J_4 [74].

Two points have to be kept in mind. First, though the exchange Hamiltonian of helium-3 looks like the fourth-order effective Hamiltonian for a half-filled Hubbard model, here the underlying physics is completely different. For this reason, it need not be true that J_4 is much smaller than J_2; in fact, they are roughly equal. Second, the energy scale is determined by the overlap of nuclear wave functions, rather than of electron wave functions, thus it is much smaller. The ordering temperature of ^{3}He is $\sim$1mK.

Remarks. Let us recall our discussion of the exchange process in Sec. 2.2 and, in particular, footnote 9. There we argued that the essential condition for exchange is that the particles be able to visit the same region of space but they may do so while keeping away from each other all the time. This is how the exchange of ^{3}He atoms should be understood: it would require very high energies to press two $1s^2$ shells into a single ball, so the two nuclei certainly never see each other! Let us describe the motion of an atom about its average position by $\phi_a(\mathbf{r}) = \phi(\mathbf{r} - \mathbf{R}_a)$, and its neighbour by $\phi_b(\mathbf{r}) = \phi(\mathbf{r} - \mathbf{R}_b)$. In a mean-field sense, the two-nucleon system would be described by a Slater determinant formed of ϕ_a and ϕ_b (with the corresponding spin functions), and the large amplitude of the atomic motion means that there must be an appreciable "exchange charge" density $\phi_a^*(\mathbf{r})\phi_b(\mathbf{r})$. However, the relative motion of the two nuclei should rather be described by a correlated wave function $\psi(\mathbf{r}_1, \mathbf{r}_2) \sim \phi_a(\mathbf{r}_1)\phi_b(\mathbf{r}_2)g(\mathbf{r}_1-\mathbf{r}_2)$, where the correlator $g(\mathbf{r}_1-\mathbf{r}_2)$ prevents the nuclei approaching each other closer than $\sim 2a_{1s}$, where a_{1s} is the radius of the $1s^2$ shell. The interchange of the two nuclear spins should proceed by the atoms squeezing past each other in the space allotted for them, thus even the nominally pair exchange process must proceed by the nuclei moving in a loop. For this reason, the exchange coefficient must be rather smaller than what we would expect from the naively defined exchange charge. – Expressed in terms of the atomic basis functions ϕ, $\psi(\mathbf{r}_1, \mathbf{r}_2)$ would be a sum of many Slater determinants: we would be dealing with a configuration interaction problem. This is the reason why this possibility could not be discussed within the restricted Hilbert space used in Sec. 2.2.

In Hubbard-like lattice fermion models, we meet a similar difficulty when discussing kinetic exchange in the limit $U \to \infty$. The pair term $\sim t^2/U$ certainly vanishes but the possibility of ring exchange processes via intermediate empty sites remains. We will describe such a case in detail in the solution to Problem 7.T (see p. 437 and p. 488). In another interesting case, the intervening sites are anion sites in a transition metal compound. We will see shortly that the superexchange coupling contains contributions in which the Hubbard U for the cation orbitals does not appear.

5.2 Superexchange

We declare ourselves basically satisfied with the kinetic exchange mechanism of antiferromagnetism we learned about in Sec. 5.1.5, accepting that it suffices for the qualitative understanding of the majority of the large number of insulating antiferromagnets. There is, however, a stumbling block: our previous argument worked for the single-orbital Hubbard model where the source of kinetic exchange is that electrons can hop from site to site, restrained only by Coulomb repulsion. But most antiferromagnetic insulators are transition metal compounds, in which the d-electron-bearing cations are separated by large anions (a typical situation is shown in Fig. 5.1). Direct hopping between d-orbitals seems rather unlikely.

Anderson [20, 25] has shown that the concept of kinetic exchange can be extended to such situations and it is even enhanced by the covalent mixing between cation and anion orbitals, thereby explaining the large ordering temperatures of many compounds. To understand how it works, let us consider the rather idealized situation shown in Fig. 5.1. Say, the two cations can be thought of as Ti^{3+} ions with one d-electron each, and the intervening anion is an O^{2-} ion, as in $LaTiO_3$. Zeroth-

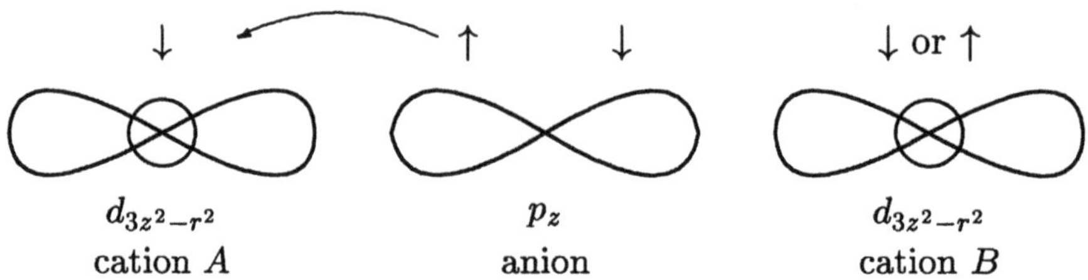

Figure 5.1: Superexchange between two cations via an intermediate anion. In zeroth order (ionic picture), the anion has two electrons (filled shell). Covalent mixing allows the anion p electrons to partially reoccupy the cation d orbitals. The degree of mixing depends on the spin configuration.

order energetics gives an ionic picture with filled oxygen p-shell: this is represented by putting two electrons into the p_z orbital we have shown (we show only the orbitals for which a strong mixing is expected). The overlap between $d_{3z^2-r^2}$ and p_z orbitals gives rise to a covalent mixing which allows the nominally p-electrons to partially reoccupy the cations, leading to fractional charges. This rearrangement depends on the relative orientation of the cation d-spins, thus the process transmits a d–d interaction which can be formulated as a spin Hamiltonian.

We merely sketch the derivation which follows [349]. The energetics is treated from the point of view of the p-electrons, for the two cases when the d-spins are either in the $\downarrow\downarrow$, or the $\downarrow\uparrow$ configuration.

As shown if Fig. 5.1, the $\uparrow$-spin p-electron can hop to the left, thus the corresponding p-orbital acquires a covalent admixture of d-character:

$$|p_{z\uparrow}\rangle' \sim \frac{|p_{z\uparrow}\rangle + b|d_{A\uparrow}\rangle}{\sqrt{1+b^2}} \tag{5.54}$$

where the mixing amplitude b is, from perturbation theory

$$b \sim \frac{\langle p_z|\mathcal{H}|d_A\rangle}{\epsilon_p - \epsilon_d}. \tag{5.55}$$

The atomic levels are understood to arise from the solution of the many-electron problem at the ionic level. We assume that $\epsilon_p < \epsilon_d$.

Now, if both d-spins are $\downarrow$, the $\uparrow$-spin p-electron can similarly extend to the cation on the right, while the $\downarrow$-spin p-electron must stay at the anion. The energy of the two p-electrons is approximately

$$\begin{aligned}\mathcal{E}_{\downarrow\downarrow} &\sim \frac{1}{1+2b^2}\cdot\left(\langle p_{z\uparrow}| + b\langle d_{A\uparrow}| + b\langle d_{B\uparrow}|\right) \\ &\quad \cdot\langle p_{z\downarrow}|\mathcal{H}|p_{z\downarrow}\rangle\left(|p_{z\uparrow}\rangle + b|d_{A\uparrow}\rangle + b|d_{B\uparrow}\rangle\right) \\ &= 2\epsilon_p + \frac{2b^2}{1+2b^2}(\epsilon_p - \epsilon_d)\,. \end{aligned} \tag{5.56}$$

If the d-electron spin configuration is $\downarrow\uparrow$, the $\downarrow$-spin p-electron can extend to the B cation

$$|p_{z\downarrow}\rangle' \sim \frac{|p_{z\downarrow}\rangle + b|d_{B\downarrow}\rangle}{\sqrt{1+b^2}} \tag{5.57}$$

and the total p-electron energy contribution is estimated as

$$\mathcal{E}_{\downarrow\uparrow} \sim {}'\langle p_{z\downarrow}|{}'\langle p_{z\uparrow}|\mathcal{H}|p_{z\uparrow}\rangle'|p_{z\downarrow}\rangle' = 2\epsilon_p + \frac{2b^2}{1+b^2}(\epsilon_p - \epsilon_d)\,. \tag{5.58}$$

The exchange splitting is approximately

$$\mathcal{E}_{\downarrow\downarrow} - \mathcal{E}_{\downarrow\uparrow} \sim 2b^4(\epsilon_d - \epsilon_p) \sim J_{\text{eff}}^{(1)}\,. \tag{5.59}$$

The parallel configuration costs more energy, thus the effective exchange is *antiferromagnetic*. The anion-mediated exchange described here is known as *superexchange*.

In terms of perturbation theory, $J_{\text{eff}}^{(1)}$ arises from a fourth-order process, with two electrons jumping away from the central anion, and two coming back. There is another fourth-order process which also contributes to superexchange. Namely, covalent mixing changes the character of the d-states as well, giving new d-like states of the kind $|d_A\rangle' \sim |d_A\rangle - b|p_z\rangle$, etc. Thus even if the original d-orbitals were strictly non-overlapping, the new d-orbitals strongly overlap at the anion site, the overlap being essentially $\sim b^2(\epsilon_d - \epsilon_p)$. The same can be our estimate for an effective hopping amplitude between the two d-like orbitals $t^*_{AB} \sim b^2(\epsilon_d - \epsilon_p)$. Thus we can envisage the process of a $\downarrow$-spin electron jumping from cation A to the anion, from there to the cation B, and an $\uparrow$-spin electron coming back the opposite way. This is quite literally an exchange of spins, with the coupling

$$J_{\text{eff}}^{(2)} \sim 2b^4\frac{(\epsilon_d - \epsilon_p)^2}{U} \sim \frac{(t^*_{AB})^2}{U}\,, \tag{5.60}$$

where U appears from the intermediate state when two electrons sit on the same cation. Now it is apparent that the superexchange process is quite analogous to the kinetic exchange we have found for the Hubbard model [20]. The total superexchange coupling is $J_{\text{eff}} = J_{\text{eff}}^{(1)} + J_{\text{eff}}^{(2)}$.

The present discussion should make us aware of a difficulty in identifying the ionic moments. The starting configuration in Fig. 5.1 is quite clear: there is an $\downarrow$-electron at A, the anion is non-magnetic, etc. Covalent mixing removes some spin-polarization from the d-site, leaving it with a reduced, fractional moment. Naturally, the electron spin cannot have become less than 1/2; only it is not localized completely to a cation[3]. The spin-operators figuring in effective spin Hamiltonians act on these "complete" spins; but they may belong to rather extended spin clouds. Neutron scattering measurements of the magnetic form factors should tell us where the spins really are.

Finally, the reader should be cautioned not to believe that the above naive estimate is anything like calculating the coupling constant of superexchange. A large number of higher-lying orbital states for all the ions have to be taken into account [465], resulting in a bewildering array of intermediate states; it is difficult to judge the reliability of these complicated calculations[4]. The direct exchange contributions we discuss in Ch. 2 and in Sec. 5.3 should not be forgotten, either. Extensive work on the subject resulted in the understanding that the geometry of the environment, in particular the bond angles, play an important role, and can even decide the sign of the interaction (the Goodenough–Kanamori rules; see, e.g., [20] and [465]). We need not be worried to learn that a few insulating substances are ferromagnetic: superexchange theory can (in principle) describe those as well. Besides, taking the orbital degeneracy into account provides us with a further, quite transparent reason why some magnetic insulators may order ferromagnetically.

---

[3] By the same token, some spin polarization may be seen on the nominally non-magnetic anion.

[4] Though it describes a story from an epoch when computational power was much less than nowadays, it is educative to read the ups and downs of an effort to do the calculation for the seemingly simple case of $KNiF_3$ [25].

5.3 The Extended Hubbard Model

Most of the literature deals with the standard form (4.10) of the Hubbard model, where only the on-site interaction U is kept. Adopting this simplification was based on the hope that, qualitatively speaking, the same kind of physics can be derived from the Hubbard model as from fuller forms of the Coulomb interaction. Naturally, it was understood that the explanation of charge ordering in systems with less-than-half-filled bands requires the inclusion of intersite interactions (Sec. 1.3.3). However, it was believed that all the major kinds of magnetic ordering can be understood on the basis of (4.10).

Beginning from the nineties, this view has undergone an essential revision. The main reason is that it has turned out that ferromagnetism is not a generic feature of the single-orbital Hubbard model (see Ch. 8 for details). Searching for the mechanism for ferromagnetic ordering has led to renewed interest in the habitually neglected terms of the Coulomb interaction (let us recall that direct exchange which we demonstrated to be ferromagnetic in Sec. 2.2, is omitted in the Hubbard model). At the same time, one was looking for terms which might cause superconductivity.

Starting his monumental series of papers [174], Hubbard himself discussed such terms, only to discard them because they are numerically weaker than *the* Hubbard term. The procedure has been reconsidered many times; our discussion is largely based on [222]. We assume a single, non-degenerate band from the outset (see, however, later remarks about the nature of the single-band model).

The band energy is diagonal in the Bloch basis

$$\mathcal{H}_{\text{band}} = \sum_{\mathbf{k}} \sum_{\sigma} \epsilon_{\mathbf{k}} \, \hat{n}_{\mathbf{k}\sigma} \, . \tag{5.61}$$

In contrast, the Coulomb interaction would be diagonal in the **r**-representation[5]

$$\mathcal{H}_{\text{Coulomb}} = \int d\mathbf{r}_1 \int d\mathbf{r}_2 \sum_{\sigma_1} \sum_{\sigma_2} \Psi^\dagger_{\sigma_1}(\mathbf{r}_1)\Psi^\dagger_{\sigma_2}(\mathbf{r}_2)\frac{e^2}{|\mathbf{r}_1 - \mathbf{r}_2|}\Psi_{\sigma_2}(\mathbf{r}_2)\Psi_{\sigma_1}(\mathbf{r}_1) \tag{5.62}$$

but it is given[6] by a multiple sum in the **k**-representation

$$\mathcal{H}_{\text{Coulomb}} = \frac{1}{2}\sum_{\mathbf{k}_1}\sum_{\mathbf{k}_2}\sum_{\mathbf{k}_3}\sum_{\mathbf{k}_4}\sum_{\mathbf{G}}\sum_{\sigma_1}\sum_{\sigma_2} \delta(\mathbf{k}_1 + \mathbf{k}_2 - \mathbf{k}_3 - \mathbf{k}_4 + \mathbf{G}) \\ c^\dagger_{\mathbf{k}_1\sigma_1} c^\dagger_{\mathbf{k}_2\sigma_2} \mathcal{V}(\mathbf{k}_1, \mathbf{k}_2, \mathbf{k}_3, \mathbf{k}_4)\, c_{\mathbf{k}_4\sigma_2} c_{\mathbf{k}_3\sigma_1} \tag{5.63}$$

where the **G**s are reciprocal lattice vectors.

It is more convenient to express $\mathcal{H}_{\text{Coulomb}}$ in the Wannier representation because then we can exploit that the Coulomb force decreases with increasing distance. Doing a Fourier transform, we get

$$\mathcal{H}_{\text{Coulomb}} = \frac{1}{2}\sum_{\mathbf{g}_1}\sum_{\mathbf{g}_2}\sum_{\mathbf{g}_3}\sum_{\mathbf{g}_4}\sum_{\sigma_1}\sum_{\sigma_2} c^\dagger_{\mathbf{g}_1\sigma_1} c^\dagger_{\mathbf{g}_2\sigma_2} \mathcal{V}(\mathbf{g}_1, \mathbf{g}_2, \mathbf{g}_3, \mathbf{g}_4)\, c_{\mathbf{g}_4\sigma_2} c_{\mathbf{g}_3\sigma_1} \tag{5.64}$$

where $\mathbf{g}_1$,... are lattice sites, and the matrix elements are expressed with the corresponding Wannier functions $\varphi(\mathbf{r} - \mathbf{g})$ as

$$\mathcal{V}(\mathbf{g}_1, \mathbf{g}_2, \mathbf{g}_3, \mathbf{g}_4) = \int d\mathbf{r}_1 \int d\mathbf{r}_2\, \varphi^*(\mathbf{r}_1 - \mathbf{g}_1)\varphi^*(\mathbf{r}_2 - \mathbf{g}_2) \\ \cdot \frac{e^2}{|\mathbf{r}_1 - \mathbf{r}_2|}\varphi(\mathbf{r}_1 - \mathbf{g}_3)\varphi(\mathbf{r}_2 - \mathbf{g}_4)\,. \tag{5.65}$$

Though three- and four-center terms exist, it is reasonable to confine the discussion to the on-site term, and pair terms. It is clear that $\mathcal{V}$ depends only on the separation $|\mathbf{i} - \mathbf{j}|$ of the sites, but not on the orientation of the pair. We keep only the nearest-neighbour terms. The problem of the

[5]Meaning the representation in the δ-function basis. As we are going to discuss, the interaction is not diagonal in the Wannier representation.

[6]Actually, if the band structure is the result of a Hartree–Fock calculation, then the nominally one-electron term $\mathcal{H}_{\text{band}}$ contains Coulomb contributions which have to be subtracted to avoid overcounting. Since we do not aim at carrying out a realistic calculation, we do not discuss this correction.

classification of all the Coulomb matrix elements was solved in Problem 2.2 (see p. 68), so here it is sufficient to quote the results.

The largest matrix element is the familiar Hubbard $U \equiv \mathcal{V}(i,i,i,i)$. $\mathcal{V}(i,j,i,j) \equiv V$ gives the strength of Coulomb repulsion at nearest-neighbour distance. The corresponding density–density interaction term

$$\mathcal{H}_V = V \sum_{\langle \mathrm{i,j} \rangle} \sum_{\sigma_1} \sum_{\sigma_2} \hat{n}_{\mathrm{i}\sigma_1} \hat{n}_{\mathrm{j}\sigma_2} \tag{5.66}$$

is analogous to what we introduced in (1.5).

The coupling constant of *direct exchange* $F \equiv \mathcal{V}(i,j,j,i)$ is readily seen to be the same kind of matrix element which we introduced in (2.48). The corresponding term in (5.64) is

$$\mathcal{H}_F = F \sum_{\langle \mathrm{i,j} \rangle} \sum_{\sigma_1} \sum_{\sigma_2} c^\dagger_{\mathrm{i}\sigma_1} c^\dagger_{\mathrm{j}\sigma_2} c_{\mathrm{i}\sigma_2} c_{\mathrm{j}\sigma_1} = -2F \sum_{\langle \mathrm{i,j} \rangle} \left(\mathbf{S}_\mathrm{i} \cdot \mathbf{S}_\mathrm{j} + \frac{1}{4} \hat{n}_\mathrm{i} \hat{n}_\mathrm{j} \right) . \tag{5.67}$$

There are two more curious terms. $X \equiv \mathcal{V}(i,i,i,j)$ is the amplitude of a *Coulomb-assisted hopping* process

$$\mathcal{H}_X = X \sum_{\langle \mathrm{i,j} \rangle} \sum_{\sigma} (c^\dagger_{\mathrm{i}\sigma} c_{\mathrm{j}\sigma} + \mathrm{H.c.})(\hat{n}_{\mathrm{i}-\sigma} + \hat{n}_{\mathrm{j}-\sigma}) . \tag{5.68}$$

It is also referred to as density-dependent hopping. An interesting interpretation can be given if we recall (Sec. 2.5) that the molecular orbital picture for the bond between two sites envisages electrons in the ground state of the hopping term

$$\hat{B}_{\mathrm{i,j}} = \sum_{\sigma} (c^\dagger_{\mathrm{i}\sigma} c_{\mathrm{j}\sigma} + c^\dagger_{\mathrm{j}\sigma} c_{\mathrm{i}\sigma}) . \tag{5.69}$$

For this reason, we may call $\hat{B}_{\mathrm{i,j}}$ the bond-charge operator. Correspondingly, $\mathcal{H}_X$ can be looked upon as a piece of the "bond-charge–site-charge repulsion". Its role has been discussed[7] in connection with the phase

[7] An alternative definition of $\mathcal{H}_X$ differs from (5.68) by using the total charge $\sum_\sigma (\hat{n}_{\mathrm{i}\sigma} + \hat{n}_{\mathrm{j}\sigma})$ rather than $\hat{n}_{\mathrm{i}-\sigma} + \hat{n}_{\mathrm{j}-\sigma}$ as the site charge factor, thus $\mathcal{H}_X$ becomes the full bond-site repulsion [56]. Similarly, $\mathcal{H}_F$ and $\mathcal{H}_Y$ can be summarized in a bond-charge–bond-charge repulsion term. It could be argued that this convention is more physical because it keeps fuller memory of the fact that the Coulomb interaction is a spin-independent density–density interaction. However, these alternative definitions differ merely in rearranging pieces of the total Hamiltonian, and we find the convention followed in [174] and [222] more convenient.

diagram of quasi-one-dimensional systems like polyacetylene [56].

Finally, $\mathcal{V}(i,i,j,j) \equiv Y$ is the amplitude for the simultaneous hopping of two electrons

$$\mathcal{H}_Y = Y \sum_{\langle \mathbf{i},\mathbf{j} \rangle} (c^\dagger_{\mathbf{i}\uparrow} c^\dagger_{\mathbf{i}\downarrow} c_{\mathbf{j}\downarrow} c_{\mathbf{j}\uparrow} + \text{H.c.}) \; . \tag{5.70}$$

Note that though the Coulomb interaction is diagonal in the **r**-representation, it is not diagonal in the Wannier representation. Therefore the electron–electron interaction can, to some extent, act like kinetic energy: it causes electron motion from site to site.

We have seen that $F = Y$ if the Wannier functions are real. The equality of these couplings is apparent in the numerical estimates given below. Still, we keep different notations for two reasons. First, the corresponding physical processes (direct exchange and double hopping, respectively) are very different. Second, all versions of the Hubbard model had better be thought of as effective Hamiltonians where the coupling constants arise from complicated processes involving many bands, rather than following the literal one-band definition (5.65). In this interpretation, F and Y may be different.

The full pair Hamiltonian is obtained by augmenting the Hubbard model with the new terms we have just found

$$\mathcal{H}_{\text{pair}} = \mathcal{H}_{\text{Hubbard}} + \mathcal{H}_V + \mathcal{H}_F + \mathcal{H}_X + \mathcal{H}_Y \; . \tag{5.71}$$

But how important are these additional terms? Hubbard [174] tried to estimate the couplings for transition metals, and he came up with $U \approx 10$eV, $V \approx 2-3$eV, $X \approx 1$eV, and $F, Y \approx 1/40$eV. At the same time, the hopping amplitude may be $t \approx 0.5 - 1.5$eV. It is also usual to cite the example of an isolated benzene molecule, for which $U \approx 17$eV, $V \approx 9$eV, $X \approx 3$eV, and $F, Y \approx 1/2$eV [56]. Thus purely on numerical grounds, introducing the Hubbard model is quite a reasonable approximation. However, the new terms are certainly important when they introduce energy splittings between states which would be (nearly) degenerate in terms of the original Hubbard model. In any case, let us keep in mind that not only V is quite substantial, but also the non-diagonal Coulomb matrix element X.

The direct exchange coupling F (and with it, Y) looks small in any comparison, so we might consider neglecting at least $\mathcal{H}_F$ and $\mathcal{H}_Y$ in "reasonable" extended Hubbard models. There is, however, a convincing counter-argument against omitting $\mathcal{H}_F$. The limit of strong-coupling Mott insulators is a physically relevant case. In zeroth order, all the neutral configurations have the same energy contribution from the diagonal Coulomb terms, and all hopping processes are blocked. The exchange term $\mathcal{H}_F$ splits the degeneracy, favouring parallel spin alignment. Naturally, we understand that this cannot be the full story, otherwise Mott insulators would be ferromagnetic! The solution to Problem 5.5 (see p. 254) should convince us that the effective exchange coupling is $J_{\text{eff}} \sim -F + (t - X)^2/U$. Thus F should not be compared to the large energies U and V, but rather to the perturbatively small $(t - X)^2/U$. In this context, there is no a priori reason to neglect $\mathcal{H}_F$.

We do not attempt even to sketch the phase diagram of the extended model $\mathcal{H}_{\text{pair}}$, apart from reconsidering the question of ferromagnetism in Ch. 8. Let us, however, make one general remark. We learned in Sec. 4.5.2 that the phase diagram of the simple Hubbard model is symmetrical about half-filling for bipartite lattices. What about the extended model? We find that the argument using the electron–hole transformation (4.45) goes through for most of the new terms, but not for $\mathcal{H}_X$. The assisted hopping term (5.68) breaks the electron–hole symmetry even for bipartite lattices, and the resulting phase diagram must be asymmetric.

There is a further point to consider, which may make the interpretation of the one-band Hubbard model really complicated. Strictly speaking, the band structure consists of (infinitely) many bands. The Coulomb interaction is not diagonal in the band index, thus it will mix the bands. Treating the partially occupied Hubbard band in isolation is an acceptable aprroximation only if this band is separated by a sufficiently wide gap from the other bands. If this condition is not satisfied, then we have to be careful and instead of just cutting off the rest of the band structure, we have to project the Coulomb interaction onto a single band [222]. This procedure may give rise to parameters which are rather different from what one would expect for an isolated band.

Problem 5.3 What kind of correlations does the X-term "want" to induce in the system? You can try to find this out by considering the eigenstates of $N = 2$ electrons on a linear chain, described by the truncated operator $\mathcal{H}_U + \mathcal{H}_X$.

Problem 5.4. Use the Hubbard operators to express the non-diagonal terms of the generalized Hubbard model $\mathcal{H}_{\text{pair}}$ given in (5.71). Is there anything special about the point $X = t$?

Problem 5.5. Derive the effective Hamiltonian for the half-filled extended Hubbard model! Take $\mathcal{H}_U + \mathcal{H}_V$ as the large term of the Hamiltonian. Discuss the condition for a ferromagnetic effective exchange coupling!

5.4 Orbital Degeneracy, Orbital Ordering

The orbital degeneracy of the d-shell makes the physics of transition metal compounds much richer than our previous experience with the non-degenerate Hubbard model would lead us to guess. At the same time, our understanding of these complex phenomena is less detailed than what we described in the foregoing text. We will confine ourselves to simple arguments to identify pieces of an effective Hamiltonian, and to characterize the ensuing novel possibilities of ordering. Our discussion is based on [76] and [206], and on the short review [207]. A more complete calculation of the same kind (applied to V_2O_3) was given in [60].

For the sake of simplicity, let us consider two sites, each with a two-fold degenerate orbital (we may think of e_g states). The orbitals are denoted by φ_a and φ_b, and the corresponding Wannier operators are $c^\dagger_{ja\sigma}$ and $c^\dagger_{jb\sigma}$, where $j = 1, 2$ is the site index and σ the spin. Assuming that the hopping does not mix φ_a with φ_b, the one-electron part of the Hamiltonian is

$$\mathcal{H}_{\text{hop}} = -t_a \sum_\sigma (c^\dagger_{1a\sigma} c_{2a\sigma} + \text{H.c.}) - t_b \sum_\sigma (c^\dagger_{1b\sigma} c_{2b\sigma} + \text{H.c.}) . \qquad (5.72)$$

Putting two electrons on the same site costs a Hubbard-like repulsion energy which depends on which orbitals are occupied; these will be denoted by U_a, U_b, and U_{ab}. Furthermore, if one electron is in the a state, and the other in the b state, they can form either a spin singlet or a spin triplet. Hund's first rule says that the triplet level is lower. Let the (ferromagnetic) intra-atomic exchange constant be $2J$. The diagonal interaction terms are collected in

$$\begin{aligned}\mathcal{H}_{\text{int}} &= U_a \sum_{j=1}^{2} \hat{n}_{ja\uparrow}\hat{n}_{ja\downarrow} + U_b \sum_{j=1}^{2} \hat{n}_{jb\uparrow}\hat{n}_{jb\downarrow} + U_{ab} \sum_{j=1}^{2} \sum_{\sigma\sigma'} \hat{n}_{ja\sigma}\hat{n}_{jb\sigma'} \\ &\quad -2J \sum_{j=1}^{2} \left(\hat{S}^z_{ja}\hat{S}^z_{jb} + \frac{1}{4} \right), \qquad (5.73)\end{aligned}$$

where $\hat{S}^z_{1a} = (\hat{n}_{1a\uparrow} - \hat{n}_{1a\downarrow})/2$, etc. Finally, there are the spin-flip terms of the intra-atomic exchange

$$\mathcal{H}_{\text{flip}} = -J \sum_{j=1}^{2} (c^\dagger_{ja\uparrow} c_{ja\downarrow} c^\dagger_{jb\downarrow} c_{jb\uparrow} + c^\dagger_{ja\downarrow} c_{ja\uparrow} c^\dagger_{jb\uparrow} c_{jb\downarrow}) \,. \qquad (5.74)$$

The full Hamiltonian of our "e_g molecule" is

$$\mathcal{H} = \mathcal{H}_{\text{hop}} + \mathcal{H}_{\text{int}} + \mathcal{H}_{\text{flip}} \,. \qquad (5.75)$$

We introduced different ts (and Us) for a and b primarily for book-keeping purposes: it makes it easier to trace the origin of various terms. At the same time, it is not unrealistic to use this distinction. Thinking of $(x^2 - y^2)$-like and $(3z^2 - r^2)$-like orbitals, and taking z as the molecule axis, it is clear that the two hopping amplitudes will be quite different.

Assuming that there are $N = 2$ electrons and that the Us and J are large compared to the ts, we may try to emulate our procedure in Sec. 2.2 to derive an effective (exchange) Hamiltonian by calculating how $\mathcal{H}$ acts in the subspace of the low-energy configurations. However, while for the H_2 molecule there are only 4 low-energy states which conveniently separate into a singlet and a triplet, for the e_g molecule we find 16 low-lying states, organized into four triplets and four singlets. Obviously, we are going to need a rather complicated effective Hamiltonian to describe the eight low-lying energy levels. Leaving the detailed argument to Problem 5.6, here we look at just a few possibilities to convince ourselves that the ground state is a triplet.

Start with an a-type singlet. The sequence of hoppings shown below in Process I

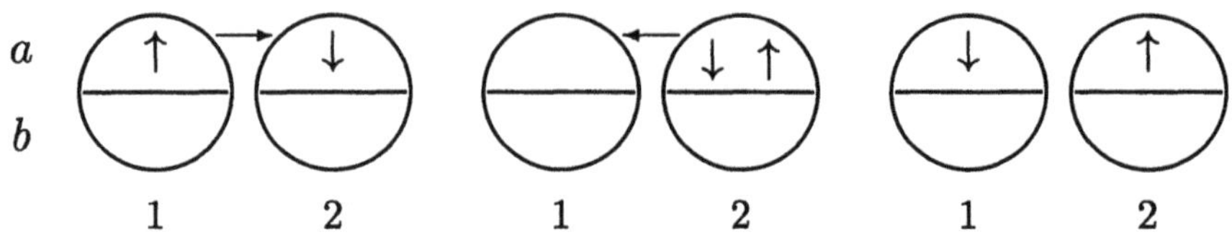

shows that there must be a singlet built of φ_a orbitals, with the energy

$$E_{a,\mathrm{si}} \approx -\frac{4t_a^2}{U_a}\,. \tag{5.76}$$

This is analogous to the antiferromagnetic kinetic exchange processes we have met in (2.135) and (5.36). Similarly, there must be a b-type singlet with the energy

$$E_{b,\mathrm{si}} \approx -\frac{4t_b^2}{U_b}\,. \tag{5.77}$$

Starting out with parallel-spin electrons, no hopping can take place, and thus the analogous triplet levels are

$$E_{a,\mathrm{tr}} = E_{b,\mathrm{tr}} = 0\,. \tag{5.78}$$

The remaining two singlets and two triplets have mixed orbital character. Looking for a low-lying triplet, consider Process II

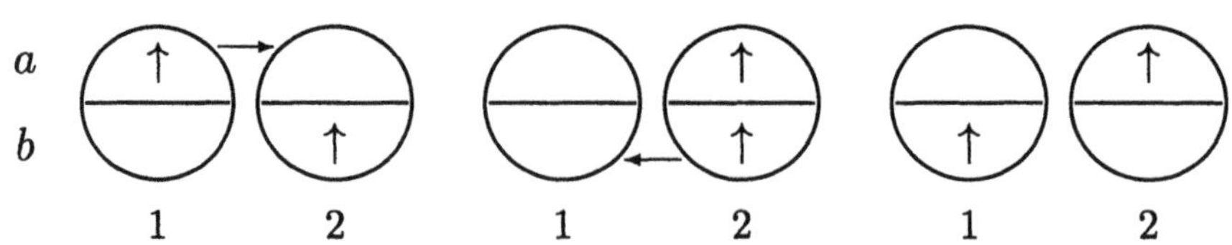

The energy of the intermediate state is $U_{ab} - J$. It is plausible to assume that the inter-orbital U_{ab} is smaller than either U_a or U_b (actually, for two e_g electrons one finds $U_a = U_b = U$, $U_{ab} = U - 2J$ [379]). Furthermore, the energy is further reduced by the Hund's rule J. Thus we have found the smallest energy denominator, and thereby also the strongest coupling. It favours a *triplet ground state*.

According to the previous picture, the energy numerator would be $t_a t_b$ but there are other possibilities: we could have let the a-electron jump back, etc. Furthermore, the picture serves to the generation of two different triplets: one even, and one odd with respect to reflection. The energy of the former is

$$E_{e,\mathrm{tr}} \approx -\frac{(t_a + t_b)^2}{U_{ab} - J} \,. \tag{5.79}$$

This is the approximate ground state energy. We see now how *the intraatomic Hund's rule coupling can give rise to intersite ferromagnetic exchange*: it works by reducing certain energy denominators.

It is clear that spin ordering is only one side of the story. Let us observe that for the arising of ferromagnetic exchange coupling, it was important that the electrons occupy different orbitals. Loosely speaking, *spin ferromagnetism must be accompanied by orbital antiferromagnetism*[8], and vice versa. We may envisage the alternating orbital order shown in the left panel of Fig. 5.2.

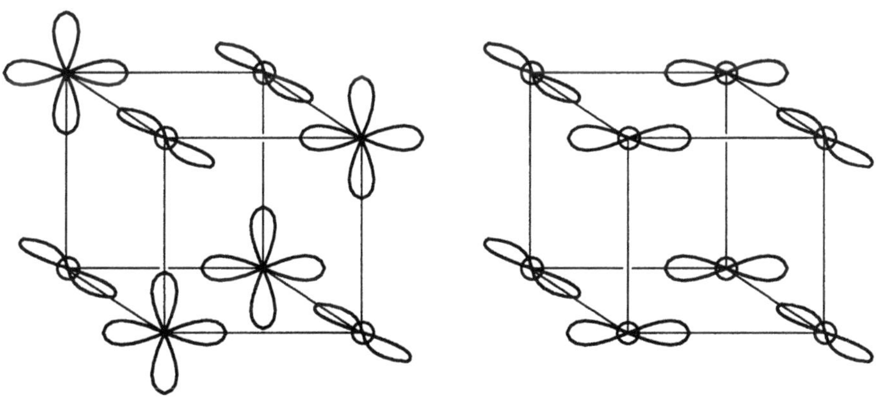

Figure 5.2: *Left*: the kinetic exchange of e_g electrons would favour the alternation of $d_{x^2-y^2}$ and $d_{3z^2-r^2}$ orbitals. *Right*: the in-plane orbital order in $LaMnO_3$ is approximately $d_{3x^2-r^2}/d_{3y^2-r^2}$.

Using a realistic description of d-orbitals, orbital ordering could be reinterpreted as quadrupolar (and possibly octupolar) ordering [362],

[8] Just to exclude the possibility of misunderstanding: "orbital antiferromagnetism" does not imply antiferromagnetism due to the orbital magnetic moments! It merely signifies an alternating order in the occupation of the two Wannier states. It is perhaps more exactly descriptive to speak about staggered orbital order.

establishing connection with our previous discussion of CeB_6, URu_2Si_2, etc. However, we do not follow this line, because it would require a lengthy formal development.

Let us recall that in Sec. 3.6 we found a different reason for the arising of an orbital superlattice: the cooperative Jahn–Teller effect. The Jahn–Teller effect does not necessarily "want" the same kind of order as the kinetic exchange processes described here. The resulting competition between the electron–electron and electron–lattice interaction[9] gives rise to complicated physics. As for $LaMnO_3$, the resulting orbital order can be described approximately as the in-plane alternation of $d_{3x^2-r^2}$ and $d_{3y^2-r^2}$ orbitals (Fig. 5.2, right). It is interesting to notice that there is no alternation along the c axis, corresponding to the fact that in that direction, the spin order alternates (compare to Fig. 5.3, right). We explain this finding shortly.

To describe orbital ordering, it is useful to introduce a pseudospin variable: let $\tau_j^z = 1/2$ if φ_a is occupied on site j, while $\tau_j^z = -1/2$ belongs to the occupation of φ_b. Process II did not change the spin but it did flip the pseudospin; it is described by

$$\sim -\frac{t_a t_b}{U_{ab} - J}\left(2S_1^z S_2^z + \frac{1}{2}\right)\tau_1^- \tau_2^+ \,.$$

In contrast, Process I belongs to

$$\sim \frac{t_a^2}{U_a}\left(\tau_1^z + \frac{1}{2}\right)\left(\tau_2^z + \frac{1}{2}\right) S_1^- S_2^+ \,.$$

Many terms like these have to be collected. The full effective Hamiltonian contains, in addition to the usual spin–spin coupling, also orbital couplings $\sim \tau_1\tau_2$, as well as mixed terms of the sort $\sim (S_1S_2)(\tau_1\tau_2)$. Fig. 5.2 shows the alternating orbital order which we would expect to be accompanied by parallel spin alignment. It is interesting that a mean field calculation based on the assumption of both spin and orbital ordering finds that the orbital superlattice may survive to higher temperatures than the spin magnetism [76]. This case seeems to be realized in $La_{0.5}Sr_{1.5}MnO_4$ [301] (see p 480).

[9]Here we may lump together all the lattice strain–electric multipole and multipole–multipole interaction terms which do not contain the magnetic degrees of freedom explicitly.

Note that the conclusion about the nature of the ground state depends on what we think about t_a, t_b, U_a, etc. In any case, our two-site model has a triplet ground state if we assume $t_a = t_b$, $U_a = U_b \geq U_{ab}$, and $J > 0$. This is fine for an exercise but what about real materials?

We cite two examples for Mott insulators[10] with one e_g electron (or one e_g hole) per magnetic ion: K_2CuF_4 and $LaMnO_3$. K_2CuF_4 is indeed ferromagnetic and it is considered as a successful realization of the idea explained above [206]. On the other hand, $LaMnO_3$ is an antiferromagnet. What has gone wrong with our argument?

One can think of several complications which can turn our would-be ferromagnet into an antiferromagnet: the presence of the other 3*d* electrons, the anisotropy of the hopping, and even the Jahn–Teller effect.

First, let us recall that Mn^{3+} has the configuration $t_{2g}^3 e_g^1$, and the ionic spin $S = 2$ arises mainly from the t_{2g} electrons. The fully polarized t_{2g}^3 "core" has no orbital degree of freedom, and the usual kinetic exchange of the t_{2g} electrons gives a strong antiferromagnetic contribution to the overall coupling. In fact, we can find out what the t_{2g} electrons alone would be doing, by looking at the magnetic structure of $CaMnO_3$ which contains $Mn^{4+} \leadsto t_{2g}^3$ ions. $CaMnO_3$ is a straightforward alternating antiferromagnet, with all nearest-neighbour pairs antiparallel (Fig. 5.3, left). In contrast, the structure of $LaMnO_3$ can be envisaged as the up–down alternation of ferromagnetic layers (Fig. 5.3, right). $LaMnO_3$ is a half-hearted antiferromagnet in which two-thirds of the nearest neighbour spin pairs are parallel. It is reckoned that the in-plane ferromagnetic coupling $J \approx -10$K is somewhat stronger than the interlayer antiferromagnetic $J' \approx 7$K [277]. In view of this, we might say that the case of $LaMnO_3$ is not all that different from that of K_2CuF_4. In K_2CuF_4, too, the primary phenomenon is strong intra-layer ferromagnetism and it happens more-or-less by chance that the much weaker $J' \sim 0.003J$ is also ferromagnetic. There are many quasi-two-dimensional Cu-based magnets which are built up of ferromagnetic Cu^{2+} layers; in $(CH_3NH_3)_2CuCl_4$ the inter-layer coupling is ferromagnetic, while in many other $(C_nH_{2n+1}NH_3)_2CuCl_4$ systems it is antiferromagnetic. In any case, $|J'/J| \sim 10^{-4}$–10^{-5}, thus the two-

[10]They may be charge transfer insulators in the sense of a finer classification which we describe in Sec. 5.5.

dimensional ferromagnetic character predominates [84].

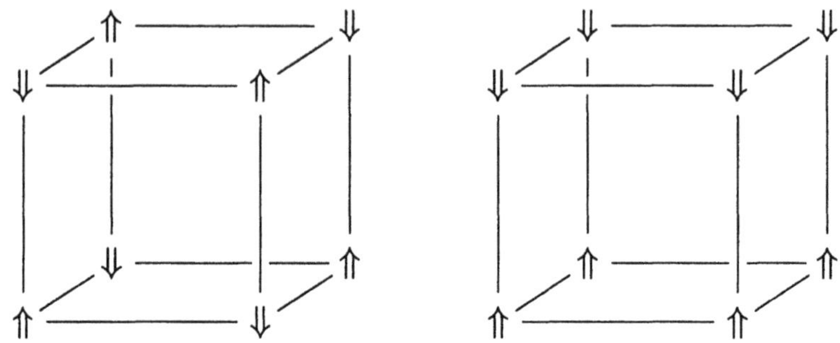

Figure 5.3: Antiferromagnetic order in $CaMnO_3$ (left), and in $LaMnO_3$ (right). Only the Mn ions are shown. Note the presence of ferromagnetic planes in $LaMnO_3$.

The extremely anisotropic character of the $(C_nH_{2n+1}NH_3)_2CuCl_4$ compounds arises from the crystal structure: the big organic molecules keep the Cu planes widely separated. However, the directional character of the e_g orbitals can give rise to strongly anisotropic[11] magnetic behaviour even in systems whose crystal structure is nearly cubic. It is educative to compare the perovskites $KCuF_3$ and $LaMnO_3$.

Recalling the orbital order of $KCuF_3$ (see Fig. 3.8 on p. 104), we find that it has the same spatial pattern as that of $LaMnO_3$, only the orbitals are different: $LaMnO_3$ has a $d_{3x^2-r^2}/d_{3y^2-r^2}$ order, while $KCuF_3$ a $d_{z^2-x^2}/d_{z^2-y^2}$ order. For $KCuF_3$, it is apparent that the hopping amplitude (and thus the kinetic exchange) is large along the vertical chains, while in perpendicular directions, there is practically no hopping. The substance is a quasi-one-dimensional antiferromagnet with weak ($J'/J \sim -0.03$) ferromagnetic interchain couplings [165]. In contrast, the $d_{3x^2-r^2}/d_{3y^2-r^2}$ orbitals of $LaMnO_3$ have a relatively small overlap in the z direction. The corresponding exchange coupling is antiferromagnetic since the orbital order in this direction is of the "ferro" type (Fig. 5.3, right). The intralayer (ferromagnetic) spin–spin coupling turns out to be somewhat larger, and the system can be looked upon as

[11]Note that in the present discussion, "anisotropic" means different coupling strengths in different crystallographic directions. The interaction is isotropic in spin space.

a sequence of ferromagnetic planes coupled by (not particularly weak) antiferromagnetic interactions. The arising of these two very different magnetic structures can be traced back to differing orbital orders on the background of the same nearly cubic perovskite structure.

It has been claimed [282] that with kinetic exchange alone, $LaMnO_3$ would be a ferromagnet, and it is just the Jahn–Teller effect what changes the character of the orbital super-lattice and stabilizes antiferromagnetism. Then it would follow that if we could somehow weaken the Jahn–Teller mechanism, we could resurrect ferromagnetism. Indeed, this is one of the ways to interpret the appearance of ferromagnetism in the $La_{1-x}(Ca/Sr)_xMnO_3$ system (see Sec. 8.5.3 for a discussion for the multi-faceted behaviour of these materials).

The intimate relationship between lattice structure and magnetism is highlighted by comparing La_2CuO_4 and K_2CuF_4. Ideally (neglecting distortions) both substances have the same K_2NiF_4 structure (see p. 79), and thus contain Cu^{2+} ions in geometrically equivalent environments. However, K_2CuF_4 is antiferrodistortive and ferromagnetic, with a Curie temperature $T_C \approx 6.25$K which is ascribed to a $J/k_B \approx 11$K between in-plane nearest neighbours. In contrast, La_2CuO_4 is ferrodistortive and antiferromagnetic, with an exceptionally strong in-plane exchange constant ($|J|/k_B$ values in the range ~1200–1600K have been deduced from fits to a variety of measurements [260]). Preliminary experimental results indicate that under sufficiently high pressure, the behaviour of K_2CuF_4 changes from an antiferrodistortive ferromagnet to a ferrodistortive antiferromagnet [181].

How can we detect orbital ordering? It is not easy. If we find a slight lattice distortion, it has to be decided whether it is due to the kinetic exchange mechanism described here, or due to some other reason. It could be due to the Jahn–Teller effect. Besides, we should not forget that any spin order induces some lattice change via the spin–orbit coupling (magnetoelasticity). Thus the evidence from lattice structure is rather indirect. However, remembering the systems discussed before, we could argue that the most sensitive measure of orbital order is just the sign and size of the effective spin–spin interaction, and the associated magnetic order. We can envisage a situation that the spin–spin coupling is ferromagnetic in the presence of orbital order, while above the orbital ordering temperature it reverts to the usual antiferromagnetism (see Problem 5.7). Such a drastic temperature dependence of the Heisenberg Hamiltonian would be difficult to explain by any other mechanism. It seems likely that this scenario is part of the story of the V_2O_3 system which we discuss in the next Section.

Problem 5.6. Determine all the low-lying levels of the "e_g molecule" specified in (5.72)–(5.74), for $N = 2$ electrons. Second-order perturbation theory is sufficient. Use the pseudospin variable τ to express the low-energy effective Hamiltonian for the symmetrical model $t_a = t_b = t$, $U_a = U_b = U$, $U_{ab} = U - J$.

Problem 5.7 The following kind of situation may be realized in V_2O_3: there is orbital ordering at $T = 0$, but at a relatively low temperature, the system undergoes a first-order transition into a state without orbital order. How does the spin–spin coupling change at the transition? For the sake of simplicity, you may assume that the system is described by the effective Hamiltonian (5.120) derived in the previous problem.

Problem 5.8. *The derivation of an $S = 1$ exchange Hamiltonian.* Consider the e_g molecule with $N = 4$ electrons in the strongly interacting limit when $U_a = U_b = U$, U_{ab}, and even J are considerably larger than $t_a = t_b = t$. The low-energy subspace is spanned by the on-site triplets. Express the effective Hamiltonian in terms of the $S = 1$ spin operators!

Compare the results of second-order and fourth-order perturbation theory. Does the form of the Hamiltonian change in an essential manner if you include the fourth-order corrections? Discuss the result, assuming that $U - U_{ab} \sim J$.

5.5 Correlated Insulators

The realization that the insulating nature of certain transition metal oxides is an electron correlation effect was the starting point of the study of "Mott phenomena" [24] a long time ago, and correlated insulators have remained a major field of investigation ever since. In the meantime, we have discovered that there are several different classes of correlated insulators. In Ch. 4, we started to discuss the most important class: systems which would be metallic if it were not for the opening of a Mott–Hubbard gap between the Hubbard subbands. In this sense, all these (usually transition-metal-based) systems are Mott insulators but shortly, we will introduce a further criterion to classify them as Mott insulators in a narrower sense, and charge transfer insulators. In any case, their low-energy effective Hamiltonian is some kind of Heisenberg model, with predominantly antiferromagnetic couplings, so the typical outcome is that at low temperatures, the system is an ordered insulating antiferromagnet. Neglecting magnetic anisotropy, the low-energy spin excitations are gapless antiferromagnetic spin waves. *Mott insulators have a large charge gap, and no spin gap*[12]. This behaviour should be contrasted with that of conventional band insulators like diamond: those are non-magnetic, and have both a charge gap and a spin gap which can be taken as equal.

There is an interesting little subclass represented by Y_2BaNiO_5. It is a Mott insulator which has a spin gap, for very special reasons. The Ni^{++} ions carry $S = 1$ spins, and the system is magnetically quasi-one-dimensional, with essentially non-communicating Ni-chains. It turns out that the singlet ground state of the isotropic $S = 1$ Heisenberg chain is characterized by an exponential fall-off of spin correlations, and a spin gap. The charge gap is large (of the order of ~1eV), while the spin gap is small (of the order of the Heisenberg coupling, i.e., ~0.01eV). More about this on pp. 303–305.

The second major class of correlated insulators is that of mixed valent and Kondo insulators. They have a spin gap as well as a charge gap, and both tend to be small but they need not be equal. We discuss them briefly in Ch. 11.

5.5.1 Mott Transitions in Transition Metal Oxides

The vanadium sesquioxide V_2O_3 has long been renowned as *the* model system for a variety of Mott phenomena [293]. However, it turned out that it is far from being a straighforward realization of the single-band Hubbard model. We will try to describe those features first, which can

[12]The spin gap is the energy of the lowest excitation which changes the spin character of the system. Spin waves do that, and (neglecting anisotropy and dipole–dipole interactions) their dispersion starts from zero.

Since there are systems which have a charge gap but no spin gap, the reader may ask whether the reverse case is also possible. Purely formally, we can readily say yes. The strong-coupling negative-U Hubbard model at fractional fillings should be a liquid of doubly occupied sites, i.e., a conductor with a spin gap $\sim U$. For more general interactions, we may envisage a (normal) quantum liquid of mobile singlet bonds which extend over several lattice sites. We do not wish to join the dispute as to whether such model results are pertinent to the behaviour of the strange normal metallic phase of certain high-T_C superconductors.

be qualitatively understood on the basis of the simple picture developed in Ch. 4. Afterwards, we discuss some complications.

V_2O_3 has the so-called corundum structure. For our purposes, it suffices to think of the structure as the sequence of puckered honeycomb lattices of vanadium sites. We will call these the *ab* planes, and the orthogonal direction the *c*-axis. The local environment is such that the *d*-level splits[13] into a non-degenerate a_{1g} level and two twofold degenerate e_g levels. Correspondingly, we envisage a tight binding band structure consisting of non-degenerate a_{1g} and twofold degenerate e_g bands.

The band structure of a complicated material like this is subject to recurring revisions, thus we rather prefer to take clues from experiments. Ti_2O_3 is another interesting transition metal oxide, and it also has the corundum structure. At low temperatures, it is a non-magnetic insulator. People often use a real-space picture, in which each Ti ion donates its 3*d* electron to form a singlet bond with its nearest neighbour along the *c*-direction. This also implies a doubling of the unit cell along the *c*-axis. In band language, the nature of the ground state can be understood to arise from the filling of a non-degenerate a_{1g} band. We assume that the sequence of the bands is the same in V_2O_3, thus the lowest-lying a_{1g} band is filled. The electrical and magnetic properties are then ascribed to the remaining one electron per V site which become Mott-localized. This would suggest that though a free V^{3+} ion has $S = 1$, in V_2O_3 the localized moments belong rather to $S = 1/2$, possibly augmented by some orbital contributions[14].

Nominally pure and stoichiometric V_2O_3 is an antiferromagnetic insulator below $T_N \sim 160$K. The ordered moment is $\sim 1.2\mu_B$. It is somewhat larger than the spin-only value for $S = 1/2$.

The antiferromagnetic–paramagnetic transition at the Néel temperature T_N is at the same time an insulator-to-metal transition of strongly first-order character (Fig. 5.4, left): the resistivity drops by about six orders of magnitude [115, 255]. We identify this transition from an anti-

[13]The six oxygen ions surrounding the V site form approximately an octahedron. The insulating phase shows a marked trigonal distortion, and this gives rise to the a_{1g}–e_g splitting.

[14]NMR measurements [388] confirm the presence of an orbital contribution.

ferromagnetic insulator (AFI) to a paramagnetic metal (PM) as a Mott transition. Since the transition is accompanied by substantial changes in the structure and volume, it is plausible to assume that the change of the bandwidth W is sufficient to take us from $(U/W) > (U/W)_{cr}$ to $(U/W) < (U/W)_{cr}$.

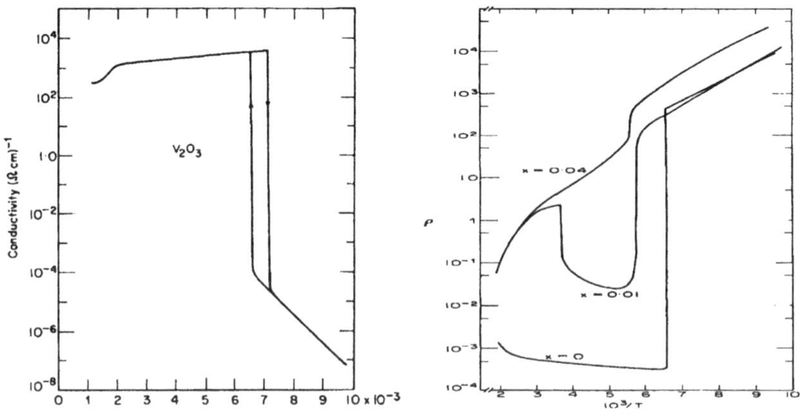

Figure 5.4: *Left*: The conductivity of V_2O_3 as a function of the inverse temperature [115]. A metal–insulator transition occurs at $T \sim 160$K. *Right*: The resistivity of $(V_{1-x}Cr_x)_2O_3$ in Ωcm as a function of inverse temperature, for $x = 0$, 0.01, and 0.04 [255].

An obvious way to decrease U/W is to apply pressure which, by compressing the lattice, widens the bands. Indeed one finds that T_N decreases with increasing pressure, and the AFI phase is completely suppressed at $P \sim 24$kbar (Fig. 5.5, left). We may also be wondering what happens at larger interatomic distances, and feel sorry that we cannot apply negative pressure! However, in some sense, we can. Namely, we can argue that the primary effect of introducing impurities which are bigger than the V ions is the expansion of the lattice. This idea seems to work wonderfully with Cr-doping: substituting several per cent of V with Cr produces the effects which we expect from an increasing U/W [256]. Naturally, we can check the correctness of this approach by substituting also smaller ions (such as Ti), and look if this acts like positive pressure. It seems that it does and one can actually determine which pressure is equivalent to a given degree of doping. Fig. 5.5 (left) shows an experimental phase diagram, in which a unified scale is used for pressure, and Ti- and Cr-substitution.

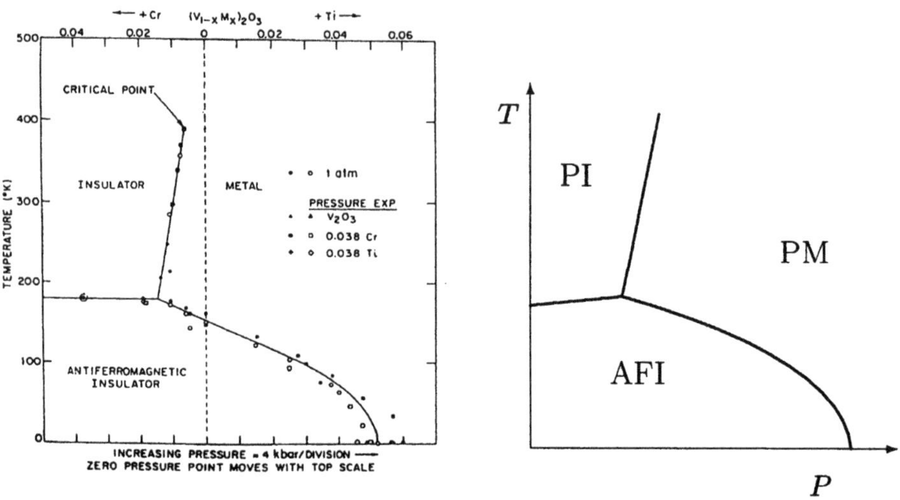

Figure 5.5: *Left*: Generalized phase diagram of the $(V_{1-x}Cr_x)_2O_3$ system, based on experiments by McWhan *et al.* [256]. *Right*: Schematic phase diagram of a non-nesting, half-filled Hubbard model in the T–P plane. The phases: AFI: antiferromagnetic insulator, PI: paramagnetic insulator, PM: paramagnetic metal. It is assumed that at $P = 0$, t/U is still fairly large.

Studying the $(V_{1-x}Cr_x)_2O_3$ system, we become aware of the existence of *two* paramagnetic phases: there is a paramagnetic insulator (PI) in addition the PM phase which we met at $x = 0$. Actually, this was to be expected: in Sec. 4.2, we were forcefully arguing that the insulating nature of a Mott insulator is not due to magnetic long-range order! The canonical behaviour of a Mott insulator is to undergo an AFI–PI transition at the Néel temperature, and this is what we find in the "negative pressure" regime (say, at x=0.04). On the other hand, the high-P end of the AFI domain is apparently not so strongly correlated, and the gap is sustained (at least partially) by the AF order[15].

[15]At this point, it is becoming debatable whether we are speaking about a Mott insulator or a spin density wave (SDW) insulator (the latter is also called a Slater insulator). If the AFI–PM transition were second-order, we should not hesitate to identify it as a SDW transition, but since it is rather strongly first-order, we think

It is a different question that at finite temperatures, we cannot really tell a (poor) metal from an insulator: they both have finite conductivities and need not differ in their symmetry properties. The change from PI to PM could be perfectly continuous, and in fact, at high enough temperatures it is just that. The PM–PI phase boundary in Fig. 5.5 is a line of first-order transitions, and it ends at a critical point. It reminds us of the well-known liquid–gas phase boundary. In both cases, the transition occurs without a change in global symmetries, being associated with a jump in density-like quantities, and in the entropy.

We can even think of a reason why there should be a first-order transition from a low-temperature correlated PM phase to a high-temperature PI phase. The conduction electron entropy of a metal is $S \sim k_{\rm B}T/T_{\rm F}$, where $T_{\rm F}$ is the Fermi temperature. On the other hand, the electrons are localized in the paramagnetic insulator, and therefore they have the spin disorder entropy $S \sim k_{\rm B} \ln 2$. Making the plausible assumption that $\ln 2 \gg T/T_{\rm F}$, we predict that at some finite temperature, the free energy $\mathcal{F} = U - TS$ of the insulator becomes lower than that of the metal. This explains why the PM–PI phase boundary in Fig. 5.5 is slanting to the right[16]. We mention that the first-order PM–PI transition is also called a Mott transition, and rightly: it takes from a metallic state to an insulator in which the electrons are collectively localized. Some people would say that only this is a true Mott transition, since it is not associated with magnetic ordering.

The schematic phase diagram shown in the right-hand panel of Fig. 5.5 sums up our expectations about the typical behaviour of a half-filled non-degenerate Hubbard model[17]. Essentially the same phase diagram is obtained from studies on non-nesting variants of the infinite

---

that it keeps some of its Mott-character. SDW's are discussed in Sec. 7.6.

[16]This feature makes it possible that at certain pressures (or Cr concentrations) *two* consecutive transitions can be observed: first from AFI to PM, and at a higher T, from PM to PI (see the curve $x = 0.01$ in Fig. 5.4, right).

[17]"Typical" implies that the band structure does not have the perfect nesting property which would require a fine-tuning of the parameters. The schematic phase diagram for a nested Fermi surface is shown in Fig. 7.6. Furthermore, note that since U/W is increasing to the left, the antiferromagnetic exchange coupling $J \sim W^2/U$ should be decreasing to the left. That's why we were careful to draw a gently sloping AFI–PI boundary in the schematic phase diagram. This feature does not become apparent in the experimental phase boundary.

dimensional Hubbard model [130]. Apparently, the overall features of the Mott transition are the same in the half-filled single-band model, and in degenerate models with $n = 1$ [225]. The only feature we have to be cautious about is the character of the PM–PI phase boundary. Though we brought a qualitative argument in favour of a line of first-order transitions, it is possible that the pure Hubbard model does not provide a mechanism for discontinuous behaviour above T_N, and the fact that real systems do show first-order transitions, has to be ascribed to the long-range part of the Coulomb interaction or/and to electron–lattice coupling. It remains possible that the sharp PM–PI boundary has to be replaced by a broad range of "semimetallic" behaviour (see Fig. 10.18).

A good 25 years after the discovery of the metal–insulator transition [115], V_2O_3 research had a flourishing period in the early seventies when the phase diagram shown in Fig. 5.5 was established. Another 20 years later, the production of $V_{2-y}O_3$ samples with a carefully controlled oxygen content led to the discovery a novel metallic spin density wave (SDW) phase [39]. Fig. 5.6 shows the combined phase diagram of the $V_{2-y}O_3$ and $(V_{1-x}Cr_x)_2O_3$ systems. It appears that the new findings require a serious consideration of orbital degeneracy.

The metallic SDW phase appears in vanadium deficient (hole-doped) samples at temperatures $T_{SDW} \sim 9$K. Like the Néel temperature T_N of the AFI, T_{SDW} decreases under pressure (Fig. 5.6). The order is not far from an alternating AF order (Fig. 5.7, left); the angle between nearest-neighbour spins in the *ab* planes is $\sim 0.87\pi$. All this would be just what one expected on the basis of a single-band model. However, we would also expect that the AFI state arises from the continuation of the SDW behaviour and that is certainly *not* the case. The honeycomb lattice is bipartite and the usual kinetic exchange should lead to two-sublattice antiferromagnetism, the apparent limiting behaviour of the SDW (this would hold also for the *c*-axis order). Look, however, what the AFI is doing (Fig. 5.7, right): one third of the nearest neighbour spins is parallel!

It is clear that the AFI phase is governed by a spin Hamiltonian in which a fraction of the bonds is ferromagnetic. Actually, we can think of a way how this may come about: due to orbital ordering. The

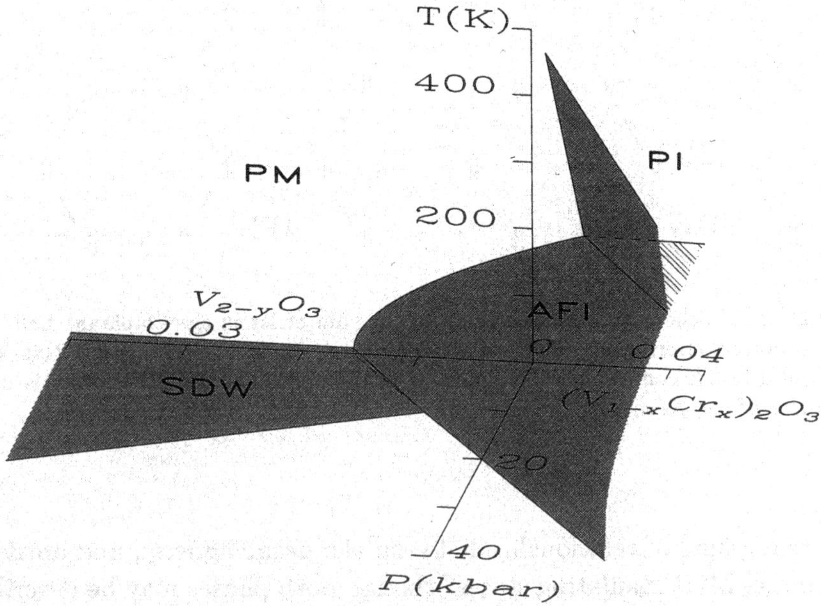

Figure 5.6: Unified phase diagram of the $V_{2-y}O_3$ and $(V_{1-x}Cr_x)_2O_3$ systems, after [38]. There are two ordered phases: the antiferromagnetic Mott insulator (AFI), and a metallic spin density wave (SDW). The PM and PI phases are separated by a surface of first-order transitions, cut off by a critical line. (By courtesy of G. Aeppli).

unpaired electrons of V_2O_3 reside in twofold degenerate e_g shells, and we have seen in Sec. 5.4 that the alternating occupation of the two orbitals induces ferromagnetic coupling. The AFI must be an orbitally ordered state. It has been suggested that *orbital antiferromagnetism is the primary order* of the AFI, and that the strongly first-order transitions leading from the AFI to any other state are associated with the disappearance of orbital order [40]. We have seen in Problem 5.7 (see p. 257) that the collapse of orbital order causes a dramatic switching of the exchange coupling from ferromagnetic to antiferromagnetic. This would also explain that the short-range order seen in the PM and PI phases is of the same nature as the order of the SDW state, but quite unlike the AFI order. The postulated primacy of orbital order implies a

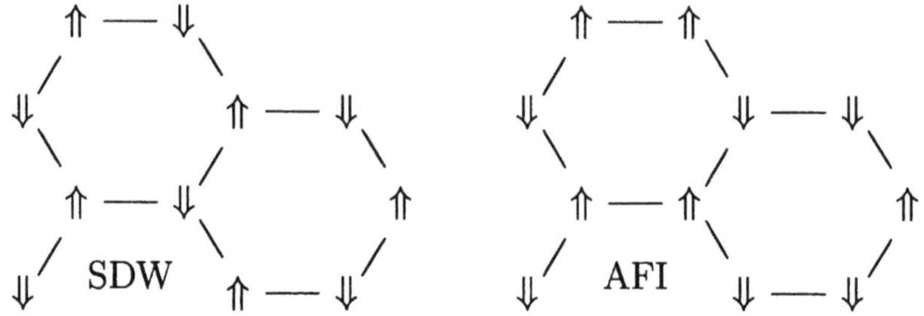

Figure 5.7: **Schematic representation of the spin order in the *ab* plane. *Left*: the approximately alternating order of the SDW state. *Right*: the AFI phase. Note that horizontal bonds connect parallel spins. It is apparent that the SDW order is not a precursor of the AF order.**

surprising lack of relationship between the paramagnetic, and antiferomagnetic, Mott insulating states: though both phases may be described by effective (spin) Heisenberg models, the two Heisenberg models are completely different!

A cautionary remark: the picture we sketched is not necessarily the correct one for V_2O_3. We relied on the concept of a filled a_{1g} band and a quarter-filled e_g band. In the literature, the reader will find claims that the a_{1g} band is empty, or that it is not empty but it is spin-polarized. It cannot be our task here to decide which view is correct. Rather, the richness of the material world should be taken as a challenge to exercise ourselves in the use of the concepts we learned. The insight that orbital ordering must be playing an important role, is likely to be of lasting value, whatever the details of the band structure will turn out to be.

V_2O_3 is only one of a substantial number of transition metal compounds which show metal–insulator transitions, and of those, many are arguably Mott transitions, or have such an aspect. We cannot attempt to review them. However, we mention that the $La_{1-x}Sr_xTiO_3$ system gives a beautiful example of what happens when the Mott insulating phase is approached by changing the band filling. This will be described in Sec. 10.5.

5.5.2 Mott Insulators versus Charge Transfer Insulators

Materials like CoO, or La_2CuO_4, are insulators because an electron correlation effect splits the partially filled d-band into a set of either completely filled, or empty, Hubbard subbands. However, these subbands are imbedded into a sequence of wider bands which are derived from the oxygen $2p$ and the transition metal[18] $4s$ states. The character of the insulator is decided by the lowest excitation energy which is sufficient for creating a charge carrier: this will appear as the activation energy in the Arrhenius law for the electrical conductivity. The Mott–Hubbard gap is not necessarily the smallest of the gaps; it is possible that the relevant excitation is a charge transfer process between a Hubbard subband and a wide band.

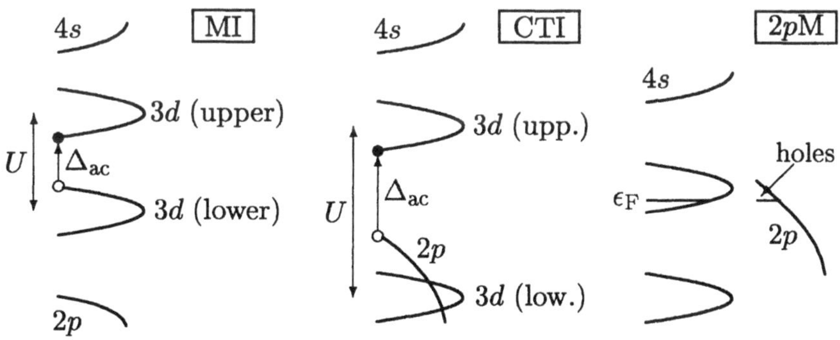

Figure 5.8: Shifting the relative positions of the Hubbard subbands and the $2p$ band gives rise to different kinds of systems. In a Mott insulator (MI), the smallest gap is between two Hubbard subbands. In a charge transfer insulator (CTI), the lowest-energy process is the excitation of an electron from the $2p$ band to the upper Hubbard subband. In both cases, the activation energy Δ_{ac} is marked. In the right-hand panel, the overlapping of the $2p$ band with the upper Hubbard subband gives a metal ($2p$M) in which the dominant charge carriers are $2p$ holes. The horizontal lines indicate the Fermi level ϵ_F.

Fig. 5.8 illustrates several of the possibilities, by inserting the Hubbard subbands into the midst of the wider bands. We should remember

[18]We do not bother to distinguish the noble metal Cu from transition metals; Cu^{2+} is like the transition metal ions.

that showing just two Hubbard subbands is an oversimplification: crystal field splittings and the actual degeneracy of the d-band give rise to a number of subbands. For the present argument, we assume that there are enough electrons to fill the $2p$ valence band and the lower Hubbard subband completely, while the upper Hubbard subband and the $4s$ band remain empty. The canonical case of Mott insulators is shown on the left: the $3d$ levels lie in between the wider bands, and the lowest-energy charge excitation is from the lower to the upper Hubbard subband. V_2O_3 and $LaTiO_3$ may be insulators of this sort. Though the most abundant charge carriers are d-like, it may turn out that substantial electrical current is carried by the $4s$-electrons: they require a higher activation energy but are much more mobile than the d-electrons.

The insulating character survives if the filled bands overlap, with the top of the valence band lying above the lower Hubbard subband (Fig. 5.8, middle picture). However, the nature of the gap changes: now the most relevant excitation process removes an electron from the $2p$ band, placing it into the upper Hubbard subband. A substance showing this kind of behaviour is called a *charge transfer insulator* [462, 348]. The excitation process creates a d-like quasiparticle and a p-hole; we expect that the outcome is hole-type conduction, because of the higher mobility of the $2p$ holes. La_2CuO_4 is one of the materials which we classify as charge transfer insulators [338].

We should understand clearly that the reason for the insulating nature is in both cases the same: the Mott localization, and the resulting splitting of the d-band into Hubbard subbands. Charge transfer insulators are *not* straightforward band insulators, even though the coexistence of wide bands with narrower, strongly correlated bands is essential for their definition. In a broader sense, both V_2O_3 and La_2CuO_4 are Mott insulators. The presently discussed subdivision into Mott insulators in a narrower sense, and charge transfer insulators is, however, useful because it helps to bring order into the bewildering variety of materials and phenomena.

The right panel of Fig. 5.8 shows the extreme case when the $2p$ band overlaps the upper Hubbard subband. In this situation, a number of electrons is transferred from the top of the valence band into d-type states; the Fermi level ϵ_F lies inside the $2p$ band (and the upper Hubbard subband), and we have a p-metal with hole-type conduction. CuS gives an example for such a behaviour. Here it is interesting to notice that metallic character can arise in spite of the presence of a large Mott–Hubbard gap. Let us observe that this is associated with the coexistence of the d^n and d^{n+1} configurations. We will systematically investigate this situation in the chapter on mixed valence.

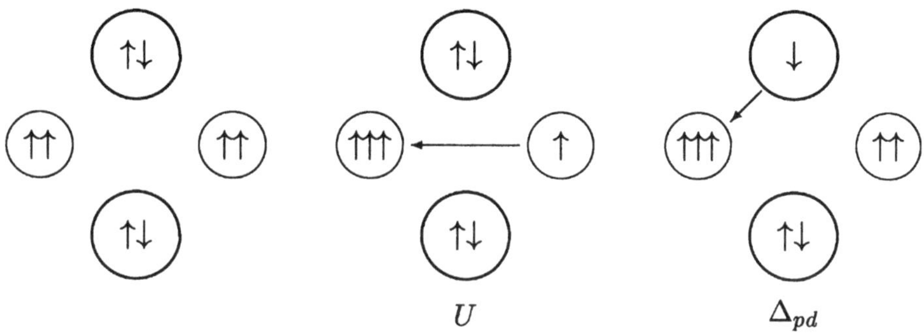

Figure 5.9: Excitation processes in a transition metal oxide (small circles: cations, large circles: anions). Middle: the cation–cation transfer of a d-electron costs an energy U. Right: moving an electron from an anionic p-shell to a cation requires the charge transfer energy Δ_{pd}.

Fig. 5.9 shows the two kinds of charge excitations as real-space processes (for illustration, we have chosen the d-band filling $n = 2$). Excitation from the lower to the upper Hubbard subband corresponds to the process

$$d^n + d^n \rightarrow d^{n+1} + d^{n-1} \,. \tag{5.80}$$

Writing things like d^{n+1} is not unambiguous at all: with our d^3 it could mean either a low-spin ($S = 1/2$), or a high-spin ($S = 3/2$) state, and further splittings arise in a crystal field. Here, and in similar arguments, we will assume that each of the ions is in its Hund's rule ground state (the middle panel of Fig. 5.9 illustrates the point). With such restrictions, the d-electron hopping becomes a difficult-to-describe, correlated event like the ones we studied in Sec. 5.1.1. The corresponding activation energy can be estimated as

$$\Delta_{\mathrm{ac}} = U - W_d \,. \tag{5.81}$$

The argument for subtracting the d-bandwidth W_d is the same that we used for Fig. 4.8: we have created a quasi-particle (d^{n+1}) and a quasi-hole (d^{n-1}) which can propagate on the background of d^n ions with the d-electron hopping amplitudes, broadening the levels into subbands.

The $p-d$ charge transfer process (Fig. 5.9, right) costs primarily $\Delta_{pd} = \tilde{\epsilon}_d - \epsilon_p$, where $\tilde{\epsilon}_d$ is the center of the upper Hubbard subband. To get the activation energy, this has to be corrected for the broadening of the quasi-particle, and quasi-hole, levels into bands:

$$\Delta_{\rm ac} = \Delta_{pd} - \frac{W_p + W_d}{2} \tag{5.82}$$

whereby the contribution due to the large p-bandwidth is essential.

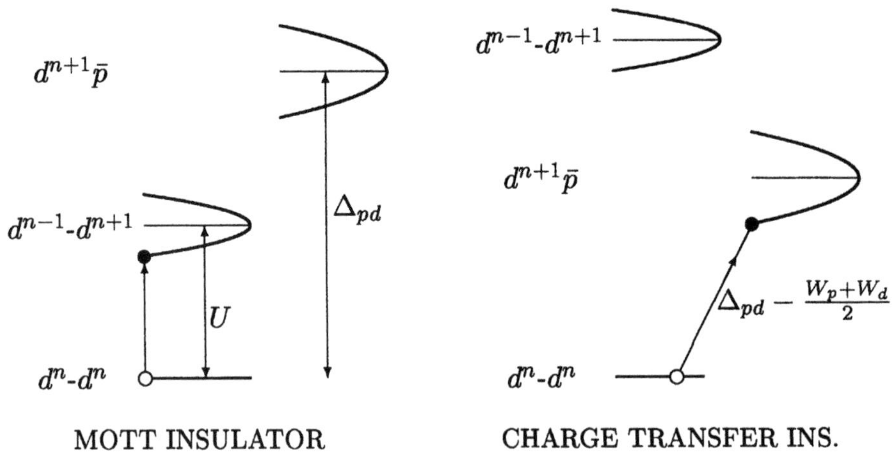

Figure 5.10: Excitation spectra of $d-d$ and $p-d$ type charge transfer processes in Mott insulators (left), and charge transfer insulators (right). $\bar{p}$ signifies a p-hole.

Though Fig. 5.8 looks as if it were about ordinary bands, it has to be used with great caution because what it really shows are spectra of itinerant many-electron excitations. If we prefer to be more circumspect [348], we have to redraw the diagrams as shown in Fig. 5.10. In contrast to 5.8 with its filled or empty bands, the ground state is represented here as the dispersionless level d^n–d^n. This corresponds to our intuitive notion of a rigid Mott-localized state. It is a quite different matter that excited particles (and holes) propagate and have a finite dispersion. In Fig. 5.10, band broadening is ascribed to the quasiparticle excitations but not to the strongly correlated ground state.

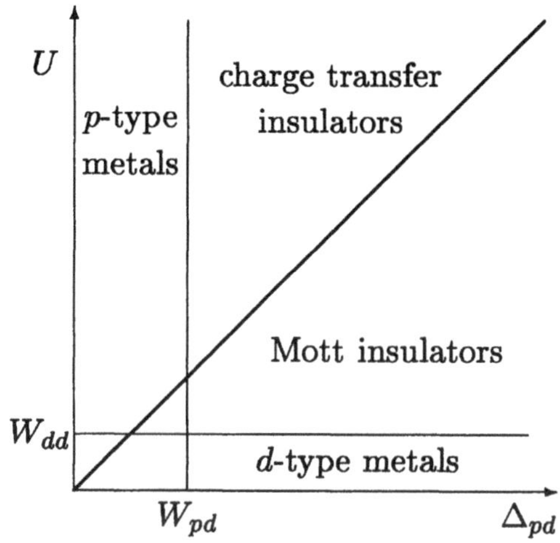

Figure 5.11: The classification of correlated insulators in the parameter plane of the p-d two-band model. For $U > \Delta_{pd}$ a charge transfer insulator, while for $\Delta_{pd} > U$ a Mott insulator is expected. Weakly correlated systems ($U < W_{dd}$) are d-type metals. A high-lying, broad p-band ($W_{pd} > \Delta_{pd}$) overlaps the upper Hubbard subband, resulting in a metal with p-type holes.

The diagram shown in Fig. 5.11 serves to classify the transition metal compounds into the broad classes of Mott insulators, and charge transfer insulators [462, 348]. Formally, both activation energies (5.81) and (5.82) can become negative, marking the transition to a weakly correlated d-metal, or a correlated p-hole metal, respectively.

The knowledge about the relative sizes of Δ_{pd} and U is important in superexchange theory (Sec. 5.2), and for understanding the results of various spectroscopies. A rule of thumb says that the early transition metals (Ti, V, Cr) tend to form Mott insulators, while towards the end of the $3d$ series (Ni, Cu) we are likely to find charge transfer insulators. Apparently, CoO belongs to the intermediate regime where $\Delta_{pd} \sim U$. The sharp insulator–insulator boundary shown if Fig. 5.11 is clearly artificial (it is not a phase boundary!); we had better speak about a broad transition region between the two kinds of insulator [462]. In contrast, the (very schematically drawn) metal–insulator boundaries belong to ground state phase transitions and have to be sharp.

The basic similarity of the two classes of insulators is also evident in the fact that the dominating term of their low-energy effective Hamiltonians is the same: it is the (most often antiferromagnetic) nearest-neighbour Heisenberg coupling. The difference is merely that in a Mott insulator the superexchange process (5.60), while in a charge transfer insulator the process (5.59) is more important.

Hole Doping

In our model studies, the ground state is predicted to be metallic whenever the d-band filling deviates from its stoichiometric (integer) values. The experimental tool for achieving this is to dope the material with cations (or anions) of a different valence. For instance, replacing a fraction of the trivalent La ions in La_2CuO_4 (or $LaTiO_3$) by divalent Sr ions gives the hole-doped metals $La_{2-x}Sr_xCuO_4$ (or $La_{1-x}Sr_xTiO_3$). Though both undoped systems are antiferromagnetic insulators with $S = 1/2$ localized moments, Sr doping produces very different results: $x \approx 0.125$ is the optimum doping for making $La_{2-x}Sr_xCuO_4$ a high-temperature superconductor, while $La_{1-x}Sr_xTiO_3$ is a non-magnetic heavy Fermi liquid[19].

Though it cannot give any immediate insight into why superconductivity appears in one case but not in the other, let us point out one important difference between the two systems: we classified $LaTiO_3$ as a Mott insulator, while La_2CuO_4 as a charge transfer insulator. We can use Fig. 5.8 to predict where the holes are going. In a Mott insulator, the highest occupied states are d-like, thus we describe doped $LaTiO_3$ as $La_{1-x}Sr_xTi^{3+}_{1-x}Ti^{4+}_xO_3$. The Fermi level is lying in the lower Hubbard subband. – In contrast, in La_2CuO_4 the highest occupied states are p-like, thus Sr doping creates holes at the top of the $2p$ valence band. $La_{2-x}Sr_xCuO_4$ might have turned out to be a relatively straightforward p-type metal. However, it is anything but straightforward: either it is a high-T_c superconductor, or a very strange "normal" conductor.

Considering the Sr-doping of La_2CuO_4 highlights a peculiarity of doping Mott insulators: we not only introduce an extra charge into a

---

[19]The $La_{1-x}Sr_xTiO_3$ system will be discussed in detail in Sec. 10.5.

non-conductor, but also an extra spin into an ordered magnet. Namely, a p-hole amounts to leaving one of the p-lobes of an oxygen ion with an uncompensated spin. This spin is sorrounded by Cu-site spins, and p–d hybridization gives rise to an antiferromagnetic Cu^{2+}–O^- exchange interaction which consists of contributions $\sim t_{pd}^2/\epsilon_p$ and $\sim t_{pd}^2/(U_d - \epsilon_p)$. Actually, it is best to start by considering the four oxygen sites surrounding a central copper site, and let the p-hole spread out to the four sites occupying an orbital which has the same symmetry as the spin-carrying d-orbital at the central Cu site. This p-state binds optimally to the central d-spin, forming a *Zhang–Rice singlet* [468]. Low-level Sr doping of La_2CuO_4 can be envisaged as giving rise to a dilute gas of Zhang–Rice singlets which, though they are fairly complex, spread-out objects[20], still can propagate[21].

The case of electron doping is different in the sense that it does not distinguish between the two classes of insulators: in either case, added electrons go into the upper Hubbard subband[22]. This means that electron doping and hole doping play asymmetric roles in charge transfer insulators: for instance, electron-type La_2CuO_4-like superconductors (of which $Nd_{2-x}Ce_xCuO_4$ provides an example) are very different from the hole-type ones. In contrast, for a Mott insulator, the effects of electron-doping and hole-doping can be interpreted in a unified scheme.

Doping is often not intentional at all. Apart from the unavoidable impurities, transition metal oxides just do not want to grow stoichiometrically: it is not unusual to find several per cent deviation of the oxygen content from its nominal integer value[23]. This is a serious problem:

---

[20]Looking from far away, we see three charges missing from the immediate environment of a Cu atom, so we may think it is the low-spin state of Cu^{3+}. At a closer look, into resolves into an ordinary Cu^{2+} bound to a surrounding p-hole complex. It is a not unfrequent experience that uncommonly looking higher oxidation states of metal ions had better be envisaged as the ion in a lower oxidation state, bound by exchange interaction to a surrounding ligand hole.

[21]However, consideration of the long-range Coulomb force may lead to the conclusion that they localize into thin domain walls. See p. 568.

[22]We might have thought that sometimes they go into the 4s band. However, this case does not seem to be realized.

[23]In spite of this fact, CoO or magnetite are not metallic. The reason is that, because of the narrowness of the 3d-band, polaronic effects come into play. Small polarons are heavy quasi-particles which are readily localized by the random potential

it is clear that in the vicinity of the metal–insulator phase boundary, all properties depend very sensitively on the degree of stoichiometry! Furthermore, we should be conscious of the fact that this makes it difficult to judge the relevance of particular experimental results. It is not enough to say: "measurements on magnetite (or on V_2O_3, etc.) found this and this"; we have to check how the sample was grown, what was the nominal composition, the likely concentration of vacancies, sample homogeneity, etc. This is the reason why a few well-characterized samples of various substances are held in great esteem for a while. Let us recall that when, thanks to a great experimental effort, it became possible to control the oxygen content of highly homogeneous V_2O_3 samples to three decimal places, a new ordered phase was discovered [39].

Solutions to the Problems

Problem 5.1 In the limit $t/U \ll 1$, second-order perturbation theory captures only pair processes, thus it is justified to confine our interest to a nearest-neighbour pair of sites 1 and 2. The zeroth-order low-energy states are the pure spin states $|\uparrow\uparrow\rangle = c^\dagger_{1\uparrow}c^\dagger_{2\uparrow}|0\rangle$, $|\uparrow\downarrow\rangle = c^\dagger_{1\uparrow}c^\dagger_{2\downarrow}|0\rangle$, $|\downarrow\uparrow\rangle = c^\dagger_{1\downarrow}c^\dagger_{2\uparrow}|0\rangle$, and $|\downarrow\downarrow\rangle = c^\dagger_{1\downarrow}c^\dagger_{2\downarrow}|0\rangle$. In first order of t, we get the new ("rotated") basis states

$$\begin{aligned} |\uparrow\downarrow\rangle' &= |\uparrow\downarrow\rangle + \frac{t}{U}(|0d\rangle + |d0\rangle) \\ |\downarrow\uparrow\rangle' &= |\downarrow\uparrow\rangle - \frac{t}{U}(|0d\rangle + |d0\rangle)\,. \end{aligned} \tag{5.83}$$

Though the perturbed states contain an admixture of ionized configurations, they can still be indexed unambiguously by their parent spin configurations. The parallel-spin states remain unaltered: $|\uparrow\uparrow\rangle' = |\uparrow\uparrow\rangle$, and $|\downarrow\downarrow\rangle' = |\downarrow\downarrow\rangle$. In the rotated basis, the matrix elements are

$$\begin{aligned} {}'\langle\uparrow\downarrow|\mathcal{H}_{\text{band}}|\uparrow\downarrow\rangle' &= -4t^2/U \\ {}'\langle\uparrow\downarrow|\mathcal{H}_{U}|\uparrow\downarrow\rangle' &= 2t^2/U \end{aligned} \tag{5.84}$$

thus

$${}'\langle\uparrow\downarrow|\mathcal{H}_{\text{Hubbard}}|\uparrow\downarrow\rangle' = -2t^2/U\,, \tag{5.85}$$

---

of charged impurities or vacancies [5, 293].

and similarly for $|\downarrow\uparrow\rangle'$. The diagonal matrix elements of $\mathcal{H}_{\text{Hub.}}$ are those of the Ising-like term

$$\mathcal{H}_{\text{eff}}^{(1)} = -\frac{2t^2}{U}\left(\frac{1}{2} - 2S_1^z S_2^z\right) . \tag{5.86}$$

We find also non-diagonal matrix elements:

$$\begin{aligned} {}'\langle\uparrow\downarrow|\mathcal{H}_{\text{band}}|\downarrow\uparrow\rangle' &= 4t^2/U \\ {}'\langle\uparrow\downarrow|\mathcal{H}_U|\downarrow\uparrow\rangle' &= -2t^2/U \end{aligned} \tag{5.87}$$

which add up to

$$ {}'\langle\uparrow\downarrow|\mathcal{H}_{\text{Hubbard}}|\downarrow\uparrow\rangle' = 2t^2/U . \tag{5.88}$$

These can be expressed by the spin-flip (or x–y) term

$$\mathcal{H}_{\text{eff}}^{(2)} = \frac{2t^2}{U}\left(S_1^+ S_2^- + S_2^+ S_1^-\right) . \tag{5.89}$$

All in all, in the rotated basis the Hubbard Hamiltonian acts like the isotropic Heisenberg model

$$\mathcal{H}_{\text{eff}} = \mathcal{H}_{\text{eff}}^{(1)} + \mathcal{H}_{\text{eff}}^{(2)} \rightsquigarrow \frac{4t^2}{U}\mathbf{S}_1 \cdot \mathbf{S}_2 . \tag{5.90}$$

The partitioning of the matrix elements into hopping and interaction contributions makes it clear that the "driving force" of the antiferromagnetic Heisenberg coupling is the band term: it leads to a small ($\sim t/U$) delocalization of the electrons, and the resulting gain in kinetic energy exceeds the Coulomb energy cost.

Problem 5.2 Expressing $\hat{n}_{\mathbf{k}\sigma}$ in terms of Wannier operators

$$\langle\Psi|c_{\mathbf{k}\sigma}^\dagger c_{\mathbf{k}\sigma}|\Psi\rangle = \frac{1}{2} + \frac{1}{L}\sum_m \sum_{n\neq m} e^{i\mathbf{k}(\mathbf{m}-\mathbf{n})}\langle\Psi|c_{\mathbf{m}\sigma}^\dagger c_{\mathbf{n}\sigma}|\Psi\rangle . \tag{5.91}$$

To evaluate the $\mathbf{m} \neq \mathbf{n}$ terms let us make the canonical transformation to the state $|\tilde{\Psi}\rangle$ which contains no doubly occupied or empty sites

$$\begin{aligned} \langle\Psi|c_{\mathbf{m}\uparrow}^\dagger c_{\mathbf{n}\uparrow}|\Psi\rangle &= \langle\Psi|e^{-iS}e^{iS}c_{\mathbf{m}\uparrow}^\dagger c_{\mathbf{n}\uparrow}e^{-iS}e^{iS}|\Psi\rangle = \langle\tilde{\Psi}|e^{iS}c_{\mathbf{m}\uparrow}^\dagger c_{\mathbf{n}\uparrow}e^{-iS}|\tilde{\Psi}\rangle \\ &\approx i\langle\tilde{\Psi}|[S, c_{\mathbf{m}\uparrow}^\dagger c_{\mathbf{n}\uparrow}|\tilde{\Psi}]\rangle . \end{aligned} \tag{5.92}$$

Replacing $S = -(i/U)(\mathcal{H}_t^+ - \mathcal{H}_t^-)$, and remembering that the action of $\mathcal{H}_t^-$ on $|\tilde{\Psi}\rangle$ gives zero

$$\langle\Psi|c_{\mathbf{m}\uparrow}^\dagger c_{\mathbf{n}\uparrow}|\Psi\rangle = -\frac{1}{U}\langle\tilde{\Psi}|\mathcal{H}_t^- c_{\mathbf{m}\uparrow}^\dagger c_{\mathbf{n}\uparrow}|\tilde{\Psi}\rangle - \frac{1}{U}\langle\tilde{\Psi}|c_{\mathbf{m}\uparrow}^\dagger c_{\mathbf{n}\uparrow}\mathcal{H}_t^+|\tilde{\Psi}\rangle . \tag{5.93}$$

The neutrality of the ground state $|\tilde{\Psi}\rangle$ requires that $\mathcal{H}_t^+$ ($\mathcal{H}_t^-$) must be acting on the same pair $\mathbf{m}$, $\mathbf{n}$ (there are no three-site contributions). It follows that the above expectation value is non-vanishing only if $\mathbf{m}$ and $\mathbf{n}$ are nearest neighbours. The subsequent calculation is easiest if we express everything with Hubbard operators

$$\begin{aligned}\langle\Psi|c^{\dagger}_{\mathbf{m}\uparrow}c_{\mathbf{n}\uparrow}|\Psi\rangle \quad &\rightsquigarrow \quad \frac{t}{U}\langle\tilde{\Psi}|(X^{\downarrow\leftarrow d}_{\mathbf{m}}X^{\uparrow\leftarrow 0}_{\mathbf{n}} - X^{\uparrow\leftarrow d}_{\mathbf{m}}X^{\downarrow\leftarrow 0}_{\mathbf{n}})X^{d\leftarrow\downarrow}_{\mathbf{m}}X^{0\leftarrow\uparrow}_{\mathbf{n}}|\tilde{\Psi}\rangle \\ &\quad + \frac{t}{U}\langle\tilde{\Psi}|X^{\uparrow\leftarrow 0}_{\mathbf{m}}X^{\downarrow\leftarrow d}_{\mathbf{n}}(X^{d\leftarrow\downarrow}_{\mathbf{n}}X^{0\leftarrow\uparrow}_{\mathbf{m}} - X^{d\leftarrow\uparrow}_{\mathbf{n}}X^{0\leftarrow\downarrow}_{\mathbf{m}})|\tilde{\Psi}\rangle \\ &= \frac{2t}{U}\langle\tilde{\Psi}|(X^{\uparrow\leftarrow\uparrow}_{\mathbf{m}}X^{\downarrow\leftarrow\downarrow}_{\mathbf{n}} - X^{\uparrow\leftarrow\downarrow}_{\mathbf{m}}X^{\downarrow\leftarrow\uparrow}_{\mathbf{n}})|\tilde{\Psi}\rangle. \qquad (5.94)\end{aligned}$$

In the last line, the first term expresses z-z type spin correlations

$$\begin{aligned}\langle\tilde{\Psi}|X^{\downarrow\leftarrow\downarrow}_{\mathbf{m}}X^{\uparrow\leftarrow\uparrow}_{\mathbf{n}}|\tilde{\Psi}\rangle &= \langle\tilde{\Psi}|\left(\frac{1}{2}+S^z_{\mathbf{m}}\right)\left(\frac{1}{2}-S^z_{\mathbf{n}}\right)|\tilde{\Psi}\rangle \qquad (5.95)\\ &= \frac{1}{4}+\frac{1}{2}\langle\tilde{\Psi}|S^z_{\mathbf{m}}|\tilde{\Psi}\rangle - \frac{1}{2}\langle\tilde{\Psi}|S^z_{\mathbf{n}}|\tilde{\Psi}\rangle - \langle\tilde{\Psi}|S^z_{\mathbf{m}}S^z_{\mathbf{n}}|\tilde{\Psi}\rangle.\end{aligned}$$

The second term is essentially a spin-flip term

$$\langle\tilde{\Psi}|X^{\uparrow\leftarrow\downarrow}_{\mathbf{m}}X^{\downarrow\leftarrow\uparrow}_{\mathbf{n}}|\tilde{\Psi}\rangle = \langle\tilde{\Psi}|S^+_{\mathbf{m}}S^-_{\mathbf{n}}|\tilde{\Psi}\rangle. \qquad (5.96)$$

A similar derivation (paying due attention to the negative sign appearing in (5.24)) can be done for the hopping amplitude of the $\downarrow$-spin electrons, yielding

$$\begin{aligned}\langle\Psi|c^{\dagger}_{\mathbf{m}\downarrow}c_{\mathbf{n}\downarrow}|\Psi\rangle &= \frac{2t}{U}\left(\frac{1}{4}+\frac{1}{2}\langle\tilde{\Psi}|S^z_{\mathbf{n}}|\tilde{\Psi}\rangle - \frac{1}{2}\langle\tilde{\Psi}|S^z_{\mathbf{m}}|\tilde{\Psi}\rangle - \langle\tilde{\Psi}|S^z_{\mathbf{m}}S^z_{\mathbf{n}}|\tilde{\Psi}\rangle\right. \\ &\quad \left. - \langle\tilde{\Psi}|S^-_{\mathbf{m}}S^+_{\mathbf{n}}|\tilde{\Psi}\rangle\right). \qquad (5.97)\end{aligned}$$

Replacing these results in (5.91), we arrive at the remarkable result cited in (5.45), namely that the leading correction to $n_{\mathbf{k}}$ is related to the nearest-neighbour spin–spin correlation function [385]

$$\langle\Psi|\hat{n}_{\mathbf{k}}|\Psi\rangle = \langle\Psi|\hat{n}_{\mathbf{k}\uparrow}+\hat{n}_{\mathbf{k}\downarrow}|\Psi\rangle \approx 1 - \frac{\epsilon(\mathbf{k})}{U}\left(1-4\langle\tilde{\Psi}|\mathbf{S}_1\cdot\mathbf{S}_2|\tilde{\Psi}\rangle\right) \qquad (5.98)$$

where $\epsilon(\mathbf{k})$ is the tight binding band energy.

Note that terms containing the spin density $\langle\tilde{\Psi}|S^z_{\mathbf{m}}|\tilde{\Psi}\rangle$ dropped out of the expression. (5.98) is equally valid for a paramagnetic ground state where $\langle\tilde{\Psi}|S^z_{\mathbf{m}}|\tilde{\Psi}\rangle = \langle\tilde{\Psi}|S^z_{\mathbf{n}}|\tilde{\Psi}\rangle = 0$, and for a magnetically ordered ground state with $\langle\tilde{\Psi}|S^z_{\mathbf{n}}|\tilde{\Psi}\rangle \neq 0$.

The one-dimensional Hubbard model has a ground state without magnetic long-range-order. It is known [175] that

$$\langle\tilde{\Psi}|\mathbf{S}_1 \cdot \mathbf{S}_2|\tilde{\Psi}\rangle = \frac{1}{4} - \ln 2\,. \tag{5.99}$$

Replacing this in (5.45), we get

$$\langle\Psi|\hat{n}_{k\sigma}|\Psi\rangle \approx \frac{1}{2}\left[1 + \frac{8t}{U}\cdot \ln 2 \cdot \cos k\right]. \tag{5.100}$$

This gives a characteristic example for the smooth k-dependence of the band state occupation number in a correlated insulator.

Problem 5.3. Let us consider first two electrons described by $\mathcal{H}_X$. It is clear that the electrons cannot hop unless they are at nearest neighbour sites, or on the same site. Since the hopping creates or destroys double occupation, only the singlet state is mobile. We can envisage two-electron propagation in the following manner

$$|000 \uparrow\downarrow 000\rangle \rightsquigarrow |0000d000\rangle \rightsquigarrow |0000 \downarrow\uparrow 00\rangle \ldots$$

Thus $\mathcal{H}_X$ "wants" to create mobile bound singlets which move with hopping amplitude X. One may wish to discuss this observation in connection with superconductivity [167, 43].

Switching on U, the pair propagation rate is slowed down to $\sim X^2/U$.

Problem 5.4. $\mathcal{H}_Y$ can be rewritten as

$$\mathcal{H}_Y = Y \sum_{\langle \mathbf{i},\mathbf{j}\rangle} (X_\mathbf{i}^{d\leftarrow 0} X_\mathbf{j}^{0\leftarrow d} + X_\mathbf{j}^{d\leftarrow 0} X_\mathbf{i}^{0\leftarrow d})\,. \tag{5.101}$$

As for $\mathcal{H}_X$, we decompose it like we did with $\mathcal{H}_{\text{band}}$ in (5.9). Let us gather the processes which increase the number of doubly occupied sites by creating one at $\mathbf{i}$. They arise from

$$\begin{aligned}
& X(c_{\mathbf{i}\uparrow}^\dagger c_{\mathbf{j}\uparrow}\hat{n}_{\mathbf{i}\downarrow} + c_{\mathbf{i}\downarrow}^\dagger c_{\mathbf{j}\downarrow}\hat{n}_{\mathbf{i}\uparrow}) \\
= \; & X[(X_\mathbf{i}^{\uparrow\leftarrow 0} + X_\mathbf{i}^{d\leftarrow\downarrow})(X_\mathbf{i}^{\downarrow\leftarrow\downarrow} + X_\mathbf{i}^{d\leftarrow d})(X_\mathbf{j}^{0\leftarrow\uparrow} + X_\mathbf{j}^{\downarrow\leftarrow d}) \\
& +(X_\mathbf{i}^{\downarrow\leftarrow 0} - X_\mathbf{i}^{d\leftarrow\uparrow})(X_\mathbf{i}^{\uparrow\leftarrow\uparrow} + X_\mathbf{i}^{d\leftarrow d})(X_\mathbf{j}^{0\leftarrow\downarrow} - X_\mathbf{j}^{\uparrow\leftarrow d}) \\
\rightsquigarrow \; & X(X_\mathbf{i}^{d\leftarrow\downarrow} X_\mathbf{j}^{0\leftarrow\uparrow} - X_\mathbf{i}^{d\leftarrow\uparrow} X_\mathbf{j}^{\uparrow\leftarrow d})\,.
\end{aligned}$$

These are the same processes as in $\mathcal{H}_t^+$, only $-t$ has to be replaced by X. The same holds for $\mathcal{H}_t^-$. Thus the decomposition of $\mathcal{H}_X$ is

$$\mathcal{H}_X = -\frac{X}{t}(\mathcal{H}_t^+ + \mathcal{H}_t^-) + \mathcal{H}_X^0\,, \tag{5.102}$$

where the last term can be written as

$$\mathcal{H}_X^0 = 2X \sum_{\langle \mathbf{i},\mathbf{j} \rangle} \sum_{\sigma} (X_{\mathbf{i}}^{d\leftarrow\sigma} X_{\mathbf{j}}^{\sigma\leftarrow d} + \text{H.c.}) . \tag{5.103}$$

It follows that at $X = t$ the extended Hubbard model acquires an additional conservation law: *it preserves the number of doubly occupied sites.*

Problem 5.5. At large U, the low-energy states of a half-filled band are essentially neutral, and the lowest-order effective Hamiltonian is of the Heisenberg form (5.42). The direct exchange term $\mathcal{H}_F$ is of this form to begin with.

Our derivation follows that in Sec. 5.1. We divide the Hamilton operator (5.71) into three terms

$$\mathcal{H}_{\text{pair}} = \mathcal{H}^0 + \mathcal{H}^+ + \mathcal{H}^- , \tag{5.104}$$

where $\mathcal{H}^+$ ($\mathcal{H}^-$) describe processes which increase (decrease) the number of doubly occupied sites by one. We know from the solution of Problem 5.4 that the operators $\mathcal{H}^+$ and $\mathcal{H}^-$ are of the same form as in the decomposition of the band term, only $-t$ has to be replaced by $X - t$. We want to find an operator S satisfying

$$i[S, \mathcal{H}_U + \mathcal{H}_V] = -(\mathcal{H}^+ + \mathcal{H}^-) \tag{5.105}$$

because then this S can be used in the canonical transformation (5.11) which leads to the effective Hamiltonian. At half-filling, the relevant piece of $\mathcal{H}_V$ can be written as

$$\mathcal{H}_V = V \sum_{\sigma} \sum_{\sigma'} X_{\mathbf{i}}^{\sigma\leftarrow\sigma} X_{\mathbf{j}}^{\sigma'\leftarrow\sigma'} \tag{5.106}$$

from which, analogously to (5.13), we can deduce that

$$[\mathcal{H}^+, \mathcal{H}_V] = V\mathcal{H}^+ . \tag{5.107}$$

All in all, we have learned that in the familiar derivation of the effective Hamiltonian, $-t$ has to be replaced by $X - t$, and U is replaced by $U - V$. Repeating the steps which led to (5.36), and including also the direct exchange term, we obtain

$$\mathcal{H}_{\text{eff}} = \left(\frac{4(t-X)^2}{U-V} - 2F \right) \sum_{\langle \mathbf{i},\mathbf{j} \rangle} \left[\mathbf{S}_{\mathbf{i}} \cdot \mathbf{S}_{\mathbf{j}} - \frac{\hat{n}_{\mathbf{i}} \hat{n}_{\mathbf{j}}}{4} \right] . \tag{5.108}$$

(The terms included in $\mathcal{H}^0$ are blocked at half-filling). It is interesting that V and X are also playing a role in determining the threshold value F_{cr} where the effective exchange becomes ferromagnetic. In particular, the ground state is ferromagnetic for any $F > 0$ if $X = t$ [378]. In the strongly correlated limit, we found $F_{\text{cr}} \sim (t - X)^2/(U - V)$, in agreement with the conclusions of [56].

Non-perturbative results are derived in [378].

Problem 5.6. We will call a configuration neutral if there is one electron at each site. There are sixteen neutral configurations. These are conveniently labelled by the total spin S (pseudospin τ), and its z-component S^z (τ^z). Let us recall that the pseudospin was introduced by assigning $\tau^z = 1/2$ to the orbital a, and $\tau^z = -1/2$ to the orbital b. We find that there is a subspace with $S = 1$, $\tau = 1$ (9 states), another with $S = 1$, $\tau = 0$ (3 states), still another with $S = 0$, $\tau = 1$ (3 states), and finally a state with $S = 0$, $\tau = 0$. The basis functions will be denoted by $|S, S^z, \tau, \tau^z\rangle$. The degeneracy of the three components of spin-triplets is evident, but the Hamiltonian defined in (5.72)–(5.74) explicitely breaks pseudospin-rotational invariance, thus states with differing τ^z have to be examined separately. We will, however, find that the choice $t_a = t_b = t$, $U_a = U_b = U$, $U_{ab} = U - J$ assures pseudospin isotropy, and thus the corresponding effective Hamiltonian describes a four-level system.

Let us start with the state with $S^z = 1$, and $\tau^z = 1$; it is

$$|1,1,1,1\rangle = c^\dagger_{1a\uparrow} c^\dagger_{2a\uparrow}|0\rangle\,. \tag{5.109}$$

It is clear that this is an eigenstate with energy 0, and similarly if we replace a by b. Thus there is a sixfold degenerate level at

$$E(S = 1, \tau = 1, \tau^z = \pm 1) = 0\,. \tag{5.110}$$

The case of the $\tau^z = 0$ component of the pseudospin triplet is different. We can introduce the pseudospin lowering operator

$$\tau^- = \sum_{j\sigma} c^\dagger_{jb\sigma} c_{ja\sigma} \tag{5.111}$$

to generate $|1,1,1,0\rangle$ from $|1,1,1,1\rangle$. Acting with $\mathcal{H}_{\text{hop}}$ on the former, we find that it is connected with the matrix element $t_a - t_b$ to the inonized configuration

$$\frac{1}{\sqrt{2}}(c^\dagger_{1a\uparrow} c^\dagger_{1b\uparrow} - c^\dagger_{2a\uparrow} c^\dagger_{2b\uparrow})|0\rangle\,,$$

thus the corresponding (threefold degenerate) energy level is

$$E(S = 1, \tau = 1, \tau^z = 0) \approx -\frac{(t_a - t_b)^2}{U_{ab} - J}\,. \tag{5.112}$$

Proceeding similarly with

$$|1,1,0,0\rangle = \frac{1}{\sqrt{2}}(c^\dagger_{1a\uparrow} c^\dagger_{2b\uparrow} - c^\dagger_{1b\uparrow} c^\dagger_{2a\uparrow})|0\rangle \tag{5.113}$$

we find that it has the matrix element $-(t_a + t_b)$ to the intermediate state

$$\frac{1}{\sqrt{2}}(c^\dagger_{1a\uparrow}c^\dagger_{1b\uparrow} + c^\dagger_{2a\uparrow}c^\dagger_{2b\uparrow})|0\rangle ,$$

thus we found another threefold degenerate energy level at

$$E(S=1, \tau=0, \tau^z=0) \approx -\frac{(t_a+t_b)^2}{U_{ab}-J} . \tag{5.114}$$

The $S = 0$, $\tau^z = \pm 1$ cases are really the same what we saw in the spin-only case, so it is enough to quote

$$E(S=0, \tau=1, \tau^z=1) \approx -\frac{4t_a^2}{U_a} , \tag{5.115}$$

and

$$E(S=0, \tau=1, \tau^z=-1) \approx -\frac{4t_b^2}{U_b} . \tag{5.116}$$

The $\tau^z = 0$ component is the space-even combination of two inter-site singlets

$$|0,0,1,0\rangle = \frac{1}{2}(c^\dagger_{1a\uparrow}c^\dagger_{2b\downarrow} - c^\dagger_{1a\downarrow}c^\dagger_{2b\uparrow} + c^\dagger_{1b\uparrow}c^\dagger_{2a\downarrow} - c^\dagger_{1b\downarrow}c^\dagger_{2a\uparrow})|0\rangle . \tag{5.117}$$

It is connected with the matrix element $-(t_a + t_b)$ to a similar combination of two on-site inter-orbital singlets

$$\frac{1}{2}(c^\dagger_{1a\uparrow}c^\dagger_{1b\downarrow} - c^\dagger_{1a\downarrow}c^\dagger_{1b\uparrow} + c^\dagger_{2a\uparrow}c^\dagger_{2b\downarrow} - c^\dagger_{2a\downarrow}c^\dagger_{2b\uparrow})|0\rangle .$$

Notice that this state has a diagonal energy J from the spin-flip term. We deduce

$$E(S=0, \tau=1, \tau^z=0) \approx -\frac{(t_a+t_b)^2}{U_{ab}+J} . \tag{5.118}$$

Finally, $|0,0,0,0\rangle$ is connected with $t_a - t_b$ to the space-odd linear combination of the on-site inter-orbital singlets, giving rise to

$$E(S=0, \tau=0, \tau^z=0) \approx -\frac{(t_a-t_b)^2}{U_{ab}+J} . \tag{5.119}$$

Introducing the symmetrical model $t_a = t_b = t$, $U_a = U_b = U$, and $U_{ab} = U - J$, we find[24] that the energies do not depend on τ^z, only on τ. The

[24] With this choice, the three on-site singlets have the same energy U, thus it is reasonable that it is associated with pseudospin-rotational symmetry.

ground state is spin-triplet and orbital singlet. The effective Hamiltonian can be written as

$$\begin{aligned}\mathcal{H}_{\text{eff}} &= -\frac{4t^2}{U-2J}\left(\vec{S}_1\cdot\vec{S}_2+\frac{3}{4}\right)\left(\frac{1}{4}-\vec{\tau}_1\cdot\vec{\tau}_2\right)\\ &\quad -\frac{4t^2}{U}\left(\frac{1}{4}-\vec{S}_1\cdot\vec{S}_2\right)\left(\frac{3}{4}+\vec{\tau}_1\cdot\vec{\tau}_2\right)\,. \qquad (5.120)\end{aligned}$$

It might seem that $\vec{S}$ and $\vec{\tau}$ are playing perfectly symmetrical roles but that would be the wrong conclusion. We should remember that the pesudospin-rotational symmetry is not required by physics; it is merely a consequence of the above choice of the parameters. The generic case is described by an effective Hamiltonian which is anisotropic in pseudospin space and looks much more complicated than our result [76, 206].

Problem 5.7. Using the mean field approximation, we calculate the effective spin-spin coupling by replacing $\vec{\tau}_1\cdot\vec{\tau}_2$ by its expectation value in (5.120). At $T = 0$ the orbital order is complete, $\langle\vec{\tau}_1\cdot\vec{\tau}_2\rangle = -3/4$, and we find the ferromagnetic

$$J_{\text{eff}} = -\frac{4t^2}{U-2J} \qquad (5.121)$$

familiar from Sec. 5.4. If the frirst-order transition sets is at a relatively low temperature, we may keep the above J_{eff} up to the transition. In the disordered phase $\langle\vec{\tau}_1\cdot\vec{\tau}_2\rangle = 0$ and we find

$$J_{\text{eff}} = \frac{2t^2}{U}\left(1-\frac{J}{U-2J}\right)\,, \qquad (5.122)$$

an antiferromagnetic coupling with the usual order of magnitude of the kinetic exchange.

In a more realistic calculation the thermal expectation value would change gradually in the ordered phase, and would do so even in the orbitally disordered phase, if we allowed for the presence of short range order. It may be found that the coupling is different for different pairs. In any case, it should not be forgotten that the pseudospin-rotationally invariant form (5.120) resulted from a simplifying assumption, and the true expression is pseudospin-anisotropic and more complicated. However, the general lesson would remain the same: the inclusion of orbital degeneracy allows for the possibility that the low-temperature order is described by a spin Hamiltonian which is quite different from the one which characterizes the high-temperature paramagnetic phase. A similar scenario was suggested for the explanation of certain properties of V_2O_3 [40].

Problem 5.8. Let us first identify the low-energy subspace. Since the Coulomb repulsion is strong, this corresponds to the uniform distribution of charge, i.e., to two electrons at each site (we assume that this assures local neutrality). Furthermore, because of the large intrasite exchange interaction, the two electrons at a site must have parallel spins: $|\mathbf{S}_1| = |\mathbf{S}_2| = 1$. If we neglect the hopping processes, the two sites are independent, and the local triplets can be combined arbitrarily. The subspace of the low-lying states is $3^2 = 9$-dimensional. Allowing for non-zero t leads to small splittings between the nine states, which can be described by an effective Hamiltonian acting on the $S = 1$ spins.

The Hamiltonian is spin-rotationally invariant, thus the total spin $\mathcal{S}$, and its z-component $\mathcal{S}^z$, are good quantum numbers. The eigenstates can be denoted by $|\mathcal{S}, \mathcal{S}^z\rangle$. Two $S = 1$ spins can be combined into a total $\mathcal{S}$=2, 1, or 0, and the corresponding levels are 5-, and 3-fold degenerate, and non-degenerate, respectively, accounting for all the nine states. To determine the corresponding energies, it is sufficient to consider the $\mathcal{S}^z = \mathcal{S}$ states $|2,2\rangle$, $|1,1\rangle$, and $|0,0\rangle$.

In second quantization

$$|2,2\rangle = c^\dagger_{1a\uparrow} c^\dagger_{1b\uparrow} c^\dagger_{2a\uparrow} c^\dagger_{2b\uparrow} |0\rangle \,. \tag{5.123}$$

Starting with $|2,2\rangle$, the Pauli principle prevents any hopping or spin-flip events, thus the energy of the $\mathcal{S} = 2$ states is fixed at $2U_{ab} - J$. From now on, we choose this as the new energy zero for the four-electron states:

$$E_{\mathcal{S}=2} = 0. \tag{5.124}$$

To find $|1,1\rangle$, let us first consider $|2,1\rangle$ which can be obtained from (5.123) by stepping the electron spins down in a symmetrical fashion

$$\begin{aligned} |2,1\rangle &= \mathcal{S}^-|2,2\rangle = (c^\dagger_{1a\downarrow}c_{1a\uparrow} + c^\dagger_{1b\downarrow}c_{1b\uparrow} + c^\dagger_{2a\downarrow}c_{2a\uparrow} + c^\dagger_{2b\downarrow}c_{2b\uparrow})|2,2\rangle \qquad (5.125) \\ &= \frac{1}{2}\left[(c^\dagger_{1a\downarrow}c^\dagger_{1b\uparrow} + c^\dagger_{1a\uparrow}c^\dagger_{1b\downarrow})c^\dagger_{2a\uparrow}c^\dagger_{2b\uparrow} + c^\dagger_{1a\uparrow}c^\dagger_{1b\uparrow}(c^\dagger_{2a\downarrow}c^\dagger_{2b\uparrow} + c^\dagger_{2a\uparrow}c^\dagger_{2b\downarrow})\right]|0\rangle \end{aligned}$$

(the factor $1/2$ was inserted for normalization). The + sign within the brackets corresponds to having the $S^z = 0$ component of a local triplet, and cannot be changed. However, the + between the two brackets can be changed to a minus sign; it leads to the only other $\mathcal{S}^z = 1$ state composed of two total triplets. Consequently,

$$|1,1\rangle = \frac{1}{2}\left[(c^\dagger_{1a\downarrow}c^\dagger_{1b\uparrow} + c^\dagger_{1a\uparrow}c^\dagger_{1b\downarrow})c^\dagger_{2a\uparrow}c^\dagger_{2b\uparrow} - c^\dagger_{1a\uparrow}c^\dagger_{1b\uparrow}(c^\dagger_{2a\downarrow}c^\dagger_{2b\uparrow} + c^\dagger_{2a\uparrow}c^\dagger_{2b\downarrow})\right]|0\rangle. \tag{5.126}$$

We do not give the details of how to construct $|0,0\rangle$. Stepping down the electron spins one by one, we can derive $|2,0\rangle$ from $|2,1\rangle$, and $|1,0\rangle$ from $|1,1\rangle$. Demanding that $|0,0\rangle$ be composed of local triplets, and that it should be orthogonal to both $|2,0\rangle$ and $|1,0\rangle$, we arrive at

$$\begin{aligned} |0,0\rangle &= \frac{1}{\sqrt{3}}\Big[c^\dagger_{1a\uparrow}c^\dagger_{1b\uparrow}c^\dagger_{2a\downarrow}c^\dagger_{2b\downarrow} + c^\dagger_{1a\downarrow}c^\dagger_{1b\downarrow}c^\dagger_{2a\uparrow}c^\dagger_{2b\uparrow} \\ &\quad -\frac{1}{2}\left(c^\dagger_{1a\downarrow}c^\dagger_{1b\uparrow} + c^\dagger_{1a\uparrow}c^\dagger_{1b\downarrow}\right)\left(c^\dagger_{2a\downarrow}c^\dagger_{2b\uparrow} + c^\dagger_{2a\uparrow}c^\dagger_{2b\downarrow}\right)\Big]|0\rangle \quad (5.127) \end{aligned}$$

a) *Second-order perturbation theory.* The application of $\mathcal{H}_{\text{hop}}$ to the low-lying (neutral) states leads to high-energy (ionized) states. We find

$$\begin{aligned} \mathcal{H}_{\text{hop}}|1,1\rangle &= 2t\cdot\frac{1}{2}\Big[c^\dagger_{1a\uparrow}c^\dagger_{2a\uparrow}(c^\dagger_{1b\uparrow}c^\dagger_{1b\downarrow} + c^\dagger_{2b\uparrow}c^\dagger_{2b\downarrow}) \\ &\quad + c^\dagger_{1b\uparrow}c^\dagger_{2b\uparrow}(c^\dagger_{1a\uparrow}c^\dagger_{1a\downarrow} + c^\dagger_{2a\uparrow}c^\dagger_{2a\downarrow})\Big]|0\rangle. \quad (5.128) \end{aligned}$$

The zeroth-order neutral state $|1,1\rangle$ gets modified by the small admixture of a higher-lying state, to which it is connected by the matrix element $2t$. The excited state is a charge excitation since each of its components belongs to an ionized configuration, with one of the sites containing three electrons. The excitation energy is $U+J$. The energy shift of the $\mathcal{S}=1$ overall triplet is obtained from second-order perturbation theory as

$$E_{\mathcal{S}=1} \approx -\frac{4t^2}{U+J}\,. \quad (5.129)$$

Similarly

$$\begin{aligned} \mathcal{H}_{\text{hop}}|0,0\rangle &= \sqrt{6}t\cdot\frac{1}{\sqrt{8}}\Big[(c^\dagger_{1a\uparrow}c^\dagger_{2a\downarrow} - c^\dagger_{1a\downarrow}c^\dagger_{2a\uparrow})(c^\dagger_{1b\uparrow}c^\dagger_{1b\downarrow} + c^\dagger_{2b\uparrow}c^\dagger_{2b\downarrow}) \\ &\quad + (c^\dagger_{1b\uparrow}c^\dagger_{2b\downarrow} - c^\dagger_{1b\downarrow}c^\dagger_{2b\uparrow})(c^\dagger_{1a\uparrow}c^\dagger_{1a\downarrow} + c^\dagger_{2a\uparrow}c^\dagger_{2a\downarrow})\Big]|0\rangle. \quad (5.130) \end{aligned}$$

The excited state is again composed of singly ionized configurations, thus second-order perturbation theory yields

$$E_{\mathcal{S}=0} \approx -\frac{6t^2}{U+J}\,. \quad (5.131)$$

Verifying that for $S=1$ spins

$$2\mathbf{S}_1\cdot\mathbf{S}_2 - 2 = \begin{cases} -6 & \text{for } \mathcal{S}=0 \\ -4 & \text{for } \mathcal{S}=1 \\ 0 & \text{for } \mathcal{S}=2 \end{cases} \quad (5.132)$$

the energies (5.124), (5.129), and (5.131) can be thought of as the eigenenergies of the effective (spin) Hamiltonian

$$\mathcal{H}^{(2)}_{\text{eff}} = \frac{2t^2}{U+J}\mathbf{S}_1\cdot\mathbf{S}_2\,. \quad (5.133)$$

Thereby we have shown that (similarly to the $S = 1/2$ case studied earlier) the antiferromagnetic $S = 1$ Heisenberg model can also be derived as an effective Hamiltonian describing the low-energy states of an interacting electron system. The origin of the coupling is the same kinetic exchange that we learned about in Sec. 2.5.2, and in Sec. 5.1.5.

Fourth-order processes. We would be right to feel surprised that our first result (5.133) for the effective Hamiltonian has the same simple form as what we found for $S = 1/2$ spins: could $S = 1$ spins not do something more complicated? In fact they can, as we are going to find by examining the fourth-order contributions to the energy levels.

(5.124) remains true. We can convince ourselves that in the $\mathcal{S} = 1$, $\mathcal{S}^z = 1$ subspace only the singly ionized state found in (5.128) can be reached from $|1,1\rangle$, thus there is a 2×2 problem to solve. Expanding the relevant eigenvalue to fourth order, we find

$$E'_{\mathcal{S}=1} \approx -\frac{4t^2}{U+J} + \frac{16t^4}{(U+J)^3} \,. \tag{5.134}$$

The singlet subspace is larger: we find two more excited states which can be reached from $|0,0\rangle$. One of them is the doubly ionized state

$$\frac{1}{\sqrt{2}}(c^\dagger_{1a\uparrow}c^\dagger_{1a\downarrow}c^\dagger_{1b\uparrow}c^\dagger_{1b\downarrow} + c^\dagger_{2a\uparrow}c^\dagger_{2a\downarrow}c^\dagger_{2b\uparrow}c^\dagger_{2b\downarrow}) \tag{5.135}$$

with excitation energy $2(U + U_{ab}) + J$. The other one is composed of local singlets:

$$\frac{1}{\sqrt{2}}(c^\dagger_{1a\uparrow}c^\dagger_{1a\downarrow}c^\dagger_{2b\uparrow}c^\dagger_{2b\downarrow} + c^\dagger_{1b\uparrow}c^\dagger_{1b\downarrow}c^\dagger_{2a\uparrow}c^\dagger_{2a\downarrow}) \,. \tag{5.136}$$

This state has the low excitation energy $2(U - U_{ab}) + J$: it violates Hund's rules but preserves local neutrality. The matrix of the 4×4 eigenvalue problem is

$$\begin{pmatrix} 0 & t\sqrt{6} & 0 & 0 \\ t\sqrt{6} & U+J & 2t & 2t \\ 0 & 2t & 2(U+U_{ab})+J & 0 \\ 0 & 2t & 0 & 2(U-U_{ab})+J \end{pmatrix} . \tag{5.137}$$

The singlet level is now

$$E'_{\mathcal{S}=0} \approx -\frac{6t^2}{U+J} + \frac{12t^4}{(U+J)^2}\left(\frac{3}{U+J} - \frac{2}{2(U+U_{ab})+J} - \frac{2}{2(U-U_{ab})+J}\right) . \tag{5.138}$$

Again, we wish to associate the energy levels (5.124), (5.134), and (5.138) with an effective Hamiltonian expressed in terms of $S = 1$ spin operators. It cannot be of the simple (bilinear) Heisenberg form (5.133) because we know from (5.132) that that gives the ratio 1:2 for consecutive energy splittings, and it is no longer the case. To describe the additional shift of the singlet level, we need a spin operator which acts only on the overall singlet. Such is the *projection operator*

$$\hat{P}_{\mathcal{S}=0} = \frac{1}{12}[(\mathbf{S}_1 + \mathbf{S}_2)^2 - 2][(\mathbf{S}_1 + \mathbf{S}_2)^2 - 6] \,. \tag{5.139}$$

It is easy to check that $\hat{P}_{\mathcal{S}=0}$ gives zero if applied on $\mathcal{S} = 1$ or $\mathcal{S} = 2$ states, and it gives a factor 1 when applied to an $\mathcal{S} = 0$ state. In other words, $\hat{P}_{\mathcal{S}=0}$ projects onto the singlet subspace.

The new effective Hamiltonian can be written as

$$\begin{aligned} \mathcal{H}_{\text{eff}}^{(4)} &= \left(\frac{2t^2}{U+J} - \frac{8t^4}{(U+J)^3}\right) \mathbf{S}_1 \cdot \mathbf{S}_2 \\ &+ \frac{12t^4}{(U+J)^2} \left(\frac{1}{U+J} - \frac{2}{2(U+U_{ab})+J} - \frac{2}{2(U-U_{ab})+J}\right) \hat{P}_{\mathcal{S}=0} \,. \end{aligned} \tag{5.140}$$

It is useful to rewrite the projection operator in the form

$$\hat{P}_{\mathcal{S}=0} = \frac{1}{3} \left[(\mathbf{S}_1 \cdot \mathbf{S}_2)^2 - 1\right]. \tag{5.141}$$

Replacing it in (5.140), we get the *bilinear–biquadratic exchange Hamiltonian*

$$\mathcal{H}_{\text{eff}}^{(4)} = J_{12}\, \mathbf{S}_1 \cdot \mathbf{S}_2 + Q\, (\mathbf{S}_1 \cdot \mathbf{S}_2)^2 + \text{const.} \tag{5.142}$$

where the biquadratic coupling is given by

$$Q = \frac{12t^4}{(U+J)^2} \left(\frac{1}{U+J} - \frac{2}{2(U+U_{ab})+J} - \frac{2}{2(U-U_{ab})+J}\right) . \tag{5.143}$$

Q is dominated by the term with the smallest denominator $2(U - U_{ab}) + J$. It is not unnatural to assume that $U - U_{ab}$ is rather smaller than U, and it is perhaps of the same order as J. For a rough estimate, let us adopt our previous $U - U_{ab} = 3J/2$. We find

$$Q \sim \frac{6t^4}{U^3} - \frac{6t^4}{U^2 J} \tag{5.144}$$

where the negative term dominates.

Did we miss further, still more complicated coupling terms? In the present case, no: It is easy to convince ourselves that $J(\mathbf{S}_1 \cdot \mathbf{S}_2) + Q(\mathbf{S}_1 \cdot \mathbf{S}_2)^2$ is the most general form of a spin-rotationally invariant two-spin coupling for $S = 1$

spins. As we discussed previously, there are three energy levels, corresponding to total spin $\mathcal{S} = 2$, 1, and 0. Setting one of these energies arbitrarily to zero, the effective Hamiltonian has to describe two energy differences, which can be achieved by the proper choice of the two coupling constants.

By the same token, the general form of the isotropic coupling between two spins of size S looks like

$$\mathcal{H} = J_1 \mathbf{S}_1 \cdot \mathbf{S}_2 + J_2 (\mathbf{S}_1 \cdot \mathbf{S}_2)^2 + \ldots + J_{2S} (\mathbf{S}_1 \cdot \mathbf{S}_2)^{2S} . \tag{5.145}$$

Since higher powers of $(\mathbf{S}_1 \cdot \mathbf{S}_2)$ arise in higher orders of perturbation theory, the biquadratic term is the most essential correction to the standard Heisenberg model for any S. In fact, the first experimental evidence of a biquadratic term was found for $S = 5/2$ Mn^{++} ions (in antiferromagnetic MnO [339], as well as in Mn-doped MgO [158]). The measured ratio $Q/J \sim -0.05$ was realized to be in accordance with the predictions of Anderson's theory of superexchange [20]. It is also quite compatible with our (5.144): our Q arises from the competition of two terms of which $-6t^4/(U^2 J)$ is the stronger.

There is another, more formal justification for introducing (5.145). It relies on showing that the higher powers of the scalar product $(\mathbf{S}_1 \cdot \mathbf{S}_2)^{2S+1}, \ldots$, are not independent of the first $2S$ operators of this kind, and therefore introducing them would not really change the form of the Hamiltonian. This must be a property of the algebra of the spin operators. For an illustration, let us consider $S = 1/2$ where we should find that $(\mathbf{S}_1 \cdot \mathbf{S}_2)^2$ can be expressed with $\mathbf{S}_1 \cdot \mathbf{S}_2$. Let us use Pauli matrices instead of the components of $\mathbf{S}$: $2S^x = \sigma^x$, etc., and exploit the wellknown property that the square of any of the Pauli matrices is the 2×2 unit matrix. Then

$$(\vec{\sigma}_1 \cdot \vec{\sigma}_2)^2 = (\sigma_1^x \sigma_2^x + \sigma_1^y \sigma_2^y + \sigma_1^z \sigma_2^z)^2 = 3 - 2\vec{\sigma}_1 \cdot \vec{\sigma}_2 \tag{5.146}$$

where we used the identities $\sigma^x \sigma^y = i\sigma^z$, $\sigma^y \sigma^z = i\sigma^x$, and $\sigma^z \sigma^x = i\sigma^y$. In terms of the $S = 1/2$ spin operators

$$(\mathbf{S}_1 \cdot \mathbf{S}_2)^2 = \frac{3}{16} - \frac{1}{2} \mathbf{S}_1 \cdot \mathbf{S}_2 . \tag{5.147}$$

It follows that for $S = 1/2$, any isotropic (pairwise) coupling can be brought to the well-known Heisenberg form. A similar proof should be possible to justify (5.145) for general S.

Chapter 6

Heisenberg Magnets

We have seen that the low-energy effective Hamiltonian of the half-filled non-degenerate Hubbard model is the $S = 1/2$ antiferromagnetic Heisenberg model (5.42) with nearest neighbour interactions

$$\mathcal{H} = J \sum_{\mathbf{i},\mathbf{j}} \mathbf{S}_{\mathbf{i}} \cdot \mathbf{S}_{\mathbf{j}} \,. \tag{6.1}$$

More precisely, this is only the leading term of the effective Hamiltonian; going to higher orders, we find also ring exchange processes (5.50), second-neighbour interactions (5.51), etc. Similarly, we get $S = 1/2$ Heisenberg models if we consider a Hubbard model with orbital degeneracy N, and the band filling is $n = 1$ (one electron per site), or $n = 2N - 1$ (one electron per site less than a filled band). However, interplay with orbital ordering may lead to a ferromagnetic spin–spin coupling (see Sec. 5.4). Furthermore, including spin–orbit coupling in the electron Hamiltonian can yield a Heisenberg model with exchange anisotropy [454]. Thus (6.1) and its more complicated versions can describe the low-energy, low-temperature physics of Mott insulators like $LaTiO_3$ ($n = 1$), or La_2CuO_4 ($n = 9$, antiferromagnet), or K_2CuF_4 ($n = 9$, ferromagnet).

Considering a degenerate Hubbard model with $n = 2$, and a strong Hund's rule exchange, we convinced ourselves that the leading-order effective Hamiltonian is the isotropic $S = 1$ Heisenberg antiferromagnet, and going to next order, we find also a biquadratic coupling (5.142).

Spin–orbit coupling can give rise to a single-ion anisotropy term like (3.34), so we may consider a class of $S=1$ spin models of the kind

$$\mathcal{H} = \sum_{\mathbf{i},\mathbf{j}} \left[J\mathbf{S}_\mathbf{i} \cdot \mathbf{S}_\mathbf{j} + Q(\mathbf{S}_\mathbf{i} \cdot \mathbf{S}_\mathbf{j})^2 \right] + \Delta \sum_{\mathbf{j}} \left(S_\mathbf{j}^z \right)^2 . \qquad (6.2)$$

For $n = 3$, etc., we may a find an effective Hamiltonian of considerable complexity. However, for transition metal compounds, the leading term is likely to be an antiferromagnetic Heisenberg model for spins, where S follows from Hund's first rule. Occasionally, we may find ferromagnetic coupling, but then we probably have to treat coexisting spin and orbital order. It is a quite different question whether the ordering moment has both spin and orbital components, due to an incomplete quenching of the orbital moment.

Though spin Hamiltonians may be quite complicated, the basic fact remains that the possibility of neglecting the charge degrees of freedom brings great simplification in the treatment of the underlying problem of interacting fermions. For a non-degenerate model, the dimensionality of the local Hilbert space is reduced from 4 to 2. With comparable effort, we can usually go much farther with a pure spin problem, than with an electron problem. Naturally, we have to remember that we can treat only low-energy, low-temperature phenomena, up to energies $\sim t^2/U$. Investing an excitation energy $\sim U$ sets the electrons in motion, and requires returning to the original Hubbard model.

In rare earth compounds, **L** and **S** are strongly coupled, and effective Hamiltonians are usually written in terms of **J**, or the equivalent Stevens operators (Sec. 3.10). Note that – in contrast to the case of $3d$ compounds – such a description can be used for metals as well as for insulators, because in many rare earth systems, we may speak about a Mott localization of the subsystem of $4f$ electrons.

6.1 Ferromagnetic Heisenberg Model

The simplest problem in the theory of lattice spin systems is posed by the ferromagnetic (FM) Heisenberg model

$$\mathcal{H} = J \sum_{\langle \mathbf{i},\mathbf{j} \rangle} \mathbf{S}_\mathbf{i} \cdot \mathbf{S}_\mathbf{j} \qquad (6.3)$$

where $J < 0$, and the summation extends over nearest neighbour pairs. The uniquely convenient feature of the model is that it has an exactly known ground state, and there can be no question about the existence of long range order at $T = 0$. The spectrum of single-particle excitations, and thermal properties at low temperatures, can be worked out with relative ease.

The model can be thought to describe ferromagnetic Mott insulators like EuO or K_2CuF_4. From our previous studies, we are aware that the real difficulty lies in arriving at (6.3) as the effective Hamiltonian of an interacting electron system: FM superexchange is relatively rare, and FM kinetic exchange arises from an interplay of spin and orbital degrees of freedom (see Sec. 5.4).

6.1.1 Ground State: Symmetry Breaking

Let us consider (6.3) for a general spin $|\mathbf{S_j}| = S$. The total energy is the sum of the bond energies $E_{\mathbf{ij}}$ ($\mathbf{i}$ and $\mathbf{j}$ are nearest neighbours)

$$\begin{aligned} E_{\mathbf{ij}} &= J\mathbf{S_i}\cdot\mathbf{S_j} = \frac{J}{2}[(\mathbf{S_i}+\mathbf{S_j})^2 - \mathbf{S_i^2} - \mathbf{S_j^2}] \\ &= |J|S(S+1) - \frac{|J|}{2}(\mathbf{S_i}+\mathbf{S_j})^2 \end{aligned} \tag{6.4}$$

which can be minimized by letting the two spins be parallel: $|\mathbf{S_i}+\mathbf{S_j}| = 2S$. Thus the minimum of the bond energy is

$$(E_{\mathbf{ij}})_{\text{min.}} = |J|[S(S+1) - S(2S+1)] = -|J|S^2. \tag{6.5}$$

This energy is obtained if both spins are "pointing up": $S_{\mathbf{i}}^z = S_{\mathbf{j}}^z = S$.

Since the Hamiltonian is the sum of bond terms, it is clear that for the ground state energy of the lattice

$$\mathcal{E}_{\text{gr.st.}}^{\text{FM}} \geq \sum_{\langle\mathbf{i},\mathbf{j}\rangle}(E_{\mathbf{ij}})_{\text{min.}} = -\frac{1}{2}L|J|zS^2 \tag{6.6}$$

holds, where L is the number of the lattice sites, and z is the coordination number. If we find a state for which the equality holds then we have found the exact ground state – or, at least, one of the exact ground states. Now, the essential point is that all bond energies can be *simultaneously* minimized if all spins are pointing up: $S^z_{\mathbf{j}} = S$ for all $\mathbf{j}$. It follows that the ground state energy of (6.3) is

$$\mathcal{E}^{\text{FM}}_{\text{gr.st.}} = -\frac{1}{2}L|J|zS^2 \ . \tag{6.7}$$

The ground state is not unique. We have just found that the system has the maximum value of the total spin $S_{\text{tot}} = LS$. $S^z_{\text{tot}} = LS$ state which is maximally polarized in the z-direction. However, the Hamiltonian is spin-rotationally invariant, hence turning the total spin in another direction does not change the energy: *the ground state must be* $(2LS+1)$*-fold degenerate.*

Quite generally, *ground state degeneracy indicates the breaking of a global symmetry.* The term "symmetry breaking" has here a very simple meaning: $\mathcal{H}$ is spin-rotationally invariant while $\mathbf{S}_{\text{tot}}$ is pointing into a preferred direction. If we find a ground state which does not have the full symmetry of the Hamilton operator, then applying to it the symmetry operations of $\mathcal{H}$ generates other ground states.

We are now confronted with an interesting conceptual question: how to do statistical physics in the presence of symmetry breaking? Remember that the partition function is defined as

$$\mathcal{Z} = \sum_{\text{states}} e^{-\epsilon_i/k_{\text{B}}T} \ .$$

Should we include all the ground states in the definition of the partition function? If we did so, statistical averaging would automatically average over all the possible orientations of the total spin, giving zero expectation value for the total magnetization of a ferromagnet, which is clearly a nonsensical result. The proper recipe[1] for the construction of the partition function is this: Restrict your Hilbert space by picking out one of the ground states, and consider the finite excitations above that ground state.

Naturally, this makes sense only if it is possible to say about an excited state that "this is a finite excitation above this particular ground state but not above the other ground states". Here the key word is

---

[1]For a macroscopically large system.

"finite". Let us consider two ground states: let $|\Psi_I\rangle$ be polarized in the z-direction, and $|\Psi_{II}\rangle$ in a direction subtending an angle ϑ with the z-axis. A finite excitation above $|\Psi_I\rangle$ can be visualized as $n \ll L$ spins turned over while $L-n$ spins are still pointing in the z-direction. The scalar product of this state with $|\Psi_{II}\rangle$ cannot be larger than the classically estimated overlap $(\cos\vartheta)^{L-n}$ which approaches 0 as $L \to \infty$. Similarly, in the limit $L \to \infty$ any finite excitation above $|\Psi_I\rangle$ becomes orthogonal to any finite excitation above $|\Psi_{II}\rangle$. Consequently, in the thermodynamic limit not only the ground states are mutually orthogonal but also the Hilbert spaces built upon them. Let us notice that the two Hilbert spaces differ in a c-number charateristic. All states from the Hilbert space of $|\Psi_I\rangle$ have the spin density $\langle S^z\rangle = S$, while for the states from the Hilbert space derived from $|\Psi_{II}\rangle$, $\langle S^z\rangle = S\cos\vartheta$. This means that the finite excitations above one of the ground states are *classically distinguishable* from the states derived from the other ground states. It is thus easy to accept that the proper definition of the (quantum-mechanical) partition function should include only the states from one of the Hilbert spaces.

It is conventional to choose the $S^z_{\text{tot}} = LS$ state as the ground state of (6.3) and construct the Hilbert space by turning over a finite number of spins. The formal device for picking this Hilbert space is a trick devised by Bogolyubov: let us introduce an infinitesimal auxiliary field acting in the z-direction

$$\mathcal{H} \longrightarrow \mathcal{H}' = J\sum_{\langle \mathrm{i,j}\rangle} \mathbf{S}_\mathrm{i}\cdot\mathbf{S}_\mathrm{j} - h\sum_{\mathrm{j}} S^z_\mathrm{j} \,. \tag{6.8}$$

The classification of the eigenstates according to S^z_{tot} remains valid but the ground state degeneracy is lifted: the state with all spins pointing up will be the unique ground state. Thermodynamic quantities can be evaluated, and at the end of the calculation, we let $h \to 0$. This leaves us with a well-defined *order parameter* $M^z = \lim_{h\to 0}\langle S^z_{\text{tot}}\rangle$. $M^z \neq 0$ below the Curie temperature.

In practice, we need not perform the calculations with $\mathcal{H}'$, since the same results can be obtained by using $\mathcal{H}$ and sticking to the $S^z_{\text{tot}} = LS$ state as *the* ground state.

We remark that in most other cases of symmetry breaking, the ground state degeneracy arises only asymptotically, in the thermodynamic limit $L \to \infty$. What is special about ferromagnetic ordering is that here the order parameter is a conserved quantity: its operator S^z_{tot} commutes with the Hamiltonian. The ground state of (6.3) is degenerate for any L. The contrast with the more usual case when the order parameter is not a conserved quantity, will become clear from our study of antiferromagnetism.

6.1.2 Excitations: Spin Waves

Since the ground state is maximally polarized: $S^z_{\text{tot}} = LS$, it is reasonable to expect that the low-lying excitations can be created by turning a spin one step down, from $S^z_{\mathbf{j}} = S$ to $S^z_{\mathbf{j}} = S - 1$, and then letting this disturbance propagate in the crystal. The arising excitations are spin waves.

Exciting a spin wave means creating a quasiparticle called *magnon*. Magnons are quantized magnetization density waves, just as phonons are quantized lattice waves. Like the phonons, the magnons are *bosonic quasiparticles*.

Now we wish to express the magnon creation and destruction operators in terms of the spin operators. For each site, we have three spin operators: $S^x_{\mathbf{j}}$, $S^y_{\mathbf{j}}$, and $S^z_{\mathbf{j}}$ but these are not independent: they are coupled by the identity $(S^x_{\mathbf{j}})^2 + (S^y_{\mathbf{j}})^2 + (S^z_{\mathbf{j}})^2 = S(S+1)$. It is advantageous to choose the two independent operators as the boson creation and destruction operators $a^\dagger_{\mathbf{j}}$, and $a_{\mathbf{j}}$, satisfying the canonical commutation relations

$$[a_{\mathbf{j}}, a^\dagger_{\mathbf{l}}] = \delta_{\mathbf{j}\,\mathbf{l}} \qquad \text{and} \qquad [a_{\mathbf{j}}, a_{\mathbf{l}}] = [a^\dagger_{\mathbf{j}}, a^\dagger_{\mathbf{l}}] = 0\,. \tag{6.9}$$

We still have to find the relationship between the boson operators and the spin operators. The basic idea is that stepping the spin down by one step amounts to creating a boson. However, we cannot simply equate $a^\dagger_{\mathbf{j}}$ with $S^-_{\mathbf{j}}$, for the following reason: boson occupation numbers can be arbitrarily large but a spin can be turned down only $2S$ times, to $S^z_{\mathbf{j}} = -S$, and then the process must stop. Somehow, we have to make sure that starting from the zero-boson state (i.e., with $S^z_{\mathbf{j}} = S$) we cannot reach the non-physical part of the bosonic Hilbert space where $\langle a^\dagger_{\mathbf{j}} a_{\mathbf{j}} \rangle > 2S$. This is guaranteed if the bosonic operators are related to the spin raising and lowering operators via the *Holstein–Primakoff transformation* [173, 210]

$$
\begin{aligned}
S_{\mathbf{j}}^{+} &= \sqrt{2S}\left(1-\frac{a_{\mathbf{j}}^{\dagger}a_{\mathbf{j}}}{2S}\right)^{1/2} a_{\mathbf{j}} \\
S_{\mathbf{j}}^{-} &= \left(S_{\mathbf{j}}^{+}\right)^{\dagger} = \sqrt{2S}\, a_{\mathbf{j}}^{\dagger}\left(1-\frac{a_{\mathbf{j}}^{\dagger}a_{\mathbf{j}}}{2S}\right)^{1/2} .
\end{aligned}
\tag{6.10}
$$

If we are trying to act with $S_{\mathbf{j}}^{-}$ on the state $S_{\mathbf{j}}^{z} = -S$, the factor $(1 - a^{\dagger}a/2S)^{1/2}$ gives zero[2]. It can be checked that the standard spin commutation relation $[S_{\mathbf{j}}^{+}, S_{\mathbf{j}}^{-}] = 2S_{\mathbf{j}}^{z}$ is consistent with (6.9).

In order to express $S_{\mathbf{j}}^{z}$ with $a_{\mathbf{j}}^{\dagger}$ and $a_{\mathbf{j}}$, let us consider

$$
\begin{aligned}
(S_{\mathbf{j}}^{z})^2 &= S(S+1) - \frac{1}{2}(S_{\mathbf{j}}^{+}S_{\mathbf{j}}^{-} + S_{\mathbf{j}}^{-}S_{\mathbf{j}}^{+}) = S(S+1) \\
&\quad -S\left[a_{\mathbf{j}}^{\dagger}\left(1-\frac{a_{\mathbf{j}}^{\dagger}a_{\mathbf{j}}}{2S}\right)a_{\mathbf{j}} + \left(1-\frac{a_{\mathbf{j}}^{\dagger}a_{\mathbf{j}}}{2S}\right)^{1/2} a_{\mathbf{j}}a_{\mathbf{j}}^{\dagger}\left(1-\frac{a_{\mathbf{j}}^{\dagger}a_{\mathbf{j}}}{2S}\right)^{1/2}\right] \\
&= S(S+1) - [S + 2Sa_{\mathbf{j}}^{\dagger}a_{\mathbf{j}} - (a_{\mathbf{j}}^{\dagger}a_{\mathbf{j}})^2] = (S - a_{\mathbf{j}}^{\dagger}a_{\mathbf{j}})^2
\end{aligned}
\tag{6.11}
$$

where we used (6.9) repeatedly. From the last line we conclude that

$$S_{\mathbf{j}}^{z} = S - a_{\mathbf{j}}^{\dagger}a_{\mathbf{j}} . \tag{6.12}$$

This corresponds to our original notion that the number of bosons is equal to the number of times the spin has been stepped down.

In order to be able to describe propagating excitations, let us introduce the $\mathbf{k}$-space operators $b_{\mathbf{k}}$, $b_{\mathbf{k}}^{\dagger}$

$$a_{\mathbf{j}} = \frac{1}{\sqrt{L}}\sum_{\mathbf{j}} e^{-i\mathbf{k}\mathbf{j}} b_{\mathbf{k}} \qquad a_{\mathbf{j}}^{\dagger} = \frac{1}{\sqrt{L}}\sum_{\mathbf{j}} e^{i\mathbf{k}\mathbf{j}} b_{\mathbf{k}}^{\dagger} \tag{6.13}$$

---

[2]This is, in fact, not enough to ensure that the spin operators are represented faithfully, because the operators (6.10) have matrix elements between the unphysical states as well as between physical states. We had better stick to the rule that (6.10) is used for states with $n_{\mathbf{j}} \leq 2S$.

$$b_{\mathbf{k}} = \frac{1}{\sqrt{L}} \sum_{\mathbf{j}} e^{i\mathbf{k}\mathbf{j}} a_{\mathbf{j}} \qquad b_{\mathbf{k}}^{\dagger} = \frac{1}{\sqrt{L}} \sum_{\mathbf{j}} e^{-i\mathbf{k}\mathbf{j}} a_{\mathbf{j}}^{\dagger} . \tag{6.14}$$

These also satisfy the boson commutation relations

$$[b_{\mathbf{k}}, b_{\mathbf{k}'}^{\dagger}]_{-} = \frac{1}{L} \sum_{\mathbf{j}} \sum_{\mathbf{l}} e^{i\mathbf{k}\mathbf{j} - i\mathbf{k}'\mathbf{l}} [a_{\mathbf{j}}, a_{\mathbf{l}}^{\dagger}]_{-} = \frac{1}{L} \sum_{\mathbf{j}} e^{i\mathbf{j}(\mathbf{k}-\mathbf{k}')} = \delta_{\mathbf{k}\mathbf{k}'} \tag{6.15}$$

$$[b_{\mathbf{k}}, b_{\mathbf{k}'}]_{-} = [b_{\mathbf{k}}^{\dagger}, b_{\mathbf{k}'}^{\dagger}]_{-} = 0. \tag{6.16}$$

$b_{\mathbf{k}}^{\dagger}$ creates ($b_{\mathbf{k}}$ annihilates) a magnon with wavevector $\mathbf{k}$.

We wish to express $S_{\mathbf{j}}^{+}$, $S_{\mathbf{j}}^{-}$, and $S_{\mathbf{j}}^{z}$ with the magnon operators. Let us assume that the temperature is sufficiently low so that only a few magnons are excited, and the relative spin deviation from the ground state saturation value is small

$$\frac{\langle a_{\mathbf{j}}^{\dagger} a_{\mathbf{j}} \rangle}{S} = \frac{\langle \hat{n}_{\mathbf{j}} \rangle}{S} \ll 1. \tag{6.17}$$

This permits us to expand the square roots in (6.10)

$$\begin{aligned} S_{\mathbf{j}}^{+} &= \sqrt{2S} \left(1 - \frac{a_{\mathbf{j}}^{\dagger} a_{\mathbf{j}}}{2S} \right)^{1/2} a_{\mathbf{j}} \approx \sqrt{2S} \left(a_{\mathbf{j}} - \frac{a_{\mathbf{j}}^{\dagger} a_{\mathbf{j}} a_{\mathbf{j}}}{4S} \right) \\ &= \sqrt{\frac{2S}{L}} \left[\sum_{\mathbf{k}} e^{-i\mathbf{k}\mathbf{j}}\, b_{\mathbf{k}} - \frac{1}{4SL} \sum_{\mathbf{k}_1} \sum_{\mathbf{k}_2} \sum_{\mathbf{k}_3} e^{i(\mathbf{k}_1 - \mathbf{k}_2 - \mathbf{k}_3)\mathbf{j}}\, b_{\mathbf{k}_1}^{\dagger} b_{\mathbf{k}_2} b_{\mathbf{k}_3} \right] \end{aligned} \tag{6.18}$$

and similarly

$$S_{\mathbf{j}}^{-} \approx \sqrt{\frac{2S}{L}} \left[\sum_{\mathbf{k}} e^{i\mathbf{k}\mathbf{j}}\, b_{\mathbf{k}}^{\dagger} - \frac{1}{4SL} \sum_{\mathbf{k}_1} \sum_{\mathbf{k}_2} \sum_{\mathbf{k}_3} e^{i(\mathbf{k}_1 + \mathbf{k}_2 - \mathbf{k}_3)\mathbf{j}}\, b_{\mathbf{k}_1}^{\dagger} b_{\mathbf{k}_2}^{\dagger} b_{\mathbf{k}_3} \right] . \tag{6.19}$$

(6.12) is rewritten as

$$S_{\mathbf{j}}^{z} = S - a_{\mathbf{j}}^{\dagger} a_{\mathbf{j}} = S - \frac{1}{L} \sum_{\mathbf{k}_1} \sum_{\mathbf{k}_2} e^{i(\mathbf{k}_1 - \mathbf{k}_2)\mathbf{j}}\, b_{\mathbf{k}_1}^{\dagger} b_{\mathbf{k}_2} \tag{6.20}$$

so the total spin z-component is

$$\begin{aligned} S^z_{\rm tot} &= \sum_{\mathbf{j}} S^z_{\mathbf{j}} = LS - \frac{1}{L}\sum_{\mathbf{k}_1}\sum_{\mathbf{k}_2}\left(\sum_{\mathbf{j}} e^{i(\mathbf{k}_1-\mathbf{k}_2)\mathbf{j}}\right) b^\dagger_{\mathbf{k}_1} b_{\mathbf{k}_2} \\ &= LS - \sum_{\mathbf{k}} b^\dagger_{\mathbf{k}} b_{\mathbf{k}}. \end{aligned} \tag{6.21}$$

It should be remarked that, in contrast to the approximate nature of (6.18) and (6.19), (6.21) is exact. It expresses the fact that the deviation of the total spin from its saturation value corresponds to the total number of the magnons.

Now we are ready to express the Heisenberg Hamiltonian in terms of the magnon operators, by replacing (6.18), (6.19), and (6.20) in (6.3). The structure of the Hamiltonian turns out to be

$$\mathcal{H} = -\frac{1}{2}|J|LzS^2 + \mathcal{H}_0 + \mathcal{H}_1 \tag{6.22}$$

where the first term is the ground state energy. $\mathcal{H}_0$ is bilinear in the magnon operators, and gives the sum of the energies of the magnons which are present in the system. $\mathcal{H}_1$ contains higher-order terms which describe magnon–magnon interactions. We will be interested mainly in $\mathcal{H}_0$.

Let us begin with the evaluation of the z–z term

$$\begin{aligned} S^z_{\mathbf{j}} S^z_{\mathbf{j}+\boldsymbol{\delta}} = & \left\{ S - \frac{1}{L}\sum_{\mathbf{k}_1}\sum_{\mathbf{k}_2} e^{i(\mathbf{k}_1-\mathbf{k}_2)\mathbf{j}}\, b^\dagger_{\mathbf{k}_1} b_{\mathbf{k}_2} \right\} \\ & \cdot \left\{ S - \frac{1}{L}\sum_{\mathbf{k}_3}\sum_{\mathbf{k}_4} e^{i(\mathbf{k}_3-\mathbf{k}_4)(\mathbf{j}+\boldsymbol{\delta})}\, b^\dagger_{\mathbf{k}_3} b_{\mathbf{k}_4} \right\} \end{aligned} \tag{6.23}$$

where $\boldsymbol{\delta}$ is the vector pointing to a nearest-neighbour site. S^2 corresponds to the contribution of this bond to the ground state energy. The bilinear terms are

$$-\frac{S}{L}\sum_{\mathbf{k}_1}\sum_{\mathbf{k}_2} e^{i(\mathbf{k}_1-\mathbf{k}_2)\mathbf{j}}\,[1+e^{i\boldsymbol{\delta}(\mathbf{k}_1-\mathbf{k}_2)}] b^\dagger_{\mathbf{k}_1} b_{\mathbf{k}_2}. \tag{6.24}$$

Summing the exponential factor over $\mathbf{j}$ brings $L\delta_{\mathbf{k}_1\mathbf{k}_2}$. The remaining expression does not depend on $\boldsymbol{\delta}$, thus summing over $\boldsymbol{\delta}$ brings merely a factor z. The z–z contribution to $\mathcal{H}_0$ is thus

$$\mathcal{H}_0^{zz} = z|J|S \sum_{\mathbf{k}} b_{\mathbf{k}}^\dagger b_{\mathbf{k}} . \tag{6.25}$$

Now let us consider the contribution coming from the spin-flip terms. Using (6.18) and (6.19), we find

$$\begin{aligned} \mathcal{H}_0^{+-} &= -\frac{|J|}{4} \sum_{\mathbf{j}} \sum_{\boldsymbol{\delta}} S_{\mathbf{j}}^+ S_{\mathbf{j}+\boldsymbol{\delta}}^- \\ &= -\frac{S|J|}{2L} \sum_{\mathbf{j}} \sum_{\boldsymbol{\delta}} \sum_{\mathbf{k}_1} e^{-\mathbf{k}_1\mathbf{j}} b_{\mathbf{k}_1} \sum_{\mathbf{k}_2} e^{i\mathbf{k}_2(\mathbf{j}+\boldsymbol{\delta})} b_{\mathbf{k}_2}^\dagger \\ &= -\frac{S|J|z}{2} \sum_{\mathbf{k}} \gamma_{\mathbf{k}} b_{\mathbf{k}} b_{\mathbf{k}}^\dagger \end{aligned} \tag{6.26}$$

where we introduced the geometrical factor

$$\gamma_{\mathbf{k}} = \frac{1}{z} \sum_{\boldsymbol{\delta}} e^{i\mathbf{k}\boldsymbol{\delta}} . \tag{6.27}$$

Similarly,

$$\mathcal{H}_0^{-+} = -\frac{|J|}{4} \sum_{\mathbf{j}} \sum_{\boldsymbol{\delta}} S_{\mathbf{j}}^- S_{\mathbf{j}+\boldsymbol{\delta}}^+ = -\frac{S|J|z}{2} \sum_{\mathbf{k}} \gamma_{-\mathbf{k}} b_{\mathbf{k}}^\dagger b_{\mathbf{k}} . \tag{6.28}$$

Thus, up to bilinear terms the Hamiltonian is

$$\begin{aligned} \mathcal{H} &\approx -\frac{1}{2}|J|zS^2L + \mathcal{H}_0^{zz} + \mathcal{H}_0^{+-} + \mathcal{H}_0^{-+} \\ &= -\frac{1}{2}|J|zS^2L + \frac{|J|zS}{2} \sum_{\mathbf{k}} [2b_{\mathbf{k}}^\dagger b_{\mathbf{k}} - \gamma_{\mathbf{k}} b_{\mathbf{k}} b_{\mathbf{k}}^\dagger - \gamma_{-\mathbf{k}} b_{\mathbf{k}}^\dagger b_{\mathbf{k}}] . \end{aligned} \tag{6.29}$$

The expression can be simplified if the structure has inversion symmetry because then $\gamma_{-\mathbf{k}} = \gamma_{\mathbf{k}}$. Using $b_{\mathbf{k}} b_{\mathbf{k}}^\dagger = 1 + b_{\mathbf{k}}^\dagger b_{\mathbf{k}}$, and noticing that $\sum_{\mathbf{k}} \gamma_{\mathbf{k}} = 0$, the lowest-order spin wave Hamiltonian is

$$\mathcal{H}_0 = |J|zS \sum_{\mathbf{k}} (1 - \gamma_{\mathbf{k}}) b_{\mathbf{k}}^\dagger b_{\mathbf{k}} = \sum_{\mathbf{k}} \omega_{\mathbf{k}} \hat{n}_{\mathbf{k}} \tag{6.30}$$

where we introduced the magnon occupation number $\hat{n}_{\mathbf{k}} = b^\dagger_{\mathbf{k}} b_{\mathbf{k}}$, and the ferromagnon dispersion relation is

$$\omega_{\mathbf{k}} = |J| z S (1 - \gamma_{\mathbf{k}}) . \tag{6.31}$$

Let us discuss the long-wavelength limit where $|\mathbf{k} \cdot \boldsymbol{\delta}| \ll 1$. Expanding

$$z(1 - \gamma_{\mathbf{k}}) \approx (1/2) \sum_{\boldsymbol{\delta}} (\mathbf{k} \cdot \boldsymbol{\delta})^2$$

gives the result that long-wavelength magnons have an approximately quadratic dispersion

$$\omega_{\mathbf{k}} \approx S|J| \, (ka)^2 \tag{6.32}$$

where a is the lattice constant.

Let us note that $\omega_{\mathbf{k}} \to 0$ as $|\mathbf{k}| \to 0$. This follows from the celebrated *Goldstone theorem* which states that if a continuous symmetry is spontaneously broken, and the forces are sufficiently short-ranged, then there must exist a branch of excitations with the property that the energy vanishes in the limit $|\mathbf{k}| \to 0$. In our case, the spin-rotational symmetry of (6.3) is spontaneously broken by the non-vanishing magnetization of the ground state. A long-wavelength magnon amounts to a nearly uniform rotation of the spins. However, we have seen in our discussion of the ground state degeneracy that turning all the spins together does not cost energy. The vanishing of the magnon energy at $|\mathbf{k}| \to 0$ is related to the fact that it is a "symmetry-restoring mode".

Since the magnon dispersion (6.31) was derived by assuming a small deviation $\Delta S/S$ from the saturation value, we might be worried that it is only of approximate validity, especially for small S. However, this would be misreading the meaning of the approximation. For a single-magnon state, (6.31) is *exact*, even for $S = 1/2$. One way to convince ourselves about it is by checking that the single-magnon states are exact eigenstates, and the eigenvalues are those given in (6.31) (see Problem 6.1). Furthermore, a glance at the form (6.41) of the fourth-order term of $\mathcal{H}$ shows that it consists of terms like $b^\dagger b^\dagger b b$, telling us that a magnon can decay only by colliding with another magnon. If the magnon is alone in the system, it is an exact eigenstate.

The good quality of the lowest-order result may seem surprising because the square roots in (6.10) give the impression as if it were an essential aspect of the Heisenberg problem that only the summing of an infinite series can yield a correct result. However, this is a misleading appearance, which is shown by the fact that equivalent (finite) polynomial expressions can also be found [264]. For instance, for $S = 1/2$

$$\sqrt{1-\hat{n}} \equiv 1-\hat{n} \tag{6.33}$$

for the physically allowed states $n = 0, 1$. $S = 1$ spins can be represented by a second-order polynomial of $\hat{n}$, because the equivalence

$$\sqrt{1-\frac{\hat{n}}{2}} \equiv 1-\left(\frac{3}{2}-\sqrt{2}\right)\hat{n}-\frac{1}{2}\left(\sqrt{2}-1\right)\hat{n}^2 \tag{6.34}$$

holds for $n = 0, 1, 2$. For any S, an equivalent polynomial of degree $2S$ can be found.

There is a different method to get rid of the square roots, which works for arbitrary S. To see how this is possible, let us consider the generalized bosonic representation

$$S_{\rm j}^{+} = \sqrt{2S}\left(1-\frac{a_{\rm j}^{\dagger}a_{\rm j}}{2S}\right)^{\alpha} a_{\rm j} \qquad S_{\rm j}^{-} = \sqrt{2S}\, a_{\rm j}^{\dagger}\left(1-\frac{a_{\rm j}^{\dagger}a_{\rm j}}{2S}\right)^{1-\alpha} \tag{6.35}$$

where $0 \le \alpha \le 1$. The reader will find it easy to check that the commutation relations $[S^+, S^-] = 2S^z$, $[S^\pm, S^z] = \mp S^\pm$ hold for arbitrary α. However, $(S^+)^\dagger \neq S^-$ unless $\alpha = 1/2$, when we recover the Holstein–Primakoff representation (6.10). For $\alpha \neq 1/2$, the transformation leading from (6.10) to (6.35) is non-unitary. In the extreme case $\alpha = 1$, we get the Dyson–Maleev transformation [94, 259, 264]

$$S_{\rm j}^{+} = \sqrt{2S}\left(1-\frac{a_{\rm j}^{\dagger}a_{\rm j}}{2S}\right) a_{\rm j} \qquad S_{\rm j}^{-} = \sqrt{2S}\, a_{\rm j}^{\dagger}\,. \tag{6.36}$$

The absence of a fractional power is an appealing feature but it has its price: the non-physical part of the Hilbert space (with $n_{\rm j} > 2S$) becomes connected to the physical part, because $S^-|-S\rangle \rightsquigarrow 0$ is not automatically ensured. While this is a principal difference from the Holstein–Primakoff version, it is not much of a difference in practice, because by using the low-order expansion of the square root, we sacrificed this feature in our previous derivation, anyway. As long as we do linear spin wave theory, the Holstein–Primakoff and Dyson–Maleev transformations give the same results. If we wish to go further, we have the choice between several complicated schemes for getting rid of the effect of unphysical states [94, 42].

Apparently, there are a number of ways to do spin wave theory which seem to be either more aesthetic-looking than the Holstein–Primakoff version, or better suited for a particular value of S. It is, however, the Holstein–Primakoff approach which becomes reasonable in the quasi-classical limit $S \gg 1$. Though we have no immediate reason for introducing this limit for ferromagnets, it is of fundamental importance in the theory of antiferromagnetism.

Effects of Anisotropy

It is easy to generalize the treatment to the Heisenberg model with easy axis anisotropy

$$\mathcal{H} = J \sum_{\mathbf{i},\mathbf{j}} (S_{\mathbf{i}}^x S_{\mathbf{j}}^x + S_{\mathbf{i}}^y S_{\mathbf{j}}^y + A S_{\mathbf{i}}^z S_{\mathbf{j}}^z) \tag{6.37}$$

where $J < 0$ and $A > 1$. The ground state is twofold degenerate ($S_{\mathrm{tot}}^z = \pm LS$). The spin wave spectrum is given by (see Problem 6.1)

$$\omega_{\mathbf{k}} = |J| z S (A - \gamma_{\mathbf{k}})\,. \tag{6.38}$$

The smallest excitation energy is $\omega(\mathbf{k} = 0) = A - 1 > 0$, i.e., the spectrum has acquired a gap. Let us recall that we interpreted the gapless nature of the spectrum (6.31) of the isotropic model by pointing it out that the $\mathbf{k} \to 0$ spin wave is a "symmetry-restoring" mode which essentially turns the total spin around, traversing over a set of ground states. In contrast, the ground state of (6.37) is merely twofold degenerate (like the ground state of the Ising model), and turning the total spin away from the easy axis gives a high-energy state.

It must be mentioned that the spin wave spectrum of real ferromagnets has a profound reason to be gapped even if the effective exchange coupling could be assumed to be perfectly isotropic. The reason is the classical dipole–dipole interaction acting between the magnetic moments $\boldsymbol{\mu}$ of the ions

$$\mathcal{H}_{\text{dip}} = \sum_{\text{i}\neq\text{j}} \frac{1}{r_{\text{ij}}^3}[\boldsymbol{\mu}_\text{i} \cdot \boldsymbol{\mu}_\text{j} - 3(\boldsymbol{\mu}_\text{i} \cdot \hat{\mathbf{r}}_{\text{ij}})(\boldsymbol{\mu}_\text{j} \cdot \hat{\mathbf{r}}_{\text{ij}})] \tag{6.39}$$

where $\hat{\mathbf{r}}_{\text{ij}}$ is the unit vector pointing in the direction of $\mathbf{r}_\text{j} - \mathbf{r}_\text{i} = \mathbf{r}_{\text{ij}}$. It is worth noticing that $\boldsymbol{\mu}$ is the total magnetic moment of the ion including the orbital contribution. Thus the effects of the dipole–dipole interaction will be relatively strong for rare earth ions, while they are relatively weak for ions with quenched orbital moments. For our present purposes we think of the $\boldsymbol{\mu}$'s as arising from spins: $\boldsymbol{\mu} \propto \mathbf{S}$.

Adding $\mathcal{H}_{\text{dip}}$ to the Heisenberg Hamiltonian leads to an interesting new problem because, even restricting ourselves to the bilinear part, new terms like $S_\text{i}^z S_\text{j}^+$, $S_\text{i}^z S_\text{j}^-$, $S_\text{i}^+ S_\text{j}^+$, and $S_\text{i}^- S_\text{j}^-$ appear. We refer to [264, 439] for the derivation of the spin wave spectrum. The result is the following: the spectrum (6.31) is essentially unaltered in most of the Brillouin zone but it becomes strongly modified in the immediate vicinity of $\mathbf{k} = \mathbf{0}$. The spin waves begin to feel the demagnetization field when $|\mathbf{k}|$ is reduced to the order of $1/l$, where l is a linear dimension of the sample, and their energy becomes dependent on the shape of the sample, and on the direction of $\mathbf{k}$. Including also the effect of an external field H (which is treated separately in Problem 6.1), we find that at $\mathbf{k} \to \mathbf{0}$, the modified spin wave energy of an ellipsoidal sample approaches that of the uniform mode of ferromagnetic resonance

$$\epsilon(\mathbf{k} = \mathbf{0}) = \sqrt{[H + (N_x - N_z)M_0][H + (N_y - N_z)M_0]} \tag{6.40}$$

where N_x, N_y, N_z are the demagnetization factors, and M_0 is the magnetization density (see also [209], Ch. 17). Generally speaking, the long-range nature of the dipole force leads to the appearance of a gap, and also of anisotropy, in the spin wave spectrum. However, these effects are absent in a spherical sample.

Magnon–Magnon Interactions

The usefulness of linear spin wave theory depends on whether the consideration of the magnon–magnon interaction $\mathcal{H}_1$ would essentially alter the conclusions derived from $\mathcal{H}_0$. The lowest-order term of $\mathcal{H}_1$ is

$$\mathcal{H}_1^{(4)} = \frac{1}{L} \sum_{\mathbf{k}_1} \sum_{\mathbf{k}_2} \sum_{\mathbf{k}_3} \sum_{\mathbf{k}_4} b_{\mathbf{k}_1}^\dagger b_{\mathbf{k}_2}^\dagger b_{\mathbf{k}_3} b_{\mathbf{k}_4} \delta(\mathbf{k}_1 + \mathbf{k}_2 - \mathbf{k}_3 - \mathbf{k}_4) F(\mathbf{k}_1, \mathbf{k}_2, \mathbf{k}_3, \mathbf{k}_4) \tag{6.41}$$

where the structure of F should be identified from (6.18), (6.19), and (6.23). Since a magnon can decay only by colliding with another magnon, we reconfirm our earlier conclusion that the single-magnon eigenstates are exact. Furthermore, it is found that for small $\mathbf{k}$ the collision probability of two magnons is $\propto k^4$ where k is the order of magnitude of both $|\mathbf{k}_1|$ and $|\mathbf{k}_2|$ [210], thus long-wavelength magnons are very sharply defined quasiparticles. For most of the two-magnon states, it would be acceptable to say that they are well approximated by combining two independent magnons. However, the magnon–magnon interaction produces also non-perturbative effects: it can give rise to *two-magnon bound states*. Linear spin wave theory cannot account for phenomena for which the existence of bound states is important.

The extent of the $\mathbf{k}$-space domain allowed for bound states depends on the dimensionality of the system [264]. It turns out that in $D = 1$ and 2, there is at least one bound state for each wavevector. It follows that bound states are playing an important role in the physics of low-dimensional systems. On the other hand, in $D = 3$ there is a substantial part of the Brillouin zone (including the long-wavelength regime) where bound states cannot exist at all. Since scattering states are well approximated by non-interacting magnons, it can be claimed that using linear spin wave theory is justifiable for the low-energy excitations of three-dimensional ferromagnets. The formation of bound states is a quantum effect: the $\mathbf{k}$-space volume allowed for bound states is largest for $S = 1/2$, and it vanishes as $S \to \infty$.

Finite Temperatures

At sufficiently low temperatures, the magnons are independent bosonic quasiparticles, and their occupation numbers are given by the Planck distribution. The total number of magnons at temperature T is

$$N(T) = \int \frac{d\mathbf{k}}{e^{\omega(\mathbf{k})/k_B T} - 1} \,. \tag{6.42}$$

At low T only low-energy magnons are excited in appreciable numbers, and we may use (6.32). For a D-dimensional system

$$N(T) \approx \zeta_D \left(\frac{k_B T}{|J|Sa^2}\right)^{D/2} \int_0^\infty \frac{x^{(D-2)/2}dx}{e^x - 1} \tag{6.43}$$

where $\zeta_D = 2\pi$, π and $1/2$ for $D = 3$, 2 and 1, and we invoke the usual argument for replacing the upper limit of integration by ∞. For $D = 3$, the integrand has an integrable singularity at $x = 0$, and we get Bloch's $T^{3/2}$ law for the temperature dependence of the magnetization

$$\frac{M(0) - M(T)}{M(0)} = \frac{\xi}{S}\left(\frac{k_B T}{|J|S}\right)^{3/2} \tag{6.44}$$

where the factor ξ depends on the lattice structure [209]. Like most results of spin wave theory, the law is supposed to give a good approximation up to $T \sim T_C/2$. At higher temperatures, the magnon–magnon interaction becomes important.

For dimensions $D = 1$ and 2, the integral picks up a divergent contribution from the neighbourhood of $x = 0$. It follows that our starting assumption about a low-temperature state with essentially complete order (which is the necessary background for defining the magnons in the first place) is not self-consistent. We may say that at any finite T, a diverging number of magnons is created which destroy the ferromagnetic state. This finding is in accordance with the Mermin–Wagner theorem which states that the isotropic Heisenberg model cannot have long range order at $T > 0$ in one and two dimensions.

The total energy of the thermally excited magnons is

$$U_m(T) = \int \frac{\omega(\mathbf{k})d\mathbf{k}}{e^{\omega(\mathbf{k})/k_B T} - 1} \rightsquigarrow \zeta_D k_B T \left(\frac{k_B T}{|J|Sa^2}\right)^{D/2} \int_0^\infty \frac{x^{D/2}dx}{e^x - 1}. \tag{6.45}$$

The integral is finite even for $D = 1$. The magnon specific heat defined as $C_m = (\partial U_m/\partial T)$ is proportional to $T^{D/2}$.

The low-temperature specific heat of a three-dimensional isotropic insulating ferromagnet is well described by $C = C_m + C_{ph} = aT^{3/2} + bT^3$, where the T^3 term is the phonon contribution. However, the $T^{3/2}$ behaviour of the magnon specific heat is a low-T approximation and it is expected to break down when T reaches about $T_C/2$. To describe a broader range of T, one uses the expansion

$$C_{\rm m} = a_0 \left(\frac{k_{\rm B}T}{|J|S}\right)^{3/2} + a_1 \left(\frac{k_{\rm B}T}{|J|S}\right)^{5/2} + a_2 \left(\frac{k_{\rm B}T}{|J|S}\right)^{7/2} + a_3 \left(\frac{k_{\rm B}T}{|J|S}\right)^{4} \quad (6.46)$$

worked out by Dyson [94], which includes terms arising from expanding $\omega(\mathbf{k})$ beyond the order k^2 as well as a contribution from the magnon–magnon interaction. It seems that a better fit to measured data is achieved primarily by including the $T^{5/2}$- and $T^{7/2}$-terms which arise from a more careful evaluation of the Brillouin-zone sum in (6.45), while the physically more interesting interaction term $\sim T^4$ is difficult to identify. We refer to [84] for a systematic survey of a large number of magnetic insulators, including the comparison of the observed behaviour with theoretical predictions.

For a two-dimensional Heisenberg ferromagnet, spin wave theory predicts a linear specific heat $C \sim T$. We may fear that this result is absolutely valueless, since we have just convinced ourselves that there is no long range order at finite temperatures. However, the low-temperature specific heat of the quasi-two-dimensional ferromagnet[3] $(CH_3NH_3)_2CuCl_4$ looks perfectly linear in T [84]. We can give the following argument: since there is perfect order at $T = 0$, the correlation length must diverge as $T \to 0$ (in fact, it is predicted to behave like $\xi \sim \sqrt{J/T} \cdot \exp(JS^2/T)$), and thus the system can support spin-wave-like long-wavelength excitations. This explains why spin wave theory gives good results for the first two terms of the low-T expansion of the free energy, and also for the specific heat [387].

Problem 6.1. Derive the spin wave dispersion (6.31) by constructing a Bloch wave of single spin flips, and proving that it is an eigenstate of the Heisenberg Hamitonian. Discuss the result a) for the isotropic model, b) for the same model in an external field and c) for the anisotropic model (6.37).

[3] Remember that we met this substance in Sec. 5.4.

6.2 Antiferromagnetic Heisenberg Model

6.2.1 Introduction

Let us consider the isotropic nearest-neighbour antiferromagnetic Heisenberg model which we write as the sum of z–z (or Ising) and spin-flip (or x–y) terms

$$\mathcal{H} = \mathcal{H}_{zz} + \mathcal{H}_{\pm} = J \sum_{\mathbf{i},\mathbf{j}} S_{\mathbf{i}}^z S_{\mathbf{j}}^z + \frac{J}{2} \sum_{\mathbf{i},\mathbf{j}} (S_{\mathbf{i}}^+ S_{\mathbf{j}}^- + S_{\mathbf{i}}^- S_{\mathbf{j}}^+) \tag{6.47}$$

with $J > 0$. In most of our preliminary discussion, we assume that the lattice is bipartite. The great difference between the ferromagnetic and antiferromagnetic cases becomes immediately clear if we follow the strategy used at the beginning of Sec. 6.1.1, and try to minimize the ground state energy on a bond-by-bond basis. A bond energy is

$$E_{\mathbf{ij}} = J\mathbf{S}_{\mathbf{i}} \cdot \mathbf{S}_{\mathbf{j}} = -JS(S+1) + \frac{J}{2}(\mathbf{S}_{\mathbf{i}} + \mathbf{S}_{\mathbf{j}})^2 \tag{6.48}$$

which is the smallest if the two spins form a singlet $|\mathbf{S}_{\mathbf{i}} + \mathbf{S}_{\mathbf{j}}| = 0$. Thus the minimum of the bond energy is

$$(E_{\mathbf{ij}})_{\text{min.}} = -JS(S+1)\,. \tag{6.49}$$

Let us observe that the minimum is *not* obtained by simply letting the two spins point in opposite directions, say, in the state $|S\rangle_{\mathbf{i}}|-S\rangle_{\mathbf{j}}$ (meaning that $S_{\mathbf{i}}^z = S$, and $S_{\mathbf{j}}^z = -S$). That would merely give the energy $-JS^2$. However, $|S\rangle_{\mathbf{i}}|-S\rangle_{\mathbf{j}}$ cannot be the ground state of the bond, because it is not even an eigenstate:

$$\mathbf{S}_{\mathbf{i}} \cdot \mathbf{S}_{\mathbf{j}} |S\rangle_{\mathbf{i}}|-S\rangle_{\mathbf{j}} \rightsquigarrow S_{\mathbf{i}}^- S_{\mathbf{j}}^+ |S\rangle_{\mathbf{i}}|-S\rangle_{\mathbf{j}} \rightsquigarrow |S-1\rangle_{\mathbf{i}}|-S+1\rangle_{\mathbf{j}}\,.$$

Strict antiparallel alignment gains energy only from the z–z part of the Heisenberg interaction. In order to get the true ground state, we have to let the spin z-components fluctuate so that the system can gain energy also from the spin-flip terms (the extra energy lowering is $-JS$). In contrast to the ferromagnetic ground state which is the simultaneous eigenstate of all the $S_{\mathbf{j}}^z$, the ground state of the antiferromagnetic Heisenberg model shows *quantum fluctuations*.

How important is the quantum-mechanical nature of the spins? We can measure this by the ratio of the quantum correction to the (classical) z–z energy: $(-JS)/(-JS^2) = 1/S$. Judging from the two-site case, quantum effects are of the order $\sim 1/S$. Going over to the lattice case, we will conclude that the relevant parameter is more like $1/zS$ for bipartite structures, and that the details of the lattice structure are important for non-bipartite structures. In any case, quantum fluctuations are the strongest for small spins, particularly for $S = 1/2$.

Now let us consider the lattice case. The knowledge of $(E_{\mathrm{ij}})_{\mathrm{min.}}$ allows to give a crude lower bound for the ground state energy:

$$\mathcal{E}^{\mathrm{AF}}_{\mathrm{gr.st.}} > \sum_{\langle \mathrm{i,j} \rangle} (E_{\mathrm{ij}})_{\mathrm{min.}} = -\frac{JLz}{2} \cdot S(S+1)\,. \tag{6.50}$$

In contrast to the analogous (6.7), (6.50) can never[4] hold as an equality. The basic reason is that if a spin participates in a singlet bond with one of its neighbours then it cannot form a singlet with another neighbour as well: *the bond energies cannot be simultaneously optimized*[5]. It is possible to take this "one bond out of z" effect into account by writing the Hamiltonian as a sum of site-Hamiltonians

$$\mathcal{H} = \frac{J}{2} \sum_{\mathrm{j}} \mathcal{H}_{\mathrm{j}} \tag{6.51}$$

where $\mathcal{H}_{\mathrm{j}}$ describes the interaction of $\mathbf{S}_{\mathrm{j}}$ with its z neighbours $\mathbf{S}_{\mathrm{m}}$

$$\mathcal{H}_{\mathrm{j}} = \mathbf{S}_j \cdot \sum_{\mathrm{m}} \mathbf{S}_{\mathrm{m}} = \frac{1}{2} \left[\left(\mathbf{S}_j + \sum_{\mathrm{m}} \mathbf{S}_{\mathrm{m}} \right)^2 - S(S+1) - \left(\sum_{\mathrm{m}} \mathbf{S}_{\mathrm{m}} \right)^2 \right]. \tag{6.52}$$

The lowest energy level E^0_{j} of $\mathcal{H}_{\mathrm{j}}$ is obtained by making the total spin of the neighbours as large as possible $(\sum \mathbf{S}_{\mathrm{m}})^2 = zS(zS+1)$, and then

---

[4] Apart from the trivial case $z = 1$ when the system consists of disjoint bonds.

[5] Let us consider $S = 1/2$ spins, and assume that both 1 and 3 are neighbours of 2. If $\mathbf{S}_1$ and $\mathbf{S}_2$ form the singlet $(|\uparrow\rangle_1 |\downarrow\rangle_2 - |\uparrow\rangle_2 |\downarrow\rangle_1)/\sqrt{2}$, then the instantaneous state of $\mathbf{S}_2$ is determined by what $\mathbf{S}_1$ is doing "just then", and cannot be similarly correlated with $\mathbf{S}_3$. An overall singlet of more than two spins can be built up of disjoint singlet pairs (see p. 307 and Problem 6.3).

minimizing the first term by making the total spin of the $(z+1)$-spin system the smallest possible: $(z-1)S$. The result is

$$E_{\mathbf{j}}^{0} = -zS^2\left(1+\frac{1}{zS}\right) . \tag{6.53}$$

Apart from the trivial case of $z = 1$, the clusters overlap, and their energies cannot attain the above-shown optimum value simultaneously. We get the following inequality for the ground state energy of $\mathcal{H}$ [16]

$$\mathcal{E}_{\text{gr.st.}}^{\text{AF}} > -\frac{JLzS}{2}\left(S+\frac{1}{z}\right) \tag{6.54}$$

which represents a significant improvement over (6.50).

Since we cannot tell readily what the ground state energy is[6], it is useful to determine an interval for $\mathcal{E}_{\text{gr.st.}}^{\text{AF}}$. We already know the lower bound (6.54), and we can find an upper bound variationally. Dividing the bipartite lattice into the alternate sublattices A and B, the *classical Néel state* Ψ_{N} is defined as the state with maximum sublattice magnetizations: all spins are $S_{\mathbf{j}}^{z} = S$ for $\mathbf{j} \in A$, and $S_{\mathbf{m}}^{z} = -S$ for $\mathbf{m} \in B$ (see Fig. 7.1). The corresponding energy

$$\langle \Psi_{\text{N}} | \mathcal{H} | \Psi_{\text{N}} \rangle = \langle \Psi_{\text{N}} | \mathcal{H}_{zz} | \Psi_{\text{N}} \rangle = -\frac{z}{2} JS^2 L \tag{6.55}$$

contains only the contribution of the z–z (Ising) term. Measuring the ground state energy density in units of the corresponding quantity for the Néel state, we get the interval

$$-1 - \frac{1}{zS} < \frac{\mathcal{E}_{\text{gr.st.}}^{\text{AF}}}{(zJS^2/2)} < -1 . \tag{6.56}$$

It is clear that there is a well-defined sense in which quantum fluctuations become unimportant for large S and/or large z (large dimensionality): the ground state energy is confined to a narrow interval, and the Néel energy (6.55) gives a good approximation. We are tempted to infer

[6]It turns out to be a very difficult problem. The Bethe Ansatz technique gives an exact solution for the $S = 1/2$ Heisenberg chain [48, 175], but there is no exact solution for higher-dimensional or higher-spin systems.

that not merely the energy is given acceptably by the classical expression but also the ground state is well approximated by Ψ_N. Though the guess turns out to go wrong in $D = 1$, it works well for $D \geq 2$. The consequences of taking the classical Néel state as the starting point are worked out in the theory of antiferromagnetic spin waves.

6.2.2 Spin Waves

Our mental picture of antiferromagnetism is based on regarding the spins as classical vectors[7]. Considering a bipartite lattice which we divide into the alternate sublattices A and B, we envisage the spins pointing up on sublattice A ($S_{\mathbf{j}}^z = S$ for $\mathbf{j} \in A$), and down on sublattice B ($S_{\mathbf{l}}^z = -S$ for $\mathbf{l} \in B$). This is the classical Néel state. The *quasiclassical theory of antiferromagnetic spin waves* is based on the idea that continuation from the classical limit to finite (and even to small) spins is possible, and that two-sublattice long range order survives in spite of the quantum fluctuations. We anticipate that the sublattice magnetization gets diminished due to quantum fluctuations: $\langle S_{\mathbf{j}}^z \rangle = -\langle S_{\mathbf{l}}^z \rangle = S - \Delta S$, but it remains finite. We want a theory for calculating quantum corrections to classical results.

Here we may let ourselves be guided by an analogy with the quantum theory of solids. The perfect crystalline state, in which the nuclei are strictly localized at the points $\mathbf{R}_{\mathbf{j}}^0$ of a periodic lattice, can be the ground state for infinitely heavy ($M \to \infty$) nuclei only. For finite M, we get quantum corrections: the nuclei oscillate about the lattice points, thus $\mathbf{R}_{\mathbf{j}}^0$ is no longer an eigenvalue of the coordinate $\mathbf{R}_{\mathbf{j}}$, but rather its expectation value $\langle \mathbf{R}_{\mathbf{j}} \rangle = \mathbf{R}_{\mathbf{j}}^0$. The deviation from the classical position is measured by $\delta a = \sqrt{\langle (\mathbf{R}_{\mathbf{j}} - \mathbf{R}_{\mathbf{j}}^0)^2 \rangle}$ which increases with decreasing M. The ground state preserves crystalline order, even though for light atoms δa may be an appreciable fraction[8] of the lattice constant a. The

[7] Spins behave as classical vectors in the limit $S \to \infty$, where contributions due to the non-commutative nature of the spin components become negligible.

[8] The Lindeman criterion for melting would be $\delta a / a \sim 0.25$. We may assume that hypothetical solid helium would have a larger $\delta a / a$, and the assumption about broken translational invariance being not self-consistent, helium "quantum-melts" and becomes a quantum fluid. This leads to the conjecture that certain antiferromagnetic spin systems may have a sufficiently large ΔS so that they quantum-melt, too, and

simplest way to describe this quantum effect is to calculate the phonon spectrum: δa arises as an integrated effect of the zero-point oscillations of the phonon modes, while the zero-point energies give a contribution to the ground state energy of the lattice.

In this section, we will attempt to do something quite similar for the spin problem: the limit $S \to \infty$ plays the role of $M \to \infty$, the classical equilibrium state is like the Néel state, and ΔS is analogous to δa. Upon making S finite, the spin system begins to oscillate about the classical Néel state, and the normal modes are the spin waves.

The theory of antiferromagnetic spin waves[9] shows many formal similarities to ferromagnetic spin wave theory, but there are also essential differences. Most important is that now we do not know the exact ground state, and one of the aims of the theory is to check the consistency of our assumption of a Néel-type ground state.

We have to define two kinds of Holstein–Primakoff bosons: on sublattice A they are defined as in (6.10)

$$S_{A\mathrm{j}}^{+} = \sqrt{2S}\left(1 - \frac{a_{\mathrm{j}}^{\dagger}a_{\mathrm{j}}}{2S}\right)^{1/2} a_{\mathrm{j}} \qquad S_{A\mathrm{j}}^{-} = \sqrt{2S}\, a_{\mathrm{j}}^{\dagger}\left(1 - \frac{a_{\mathrm{j}}^{\dagger}a_{\mathrm{j}}}{2S}\right)^{1/2} \tag{6.57}$$

while on sublattice B the creation of a boson belongs to raising S^z

$$S_{B\mathrm{l}}^{+} = \sqrt{2S}\, b_{\mathrm{l}}^{\dagger}\left(1 - \frac{b_{\mathrm{l}}^{\dagger}b_{\mathrm{l}}}{2S}\right)^{1/2} \qquad S_{B\mathrm{l}}^{-} = \sqrt{2S}\left(1 - \frac{b_{\mathrm{l}}^{\dagger}b_{\mathrm{l}}}{2S}\right)^{1/2} b_{\mathrm{l}}\,. \tag{6.58}$$

When introducing propagating states, we have to remember that the unit cell is now the magnetic supercell consisting of an A-site and a B-site[10]. The reduced Brillouin zone contains $L/2$ $\mathbf{k}$-vectors. In analogy

become "spin quantum liquids". This certainly happens in $D = 1$, but can it happen in higher dimensions as well? We return to this question in Sec. 6.3.

[9]Developed by Anderson in [17]. Our derivation follows [210]. To cite just a few sources, [33], [213], [260], and [264] review spin wave theory from a variety of points of view, and compare its results to those obtained from more sophisticated treatments.

[10]For the sake of simplicity, let us henceforth assume that we have a D-dimensional cubic lattice, for which imposing AF order leads to cell doubling. This would not be true for all bipartite lattices: e.g., the unit cell of the honeycomb lattice contains two sites, and introducing two-sublattice magnetic order does not change the unit cell.

to (6.13), let us define **k**-states propagating on sublattice A

$$c_{\mathbf{k}}^{\dagger} = \sqrt{\frac{2}{L}} \sum_{\mathbf{j} \in A} e^{-i\mathbf{k}\mathbf{j}} a_{\mathbf{j}}^{\dagger} \qquad a_{\mathbf{j}}^{\dagger} = \sqrt{\frac{2}{L}} \sum_{\mathbf{k}} e^{i\mathbf{k}\mathbf{j}} c_{\mathbf{k}}^{\dagger}, \tag{6.59}$$

and similar states for sublattice B

$$d_{\mathbf{k}}^{\dagger} = \sqrt{\frac{2}{L}} \sum_{\mathbf{l} \in B} e^{-i\mathbf{k}\mathbf{l}} b_{\mathbf{l}}^{\dagger} \qquad b_{\mathbf{l}}^{\dagger} = \sqrt{\frac{2}{L}} \sum_{\mathbf{k}} e^{i\mathbf{k}\mathbf{l}} d_{\mathbf{k}}^{\dagger}. \tag{6.60}$$

We cannot, however, expect that the effect of flipping a spin at an A-site remains confined to sublattice A, because the nearest neighbour coupling connects A- and B-sites. The proper eigenstates will be described by combining c- and d-operators.

The sublattice magnetizations are given by

$$S_A^z = \sum_{\mathbf{j}} S_{A\mathbf{j}}^z = \frac{LS}{2} - \sum_{\mathbf{k}} c_{\mathbf{k}}^{\dagger} c_{\mathbf{k}} \qquad S_B^z = \sum_{\mathbf{l}} S_{B\mathbf{l}}^z = -\frac{LS}{2} + \sum_{\mathbf{k}} d_{\mathbf{k}}^{\dagger} d_{\mathbf{k}}. \tag{6.61}$$

In what follows, we proceed as we did on pp. 270–272, replacing the expansions (6.59), (6.60) into (6.57), and (6.58), and those into the expression of the Hamiltonian. However, here we declare at the outset that we do not wish to go beyond linear spin wave theory, and neglect magnon–magnon interactions. $\mathcal{H}$ is approximated by

$$\mathcal{H} = -\frac{Lz}{2} JS^2 + \mathcal{H}_0 \tag{6.62}$$

where the first term is the classical Néel energy, and the second term is a bilinear expression of c- and d-operators.

The z–z coupling of a single pair of sites gives rise to

$$\begin{aligned} J \cdot S_{A\mathbf{j}}^z S_{B\mathbf{j}+\boldsymbol{\delta}}^z &\rightsquigarrow JS a_{\mathbf{j}}^{\dagger} a_{\mathbf{j}} + JS b_{\mathbf{j}+\boldsymbol{\delta}}^{\dagger} b_{\mathbf{j}+\boldsymbol{\delta}} \\ &= \frac{2JS}{L} \sum_{\mathbf{k}_1} \sum_{\mathbf{k}_2} e^{i(\mathbf{k}_1 - \mathbf{k}_2)\mathbf{j}} \left[c_{\mathbf{k}_1}^{\dagger} c_{\mathbf{k}_2} + e^{i(\mathbf{k}_1 - \mathbf{k}_2)\boldsymbol{\delta}} d_{\mathbf{k}_1}^{\dagger} d_{\mathbf{k}_2} \right]. \end{aligned} \tag{6.63}$$

Summing over **j** or (or **l**, whichever is more convenient) gives $(L/2)\delta_{\mathbf{k}_1\mathbf{k}_2}$, and the sum over $\boldsymbol{\delta}$ brings merely a factor of z

$$J \cdot \sum_{\mathbf{j}} \sum_{\mathbf{l}} S_{A\mathbf{j}}^z S_{B\mathbf{l}}^z \rightsquigarrow JSz \sum_{\mathbf{k}} (c_{\mathbf{k}}^{\dagger} c_{\mathbf{k}} + d_{\mathbf{k}}^{\dagger} d_{\mathbf{k}}). \tag{6.64}$$

Repeating the same procedure for the spin-flip terms, we find a structure-sensitive contribution:

$$\frac{J}{2} \cdot \sum_{\mathbf{j}} \sum_{\mathbf{l}} S^{+}_{A\mathbf{j}} S^{-}_{B\mathbf{l}} \rightsquigarrow \frac{J}{2} \cdot \sum_{\mathbf{j}} \sum_{\mathbf{l}} a_{\mathbf{j}} b_{\mathbf{l}}$$

$$= \frac{JS}{L} \cdot \sum_{\mathbf{j}} \sum_{\boldsymbol{\delta}} \sum_{\mathbf{k}_1} \sum_{\mathbf{k}_2} e^{-\mathbf{k}_1 \mathbf{j}} e^{-i\mathbf{k}_2(\mathbf{j}+\boldsymbol{\delta})} c_{\mathbf{k}_1} d_{\mathbf{k}_2}$$

$$= JSz \sum_{\mathbf{k}} \gamma_{\mathbf{k}} c_{\mathbf{k}} d_{-\mathbf{k}} \tag{6.65}$$

where $\gamma_{\mathbf{k}}$ is given by (6.27). Similarly,

$$\frac{J}{2} \cdot \sum_{\mathbf{j}} \sum_{\mathbf{l}} S^{-}_{A\mathbf{j}} S^{+}_{B\mathbf{l}} \rightsquigarrow JSz \sum_{\mathbf{k}} \gamma_{\mathbf{k}} d^{\dagger}_{\mathbf{k}} c^{\dagger}_{-\mathbf{k}} . \tag{6.66}$$

All in all

$$\mathcal{H}_0 = JzS \sum_{\mathbf{k}} [\gamma_{\mathbf{k}}(c_{\mathbf{k}} d_{-\mathbf{k}} + d^{\dagger}_{\mathbf{k}} c^{\dagger}_{-\mathbf{k}}) + (c^{\dagger}_{\mathbf{k}} c_{\mathbf{k}} + d^{\dagger}_{\mathbf{k}} d_{\mathbf{k}})] . \tag{6.67}$$

In contrast to the ferromagnetic case, Fourier transformation has not sufficed to give a diagonal form of $\mathcal{H}_0$. The basic reason is that the occupation numbers of the c- and d-bosons are not conserved quantities. For instance, $c_{\mathbf{k}}$ evolves into $d^{\dagger}_{-\mathbf{k}}$

$$\dot{c}_{\mathbf{k}} = -i[c_{\mathbf{k}}, \mathcal{H}] \rightsquigarrow c_{\mathbf{k}} + \gamma_{\mathbf{k}} d^{\dagger}_{-\mathbf{k}} . \tag{6.68}$$

Hence we infer that in a true eigenoperator, $c_{\mathbf{k}}$ and $d^{\dagger}_{-\mathbf{k}}$ are mixed. The proper combination is found by a Bogolyubov transformation.

Noticing that $\mathbf{k}$ and $-\mathbf{k}$ are playing formally dissimilar roles, it is worth rewriting $\mathcal{H}_0$ is a symmetrized form according to $\mathcal{H}_0 = \sum_{\mathbf{k}} = (\sum_{\mathbf{k}} + \sum_{-\mathbf{k}})/2$, and redistribute the terms as follows

$$\frac{2}{zJS} \mathcal{H}_0 = \sum_{\mathbf{k}} \mathcal{H}^{(1)}_{\mathbf{k}} + \sum_{\mathbf{k}} \mathcal{H}^{(2)}_{\mathbf{k}} \tag{6.69}$$

where

$$\mathcal{H}^{(1)}_{\mathbf{k}} = \gamma_{\mathbf{k}}(c^{\dagger}_{\mathbf{k}} d^{\dagger}_{-\mathbf{k}} + d_{-\mathbf{k}} c_{\mathbf{k}}) + (c^{\dagger}_{\mathbf{k}} c_{\mathbf{k}} + d^{\dagger}_{-\mathbf{k}} d_{-\mathbf{k}}) \tag{6.70}$$

and

$$\mathcal{H}_{\mathbf{k}}^{(2)} = \gamma_{\mathbf{k}}(d_{\mathbf{k}}^{\dagger}c_{-\mathbf{k}}^{\dagger} + c_{-\mathbf{k}}d_{\mathbf{k}}) + (c_{-\mathbf{k}}^{\dagger}c_{-\mathbf{k}} + d_{\mathbf{k}}^{\dagger}d_{\mathbf{k}})\,. \tag{6.71}$$

Proper eigenoperators can be found for $\mathcal{H}_{\mathbf{k}}^{(1)}$ and $\mathcal{H}_{\mathbf{k}}^{(2)}$ separately.

We wish to bring $\mathcal{H}_{\mathbf{k}}^{(1)}$ to the diagonal form $\lambda_{\mathbf{k}}\alpha_{\mathbf{k}}^{\dagger}\alpha_{\mathbf{k}}$ by introducing new field operators $\alpha_{\mathbf{k}}^{\dagger}$, $\alpha_{\mathbf{k}}$ which satisfy boson commutation relations

$$[\alpha_{\mathbf{k}}, \alpha_{\mathbf{k}'}^{\dagger}] = \delta_{\mathbf{k}\mathbf{k}'}\,. \tag{6.72}$$

The requirement for getting a diagonal form can be rephrased as

$$[\alpha_{\mathbf{k}}, \mathcal{H}_{\mathbf{k}}^{(1)}] = \lambda_{\mathbf{k}}\alpha_{\mathbf{k}} \quad \text{and} \quad [\alpha_{\mathbf{k}}^{\dagger}, \mathcal{H}_{\mathbf{k}}^{(1)}] = -\lambda_{\mathbf{k}}\alpha_{\mathbf{k}}^{\dagger} \tag{6.73}$$

where $\lambda_{\mathbf{k}}$ is to be determined.

Referring to (6.68), we seek the new operators in the form

$$\alpha_{\mathbf{k}} = u_{\mathbf{k}}c_{\mathbf{k}} - v_{\mathbf{k}}d_{-\mathbf{k}}^{\dagger} \qquad \alpha_{\mathbf{k}}^{\dagger} = u_{\mathbf{k}}c_{\mathbf{k}}^{\dagger} - v_{\mathbf{k}}d_{-\mathbf{k}}\,. \tag{6.74}$$

$u_{\mathbf{k}}$ and $v_{\mathbf{k}}$ are assumed to be real. $[\alpha_{\mathbf{k}}, \alpha_{\mathbf{k}}^{\dagger}] = 1$ holds if

$$u_{\mathbf{k}}^2 - v_{\mathbf{k}}^2 = 1\,. \tag{6.75}$$

Now we evaluate $[\alpha_{\mathbf{k}}, \mathcal{H}_{\mathbf{k}}^{(1)}]$, requiring that the result looks like (6.73)

$$\begin{aligned}[\alpha_{\mathbf{k}}, \mathcal{H}_{\mathbf{k}}^{(1)}] &= u_{\mathbf{k}}\gamma_{\mathbf{k}}d_{-\mathbf{k}}^{\dagger} + u_{\mathbf{k}}c_{\mathbf{k}} + v_{\mathbf{k}}\gamma_{\mathbf{k}}c_{\mathbf{k}} + v_{\mathbf{k}}d_{-\mathbf{k}}^{\dagger} \\ &= \lambda_{\mathbf{k}}(u_{\mathbf{k}}c_{\mathbf{k}} - v_{\mathbf{k}}d_{-\mathbf{k}}^{\dagger})\,.\end{aligned} \tag{6.76}$$

Term-by-term comparison gives the system of equations

$$\begin{aligned}u_{\mathbf{k}} + \gamma_{\mathbf{k}}v_{\mathbf{k}} &= \lambda_{\mathbf{k}}u_{\mathbf{k}} \\ v_{\mathbf{k}} + \gamma_{\mathbf{k}}u_{\mathbf{k}} &= -\lambda_{\mathbf{k}}v_{\mathbf{k}}\end{aligned} \tag{6.77}$$

which has a non-trivial solution if

$$\begin{vmatrix} 1-\lambda_{\mathbf{k}} & \gamma_{\mathbf{k}} \\ \gamma_{\mathbf{k}} & 1+\lambda_{\mathbf{k}} \end{vmatrix} = 0\,, \tag{6.78}$$

i.e.,

$$\lambda_{\mathbf{k}}^2 = 1 - \gamma_{\mathbf{k}}^2\,. \tag{6.79}$$

Repeating the previous procedure, we find that $\mathcal{H}_{\mathbf{k}}^{(2)}$ is diagonalized by a second set of bosonic operators

$$\begin{aligned} \beta_{\mathbf{k}} &= u_{\mathbf{k}} d_{\mathbf{k}} - v_{\mathbf{k}} c_{-\mathbf{k}}^{\dagger} \\ \beta_{\mathbf{k}}^{\dagger} &= -v_{\mathbf{k}} c_{-\mathbf{k}} + u_{\mathbf{k}} d_{\mathbf{k}}^{\dagger} \,. \end{aligned} \tag{6.80}$$

$[\alpha_{\mathbf{k}}, \beta_{\mathbf{k}'}] = [\alpha_{\mathbf{k}}, \beta_{\mathbf{k}'}^{\dagger}] = 0$, and $\lambda_{\mathbf{k}}$ is the same as in (6.79).

Re-expressing $\mathcal{H}_{\mathbf{k}}^{(1)}$ and $\mathcal{H}_{\mathbf{k}}^{(2)}$ with the new operators requires using the inverse formulas

$$\begin{aligned} c_{\mathbf{k}} &= u_{\mathbf{k}} \alpha_{\mathbf{k}} + v_{\mathbf{k}} \beta_{-\mathbf{k}}^{\dagger} & \qquad c_{\mathbf{k}}^{\dagger} &= u_{\mathbf{k}} \alpha_{\mathbf{k}}^{\dagger} + v_{\mathbf{k}} \beta_{-\mathbf{k}} \\ d_{\mathbf{k}} &= u_{\mathbf{k}} \beta_{\mathbf{k}} + v_{\mathbf{k}} \alpha_{-\mathbf{k}}^{\dagger} & \qquad d_{\mathbf{k}}^{\dagger} &= u_{\mathbf{k}} \beta_{\mathbf{k}}^{\dagger} + v_{\mathbf{k}} \alpha_{-\mathbf{k}} \,. \end{aligned} \tag{6.81}$$

The transformation brings $\mathcal{H}_0$ to a form which is diagonal in the occupation number representation of the α-, and β-bosons. We find

$$\mathcal{H}_0 = -\frac{JzL}{2} S(S+1) + JzS \sum_{\mathbf{k}} \sqrt{1-\gamma_{\mathbf{k}}^2} \left\{ \left(\alpha_{\mathbf{k}}^{\dagger} \alpha_{\mathbf{k}} + \frac{1}{2} \right) + \left(\beta_{\mathbf{k}}^{\dagger} \beta_{\mathbf{k}} + \frac{1}{2} \right) \right\} . \tag{6.82}$$

Since all boson energies

$$\omega_{\mathbf{k}} = JzS \sqrt{1-\gamma_{\mathbf{k}}^2} \tag{6.83}$$

are non-negative, the boson vacuum is the ground state[11].

The linearized Hamiltonian can be rewritten as

$$\mathcal{H} = E_0 + \sum_{\mathbf{k}} \omega_{\mathbf{k}} (\alpha_{\mathbf{k}}^{\dagger} \alpha_{\mathbf{k}} + \beta_{\mathbf{k}}^{\dagger} \beta_{\mathbf{k}}) \tag{6.84}$$

where the **k**-sum gives the total energy of the antiferromagnetic magnons. *The magnon spectrum is twofold degenerate.* We can understand this

[11]This is the vacuum of α- and β-bosons, not of c- and d-bosons, thus it contains many spin-flips with respect to the Néel-state! Since linearized spin wave theory does not amount to an exact implementation of the Holstein–Primakoff transformation, we have to worry about the possibility that the ground state contains components from the non-physical part of the enlarged bosonic Hilbert space. Well, it does, but it can be argued [42, 213] that in spite of this shortcoming spin wave theory gives surprisingly good results, so we leave it at that.

by checking that $\alpha^\dagger$ ($\beta^\dagger$) creates a $\Delta S^z_{\rm tot} = -1$ ($\Delta S^z_{\rm tot} = 1$) excitation. Postulating the Néel state, the system lost its isotropy but still retains axial symmetry about the spin z-axis, and this is enough to ensure the degeneracy of $\Delta S^z_{\rm tot} = \pm 1$ excitations. – In the long-wavelength limit $|\mathbf{k}| \ll a^{-1}$, $\sqrt{1-\gamma_{\mathbf{k}}^2} \approx |\mathbf{k}|a$, thus the antiferromagnon spectrum is approximately linear

$$\omega_{\mathbf{k}} \approx JzSa|\mathbf{k}| \,, \tag{6.85}$$

like the phonon spectrum. The gaplessness of the spectrum is, again, a consequence of the Goldstone theorem: the $\omega \to 0$ excitations are symmetry-restoring modes.

The ground state energy

$$E_0 = -\frac{1}{2}LzJS^2 + JzS\sum_{\mathbf{k}}(\sqrt{1-\gamma_{\mathbf{k}}^2} - 1) \tag{6.86}$$

is *lowered* due to quantum fluctuations[12]. Let us recall that the energy (6.55) of the classical Néel state is a variational bound, so the energy can only decrease if we allow the spins to deviate from the z-direction.

An often-used alternative way of writing the ground state energy indicates the extent to which E_0 falls short of reaching the lower bound given in (6.54):

$$E_0 = -\frac{1}{2}zLJS(S+\zeta) \tag{6.87}$$

where

$$\zeta = \frac{2}{L}\sum_{\mathbf{k}}(1 - \sqrt{1-\gamma_{\mathbf{k}}^2}) \,. \tag{6.88}$$

Obviously, $0 < \zeta < 1$. For the square lattice $\zeta \approx 0.158$, for the simple cubic lattice $\zeta \approx 0.097$, while for the bcc lattice $\zeta \approx 0.072$. Let us notice that the ratio of these numbers is roughly (1/4):(1/6):(1/8), thus ζ is a correction of order $1/z$. In retrospect, we see that (6.56) indicates the correct order of magnitude $1/zS$ of the leading quantum correction[13] to

[12]Observe that the quantum correction is not simply the sum of the zero-point energies of the magnons but contains also $-zJSL/2$.

[13]Though the first term is $\sim 1/zS$, it does not follow that $1/zS$ is *the* small parameter. Higher-order spin wave theory looks more like a simultaneous expansion in $1/S$ and $1/z$, thus one finds also corrections like $1/z^2S$, etc.

the ground state energy, and linear spin wave theory provides us with a structure-dependent numerical factor to get this quantity right.

The highest ζ is expected for the linear chain, where we would naively estimate $1/2$. Actually, linear spin wave theory gives $\zeta \approx 0.363$, while the exact ground state energy of the $S = 1/2$ model [175] can be parametrized by $\zeta = 2\ln 2 - 1 \approx 0.386$. It is remarkable that a relatively accurate ground state energy is obtained, notwithstanding the fact that spin wave theory will be seen to fail for the linear chain.

The specific heat contribution of antiferromagnons is calculated analogously to that of ferromagnons (see p. 278). However, since the long-wavelength dispersion (6.85) is linear, we find $C \sim T^3$ in $D = 3$. Since the phonon specific heat has the same T-dependence, it takes due care to distinguish the two contributions in measured data [84].

Sublattice Magnetization

The order parameter is the staggered magnetization density m_{st} (or, alternatively, the sublattice magnetization density). Since its operator does not commute with $\mathcal{H}$, we have to define it as the ground state expectation value (see (6.61) and (6.81))

$$\begin{aligned} m_{\mathrm{st}} &= \frac{1}{L}\left\langle \sum_{\mathbf{j}\in A} S^z_{\mathbf{j}} - \sum_{\mathbf{m}\in B} S^z_{\mathbf{m}} \right\rangle = S - \frac{1}{L}\sum_{\mathbf{k}} \left(\langle c^\dagger_{\mathbf{k}} c_{\mathbf{k}} \rangle + \langle d^\dagger_{\mathbf{k}} d_{\mathbf{k}} \rangle \right) \\ &= S - \frac{2}{L}\sum_{\mathbf{k}} v^2_{\mathbf{k}} = S - \frac{1}{L}\sum_{\mathbf{k}} \left[\frac{1}{\sqrt{1-\gamma^2_{\mathbf{k}}}} - 1 \right] \end{aligned} \tag{6.89}$$

where we exploited that off-diagonal terms like $u_{\mathbf{k}} v_{\mathbf{k}} \alpha^\dagger_{\mathbf{k}} \beta^\dagger_{-\mathbf{k}}$ do not contribute to the expectation value, and that in the ground state, $\langle \alpha^\dagger_{\mathbf{k}} \alpha_{\mathbf{k}} \rangle = 0$, and $\langle \beta_{-\mathbf{k}} \beta^\dagger_{-\mathbf{k}} \rangle = 1$. The expression for $v^2_{\mathbf{k}}$ is readily obtained from (6.75) and (6.77).

We find that quantum fluctuations reduce the staggered magnetization from its classical value S by

$$\Delta S = \frac{1}{2}\left(\int \frac{d^D\mathbf{k}}{(2\pi)^D} \frac{1}{\sqrt{1-\gamma^2_{\mathbf{k}}}} - 1 \right) . \tag{6.90}$$

ΔS is called the spin reduction. At small $|\mathbf{k}|$ the integrand diverges as $1/|\mathbf{k}|$, and it depends on the dimensionality D whether this has a drastic effect. In $D = 1$, we find a logarithmic divergency

$$\Delta S \sim \int_0^{O(1)} \frac{dk}{k} \rightsquigarrow -\ln 0 \to \infty . \tag{6.91}$$

The finding of a nominally divergent spin reduction[14] shows that our starting assumption of antiferromagnetic long-range order in the ground state is untenable. *For the one-dimensional isotropic Heisenberg model, quantum fluctuations destroy the antiferromagnetic long-range order for any finite spin S.* Though our reasoning was approximate, the above conclusion is strictly valid[15]. We conclude that the total spin of the ground state must be $S_{\text{tot}} = 0$. But what kind of singlet is it? We return to this question on pp. 299–306, and in Sec. 6.3.

The neighbourhood of $\mathbf{k} = 0$ gives a negligible contribution in D=3, and the spin reduction is small, being of the order of $1/2z$. One finds $\Delta S \approx 0.078$ for the simple cubic lattice. There are several reasons why this small correction is difficult to measure [84]. Extending spin wave theory to the anisotropic Heisenberg model shows that ΔS is very sensitive to anisotropy (even a weak uniaxial anisotropy is efficient in suppressing quantum fluctuations). Seeking a good model material, one finds the perovskite $RbMnF_3$ in which the Mn^{2+} ions have a half-filled shell, thus anisotropy and complications due to an orbital moment are negligible. However, ΔS amounts to only about 3% of $S = 5/2$. Neutron scattering or nuclear resonance methods may have the necessary resolution but we are faced with the next difficulty: covalent mixing spreads out the moment to the neighbouring anions where it gets cancelled by contributions arising from other Mn ions, and this apparent loss of moment is difficult to disentangle from spin reduction. In spite of all these difficulties, $\Delta S \approx 0.078$ can be regarded as experimentally confirmed in $RbMnF_3$ [84].

[14]Naturally, the result $\Delta S = \infty$ cannot be taken literally. A more careful argument should give $\Delta S = S$.

[15]There are general arguments to rule out the breaking of a continuous symmetry in one-dimensional quantum systems with short-range interactions [67]. However, we cannot exclude that an isotropic Heisenberg chain with short-range interactions breaks its (discrete) translational symmetry. See Sec. 6.3.2.

It would be difficult to miss observing ΔS in quasi-two-dimensional low-spin antiferromagnets. $D = 2$ is an intermediate case in the sense that both the vicinity of $\mathbf{k} = 0$ and the outer parts of the Brillouin zone give comparable contributions to the integral (6.90). Spin wave theory predicts $\Delta S \approx 0.197$ for the square lattice. That leaves us with only ~60% of the classically estimated order parameter for an $S = 1/2$ antiferromagnet! The case is of great interest because it closely corresponds to La_2CuO_4. However, let us not forget that our spin wave theory amounts to keeping the lowest-order contributions in a $1/S$ expansion, thus finding $\Delta S/S \sim 0.4$ in lowest order may have led us to suspect that in higher orders ΔS is even larger, or maybe long range order is destabilized completely. A great variety of theoretical techniques have been brought to bear upon this problem and the upshot is that the simplest estimate is basically right [260]. It is interesting that $m_{\rm st} \approx 0.3$ is essentially confirmed by experiments, thanks to the fact that in La_2CuO_4, potential disturbing effects are almost entirely absent: the interlayer coupling is very weak, and anisotropy effects apparently small. The situation is discussed in [260].

There are two-dimensional lattices with coordination number $z = 3$ and we expect that for these, fluctuation effects are even stronger. A honeycomb lattice $S = 1/2$ antiferromagnet looses ~56% of its classical sublattice magnetization: $m_{\rm st} \approx 0.22 \pm 0.03$ both from spin wave theory and Monte Carlo studies [333]. A different $z = 3$ lattice is the idealized structure of the quasi-two-dimensional $S = 1/2$ Mott insulator CaV_4O_9, which we show in Fig. 6.4 (top left) (see p. 327). Notice that inequivalent bonds have the differing coupling constants J_1 and J_2. If we take these as equal[16], spin wave theory gives $m_{\rm st} \approx 0.228$ [412]. We mention that for these lattices, the lack of inversion symmetry would make the formal development of spin wave theory slightly dissimilar to what we have done before.

We have found that even the most wildly fluctuating bipartite-lattice[17] two-dimensional antiferromagnets have sublattice polarizations $m_{\rm st} > 0.2$, while (isotropic) $D = 1$ antiferromagnets do not order at all.

[16]But in real life, they are rather different, and CaV_4O_9 is not an ordered antiferromagnet. The situation is discussed in Sec. 6.3.3.

[17]The case of non-bipartite lattices is a quite different story. See Sec. 6.3.4.

Can we find somewhere a limit of extreme quantum antiferromagnetism with arbitrarily small (but finite) $m_{\rm st}$? Indeed we can, in quasi-one-dimensional systems. Let us assume that the (antiferromagnetic) coupling J is strong along the chains, while it is the substantially weaker J' in perpendicular directions[18]. Interchain couplings can suppress the divergent fluctuations, because they increase the dimensionality of the system from 1 to 1+$(D-1)$, where $D-1$ is the number of independent directions in the subspace perpendicular to the chains. Thus we expect that long range order becomes possible at $T=0$ in 1+1 dimensions, or even at finite T in 1+2 dimensions. There is, however, a difficult problem to solve: is an arbitrarily small $|J'/J|$ sufficient to stabilize long range order, or does one need a small but finite threshold value of the interchain coupling? That the answer is not obvious at all, should become apparent from the fact that spin wave theory predicts a finite threshold but more careful reasoning shows that this conclusion is false [9, 354].

Let us consider a square lattice where the coupling is J in the x-direction, J' in the y-direction, and $J' \ll J$ [9]. Let us first try linear spin wave theory. We can use (6.90), only γ has to be redefined to describe the lattice anisotropy

$$\gamma_{\mathbf{k}} = \frac{J\cos k_x + J'\cos k_y}{J+J'}\,, \tag{6.92}$$

leading to

$$\Delta S \approx \frac{1}{2\pi}\ln\frac{J}{J'}\,. \tag{6.93}$$

Thus spin wave theory would predict that there is critical J'/J below which long-range order is impossible even in the presence of interchain coupling. From $\Delta S = S$, we get

$$J'_{\rm cr} = Je^{-2\pi S}\,. \tag{6.94}$$

For instance, for $S=1/2$ one finds $J'_{\rm cr}/J \approx 0.034$, which is not such a small value at all; there exist systems with smaller J'/J, and at least several of them are ordered.

[18]The sign of J' does not matter for the argument if the overall structure is unfrustrated.

A more sophisticated approach is to use a *chain mean field theory*, particularly for $S = 1/2$. The idea is to regard the interchain coupling as an external staggered field, and to treat the chains as accurately as one can. Though the latter step does not mean an exact solution in the literal sense, it comes very close to it for long-wavelength properties, including the question of ordering [9, 354]. The essential result is that the $T = 0$ order parameter vanishes as

$$m_{\text{st}} \sim \left(\frac{J'}{J}\right)^{1/2} . \tag{6.95}$$

Long range order survives as long as there is a non-zero interchain coupling, however small it may be. If the model is at least $(1+2)$-dimensional, there is also a finite Néel temperature. Its behaviour is not so simple as that of m_{st} but, omitting logarithmic factors, the rough proportionality $T_{\text{N}} \sim J'$ holds [354].

There are several well-studied nearly isotropic quasi-one-dimensional antiferromagnets. Let us recall that in the case of $KCuF_3$, the spatial anisotropy of the coupling does not arise from that of the crystal structure (which is almost cubic) but from the alternating occupation of different orbitals, and the resulting anisotropy of the hopping (see Fig. 3.8). Actually, the corresponding $J'/J \sim 0.01$ is not considered as particularly small, since it implies a sublattice magnetization of $O(0.1)$. The experimental value is $m_{\text{st}} \approx 0.25$, showing that $KCuF_3$ is only a little bit more fluctuating than La_2CuO_4.

The much more conspicuous magnetic quasi-one-dimensionality of the cuprates Sr_2CuO_3 and Ca_2CuO_3 originates from the interplay of an anisotropic crystal structure (consisting of well-separated CuO_2 chains), and the directional character of the superexchange path. It is estimated that $J'/J \sim 10^{-3}$. Since $J \approx 1300$K is supposed to be the same for both materials, measuring m_{st} and T_{N} allows a check of theoretical predictions. It is found that $m_{\text{st}}^{\text{Sr}} \approx 0.03$ and $T_{\text{N}}^{\text{Sr}} \approx 5$K for Sr_2CuO_3, while $m_{\text{st}}^{\text{Ca}} \approx 0.045$ and $T_{\text{N}}^{\text{Ca}} \approx 11$K for Ca_2CuO_3 [221]. These small ordered moments[19] are at the limit of detectability, thus we should not be surprised that different methods tend to give somewhat different

[19] Remember that in this discussion, m_{st} is measured in units of $g\mu_{\text{B}}$, thus the ordered moment of Sr_2CuO_3 is $0.06\mu_{\text{B}}$.

results. However, $m_{\rm st}^{\rm Sr}/m_{\rm st}^{\rm Ca} \approx \sqrt{T_{\rm N}^{\rm Sr}/T_{\rm N}^{\rm Ca}}$ seems to hold, in essential agreement with chain mean field theory.

Equation (6.95) allows the existence of arbitrarily small $T = 0$ sublattice magnetization. The finding of ordered moments of the order of $0.01\mu_{\rm B}$ in pure Heisenberg antiferromagnets[20] is an example of extreme quantum behaviour. It seems to confirm the idea that there is no intrinsic limit to the smallness of $m_{\rm st}$ in bipartite antiferromagnets, i.e., the fluctuation reduction of the sublattice magnetization can get arbitrarily near to 100%, without causing a breakdown of the quasi-classical picture. We return to this point in our discussion of the spin liquid controversy in Sec. 6.3.

To briefly review the situation at finite temperatures, let us re-evaluate the spin reduction by replacing $\langle \alpha_{\mathbf{k}}^{\dagger}\alpha_{\mathbf{k}}\rangle$, etc. by the Bose occupation numbers. Referring to (6.89), we can write

$$\begin{aligned}\Delta S(T) &= \frac{1}{L}\sum_{\mathbf{k}}\left(\langle c_{\mathbf{k}}^{\dagger}c_{\mathbf{k}}\rangle + \langle d_{\mathbf{k}}^{\dagger}d_{\mathbf{k}}\rangle\right) = \frac{1}{L}\sum_{\mathbf{k}}\Big(u_{\mathbf{k}}^2\langle \alpha_{\mathbf{k}}^{\dagger}\alpha_{\mathbf{k}}\rangle \\ &\quad + v_{\mathbf{k}}^2\langle \beta_{-\mathbf{k}}\beta_{-\mathbf{k}}^{\dagger}\rangle + u_{\mathbf{k}}^2\langle \beta_{\mathbf{k}}^{\dagger}\beta_{\mathbf{k}}\rangle + v_{\mathbf{k}}^2\langle \alpha_{-\mathbf{k}}\alpha_{-\mathbf{k}}^{\dagger}\rangle\Big) \\ &= \frac{2}{L}\sum_{\mathbf{k}}\frac{u_{\mathbf{k}}^2+v_{\mathbf{k}}^2}{e^{\omega_{\mathbf{k}}/k_{\rm B}T}-1} + \Delta S(0)\end{aligned} \tag{6.96}$$

where $\Delta S(0)$ is the quantum fluctuation contribution (6.90). The thermally induced part of the spin reduction is

$$\Delta S(T) - \Delta S(0) = \frac{1}{L}\sum_{\mathbf{k}}\frac{1}{\sqrt{1-\gamma_{\mathbf{k}}^2}}\cdot\frac{1}{e^{\omega_{\mathbf{k}}/k_{\rm B}T}-1}\,. \tag{6.97}$$

At low temperatures, we can approximate $\omega_{\mathbf{k}}$ by its long-wavelength expression (6.85), and use $\sqrt{1-\gamma_{\mathbf{k}}^2} \propto |\mathbf{k}|$. For $D = 3$, standard arguments

[20] As we learn elsewhere, there are small ordered moments in certain f-electron metals like $\rm URu_2Si_2$ (Ch. 3), and also in weak-coupling SDW materials like $\rm (TMTSF)_2PF_6$ (Ch. 7). These phenomena are entirely different from what we see in $\rm Sr_2CuO_3$. As for $\rm (TMTSF)_2PF_6$, we understand well that the small moment arises from the fact that the system is far from the large-U limit, while in $\rm URu_2Si_2$, we do not understand it at all. Although the debate about the role of the orbital degrees of freedom can reopen about any particular material at any time, we see nothing wrong with the idea that the small moment of $\rm Sr_2CuO_3$ can be explained in terms of a simple Heisenberg model.

lead to

$$\Delta S(T) - \Delta S(0) \approx \frac{1}{2\pi^2} \int \frac{k^2 dk}{k} \cdot \frac{1}{e^{Jzsk/k_\mathrm{B}T} - 1} \propto T^2 \,. \tag{6.98}$$

The temperature dependence of the order parameter follows a T^2-law.

In $D = 2$, the integral

$$\Delta S(T) - \Delta S(0) \approx \frac{1}{2\pi} \int \frac{dk}{e^{Jzsk/k_\mathrm{B}T} - 1} \tag{6.99}$$

picks up a divergent contribution from the vicinity of $\mathbf{k} = 0$, signalling that Néel-type order cannot survive at finite temperatures. This a further illustration of the Mermin–Wagner theorem. At any finite T, correlations are isotropic, and fall off exponentially with increasing distance. Asymptotically

$$\langle \mathbf{S_j} \cdot \mathbf{S_m} \rangle \approx (-1)^{||r||} \left(\frac{T}{c_1 JS} \right)^2 \frac{\xi}{r} e^{-r/\xi} \tag{6.100}$$

where ξ is the correlation length, $r = |\mathbf{j} - \mathbf{m}|$, $(-1)^{||r||}$ symbolizes the alternating character of correlations, and c_1 is a constant of $O(1)$ [260]. However, in the limit $T \to 0$ we approach a state with long range order, and this shows up in the exponential divergency of the correlation length

$$\xi \approx \frac{S\hbar c_2}{k_\mathrm{B}T} \exp\left\{ \frac{\rho_S J}{k_\mathrm{B}T} \right\} \tag{6.101}$$

where ρ_S is an S-dependent factor. For the square lattice $S = 1/2$ antiferromagnet $\rho_S = 0.1757$.

Strong intra-plane correlations are very much in evidence in layered cuprates like La_2CuO_4 [260]. A correlation length of at least 20nm has been detected by neutrons in a state without $3D$ order. The development of extensive quasi-ordered regions within the paramagnetic phase makes it understandable that an extremely weak interlayer coupling $J' \sim 10^{-5} J$ can stabilize three-dimensional antiferromagnetic order at relatively high temperatures.

Before closing this subsection, let us mention that according to spin wave theory, the $T = 0$ transverse susceptibility is reduced from its mean field value $\chi_\perp^0 = (g\mu_B)^2/(2zJ)$ [209] by the factor

$$\frac{\chi_\perp}{\chi_\perp^0} = 1 - \frac{\Delta S}{S} - \frac{\zeta}{2S} . \tag{6.102}$$

$\Delta S/S$ is the fraction of the spin which cannot be aligned by the exchange field, thus we are not surprised that it cannot be aligned by the external field, either. The term $\zeta/2S$ is related to the effect of quantum fluctuations on the ground state energy. We can understand that a correction of this order of magnitude has to arise, by the following argument: let us reinterpret (6.87) as the Néel energy with an effective coupling: $J_{\text{eff}} S^2 = JS(S+\zeta)$. This indicates a quantum fluctuation enhancement of the exchange coupling: $J_{\text{eff}} = J(1+\zeta/S)$. Since $\chi_\perp^0 \propto 1/J$, the appearance of a term $\sim -\zeta/S$ becomes plausible.

Anisotropy

Uniaxial anisotropy may result either from the anisotropy of exchange or from the single-ion mechanism we discussed on p. 98. We consider the model

$$\mathcal{H}' = J \sum_{\mathbf{i},\mathbf{j}} S_\mathbf{i}^z S_\mathbf{j}^z + \frac{J\alpha}{2} \sum_{\mathbf{i},\mathbf{j}} (S_\mathbf{i}^+ S_\mathbf{j}^- + S_\mathbf{i}^- S_\mathbf{j}^+) - \Lambda_{zz} \sum_\mathbf{i} (S_\mathbf{i}^z)^2 \tag{6.103}$$

with $\alpha \le 1$ and $\Lambda_{zz} \ge 0$. The energetical preference for the z-axis makes the starting assumption of a Néel state with sublattice magnetizations in the z-direction look perfectly natural, in contrast to the isotropic case where it was an arbitrary choice. For states sufficiently near the Néel state, the single-ion term can be decoupled[21] in the following manner

$$-\Lambda_{zz} \sum_\mathbf{i} (S_\mathbf{i}^z)^2 \rightsquigarrow -2\Lambda_{zz} \sum_\mathbf{i} \langle S_\mathbf{i}^z \rangle S_\mathbf{i}^z \approx -2\Lambda_{zz} S \left(\sum_{\mathbf{j}\in A} S_\mathbf{j}^z - \sum_{\mathbf{l}\in B} S_\mathbf{j}^z \right) . \tag{6.104}$$

[21] The apparent mean-field character of this step does not mean that a new, cruder, approximation is introduced! We did exactly the same as in deriving (6.64): we neglected the four-boson terms. The decoupling form makes it, however, clearer that the same single-ion anisotropy term can act like a uniform field in ferromagnets, and like a staggered field in collinear antiferromagnets.

The development of linearized spin wave theory is much the same as in the isotropic case. Introducing the Holstein–Primakoff bosons, the bilinear part of the Hamiltonian is similar to (6.67)

$$\mathcal{H}'_0 = zJS\sum_{\mathbf{k}}[\alpha\gamma_{\mathbf{k}}(c_{\mathbf{k}}d_{-\mathbf{k}} + d^\dagger_{\mathbf{k}}c^\dagger_{-\mathbf{k}}) + (1+\rho)(c^\dagger_{\mathbf{k}}c_{\mathbf{k}} + d^\dagger_{\mathbf{k}}d_{\mathbf{k}})] \qquad (6.105)$$

where $\rho = 2\Lambda_{zz}/Jz$. Going through the argument leading from (6.68) to (6.79), we find that the spin wave dispersion

$$\omega_{\mathbf{k}} = zJS\sqrt{(1+\rho)^2 - \alpha^2\gamma_{\mathbf{k}}^2} \qquad (6.106)$$

has acquired a gap at $\mathbf{k} = 0$: $\omega(0) = JzS\sqrt{(1+\rho)^2 - \alpha^2} > 0$ if at least one of the inequalities $\alpha < 1$, $\Lambda_{zz} > 0$ is valid.

The experimentally measured gap is often discussed in terms of a formula in which energies are characterized by effective magnetic fields:

$$\omega(0) = g\mu_{\mathrm{B}}\sqrt{(H_{\mathrm{e}} + H_{\mathrm{a}})^2 - \alpha^2 H_{\mathrm{e}}^2} \qquad (6.107)$$

where $H_{\mathrm{e}} = zJS/g\mu_{\mathrm{B}}$ is the exchange field, and $H_{\mathrm{a}} = 2\Lambda_{zz}S/g\mu_{\mathrm{B}}$ is the anisotropy field. It would be usual to find that $H_{\mathrm{a}}/H_{\mathrm{e}} \ll 1$. Assuming furthermore that $\alpha = 1$, we may write

$$\omega(0) \approx g\mu_{\mathrm{B}}\sqrt{2H_{\mathrm{e}}H_{\mathrm{a}}}\,. \qquad (6.108)$$

This teaches us that anisotropy has a more drastic effect in antiferromagnets than in ferromagnets. While the magnon gap of a ferromagnet is of the order of the anisotropy field[22] (which is small), the gap in the antiferromagnon spectrum is of the order of the geometric mean of the exchange and anisotropy fields (which is larger). Taking the quasi-two-dimensional antiferromagnet K_2NiF_4 as an example, neutron scattering data reveal a gap of $\sim$2.2meV in the magnon spectrum whose overall width is $\sim$38meV, in spite of the smallness of $H_{\mathrm{a}}/H_{\mathrm{e}} \approx 2\cdot 10^{-3}$ [84]. On the other hand, experiments show that the gap must be very small for La_2CuO_4, and it is concluded that the exchange anisotropy cannot exceed 0.01% [260].

[22] Essentially this result was found in Problem 6.1, since for a ferromagnet, a decoupling like (6.104) makes the anisotropy act like a uniform magnetic field.

The introduction of easy-axis anisotropy helps to stabilize the Néel-type order. In the limit $\alpha \to 0$ we get the Ising model with a rigidly ordered ground state. One of the motivations for introducing the anisotropic Heisenberg model is purely theoretical: for a clearer understanding of the antiferromagnetism of the isotropic model (6.47), we may wish to derive it by continuation from the well-controlled Ising limit. We can expand in terms of α (Ising expansion), and then let $\alpha \to 1$. This method has the advantage that it does not require taking the quasi-classical limit $S \gg 1$ but works for arbitrary spin S, including $S = 1/2$ where the $1/S$-expansion is on shaky grounds.

A Glimpse at the Strange World of $D = 1$

The spectrum of antiferromagnons (6.83) becomes particularly simple for a one-dimensional system:

$$\omega_k = JSz|\sin(ka)| \,. \tag{6.109}$$

We have good reason to fear that this attractive result has to be discarded because it is derived by assuming two-sublattice long range order, and we know that that assumption fails. Nevertheless, (6.109) is remarkably similar to what is observed in many quasi-one-dimensional materials, for instance, in the nearly isotropic $S = 5/2$ salt $CsMnCl_3 \cdot 2H_2O$ [84]. The formula works so well because it is expected that the exact spectrum of isotropic $S =$ half-integer Heisenberg chains shows the same functional dependence. Though we do not have an analytic result for $S = 5/2$, we can quote the des Cloizeaux–Pearson spectrum of triplet excitations [86] which is derived for $S = 1/2$

$$\omega_k = J\frac{\pi}{2}|\sin(ka)| \,. \tag{6.110}$$

Replacing $z = 2$ and $S = 1/2$ in (6.109), we would get $\omega_k = J|\sin(ka)|$, so we may be tempted to say that the spin wave calculation misses the exact result only by a factor of $\pi/2$. However, this would be rather superficial, because there are essential dissimilarities in the nature of (6.109) and (6.110). The excited states (6.110) are threefold degenerate because they are true triplets over a non-symmetry-breaking singlet ground state. In contrast, (6.109) describes excitations whose twofold degeneracy arises from the symmetry-breaking two-sublattice Néel state. Furthermore, (6.110) holds in the entire Brillouin zone (because translational symmetry has not been broken), while (6.109) would hold only in the reduced zone (because of magnetic cell-doubling). Such qualifying statements notwithstanding, the similarity of the two formulae is striking and one may ask if there is something common in the physical mechanisms leading to them. We cite the finding [250] that the asymptotic behaviour of the $T = 0$ spin–spin correlation function is[23]

$$\langle S_i^z S_{i+j}^z \rangle \propto \frac{(-1)^{|j|}}{|j|} S^2 \tag{6.111}$$

where we set the lattice constant $a = 1$. If a spin looks around to see if its fellows line up with it, it may say that on its sublattice they tend to be parallel, while on the alternate sublattice antiparallel, only this order will eventually decay in the far distance. The situation is often described as *quasi-long-range order*, emphasizing that correlations decay very slowly (algebraically rather than exponentially). We may envisage that an excitation propagating down the lattice sees enough of alternating order to behave as if it were in a long-range-ordered system, and develops a gapless spectrum. The $\omega \to 0$ excitation is not a Goldstone mode but it is something similar.

The $1/r$ decay law shown in (6.111) has an immediate corollary. Since according to linear response theory, the static susceptibility is related to intersite correlations via

$$\chi^{zz}(\mathbf{q}) \propto \frac{1}{L} \sum_{\mathbf{m}} e^{i\mathbf{q}\mathbf{m}} \langle S_{\mathbf{j}}^z S_{\mathbf{j+m}}^z \rangle \tag{6.112}$$

the staggered susceptibility of the linear chain diverges like the harmonic series

$$\chi_{\mathrm{st}} = \chi(q = \pi) \propto \sum_{m=1}^{\infty} e^{im\pi} \frac{(-1)^m}{m} \propto \sum_{m=1}^{\infty} \frac{1}{m} \to \infty \,. \tag{6.113}$$

We may say that the ground state is fluctuating so wildly because $T = 0$ is the critical temperature of the system. The divergent response to a

[23] An accurate expression, including a multiplicative logarithmic correction, has been derived for $S = 1/2$ [8].

staggered field is the reason why the chain mean field theory mentioned on p. 294 predicts long range order for arbitrarily small interchain coupling. Spin wave theory fails to do the same because it cannot capture the algebraic decay (6.111), and therefore predicts a finite staggered susceptibility.

Our opening remark was about $CsMnCl_3{\cdot}2H_2O$ in which magnetic ions carry $S = 5/2$, and then we were citing exact results for $S = 1/2$. Is it justified to think that one-dimensional systems with higher S behave qualitatively similarly to the $S = 1/2$ chain? Spin wave theory would lead (but as it turns out, in this case mislead) us to think that the answer is yes. However, the truth is more surprising, so surprising, that after Bethe's exact solution for the $S = 1/2$ model (1931) it took more than 50 years to discover that the solution of the $S = 1$ model would be something quite different [148]. Well, we still do not know the full solution of (6.47) with $S = 1$ in the same sense as we do for $S = 1/2$, but about one thing we are quite certain: its excitation spectrum is gapped. It must be emphasized that here we are speaking about a gap in an isotropic model, which is quite different from the anisotropy-induced gap we learned about earlier.

What do we mean by a gap? In a finite system, all the levels are discrete, and the spin wave energies are spaced like $\sim J/L$. If the ground state is separated by $\Delta E \sim J/L$ from the first excited state, we do not speak about a gap, but rather interpret this as a sign of continuity because $\Delta E \to 0$ as $L \to \infty$. A true gap remains finite even for an infinite system.

Let the ground state be ψ_0, and the ground state energy E_0. Let us pose the question whether we can find a state ψ_1 such that

$$\langle \psi_1 | \mathcal{H} | \psi_1 \rangle - E_0 \sim 1/L\,.$$

The following argument does not require an explicit knowledge of ψ_0, and it is not assumed that ψ_1 is an eigenstate of $\mathcal{H}$. We merely sketch the derivation which was given in precise from by Lieb, Schultz, and Mattis [247] for $S = 1/2$, and by Affleck and Lieb [11] for higher half-integer spins.

The idea is that a low-lying state can be generated from the ground state by slowly twisting it

$$\psi_1 \sim \exp\left\{ i\frac{\pi}{l} \sum_{n=-l}^{l} (n+l) S_n^z \right\} \psi_0 \tag{6.114}$$

where (numbering the sites from $-L/2$) we may take $l = L/2$. The twisting axis is the z-axis, and the twisting angle increases from 0 to 2π as we pass along the chain. Since neighbouring spins have become misaligned by $2\pi/L$, the twisting energy is of the order $L[1 - \cos(2\pi/L)] \sim 1/L$, what we required. The same result is obtained if there are also second-neighbour, or generally, finite-range interactions.

Up to this point, the argument works for any spin S. But do we really have an excited state? To be sure about this, we want to make certain that ψ_1 is orthogonal to ψ_0. One end of the chain is twisted by an angle of 2π relative to the other end, and a 2π-rotation gives a factor of -1 for half-integer spins, thus ψ_1 has a parity opposite to that of the ground state $\leadsto$ ψ_1 and ψ_0 are orthogonal for half-integer spins. Thus we arrived at the *Lieb–Schultz–Mattis–Affleck theorem*: an isotropic Heisenberg chain of half-integer spins, with short-range interactions, can show two kinds of behaviour in the limit $L \to \infty$:

- The spectrum of excitations is gapless, or
- There are two ground states, as the result of breaking a discrete symmetry (in the present formulation, reflection symmetry about the centre of the chain).

Though we may have felt that we were proving the first statement, it has to be realized that we never excluded the second one: our ψ_1 is not an eigenstate, thus it is possible that for a true eigenstate, the energy separation is less than $O(1/L)$, maybe even exactly 0. Though such carefulness may appear as an example of pedantry taken to the extreme, nature did not overlook this possibility: it is realized in the so-called dimer phase (see later).

What about integer spins? For those, a 2π-rotation gives a factor 1, the phase of the wave function ψ_1 is the same as that of ψ_0 at both ends of the chain, thus the parity-based proof of orthogonality breaks down. We cannot be certain that ψ_1 is linearly independent of ψ_0, so we have no statement to make. Since the previous argument looks deceptively effortless, one might have thought that with a bit more work, a different method can be devised to derive similar results for $S = 1$, etc. It is not so. The breakdown of the simple proof gives the right clue: it does not work because the statement would be false. There are several different ways to find out that Heisenberg chains formed of integer spins are profoundly different from systems with half-integer spins [7]. The ground state of an $S = 1$ system is separated by the substantial gap

$$\Delta_{S=1} \approx 0.41J \tag{6.115}$$

from the first excited state. This is called the *Haldane gap* [148].

Quite generally, a gap above the ground state level is associated with short-range correlations in the ground state. In contrast to the algebraic behaviour in (6.111) which is characteristic of the half-integer case, spin–spin correlations of integer-spin models decay exponentially

$$\langle S_i^z S_{i+j}^z \rangle \propto S^2 \frac{(-1)^{|j|}}{\sqrt{|j|}} \exp\left\{-\frac{|j|}{\xi}\right\} . \tag{6.116}$$

For $S = 1$, the correlation length $\xi \approx 6.25$ is rather short. A similar behaviour is obtained for $S = 2$, but with a smaller gap $\Delta \approx 0.048J$, and larger $\xi \approx 83.3$. The gapped spectrum can be represented by

$$\omega(k) = \sqrt{(cJ)^2 \sin^2 k + \Delta^2} \tag{6.117}$$

where c depends on S. For large S, one finds [33]

$$\Delta \propto JSe^{-\pi S} \qquad \text{and} \qquad \xi \propto e^{\pi S} . \tag{6.118}$$

The vanishing of the Haldane gap for $S \to \infty$ shows that this is a purely quantum phenomenon, without a classical analogue. In the same limit, the dispersion (6.117) approaches the gapless form $\omega(k) \sim |\sin k|$ which *does* have a classical derivation [209].

There is no simple argument to demonstrate the results for the $S = 1$ model with pure Heisenberg coupling. However, we will find qualitatively the same behaviour for a closely related bilinear–biquadratic

Hamiltonian, and this will enable us to interpret the difference of integer and half-integer spin models in terms of their valence bond distributions.

All in all, we think that all (nearest neighbour coupling) half-integer spin chains behave similarly, having gapless spectra looking like (though not exactly given by) (6.109), and $T = 0$ quasi-long-range order of the kind shown in (6.111). Similarly, all S = integer chains are similar to each other[24], having their Haldane gap, and an exponential fall-off of correlations like (6.116). It must be remarked that Haldane's argument being rather indirect and applicable primarily for large S, the previous sweeping statements are referred to as "Haldane's conjecture" rather than as a "theorem". There is an on-going effort to check them for specific small values of S. The Haldane phenomenon is established beyond doubt for $S = 1$. A careful investigation of the $S = 3/2$ chain [150] shows that the correlation function follows (6.111), and that the prefactor of the long-wavelength $|\sin k| \sim |k|$ dispersion exceeds the spin wave prediction (6.109) by a factor of ≈ 1.29. This is quite a bit smaller than the factor $\pi/2 \approx 1.57$ found for $S = 1/2$, showing that quantum fluctuation corrections diminish quickly with increasing S. Extending the trend to $S = 5/2$, we see that it was not such a bad idea to think of (6.109) in connection with $CsMnCl_3 \cdot 2H_2O$.

Experimental checks of Haldane's conjecture have to come from the study of quasi-one-dimensional systems, and at this point, we become aware of a further essential difference between integer spin and half-integer spin systems: they react differently to a weak interchain coupling. Let us recall our previous discussion of Sr_2CuO_3 and Ca_2CuO_3 which are $S = 1/2$ systems (see pp. 292–294). We were arguing that an arbitrarily small interchain $|J'|$ is sufficient to lead to $3D$ ordering and a finite T_N, essentially because the staggered susceptibility (6.113) of the chains diverges, which follows from the power-law correlations. However, the exponentially decaying correlations of an $S = 1$ spin chain indicate that it has a finite staggered susceptibility, and we may surmise that in this case, $|J'|$ has to exceed a finite threshold value to

[24] To be more precise: the existence of a gap is the generic behaviour of, say, $S = 1$ isotropic antiferromagnets with a general short-range interaction. As we are going to see in our discussion of the $S = 1$ bilinear–biquadratic model (6.148), this does not exclude that the gap vanishes at a critical point between two gapful phases.

overcome the Haldane gap, and induce $3D$ ordering. Thus even the three-dimensional system can have a disordered ground state[25] if $|J'|$ is small enough, while it shows antiferromagnetic ordering below a small Néel temperature, if $|J'|$ is larger [224, 391]. However, $3D$ ordering is a weak effect, and the Haldane gap becomes visible at $T > T_{\rm N}$.

There are experimental examples of both kinds of behaviour, notably in nickel compounds. Let us recall that in an octahedral environment, the ion $Ni^{2+} \rightsquigarrow 3d^8$ has the closed subshell configuration t_{2g}^3 with $S = 1$, and no orbital degeneracy, providing good chances for finding a purely spin Heisenberg model with relatively weak anisotropy[26]. The same remains true if the environment is distorted to a lower-symmetry configuration. We mention two materials. Y_2BaNiO_5 is a nearly ideal Haldane gap system with $J \sim$ 21meV [449]. It is estimated that $|J'/J| < 5{\cdot}10^{-4}$, well below the critical ratio $\sim 10^{-2}$, thus Y_2BaNiO_5 does not order at any temperature, and the average size $\sim$8.5meV of the weakly anisotropic Haldane gap fits (6.115) well. In contrast, in the compound $CsNiCl_3$ (which can be pictured as a triangular lattice of weakly coupled Ni-chains), $|J'/J| \approx 1.8{\cdot}10^{-2}$ which is sufficiently large to cause three-dimensional ordering below $T_{\rm N} = 4.8$K. At $T < T_{\rm N}$, the spectrum is gapless as in any ordinary antiferromagnet, but the Haldane gap becomes detectable by neutron scattering at $T > T_{\rm N}$ [53].

The Haldane gap is a *spin gap*, and it should show up in the thermally activated behaviour[27] of the spin susceptibility

$$\chi(T) \propto T^{-1/2} \exp\left(-\Delta/k_{\rm B}T\right) \tag{6.119}$$

which is analogous to the Arrhenius law of the conductivity which follows from the existence of a charge gap. We mention in passing that the Haldane phenomenon is only one of the possible mechanisms leading to

---

[25] We can call it a quantum paramagnet to emphasize that we are speaking about the unique ground state of an interacting system in which long-range correlations are disrupted by quantum fluctuations.

[26] It appears that in the substances we mention, one finds very little exchange anisotropy, and a moderate single-ion anisotropy. We do not go into this, but it has been shown that the Haldane phase extends over a finite range of anisotropy.

[27] We can get the T-dependent prefactor by postulating that χ is proportional to the number of thermally excited triplets, and expanding (6.117) about the bottom of the band. In D-dimensions, the prefactor would be $\propto T^{(D-2)/2}$.

the appearance of a spin gap in a system with isotropic interactions between localized moments. Spin gaps are found also in dimerized $S = 1/2$ systems like $CuGeO_3$ (see p. 318), in "plaquette-resonating" CaV_4O_9 (see Sec. 6.3.3), in the $S = 1/2$ massive spin liquid of the pyrochlore lattice (see p. 335), and in Kondo insulators.

6.3 Néel Order versus Valence Bond States

The nature of symmetry breaking in antiferromagnets is quite different from what we find in ferromagnets. This can be demonstrated by comparing the behaviour of small systems. To be specific, let us take a four-site Heisenberg system, a single plaquette from the square lattice, and determine the ground state first for ferromagnetic (FM), and then for antiferromagnetic (AF), coupling. The nature of FM order is recognized immediately, because we can choose the ground state as fully polarized in the z-direction: $\langle S_j^z \rangle = S$ for $j = 1, \ldots, 4$. The $L = 4$ system is ordered in the same sense as a very large system is. In contrast, the ground state[28] of the isotropic AF Heisenberg model is a singlet ($S_{\rm tot} = 0$) state in which $\langle S_j^z \rangle = 0$ for any j. The ground state certainly does not show the alternating Néel order which we associate with the name "antiferromagnetism".

The situation remains the same if we go on to larger clusters. The ground state of the FM model is ferromagnetic, but the ground state of the AF model is *not* antiferromagnetic: it is a singlet[29] [261, 246]. The reason for this difference is that the order parameter of ferromagnetism ($S_{\rm tot}^z$) commutes with the Hamiltonian, while that of antiferromagnetism (the staggered magnetization $m_{\rm st} = \sum_{{\rm j}\in A} S_{\rm j}^z - \sum_{{\rm l}\in B} S_{\rm l}^z$) does not. $\mathcal{H}$ and $S_{\rm tot}^z$ can be diagonalized simultaneously, and $S_{\rm tot}^z$ happens to have a large eigenvalue in the ground state of the FM model. In contrast, $m_{\rm st}$ cannot have a non-vanishing expectation value in a $S_{\rm tot} = 0$ ground

[28] We will do the calculation for $S = 1/2$ in Problem 6.3.

[29] For even L, assuming that we let the cluster grow in a balanced manner, so that the number of sites on the two sublattices is always the same. We sketch a proof on p. 314. – For $S = 1/2$, we can regard this as a corollary of Lieb's theorem (p. 426) in the limit $U/t \gg 1$ where the Hubbard model becomes equivalent to an AFM Heisenberg model. The theorem says that the ground state spin is $S_{\rm tot} = (|A| - |B|)/2$.

state, because that would imply the existence of a preferred direction, and a singlet state has no such preference. The possibility of a Néel-type symmetry breaking has to emerge in the thermodynamic limit.

The nature of antiferromagnetism is more elusive than that of ferromagnetism. On the one hand, it should be possible to understand the Néel-type order of isotropic antiferromagnets as the limiting behaviour of singlets. On the other hand, we have to be aware that antiferromagnetic interactions do not necessarily lead to antiferromagnetic order. There are models for which the ground state remains a singlet even when $L \to \infty$. We have seen that the ground state retains the full symmetry of the Hamiltonian for the nearest-neighbour Heisenberg model in $D = 1$. It is an interesting question whether this happens in some higher dimensional systems as well. Furthermore, there are cases in which translational symmetry is broken, but spin-rotational symmetry is not: we will meet a model with a twofold degenerate singlet ground state in Sec. 6.3.2.

To discuss these questions, it is desirable to have a method by which the structure of $S_{\text{tot}} = 0$ states can be visualized. This is provided by building up the many-spin singlet states from singlet pairs.

6.3.1 Resonating Valence Bonds

What the antiferromagnetic interaction "wants", is to make neighbouring spins antiparallel but (because of the spin-rotational invariance of the coupling) in an isotropic manner. This can be achieved by forming an overall singlet with $S_{\text{tot}} = 0$ (we stick to L=even, assuming that even–odd effects become irrelevant for large L). The difficulty is that – in contrast to the ferromagnetic case where the $S_{\text{tot}} = LS$ multiplet is unique – requiring $S_{\text{tot}} = 0$ does not fix the state uniquely. For instance, for $S = 1/2$ spins, the number of singlets is (see Problem 6.2)

$$\mathcal{N}_{S_{\text{tot}}=0} = \frac{2C^L_{L/2}}{L+2} \sim \frac{2^L}{L} , \tag{6.120}$$

an exponentially large number! And what is worse, most of these singlets have a fairly complicated structure.

To get an idea of the nature of the singlet states, we assume that the system consists of $S = 1/2$ spins, and follow Hulthén's early work

[175]. The basic entity is the singlet pair state (or singlet bond)

$$[\mathbf{ij}] = \frac{1}{\sqrt{2}}[\alpha(\mathbf{i})\beta(\mathbf{j}) - \beta(\mathbf{i})\alpha(\mathbf{j})] = \quad (6.121)$$

The bond is directed because [**ij**] changes sign if **i** and **j** are interchanged. Singlet states of more than two spins are obtained by letting all the spins participate in pair bonds. For instance, the singlet basis for $L = 4$ can be chosen to consist of the two states represented by non-crossing bonds

$[12][34] =$

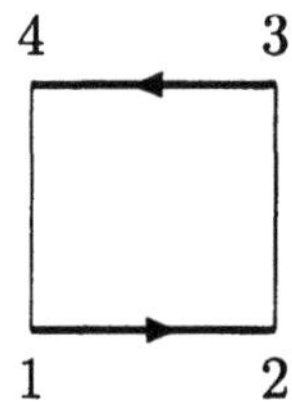

and $[23][41] =$

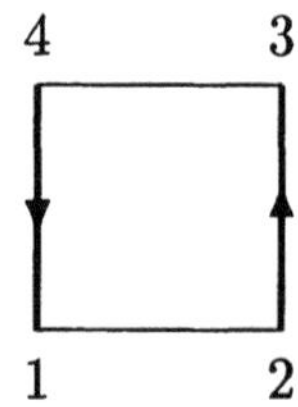

(6.122)

The diagrams shown above are called Rumer diagrams[30]. One may think that there is a third way to pair the spins into singlets: why not take [13][24] as well? However, this state would not be independent of the previous two; in fact $[13][42] = [12][34] - [23][41]$, or graphically

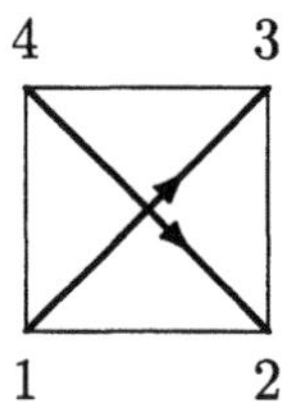

=

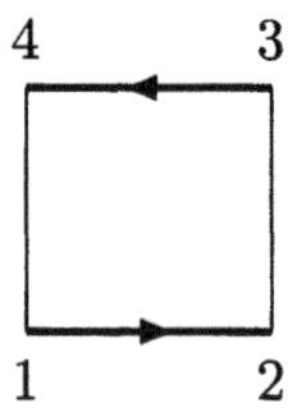

−

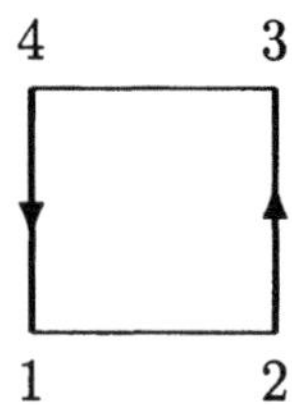

(6.123)

This is the simplest illustration of the general statement that the non-crossing Rumer diagrams[31] form a basis of the singlet subspace.

The singlet subspace is an eigensubspace of the Hamiltonian, so it is worth to work out how $\mathcal{H}$ acts on the elements of the singlet basis.

[30] The application of Rumer diagrams [343] goes back to the beginnings of quantum chemistry but using them in the present context was pioneered by Hulthén [175].

[31] In $D = 1$, the meaning of "non-crossing" is straightforward: the system can be represented as a ring, and the bonds are cords in a circle. For $D > 1$, the sites have to be numbered in some order, and the system redrawn as a circle. This implies that in order to exhaust the singlet subspace, we have to consider also bonds connecting distant lattice sites.

Let us rewrite a single term of $\mathcal{H}$ as

$$\mathbf{S_i} \cdot \mathbf{S_j} = -\frac{1}{2}\left(\frac{1}{2} - 2\mathbf{S_i} \cdot \mathbf{S_j}\right) + \frac{1}{4} = -\frac{1}{2} \cdot Q_{\mathbf{ij}} + \frac{1}{4}. \tag{6.124}$$

The point of introducing the $Q_{\mathbf{ij}}$ is that these are the operators which act directly on the singlet bonds. It is easy to check the following simple rules: $[\mathbf{ij}]$ is an eigenstate of $Q_{\mathbf{ij}}$

$$Q_{\mathbf{ij}}[\mathbf{ij}] = 2[\mathbf{ij}] \, , \tag{6.125}$$

while non-diagonal processes are governed by

$$Q_{\mathbf{jk}}[\mathbf{ij}][\mathbf{kl}] = [\mathbf{jk}][\mathbf{li}] \, . \tag{6.126}$$

Expressing with Rumer diagrams

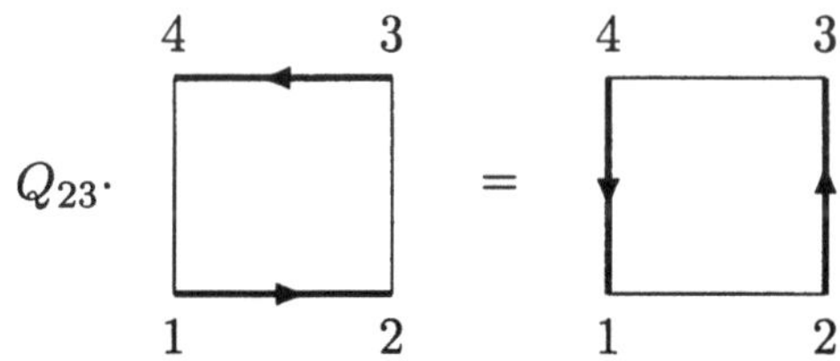

(6.127)

Thus we learn that certain terms of the Hamiltonian make the system "resonate" between different configurations of the singlet bonds. This may remind us of the resonance of the benzene molecule between the Kekule configurations (Fig. 6.1). There are two, energetically equivalent ways to distribute the three double bonds in the molecule. The ground state is a superposition of bond configurations and this fact is expressed by saying that the molecule resonates between them. This is called the *resonating valence bond* (RVB) state in chemistry [320].

For us, RVB means processes like (6.127). Suggestive as the phrase "resonating valence bonds" is, we should refrain from ascribing a direct chemical meaning to it. We are speaking about a pictorial representation of the singlet states of the $S = 1/2$ Heisenberg antiferromagnet. We will see shortly that this involves resonance to configurations with some "long bonds" which connect distant lattice sites[32], and these are certainly not bonds in the chemical sense.

[32]For an illustration, see (6.129).

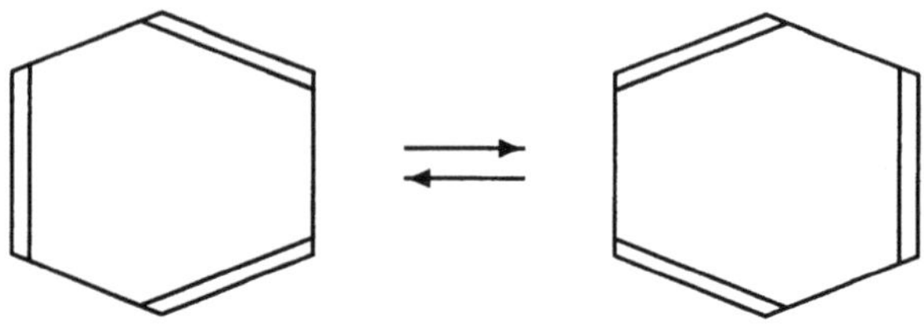

Figure 6.1: Schematic representation of the resonating valence bond state of the C_6H_6 molecule. Double bonds between the C atoms are indicated; the H atoms are not shown.

The ground state of the 4-site Heisenberg model can be sought as the linear combination of the non-crossing states [12][34] and [23][41] (see Problem 6.3). It turns out that the ground state (6.173) resonates between these two configurations

$$\psi_0 = (1/\sqrt{3})([12][34] + [23][41]) \quad \text{with energy} \quad \epsilon_0 = -2J\,. \tag{6.128}$$

This is the simplest example of an RVB ground state. The perfectly symmetrical ψ_0 is very different from the four-site Néel state $\psi_N = |\uparrow\downarrow\uparrow\downarrow\rangle$, since $\langle\psi_0|S_j^z|\psi_0\rangle = 0$, while $\langle\psi_N|S_j^z|\psi_N\rangle = (-1)^{j+1}/2$.

What about larger systems? Let us see what happens in a six-site cluster carved out from the square lattice. Starting from the reasonable expectation that the short-bond configuration [12][34][56] must be a component of the ground state, we find that under the action of the nearest-neighbour coupling $\mathbf{S}_2 \cdot \mathbf{S}_5$, it resonates to the configuration [25][61][34]:

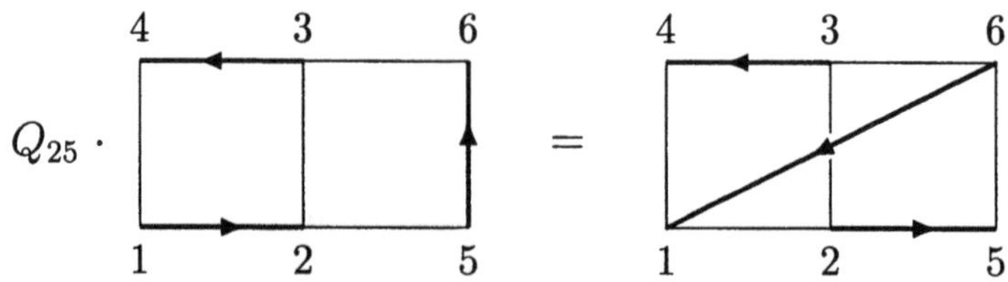

(6.129)

The long bond [61] "does not bind" (since $\mathbf{S}_1 \cdot \mathbf{S}_6$ does not occur in the Hamiltonian), thus it should be included in the ground state with a smaller amplitude than the nearest-neighbour bonds. Nevertheless, it does have some amplitude, and so do the longer bonds. Generally, the RVB ground state can be envisaged as a fluctuating criss–cross pattern of bonds with all possible lengths. It is an attractive picture, but difficult to follow with calculations [382]. It may appear that there is no systematics in the pattern of the haystack of bonds, but we may notice at least one feature: starting with a configuration with nearest-neighbour bonds, and letting it resonate, all the resulting long bonds connect an A site with a B site. Furthermore, even though one finds bonds with all possible lengths, certainly there should be a rule which says that short bonds are more likely to be seen than long bonds. One may attempt to characterize RVB states by their bond length distributions [242].

Problem 6.2. Derive (6.120). What do you think, is the number of triplets larger or smaller than the number of singlets?

Problem 6.3. Determine the spectrum of the 4-site $S = 1/2$ antiferromagnetic Heisenberg model

$$\mathcal{H}_4 = J(\mathbf{S}_1 \cdot \mathbf{S}_2 + \mathbf{S}_2 \cdot \mathbf{S}_3 + \mathbf{S}_3 \cdot \mathbf{S}_4 + \mathbf{S}_4 \cdot \mathbf{S}_1). \qquad (6.130)$$

What is the nature of the ground state? Does it show Néel-type order, or at least antiferromagnetic correlations?

What are the level degeneracies? Do they change if we include the next-nearest neighbour couplings contained in

$$\mathcal{H}_4' = \mathcal{H}_4 + J'(\mathbf{S}_1 \cdot \mathbf{S}_3 + \mathbf{S}_2 \cdot \mathbf{S}_4). \qquad (6.131)$$

Problem 6.4. *Dimer State*. Discuss the gound state of the $S = 1/2$ antiferromagnetic J_1–J_2 chain

$$\mathcal{H} = J_1 \sum_i \mathbf{S}_i \cdot \mathbf{S}_{i+1} + J_2 \sum_i \mathbf{S}_i \cdot \mathbf{S}_{i+2} \qquad (6.132)$$

for $J_2 = J_1/2 > 0$. It is useful to start from the nearest-neighbour valence bond state $[12][34][56]\ldots[N-1, N]$.

It would be impossible to guess the nature of the quantum ground state by starting from the classical limit. By the way, what is the ground state of (6.132) for classical vectors?

Digression: Symmetry Breaking (or the Lack of It)

The ground state of any even-L finite piece of the square lattice is a singlet, but there is overwhelming evidence (though no exact proof[33]) that the infinite square lattice is an antiferromagnet. The Néel state (as we know it from spin wave theory) is definitely not a singlet state, since it breaks spin-rotational invariance. This is possible only if for large L, the optimally constructed Néel-type state becomes *asymptotically degenerate* with the exact singlet ground state.

The phenomenon of *spontaneous symmetry breaking* can be envisaged in the following manner [17, 382]: as L increases, we witness an accumulation of a large number of very low-lying $S_{\text{tot}} = 1, 2, \ldots$ states. As $L \to \infty$, an infinitesimally small perturbation (such as the presence of a very weak easy-axis anisotropy) suffices to mix the low-lying states into a symmetry-breaking wave packet which is asymptotically degenerate with the singlet ground state. The Néel-type ground state is not an eigenstate, but rather a wave packet of true eigenstates with $S_{\text{tot}} = 0, 1, 2, \ldots, O(\sqrt{L})$ [17]. Generally speaking, a similar asymptotic degeneracy, and wave-packet formation, provides the mechanism of spontaneuos symmetry breaking in cases where the order parameter is not a conserved quantity.

There is an alternative, down-to-earth method to convince ourselves that the possibility of two-sublattice order is latent in the exact eigenstates of finite systems. The characteristic up–down pattern of Néel ordering arises from the bipartite structure of the square lattice. How can the singlet ground state "know about" the bipartite nature of the lattice? We can find out about this by returning to the S^z representation. The argument illuminates that the phase relationships of the unique quantum-mechanical ground state reflect a two-sublattice partitioning of the lattice, and as a corollary, we find a proof that the ground state is indeed a singlet [261].

[33] It is exactly proven that the ground state of the isotropic AF Heisenberg model shows AF order for any S on the cubic lattice, and for $S \geq 1$ on the square lattice [95]. As for the $S = 1/2$ square lattice antiferromagnet, rigorous proofs of $T = 0$ order have to assume a sufficiently large anisotropy [444]. In spite of this technical difficulty, the existence of AF order is universally accepted even for the isotropic model.

For even L, the ground state can be sought in the $S^z_{\text{tot}} = 0$ subspace. A configuration can be specified by the sites $\mathbf{g}_1, \ldots, \mathbf{g}_{L/2}$ where $\uparrow$-spins are sitting; we will use the simple notation $|\mathbf{g}_1, \ldots, \mathbf{g}_{L/2}\rangle \rightsquigarrow |\mu\rangle$. The ground state (which we assume to be normalized) can be written as

$$|\Psi\rangle = \sum_{\mu} a(\mu)|\mu\rangle \,. \tag{6.133}$$

Let us write the Hamiltonian as

$$\mathcal{H} = \mathcal{H}_{zz} + \mathcal{H}_{\pm} = J^z \sum_{\mathbf{i},\mathbf{j}} S^z_{\mathbf{i}} S^z_{\mathbf{j}} + \frac{J^{\pm}}{2} \sum_{\mathbf{i},\mathbf{j}} (S^+_{\mathbf{i}} S^-_{\mathbf{j}} + S^-_{\mathbf{i}} S^+_{\mathbf{j}}) \,. \tag{6.134}$$

The Ising term is diagonal in the present representation but $\mathcal{H}_{\pm}$ connects the configuration $|\mu\rangle$ to its "neighbours" $|\mu'(\mu)\rangle$. We assume that the lattice can be divided into alternate sublattices A and B in such a way that $\mathcal{H}_{\pm}$ connects A-sites only to B-sites, and vice versa. In general, the problem of minimizing

$$\langle\Psi|\mathcal{H}|\Psi\rangle = J^z \sum_{\mu} |a(\mu)|^2 + \frac{J^{\pm}}{2} \sum_{\mu} \sum_{\mu'(\mu)} a^*(\mu')a(\mu) \tag{6.135}$$

with respect to the coefficients $a(\mu)$ cannot be solved. However, we can solve a restricted problem: Let us assume that the $|a(\mu)|$ are fixed. What is the optimal choice of the phase factors? The relative phases do not matter for the Ising term (which contains only $|a(\mu)|^2$), but they influence the contribution of the x–y term. We want each $a^*(\mu')a(\mu)$ to be as negative as possible. It is obvious that the optimum is reached if

$$a^*(\mu')a(\mu) = -|a(\mu')a(\mu)| \tag{6.136}$$

for all μ, $\mu'(\mu)$ simultaneously. This can be achieved by imposing the *Marshall–Peierls sign rule* [261]

$$a(\mu) = (-1)^{p(\mu)}|a(\mu)| \tag{6.137}$$

where $p(\mu)$ is the number of $\uparrow$-spins on sublattice A. Namely, a spin-flip event (which is alway affecting an A-site and a B-site) is bound to change $p(\mu)$ by 1, thus $p(\mu') = p(\mu) \pm 1$, and (6.136) is fulfilled. Q.E.D.

Though we had the nearest-neighbour model in mind, it is clear that the argument applies also to models with longer-range interactions, provided that $\mathcal{H}_\pm$ connects only sites belonging to different sublattices. We may even consider the infinite-range model in which every A-spin interacts with every B-spin with the same strength

$$\begin{aligned} \mathcal{H}_\infty &= J \sum_{\mathrm{i}\in A} \sum_{\mathrm{j}\in B} \mathbf{S}_\mathrm{i}\cdot\mathbf{S}_\mathrm{j} \qquad (6.138) \\ &= \frac{J}{2} \left[S_{\mathrm{tot}}(S_{\mathrm{tot}}+1) - S_{A,\mathrm{tot}}(S_{A,\mathrm{tot}}+1) - S_{B,\mathrm{tot}}(S_{B,\mathrm{tot}}+1) \right] . \end{aligned}$$

It is easy to verify that the energy is lowered by increasing the sublattice spins $S_{A,\mathrm{tot}}$, $S_{B,\mathrm{tot}}$, and by decreasing their sum. The minimum belongs to $S_{A,\mathrm{tot}} = S_{B,\mathrm{tot}} = L/4$, and $S_{\mathrm{tot}} = 0 \leadsto$ the ground state $|\Psi\rangle_\infty$ of $\mathcal{H}_\infty$ is a singlet. Since the Marshall–Peierls sign rule is valid for both $|\Psi\rangle$ and $|\Psi\rangle_\infty$, $\langle\Psi|\Psi\rangle_\infty$ is a sum of non-negative numbers $\leadsto |\Psi\rangle$ and $|\Psi\rangle_\infty$ have non-vanishing overlap. Both $\mathcal{H}$ and $\mathcal{H}_\infty$ being isotropic, their eigenstates are simultaneously S_{tot} eigenstates, i.e, they are pure singlets, or triplets, etc. Since $|\Psi\rangle_\infty$ and $|\Psi\rangle$ are not orthogonal, $|\Psi\rangle$ must be a singlet[34].

The singlet ground states of finite systems can be described in the language of singlet pair bonds, and are thus RVB states, but this is a trivial statement. The true interest of the RVB idea lies in the question whether there are models for which the ground state remains an exact singlet even in the thermodynamic limit [21, 106]. The singlet ground state of an infinite system could be called a "spin quantum liquid", in contrast to the Néel-type state which is then seen as a "spin crystal". We know that Néel-order "quantum-melts" in $D = 1$, but are $D > 1$ systems necessarily spin crystals (ordinary antiferromagnets) ? The usage of these expressions recalls the analogous situation in the quantum theory of solids, which we briefly mentioned in footnote 8 on p. 283. The ground state of a large number of atoms of mass M is usually a crystal. However, if the atoms are light enough (like the helium isotopes), quantum fluctuations destroy the crystalline state[35], and the true ground

[34] In sketching the present simple proof of the singlet nature of the ground state, we followed [33], rather than the original work by Marshall [261]. See also [246].

[35] Putting the system under pressure restrains the atomic motion and stabilizes

state is a quantum liquid. We may envisage a quantum version of the Lindeman criterion of melting: if the amplitude of zero point oscillations δa exceeds a threshold, say $\delta a > 0.25a$, then the ground state is no longer a crystal, but something entirely different: a quantum fluid. Observe that a crystal would be still recognizable if, say, $\delta a \sim 0.35a$; nevertheless we envisage that if fluctuations in the crystalline state are strong enough, then the system goes and seeks a different ground state somewhere far away in the Hilbert space, even though the ordered state has not yet become *locally* unstable. Recognizing that S and M play similar roles, and knowing that quantum fluctuations are pretty strong in low-dimensional antiferromagnets, it was entirely reasonable to assume that a spin system chooses a spin liquid (RVB) ground state if the spin reduction ΔS in the AF state exceeds a certain fraction of S [21]. The RVB conjecture looked a particularly safe bet for systems like the $S = 1/2$ triangular antiferromagnet, because frustration enhances ΔS (see Sec. 6.3.4). For a variety of complicated reasons [214, 215], it has not come off yet; spin crystals are apparently capable of enduring a greater vexation by quantum fluctuations, than crystals of atoms are. To cite a further evidence supporting this view, let us recall our discussion of quasi-one-dimensional substances on pp. 292–295; there we found that ΔS can get arbirarily near to $S = 1/2$, without causing a collapse of AF order. Still, a large ΔS in a non-frustrated system is not quite the same thing as the combination of a large ΔS *and* frustration, and there is still hope that some wildly frustrated systems like the kagome or pyrochlore lattices, do have spin liquid ground states. We return to this question at the end of Sec. 6.3.4 (see p. 334).

It was an unexpected turn of the story that states with non-resonating singlets (the so-called *valence bond solids*) provide an entire class of interesting, and physically relevant, ground states. We discuss them in the next section.

6.3.2 Valence Bond Solids

The idea of representing an overall singlet state as consisting of singlet pairs finds some of its most useful applications in situations where the

---

crystalline order. The (super)fluid ground states are low-pressure phases.

bonds cease to resonate. Let us consider the $S = 1/2$ isotropic Heisenberg chain with first- and second-neighbour interactions

$$\mathcal{H} = J_1 \sum_j \mathbf{S}_j \cdot \mathbf{S}_{j+1} + J_2 \sum_j \mathbf{S}_j \cdot \mathbf{S}_{j+2} \tag{6.139}$$

where we assume $J_1 > 0$, $J_2 \geq 0$. In general, this presents a much more difficult problem than the nearest-neighbour model. Let us note, however, that in the special case $J_2 = J_1/2$, $\mathcal{H}$ can be rewritten as

$$\mathcal{H} = J \sum_j \mathbf{S}_j \cdot \mathbf{S}_{j+1} + (J/2) \sum_j \mathbf{S}_j \cdot \mathbf{S}_{j+2} = (J/4) \sum_j \mathcal{P}_j + \text{const} \tag{6.140}$$

where

$$\mathcal{P}_j = (\mathbf{S}_{j-1} + \mathbf{S}_j + \mathbf{S}_{j+1})^2 \tag{6.141}$$

measures the total spin of three consecutive sites. $\mathcal{P}_j$ has the eigenvalues 15/4 and 3/4 for $|\mathbf{S}_{j-1} + \mathbf{S}_j + \mathbf{S}_{j+1}| = 3/2$ and 1/2, respectively. If a contribution 3/4 is picked up from every j, we have certainly reached the possible lowest energy level; the question is whether we can find a state in which none of the three-site clusters has a spin 3/2. This is achieved by putting the spins pairwise in singlet bonds, because then of any three consecutive spins, there will be two with combined spin 0, and the three spins can only have a total spin 1/2. Thus $\Phi_1 = [12][34]\dots[N-1, N]$, or pictorially

$\Phi_1 =$

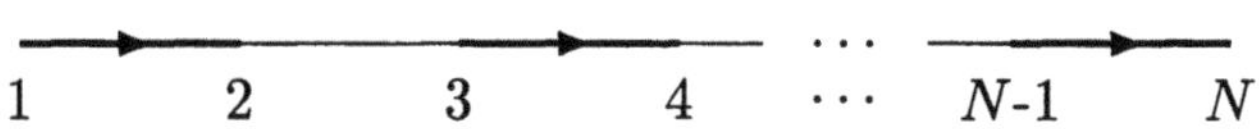

(6.142)

is definitely a ground state. Assuming the periodic boundary condition $N{+}1 \equiv 1$, the state obtained by translating the bond pattern by one step gives another ground state: $\Phi_2 = [23][45]\dots[N, 1]$. There are no other ground states. It has been shown that the twofold degenerate ground state is separated from the excited states by a gap [10]. Symmetry breaking has occured, but in a manner which we may find unexpected from a spin system: spin-rotational invariance remains intact, but the (discrete) translational symmetry is broken.

Φ_1 is represented by a frozen pattern of bonds, giving an example of *valence bonds solids* (in contrast to a resonating state which is thought of as a quantum liquid of valence bonds). It is often called the Majumdar–Ghosh state [258]. One may also speak about a *dimer state*, emphasizing the order peculiar to Φ_1. We can choose the order parameter of dimerization as

$$\mathcal{D}_j = \frac{\langle \mathbf{S}_{j-1} \cdot \mathbf{S}_j - \mathbf{S}_j \cdot \mathbf{S}_{j+1} \rangle}{(-3J/4)} \tag{6.143}$$

which measures whether the bonding strength alternates from bond to bond. Let us note that $\mathcal{D}_j = 0$ for the homogeneous ground state of the usual $J_2 = 0$ Heisenberg chain. In contrast, the Majumdar-Ghosh states show the greatest possible degree of alternation: we find $\mathcal{D}_j = (-1)^{j+1}$ for Φ_1, and $\mathcal{D}_j = (-1)^j$ for Φ_2. The twofold degeneracy of the ground state is ascribed to the two possible ordering patterns of a scalar order parameter, like in the AF Ising model. Let us remark that though there are general arguments to exclude the breaking of a continuous symmetry[36] in linear chain models with short-range interactions, there is no theorem which would forbid the $T = 0$ breaking of a discrete symmetry.

We should be wondering how Φ_1 manages to remain frozen when both first- and second-neighbour interactions couple it to other configurations by events like (6.126). The solution to Problem 6.4 shows that these off-diagonal processes cancel each other exactly at $J_2 = J_1/2$, but at any other ratio, the ground state resonates away from Φ_1. Can we still speak about a dimer state? Yes, we can: it is found that the dimer order parameter is non-vanishing in a finite interval of J_2/J_1 [147]. In the parameter range of interest to us, the J_1–J_2 model has two phases: the gapless, homogeneous spin liquid (or RVB) state at $J_2/J_1 < (J_2/J_1)_{\rm cr}$, and the dimerized, gapfull phase[37] at $J_2/J_1 > (J_2/J_1)_{\rm cr}$. It is estimated that $(J_2/J_1)_{\rm cr} \approx 0.24$ [313, 61]. The gapless phase has algebraic spin correlations of the type (6.111), while the gapfull state has a finite correlation length. In fact, exactly at $J_2 = J_1/2$ we find ultrashort-range correlations, since

$$\langle \Phi_1 | S_j^z S_{j+m}^z | \Phi_1 \rangle = 0 \qquad \text{for } m > 1 \ . \tag{6.144}$$

[36] Except when the order parameter is a conserved quantity as for ferromagnetism.

[37] One often calls it a massive phase, emphasizing that excitations have a finite rest mass.

In terms of bond patterns, we can envisage that in the dimer phase, the short-bond configurations dominate, and though some resonance reaching beyond nearest neighbours certainly evolves, it remains a secondary feature. In contrast, the gapless phase is dominated by long-bond configurations. All in all, we find that the two options allowed by the Lieb–Schulz–Mattis–Affleck theorem (see p. 302) are realized in separate phases of the J_1–J_2 Heisenberg model[38].

Can dimer order be detected? We should bear in mind that though (6.142) gives a clear picture of bond alternation, we are not speaking about dimerization in the chemical sense. The underlying Hamiltonian (6.140) does not include a coupling to the lattice, and therefore all bond lengths remain equal, even though there is alternation in the sense of the parameter (6.143). We may say that (6.140) gives a purely electronic mechanism for the spin-Peierls transition[39]. However, if we allow a weak coupling to the phonons, the lattice will surely find it easy to follow the "electronically preformed" dimerization, and distorts accordingly. We get an alternation of short and long bonds, and this can be seen by diffraction measurements.

The question is whether the above scenario stands a chance to be realized at all. After all, we require a fairly strong ($J_2 > 0.24J_1$) second-neighbour coupling, and this should have an explanation in terms of a higher-order superexchange process which is reaching several steps beyond the first neighbour magnetic ion. The reasonable expectation is that J_1 is much bigger than J_2. However, it has been found that the geometry of $CuGeO_3$ is special, because the $2p$ lobes of an oxygen ion cannot point towards two neighbouring Cu^{2+} ions simultaneously, and therefore nearest-neighbour superexchange is anomalously weak [262, 61]. Indeed, fits to various measurements indicate J_1 ~140–180K, which is exceptionally low (let us recall that in Sr_2CuO_3, $J_1 \approx 1300$K). And

[38]The model can be understood as the large-U effective Hamiltonian of the half-filled t_1–t_2 Hubbard chain. Away from half-filling, the same model has an extended domain of ferromagnetism (see Ch. 8). The apparently extraordinarily rich phase diagram of the t_1–t_2 Hubbard model is the subject of continuing speculations.

[39]The ordinary Peierls transition is a spontaneous lattice distortion which lowers the energy of a one-dimensional electron system by opening a gap at the Fermi level [138]. One may envisage an analogous phenomenon when the electrons are so strongly correlated that they have to be described by an effective spin model.

$CuGeO_3$ *does* show a phase transition which is customarily referred to as the spin-Peierls transition. It can be deduced from the temperature dependence of the susceptibility that a spin gap appears at $\sim$ 14K [159]. The transition is accompanied by a weak lattice distortion. Expert opinion diverges concerning the value of J_2/J_1; early estimates indicated that it may be marginally subthreshold [61], but lately, it is becoming accepted that it must be well above the critical value [455, 100]. In any case, it can be asserted that J_2/J_1 is substantial, and that it is the driving force of the phenomenon.

The experience of having built a non-resonating singlet ground state of $S = 1/2$ spins is all the more valuable because familiarity with the idea of valence bond solids helps us to understand the nature of the Haldane state of the $S = 1$ chain as well. Physically, the $S = 1$ spins of, say, Ni^{2+} ions are composed of the $S = 1/2$ spins of the constituent $3d$ electrons, thus whatever the $S = 1$ spins are doing, we should be able to interpret it as a state of twice as many $S = 1/2$ spins. We are led to model $S = 1$ systems by putting two strongly coupled $S = 1/2$ spins onto each lattice site. Each of the constituent spins can participate in a singlet bond with a different neighbour, and this state can be continued all through the chain (we assume an even L, and periodic boundary condition), to give the *valence bond solid state* of the $S = 1$ Heisenberg chain [10]

$|\text{VBS}\rangle =$ 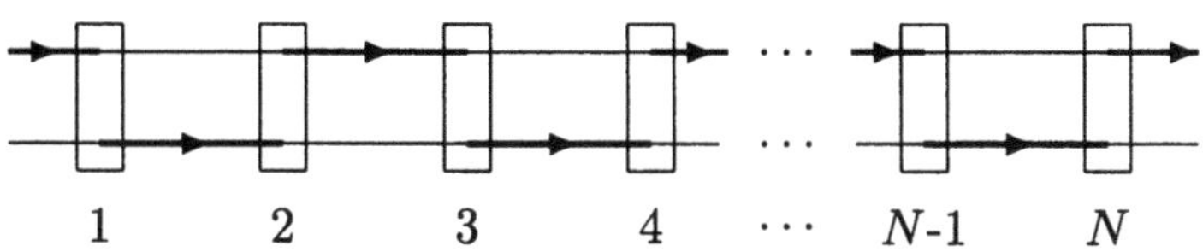

(6.145)

Forming the singlet bonds indicated by arrows is not enough: it would allow the two spins at a site to combine into either a singlet, or a triplet. However, we insist that we want to construct a state of the $S = 1$ chain, and to this end, we act with a symmetrizing projector at each of the sites, which lets only the local $|\mathbf{S}_j| = 1$ components survive: this is symbolized by the vertical boxes. We do not write this down formally; it is enough to know that the procedure leads to a well-defined state. Since $|\text{VBS}\rangle$ is composed of nearest-neighbour bonds, it is easy to accept that it has short-range correlations, and the spectrum is gapped.

The only trouble is that it can be checked that $|\mathrm{VBS}\rangle$ is *not* the ground state of the $S = 1$ AF Heisenberg model. What good is it then? Well, once one finds such a beautiful state, one is not willing to discard it, but instead one finds the problem for which it *is* the solution. We construct a Hamiltonian for which $|\mathrm{VBS}\rangle$ is the ground state, following the same strategy by which we proved that the dimer states are the ground states of (6.140). In general, two $S = 1$ spins can be combined into $|\mathbf{S}_1 + \mathbf{S}_2| = 2$, 1, and 0. Let us, however, observe that in $|\mathrm{VBS}\rangle$, the four $S = 1/2$ spins making up two neighbouring $S = 1$ spins cannot add up to 2, because two of them form a singlet, and thus the rest can make only a singlet or a triplet.

Now consider the following expression of two $S = 1$ operators

$$\mathcal{P} = \mathbf{S}_1 \cdot \mathbf{S}_2 + \frac{1}{3}(\mathbf{S}_1 \cdot \mathbf{S}_2)^2 + \frac{2}{3} \, . \tag{6.146}$$

It is easy to check that $\mathcal{P}$ projects onto the subspace $|\mathbf{S}_1 + \mathbf{S}_2| = 2$. It is generally true that the smallest eigenvalue of a projection operator is 0. Consequently, the ground states of $\mathcal{P}$ are those which do not contain a $|\mathbf{S}_1 + \mathbf{S}_2| = 2$ component. Generalizing for the lattice, $|\mathrm{VBS}\rangle$ is a ground state of the VBS-point bilinear–biquadratic Hamiltonian

$$\mathcal{H}_{\mathrm{VBS}} = \sum_j \left\{ \mathbf{S}_j \cdot \mathbf{S}_{j+1} + \frac{1}{3}(\mathbf{S}_j \cdot \mathbf{S}_{j+1})^2 \right\} \tag{6.147}$$

(we omitted the additive constant). It can be shown that the ground state is unique[40], if we use periodic boundary condition. The effect of free chain ends will be discussed later.

The lesson is that though $|\mathrm{VBS}\rangle$ is not the ground state of the Heisenberg model, it is the ground state of a related bilinear–biquadratic model. Both the Heisenberg model and $\mathcal{H}_{\mathrm{VBS}}$ can be regarded as special cases of the general bilinear–biquadratic model

$$\mathcal{H} = \sum_j \left[J\mathbf{S}_j \cdot \mathbf{S}_{j+1} + Q(\mathbf{S}_j \cdot \mathbf{S}_{j+1})^2 \right] \, . \tag{6.148}$$

[40]Looking at the graphical representation (6.145), we might have thought that there are two degenerate ground states, because the starting bond [12] can be placed in either the lower, or the upper, row. However, it follows from the process of symmetrization that we do not distinguish the bonds emanating from a site [10].

which has several distinct $T = 0$ phases [102]. It turns out that the pure Heisenberg model ($Q = 0$), and the VBS model ($Q = J/3$) are indeed closely related, in the sense that they belong to the same Haldane phase which stretches over the interval $-J < Q < J$. The Haldane phase is gapped, and homogeneous: a glance at (6.145) shows that the bonds are energetically equivalent.

Is the valence bond state completely disordered? In the usual sense of the word, it is: it does not break any global symmetry, and its spin–spin correlation function decays exponentially. Still, we would expect that the ground state of an interacting system should show some interesting coherence property, and such is indeed the case. The eigenstates of S_j^z are 1, 0, and -1; let us denote the corresponding states by 1, 0, and $\bar{1}$. For instance, the state 1 arises from both constituent 1/2-spins being $\uparrow$. If the ground state were completely featureless, a "snapshot" would give an arbitrary sequence of these species, say $10\bar{1}1001\bar{1}0\ldots$. In particular, a 1 site could be followed either by another 1, or by $\bar{1}$, or a 0. Let us, however, consider the construction of the VBS state (6.145), and assume that site $j = 2$ is in the state 1. This means that from both bonds ending at 2, we used up the $\uparrow$-electron. Consequently, at the site $j = 3$, we can use only the $\downarrow$-electron from the bond [23], and thus site 3 cannot be in the state 1. It follows that in the VBS state, 1 can be followed only by 0 or $\bar{1}$, but not by 1, and similarly, $\bar{1}$ can be followed only by 1 or 0, but not by $\bar{1}$. If we contract the state by leaving out the zeros, we get the ideal antiferromagnetic sequence $1\bar{1}1\bar{1}1\bar{1}\ldots$. However, this feature is not manifested in AFM long-range order, not even in quasi-long-range order, because between any two 1 and $\bar{1}$ sites, an arbitrary number of 0 sites can be inserted. Therefore one speaks of *hidden order*. What we have just met is saturated hidden order in the sense that, after leaving out the zeros, a 1 is always followed by a $\bar{1}$, and vice versa. We may also envisage hidden order reduced by quantum fluctuations, so that following 1 by $\bar{1}$ is no longer certain but still has ¿50% probability. We have convinced ourselves that hidden order is saturated in the VBS point. It can be shown that it is non-saturated but non-vanishing in the entire Haldane phase[41].

---

[41] The existence and degree of Néel order could be detected by calculating $\lim_{m\to\infty}\langle S_j^z S_{j+m}^z\rangle$. Quantifying the hidden order is not so simple: it requires check-

There is a further charateristic of the Haldane phase which can be understood in simple terms: the effect of chain ends. Let us consider the same N-site chain as in (6.145), but with free boundary conditions (Fig. 6.2). Since now the bond connecting the sites N and 1 is missing, we get one unbound spin $S = 1/2$ at each chain end (it was arbitrary to draw them as $\uparrow$ and $\downarrow$; the spin directions can be chosen independently). It follows that the VBS ground state of a chain with free ends is fourfold degenerate (in contrast to the case of periodic boundary condition, where the ground state is unique). If we come away from the VBS point, but remain in the Haldane phase, the ensuing valence bond resonance couples the end spins with an effective interaction which decreases exponentially with chain length. Even though the singlet and triplet combinations of the end spins are no longer exactly degenerate, they are separated by a very small energy, thus we can at least speak about quasi-degeneracy: the $S = 1/2$ entities become liberated at very low temperatures.

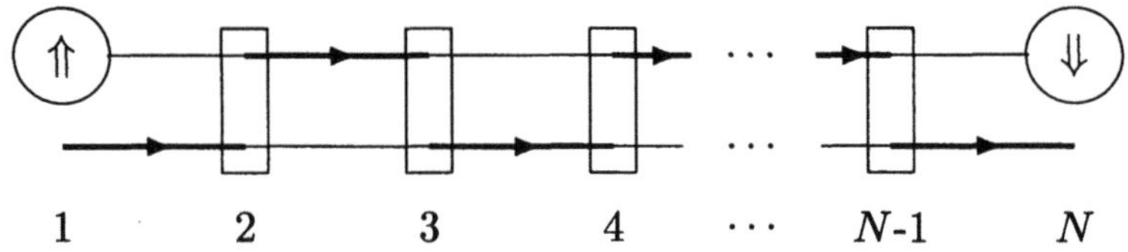

Figure 6.2: The VBS construction gives rise to free spins at the ends of the chain.

The above consideration helps in understanding the following observation: the susceptibility of $CsNiCl_3$ *increases* if we replace 2% of the Ni atoms with Mg [31]. This result is curious because we are replacing magnetic atoms with non-magnetic ones. The explanation is that the pure system of magnetic ions has a spin gap, and has therefore a very low susceptibility, while disrupting the chains with Mg atoms creates chain ends with essentially free $S = 1/2$ spins which make a substantial Curie contribution to the susceptibility.

We have learned about two kinds of VBS states: (6.142), and (6.145).

ing the state of all the sites in the string which leads from j to $j + m$. This can be achieved by introducing a non-local, or string, order parameter [85, 205].

Are they in the same class? We should say no: the dimer state (6.142) breaks translational invariance, and perforce has discrete degeneracy, while (6.145) is homogeneous and unique. It is a characteristic property of integer-spin chains that they allow the construction of homogeneous VBS states, and this fact illuminates the essential difference between integer and half-integer cases [7]. However, this does not mean that we could not construct symmetry-breaking VBS states for integer spins! It is easy to devise a dimer state of the $S = 1$ chain, in which sites $2j + 1$ and $2j + 2$ are connected by a double bond, while $2j + 2$ and $2j + 3$ are unbound:

$|S = 1, \mathrm{dim}\rangle =$ 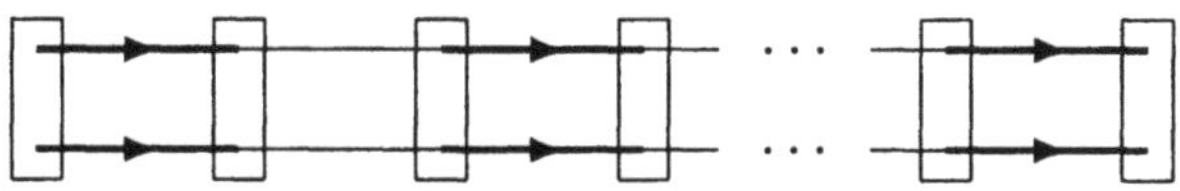

(6.149)

Assuming periodic boundary condition, the dimer ground state is twofold degenerate in the same sense as the Majumdar–Ghosh state is.

Is it physically less motivated to introduce $|S = 1, \mathrm{dim}\rangle$ than $|\mathrm{VBS}\rangle$? The answer is no; it depends on the ratio Q/J, which of the two states is nearer to the true ground state. It turns out that each of them represents one of the singlet phases of the model (6.148). At the simplest level, the competition of the two phases can be discussed by using $|\mathrm{VBS}\rangle$ and $|S = 1, \mathrm{dim}\rangle$ as variational trial states. One finds that [10]

$$\langle \mathrm{VBS}|\mathcal{H}|\mathrm{VBS}\rangle = -\frac{4J}{3} + 2Q \tag{6.150}$$

and

$$\langle S = 1, \mathrm{dim}|\mathcal{H}|S = 1, \mathrm{dim}\rangle = -J + \frac{8Q}{3} \,. \tag{6.151}$$

We can offer a handwaving argument for these results. In the VBS state, neighbouring spins can only form a singlet with

$$\langle \mathbf{S}_j \cdot \mathbf{S}_{j+1}\rangle = \frac{1}{2}[(\mathbf{S}_j + \mathbf{S}_{j+1})^2 - \mathbf{S}_j^2 - \mathbf{S}_{j+1}^2] \rightsquigarrow \frac{0-2-2}{2} = -2\,,$$

$\langle (\mathbf{S}_j \cdot \mathbf{S}_{j+1})^2\rangle = 4$, or a triplet with $\langle \mathbf{S}_j \cdot \mathbf{S}_{j+1}\rangle = -1$ and $\langle (\mathbf{S}_j \cdot \mathbf{S}_{j+1})^2\rangle = 1$. The question is what is the probability of finding a singlet (triplet). Let us recall our discussion of hidden order, where we found that in the S^z basis, possible sequences are 10, $1\bar{1}$, 00, etc. but 11 and $\bar{1}\bar{1}$ are excluded. Writing down the configurations in terms of their constituent 1/2-spins, and counting those which are compatible with an intersite triplet, we find 10, $1\bar{1}$, 00 (twice), $0\bar{1}$, 01, $1\bar{1}$, and $\bar{1}1$, eight configurations. Similarly, four configurations are compatible with an intersite singlet. Since the VBS Hamiltonian does not introduce a bias between triplets and singlets, we may assume that the ratio of their relative abundances is 8:4. Taking the singlet contribution with weight 1/3, and the triplet contribution with weight 2/3, we arrive at (6.150). For the pure Heisenberg model

$$\langle \mathrm{VBS}|\mathbf{S}_j\cdot\mathbf{S}_{j+1}|\mathrm{VBS}\rangle = -4/3 \tag{6.152}$$

which is only 5% higher than the numerically estimated exact ≈ -1.4. If we wish, we can say that the VBS state is quite a good approximation for the Heisenberg ground state. However, we have to be cautious with such numerical coincidences: it is often seen that trial ground states which are quite far away in Hilbert space, give nearly the same energy. Regarding the VBS state as closely related to the Heisenberg ground state relies on a variety of evidence, not merely on the nearness of the energy estimates.

As for the dimer state, half of the first neighbour pairs form intersite singlets; they contribute $-J + 2Q$ to the energy density. On the other hand, in an unbound pair (like in the pair of sites 2 and 3 in (6.149)), the relative orientation of the two spins is quite arbitrary, thus the probabilities of finding an intersite singlet, triplet, and quintiplet are in the ratios 1:3:5, the numbers of corresponding basis states in the S^z basis. The contribution from the bilinear term $(1/9)(-2)+(3/9)(-1)+(5/9)\cdot 1 = 0$ vanishes, as expected. However, there is a contribution to the biquadratic energy term: $(1/9)\cdot 4+(3/9)+(5/9) = 4/3$. Only half of the pairs is of this kind, thus we found $2Q/3$, whereby (6.151) is derived.

Comparing (6.150) and (6.151), we predict that a *phase transition between the two singlet phases* is taking place at $Q = -J/2$ $(J > 0)$. We expect that the ground state is homogeneous, and the spectrum gapped for $Q > -J/2$: this is the *Haldane phase.* On the other hand, the ground state breaks translational invariance, and becomes twofold degenerate in the *dimer phase* which occupies the regime $Q < -J/2$. The dimer phase is also gapped. Since our naive argument works with inflexible trial states, it is inevitable that it indicates a first-order transition.

More sophisticated approaches confirm that the above predictions are basically right: the Haldane phase and the dimer phase are indeed there, only the position of the phase boundary was misjudged: it is really at $Q = -J$ [10]. It is a critical point where the gap collapses, and the correlations follow a power law. The Haldane phase is stretching over the region $-J < Q < J$, thus it includes $Q = 0$ (the pure AF Heisenberg model), and $Q = J/3$ (the VBS-point Hamiltonian (6.147)). The dimer state is superior to any other state if the negative biquadratic coupling is strong enough ($-Q > J$ for AF bilinear coupling). It is interesting that the pure $Q < 0$ biquadratic model can be solved exactly [216], confirming our idea of a singlet state with cell doubling, a gap, and exponential decay of correlations.

The ensuing picture is that of two extended singlet phases, with their respective spin gaps, and a gapless critical point in between. This confirms that the gapfullness is the generic behaviour of $S = 1$ chains, and the collapse of the gap can occur only at isolated points.

Fig. 6.3 summarizes what we learned about the contrasting behaviour of isotropic $S = 1/2$ and $S = 1$ spin chains. The $S = 1/2$ model has two canonical phases: a gapless and a massive phase. Correlations in the gapless state decay according to power laws, so we may speak about a critical phase (a whole region of critical points). Entering the massive (dimer) phase, the gap is initially exponentially small [147]. In contrast, the canonical behaviour of the $S = 1$ chain is gapfull, and gaplessness can occur only at a critical point between two massive phases. The spin gap may rise linearly on both sides of the critical point [369].

Exploring the domains $Q > J > 0$, and $J < 0$, we find further phases of (6.151), including a ferromagnetic one. We refer to the literature for current ideas about the complete phase diagram [102].

Finally, a word about the practical relevance of the various phases of the bilinear–biquadratic model (6.151). We have discussed that there is indeed an electron mechanism giving rise to biquadratic coupling (see pp. 261–262), but it is a higher-order effect, and inherently weaker than the bilinear coupling. Thus we would say that it is alright to consider the pure AF Heisenberg model, and its immediate surroundings with

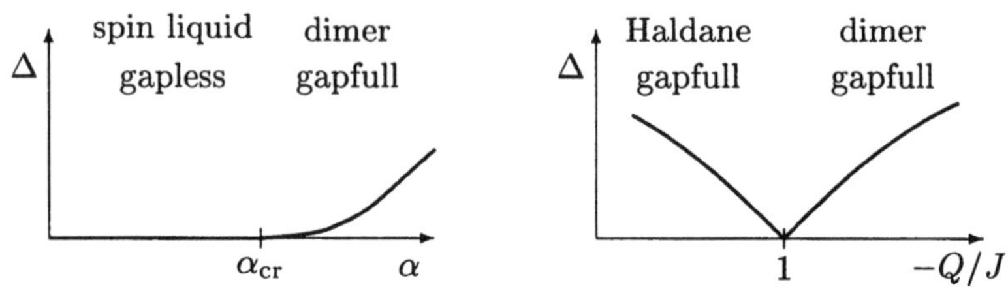

Figure 6.3: A schematic representation of $T = 0$ phase transitions in isotropic spin chains shows a marked difference in the behaviour of half-integer and integer spin systems. *Left*: The $S = 1/2$ J_1–J_2 model (6.139) passes from a critical (gapless) phase to a massive phase at α_{cr} ($\alpha = J_2/J_1$). Initially, the spin gap Δ is exponentially small. – *Right*: The $S = 1$ bilinear–biquadratic model (6.147) has a critical point between too massive singlet phases.

a small $|Q/J|$; this is well inside the Haldane phase and we have seen experimental evidence for the Haldane gap. More exotic phases, demanding $|Q/J| \sim 1$ (or larger) are unlikely to be realized.

6.3.3 The CaV_4O_9 Story

The remarkable compound CaV_4O_9 offers the chance to witness the competition of the three qualitatively different kinds of behavior of an isotropic Heisenberg antiferromagnet: Néel order, valence bond solid, and resonating singlet bonds. The V ions have a single $3d$-electron, so we are dealing with $S = 1/2$ spins[42]. The material is quasi-two-dimensional, and the arrangement of the V ions in a plane is shown in Fig. 6.4 (upper left). The structure can be understood as a network of square plaquettes interconnected by bonds of strength J_2, while the coupling within the plaquettes is J_1. We assume that neighbouring spins interact like $J_1 \mathbf{S_m} \cdot \mathbf{S_n}$, or $J_2 \mathbf{S_m} \cdot \mathbf{S_n}$, whichever is appropriate.

The lattice is bipartite with coordination number $z = 3$, thus it is well suited for antiferromagnetic order (Fig. 6.4, lower right). Two-sublattice Néel order, which does not distinguish between the two kinds of bonds, is likely if $J_1 \approx J_2$. However, if one of the bond strengths is significantly greater than the other, the system will surely prefer to

[42]The role of the orbital degrees of freedom is a matter of continuing debate.

optimize the energy gain from the strong bonds by adopting a suitable valence bond state. If J_1 is strong enough, we envisage resonance within the essentially independent plaquettes (Fig. 6.4, upper right). On the other hand, if J_2 sufficiently larger than J_1, it is best to form non-resonating singlet bonds (dimers), as shown in Fig. 6.4, lower left.

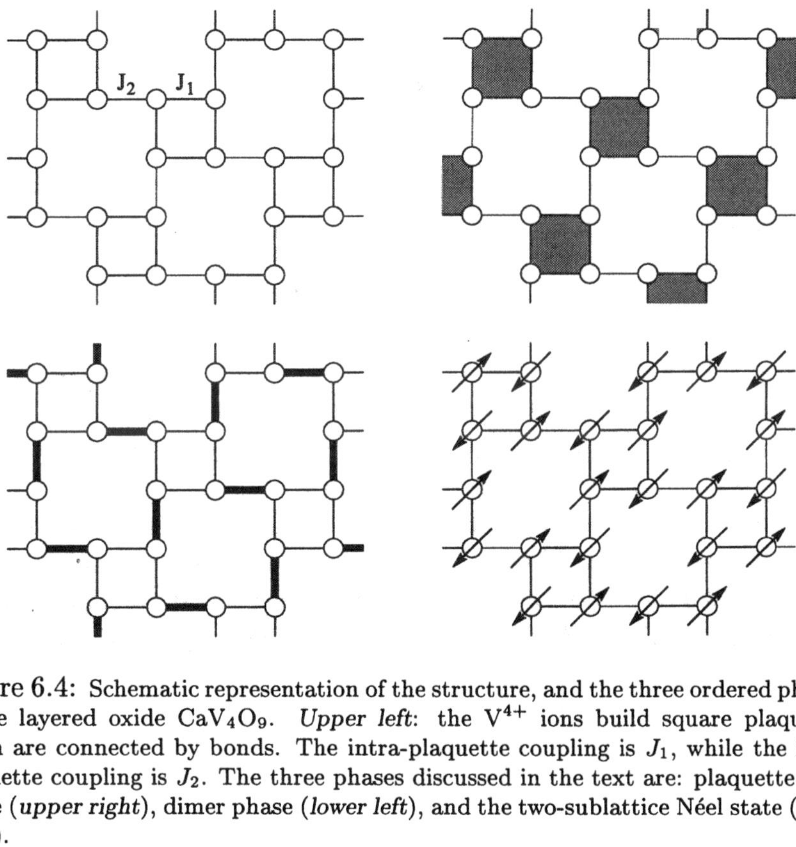

Figure 6.4: Schematic representation of the structure, and the three ordered phases, of the layered oxide CaV_4O_9. *Upper left*: the V^{4+} ions build square plaquettes which are connected by bonds. The intra-plaquette coupling is J_1, while the inter-plaquette coupling is J_2. The three phases discussed in the text are: plaquette RVB phase (*upper right*), dimer phase (*lower left*), and the two-sublattice Néel state (*lower right*).

We use simple energy estimates to work out a rudimentary ground state phase diagram. The dimer phase (which is a two-dimensional version of the Majumdar–Ghosh phase) is treated as an array of independent singlet pairs, thus its energy density is

$$E_{\text{dimer}} = -\frac{3J_2}{8} . \tag{6.153}$$

The plaquette RVB phase is envisaged as a system of independent squares, each plaquette resonating according to (6.128). The energy density is

$$E_{\rm RVB} = -\frac{J_1}{2} \, . \tag{6.154}$$

The energy density of the ordered antiferromagnetic state can be estimated simply but (we believe) fairly accurately by the following argument: let us recall that for a lattice with all bond strengths equal, spin wave theory gives the ground state energy density as

$$E_{\rm AF} = -\frac{zJ}{2} S \left(S + \frac{\xi}{z} \right) \, . \tag{6.155}$$

This parametrization is slightly different from (6.87); we sought to emphasize that the quantum correction is of the order $1/zS$, as it follows from the argument leading to (6.54). We use the result for the square lattice to deduce that for two-dimensional lattices $\xi \approx 0.632$. Replacing this and $z = 3$ in (6.155), we would guess that the Néel energy density of the $J_1 = J_2$ CaV_4O_9 lattice is $-0.533J_1$ which is not so bad at all, considering that the actual spin wave calculation gives $-0.5376J_1$ [412]. Let us note that this is lower than either (6.153) or (6.154), thus the two singlet phases are separated by a regime of Néel-type order. We extend the result (6.155) to the $J_1 \neq J_2$ case by observing that zJ has to be replaced by a suitably weighted sum over nearest neighbour bonds $zJ \rightsquigarrow z_1 J_1 + z_2 J_2 = 2J_1 + J_2$. This leads to the AF energy density

$$E_{\rm AF} \approx -0.355 J_1 - 0.178 J_2 \, . \tag{6.156}$$

We follow [412] by parametrizing the couplings as $J_1 = \alpha J$, and $J_2 = (1 - \alpha)J$, where $0 \leq \alpha \leq 1$. The ground state phase diagram is obtained by determining which of the energies (6.153), (6.154), and (6.156) is the lowest (see Fig. 6.5). We find that the dimer phase is stable if $\alpha < 0.357$, the RVB state prevails for $\alpha > 0.551$, while the intermediate region $0.357 < \alpha < 0.551$ is Néel-ordered. Naturally, we should be aware that the argument for (6.156) is very handwaving, and the estimates (6.153) and (6.154) are certainly crude, since they rely on neglecting the coupling between dimers, and plaquettes, respectively. Better results can be obtained by perturbation expansion in J_1/J_2 (or J_2/J_1, whichever is appropriate) [412]. However, the overall appearance

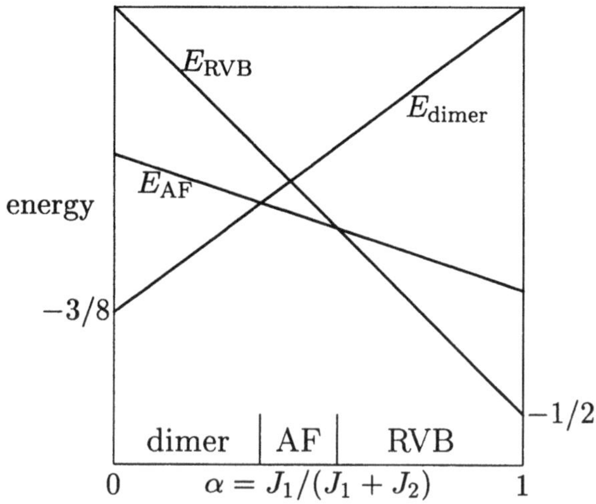

Figure 6.5: An approximate ground state phase diagram of the Heisenberg model representing CaV_4O_9 is obtained by comparing the energies (6.153), (6.154), and (6.156).

of our phase diagram is quite similar to that derived by the quantum Monte Carlo method [406], and we would like to believe that the existence of three such phases is a genuine feature of the model.

It is not unlikely that a truthful picture of CaV_4O_9 should incorporate such further ingredients as spin–phonon coupling, farther-neighbour interactions, and coupling to the orbital degrees of freedom. However, it may be asked if even our result offers some kind of explanation for the observed behaviour. CaV_4O_9 attracted great attention because the measured susceptibility shows activated behaviour[43]

$$\chi \propto \exp\left(-\Delta/k_{\mathrm{B}}T\right) \tag{6.157}$$

which can be interpreted as evidence for a spin gap $\Delta = 107\mathrm{K}$ [389]. The two singlet phases we found are gapfull phases, because a finite energy is needed for breaking up a short-range singlet object. It is likely that the observed gapped behaviour arises from plaquette resonance.

[43]The argument in footnote 27 (see p. 305) shows that there is no T-dependent prefactor in $D = 2$.

6.3.4 Frustration

The preceding discussion of bipartite lattices relied heavily on the fact that the ground state of the Ising term (or shortly, the Ising ground state) was essentially unambiguous, corresponding to an up–down alternation of the spins. However, in many interesting materials the lattice of magnetic ions is non-bipartite, as in $LiNiO_2$ (quasi-2-dim. with triangular planes), or in CoO (fcc). This makes the nature of antiferromagnetic ordering more complicated, as shown by the following simple arguments:

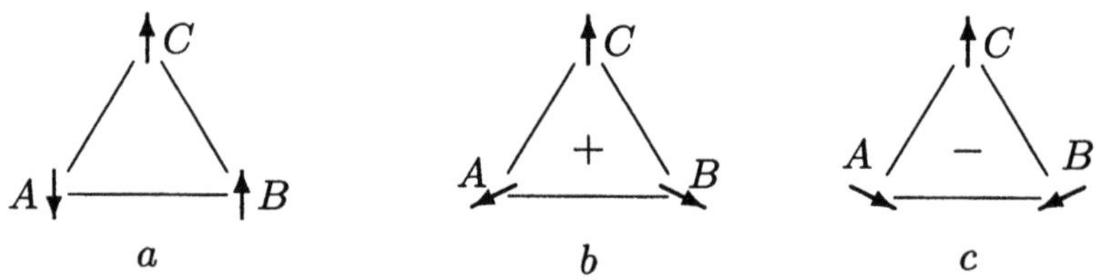

Figure 6.6: The effect of the competition of AF interactions in the triangle plaquette. *a*: One of the bonds is necessarily frustrated in the Ising model. *b* and *c* show the two non-equivalent ground states of the *XY* model.

A typical unit of many non-bipartite lattices is the triangle plaquette of sites A, B, and C (Fig. 6.6). Let us first consider the case of pure z–z interaction (Ising limit[44]), so the minimum energy of a single bond is $-JS^2$, and the corresponding lower bound for the ground state energy of the triangle is $-3JS^2$. Can this bound be reached? No, the minimum energy of the triangle is merely $-JS^2$. Putting a spin $\downarrow$ at A, we will want its neighbours at B and C to point $\uparrow$. AB and AC are low-energy bonds but the nearest-neighbour pair BC forms a high-energy bond with energy JS^2 (Fig. 6.6, left). Whatever spin arrangement we try, we find one bond which is frustrated in its effort to reach the low-energy configuration. One may speak about a *frustration effect*: seeking its ground state, the system cannot reach the state which it really desires,

[44]If we keep only $\mathcal{H}_{zz}$, we get a model in which only the z-components of the three-dimensional vector spins $\mathbf{S}_i$ are coupled. In the true Ising model $\mathcal{H} = J \sum S_i S_j$, the variables S_i are scalars which can assume the values $-S$, $-S+1$, ..., S. The two models behave identically, but the former allows the "switching on" of the x–y coupling, and can support non-collinear order.

but it can choose from several, equally unsatisfying, compromises[45]. The conclusion can be generalized to the triangular lattice Ising antiferromagnet, because each of the triangle plaquettes must contain (at least) one parallel spin pair (frustrated bond). The ground state energy is much larger than (the number of bonds)×(minimum bond energy) which would be $-3LJS^2$; it is merely $-LJS^2$. The reader can easily show (see Problem 6.5) that the number of ground states is exponentially large. One may say that the model has finite zero-point entropy[46]. Most of these ground states are non-periodic, and the triangular Ising antiferromagnet does not show long range order even at $T = 0$.

Switching on the x–y part of the Hamiltonian mixes the Ising ground states and thereby lifts the huge degeneracy. Though interesting results can be obtained by studying the effect of a weak x–y coupling on quantum spins [106, 214], we prefer to go straight to the limit of classical vectors, and an isotropic interaction. We no longer find a macroscopic degeneracy in the sense of a large number of disordered-looking ground states, but it remains true that the nature of the ground state reflects a compromise between competing interactions. For a first orientation, let us study the isotropic classical vector model for three spins. The energy

$$E(\vartheta_A, \varphi_A; \vartheta_B, \varphi_B; \vartheta_C, \varphi_C) = J(\mathbf{S}_A \cdot \mathbf{S}_B + \mathbf{S}_B \cdot \mathbf{S}_C + \mathbf{S}_C \cdot \mathbf{S}_A) \tag{6.158}$$

is a function of the polar angles (we measure ϑ from the z-direction). Collinear (Ising) configurations would only give $-JS^2$, but having the spins pointing at angles of 120 degrees yields the substantially lower $-(3/2)JS^2$. The symmetry of the small cluster makes this result easy to accept but let us note that in general, finding the non-collinear ground state of classical vectors is not straightforward. We are soon going to see that even quite simple-looking problems can have surprising solutions.

[45] The particularly fitting expression "frustration" was coined (or rather borrowed from Freudian psychoanalysis) by Toulouse in 1977 [403]. The implications of this situation had been well-known from previous studies of antiferromagnetism in close-packed structures.

[46] The entropy is proportional to the logarithm of the number of quantum states accessible to the system, so if the ground state degeneracy is $\sim e^{\alpha L}$, then the $T = 0$ entropy density is $k_B \alpha$. The model violates Nernst's theorem. This is an artificial feature: a slight modification of the model (e.g., making one bond slightly weaker than the other two in each of the plaquettes) suffices to reduce the number of ground states to a small number.

The ground state order of the triangular lattice antiferromagnet can be obtained by extending the triangle solution into a three-sublattice Néel state (see Fig. 7.1, right). The lesson is that competing interactions can give rise to non-collinear phases, leading to a great variety of magnetic ordering phenomena.

Demanding that the three vectors subtend angles $2\pi/3$ does not specify the ground state uniquely. First, there is the two-parameter continuous degeneracy which is common to all states breaking spin-rotational invariance. The two parameters can be chosen as two angles, one of them determining the orientation of the plane[47] of the spins, and the other giving the spin z-axis within the plane. There is also a discrete degeneracy, associated with the permutation of the sublattices.

The XY model (and generally, models with easy-plane anisotropy) show the curious discrete degeneracy illustrated in Fig. 6.6: the patterns b and c cannot be rotated into each other. The difference between the two plaquette solutions can be characterized by the scalar chirality [281]

$$\kappa = \frac{2}{3\sqrt{3}}[\mathbf{S}_A \times \mathbf{S}_B + \mathbf{S}_B \times \mathbf{S}_C + \mathbf{S}_C \times \mathbf{S}_A]_z = \frac{2}{3\sqrt{3}}(S_A^x S_B^y - S_A^y S_B^x + \ldots) \tag{6.159}$$

which is 1 for b and -1 for c. The manifestly Ising character of κ raises the intriguing possibility of a finite-T phase in which chirality has long range order, though the spins do not [281, 241].

What about quantum effects? Since AF interactions are doomed to be less effective in a non-collinear structure than in a collinear one, we expect that fluctuation effects become more spectacular. Such is indeed the case. Extending the spin wave calculation to the three-sublattice case, it has been found that the spin reduction is $\Delta S \approx 0.261$ [194]. This is quite a bit larger than the value for the square lattice ($\approx$0.197), and approaches the value 0.28$\pm$0.03 quoted for the honeycomb lattice. Recalling that the corresponding coordination numbers are $z = 6$, 4, and 3, we see that the simple rule that fluctuation corrections are $\sim 1/z$, ceases to hold if the structure is frustrated. We may say that frustration diminishes the effective coordination number. However, for our example it remains true that the nature of order is the same as for classical

[47]The structure is not collinear but it is still coplanar. In more complicated systems, e.g., in fcc antiferromagnets, we can find non-coplanar ordering patterns.

vectors, and only the size of the order parameter is subject to quantum corrections.

Studying close-packed structures revealed the exciting phenomenon of fluctuation-induced order, or "order through disorder". The essence of this phenomenon is that in certain situations where the classical vector model does not order even at $T = 0$, quantum fluctuations select an ordered ground state if we make the spin finite. Let us illustrate this by putting our isotropic three-site model in an external magnetic field. The energy

$$\begin{aligned} E &= J(\mathbf{S}_A \cdot \mathbf{S}_B + \mathbf{S}_B \cdot \mathbf{S}_C + \mathbf{S}_C \cdot \mathbf{S}_A) - \mathbf{H} \cdot (\mathbf{S}_A + \mathbf{S}_B + \mathbf{S}_C) \\ &= \frac{J}{2} \left(\mathbf{S}_A + \mathbf{S}_B + \mathbf{S}_C - \frac{\mathbf{H}}{2J} \right)^2 + \text{const} \end{aligned} \tag{6.160}$$

is minimum if

$$\mathbf{S}_A + \mathbf{S}_B + \mathbf{S}_C = \frac{\mathbf{H}}{2J} \,. \tag{6.161}$$

Specifying the directions of $\mathbf{S}_A$, $\mathbf{S}_B$, and $\mathbf{S}_C$ requires six angles but the vector equation (6.161) gives only three conditions. From the axial symmetry of the system, we expect that there is one angle variable on which E does not depend: the angle of the simultaneous rotation of the three spins about the field direction. However, we have just found that there must be two more angles which can be continuously varied (at least within a certain interval) without changing the energy. This is referred to as the *non-trivial continuous degeneracy*[48]. That the degeneracy deserves the name non-trivial, is illustrated by Fig. 6.7 which shows that very dissimilar-looking configurations can be simultaneous ground states. Let us remember that these are only certain extreme spin-arrangements, but they are continuously connected via other ground states.

It would be difficult to claim direct physical insight into why this should be so; it can only be found by the careful solution of the problem. It should serve as a warning that when we are faced with a problem like that of minimizing (6.160), we should not be too ready to select a

[48] In contrast, the continuous degeneracy related to global rotations may be called trivial.

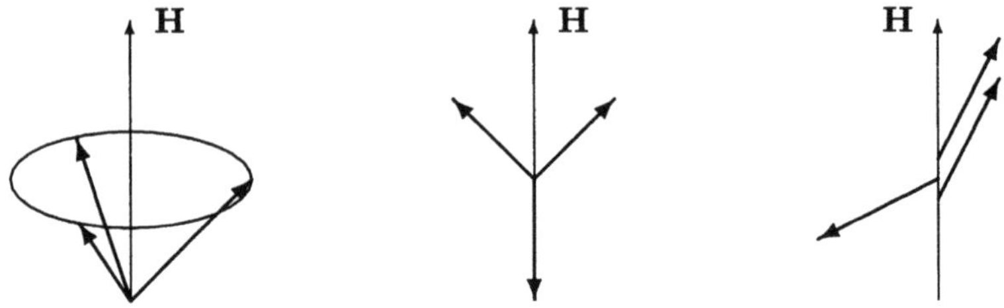

Figure 6.7: Several examples of degenerate ground states of the isotropic triangular antiferromagnet in an external field **H**. In each of the pictures, the three vectors represent the sublattice magnetizations $\mathbf{S}_A$, $\mathbf{S}_B$, and $\mathbf{S}_C$. Quite dissimilar configurations belong to the ground state set.

reasonable-looking symmetrical state, trusting that we have found *the* relevant ground state.

Let us now assume that the spins are finite, and consider quantum fluctuations. What is to be done? Is there an independent spin wave theory, and quantum ground state, for each of the classical ground states? The answer is no. The spin wave calculation yields a quantum correction to the ground state energy, and this depends on the configuration; choosing the lowest level, we get an essentially unique ground state configuration which, however, still depends on the external field [65]. The quantum fluctuations described by spin wave theory eliminate the non-trivial degeneracy, and select the $T = 0$ order.

In addition to the isotropic triangular antiferromagnet in an external field, there are other spin models which show the "order through disorder" phenomenon. Non-trivial continuous degeneracy has been found for the triangular antiferromagnet with an easy-axis anisotropy [280], and the isotropic fcc antiferromagnet in zero external field [162]. The corresponding spin wave calculations [215, 213] demonstrate the quantum selection of ground state order. We do not discuss the analogous phenomenon of selecting $T > 0$ order by thermal fluctuations.

The "order by disorder" phenomenon sheds new light on the question we posed on p. 314: why is the *spin quantum liquid* state so difficult to realize? The reason is that quantum fluctuations are playing an ambivalent role. On the one hand, they reduce the order parameter of the AF state; this is particularly noticeable at small S. On the other hand, they may force the system to pick an order even if classical spins with the same kind of interaction were non-ordering; this begins to act as soon as S is finite. In contrast to some earlier suggestions, the example of the triangular lattice has taught us that there is a continuity between the large-S and $S = 1/2$ limits [214, 215]; thus we can expect an $S = 1/2$ spin liquid only in a system which cannot order through disorder at large S. It appears that the edge-sharing of plaquettes in the triangular lattice leaves too little room for uncertainty; actually even the more open structure of the kagome lattice which is a corner-sharing network of triangles (Fig. 8.3) may settle into an ordered state.

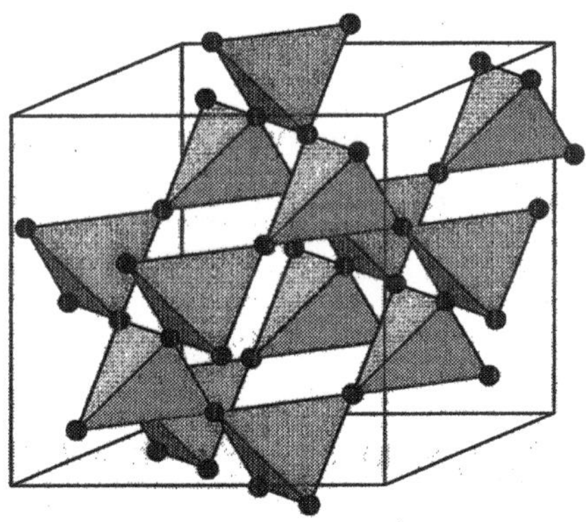

Figure 6.8: The pyrochlore lattice (after [284])

Currently the best bet for a spin liquid ground state is the pyrochlore lattice which is a corner-sharing network of tetrahedra (Fig. 6.8). We met the same lattice in a different context in Sec. 1.3.3: it is the sublattice of octahedrally coordinated sites in magnetite, and it is its highly frustrated nature what makes the finding of the Verwey order such a difficult problem. – The ground state of the classical Heisenberg model shows a high degree of non-trivial degeneracy, and thermal fluctuations do not induce order at any finite T [284]. Assuming that the same holds for quantum fluctuations, we may ask whether the $S = 1/2$ model has a spin liquid ground state. It is claimed that the answer is yes: spin–spin correlations decay exponentially, and there is a spin gap [57].

Problem 6.5. Show that the number of the ground states of the antiferro-

magnetic triangular lattice Ising model is exponentially large in L.
Hint: You are not requested to calculate the number of ground states exactly! It is enough to find a sufficiently numerous subset of ground states. You can achieve this by fixing the state of a bipartite sublattice, and checking if this leaves you with enough choice for the rest of the lattice.

Solutions to the Problems

Problem 6.1. Let us choose the state with maximum polarization $\mathcal{S}^z = LS$ as the ground state which we denote by $|0\rangle$. The subspace with $\mathcal{S}^z = LS - 1$ is spanned by the single-site spin flip states $|\mathbf{j}\rangle = S_{\mathbf{j}}^- |0\rangle$. The Bloch state with wavevector $\mathbf{k}$ is

$$|\mathbf{k}\rangle = \frac{1}{\sqrt{L}} \sum_{\mathbf{j}} e^{i\mathbf{kj}} |\mathbf{j}\rangle \,. \tag{6.162}$$

Let us consider the anisotropic Hamiltonian in an external field

$$\mathcal{H} = J \sum_{\mathbf{i},\mathbf{j}} (S_{\mathbf{i}}^x S_{\mathbf{j}}^x + S_{\mathbf{i}}^y S_{\mathbf{j}}^y + A S_{\mathbf{i}}^z S_{\mathbf{j}}^z) - H \sum_{\mathbf{j}} S_{\mathbf{j}}^z \,. \tag{6.163}$$

We show that $|\mathbf{k}\rangle$ is an eigenstate of (6.163).

The z–z contribution is

$$AJ \sum_{\mathbf{i},\mathbf{j}} S_{\mathbf{i}}^z S_{\mathbf{j}}^z |\mathbf{k}\rangle = (JASz + H + E_0)|\mathbf{k}\rangle \tag{6.164}$$

where $E_0 = -JAS^2 z$ is the ground state energy. To calculate the x–y contribution, we recall $S^{\pm}|S, S^z\rangle = \sqrt{(S \pm S^z + 1)(S \mp S^z)}|S, S^z \pm 1\rangle$, and obtain

$$\frac{J}{2} \sum_{\mathbf{i},\mathbf{j}} (S_{\mathbf{i}}^+ S_{\mathbf{j}}^- + S_{\mathbf{i}}^- S_{\mathbf{j}}^+)|\mathbf{k}\rangle = JSz\gamma(\mathbf{k})|\mathbf{k}\rangle \tag{6.165}$$

where $\gamma(\mathbf{k})$ was introduced in (6.27). Putting the above results together, the spin wave energy is

$$\omega_{\mathbf{k}} = |J|zS(A - \gamma_{\mathbf{k}}) + H \,. \tag{6.166}$$

With $H = 0$ we recover (6.38). The result (6.31) for the isotropic model belongs to $A = 1$.

The spin wave continuum is separated by a gap $|J|zS(A - 1) + H$ from the ground state. The external field and the uniaxial anisotropy act in similar ways in producing the gap.

Now we discuss the case $H = 0$. $\omega(\mathbf{k}) \geq 0$ if $A \geq 1$. This is how it should be because the condition of the stability of the ground state requires that all excitation energies be positive. Let us, however, observe that $\omega(\mathbf{k} \to 0) < 0$ for $A < 1$, indicating that for a model with easy-plane anisotropy[49], the states with maximum polarization in the z-direction are no longer ground states. In fact, the ground state is expected to lie in the subspace $\mathcal{S}^z = 0$.

Problem 6.2. The total number of states in the $S^z_{\rm tot} = 0$ subspace is $C^L_{L/2}$. Herein included are the singlets, the $S^z_{\rm tot} = 0$ components of the triplets, and the $S^z_{\rm tot} = 0$ components of all the $S_{\rm tot} > 1$ levels. Similarly, all the levels except the singlets have a component in $S^z_{\rm tot} = 1$ subspace. The number of singlets is

$$\mathcal{N}_{S_{\rm tot}=0} = C^L_{L/2} - C^L_{L/2-1} = C^L_{L/2}\frac{2}{L+2}\,. \tag{6.167}$$

Repeating the argument, the number of triplet levels is

$$\mathcal{N}_{S_{\rm tot}=1} = C^L_{L/2-1} - C^L_{L/2-2} = C^L_{L/2}\frac{6L}{(L+2)(L+4)}\,. \tag{6.168}$$

For large L, there are about three times (nine times) as many triplet levels (states) than singlet levels (states).

Problem 6.3. It is amusing that the 16-dimensional eigenvalue problem can be fully diagonalized with the help of symmetry arguments (i.e., any two eigenstates differ in at least one quantum number).

The total spin $\mathcal{S}$ and its z-component $\mathcal{S}^z$ are good quantum numbers. Eigenstates can be sought in the form $|\mathcal{S}, \mathcal{S}^z, k\rangle$, where $k = 0, \pm\pi/2, \pi$ is the Bloch wave number. For each multiplet, it is enough to consider the $\mathcal{S}^z = \mathcal{S}$ state. The eigenenergy is denoted by $\mathcal{E}_{\mathcal{S},k}$.

Since there is only one $\mathcal{S}^z = 2$ state, there is only one quintiplet. In obvious notation,

$$|2,2\rangle = |\uparrow\uparrow\uparrow\uparrow\rangle \tag{6.169}$$

and

$$\mathcal{H}_4|2,2\rangle = J|2,2\rangle. \tag{6.170}$$

The $\mathcal{E}_2 = J$ level is fivefold degenerate.

Out of the four $\mathcal{S}^z = 1$ states, one corresponds to turning down $|2,2\rangle$: it is $|2,1\rangle \sim \sum_j S^-_j|2,2\rangle$, and it has the wavenumber $k = 0$. The $k \neq 0$ states belong to $\mathcal{S} = 1$ triplets. For $k = \pi$ we find

$$\mathcal{H}_4|1,1,\pi\rangle \quad = \quad \mathcal{H}_4 \cdot \frac{1}{2}(|\downarrow\uparrow\uparrow\uparrow\rangle - |\uparrow\downarrow\uparrow\uparrow\rangle + |\uparrow\uparrow\downarrow\uparrow\rangle - |\uparrow\uparrow\uparrow\downarrow\rangle)$$

---

[49]Taking $A = 0$ gives the ferromagnetic x–y model. In spite of its apparent simplicity, it presents a more difficult problem than the isotropic Heisenberg model; we cannot even give its exact ground state in dimensions $D > 1$. One might have naively assumed that a state maximally polarized in the x-direction would do, but it is not an eigenstate of the model because S^x and S^y do not commute.

$$= -J|1,1,\pi\rangle. \tag{6.171}$$

The $\mathcal{E}_{1,\pi} = -J$ level is threefold degenerate. Since time reversal symmetry guarantees that the states $|S, S^z, k\rangle$ and $|S, -S^z, -k\rangle$ are degenerate, we conclude that the level $\mathcal{E}_{1,\pm\pi/2} = 0$ is sixfold degenerate.

The remaining two states must be singlets. Though it would be quite straightforward to deal with these using the same representation as before, we prefer to use Hulthén's technique. Re-expressing the Hamiltonian as

$$\mathcal{H}_4 = -\frac{J}{2}(Q_{12} + Q_{23} + Q_{34} + Q_{41}) + J \tag{6.172}$$

and using the rules (6.125) and (6.126), it is immediately found that the ground state is the RVB state

$$|0,0,0\rangle = \frac{1}{\sqrt{3}}([12][34] + [23][41]) = \frac{1}{\sqrt{3}} \tag{6.173}$$

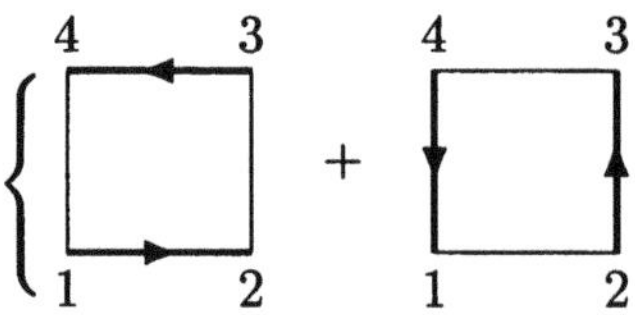

with energy $\mathcal{E}_{00} = -2J$, while the excited singlet is[50]

$$|0,0,\pi\rangle = [12][34] - [23][41] \tag{6.174}$$

with energy $\mathcal{E}_{0\pi} = 0$. We found the same energy as $\mathcal{E}_{1,\pm\pi/2}$. The degeneracy between a singlet and the triplets is not demanded by symmetry: it is an *accidental* feature of the model.

The ground state is not ordered in the Néel sense, since $\langle 0,0,0|S_j^z|0,0,0\rangle = 0$ for $j = 1, \ldots, 4$. If we wish, we can say that we see the beginnings of AF order in the correlations: $\langle 0,0,0|S_1^z S_2^z|0,0,0\rangle = -1/6$, and $\langle 0,0,0|S_1^z S_3^z|0,0,0\rangle = 1/12$. However, correlations must be isotropic in a singlet state, and indeed one finds $\langle 0,0,0|S_1^x S_2^x|0,0,0\rangle = -1/6$, etc.

The Hamiltonian $\mathcal{H}_4'$ has the same symmetries as $\mathcal{H}_4$ and therefore the only consequence of switching on the next-nearest-neighbour interaction is lifting the accidental degeneracy.

Problem 6.4. We consider a chain with an even number of sites, and periodic boundary conditions. Using the operators Q introduced in (6.124), the Hamiltonian can be rewritten as

$$\mathcal{H} = -\frac{J_1}{2}\sum_j Q_{j,j+1} - \frac{J_2}{2}\sum_j Q_{j,j+2} + \frac{3J}{8}L \tag{6.175}$$

[50]The funny-looking normalization factors arise from the non-orthogonality of the Rumer–Hulthén basis.

where $J_1 > 0$, $J_2 > 0$.

Let us start with the singlet bond configuration

$$\Phi_1 = [12][34][56]\ldots[N, N-1] , \tag{6.176}$$

and check whether acting on this with $\mathcal{H}$ causes the system to resonate to other configurations. Let us follow what happens to the bond [34]. The relevant terms of $\mathcal{H}$ are collected in

$$\mathcal{H}([34]) = -\frac{J_1}{2}(Q_{23} + Q_{34} + Q_{45}) - \frac{J_2}{2}(Q_{13} + Q_{24} + Q_{35} + Q_{46}) . \tag{6.177}$$

Applying (6.125) and (6.126), we find

$$\begin{aligned} \mathcal{H}([34])\Phi_1 &= -J_1\Phi_1 - \frac{J_1}{2}([23][41][56]\ldots + [12][45][63]\ldots) \\ &\quad -\frac{J_2}{2}(-2[13][42][56]\ldots - 2[12][35][64]\ldots) . \end{aligned} \tag{6.178}$$

The ... stand for the undisturbed part of the chain. Apparently, there is resonance: the nearest neighbour coupling induces configurations with "long bonds" like [41]. However, the configurations in the second line of the above equation contain crossed bonds and, using (6.123), they can be reexpressed in terms of the non-crossing configurations to yield

$$\mathcal{H}([34])\Phi_1 \rightsquigarrow -(J_1 + J_2)\Phi_1 - \left(\frac{J_1}{2} - J_2\right)([23][41][56]\ldots + [12][45][63]\ldots) . \tag{6.179}$$

It follows that setting $J_2 = J_1/2$ cancels the resonance completely, and Φ_1 becomes an eigenstate. It can be proved that it is the ground state. The state Φ_1 is called the *Majumdar–Ghosh state* [258], or the *dimer state.* Let us observe that Φ_1 breaks translational invariance but (being an exact singlet) it does not break spin-rotational invariance. The ground state is doubly degenerate: $\Phi_2 = [23][45]\ldots[N, 1]$ is also a ground state.

The limit of classical vectors ($S \to \infty$): Like the t_1–t_2 chain (see Fig. 4.5), the J_1–J_2 Heisenberg chain could be redrawn as a railroad trestle lattice. The elementary unit would be a triangle with bonds J_1, J_1, and J_2, wherefrom it transpires that the system is frustrated in the same sense as other lattices built of triangular plaquettes are. Applying the same argument as in Sec 6.3.4, we expect a non-collinear ground state. Let us assume that the ordering is given by the *spiral wave* $\mathbf{S}_j = (S\cos(j\alpha), S\sin(j\alpha), 0)$. Minimizing the energy density

$$E(\alpha) = J_1 S^2 \cos\alpha + J_2 S^2 \cos 2\alpha$$

with respect to α, we obtain

$$\alpha = \begin{cases} \pi & \text{if } J_2/J_1 < 1/4 \\ \arccos(-J_1/4J_2) & \text{if } J_2/J_1 \geq 1/4 \end{cases}$$

The classical ground state is a spiral wave for $J_2 > J_1/4$. As J_2/J_1 increases, α gradually decreases from π towards $\pi/2$.

The Néel-to-spiral transition predicted for the classical limit is destabilized by quantum fluctuations. For $S = 1/2$, the quantum transition is between a spin liquid and a dimer phase (see Sec. 6.3.2). Let us observe that studying the classical limit does not give any hint of dimerization; the dimer order described in Sec. 6.3.2 is a purely quantum phenomenon.

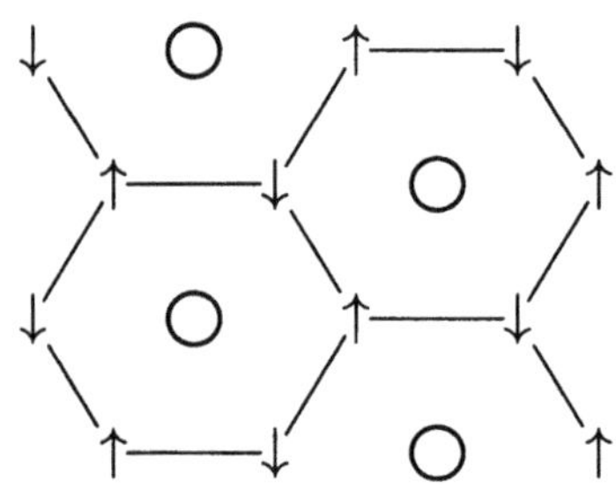

Figure 6.9: Construction of a class of Ising ground states. Two thirds of the sites form a honeycomb lattice with perfect AF order. The spin states of the remaining one third of the sites (circles) are free to choose.

Problem 6.5. A low-energy bond connects a site with $S^z = S$ to a site with $S^z = -S$; anything else is bad. In Sec. 6.3.4, we convinced ourselves that every triangle must contain at least one unsatisfied bond. A subclass of ground states with partial long range order is obtained by imposing AF order on a sublattice holding $2L/3$ sites in the form of a honeycomb lattice (Fig. 6.9). $L/3$ sites are at the hexagon centres. Each of these has three neighbours with $S^z = S$, and three with $S^z = -S$, thus whatever the spin at the centre site is, the energy remains the same. The centre spin can be chosen in $(2S+1)$ ways, thus this subclass contains $(2S+1)^{L/3}$ ground states, an exponentially large number. Naturally, the total number of the ground states is larger, and most of them show no recognizable order.

For $S = 1/2$, we constructed $2^{L/3}$ ground states, thus $(\ln 2/3)k_B = 0.231k_B$ is a lower bound for the zero point entropy density. The exact value is $0.3383k_B$ [438].

Chapter 7

Itinerant Electron Magnetism

7.1 Introduction

The Hubbard model was invented with the intention of explaining magnetism, foremost the ferromagnetism of Fe, Co, and Ni. Later on, the apparent initial successes gave rise to the bold expectation that maybe most of the major correlation phenomena are represented in the phase diagram of the simplest one-band Hubbard model. Such hopes were based on early Hartree–Fock phase diagrams (Fig. 4.11), and on the understanding that richer phase diagrams could be obtained without any difficulty, if only one postulated the existence of more complicated ordering patterns. Optimistically, one might have hoped to find the Mott transition, ferro- and antiferromagnetism, incommensurate magnetic structures, charge and spin gaps, heavy fermions, and maybe even an electronic mechanism of high-temperature superconductivity, all explained within a unified framework. Naturally we are aware that, taken in itself, the Hubbard model is far from describing any real material. However, first we wish for a qualitative understanding of the various kinds of collective behaviour: we want to identify the "driving forces" of the ordering transitions. For such a purpose, the understanding obtained from studying the Hubbard model may be sufficient. For instance, we are basically satisfied with what we learned about the (antiferromagnetic) kinetic exchange in Sec. 5.1, even though we know that it would be very difficult to calculate the superexchange coupling strength for a real compound.

Subsequent work on the Hubbard model and its cousins made us more cautious. Now we would say that there are a few important phenomena which we can understand qualitatively by studying the one-band Hubbard model; there are some which demand two- or multi-band models or an extended Hubbard model; and finally, there are phenomena for which the canonical model has not been found yet.

Foremost, let us reassert that the insight gained from deriving the antiferromagnetic Heisenberg coupling as the leading interaction term in the large-U effective Hamiltonian (5.36) is correct: the Hubbard model gives a solid basis for understanding antiferromagnetism, especially the antiferromagnetism of Mott insulators. Quite probably, we can also describe metallic states with an incommensurate magnetic order which arises by a gradual distortion from a "nearby" two-sublattice antiferromagnetic state. Thus, choosing the right parameter regime, we can in principle understand insulating antiferromagnets like NiO, and perhaps also a spin density wave metal like Cr. These phases are easily found in mean field calculations, and their existence and robustness is corroborated by a variety of more sophisticated techniques.

Then what about the ferromagnetism of iron? The $T = 0$ mean field phase diagram of the cubic lattice Hubbard model (schematically reproduced in Fig. 4.11) shows an extended ferromagnetic domain around quarter-filling, starting from intermediate coupling strengths; it looks like a good imitation of itinerant ferromagnetism, and it does not look any more fragile than the antiferromagnetic phase. Of course, one has to keep in mind that a mean field solution is obtained by neglecting fluctuations and that in an exact solution, the phase boundaries would be shifted from their Hartree–Fock positions. Still, conventional wisdom would have held that for a three-dimensional system, such shifts would not alter the overall appearance of the phase diagram and, in particular, if there is a ferromagnetic phase in the mean field approximation, then you would expect that it also exists in the exact solution. Well, it is still possible that the cubic lattice Hubbard model is ferromagnetic somewhere, but even if it is, then probably only at much larger values of

U/t, and in a narrower range of band filling, than it had been thought. It does not look like a recipe for the exceptionally robust kind of order we observe in iron. Thus we have to conclude that the seemingly simplest kind of magnetic order: ferromagnetism, is not a generic Hubbard phenomenon.

Seeing a difficulty here may look perplexing because it is known that advanced electronic structure calculations (density functional theories) provide a detailed, and often quite accurate, description of the ferromagnetic transition metals [15, 141, 145]. What more should we wish for? The point is that we would like to understand magnetism as the correlated behaviour of interacting electrons and (as we are going to discuss) it is very difficult to "translate" a density functional calculation into a simple-looking lattice fermion model. Ideally, we would like a model which explains ferromagnetism at about the same level of ease, clarity, and also confidence, as the simple Hubbard model does for antiferromagnetism. The exploration of several different routes is in progress, and apparently, there are a number of ways to get ferromagnetism. We do not know yet whether they are all realized in different systems, or we will eventually settle for a single relevant mechanism.

First, we will describe the conventional mean-field (Stoner-like) criteria of magnetic ordering. This we can do with a clear conscience even though we have just stated that the criteria are misleading when applied to the case of ferromagnetism in the simplest Hubbard model[1]. On the one hand, mean field theory can be used with more confidence for the well-understood case of antiferromagnetism: we will treat the spin density wave state in detail (Sec. 7.6). On the other hand, Stoner theory gives a convenient framework for systematizing the knowledge about band magnets. We close the chapter with a brief review of the magnetic properties of transition metals (mainly the iron group elements, and chromium), as well as of a number of nearly ferromagnetic, and weakly ferromagnetic, systems (Sec. 7.7). Further discussion of the difficult problem of ferromagnetism in lattice fermion models is relegated

[1] Single-orbital intermediate-coupling Hubbard models *can* show itinerant ferromagnetism with a Curie temperature $\sim$ 1000K for a bandwidth $\sim$ 4eV (i.e., in the parameter range relevant for Fe), but they have to be "fine-tuned" for the purpose. See the Discussion starting on p. 458.

to Ch. 8. This will necessitate a brief excursion into the realm of two- and multi-band models.

7.2 Magnetic Order

Magnetic order can be understood as a certain pattern of the local imbalance in the occupation of ↑-spin, and ↓-spin states. The z-component of the spin at site $\mathbf{j}$ is

$$\langle S^z_{\mathbf{j}} \rangle = \frac{1}{2}(\langle \hat{n}_{\mathbf{j}\uparrow} \rangle - \langle \hat{n}_{\mathbf{j}\downarrow} \rangle) . \tag{7.1}$$

Here $\langle \ \rangle$ can denote either the ground state expectation value, or the thermal expectation value at temperature T. If $\langle S^z_{\mathbf{j}} \rangle \neq 0$, we call the site $\mathbf{j}$ magnetic, since it has a non-vanishing magnetization $\langle M^z \rangle = g\mu_B \langle S^z \rangle$ in the z-direction[2]. In a magnetically ordered state either all the sites, or a certain subset of the sites, have non-vanishing magnetization.

Finding $\langle S^z_{\mathbf{j}} \rangle \neq 0$ in the ground state means that the ground state is symmetry-breaking. Let us recall that the Hubbard Hamiltonian has spin-rotational invariance. Thus all directions should be equivalent but we find that the spin of the site $\mathbf{j}$ is pointing in some specific direction. Generally, the phenomenon of magnetic ordering has to be understood as a transition to a broken-symmetry state[3]. Spin-rotational symmetry is only one of the symmetries that can get broken: certain kinds of magnetic order break translational invariance as well. Furthermore, all kinds of magnetic order break time-reversal invariance: $\mathcal{T}_t \mathbf{S} = -\mathbf{S}$ thus in a time-reversal invariant state $\langle \mathbf{S} \rangle$ should vanish.

The simplest kind of magnetic order is *ferromagnetism*: all sites are polarized in the same direction, and all magnetic moments are equal:

$$\begin{aligned} \langle \hat{n}_{\mathbf{j}\sigma} \rangle &= n_\sigma \quad \text{with} \quad n_\uparrow \neq n_\downarrow \\ \langle S^z_{\mathbf{j}} \rangle &= \frac{1}{2}(n_\uparrow - n_\downarrow) . \end{aligned} \tag{7.2}$$

[2]It is purely arbitrary that we picked out the z-direction. In general, the magnetic moment $\mathbf{M} = (M^x, M^y, M^z)$ can point in any direction.

[3]A brief general discussion of the phenomenon of spontaneous symmetry breaking will be given in the following *Digression*.

Here, full translational invariance is kept but spin-rotational invariance is broken; so is time-reversal invariance.

A widespread kind of order is *two-sublattice antiferromagnetism.* Let us assume that the lattice is bipartite, i.e., it can be divided into sublattices A and B in such a manner that all the nearest neighbours of an A-site are on sublattice B, and vice versa. We may think of the square lattice which can be subdivided in a chessboard-like fashion (Fig. 7.1, left). This is an example of symmetrical bipartite lattices: it is clear that the two sublattices play equivalent roles, and have the same number of lattice sites: $|A| = |B|$. Notice that the latter feature does not follow from the bipartite nature: the CuO_2 lattice (Fig. 3.1) is an example for non-symmetrical bipartite lattices (in this particular case, with $|B|/|A| = 2$). As we are going to see in our discussion of Lieb's ferrimagnetism, asymmetric bipartite lattices favour a special kind of magnetic order.

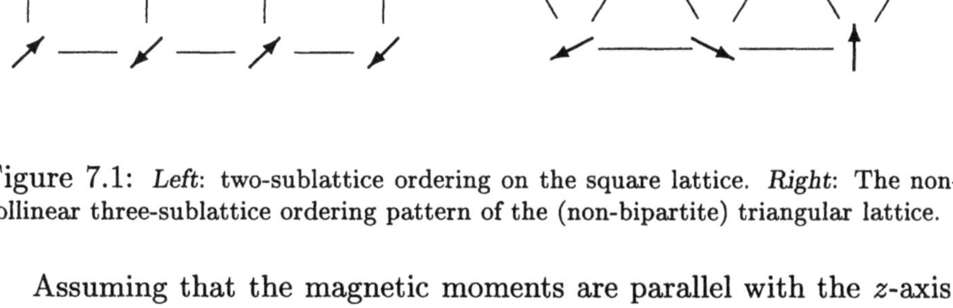

Figure 7.1: *Left*: two-sublattice ordering on the square lattice. *Right*: The non-collinear three-sublattice ordering pattern of the (non-bipartite) triangular lattice.

Assuming that the magnetic moments are parallel with the z-axis, the antiferromagnetic order of a symmetrical bipartite lattice can be described by

$$\langle \hat{n}_{\mathbf{j}\uparrow} \rangle = \begin{cases} \dfrac{n}{2} + m & \text{if } \mathbf{j} \in A \\ \dfrac{n}{2} - m & \text{if } \mathbf{j} \in B \end{cases} \qquad \langle \hat{n}_{\mathbf{j}\downarrow} \rangle = \begin{cases} \dfrac{n}{2} - m & \text{if } \mathbf{j} \in A \\ \dfrac{n}{2} + m & \text{if } \mathbf{j} \in B\,. \end{cases} \tag{7.3}$$

The antiferromagnetic order breaks, in addition to the spin-rotational and the time-reversal invariance, also the (discrete) translational invariance of the Hubbard model: in the Hamiltonian, all sites are equivalent, while in the antiferromagnetic state, a lattice site on sublattice A is inequivalent to a site on B. However, the two-sublattice state still shows translational invariance under a restricted set of translations, namely under those which take an A site into an A site, and a B site into a B site. These are the translations of the magnetic superlattice.

For a hypercubic lattice, the above conditions can be concisely rewritten as

$$\langle \hat{n}_{\mathbf{j}\sigma} \rangle = \frac{n}{2} + \eta(\sigma) m e^{i\mathbf{Q}\mathbf{j}} \tag{7.4}$$

where $\mathbf{Q} = (\pi, \pi, ..., \pi)$ is the spanning vector, and $\eta(\sigma) = \pm 1$ for $\sigma = \uparrow$, and $\downarrow$, respectively. The sublattice spin polarizations are

$$\langle S_{\mathbf{j}}^{z} \rangle = m e^{i\mathbf{Q}\mathbf{j}} = \begin{cases} m & \text{if } \mathbf{j} \in A \\ -m & \text{if } \mathbf{j} \in B. \end{cases} \tag{7.5}$$

m can be identified as the *order parameter*: it measures the sublattice magnetization.

Antiferromagnets with a closed-packed lattice (for instance, the triangular lattice in two dimensions, or the fcc lattice in three dimensions) usually have a non-collinear order with more than two sublattices (see Fig. 7.1, right).

A more complicated sort of non-collinear order is found in the *circularly polarized spin density wave*

$$\langle S_{\mathbf{j}}^{x} \rangle + i \langle S_{\mathbf{j}}^{y} \rangle = \mathcal{S} e^{i\mathbf{q}\mathbf{j}}. \tag{7.6}$$

In the general case when $\mathbf{q}$ is *incommensurate* with the reciprocal vectors of the lattice, there is no lattice translation which would map the magnetization pattern onto itself, and thus translational invariance is completely lost. In practice, it is difficult to distinguish this situation from a long-period superlattice.

Linearly polarized spin density waves are modulated collinear structures

$$\langle S_{\mathbf{j}}^{z} \rangle = \mathcal{S} \cos \mathbf{q} \cdot \mathbf{j} \,. \tag{7.7}$$

Note that both (7.2) and (7.3) can be regarded as special cases of either (7.6), or (7.7): for ferromagnetism choose $\mathbf{q} = 0$, while for antiferromagnetism, $\mathbf{q} = \mathbf{Q}$, where $\mathbf{Q}$ is the spanning vector $(\pi, \ldots, \pi)$.

7.2.1 Digression: Symmetry Breaking

Finding magnetic sites with $\langle \mathbf{S_j} \rangle \neq 0$ means that we have to do with a symmetry-breaking ground state of the Hubbard Hamiltonian. This should be understood as follows:

For finite quantum-mechanical systems, it is expected that the ground state, as well as finite-T equilibrium states, show the full symmetry of the Hamiltonian. Consequently, for a spin-rotationally invariant Hamiltonian, such as the Hubbard model, expectation values like (7.1) should vanish. In finite systems, we cannot really speak about magnetic (or any other kind of) order[4]. This may sound as if the understanding of spontaneous ordering were denied to us because strictly speaking, all systems are finite. However, macroscopic systems are so large that it can be argued that their behaviour is essentially the same as it would be for an infinitely large system. Technically speaking, one must take the *thermodynamic limit*: say, for a system with a given density, we let the particle number $N \to \infty$, and the volume $V \to \infty$, with N/V kept fixed. For lattice models, as in our present discussion, the filling has to be kept fixed: we let $N \to \infty$, and the number of lattice sites $L \to \infty$, with $n = N/L$ fixed. The important consequence of this step is that we gain a principal understanding of ordering transitions: *in the thermodynamic limit, symmetry breaking becomes possible.* The way it comes about is this: For a finite system, the free energy is an analytic function of the temperature T for all finite T, so the behaviour of the system must be perfectly continuous, there can be no phase transition. In contrast, in the thermodynamic limit the free energy can show non-analytic behaviour at some finite transition temperature $T_{\rm tr}$. This has the effect that the behaviour at $T < T_{\rm tr}$ cannot be understood by smooth continuation from $T > T_{\rm tr}$; the ordered state necessitates a new method of description. Approximate theories (such as the mean field theories we are going to discuss) are usually not sensitive to the difference between finite and infinite systems, and may predict magnetic long-range order even for small systems[5]. We understand that such a result must be spurious, and that the

[4]More precisely: the ground state belongs to one of the irreducible representations of the symmetry group of the Hamiltonian. If the ground state is non-degenerate, we expect this to be the identity representation, so the ground state is fully symmetrical; this is the case relevant for systems with a tendency to antiferromagnetic ordering. Systems inclined to become ferromagnetic can have a high-spin ground state, i.e., several degenerate ground states which can be rotated into each other. Then you might say that any of the individual ground states breaks spin-rotational invariance even for a finite system; recall the case of the O_2 molecule. However, the thermal equilibrium state of the system (including $T \to 0$ where we average over all degenerate ground states) remains fully symmetrical.

[5]We will see an example of this in Problem 7.6 (see p. 385 and p. 413).

mean-field theory of a phase transition is supposed to give an approximate description of the behaviour of macroscopically large systems.

For a macroscopic system, the fully symmetrical state is realized only at sufficiently high temperatures, i.e., above the highest ordering temperature. Upon cooling, interactions cause the system to break one, or several symmetries of the Hamiltonian[6]. *"Breaking a symmetry" means that there are symmetry operations under which the Hamiltonian is invariant but the state of the system is not.* The symmetry-breaking nature of the state is signalled by non-zero results for certain expectation values which should exactly vanish if the state were symmetrical. In our present case, the symmetry breaking is associated with the appearance of a non-vanishing magnetization $\langle \mathbf{S_j} \rangle$ in a system whose Hamilton operator is spin-rotationally invariant. Rotating all the spins by the same angle, the Hamiltonian is taken into itself, but the vector $\langle \mathbf{S_j} \rangle$ turns into a different direction, i.e., the state has been changed.

It may be mentioned that the phenomenon we are discussing is often called *spontaneous* symmetry breaking. It is "spontaneous" because we did not put into the Hamiltonian any symmetry-breaking term which could have reduced the symmetry of the energy expression to that of the ordered state.

There are cases in which even a macroscopic system retains full symmetry at all temperatures, or at least at all non-zero temperatures. We mention two basic possibilities.

First, it may seem obvious that a non-interacting system does not order, since it looks like an assembly of independent finite systems. However, if we think a bit more carefully, we realize that "non-interacting" is not quite the same as "independent": quantum statistics also induces correlations. It turns out that a free Fermi system is indeed non-ordering but a three-dimensional free Bose gas condenses at a finite transition temperature.

Second, in low dimensions, thermal and/or quantum fluctuations can prevent long-range order even in strongly interacting systems. The breaking of continuous symmetries at finite temperatures is ruled out for one- and two-dimensional systems by the powerful *Mermin–Wagner theorem*, assuming that the interaction falls off sufficiently rapidly with increasing interparticle distance [269]. The statement covers the isotropic spin models, as well as interacting electron models which do not include spin–orbit coupling or dipole–dipole interaction and thus cannot describe magnetic anisotropy. The theorem is certainly valid for the Hubbard model which has a zero-range interaction, thus we can be sure that one- and two-dimensional Hubbard models are not magnetically ordered at any finite temperature[7]. The same holds for the extended Hubbard

[6]See [22] for a thorough discussion of the concept of symmetry breaking.

[7]We have to be very cautious in interpreting this result: the non-existence of magnetic order at finite T does not necessarily mean that there cannot be a magnetic

model, and more general lattice models with arbitrary finite-range interactions, as long as they represent the Coulomb force in non-relativistic approximation[8].

Basically, the Mermin–Wagner theorem follows from the fact that in $d = 1$ and $d = 2$, a diverging number of infinitesimally low-lying excitations (spin waves) is created at any finite T, and therefore the assumption of a non-vanishing order parameter is not self-consistent (an elementary argument is presented in Ch. 6). The proof does not apply at $T = 0$, thus the ground state may be ordered. Indeed, two-dimensional ferromagnetism (or antiferromagnetism) is perfectly possible at $T = 0$. Quantum fluctuations oppose, but do not prevent, the arising of two-dimensional antiferromagnetism.

In contrast, for one-dimensional systems quantum fluctuations tend to become so strong that they prevent even ground state ordering. It is known that the ground state of the one-dimensional Hubbard model (with nearest-neighbour hopping) is a non-magnetic singlet at any band filling and U/t. This includes the large-U limit at half-filling, where the model becomes equivalent to the (isotropic) antiferromagnetic $S = 1/2$ Heisenberg chain, thus that cannot order, either (see Ch. 6). However, we emphasize that this no-go theorem holds only if the suggested order parameter is truly fluctuating, i.e., it does not commute with the Hamiltonian. The order parameter for ferromagnetism (the total spin) commutes with the Hubbard (or Heisenberg) Hamiltonian, and thus there is nothing to prevent a one-dimensional ferromagnet from becoming ordered at $T = 0$. To quote the simplest case, the fully polarized state is an eigenstate of the Heisenberg model, and it is the ground state if the coupling is ferromagnetic.

7.3 Mean Field Theories

Since the Hubbard term $\mathcal{H}_U$ must be responsible for the ordering[9], we wish to recast it in such a form that the arising of a non-vanishing $\langle S_{\mathrm{j}}^z \rangle$ appears natural. Adding and subtracting the expectation value, we find

$$\hat{n}_{\mathrm{j}\uparrow}\hat{n}_{\mathrm{j}\downarrow} = (\hat{n}_{\mathrm{j}\uparrow} - \langle \hat{n}_{\mathrm{j}\uparrow} \rangle + \langle \hat{n}_{\mathrm{j}\uparrow} \rangle)(\hat{n}_{\mathrm{j}\downarrow} - \langle \hat{n}_{\mathrm{j}\downarrow} \rangle + \langle \hat{n}_{\mathrm{j}\downarrow} \rangle)$$

phase transition at finite temperature! We may envisage a Kosterlitz–Thouless-type transition to a phase with an algebraic fall-off of correlations.

[8]We mention that it is possible to generalize the Hubbard model to include the spin–orbit coupling [454]. Such an explicitely non-isotropic model may describe a system with uniaxial anisotropy, and then – like the Ising model – it can order at finite T even in two dimensions.

[9]Assuming, of course, that the bandwidth is non-vanishing. We emphasized in Sec. 4.4 that magnetism arises from an interplay of kinetic and interaction energies.

$$
\begin{aligned}
&= (\hat{n}_{j\uparrow} - \langle \hat{n}_{j\uparrow} \rangle)\langle \hat{n}_{j\downarrow} \rangle + (\hat{n}_{j\downarrow} - \langle \hat{n}_{j\downarrow} \rangle)\langle \hat{n}_{j\uparrow} \rangle + \langle \hat{n}_{j\uparrow} \rangle \langle \hat{n}_{j\downarrow} \rangle \\
&\quad + (\hat{n}_{j\uparrow} - \langle \hat{n}_{j\uparrow} \rangle)(\hat{n}_{j\downarrow} - \langle \hat{n}_{j\downarrow} \rangle).
\end{aligned} \tag{7.8}
$$

The basic idea of the mean field approximation (MFA) is that the physical quantities show only minor fluctuations about their mean value. For instance, we will regard $\hat{n}_\uparrow - \langle \hat{n}_\uparrow \rangle$ as an inherently small quantity. Accordingly, the term in the last line of (7.8), being the product of two small quantities, is much smaller than the other terms, and can be safely neglected. In the MFA, the interaction term acts like a site- and spin-dependent external field[10]

$$
\hat{n}_{j\uparrow}\hat{n}_{j\downarrow} \rightsquigarrow \hat{n}_{j\uparrow}\langle \hat{n}_{j\downarrow} \rangle + \hat{n}_{j\downarrow}\langle \hat{n}_{j\uparrow} \rangle - \langle \hat{n}_{j\uparrow} \rangle \langle \hat{n}_{j\downarrow} \rangle \,. \tag{7.9}
$$

Let us make it apparent that this form allows both charge and spin ordering, by introducing the local charge $\hat{n}_j = \hat{n}_{j\uparrow} + \hat{n}_{j\downarrow}$, whereupon

$$
\hat{n}_{j\uparrow}\hat{n}_{j\downarrow} \rightsquigarrow \frac{1}{2}\left[\hat{n}_j\langle \hat{n}_j \rangle - 4S_j^z\langle S_j^z \rangle - \left(\frac{\langle \hat{n}_j \rangle^2}{2} - 2\langle S_j^z \rangle^2\right)\right] . \tag{7.10}
$$

We have as yet considered only expectation values which are diagonal in the occupation number representation, i.e., we used the Hartree approximation. Its insufficiency becomes clear if we remember that the Hubbard model is isotropic, and therefore spontaneous spin polarization should set in as readily in the x-direction as in the z-direction. This is not guaranteed yet. Let us, however, notice that decoupling the Hubbard term in all possible ways into the product of operator pairs with non-vanishing expectation values, we find also the Fock term

$$
\begin{aligned}
c_{j\uparrow}^\dagger c_{j\uparrow} c_{j\downarrow}^\dagger c_{j\downarrow} &= -c_{j\uparrow}^\dagger c_{j\downarrow}^\dagger c_{j\uparrow} c_{j\downarrow} \rightsquigarrow -c_{j\uparrow}^\dagger c_{j\downarrow}\langle c_{j\downarrow}^\dagger c_{j\uparrow} \rangle - c_{j\downarrow}^\dagger c_{j\uparrow}\langle c_{j\uparrow}^\dagger c_{j\downarrow} \rangle \\
&= -S_j^+\langle S_j^- \rangle - S_j^-\langle S_j^+ \rangle \,.
\end{aligned} \tag{7.11}
$$

The complete[11] Hartree–Fock approximation to the Hubbard term reads

$$
U\sum_j \hat{n}_{j\uparrow}\hat{n}_{j\downarrow} \;\rightsquigarrow\; \frac{U}{2}\sum_j \left\{\hat{n}_j(n + \langle \delta\hat{n}_j \rangle) - 4\vec{S}_j\langle \vec{S}_j \rangle\right\}
$$

[10] In the high-temperature, symmetrical phase $\langle \hat{n}_{j\sigma} \rangle$ is $\mathbf{j}$- and σ-independent: $\langle \hat{n}_{j\sigma} \rangle = n/2$. This solution is obviously self-consistent. We try to find out if there are other self-consistent solutions which give lower (free) energy.

[11] We do not define the decoupling into pair fields as part of the Hartree–Fock approximation. See, however, a later remark accompanying (7.15).

$$-\frac{U}{4}\sum_{\mathbf{j}}\left\{(n+\langle\delta\hat{n}_{\mathbf{j}}\rangle)^2-4\langle\vec{S}_{\mathbf{j}}\rangle^2\right\} \qquad (7.12)$$

where $\langle\delta\hat{n}_{\mathbf{j}}\rangle$ is the deviation of the local charge from the lattice average of the electron number $n = N/L$. Note that only the terms in the first line have operator character, the rest is just an additive constant. We can omit the constant when solving the eigenvalue problem of the mean field Hamiltonian. Its inclusion is, however, necessary for obtaining the correct total energy.

We will usually seek magnetically ordered states with a uniform distribution of charge. In other words, we exclude charge density waves, and permit only spin density waves. Setting $\langle\delta\hat{n}_{\mathbf{j}}\rangle = 0$, we find the simplified mean field scheme

$$U\sum_{\mathbf{j}}\hat{n}_{\mathbf{j}\uparrow}\hat{n}_{\mathbf{j}\downarrow} \rightsquigarrow -2U\sum_{\mathbf{j}}\vec{S}_{\mathbf{j}}\langle\vec{S}_{\mathbf{j}}\rangle + U\sum_{\mathbf{j}}\langle\vec{S}_{\mathbf{j}}\rangle^2 + L\frac{Un^2}{4}\,. \qquad (7.13)$$

Actually, there is nothing to prevent us from choosing the local quantization axis in the direction of the local magnetization, and then we recover the still simpler form (7.10) (or (7.9)). One may say that for the spin-ordering decoupling of the Hubbard term, the Hartree–Fock approximation is equivalent to the Hartree approximation. Indeed, we will use the Hartree replacement (7.9) in our study of linearly polarized spin density waves. However, it would be cumbersome to describe helicoidal structures by changing the spin quantization axis from site to site; we rather prefer to fix the spin axis, and use the Hartree–Fock decoupling (7.13) (see Problem 7.8).

There is no general principle which would tell us whether the ordering transition is of first or second order. If the transition is second-order (continuous), we can seek the phase boundary by two different strategies. We can approach the critical point from within the ordered phase by requiring that the order parameter vanishes. Or alternatively, we can approach from the disordered side where the phase transition is signalled by the divergence of the corresponding susceptibility.

Let us notice that though the decoupling scheme (7.12) appears quite natural, it is not unambiguous that this is what we should do. Recasting the Hubbard term in **k**-space representation

$$\begin{aligned}\mathcal{H}_U &= \frac{U}{L^2}\sum_{\mathbf{k}_1}\sum_{\mathbf{k}_2}\sum_{\mathbf{k}_3}\sum_{\mathbf{k}_4}\sum_{\mathbf{j}} e^{i(\mathbf{k}_1-\mathbf{k}_2+\mathbf{k}_3-\mathbf{k}_4)\mathbf{j}} c^\dagger_{\mathbf{k}_1\uparrow}c_{\mathbf{k}_2\uparrow}c^\dagger_{\mathbf{k}_3\downarrow}c_{\mathbf{k}_4\downarrow} \\ &= \frac{U}{L}\sum_{\mathbf{k}}\sum_{\mathbf{q}}\sum_{\mathbf{q}'} c^\dagger_{\mathbf{k}+\mathbf{q}/2,\uparrow}c^\dagger_{\mathbf{k}-\mathbf{q}/2,\downarrow}c_{\mathbf{k}-\mathbf{q}'/2,\downarrow}c_{\mathbf{k}+\mathbf{q}'/2,\uparrow} \qquad (7.14)\end{aligned}$$

we might feel it just as natural to perform a decoupling which points towards a superconducting state

$$\mathcal{H}_U \rightsquigarrow \frac{U}{L}\sum_{\mathbf{k}}\sum_{\mathbf{q}}(\Delta_{\mathbf{k}}c^\dagger_{\mathbf{k}+\mathbf{q}/2,\uparrow}c^\dagger_{\mathbf{k}-\mathbf{q}/2,\downarrow} + \Delta^*_{\mathbf{k}}c_{\mathbf{k}-\mathbf{q}/2,\downarrow}c_{\mathbf{k}+\mathbf{q}/2,\uparrow}) \qquad (7.15)$$

where

$$\Delta^*_{\mathbf{k}} = \sum_{\mathbf{q}}\langle c^\dagger_{\mathbf{k}+\mathbf{q}/2,\uparrow}c^\dagger_{\mathbf{k}-\mathbf{q}/2,\downarrow}\rangle \qquad (7.16)$$

is a BCS-type order parameter. Which is better: (7.12) or (7.15)? We cannot know in advance, one has to do both calculations and check which gives a better ground state energy. There might be still other, more complicated decoupling schemes which we have not thought of. In contrast to the case of classical models like the Ising model, with itinerant many-fermion models it is not even obvious how their mean field theory should be formulated. Currently it is held that this problem has found its solution in the "essentially exact" theories of the infinite-dimensional Hubbard model and other lattice fermion models[12].

In this Chapter, we confine ourselves to simple-minded cases of magnetic ordering, and use the decoupling schemes (7.13), or (7.9).

7.4 Stoner Model

Let us find the criterion for the Hubbard model becoming ferromagnetic. We place the system in a uniform magnetic field H

$$\mathcal{H} = \sum_{\mathbf{k}\sigma}\epsilon_{\mathbf{k}}\hat{n}_{\mathbf{k}\sigma} + U\sum_{\mathbf{j}}\hat{n}_{\mathbf{j}\uparrow}\hat{n}_{\mathbf{j}\downarrow} - \frac{g\mu_{\mathrm{B}}H}{2}\sum_{\mathbf{j}}(\hat{n}_{\mathbf{j}\uparrow} - \hat{n}_{\mathbf{j}\downarrow}). \qquad (7.17)$$

[12] We will briefly discuss the results of this approach in Sec. 10.6.5. As it was to be expected, the solution is not simple: in principle, there are infinitely many effective fields.

As a response, the system develops a uniform spin polarization m

$$\langle \hat{n}_{j\uparrow} \rangle = \frac{n}{2} + m \qquad \langle \hat{n}_{j\downarrow} \rangle = \frac{n}{2} - m. \tag{7.18}$$

Using the mean field decoupling (7.9), the total energy density is

$$\mathcal{E}(m) = \int_B^{\mu_\uparrow} \epsilon \rho(\epsilon) d\epsilon + \int_B^{\mu_\downarrow} \epsilon \rho(\epsilon) d\epsilon + U\left(\frac{n^2}{4} - m^2\right) - g\mu_B H m \tag{7.19}$$

where B is the bottom of the band which is filled up to μ_σ for spin σ

$$\int_B^{\mu_\sigma} \rho(\epsilon) d\epsilon = n_\sigma. \tag{7.20}$$

It is useful to regard the energy change caused by the spin polarization

$$\Delta\mathcal{E}(m) = \mathcal{E}(m) - \mathcal{E}(0) = \int_{\mu_0}^{\mu_\uparrow} \epsilon \rho(\epsilon) d\epsilon - \int_{\mu_\downarrow}^{\mu_0} \epsilon \rho(\epsilon) d\epsilon - Um^2 - g\mu_B H m \tag{7.21}$$

where μ_0 is the $H = 0$ chemical potential. The change in the interaction energy ($-Um^2$) is negative, expressing the fact that polarizing the spins makes it less likely that two electrons meet at the same site.

Increasing U from 0, magnetization sets in if for some m, $\Delta\mathcal{E}(m)$ becomes negative. This $T = 0$ paramagnetic–ferromagnetic transition may be either second-order, or first-order, depending on the band structure. A first-order transition means that the non-magnetic state becomes unstable against a finite-m state; we derive the corresponding finite-m Stoner criterion (7.111) in the solution to Problem 7.2. Here we consider the case of a second-order transition where m is increasing continuously from 0. In the pre-transition paramagnetic state, m is induced by the external field. For weak H, we may replace $\rho(\epsilon) \approx \rho(\epsilon_F)$ in the second-order expansion of the energy

$$\Delta\mathcal{E}(m) = \frac{m^2}{\rho(\epsilon_F)} - Um^2 - g\mu_B H m. \tag{7.22}$$

Minimizing with respect to m leads to the result that *the susceptibility is enhanced by the electron–electron interaction*

$$\chi = \frac{m g \mu_B}{H} = \frac{(g\mu_B)^2 \rho(\epsilon_F)}{2} \cdot \frac{1}{1 - U\rho(\epsilon_F)} = \frac{\chi_{\text{Pauli}}}{1 - U\rho(\epsilon_F)}. \tag{7.23}$$

$(1 - U\rho(\epsilon_F))^{-1}$ is often called the exchange enhancement factor. As U approaches the critical value

$$U_{cr}^{F} = \frac{1}{\rho(\epsilon_F)} \tag{7.24}$$

from below, the susceptibility diverges, i.e., even an infinitesimal field is sufficient to cause a finite magnetization. Approaching from the $U < U_{cr}^{F}$ side, we can speak about a magnetic instability: the symmetrical ground state becomes unstable against the onset of ferromagnetic ordering. (7.24) is known as the *Stoner criterion*, while the mean field approach developed in this Section is called the Stoner model[13] of band ferromagnetism.

In the present treatment m is a continuous variable and it is clear that, depending on the parameters of the problem, the equilibrium magnetization per site $m\mu_B$ can be any fraction of μ_B. Historically, the picture of localized-moment magnetism (based on the Heisenberg model) came first, which would have predicted integer multiples of μ_B. It was a triumph of the Stoner model that it could offer an explanation for the fractional saturation moments of the iron group elements.

Let us return to the instability criterion. Roughly speaking, we found $1/\rho(\epsilon_F) \sim W$, the bandwidth. Stated in the simplest terms, ferromagnetism sets in when the interaction energy reaches the order of magnitude of the kinetic energy. This holds for both (7.24), and the finite-m criterion (7.111). The difference is merely that (7.24) focusses on a narrow range of energies about ϵ_F, while (7.111) samples $\rho(\epsilon)$ over a finite interval.

Using the Stoner criterion (7.24), we conclude that it is favourable for the appearance of ferromagnetism if ϵ_F is sitting in a sharp peak of $\rho(\epsilon)$. This allows that the critical value of U be rather smaller than W, so the argument explains how some rather moderately interacting substances can satisfy the criterion by a small margin: it may happen

[13] Here we presented it as a mean field treatment of the Hubbard model. Actually, the Stoner model [376] far antedates the Hubbard model, and has a wider meaning. The Coulomb interaction need not be restricted to the Hubbard term but it is assumed that after mean field decoupling, the energy expression is similar to (7.19). The band can be degenerate so that (within its limitations) the Stoner model is directly applicable to transition metals.

that $U\rho(\epsilon_F) > 1$ but $U\rho(\epsilon_F) - 1$ is small. Since $U\rho(\epsilon_F) - 1$ sets the scale of both the ground state magnetization $m(T = 0)$ and the Curie temperature T_C , we can use an expansion in powers of $U\rho(\epsilon_F) - 1$ (see Problem 7.3) to describe the properties of the so-called *weak itinerant ferromagnets.* In this context, "weak" means that $m(T = 0) \ll n$, i.e., the magnetization is much smaller than the largest value[14] which the given number of electrons could support. It also implies $T_C \ll T_F$, T_F being the Fermi temperature. A good example is the ordered alloy $ZrZn_2$ (see p. 396) which is all the more remarkable because neither of its constituents is, in itself, magnetic. The more strongly interacting iron group elements should be thought of in terms of the finite-m criterion for which the precise location of ϵ_F is not so relevant.

The mean field theory can be extended to finite temperatures. For $U > U_{cr}^F$, one gets[15] the following criterion for the Curie temperature

$$U \int d\epsilon \frac{\partial f(T_C)}{\partial \epsilon} \rho(\epsilon) + 1 = 0\,. \tag{7.25}$$

If we choose parameters which give the right size of the $T = 0$ magnetic moment for, say, Fe then we find that T_C is overestimated by a factor of ~ 5. The essential reason for the discrepancy is that mean field theory grossly underestimates the entropy of the paramagnetic metal: it considers only the entropy coming from electron–hole excitations but neglects the fact that disordered moments exist also above T_C and this is associated with a high spin entropy. For the low-T_C weak itinerant ferromagnets, the discrepancy should be smaller.

If U is somewhat smaller than U_{cr}^F then the exchange enhancement factor is large and we expect to see a *nearly ferromagnetic metal* with

[14]For less than half filling, the largest value belongs to full polarization. For a more than half-filled band, the largest magnetization is obtained by filling the majority band completely, and putting the rest of the electrons in the minority band. We can say that the polarization of a weak ferromagnet is limited by the weakness of the interaction; in contrast, the polarization of a strong ferromagnet (such as the iron group elements) is limited by the band filling. – This is one of the ways in which the terms "weak" and "strong" are used. There is also a different usage: one may call a ferromagnet strong, if the majority band is full. In this latter sense, iron is a weak ferromagnet because $n_{d\uparrow} \sim 4.37$ and $n_{d\downarrow} \sim 2.14$.

[15]It will be derived as (7.46).

a large spin susceptibility. It is thought that the anomalously large susceptibility of Pd is basically of such a nature.

Actually, we can use (7.22) also for asserting that the order parameter must be finite if U is large enough. Let us set the external field $H = 0$. Then the total energy

$$\Delta\mathcal{E}(m) = m^2 \left(\frac{1}{\rho(\epsilon_F)} - U \right)$$

consists of two terms: spin polarization costs band energy which goes as m^2; and it gains interaction energy which is also proportional to m^2. The energy is minimized by $m = 0$ if $U < U_{cr}^F$, while for $U > U_{cr}^F$, the energy is lowered by making m finite. Just how large m will be, cannot be decided unless we return to the full energy expression (7.21) (see Problem 7.2).

It is enlightening to replace (7.18) back into the decoupled form of the Hamiltonian. Using the identity $\sum_{\mathbf{j}} \hat{n}_{\mathbf{j}\sigma} = \sum_{\mathbf{k}} \hat{n}_{\mathbf{k}\sigma}$, we get

$$\mathcal{H} \rightsquigarrow \mathcal{H}_{\text{MFA}} = \sum_{\mathbf{k}} [\tilde{\epsilon}_{\mathbf{k}\uparrow}\hat{n}_{\mathbf{k}\uparrow} + \tilde{\epsilon}_{\mathbf{k}\downarrow}\hat{n}_{\mathbf{k}\downarrow}] - U\left(\frac{n^2}{4} - m^2\right) \tag{7.26}$$

where (for $H = 0$)

$$\begin{aligned} \tilde{\epsilon}_{\mathbf{k}\uparrow} &= \epsilon_{\mathbf{k}} + \frac{Un}{2} - Um \\ \tilde{\epsilon}_{\mathbf{k}\downarrow} &= \epsilon_{\mathbf{k}} + \frac{Un}{2} + Um\,. \end{aligned} \tag{7.27}$$

The spin-$\uparrow$ and spin-$\downarrow$ bands are split by an energy which is proportional to the magnetization. This so-called *exchange splitting*[16] is an important concept in theories of band ferromagnetism.

To conclude: at which values of the band filling can we expect ferromagnetism? As long as we have nothing but the Stoner criterion, we must answer: at all fillings, because for any $\rho(\epsilon_F) \neq 0$, (7.24) predicts

[16]It is traditional terminology and it should not worry us that in the present treatment, U is a direct Coulomb matrix element, not an exchange matrix element. In any case, when the Stoner model is applied to complicated systems like transition metals, the U identifiable from the exchange splitting is an effective coupling which contains contributions from a variety of sources.

ferromagnetism at sufficiently large U. Clearly, this cannot be the whole story because we have already convinced ourselves that near half-filling, the Hubbard model prefers antiferromagnetism. We have to go on to formulate instability criteria for the other kinds of magnetic ordering.

Problem 7.1. Temperature effects in the spin susceptibility:

a) calculate the lowest-order finite-temperature correction to the Pauli susceptibility. Does it decrease or increase as you increase the temperature from $T = 0$?

b) Discuss the large-T behaviour of the spin susceptibility for a half-filled band with a symmetrical density of states.

c) Devise a DOS form for which the susceptibility has a marked maximum as a function of temperature. Discuss the effect of switching on an interaction.

Problem 7.2. Finite-m predictions of the Stoner theory:

a) Derive the $T = 0$ Stoner criterion for finite m.

b) Consider the double-peak model DOS used in the previous problem (see p. 407). Increasing U from zero, which kind of Stoner instability do you find: is m increasing continuously from $m = 0$, or does it jump immediately to a finite value?

Problem 7.3 Stoner theory should work best in the limit of very weak itinerant ferromagnetism where $T_C \ll T_F$ ($ZrZn_2$ is thought to realize this limit; see p. 396 for discussion). Here we work out some of the finite-tempereture results:

a) Use the result of Problem 7.1 to calculate the Curie temperature.

b) Assume that $\rho(\epsilon) = \rho_0 - a\epsilon^2$, and the band is half filled. Derive the temperature dependence of the order parameter.

c) For the same model, derive the susceptibility for $T > T_C$.

7.5 Generalized Susceptibility

Now we want to find out whether the Hubbard model is unstable against ordering into a spin density wave of the form (7.7). The instability is signalled by a divergent response to an external magnetic field with wavevector $\mathbf{q}$. The full Hamiltonian reads

$$\mathcal{H} = \mathcal{H}_{\text{band}} + \mathcal{H}_U + \mathcal{H}_{\text{field}}\,. \tag{7.28}$$

In first-quantized form, the coupling of the spins to the field is

$$\mathcal{H}_{\text{field}} = -\int d\mathbf{r}\mathbf{H}(\mathbf{r})\cdot\mathbf{M}(\mathbf{r}) \rightsquigarrow -g\mu_B \int d\mathbf{r}\mathbf{S}(\mathbf{r})\mathbf{H}_{\mathbf{q}}\cos(\mathbf{q}\cdot\mathbf{r}) \tag{7.29}$$

where the spin density is given as a sum over all electrons

$$\mathbf{S}(\mathbf{r}) = \sum_j \delta(\mathbf{r} - \mathbf{r}_j) \cdot \mathbf{S}_j \, . \tag{7.30}$$

It is convenient to introduce the Fourier components

$$\mathbf{S}(\mathbf{q}) = \int d\mathbf{r} e^{-i\mathbf{q}\cdot\mathbf{r}} \mathbf{S}(\mathbf{r}) = \sum_j e^{-i\mathbf{q}\cdot\mathbf{r}_j} \mathbf{S}_j \, . \tag{7.31}$$

Going over to second quantization, we can express $\mathcal{H}_{\text{field}}$ in terms of fermion operators.

$$\mathbf{S}(\mathbf{q}) = \frac{1}{2} \sum_{\mathbf{p}} \sum_{\alpha\beta} c^\dagger_{\mathbf{p}+\mathbf{q},\alpha} \vec{\sigma}_{\alpha\beta} c_{\mathbf{p},\beta} \tag{7.32}$$

where $\vec{\sigma}$ is the vector of the Pauli matrices. Instead of S^x and S^y, it is convenient to use

$$S^+(\mathbf{q}) = \sum_{\mathbf{p}} c^\dagger_{\mathbf{p}+\mathbf{q},\uparrow} c_{\mathbf{p},\downarrow} \tag{7.33}$$

and

$$S^-(\mathbf{q}) = \sum_{\mathbf{p}} c^\dagger_{\mathbf{p}+\mathbf{q},\downarrow} c_{\mathbf{p},\uparrow} \, . \tag{7.34}$$

The Hubbard model is spin-rotationally invariant, so its susceptibility must be isotropic. For reasons of convenience, we choose an external field pointing in the x-direction

$$\begin{aligned}
\mathcal{H}_{\text{field}} &= -\frac{g\mu_{\text{B}}H^x}{4} \sum_{\mathbf{p}} (c^\dagger_{\mathbf{p}+\mathbf{q},\uparrow} c_{\mathbf{p},\downarrow} + c^\dagger_{\mathbf{p}+\mathbf{q},\downarrow} c_{\mathbf{p},\uparrow} + c^\dagger_{\mathbf{p}-\mathbf{q},\uparrow} c_{\mathbf{p},\downarrow} + c^\dagger_{\mathbf{p}-\mathbf{q},\downarrow} c_{\mathbf{p},\uparrow}) \\
&= -\frac{g\mu_{\text{B}}H^x}{2} [S^x(\mathbf{q}) + S^x(-\mathbf{q})].
\end{aligned} \tag{7.35}$$

Now let us consider $\mathcal{H}_U$. Let us use (4.36) and (4.37) to write

$$\mathcal{H}_U = \frac{NU}{2} - U \sum_{\mathbf{j}} [(S^x_{\mathbf{j}})^2 + (S^y_{\mathbf{j}})^2]. \tag{7.36}$$

We make a mean-field decoupling of the form

$$S^x_{\mathbf{j}} \cdot S^x_{\mathbf{j}} \rightsquigarrow 2S^x_{\mathbf{j}} \cdot \langle S^x_{\mathbf{j}} \rangle - \langle S^x_{\mathbf{j}} \rangle^2. \tag{7.37}$$

Assuming that the only non-vanishing averages correspond to the spin density wave induced by the external field: $\langle S_{\mathbf{j}}^x \rangle = \mathcal{S} \cos(\mathbf{q} \cdot \mathbf{j})$, i.e., $\langle S^x(\mathbf{q}) \rangle = \langle S^x(-\mathbf{q}) \rangle = \mathcal{S}$. $\mathcal{H}_U$ and $\mathcal{H}_{\text{field}}$ can be summarized in

$$\mathcal{H}' = -\left(\frac{g\mu_B H^x}{2} + U\mathcal{S}\right)[S^x(\mathbf{q}) + S^x(-\mathbf{q})]. \tag{7.38}$$

Now let us assume that the spin density is small $\mathcal{S} \ll 1$. Then $\mathcal{H}'$ is a small perturbing term in comparison to $\mathcal{H}_{\text{band}}$ whose ground state is the Fermi sea $|\text{FS}\rangle$. (We make a $T = 0$ calculation.) Since the expectation values of spinflip-type operators like $c_\uparrow^\dagger c_\downarrow$ with the eigenstates of $\mathcal{H}_{\text{band}}$ vanish, we can go straight to the relevant formulae of first-order perturbation theory. The perturbed ground state is

$$\begin{aligned} |\Psi'\rangle \;&=\; |\text{FS}\rangle - \left(\frac{g\mu_B H^x}{2} + U\mathcal{S}\right) \\ &\cdot \sum_{\mathbf{p}\sigma} \left[\frac{c_{\mathbf{p}+\mathbf{q},\sigma}^\dagger c_{\mathbf{p},-\sigma}|\text{FS}\rangle}{\epsilon_{\mathbf{p}} - \epsilon_{\mathbf{p}+\mathbf{q}}} + \frac{c_{\mathbf{p}-\mathbf{q},\sigma}^\dagger c_{\mathbf{p},-\sigma}|\text{FS}\rangle}{\epsilon_{\mathbf{p}} - \epsilon_{\mathbf{p}-\mathbf{q}}}\right]. \end{aligned} \tag{7.39}$$

It is easy to check that to lowest order

$$\langle \Psi' | c_{\mathbf{p}+\mathbf{q},\sigma}^\dagger c_{\mathbf{p},-\sigma} | \Psi' \rangle = \left(\frac{g\mu_B H^x}{2} + U\mathcal{S}\right) \cdot \frac{f_{\mathbf{p}} - f_{\mathbf{p}+\mathbf{q}}}{\epsilon_{\mathbf{p}} - \epsilon_{\mathbf{p}+\mathbf{q}}} \tag{7.40}$$

where the $f_{\mathbf{p}} = \Theta(\epsilon_F - \epsilon_{\mathbf{p}})$ are the (spin-independent) occupation numbers in the unperturbed ground state $|\text{FS}\rangle$. It follows that

$$\langle \Psi' | S^x(\mathbf{q}) | \Psi' \rangle = \left(\frac{g\mu_B H^x}{2} + U\mathcal{S}\right) \cdot \chi^{(0)}(\mathbf{q}) \tag{7.41}$$

where

$$\chi^{(0)}(\mathbf{q}) = \sum_{\mathbf{p}} \frac{f_{\mathbf{p}} - f_{\mathbf{p}+\mathbf{q}}}{\epsilon_{\mathbf{p}+\mathbf{q}} - \epsilon_{\mathbf{p}}}. \tag{7.42}$$

It is easy to see that the same form applies at finite temperatures where $f_{\mathbf{p}}$ is the Fermi distribution.

The self-consistency requirement $\langle \Psi' | S^x(\mathbf{q}) | \Psi' \rangle = \mathcal{S}$ leads to

$$\mathcal{S} = \frac{g\mu_B H^x}{2} \cdot \frac{\chi^{(0)}(\mathbf{q})}{1 - U\chi^{(0)}(\mathbf{q})}. \tag{7.43}$$

Recalling (7.35), we find that the generalized **q**-dependent static susceptibility is

$$\chi(\mathbf{q}) = (g\mu_B)^2 \cdot \frac{\chi^{(0)}(\mathbf{q})}{1 - U\chi^{(0)}(\mathbf{q})}. \tag{7.44}$$

Our derivation led to the same result as the so-called random phase approximation (RPA) [90]. Occasionally we will refer to (7.44) as the RPA susceptibility. We observe that $\chi(\mathbf{q})$ differs by the presence of an enhancement factor from the polarizability $\chi^{(0)}(\mathbf{q})$ of non-interacting band electrons, just as the uniform susceptibility (7.23) is the enhanced Pauli susceptibility.

7.5.1 Criteria for q $\neq 0$ Instabilities

The generalized susceptibility $\chi(\mathbf{q})$ diverges when

$$\chi^{(0)}(\mathbf{q}) \cdot U_{\rm cr}^{\mathbf{q}} = 1 . \tag{7.45}$$

This is the *generalized Stoner criterion.* The Stoner criterion for ferromagnetism is easily recovered by taking $\mathbf{q} \to 0$

$$\frac{1}{U} \approx -\sum_{\mathbf{p}} \frac{\frac{\partial f}{\partial \epsilon} \cdot \frac{\partial \epsilon}{\partial \mathbf{p}} \cdot \mathbf{q}}{\frac{\partial \epsilon}{\partial \mathbf{p}} \cdot \mathbf{q}} \rightsquigarrow - \int \frac{\partial f}{\partial \epsilon} \rho(\epsilon) d\epsilon. \tag{7.46}$$

At finite temperatures, this is just (7.25). As $T \to 0$, $\partial f/\partial\epsilon$ can be replaced by $-\delta(\epsilon - \epsilon_F)$, and we recover (7.24).

Our conclusion is similar to that concerning the appearance of ferromagnetism. At arbitrary band filling, the system becomes unstable against the formation of a spin density wave with arbitrary **q**, if U is large enough. Naturally, only one of the instabilities can be real: that which is predicted to occur at the lowest critical U

$$U_{\rm cr} = \min_{\mathbf{q}} U_{\rm cr}^{\mathbf{q}} = \left(\max_{\mathbf{q}} \chi^{(0)}(\mathbf{q}) \right)^{-1} . \tag{7.47}$$

The minimum criterion picks out the ordering wavevector $\mathbf{q}_0$ (which depends on the band filling). As U is increased beyond $U_{\rm cr}$, the amplitude of the $\mathbf{q}_0$ spin density wave becomes finite, and this invalidates the instability criteria for the other modes of ordering. What often happens is that with a further increase of U, the amplitude of the $\mathbf{q}_0$ order increases towards saturation. There are, however, cases when at a higher U, a first-order transition to a different kind of fully developed order is taking place.

In order to make use of (7.47), we should know $\chi^{(0)}(\mathbf{q})$. $\chi^{(0)}(\mathbf{q})$ depends on the details of the band structure and generally, it has to be determined by numerical integration. There are, however, important special cases where analytical results are available.

For free electrons in three-dimension, $\chi^{(0)}(\mathbf{q})$ is the well-known Lindhard function [32] which has its maximum at $\mathbf{q} = 0$. Thus the leading (magnetic) instability of the three-dimensional electron gas would be[17] the onset of ferromagnetism at $U \sim \epsilon_F$. In one-dimension, $\chi^{(0)}(\mathbf{q})$ has a logarithmic divergence at the Fermi diameter $q = 2k_F$ (see Problem 7.4). We may want to conclude that a quasi-one-dimensional substance should have a $q = 2k_F$ spin density wave ground state, except that the dielectric susceptibility diverges at the same q, so we have to worry about the competition between spin density wave, and charge density wave, types of ordering. The physics of quasi-one-dimensional substances is an interesting, and highly developed, branch of solid state physics, which we cannot discuss here (see [138] for details).

In our model studies, an important role is played by the D-dimensional cubic lattices which have the tight binding dispersion relation

$$\epsilon(\mathbf{k}) = -2t \sum_{i=1}^{D} \cos k_i. \tag{7.48}$$

Let us observe that these satisfy the *perfect nesting condition*

$$\epsilon(\mathbf{k} + \mathbf{Q}) = -\epsilon(\mathbf{k}) \qquad \text{for all } \mathbf{k} \tag{7.49}$$

where $\mathbf{Q} = (\pi, \pi, \ldots, \pi)$ is the so-called spanning vector (we took the lattice constant $a = 1$). We have met $\mathbf{Q}$ as the wavevector of the two-sublattice antiferromagnetic order (7.4). At half-filling $\epsilon_F = 0$, and at

[17] It has to be mentioned at once that further studies show that this mean field prediction is false. The free-electron dispersion might be used at the bottom of a band, i.e., at low fillings. However, there are general arguments to rule out ferromagnetism in a low-density electron system (see Ch. 8). At larger band fillings, the response is determined by lattice effects. We will associate ferromagnetism with marked deviations from the free-electron behaviour.

$T = 0$ (7.42) can be written as

$$\chi^{(0)}(\mathbf{Q}) = \int_0^{W/2} \frac{\rho(\epsilon)}{2\epsilon} d\epsilon\,. \tag{7.50}$$

$\chi^{(0)}(\mathbf{Q})$ diverges logarithmically unless $\rho(\epsilon) \to 0$ as $\epsilon \to 0$. It follows that $U_{\rm cr}^{\mathbf{Q}} = 0$, i.e., *for half-filled bands with the perfect nesting property, an arbitrarily small interaction $U > 0$ causes a transition to a two-sublattice antiferromagnetic state.* We are going to see that the antiferromagnetic state is insulating. Thus in these circumstances, *the metal–insulator transition occurs at $U = 0$.*

When making the above statement, we were taking it for granted that the $\mathbf{q} \neq \mathbf{Q}$ susceptibilities are finite so that it is clear the $\mathbf{q} = \mathbf{Q}$ ordering is the dominant instability. A curious case is presented by the half-filled, square lattice tight binding band. Here not only $\chi(\mathbf{Q})$ diverges but, because of the logarithmic divergence of the density of states (4.31), $\chi(\mathbf{q} = 0)$ too. We have a case of *competing instabilities*: it appears that at an infinitesimal $U > 0$, the system might decide to go either ferromagnetic, or antiferromagnetic. Which instability is stronger? Strictly at $T = 0$, we should compare two infinite susceptibilities but coming from finite temperatures, we can easily convince ourselves that antiferromagnetism prevails (Problem 7.5).

Though (7.47) often gives a good idea of what can be expected to happen, we should not forget that it was derived in a simple mean field theory, and it is certainly not exact. We are speaking about Hartree–Fock instabilities. Hartree–Fock theory is notorious for underestimating the stability of the non-ordered, symmetrical ground state. We should be especially cautious about ordering away from half-filling. It turns out, however, that the prediction of the $U = 0$ metal–insulator transition for half-filled, perfectly nested bands is basically correct.

Problem 7.4 Derive $\chi^{(0)}(q)$ for the one-dimensional tight binding band. Compare the divergence at $q = 2k_{\rm F}$ to that obtained for free electrons.

Problem 7.5 Competing instabilities: For the half-filled square lattice tight binding band at $T = 0$, $\chi^{(0)}(\mathbf{Q})$ and $\chi^{(0)}(\mathbf{0})$ diverge simultaneously. Decide which divergence is stronger by doing a calculation at finite temperatures, and taking the limit $T \to 0$. (*Hint*: you can simplify the calculation by approximating the Fermi distribution by a piecewise linear function).

Compare the weak-coupling mean-field Néel temperatures for $D = 2$ dimensions, and $D \neq 2$.

7.6 Spin Density Waves

For the Hubbard model, the most robust kind of ordering is antiferromagnetism. We certainly expect it at large U at half-filling, because the effective Hamiltonian is the antiferromagnetic Heisenberg model. However, for models with the perfect nesting property (7.49), the same kind of order is present for all $U > 0$. We expect that the $T = 0$ order parameter approaches 0 continuously as $U \to 0$, and it is finite for finite U. Previously, we used the mean field theory in an infinitesimal neighbourhood of the phase boundary; now we wish to extend it to the whole ordered domain.

Let us assume that the lattice is bipartite so that we have two-sublattice antiferromagnetism. The collinear order can be assumed to arise from spin-$\uparrow$ electrons being the majority species on sublattice A, and spin-$\downarrow$ electrons being in majority on sublattice B. We quote (7.4)

$$\langle \hat{n}_{\mathbf{j}\sigma} \rangle = \frac{n}{2} + \eta(\sigma) m e^{i\mathbf{Q}\mathbf{j}}. \tag{7.51}$$

For small U, the modulation of the spin density

$$\langle S^z_{\mathbf{j}} \rangle = m e^{i\mathbf{Q}\mathbf{j}} \tag{7.52}$$

is weak; it is very far from the up-down alternation of localized spins which we find in the large-U limit and which we associate with the name "antiferromagnet". Hence we prefer to call the order specified in (7.51) and (7.52) a spin density wave (SDW). At large U, the amplitude of the spin density wave tends to the saturation value 1/2, and we have a mean field description of local moment antiferromagnetism. The change from a weak to strong SDW is perfectly continuous.

We replace (7.51) into (7.9) to get the mean field Hamiltonian

$$\mathcal{H}_{\text{SDW}} = \mathcal{H}_{\text{band}} + U \sum_{\mathbf{j}} \sum_{\sigma} \hat{n}_{\mathbf{j}\sigma} \langle \hat{n}_{\mathbf{j}-\sigma} \rangle - U \sum_{\mathbf{j}} \langle \hat{n}_{\mathbf{j}\uparrow} \rangle \langle \hat{n}_{\mathbf{j}\downarrow} \rangle \,. \tag{7.53}$$

The potential seen by a propagating $\uparrow$-spin electron is the superposition of the lattice potential, and the average field $U\langle \hat{n}_{\mathbf{j}\downarrow} \rangle$ of the $\downarrow$-electrons.

The one-dimensional case is easiest to visualize (Fig. 7.2). The electrons see a potential which is periodic only under translations by

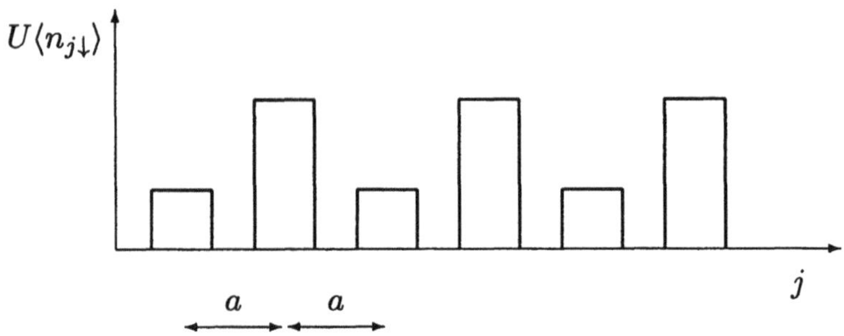

Figure 7.2: The effective field seen by a $\uparrow$–spin electron. Since the new repeat distance is $2a$, the magnetic supercell is twice as large as the crystal cell.

integer multiples of $2a$, where a is the lattice constant of the underlying periodic lattice potential which gave rise to the tight binding band $\epsilon(\mathbf{k})$. The effective field has a lower translational symmetry than the starting Hamiltonian. The new reciprocal lattice vectors are integral multiples of $2\pi/2a = \pi/a$, and the Brillouin zone is halved. To keep track of all the states, the part of the original band (dashed line) which is lying outside the new first Brillouin zone (NBZ) has to be "folded back" into it (dotted line in Fig. 7.3), so that we have a two-band situation to deal with.

The folded-back bands are modified by the extra Bragg reflection on the self-consistent potential $U\langle\hat{n}_{\mathbf{j}-\sigma}\rangle$. This is particularly relevant where it mixes nearly degenerate states, i.e., near the points $k = \pm\pi/2a$. As a consequence, gaps are opened at the new Brillouin zone boundaries. We can also say that $U\langle\hat{n}_{\mathbf{j}\downarrow}\rangle$ hybridizes the folded-back bands, leading to a splitting between the lower and upper hybridized bands. These are shown with continuous lines in Fig. 7.3.

At half-filling, with one electron per site, there are two electrons per magnetic supercell, just enough to fill the lower band, while leaving the upper band empty. The opening of the gap pushed the occupied states down, and raised the empty states. This is the energetical motivation[18] for the symmetry breaking (cell doubling due to magnetic order).

[18]Similar would be the energetical motivation for creating a supercell by redistributing the charge rather than the spin (Peierls distortion).

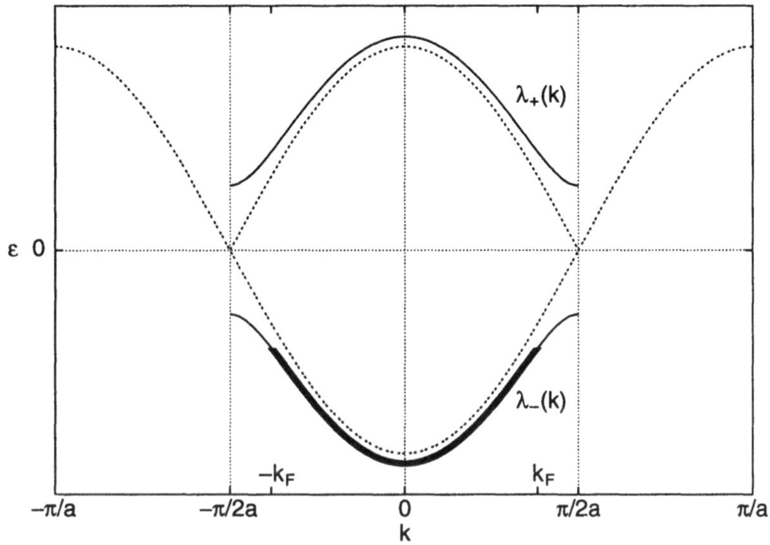

Figure 7.3: Illustration of the mean field treatment of two-sublattice SDW order. The effective field doubles the periodicity and therefore halves the Brillouin zone. The tight binding band of non-interacting electrons (dashed line) has to be folded back into a new zone, giving rise to an upper band (dotted line). The effective field acts like hybridization, producing the lower and upper SDW bands (continuous lines), separated by the SDW gap. We can envisage a similar scenario for less than half filling when k_F lies inside the new zone, and only the corresponding part of the lower SDW band is occupied (thick line; see Sec. 7.6.4).

All this was easy to visualize in the one-dimensional case, but the analogous statements hold for a D-dimensional hypercubic lattice.

These ideas are embodied in $\mathcal{H}_{\rm SDW}$ which is bilinear in the fermion operators, thus easily diagonalizable. The last obstacle is that part of it is in direct-space, and part of it in momentum-space representation. We find it convenient to rewrite the entire Hamiltonian in terms of **k**-space operators.

For each $\mathbf{k} \in$ NBZ, $\mathbf{k}+\mathbf{Q}$ is outside the NBZ. In the spirit of the reduced zone scheme, we ascribe the state created by $c^{\dagger}_{\mathbf{k}+\mathbf{Q}}$ also to **k**, by introducing the new fermion creation operator $d^{\dagger}_{\mathbf{k}} = c^{\dagger}_{\mathbf{k}+\mathbf{Q}}$. The folded-back tight binding band structure is described by

$$\mathcal{H}_{\text{band}} \rightsquigarrow \sum_{\mathbf{k}}^{\text{NBZ}} \sum_{\sigma} (\epsilon_{\mathbf{k}} c^{\dagger}_{\mathbf{k}\sigma} c_{\mathbf{k}\sigma} + \epsilon_{\mathbf{k}+\mathbf{Q}} d^{\dagger}_{\mathbf{k}\sigma} d_{\mathbf{k}\sigma}). \tag{7.54}$$

By making use of (7.51), the second term on the right-hand side of (7.9) can be conveniently expressed in terms of Bloch-operators

$$\begin{aligned} & U \sum_{\mathbf{j}} \sum_{\sigma} \hat{n}_{\mathbf{j}\sigma} \langle \hat{n}_{\mathbf{j}-\sigma} \rangle \\ & \rightsquigarrow U \sum_{\mathbf{j}} \sum_{\sigma} \frac{1}{L} \sum_{\mathbf{k}_1 \mathbf{k}_2} e^{i\mathbf{j}(\mathbf{k}_1 - \mathbf{k}_2)} c^{\dagger}_{\mathbf{k}_1\sigma} c_{\mathbf{k}_2\sigma} \left(\frac{n}{2} - m\eta(\sigma) e^{i\mathbf{Q}\mathbf{j}} \right) \\ & = \frac{Un}{2} \sum_{\mathbf{k}} \sum_{\sigma} \hat{n}_{\mathbf{k}\sigma} - Um \sum_{\mathbf{k}} \sum_{\sigma} \eta(\sigma) (c^{\dagger}_{\mathbf{k}\sigma} d_{\mathbf{k}\sigma} + d^{\dagger}_{\mathbf{k}\sigma} c_{\mathbf{k}\sigma}). \end{aligned} \tag{7.55}$$

It is apparent that the effective field hybridizes the c- and d-bands. Putting all the above contributions together, the mean-field Hamiltonian becomes block-diagonal

$$\mathcal{H}_{\text{SDW}} = \sum_{\mathbf{k}}^{\text{NBZ}} \sum_{\sigma} \mathcal{H}_{\mathbf{k}\sigma} \tag{7.56}$$

with

$$\mathcal{H}_{\mathbf{k}\sigma} = \tilde{\epsilon}_{\mathbf{k}} c^{\dagger}_{\mathbf{k}\sigma} c_{\mathbf{k}\sigma} + \tilde{\epsilon}_{\mathbf{k}+\mathbf{Q}} d^{\dagger}_{\mathbf{k}\sigma} d_{\mathbf{k}\sigma} - Um\eta(\sigma)(c^{\dagger}_{\mathbf{k}\sigma} d_{\mathbf{k}\sigma} + d^{\dagger}_{\mathbf{k}\sigma} c_{\mathbf{k}\sigma}) \tag{7.57}$$

where $\tilde{\epsilon}_{\mathbf{k}} = \epsilon_{\mathbf{k}} + (Un/2)$, and $\tilde{\epsilon}_{\mathbf{k}+\mathbf{Q}} = \epsilon_{\mathbf{k}+\mathbf{Q}} + (Un/2)$.

The present MFA reduced the problem of the Hubbard model to that of the hybridization of two bands of non-interacting fermions. The eigenvalue problem is block-diagonal in $\mathbf{k}$ and σ, and all that is left is to diagonalize the 2×2 matrices

$$\begin{pmatrix} \tilde{\epsilon}_{\mathbf{k}} & -Um\eta(\sigma) \\ -Um\eta(\sigma) & \tilde{\epsilon}_{\mathbf{k}+\mathbf{Q}} \end{pmatrix}. \tag{7.58}$$

The solution appears in a pleasingly simple form if we assume that the perfect nesting property $\epsilon_{\mathbf{k}+\mathbf{Q}} = -\epsilon_{\mathbf{k}}$ holds. The eigenvalues are

$$\lambda^{\pm}(\mathbf{k}) = \frac{Un}{2} \pm \sqrt{\epsilon_{\mathbf{k}}^2 + U^2 m^2}. \tag{7.59}$$

$\lambda^+(\mathbf{k})$ and $\lambda^-(\mathbf{k})$ correspond to the two hybridized bands shown in Fig. 7.3. Note that though the matrix was σ-dependent, the eigenvalues are not. This reflects the fact that in building up an antiferromagnetic state, the $\uparrow$-spin, and $\downarrow$-spin electrons play symmetrical roles.

The ground state energy density can be expressed as

$$\begin{aligned} E(m) &= \frac{2}{L}\sum_{\mathbf{k}}^{\text{occ.}}\left[\frac{Un}{2} - \sqrt{\epsilon_{\mathbf{k}}^2 + U^2m^2}\right] - U\sum_{\mathbf{j}}\langle\hat{n}_{\mathbf{j}\uparrow}\rangle\langle\hat{n}_{\mathbf{j}\downarrow}\rangle \\ &= U\left(\frac{n^2}{4} + m^2\right) - \frac{2}{L}\sum_{\mathbf{k}}^{\text{occ.}}\sqrt{\epsilon_{\mathbf{k}}^2 + U^2m^2} \end{aligned} \tag{7.60}$$

where the sum over $\mathbf{k}$ extends over the N lowest-lying eigenvalues $\lambda^-(\mathbf{k})$ (or $\epsilon_{\mathbf{k}}$); these levels are occupied in the ground state.

7.6.1 Gap Equation

We have to remember that (7.59) is just a formal solution in terms of the as yet unknown quantity m. In the literal sense of the mean field procedure (which we followed in Sec. 7.5), we should determine also the eigenstates corresponding to $\lambda^\pm(\mathbf{k})$, construct the ground state by filling the lowest one-electron states with as many electrons as we have, and calculate $\langle\hat{n}_{\mathbf{j}\sigma}\rangle$ therefrom. A comparison with (7.51) would yield a self-consistency equation for m, which we then would have to solve.

Here we follow another, much simpler route which leads to the same result[19]. We can regard m as a variational parameter, and determine it by requiring that $E(m)$ be minimum

$$\frac{\partial E(m)}{\partial m} = 0\,. \tag{7.61}$$

There are two solutions: either

$$m = 0\,, \tag{7.62}$$

or m is determined by the gap equation

$$1 = \frac{1}{L}\sum_{\mathbf{k}}^{\text{occ.}}\frac{U}{\sqrt{\epsilon_{\mathbf{k}}^2 + \Delta^2}} \tag{7.63}$$

[19] You'll have a chance to check the equivalence of the two methods by solving Problem 7.6.

where the notation

$$\Delta = Um \tag{7.64}$$

has been introduced.

The nonmagnetic solution $m = 0$ is always there, only it need not correspond to the actual ground state. We are, of course, interested in the case when the true minimum corresponds to some $m \neq 0$, the solution of the gap equation. At the new BZ boundary, $\epsilon_{\mathbf{k}} = 0$, and

$$\lambda^{\pm} = \frac{Un}{2} \pm \Delta\,. \tag{7.65}$$

The lower and upper SDW-split bands are separated by a gap of the size 2Δ.

At exact half-filling, where the lower band is full, and the upper band is empty, the SDW state is insulating. For lattices with perfect nesting, the ground state of the half-filled Hubbard model is predicted to be an antiferromagnetic insulator for any $U > 0$.

Let us discuss the weak-coupling ($U \ll W$) behaviour of Δ, and m. Using the symmetry of $\rho(\epsilon)$ about the centre of the band, (7.63) can be rewritten as

$$1 = U \int_0^{W/2} \frac{d\epsilon \rho(\epsilon)}{\sqrt{\epsilon^2 + \Delta^2}}\,. \tag{7.66}$$

If $\rho(0)$ is finite (as in $1D$ or $3D$), then for $U \ll W$ we expect $\Delta \ll W$. Dividing the integral in two parts

$$1 \sim U \int_0^{\Delta} \frac{d\epsilon \rho(0)}{\Delta} + U \int_{\Delta}^{W/2} \frac{d\epsilon \rho(\epsilon)}{\epsilon} \sim U\rho(0) + U\rho(0) \ln \frac{W}{2\Delta}. \tag{7.67}$$

(The estimate for the first term is obvious. The extreme smallness of Δ permits us to assume that the second integral is likewise dominated by the contribution from a small neighbourhood of $\epsilon = 0$ where $1/\epsilon$ is large.) The second term dominates, giving

$$\Delta \sim \frac{W}{2} \exp\left[-\frac{1}{U\rho(0)}\right] \sim t \exp\left[-\frac{t}{U}\right]. \tag{7.68}$$

The U-dependence is given by the extremely weak, non-analytic $e^{-t/U}$. (The precise large-D result is quoted in (10.82) on p. 588). The sublattice polarization

$$m \sim \frac{t}{U} \exp\left[-\frac{t}{U}\right] \tag{7.69}$$

is also very small.

The case of the square lattice is somewhat different because of the logarithmic divergence of $\rho(\epsilon)$ as $\epsilon \to 0$. Still we can make a derivation similar to (7.67)

$$1 \sim U \int_0^{\Delta} \frac{d\epsilon}{t\Delta} \ln \frac{t}{\epsilon} + U \int_{\Delta}^{t} \frac{d\epsilon}{\epsilon t} \ln \frac{t}{\epsilon} \sim \frac{U}{t} \ln \frac{t}{\Delta} + \frac{U}{t} \ln^2 \frac{t}{\Delta}. \tag{7.70}$$

which leads to the weak-coupling result[20]

$$\Delta \sim t \exp\left[-\sqrt{\frac{t}{U}}\right] \tag{7.71}$$

(numerical factors are omitted even in the exponent, we are interested only in the form of the function). (7.71) being in an exponential sense larger than (7.68), we may conclude that the square lattice Hubbard model has an anomalously strong tendency towards AF ordering in the weak-coupling regime.

It is educative to study the case of a band with a constant density of states: the gap equation can be solved in closed form, and the development of the SDW from weak coupling (small U) to strong coupling limit (large U) can be followed in detail (this will be discussed further in Sec. 7.6.3).

Assuming

$$\rho(\epsilon) = \begin{cases} 1/W & \text{if } -\frac{W}{2} \le \epsilon \le \frac{W}{2} \\ 0 & \text{otherwise} \end{cases} \tag{7.72}$$

and a half-filled band, (7.63) is solved by

$$\Delta = \frac{W}{2 \sinh \dfrac{W}{U}} \tag{7.73}$$

and

$$m = \frac{W}{2U} \frac{1}{\sinh \dfrac{W}{U}}. \tag{7.74}$$

---

[20] We found the same from the Stoner criterion in Problem 7.5.

The energy expression (7.60) turns into

$$E(U) = U\left(\frac{1}{4}+m^2\right) - \frac{1}{2}\sqrt{\frac{W^2}{4}+\Delta^2} - \frac{\Delta^2}{W}\cdot\ln\frac{\frac{W}{2}+\sqrt{\frac{W^2}{4}+\Delta^2}}{\Delta}. \tag{7.75}$$

For $U \ll W$, the already familiar $e^{-W/U}$ is seen to arise in the expression for Δ. Note that the exponent appearing in the expression for the ground state energy lowering is twice as large: $\sim -We^{-2W/U}$. You can think of this as the result of $\sim \rho(0)\Delta$ electrons having lowered their energies by $\sim \Delta$. All this is perfectly analogous to the BCS weak coupling results for the gap, and the ground state energy. The strong-coupling behaviour will be discuseed in Sec. 7.6.3.

The SDW ground state energy is only an approximation to the true ground state energy, so it is useful to know at least whether it is lying above, or below, the exact value. We will learn from the solution to Problem 7.9 that the $T = 0$ mean field theory can be reformulated as a variational approach to the ground state, thus our present result is an upper bound to the true ground state energy.

7.6.2 Finite Temperatures

There is no principal difficulty in extending the previous argument to finite temperatures. The mean field decoupling led to an independent-fermion problem, where thermal effects can be described by expressing the occupation numbers with the Fermi distribution f.

Coming from the high-temperature side, the onset of the SDW is signalled by the divergence of the generalized susceptibility

$$U\chi^{(0)}(\mathbf{Q}, T_{\text{cr}}) = 1\,. \tag{7.76}$$

Let us recall (7.42), exploit that with perfect nesting $f(\epsilon_{\mathbf{k}+\mathbf{Q}}) = 1 - f(\epsilon_{\mathbf{k}})$, and assume a constant DOS. Integrating by parts

$$\begin{aligned}\frac{W}{U} &= \int_{-W/2}^{W/2} \tanh\left(\frac{\epsilon}{2k_{\text{B}}T_{\text{cr}}}\right)\frac{d\epsilon}{\epsilon} = \int_0^{W/4k_{\text{B}}T_{\text{cr}}} \tanh x\frac{dx}{x} \\ &= \ln\frac{W}{4k_{\text{B}}T_{\text{cr}}}\tanh\frac{W}{4k_{\text{B}}T_{\text{cr}}} - \int_0^{W/4k_{\text{B}}T_{\text{cr}}} \frac{\ln x\,dx}{\cosh^2 x}\,.\end{aligned} \tag{7.77}$$

In the weak coupling limit, we may assume $k_{\rm B}T_{\rm cr} \ll W$, thus the tanh can be replaced by 1 in the first term, while in the second term, the integration can be extended to ∞, yielding $-\ln(4\gamma/\pi)$ where γ is related to Euler's constant $\gamma_{\rm E}$ [4] via $\gamma = \exp(\gamma_{\rm E})$. We find[21]

$$k_{\rm B}T_{\rm cr} = W\frac{\gamma}{\pi}e^{-W/U}\,. \tag{7.78}$$

The relationship of the critical temperature to the $T=0$ gap

$$2\Delta(T=0) \approx 3.53 k_{\rm B}T_{\rm cr} \tag{7.79}$$

is the same as in the BCS theory of weak coupling superconductivity.

The temperature dependence of the order parameter is decided by two competing contributions to the free energy (or the grand potential): the energy $\mathcal{U}$, and the entropy term $-T\mathcal{S}$. At $T=0$, there is only $\mathcal{U}$ to think about: it is minimized by filling the lower SDW band, which gives an energy lowering $\sim \Delta^2$. At higher T, it is more advantageous to reduce the gap which makes the process of exciting electrons into the upper SDW band easier, and thereby improves the free energy contribution from the entropy term. The disappearance of the gap at $T_{\rm cr}$ is an entropy-driven process.

The finite-T theory of the ordered state works by minimizing the grand potential density

$$\frac{1}{L}\Omega(T,\mu,\Delta) = -\frac{U}{4} + \frac{\Delta^2}{U} - \frac{k_{\rm B}T}{L}\ln\left\{\prod_{\mathbf{k}\sigma\alpha}\left[1+\exp\left(-\frac{\lambda_\sigma^\alpha(\mathbf{k})-\mu}{k_{\rm B}T}\right)\right]\right\}. \tag{7.80}$$

It is an essential simplification that the grand canonical sum factorizes in the independent-particle picture. The additional constants arise from the mean field decoupling, and are the same as in (7.60). The band index α inludes both branches: $\alpha = +,-$. Because of the symmetrical DOS, the chemical potential is pinned at $\mu = U/2$, and we can write

$$\frac{1}{L}\Omega(T,\mu,\Delta) = \frac{U}{4} + \frac{\Delta^2}{U} - \frac{k_{\rm B}T}{L}\sum_{\mathbf{k}}\ln\left\{4\cosh^2\frac{\sqrt{\epsilon^2(\mathbf{k})+\Delta^2}}{2k_{\rm B}T}\right\}. \tag{7.81}$$

---

[21] The simple method used in the solution to Problem 7.5 gives the functional form but not the numerical coefficient. It can be shown similarly that the weak-coupling critical temperature of the square lattice Hubbard model behaves like $\sim W\exp(-\sqrt{W/U})$ (see (7.132).

Minimizing w.r.t. Δ, we get the finite-T gap equation

$$\frac{1}{L}\sum_{\mathbf{k}} \frac{U}{\sqrt{\epsilon^2(\mathbf{k})+\Delta^2}} \tanh \frac{\sqrt{\epsilon^2(\mathbf{k})+\Delta^2}}{2k_\mathrm{B}T} = 1\,. \tag{7.82}$$

At $T=0$, we recover (7.63). Now we are interested in the law governing the collapse of the gap as $T \to T_\mathrm{cr} - 0$. Assuming again a constant DOS, (7.82) can be rewritten as

$$\frac{W}{U} = \int_0^{W/2} \frac{d\epsilon}{\sqrt{\epsilon^2+\Delta^2}} \tanh \frac{\sqrt{\epsilon^2+\Delta^2}}{2k_\mathrm{B}T}\,. \tag{7.83}$$

The weak-coupling T_cr (7.78) can be rederived from the above equation by setting $\Delta = 0$: $(W/U) = \ln(\gamma W/\pi k_\mathrm{B} T_\mathrm{cr})$. Now assume that $T_\mathrm{cr} - T \ll T_\mathrm{cr}$, so that $\Delta \ll W$. We expand the integral on the right hand side to order Δ^2. The $\Delta = 0$ value is $\ln(\gamma W/\pi k_\mathrm{B} T)$, while the coefficient of the term $\sim \Delta^2$ is expressed by some definite integrals. All in all, we find the relationship

$$\ln\frac{T}{T_\mathrm{cr}} = \ln\left(1 - \frac{T_\mathrm{cr}-T}{T_\mathrm{cr}}\right) \approx \frac{T_\mathrm{cr}-T}{T_\mathrm{cr}} = -A\Delta^2\,.$$

Close to the critical point, Δ follows a square root law which is a characteristic outcome of mean field calculations

$$\Delta(T) \approx \sqrt{\frac{8\pi^2}{7\zeta(3)}} k_\mathrm{B} T_\mathrm{cr} \sqrt{1-\frac{T}{T_\mathrm{cr}}} \approx 3.06 k_\mathrm{B}\sqrt{T_\mathrm{cr}-T}\,. \tag{7.84}$$

The increment of Ω near T_cr can be approximated as

$$\begin{aligned}
\Omega(\Delta) - \Omega(\Delta=0) &\approx \frac{\Delta^2}{U} - \frac{2\Delta^2}{W}\int_0^{W/2} \tanh\left(\frac{\epsilon}{2k_\mathrm{B}T_\mathrm{cr}}\right)\frac{d\epsilon}{\epsilon} \\
&= \Delta^2\left(\frac{1}{U} - \frac{2}{W}\ln\frac{\gamma W}{\pi k_\mathrm{B} T}\right) \qquad (7.85)\\
&\approx \Delta^2\left\{\frac{1}{U} - \frac{2}{W}\ln\frac{\gamma W}{\pi k_\mathrm{B} T_\mathrm{cr}} - \frac{2}{W}\frac{T_\mathrm{cr}-T}{T_\mathrm{cr}}\right\} \\
&= -\frac{2\Delta^2}{W}\frac{T_\mathrm{cr}-T}{T_\mathrm{cr}} = -\frac{8\pi^2}{7\zeta(3)}\frac{k_\mathrm{B}^2}{W}(T_\mathrm{cr}-T)^2\,.
\end{aligned}$$

The SDW order makes a finite contribution to the specific heat $C_V = -T(\partial^2\Omega/\partial T^2)$ up to $T_{\rm cr}$, where it vanishes in a step-like fashion. The prediction of such a lambda-anomaly is a hallmark of weak coupling mean-field calculations. The discontinuity at $T_{\rm cr}$ is

$$C_V(T = T_{\rm cr} - 0) - C_V(T = T_{\rm cr} + 0) = \frac{16\pi^2}{7\zeta(3)} k_{\rm B}^2 \rho(\epsilon_{\rm F}) T_{\rm cr} \tag{7.86}$$

where we recalled that $1/W$ corresponds to $\rho(\epsilon_{\rm F})$. Since the specific heat above $T_{\rm cr}$ must coincide with the non-interacting $C_V(T = T_{\rm cr} + 0) = (2\pi^2/3)k_{\rm B}^2\rho(\epsilon_{\rm F})T_{\rm cr}$, we deduce a universal weak-coupling ratio between the peak value of the lambda anomaly, and the normal-state value

$$\frac{C_V(T = T_{\rm cr} - 0) - C_V(T = T_{\rm cr} + 0)}{C_V(T = T_{\rm cr} + 0)} = \frac{12}{7\zeta(3)} \approx 1.43 \,. \tag{7.87}$$

The weak-coupling results (7.78)–(7.86) describe the observed behaviour of a number of systems quite well, just as the analogous BCS theory works beautifully for weak-coupling superconductors. This needs some explanation. The ubiquitous BCS features have their foundation in the truly universal instability[22] of the (three-dimensional) Fermi sea against Cooper pair formation, and in the fact that the phonon-induced effective attraction is weak on the scale of the Fermi energy. In contrast, the condition for the applicability of weak-coupling SDW theory is the divergence of the non-interacting $\chi^{(0)}(\mathbf{q})$, which we derived from the perfect nesting property (7.49). For three-dimensional systems, this is an artificial assumption, and we cannot expect that it is ever exactly fulfilled: a weak hopping to next-to-nearest neighbours is enough to destroy it.

Let us, however, observe that it is not really necessary to require the fulfillment of $\epsilon(\mathbf{k} + \mathbf{Q}) = -\epsilon(\mathbf{k})$ in the whole Brillouin zone if all what we want is a weak SDW. It is enough if it holds in a narrow range about the Fermi surface where we can linearize the spectrum:

---

[22]The Fermi sea is unstable against Cooper pair formation in the presence of an (effective) attractive electron–electron interaction which can be traced back to the electron–phonon interaction. Less obviously, there can be a similar instability also with a net repulsive interaction [219]. It is believed that even the apparently non-ordering (i.e., non-magnetic and non-superconducting) metals should eventually become unstable against some kind of pair condensation, though maybe only at unobservably low temperatures.

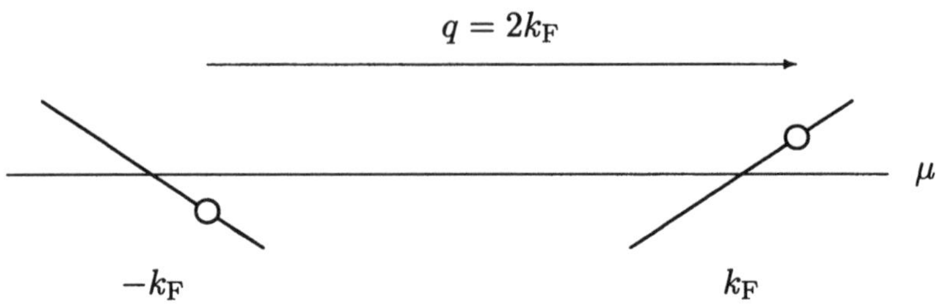

Figure 7.4: The nesting condition $\epsilon(k+q) = -\epsilon(k)$ is fulfilled in the energy range where we can linearize the spectrum about the Fermi points $\pm k_F$.

It is clear from Fig. 7.4 that in one-dimension, an approximate nesting situation is present for arbitrary band filling, and we can use our theory to describe a weak SDW with $q = 2k_F$. The effective field contains a Fourier component with $q = 2q_F$, and it opens a gap at the Fermi points. It is remarkable that a metal-insulator transition is predicted to happen at arbitrary band filling. Notice, however, that this statement does not hold for higher dimensions.

The best candidates for the observation of SDWs are quasi-one-dimensional systems which consist of weakly coupled conducting chains. To cite a well-studied example, consider the organic charge transfer salt $(TMTSF)_2PF_6$ [138]. It can be regarded as an array of approximately quarter-filled Hubbard chains, thus it is a metal though it does not contain metal atoms at all. The substance develops SDW order at $T_N = 11.5$K and at the same time, it undergoes a metal–insulator transition. Both the low critical temperature, and the small value of the $T = 0$ ordered moment $\sim 0.08\mu_B$ show that we have to do with a weak-coupling SDW. Can one observe the order directly? The straightforward method would magnetic neutron scattering, but the small saturation moment, and the small sample size make this option look pretty hopeless. Fortunately, upon the application of an external magnetic field H_0, the SDW gives rise to a modulation $\sim H_0 \cos \mathbf{qr}$ of the internal field, and this can be detected by the sensitive techniques of nuclar magnetic resonance (NMR) and muon spin resonance (μSR) [138]. The temperature dependence of the order parameter is shown in Fig. 7.5. Considering that $m(T) \propto \Delta(T)$, it is comforting to see that the measured behaviour is not far from the (BCS-like) mean field prediction which follows from (7.83).

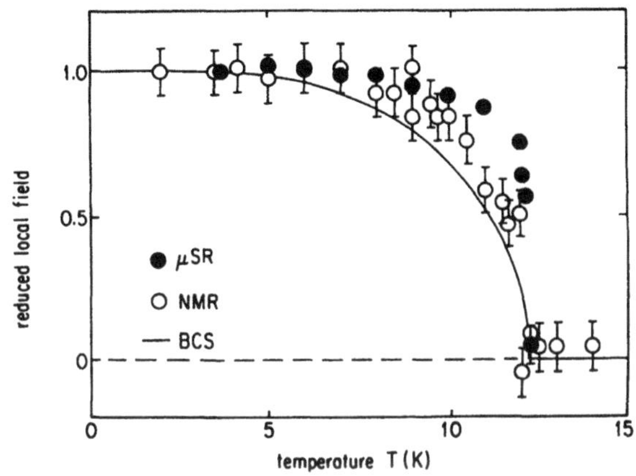

Figure 7.5: The SDW order parameter of $(TMTSF)_2PF_6$ measured by NMR (after [238]). The experimental data (circles) follow approximately the mean field curve (line).

Naturally, we have to remember that a genuinely one-dimensional system would not order. However, we can invoke the usual argument that the weak interchain coupling helps to stabilize the order which the individual chains would like to develop, and thus mean field theory gives an acceptable description[23].

In three (two) dimensions it may happen that the Fermi surface shows approximately parallel faces (line segments) connected by a spanning vector $\mathbf{q}$. Usually there are several pairs of faces whose spanning vectors are connected by a symmetry of the high-temperature phase. The nesting condition $\epsilon(\mathbf{k}+\mathbf{q}) = -\epsilon(\mathbf{k})$ is approximately satisfied over parts of the Fermi surface with one or another of the $\mathbf{q}$-vectors. This state of affairs can give rise to a spin density wave transition, with the low-temperature order built with a single Fourier component with one of the $\mathbf{q}$s (single SDW), or with several $\mathbf{q}$s (multiple SDW). A single

[23] The best argument is provided by the coincidence of SDW ordering with a metal-insulator transition. Our mean field theory predicts an insulating ground state at arbitrary filling. The exact treatment of an isolated Hubbard chain would predict that it is a conductor except at exact half-filling.

SDW can open a gap only over one pair of nested faces, but even with a multiple SDW, the typical outcome is that only a part of the Fermi surface gets gapped away. For three-dimensional, not perfectly nested systems, the SDW transition is a metal–metal transition in which the number of charge carriers becomes smaller but does not vanish. Cr and $FeGe_2$ offer examples of such a behaviour (Sec. 7.7).

7.6.3 Strong Coupling Limit

The unambiguously successful part of the SDW story is the weak coupling solution. At half filling, we should be able to proceed smoothly from the weak-coupling SDW state to the Mott insulator by increasing U/W. But do we really manage it? Looking at the strong coupling solution at $T = 0$, we get a mixed impression: there are results which we may be ready to accept, but there are also warning signs that not all is well. Going to finite temperatures, it becomes clear that the SDW mean field theory fails to capture the essential physics of Mott insulators.

For large U, bands with different DOS behave very similarly, so it is sufficient to consider the case of a constant DOS. Let us return to our $T = 0$ results (7.73)–(7.75). Expanding in terms of W/U, the leading term of the ground state energy is

$$E(U) \sim -\frac{W^2}{2U} \,. \tag{7.88}$$

Apparently, we have managed to describe, at least as far as the energy scale is concerned, the local moment antiferromagnetism expected in the large-U limit of the Hubbard model, which is ruled by the Heisenberg coupling $J \sim W^2/U$. The reason why SDW mean field is doing relatively well is that it gives a reasonable estimate for the quantity controlling the ground state energy: the density of doubly occupied sites

$$\langle \hat{n}_{\mathrm{j}\uparrow}\hat{n}_{\mathrm{j}\downarrow}\rangle = \frac{n^2}{4} - m^2 \sim \frac{1}{12}\left(\frac{W}{U}\right)^2 \,. \tag{7.89}$$

The order of magnitude is just what you would expect from perturbation theory. (7.89) and (7.88) are very similar to what we were arguing about in our qualitative discussion (Sec. 4.2) of the Mott transition. Let us note, however, that (7.89) is not achieved by dynamically correlating $\uparrow$- and $\downarrow$-electrons but by gradually localizing them on different sublattices. The electrons do not have to develop short-range correlations to avoid each other, because they are kept apart by the antiferromagnetic long range order. The mean field answer to the problem of strong electron–electron interaction works without introducing correlations. It is not what we would really like but for certain purposes, it may do quite well. However, we are bound to discover that the underlying picture has serious limitations.

The result for the sublattice polarization

$$m \approx \frac{1}{2} - \frac{1}{12}\left(\frac{W}{U}\right)^2 \tag{7.90}$$

is too near to the classical limit 1/2, and it is *not* the result which we derive from the corresponding antiferromagnetic Heisenberg model (cf. Sec. 6.2.2). The Heisenberg model's ground state contains a lot of quantum fluctuations, especially in low dimensions, and these are evidently missed in the SDW Hartree–Fock approximation. Observe that in the above expression, the correction to the saturation value 1/2 is U-dependent, and it expresses merely that a fraction of the sites is non-magnetic. In contrast, the result for the spin reduction (10.26) in AF spin wave theory relies on the concept that all sites are magnetic, and it depends strongly on the geometry.

The gap 2Δ is approximately

$$2\Delta(U) \sim U - \frac{W^2}{6U}\,. \tag{7.91}$$

At large U, the SDW bands are separated by a forbidden gap $\sim U$, and this may make you think about the Hubbard subbands which have the same separation at large U. It should be emphasized that our present SDW scenario is physically very different from the correlation picture of the Mott transition which we presented in Sec. 4.2. There we envisaged a metal-to-insulator transition due to local correlations, which was taking place (without symmetry breaking) within the paramagnetic state. As a secondary effect, the low-temperature state of a Mott-insulator may be (and usually is) found to be an antiferromagnet, but that is not relevant for the existence of the gap. In strongly correlated substances like CoO, the gap survives the collapse of antiferromagnetic long range order, remaining essentially unchanged when the temperature is raised above the Néel temperature. In contrast, the SDW ground state breaks both translational symmetry and spin-rotational invariance. The gap exists only because there is symmetry breaking, and it collapses when the temperature reaches the critical temperature $T_{\rm cr}$.

A further, more formal, difference is that the Hubbard subbands extend over the entire Brillouin zone, and can hold only one electron for each **k**-vector, while the SDW bands are defined over a shrunken Brillouin zone, and can hold two electrons for each **k**-vector.

Though we must be aware of the shortcomings of the SDW mean field theory in the strong scoupling regime, it has to be emphasized that it remains a very useful theory for the weak-to-intermediate coupling range, providing a mechanism for the appearance of antiferromagnetic insulators, which is distinct from the Mott mechanism. In Chapter 4, we were arguing that the Hubbard model should show a metal–insulator (Mott) transition at $U \sim$ bandwidth, even if one disregards magnetism. However, (either perfect or approximate) nesting can push the metal–insulator transition down to smaller values of U (maybe to $U = +0$), making it coincide with the SDW ordering transition. We would not like to call this transition a Mott transition because it leads to a weakly (or not particularly strongly) correlated insulator. If an experimentally observed metal–insulator transition coincides with a magnetic ordering transition, we are usually prepared to interpret it in the SDW framework, and speak about a *Slater insulator*, rather than a Mott insulator. This certainly applies to $(\mathrm{TMTSF})_2\mathrm{PF}_6$, and maybe even to V_2O_3 under pressure (see Fig. 5.5, left). Remember, however, that while nesting is the rule in $D = 1$, it becomes an exception in $D = 3$.

What about the U-dependence of the SDW critical temperature $T_{\rm cr}$? At weak coupling, $T_{\rm cr}$ is the exponentially small (7.78). Assuming that $T_{\rm cr} \sim \Delta$ continues to hold, we are led to guess that at large U, $T_{\rm cr} \sim U$. Indeed we get such a behaviour by applying the Stoner criterion. At large T, $\chi^{(0)}(\mathbf{Q}) \sim 1/T$, leading to $T_{\rm cr} \sim U$. But this is quite wrong. In the strong coupling limit, $T_{\rm cr}$ must be proportional to the Heisenberg coupling, so the correct result is $T_{\rm cr} \sim t^2/U$. Thus for large U, $T_{\rm cr}$

becomes small again, having passed through a maximum somewhere in the range $U \sim t$ (Fig. 7.6). Unfortunately, our simple mean field approach cannot[24] give us the proper $T_{cr} \sim t^2/U$.

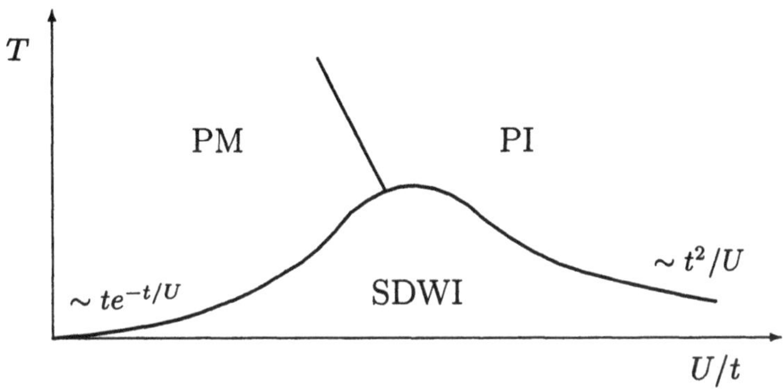

Figure 7.6: Schematic representation of the phase diagram of the Hubbard model with nesting (T denotes temperature). There are three phases: paramagnetic metal (PM), paramagnetic insulator (PI), and the spin density wave insulator (SDWI). The PM–PI phase boundary ends at a critical point. The weak-coupling, and strong-coupling expressions of the critical temperature are indicated.

In summary, the SDW mean-field description seems to have got the low-temperature energetics of the antiferromagnetic Mott-insulating state right, but it did not get the subtler features of the Mott phenomenon right. At low temperatures, it would be very difficult to distinguish an antiferromagnetic Mott insulator from a large-U SDW insulator, but the higher the temperature, the easier to tell the difference, and in the paramagnetic state it is already a difference between a metal and an insulator.

It is useful to compare the high-temperature paramagnetic phase of an $n = 1$ SDW system with that of an antiferromagnetic Mott insulator (AFMI). At the Néel temperature of the AFMI the localized moments disorder but the individual sites continue to possess a magnetic moment, and the system shows a Curie–Weiss type susceptibility. Restricting our-

[24]Remember that we have already done the derivation assuming $T_{cr} \ll t$; it can yield only the weak-coupling (7.78).

selves to the single-orbital Hubbard model, it is justified to use the same effective Heisenberg model on both sides of the phase boundary, and the phase transition of the electron system can be modelled as the phase transition of a spin system[25]. This approach can be used with confidence for systems which are wildly Mott insulating like CoO. In contrast, the collapse of a weak-coupling SDW order leaves behind a weakly correlated Fermi sea. At intermediate couplings, we would predict a Fermi sea whose susceptibilities (especially the staggered susceptibility) are (perhaps strongly) enhanced. All the same, the SDW-to-paramagnetic transition is still an insulator-to-metal transition. At a still somewhat bigger U (around $U \sim W$) the scenario changes because we get on the other side of the Mott transition: the PM phase becomes insulating. Thus there must be a finite-temperature metal–insulator phase boundary within the paramagnetic state, which meets the SDW critical line in the intermediate coupling regime (Fig. 7.6).

We were arguing that in the ground state, the perfect nesting property causes a unified behaviour from $U = +0$ to $U = \infty$; there is continual change but no sharp boundary. We feel that the weak-U system is merely a weak SDW, while the large-U system must be really a Mott insulator but remaining at $T \le T_{\rm N}$, we cannot pinpoint where the transition occurs. The situation is quite different at temperatures $T > T_{\rm N}$ where there is a sharp PM–PI phase boundary. This is a line of first-order phase transitions[26]. The PM and PI states have the same spatial symmetry, do not break the spin-rotational and time-reversal invariance, thus they can differ only[27] in density-like characteristics: the

[25]Allowing for degenerate models may bring extra complications due to the orbital degrees of freedom. We have seen in Sec. 5.4 that the interplay with orbital ordering may lead to situations where *different* Heisenberg models have to be used for the AF and the PM phases.

[26]We certainly would expect such a boundary for realistic models which incorporate long-range Coulomb interaction, and/or electron–phonon coupling. For the pure Hubbard model, we have to repeat the warning given in the discussion of Fig. 5.5: it is possible that the model cannot yield discontinuous behaviour.

[27]If we mean it as a general statement, it is not true that if a transition does not change symmetries then it cannot be a higher-order transition; the Kosterlitz–Thouless transition is a counterexample, and even apart from that, we have to worry about the possibility of hidden characteristics which are not macroscopic order parameters in the conventional sense. However, our simple-minded argument is good

volume of the unit cell, the spin susceptibility, and the conduction electron entropy. The PM–PI transition is often likened to the liquid–gas transition and like that, it can also proceed completely smoothly by going around the critical point where the first-order phase boundary ends. For a simple cubic Hubbard model with perfect nesting, we expect the phase diagram shown in Fig. 7.6. The overall features are confirmed by dynamical mean field theories for the infinite-dimensional Hubbard model.

7.6.4 Away from Half-Filling

Half-filling plays a distinguished role because the assured existence of a Mott transition at some critical value of U/t makes it certain that the ground state is insulating in the strong coupling limit[28]. The insulating domain can be enlarged by magnetic long range order which adds further Fourier components to the effective field, opening gaps in certain parts of the Brillouin zone. Perfect nesting can make the system insulating even in the weak coupling regime. Though we have seen signs that the SDW mean field theory is not completely trustworthy beyond the weak coupling regime, we do not see any reason to doubt its essential conclusion that the weak-coupling and strong-coupling insulating regimes are connected, and the antiferromagnetic Mott insulator evolves continuously from the SDW state.

The situation is rather different away from half-filling where an elementary discussion of local correlation effects led to the conclusion that the system remains metallic even at $U \to \infty$. If the system turns out to be insulating, it can only[29] be due to the formation of a spin density wave which is a weak-to-intermediate coupling phenomenon (at any rate, the kinetic exchange strength $\sim t^2/U$ diminishes as $U \to \infty$).

Here we have to discuss one-dimensional and higher-dimensional systems separately. We have pointed it out that any one-dimensional band

enough for most cases of practical interest.

[28] For degenerate Hubbard models, the same is true for any integral filling.

[29] Assuming that the system can be described by the Hubbard model. Intersite Coulomb interaction can make a quarter-filled band insulating, turning it into a Wigner crystal. Coupling to the lattice can also open gaps due to a Peierls distortion, and the accompanying charge density waves [138].

structure is approximately nesting (Fig. 7.4), and that is good enough as an argument for a weak-coupling SDW. Postulating a $q = 2k_{\rm F}$ modulation $n_{\sigma j} \sim m\eta(\sigma)\cos(qj)$, we get $\pm 2k_{\rm F}$ components of the effective field, and the subsequent derivation is very much the same as in the half-filled case. The results are qualitatively applicable to many quasi-one-dimensional systems [138].

For higher-dimensional systems, at less than half filling, the most that can be expected is that there is approximate nesting over certain regions of the Fermi surface. A weak SDW can gap away only the flat parts of the Fermi surface, leaving behind a metal with a reduced number of free carriers (we will describe some of the experimental consequences in our discussion of chromium in Sec. 7.7). We need a treatment of metallic SDW systems.

There is no reason why the simplest two-sublattice order specified in (7.51) could not be tried at less-than (or more-than) half-filling. Sufficiently close to half-filling, the spin order is expected to be approximately alternating from site to site. Furthermore, some treatments indicate that the two-sublattice state may survive inside a finite range of band filling [118, 419, 308]. It has been argued that in the weak coupling limit, the two-sublattice state is *the* stable homogeneous ground state[30] in $D \geq 3$. The one-dimensional illustration in Fig. 7.3 gives an idea of the state of affairs in any dimension. For $n < 1/2$, k_F lies inside the new Brillouin zone, and the lower SDW band is only partially filled (see the part drawn in thick line). For $n > 1/2$, one has to start filling up the upper SDW band. The two-sublattice antiferromagnetic order is not quite so stable as at half-filling, either because we do not utilize all the states which have an energy lowering, or because we begin to fill also states which have been raised. One consequence of this is that the ordering is no longer an infinitesimal-U instability but one needs a finite threshold value $U_{\rm cr}$ to get a spin density wave. For simple lattice structures, the transition is second-order, and the critical $U_{\rm cr}$ is given by the generalized Stoner criterion (7.45). Away from half-filling, all states are metallic: we work in the band picture with partially filled bands.

The expectation of a two-sublattice ordering is well-founded for bi-

[30] However, phase segregation into an insulating antiferromagnet and a metallic paramagnet gives a lower total energy. See p. 568 and p. 589.

partite lattices at half-filling but in any other case, we should suspect that the system prefers a different ordering pattern. This is quite clear in $D = 1$ where we see that $q = 2k_F$ is the best choice. For higher dimensions, the situation is not so clear-cut and in any case, the resulting states are much more difficult to handle. We refer to the literature for a sample of complicated mean field phase diagrams [179, 353].

Sufficiently far from half-filling, even the ferromagnetic ($\mathbf{q}$=0) solution becomes preferable to the antiferromagnetic solution. The simplest mean-field diagrams (whose general appearance we sketched in Fig. 4.11) show only these two kinds of ordering. For the simple cubic lattice, the corresponding phase diagram was worked out by Penn [324], showing a central region of antiferromagnetism flanked by two domains of ferromagnetism (Fig. 4.11). This gave the hope that itinerant ferromagnetism can be understood as the typical intermediate-U, intermediate-filling behaviour of the Hubbard model. Note that at large U, the ferromagnetic domain begins to outflank the AF region, reaching the $n = 1$ axis at $U \to \infty$. We have already remarked that the energy scale sustaining AF or SDW behaviour must be suppressed at large enough U.

What is the status of the ordered states predicted by Hartee–Fock theories? We understand that such theories are only approximate, and that they usually overestimate the ordering tendencies because they force the system into the straightjacket of long range order, thus even what would be really short range order effects, are compelled to appear in the disguise of long range order. But is it possible that the very prediction of a state with SDW long range order is false?

We discuss first ordering at $T = 0$. It is understood that the prediction of an antiferromagnetic insulator for (perfectly nested) cubic lattices at half-filling is a solid result for dimensions $d \geq 2$. The statement is not exactly proven in a mathematical sense, but it is supported by such a variety of techniques that there is no room for reasonable doubt. In fact, the corresponding result is "almost true" even in $D = 1$: the ground state *is* insulating, with a gap whose U-dependence is not unlike the SDW result. True that there is no long range order but it is replaced by antiferromagnetic quasi-long-range order; the correlations fall off algebraically $\langle S^z_m S^z_n \rangle \sim (-1)^{m-n}/|m-n|$, and the staggered susceptibility diverges (the system is at the critical point). At large U, the system is clearly a Mott insulator: the gap is $\sim U$, and the spin quasi-long-range order is the same as for the equivalent $S = 1/2$ Heisenberg model.

The situation is quite different away from half-filling, inasmuch as the existence of magnetic long-range order is more doubtful. The one-dimensional Hubbard model is known not to order at all[31], so let us take the example of the square lattice. Keeping only the ferromagnetic, and two-sublattice antiferromagnetic, types of order, the mean field phase diagram [166] is rather like Fig. 4.11. Allowing fancier ordering patterns, rich Hartree–Fock phase diagrams are obtained [179, 353]. But are all these phases real? The square lattice is simple enough to allow trying to check the predicted phase diagram by comparing to numerical results obtained for finite systems. The result was rather shocking: Hirsch [166] felt compelled to conjecture that any ordering away from half-filling is merely an artefact of the mean field approximation. But remember that though the numerical results gave sufficient reason to question the mean-field results, the counter-suggestion was not a rigorous statement either! Here we do not wish to discuss the difficult problem whether the square lattice Hubbard model has any kind of order (and particularly, whether it is superconducting) away from half-filling. Let us rather soften the criticism of mean field approaches by two general remarks. First, even if it turns out that a predicted phase is not realized at all, the Hartree–Fock studies are not a waste of effort. One can always argue that the hypothetical long-range order is replaced by a well-developed short-range order of much the same kind. Second, the remarkable fragility of magnetic order in the Hubbard model (which we are going to discuss at length in Ch. 8) may be due to the fact that the model contains only on-site interaction. An honest mean field should arise from the neighbours! We surmise that augmenting the Hubbard model with intersite interactions would not only take us nearer to physical reality, but would at the same time make the behaviour of the model

---

[31] We find a ferromagnetic phase in the ground state phase diagram of a generalized Hubbard chain with first- and second neighbour hoppings (see Ch. 8). For a quite different parameter regime, it may also show dimer order. It remains true that it does not show magnetic order with $q \neq 0$.

more mean-field-like.

What about three-dimensional systems (including quasi-one-dimensional and quasi-two-dimensional ones) ? We feel now on safer ground; after all, we do have chromium and $(TMTSF)_2PF_6$. The likely result is that two-sublattice antferromagnetism prevails in a finite range of n on both sides of $n = 1$, going over into incommensurate SDW states at smaller (larger) band fillings, and eventually into a featureless paramagnetic state [118, 383]. The status of metallic ferromagnetism is a debated matter, and we take up its discussion in Ch. 8.

Finally, what about finite temperatures? The Mermin–Wagner theorem prohibits the development of magnetic order in one- and two-dimensional systems. If a three-dimensional system shows SDW order, the U/t-dependence of its critical temperature resembles that shown in Fig. 7.6, with one essential modification: ordering sets in at a finite threshold value of U/t (see also Fig. 5.5). Remember, however, that the large-U part of Fig. 7.6 is *not* a Hartree–Fock result; we obtained it from a plausibility argument, using the equivalence to an equivalent Heisenberg model. Away from half-filling, we should invoke the equivalent t–J model. It remains, however, true that the energy (and temperature) scale of magnetic ordering is given by $J \sim 4t^2/U$.

Problem 7.6. *Mean field transition in a finite system.* Do the SDW mean field calculation for the two-site Hubbard molecule (with two electrons) described in Sec. 2.5.2. Compare the result a) to the exact solution, and b) to the MF solution of the infinite system.

Problem 7.7. *SDW in the absence of nesting.* In the absence of nesting, SDW appears at a finite critical U_{cr}. Derive the gap at $T = 0$, assuming that $U - U_{\text{cr}}$ is small.

Problem 7.8. *Linear versus circular polarization.* Discuss the nature of the gap opened by a circularly polarized SDW. Is it true that a linearly polarized SDW is energetically preferable to a circularly polarized SDW?

Problem 7.9. *The SDW trial state.* The $T = 0$ mean field theory can be reformulated as a variational theory, in which a factorization property of the wave function plays the role of the decoupling procedure. For the sake of simplicity, assume perfect nesting, and a half-filled band. Devise the variational trial state, and carry out the minimization procedure (the rederivation of the gap equation is sufficient).

7.7 Transition Metals and Alloys

7.7.1 Introduction

The understanding of the magnetic properties of transition metals is one of the chief aims of the theory of magnetism (after all, iron has given its name to *ferro*magnetism!). It happens to be a goal which is much more difficult to attain than it was once thought. It appears that transition metals (TM) should be classified as systems where correlations are of intermediate strength: it is no good to neglect them but methods developed for the limiting case of strong correlation are not particularly useful, either.

In order to develop a model of the TM, first we should enquire which kind of electronic states are lying near the Fermi energy $\epsilon_{\rm F}$. For all the transition metals, $\epsilon_{\rm F}$ is in the energy range where a moderately wide d-band overlaps a wide s-band. For instance, in iron, the $3d$ band of width $\sim 4.6e$V (which can hold 10 electrons) overlaps the $4s$ band of width $\sim 24e$V (which can hold 2 electrons). If we approximate the density of states as consisting of $3d$ and $4s$ parts

$$\rho(\epsilon_{\rm F}) \approx \rho_{3d}(\epsilon_{\rm F}) + \rho_{4s}(\epsilon_{\rm F})$$

it is clear that $\rho_{3d}(\epsilon_{\rm F})$ predominates. Should we consider a one-band model (with only $3d$ electrons), or keep the full two-band model?

The answer depends very much on which physical quantity we are interested in. For the electronic specific heat, it is a good approximation to keep only the d-electrons. On the other hand, the mobile s-electrons make an essential contribution to the electrical conductivity. Our present interest is in magnetism which is predominantly an affair of the d-electrons, thus we may try to think of the TM in terms of a degenerate Hubbard model which imitates the subsystem of d-electrons. However, certain parameters of this effective Hubbard model are affected by the presence of the s-band. Foremost the band filling: since the $3d$ and $4s$ bands overlap, they both have non-integer fillings. For instance, it is reckoned that in paramagnetic Fe, the $3d$ band holds ~ 6.5 electrons, while the $4s$ band has ~ 1.5 electrons. Furthermore, the effective Hubbard U is diminished because the light s-electrons screen the electron–electron interaction effectively.

Introducing lattice models of interacting fermions is the standard approach in this book but at the present, it is certainly not the most promising way of thinking about transition metals. Another possibility is to refer to the results of up-to-date electronic structure calculations. Density functional theory [92], or more specifically: the application of the local spin density approximation (LSDA) in band theory, has been remarkably successful in explaining the properties of TM elements, and their alloys [15]. We discuss the relationship of LSDA to lattice fermion models in Sec. 7.7.2. In spite of its many successes in describing weakly or moderately correlated systems, LSDA is not a proper theory of electron correlations; for instance, it often fails to find the Mott–Hubbard gap of Mott insulators. However, it works very well for the ferromagnetic iron group elements. The total energy given by LSDA can (with some further approximations) be brought to the form (7.26), confirming the picture of exchange-split bands. In principle we expect the energy difference $\tilde{\epsilon}_{\mathbf{k}\uparrow} - \tilde{\epsilon}_{\mathbf{k}\downarrow}$ to be $\mathbf{k}$-dependent but the finding [141] that the splitting is, in fact, more-or-less constant allows to use the parametrization (7.27), and to calculate the Stoner parameter I from first principles[32]. I is found to lie in the range 0.7eV-1eV for the $3d$ series (increasing towards Ni). Somewhat smaller values are found for the $4d$ and $5d$ series. This shows that the electron–electron interaction is of intermediate strength.

In view of the relative constancy of I, going from element to element in the $3d$ series amounts, roughly speaking, to changing the d-band filling. Varied behaviour is found [442]. Ti and V are paramagnetic metals. Cr has an interesting spin density wave phase which we discuss

[32]The LSDA calculations treat the true degenerate d-band which is hybridized with a wide s-band, and the interaction is the full Coulomb interaction. Thus the Stoner I should be regarded as an effective coupling constant which no longer has the simple meaning of the Hubbard U [357]. $I \sim 1e$V may well imply a Hubbard U of several eV, which confirms that we have to do with an intermediate coupling problem.

in more detail later. α-Mn is an antiferromagnet below $T_N = 95K$ but a horrendously complicated one: even the non-magnetic cubic unit cell contains 58 atoms. Fe, Co, and Ni are ferromagnetic metals, to which we return in Sec. 7.7.4. Neither of the $4d$ or $5d$ TM elements orders magnetically but Pd comes close to it.

The alloys of TM with each other, and with other elements, show a tremendous variety of behaviour [442] which we cannot endevour to review here. For illustrating various points, we will mention $ZrZn_2$, the YCo_2 system, and some of the TM silicides.

When describing the experimental findings, or the results of electronic structure calculations, we will be using the language we developed in the mean field treatment of the simplest[33], one-band Hubbard model: Stoner criterion, spin density waves, etc. We do so in the hope that these concepts turn out to be applicable also to more complicated models, including such that will, eventually, be accepted as useful models of Fe, Cr, $La_{1-x}Sr_xMnO_3$, etc. To proceed in this spirit presupposes that the mean field results are not grossly misleading, i.e., that the mean field behaviour is qualitatively the same as the exact one. Here lies a difficulty which, at the time of writing, presents one of the outstanding problems of many-body theory: the status of ferromagnetism. It appears that while the mean field description of two-sublattice antiferromagnetism and of the (in general, incommensurate) spin density wave state is in qualitative agreement with the expected exact results, the prediction of an essentially structure-insensitive, intermediate-U ferromagnetic phase is an artefact of the mean field approximation. This is merely to say that well-developed itinerant ferromagnetism does not seem to be a generic feature of the Hubbard model. It is perfectly possible that a well-chosen specific Hubbard model, or some more sophisticated lattice fermion model, will provide a satisfactory understanding of the phenomenon of itinerant ferromagnetism. We discuss this question briefly at the end of the next section, and in more detail in Ch. 8.

[33]Meaning the nearest-neighbour hopping model on a hypercubic lattice.

7.7.2 LSDA versus Lattice Fermion Models

Our classification of the solids as weakly or strongly correlated systems rests on the assumption of a small or large U/t (or the analogous quantities in multiband models). U/t is not easy to estimate, thus we work with a phenomenological input; postulating Hubbard-like models is not "first-principles" theory. A truly first-principles approach would first determine how strongly correlated a system is, and proceed then to derive the consequences of strong correlation – if indeed these two stages can be separated at all. But would such a complete treatment be feasible? Density functional theories boldly assert that the answer is yes[34]. After all, there is only one coupling constant: e^2, and only one interaction: the Coulomb interaction. Somehow, everything else must follow. Whether the outcome of the very same Coulomb interaction is a nearly-free-electron system like sodium, or a strongly correlated system like magnetite, must be associated with the distribution of the electrons in the field of the nuclei.

Density functional theory rests on a powerful theorem by Hohenberg and Kohn [172] which states that the ground state energy is uniquely associated with the ground state electron density. The theorem refers to the $T = 0$ properties of a non-relativistic, non-spin-polarized interacting electron system which is subject to the external potential $V_{\text{ext}}(\mathbf{r})$. It states that there exists a unique energy functional $E[n(\mathbf{r})]$ of the charge density $n(\mathbf{r})$ whose minimum $E_0[n_0(\mathbf{r})]$ is the ground state energy, and $n_0(\mathbf{r})$ is the ground state charge density. Moreover, the energy functional is of the form

$$E[n(\mathbf{r})] = \int d\mathbf{r}\, n(\mathbf{r}) V_{\text{ext}}(\mathbf{r}) + F[n(\mathbf{r})] \tag{7.92}$$

$V_{\text{ext}}(\mathbf{r})$ is the field of an arbitrary arrangement of nuclei. The universal functional $F[n(\mathbf{r})]$ does not depend on the external potential $V_{\text{ext}}(\mathbf{r})$, i.e., it is independent of which solid (or molecule, or atom) we are considering; it describes the universal properties of the "electron glue" which holds the nuclei together.

---

[34] There are many systematic treatises on the subject [92]. See also the insightful brief review [144].

The ground state is found by minimizing E with respect to arbitrary variations $\delta n(\mathbf{r})$ which are constrained only by the requirement that the total number of particles $N = \int d\mathbf{r} n(\mathbf{r})$ is fixed. At some stage, the form of $F[n(\mathbf{r})]$ must become explicit. An important step is to write

$$F[n(\mathbf{r})] = K_0[n(\mathbf{r})] + E_{\mathrm{H}}[n(\mathbf{r})] + E_{\mathrm{xc}}[n(\mathbf{r})] \tag{7.93}$$

where $K_0[n(\mathbf{r})]$ is the kinetic energy of a system with density $n(\mathbf{r})$,

$$E_{\mathrm{H}}[n(\mathbf{r})] = \frac{e^2}{2} \int d\mathbf{r} \int d\mathbf{r}' \frac{n(\mathbf{r})n(\mathbf{r}')}{|\mathbf{r} - \mathbf{r}'|} \tag{7.94}$$

is the Hartree term, and $E_{\mathrm{xc}}[n(\mathbf{r})]$ is the as yet unknown *exchange–correlation functional.* If we assume that it is given, the rest of the problem is a numerically well-tractable self-consistency calculation: the charge density arises from electrons occupying a set of orbitals ψ_ν

$$n(\mathbf{r}) = \sum_\nu |\psi_\nu(\mathbf{r})|^2 \Theta(\mu - \epsilon_\nu) \tag{7.95}$$

where μ is the chemical potential, and the ϵ_ν are the eigenvalues of Schrödinger-like equations

$$\left(-\frac{\hbar^2}{2m}\nabla^2 + V_{\mathrm{eff}}(\mathbf{r}, [n])\right) \psi_\nu(\mathbf{r}) = \epsilon_\nu \psi_\nu(\mathbf{r}) \, . \tag{7.96}$$

The effective potential

$$V_{\mathrm{eff}}(\mathbf{r}, [n]) = V_{\mathrm{ext}}(\mathbf{r}) + e^2 \int d\mathbf{r}' \frac{n(\mathbf{r}')}{|\mathbf{r} - \mathbf{r}'|} + V_{\mathrm{xc}}(\mathbf{r}, [n]) \tag{7.97}$$

contains, in addition to the external field, and the average Coulomb field of the electrons, the all-important exchange-correlation potential

$$V_{\mathrm{xc}}(\mathbf{r}, [n]) = \frac{\delta E_{\mathrm{xc}}}{\delta n(\mathbf{r})} \, . \tag{7.98}$$

Solving the equations (7.96) proceeds by iteration. Once this is done, the ground state energy can be expressed by the Kohn–Sham formula

[220]

$$
\begin{aligned}
E_0 &= \sum_\nu \epsilon_\nu \Theta(\mu - \epsilon_\nu) - \frac{e^2}{2}\int d\mathbf{r}\int d\mathbf{r}' \frac{n(\mathbf{r})n(\mathbf{r}')}{|\mathbf{r}-\mathbf{r}'|} \\
&+ E_{\mathrm{xc}}[n(\mathbf{r})] - \int V_{\mathrm{xc}}(\mathbf{r},[n])n(\mathbf{r})d\mathbf{r}\,. \qquad (7.99)
\end{aligned}
$$

Apparently, all the difficulties of many-body theory have been reduced to determining the exchange–correlation functional. Unfortunately, there is no systematic procedure for constructing successively improving approximations to $V_{\mathrm{xc}}(\mathbf{r},[n])$. A great part of the popularity of the density functional approach is due to the remarkable fact that the first reasonable guess for $V_{\mathrm{xc}}(\mathbf{r},[n])$, the so-called local density approximation (LDA), gives very good results. LDA postulates that $V_{\mathrm{xc}}(\mathbf{r},[n])$ depends only on the charge density at $\mathbf{r}$:

$$
V_{\mathrm{xc}}(\mathbf{r},[n]) \approx V_{\mathrm{xc}}^{\mathrm{LDA}}(\mathbf{r},n(\mathbf{r}))\,. \qquad (7.100)
$$

This opens the way to finding an expression for $V_{\mathrm{xc}}^{\mathrm{LDA}}(\mathbf{r},n(\mathbf{r}))$: it can be identified with the solution we find for a homogeneous (sufficiently dense) electron liquid on a uniform neutralizing background, the so-called jellium model. Naturally, we then go on to use the formalism to describe the non-uniform distribution of electrons in a solid. The idea is that if at a point $\mathbf{r}$ the density is $n(\mathbf{r})$, then the electron feels the same exchange–correlation potential as if it were in a medium with the same uniform density. There are deeper reasons [144] why the approach does not break down easily even in patently non-uniform systems, e.g., at surfaces, or for isolated molecules. LDA has had many sucesses as a parameter-free theory of a large number of non-magnetic molecules and solids.

Magnetically ordered ground states can be described by a spin-polarized version of LDA: the local spin density approximation (LSDA). It relies on generalizing the Hohenberg–Kohn theorem to the lowest-energy state within a subspace with a fixed spin polarization. LSDA has been eminently successful in describing the magnetism of transition metals and their alloys [15, 141]; we briefly mention some of these achievements later. It has also been possible to extend the treatment to finite temperatures, which yields fairly good estimates of the critical temperature, and allows even the derivation of a Curie–Weiss-like susceptibility in the paramagnetic state of certain band ferromagnets like bcc Fe [374]. The relativistic version of LSDA gives promising results for magnetic anisotropy.

Since LSDA seems to be a theory of almost everything, why do not we simply regard it as *the* theory of magnetism? The answer is twofold. First, LSDA (at least, in its conventional form) tends to break down for strongly correlated systems: it predicts that Mott insulators like CoO or La_2CuO_4 should be metals [306], and it needs a phenomenological input parameter to describe the large effective mass of heavy fermion compounds. Since our main interest is in Mott phenomena, we would not wish to rely on density functional theory alone[35].

However, this is only the lesser of our reasons to stick to the Hubbard model and its likes. The rapid development of density functional theory may soon make our first objection unfounded. Still, we can argue that even when we will have a complete, *ab initio* description of, say V_2O_3, giving its structure, lattice constants, orbital and spin order all coming out from a single calculation, theories of simple Hubbard-like models will not become obsolete. The point is the following: in LSDA theory, the primary quantity is the electron density, and orbitals play a merely auxiliary role. It turns out that for a sufficiently accurate expansion of the ground state density $n_0(\mathbf{r})$, one may need dozens of orbitals per lattice site, well beyond the number which is "reasonably expected" in a tight binding scheme. Indeed, the local basis one uses is not the tight binding basis at all, but a set of functions chosen for quick convergence of the calculation; the local basis functions have no immediate physical meaning. Correspondingly, if we wanted to represent the Coulomb interaction in the local basis, there would be a great many intrasite matrix elements, and we still would have the long-range part to describe! Reformulated in terms of localized orbitals, LSDA would be a many-parameter theory[36], with a large local basis. Apparently, "the

---

[35]We do not attempt to do justice to later developments, such as the LDA+U approach, which are supposed to cover the Mott–Hubbard gap and other strong correlation features.

[36]Remember that those many parameters would not be arbitrarily chosen but

'electron glue' is such a complex many-body system that it cannot be usefully parametrized" [144] – not, if we are so ambitious as to require a unified description of low- and high-energy phenomena, ranging from 0.01eV to 100eV.

Though the above statement does not exclude the possibility of finding a relatively simple low-energy effective Hamiltonian, the fact remains that it is quite difficult to establish the relationship between an LSDA calculation, and the study of a model Hamiltonian with a small local basis, and a few interaction parameters. Let us recall that the Hubbard model was constructed with a single local orbital, and a single interaction parameter! One studies such, admittedly oversimplified, models in the hope that they allow the identification of the "driving force" of certain types of ordering. The approach is useful only if we have some reason to believe that, after including many more terms into the Hamiltonian, it becomes realistic enough to describe a system, but the nature of ordering is still basically the same as in the simple model. Assuming that this is the case, the understanding gained by working with the simple model is of great value.

The problem of the connection between LSDA and Hubbard-model physics has been brought into the focus of attention by the reviving controversy about itinerant ferromagnetism. LSDA has been able to describe many properties of the ferromagnetic iron group elements in impressive detail (Sec. 7.7.4). On the other hand, it is not clear how to account for band ferromagnetism starting from the Hubbard model (Ch. 8). Thus the question can be posed: which is the simplest lattice fermion model which would have the same qualitative features as the LSDA description of iron?

We return to a detailed discussion of this question in Ch. 8. There are indications that even the one-band Hubbard model is capable of ferromagnetic ordering in a physically reasonable parameter range if we "fine-tune" the kinetic energy by choosing certain non-bipartite lattices or/and farther-neighbour hopping. Another relevant extension is to degenerate models[37], or to many-band models. Time will show which

should be derived from the underlying Coulomb force. LSDA has no adjustable interaction parameters.

[37]These allow to bring into play the intraatomic exhange (responsible for Hund's

of these developments is relevant for the understanding of iron. From the point of view of many-body theory, we would like to have a lattice model of interacting fermions, which resembles iron in essential respects, shows ferromagnetism in a reasonable parameter range *and* it is simple enough so that we can go beyond the mean field level and make sure that ferromagnetism is a true feature of the model. The finding, and solving, of such a model will provide the missing link between the LSDA calculations, and the theory of correlation phenomena.

7.7.3 Nearly Ferromagnetic Metals

Materials for which $I\rho(\epsilon_F)$ is not much smaller than 1 are near to satisfying the Stoner criterion[38] (7.24) and can be called nearly ferromagnetic metals (NFM). It means that the enhancement factor $(1 - I\rho(\epsilon_F))^{-1}$ appearing in the expression (7.23) for the spin susceptibility has a largish value thus we can alternatively speak about exchange enhanced paramagnets (EEP) with a large enhancement factor[39].

A well-known NFM is palladium. The temperature and pressure dependence of the susceptibility is shown in Fig. 7.7. First of all, let us observe that χ is about an order of magnitude larger than that of simple metals like potassium. LSDA gives $I\rho(\epsilon_F) \sim .9$ and the calculated and experimental values of $\chi(T = 0)$ are fairly close [15]. Another notable feature is that $\chi(T)$ goes through a broad maximum around $T \sim 80$K. We learn in Problem 7.1 (p. 406) that this can be accounted for as a band structure effect: it arises if $\rho''(\epsilon_F)$ is large enough, and the feature is amplified by exchange enhancement.

Once a parameter is near a critical value, we are always interested in possibilities of fine-tuning it. Applying pressure drives $I\rho(\epsilon_F)$ away from 1: increased overlap widens the bands, and diminishes $\rho(\epsilon_F)$. Thus

---

rules) which was long suspected to be essential for ferromagnetism.

[38] Our derivation was formulated for the Hubbard model, and led to the enhancement factor $(1 - U\rho(\epsilon_F))^{-1}$. We understand, however, that the Stoner mean field theory applies to a wider class of models. Since in the present discussion, the Stoner parameter usually characterizes a material rather than a model, we replace U by the Stoner parameter I.

[39] Such names always have some ambiguity. We would say that W with its enhancement factor ~ 5 is merely an EEP, not "nearly ferromagnetic".

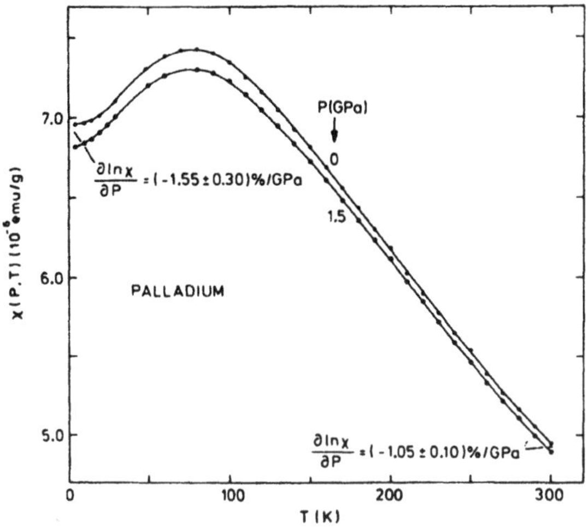

Figure 7.7: The temperature dependence of the susceptibility of Pd, under atmospheric pressure (upper curve), and a pressure of 1.5GPa (lower curve), after [131].

we are not surprised to learn that χ decreases under pressure [131]. A way to help Pd nearer to the ferromagnetic instability is to alloy it with moment-carrying atoms, say, iron. At small concentrations, "giant moments" arise: the host responds to the exchange fields of the impurities by developing huge polarization clouds which make the nominal moments of the Fe atoms look much bigger than what an atom could really carry. A concentration of a mere 0.1–0.2 at% of Fe or Co is sufficient to turn Pd into a ferromagnet.

The ordered alloys $Y(TM)_2$ form an interesting class of materials which allows to study the immediate neighbourhood of the Stoner instability in beautiful detail. YFe_2 is a large-moment ferromagnet while YCo_2 with $I\rho(\epsilon_F) \sim .85$ is a strongly exchange enhanced paramagnet [75]. The Stoner instability can be gradually approached by changing the Al-concentration in $Y(Co_{1-x}Al_x)_2$ alloys; the threshold for weak ferromagnetism is reached at $x_c = 0.13$ [344]. For a rnage of x, the susceptibility per formula unit is very much bigger than in Pd; note in particular the truly impressive susceptibility of $Y(Co_{.89}Al_{.11})_2$, for which we estimate $I\rho(\epsilon_F) \geq .98$ (Fig. 7.8, left). It has been argued [445] that there is a general relationship between a maximum seen in $\chi(T)$, and the possibility of a metamagnetic transition. Problem 7.2 demonstrates this for a specific choice of $\rho(\epsilon)$. Though the nature of such a correspondence may not be so straightforward at all, we observe that $Y(Co_{1-x}Al_x)_2$ shows both phenomena in the range $0.03 \leq x \leq 0.11$. Note that the fields in Fig. 7.8 (right) reach up to 100T and, for $H > 40$T, were produced by pulsed magnets. The width of the higher-field transitions is not inherent, but should rather be ascribed to the shortness of the pulse.

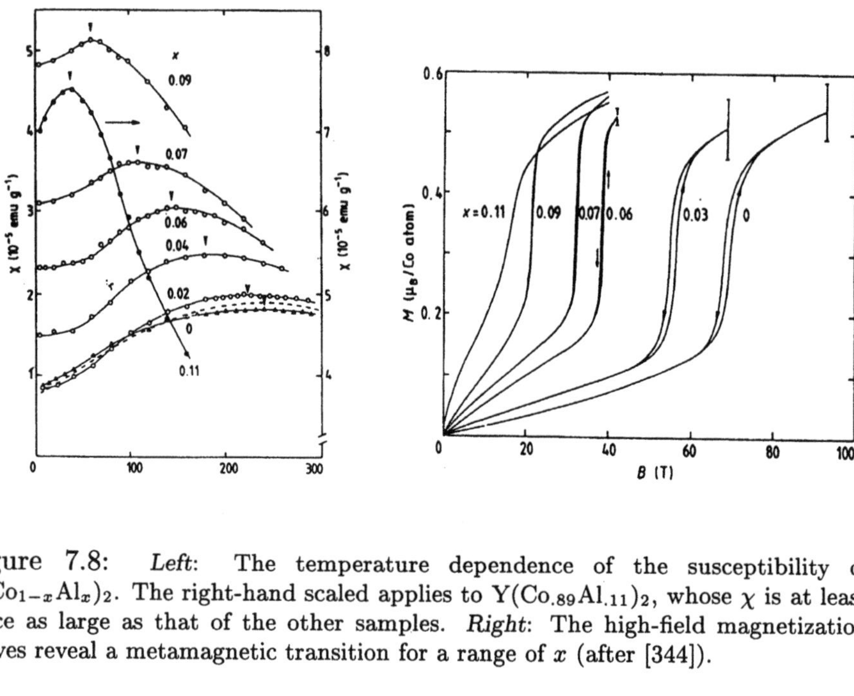

Figure 7.8: *Left:* The temperature dependence of the susceptibility of $Y(Co_{1-x}Al_x)_2$. The right-hand scaled applies to $Y(Co_{.89}Al_{.11})_2$, whose χ is at least twice as large as that of the other samples. *Right:* The high-field magnetization curves reveal a metamagnetic transition for a range of x (after [344]).

7.7.4 Ferromagnetic Metals

Weak Itinerant Ferromagnets

For us, the principal interest of the $ZrZn_2$ system lies in the fact that it provides a good testing ground for the Stoner model. Mean field theory is remarkably successful in describing the properties of the ordered phase. Its failure at $T > T_C$ highlights the essential limitations of Stoner theory.

The Stoner criterion (7.24) predicts that if ϵ_F lies in a strong maximum of the density of states then a relatively small I (or U) may be enough to drive the system ferromagnetic. It is a pure band structure effect: there is no need for the system to contain atoms which we ordinarily call magnetic (e.g., iron). A case in point is $ZrZn_2$, the ordered alloy of two nonmagnetic metals, which becomes ferromagnetic at a T_C which ranges (for various samples) from $\sim$ 21K to $\sim$ 27K. Another example would be $TiBe_2$ [442].

Very weak itinerant ferromagnets are systems for which the Stoner criterion is only marginally satisfied so that $I\rho(\epsilon_F)-1$ is a small parameter. The details of the corresponding mean field theory are relegated to Problem 7.3 (see p. 409); here we merely quote the results. We find that both T_C, and the $T=0$ band polarization $m(0)$ are small, being proportional to $\sqrt{I\rho(\epsilon_F)-1}$. In the absence of an external field, the order parameter follows

$$\frac{m^2(T)}{m^2(0)} = 1 - \left(\frac{T}{T_C}\right)^2. \tag{7.101}$$

In an external field H, the behaviour is governed by the Edwards–Wohlfarth [97] equation of the magnetic isotherms

$$\frac{m^2(T,H)}{m^2(0,0)} = 2\tilde{\chi}\frac{H}{m(T,H)} + 1 - \left(\frac{T}{T_C}\right)^2. \tag{7.102}$$

It was suggested [217] that for $ZrZn_2$, $I\rho(\epsilon_F)-1 \approx .015$, a very small value indeed. The measured temperature dependence of the magnetization (see Fig. 7.9) follows (7.101) beautifully in the entire range $T \leq T_C$. Magnetization measurements in pulsed fields [266] demonstrated that (7.102) is also well obeyed.

We contend that the weak ferromagnetism of $ZrZn_2$ is an accidental consequence of its band structure: apparently, ϵ_F is located near a sharp maximum in the density of states. If this picture is correct, then it should follow that the analogous $5d$ compound $HfZn_2$ which has the same crystal structure, and number of d-electrons, behaves similarly. Indeed, experiments show that $HfZn_2$ is an exchange enhanced paramagnet, with an estimated $I\rho(\epsilon_F) \sim 0.81$ [218]. Though the material does not manage to reach the ferromagnetic state, its $\chi(T=0)$ is uniquely

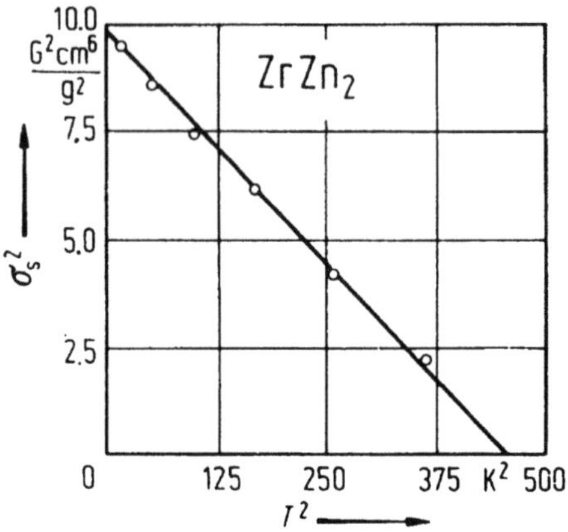

Figure 7.9: The measured temperature dependence of the magnetization of $ZrZn_2$ (circles), and the mean field prediction (7.101) (line) (after [311], reproduced in [442]).

high among $5d$ systems; in fact, it is about the same as the susceptibility of the nearly ferromagnetic $4d$ element Pd.

So far the successes of Stoner theory; now to the failures. We find that, according to (7.124), in the paramagnetic phase the inverse susceptibility should have a quadratic temperature dependence. As far as we know, such a behaviour is never observed. In fact, the susceptibility of $ZrZn_2$ is better approximated by a Curie–Weiss law in the range $T_C \leq T \leq 3T_C$.

The finding that the paramagnetic susceptibility of a supposedly itinerant magnet can be fitted by a Curie–Weiss law, is most puzzling. Sure enough, we have convinced ourselves (Problem 7.1) that at high enough temperatures, all band electron spin susceptibilities should follow a Curie–Weiss law. But that holds at $T \sim T_F$; it cannot conceivably explain the behaviour of $ZrZn_2$ at $T \sim 30$K! This is a major problem and we had better admit it at once that it cannot be tackled within any of the simple theories which we review here. For more sophisticated approaches, see [290] and [374].

It should be mentioned that not all ferromagnets formed of individually non-magnetic elements are necessarily weak. The ordered alloy VAu_4 is reported to have the quite respectable $T = 0$ moment $\sim 1\mu_B$ per formula unit [442].

Before leaving this subject of weak ferromagnetism, let us note that the smallness of the parameter $I\rho(\epsilon_F) - 1$ implies that it should be possible to drive the system through a ferromagnetic–paramagnetic transition by applying pressure. The essence of the mechanism is that bands get wider with decreasing volume, and thus $\rho(\epsilon_F)$ decreases. If we assume that $I\rho(\epsilon_F)$ depends linearly on the applied pressure P, and that $I\rho(\epsilon_F) = 1$ at a critical pressure P_{cr}, then (7.113) implies that

$$T_C \sim \sqrt{P_{cr} - P} \tag{7.103}$$

which is one of the notable predictions of Stoner theory. It is in this context that we mention the case of MnSi, for which $I\rho(\epsilon_F) = 1.03$. It is habitual to consider MnSi as a weak itinerant ferromagnet though, in fact, it has a $\mathbf{q} \neq 0$ helicoidal ground state. It is found that T_C decreases with increasing pressure, and disappears at $P_{cr} \sim 15$kbar, with the P-dependence following (7.103) in the neighbourhood of P_{cr} [395]. A further reason for mentioning MnSi is that since $T_C \to 0$ in a well accessible pressure range, it may be a good model system to study quantum critical behaviour [276].

Iron Group Elements

Whether a system at $T = 0$ is magnetic or not, is decided by the minimum condition for the ground state energy of the interacting electron system. Including the coupling to the lattice degrees of freedom, the same quantity decides about the equilibrium lattice structure and lattice constant. Consequently, there must be an intimate relationship between the structural and magnetic properties of solids. For instance, applying pressure or mechanical stress may cause a significant change in the magnetic properties. Conversely, applying a magnetic field nay drive a structural phase transition (see p. 478 for an example).

It is an outstanding achievement of LSDA theory that it offers a detailed understanding of the low-temperature properties of the ferromagnetic transition metals [141, 15]. Some of the properties of the elements are listed in the upper part of Table 7.1. What we know as ferromagnetic iron is the low-T, low-pressure phase α-Fe which has the bcc structure. Applying a pressure $P \sim 100$ kbar, there is a phase transition into the hexagonal ϵ-Fe phase which is apparently non-magnetic [443]. LSDA theory predicts that while bcc iron is magnetic for a broad range of the lattice constant, for the hexagonal phase the Stoner instability should be a first-order transition, with the magnetization appearing abruptly, at a lattice constant which is somewhat larger than what one actually finds in ϵ-Fe. It is a significant achievement that LSDA consistently predicts ferromagnetic TM to be ferromagnetic, and non-magnetic ones to be non-magnetic, even in borderline cases like that of ϵ-Fe. Also the $T = 0$ magnetization (which is shown as p_{sat} in the Table 7.1) is given accurately by the calculations. – The strongly ferromagnetic Co and Ni have $I\rho(\epsilon_F)$ well above 1, while Pd somewhat below.

substance	p_{sat} (μ_B)	$I\rho(\epsilon_F)$	T_C (K)	p_{par}/p_{sat}	struct.	type
α–Fe	2.12	1.5–1.7	1044	1.01	bcc	FM
ϵ–Fe	0	< 1	—	—	hcp	EEP
Co	1.66	1.6–1.8	1390	1.28	hcp	FM
Ni	0.54	2.1	624	1.375	fcc	FM
Pd	0	0.9	–	–	fcc	EEP
YFe_2	1.48	1.4	542		cubic	FM
YCo_2		0.85	–	–	cubic	EEP
$ZrZn_2$	0.12	1.015	21–27	4		WIF
MnSi	0.4	1.03	29	5.3	cubic	WIF

Table 7.1: The magnetic properties of some TM elements, and of several intermetallic compounds discussed in the text. The types of magnetism: FM: ferromagnet with strong moments, WIF: weak itinerant ferromagnet, EEP: exchange-enhanced paramagnet. p_{sat} is the $T = 0$ magnetic moment per atom (elements), or per formula unit (compounds). Experimental data are taken from [442] (elements), [290] (WIF), [75] (FeY_2 and CoY_2) while theoretical estimates are from sources cited in the text.

It is clear that $I\rho(\epsilon_F)$ is useful for systematizing the results, but we have to add a remark as to its meaning in the Table above. We will learn in Ch. 8 (see particularly the Discussion on p. 458) that the Hartree–Fock criterion (7.24), and also its generalization (8.37) for degenerate bands, is grossly misleading, thus I cannot be the bare U. Then what do we mean by claiming an excellent agreement between the observed behaviour and the Stoner condition? As explained previously, I (and, of course, also $\rho(\epsilon_F)$) are derived from an LSDA total energy calculation. In a ferromagnetic metal, I may appear as the average exchange splitting between spin-$\uparrow$ and spin-$\downarrow$ bands, and at the same time, we find $I\rho(\epsilon_F) > 1$, and a reasonable value of m_{sat}. Thus we may speak about a valid Stoner theory in the sense that it gives a consistent language in which the gross features of the results of a full-scale LSDA calculations can be described. However, we are not able to rederive these same results by postulating a simple lattice model, and using the mean field approximation[40], so the Stoner theory is not valid in its original naive sense. It is a future task to bridge the gap between these approaches.

Paramagnetic Susceptibility

The susceptibility of most metallic ferromagnets can be fitted to a Curie–Weiss law in a temperature range above T_C. This allows to identify a high-temperature paramagnetic moment which is regarded as an important measure of the degree of itinerancy of the magnetism. This should be understood as follows: Let us assume that the ground state saturation magnetization $p_{sat}\mu_B$ can be ascribed to localized spins; using $g = 2$, this would require the fictitious (in general, fractional) spin $S_{T=0} = p_{sat}/2$. Assuming that the high-T paramagnetic susceptibility corresponds to these localized spins, we get the estimate

$$\chi_{est} = \frac{\mu_B^2}{3k_B(T-\theta)} \cdot p_{sat}(2 + p_{sat}). \tag{7.104}$$

As a matter of fact, the observed susceptibility often follows a Curie–Weiss law, but with a different Curie constant

$$\chi_{obs} = \frac{\mu_B^2}{3k_B(T-\theta)} \cdot p_{par}(2 + p_{par}) \tag{7.105}$$

which allows us to measure the paramagnetic Bohr magneton number p_{par}. The parameter p_{par}/p_{sat} is listed in Table 7.1. For localized spins, it would have the value 1. However, for itinerant (band) ferromagnets,

---

[40]However, the overall features of an essentially exact solution for such models may closely resemble the LSDA results, and may give an insight into the role of the shape of $\rho(\epsilon)$ (p. 458), and intraatomic exchange (p. 470).

the saturation moment arises from the different partial occupations of the exchange-split spin-↑ and spin-↓ bands. In the simplest mean field treatment, the exchange splitting vanishes for $T > T_C$, so we have no right to expect that p_{sat} has any relevance for the paramagnetic susceptibility. Thus it should come as a surprise that for iron, $p_{par}/p_{sat} \sim 1$, indicating that the magnetic moments are "almost localized", and survive essentially intact into the disordered phase. This is a local correlation effect, completely beyond the scope of the simple Stoner theory. Let us note, however, that there exists a finite-T version of LSDA which gives a fairly accurate description of paramagnetic iron [374]. The same theory predicts that Ni is a much more itinerant (essentially Stoner-type) ferromagnet.

For other substances p_{par}/p_{sat} exceeds 1 the more, the more itinerant the nature of the ferromagnetic state is. A large value of p_{par}/p_{sat} (ranging up to ~ 15 for some iron–cobalt silicides) is typically associated with a low p_{sat}, and a low T_C. As we noted before, the important subject of apparent local moment behaviour in the paramagnetic phase of metallic ferromagnets lies beyond the scope of the present text.

7.7.5 Chromium

At $T_N = 311$K, structurally perfect[41], single-crystal Cr samples undergo a phase transition into an incommensurate, linearly polarized spin density wave state. An extensive review of the properties of this fascinating system is given in [103]. We learn on the example of Cr, what kind of behaviour is expected if the SDW does not open a gap over the entire Fermi surface, thus the system remains metallic below T_N.

Paramagnetic Cr has the bcc structure. It would be easy to envisage a two-sublattice antiferromagnetic structure, with the moments of the corner site atoms of the cubic unit cells pointing antiparallel to those of the center sites. This structure could be described as alternating from plane to plane, as we proceed along (say) the z-axis, thus it would have

[41] The sensitive dependence of the magnetic order on the mechanical quality of the sample was the source of a long-standing difficulty in understanding the nature of Cr. Cold-worked polycrystalline samples show a transition into a different, commensurate magnetic phase around 475K, providing an example of "technical antiferromagnetism" [103].

the wave vector $(0, 0, 1)a^*$, where $a^* = (2\pi/a)$. The actual structure of the SDW appearing below T_N differs from this inasmuch as we find a long-wavelength modulation of the sublattice magnetization: the magnetization at the corner site points opposite to that at the centre site but it is of a slightly different magnitude (Fig. 7.10). Magnetic neutron scattering indicates a linearly polarized transverse wave, with wave vector $\mathbf{Q}_z = (0, 0, 1 - \delta)a^*$. δ changes gradually from 0.037 to 0.048 as the temperature decreases from T_N to zero. We can speak about a nearly commensurate structure which is near to the two-sublattice structure described above. In fact, alloying Cr with a few per cent of Fe, Mn, Ru, etc. causes a transition into the commensurate state.

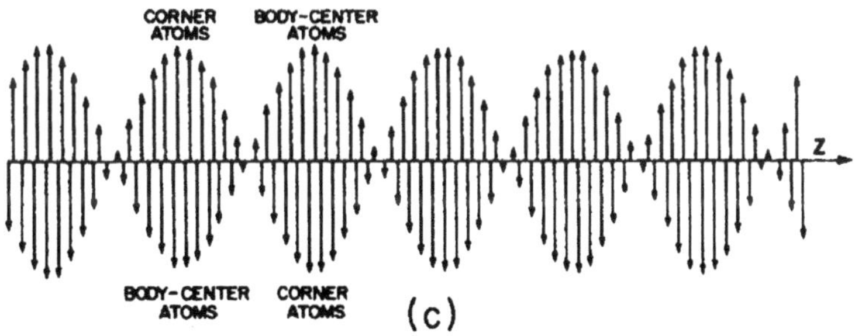

Figure 7.10: Schematic representation of the structure of the SDW in Cr at temperatures 123K < T < 311K (after [103]). z is one of the principal axes of the bcc paramagnetic phase. The magnetic moment of an atom at the center of the cubic unit cell is antiparallel to that at a corner site.

Cr is an itinerant magnet: its incommensurate SDW is well explained by its band structure. It is found that there is approximate nesting over substantial parts of the Fermi surface. The vector $\mathbf{Q}_z$ mentioned above can be identified as the spanning vector connecting electron-like and hole-like sheets of the Fermi surface. In view of the cubic symmetry of the paramagnetic phase, there must be equivalent $\mathbf{Q}_x$, and $\mathbf{Q}_y$ spanning vectors as well. We have to ask whether the SDW belongs to a single $\mathbf{Q}$, or we have a triple-$\mathbf{Q}$ structure. It turns out that the shape of the Fermi surface favours the single-$\mathbf{Q}$ SDW. Samples cooled through T_N in the absence of a magnetic field appear macroscopically cubic, since they have a polydomain structure, the individual domains corresponding with equal probability to either $\mathbf{Q}_x$, $\mathbf{Q}_y$, or $\mathbf{Q}_z$. Cooling in an external field $H \geq 3\text{T}$ produces a single-$\mathbf{Q}$ sample with tetragonal symmetry. It is easy to understand that $\mathbf{Q}$ must point in the direction of the field. The susceptibility of the SDW system is like that of an antiferromagnet: $\chi_{\perp} > \chi_{\|}$, therefore the moments tend to align in a direction which is perpendicular to the field. At least, this holds for $123\text{K} < T < 311\text{K}$. Curiously, Cr has also a second magnetic phase transition: a spin-flip transition is observed at $T_{\text{sf}} = 123\text{K}$ where the character of the linearly polarized SDW changes abruptly from transverse to longitudinal.

The shape of the Fermi surface explains the choice of the $\mathbf{Q}$ vectors, but it is not sufficient to justify the existence of the SDW. Mo and W (the $4d$ and $5d$ element in the same column of the periodic table) have rather similar Fermi surfaces but no SDW. If nesting is only approximate, we need a threshold value of the interaction to permit an SDW state; apparently, this is given only in Cr. It is a general rule that $4d$ and $5d$ transition metals show weaker interaction effects than their $3d$ counterparts.

Actually, even Cr is only marginally magnetic[42]. ϵ_{F} is sitting in a deep valley of the density of states, with an estimated $\rho(\epsilon_{\text{F}}) \approx 0.603\text{eV}^{-1}$, thus though the Stoner $I \approx 0.82\text{eV}$ is about as high as for other transition metals, $I\rho(\epsilon_{\text{F}}) \approx 0.5$ is far below 1, the critical value for ferromagnetism. Naturally, here we should rather use the generalized Stoner criterion (7.45) for SDW which contains $\chi^{(0)}(\mathbf{Q})$, where $\mathbf{Q}$ is the observed wave vector. Approximating $\mathbf{Q}$ with the nearby commensurate vector (0,0,1), it was estimated that $I\chi^{(0)}(\mathbf{Q}) \approx 1.025$, only slightly above the threshold value.

Can we say why the SDW is linearly rather than circularly polarized? We might have envisaged a helicoidal wave of the kind

$$\mathbf{S}(\mathbf{r}) = (\cos(Qz), \sin(Qz), 0) . \qquad (7.106)$$

However, there is a simple argument to show that the helicoidal wave is less effective in opening gaps than the linearly polarized wave (see Problem 7.8). The tendency to form a linearly polarized SDW is char-

[42] We use the estimates of [366], as cited in [103].

acteristic of itinerant magnets. In contrast, localized-spin systems tend to form helicoidal patterns.

The SDW amplitude in Cr is moderately large: $0.62\mu_B$. Actually, it is not really proper to speak about a spin density wave, since ~60% of the magnetization is orbital moment, and only ~40% is due to spins. The experimental observation that the susceptibility shows only a relatively small decrease due to the SDW transition, can also be explained by arguing that only the spin susceptibility is strongly affected by the SDW gaps.

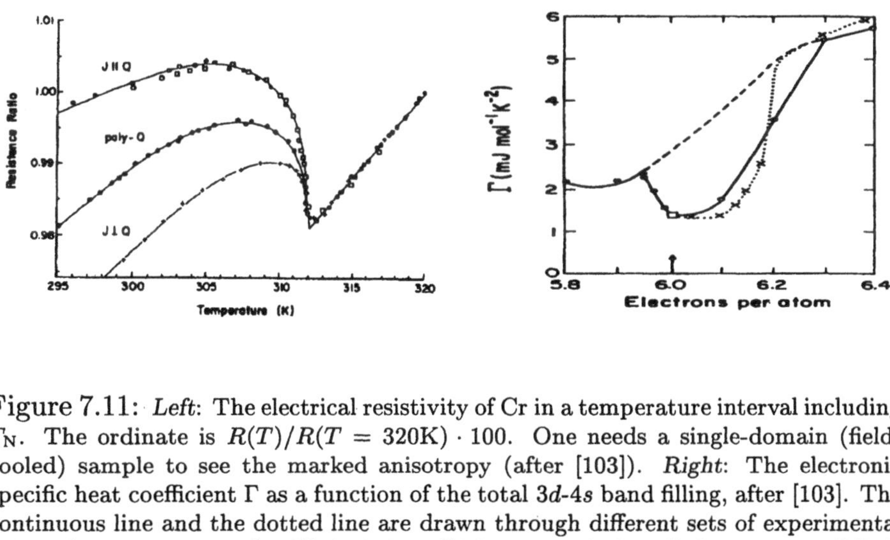

Figure 7.11: *Left*: The electrical resistivity of Cr in a temperature interval including T_N. The ordinate is $R(T)/R(T = 320\text{K}) \cdot 100$. One needs a single-domain (field-cooled) sample to see the marked anisotropy (after [103]). *Right*: The electronic specific heat coefficient Γ as a function of the total $3d$-$4s$ band filling, after [103]. The continuous line and the dotted line are drawn through different sets of experimental points (empty square: Cr, filled circles: CrV, empty circles: CrOs, crosses: CrRe), but in either case, SDW formation causes a marked dip from the hypothetical non-magnetic state (dashed line).

The effect of the opening of the SDW gaps is most clearly seen in the electronic specific heat, and the electrical resistivity. Fig. 7.11 (right) shows the electronic specific heat coefficient Γ as a function of the electron concentration. $n = 6.0$ represents Cr, and the surrounding range can be realized by alloying Cr with various transition metals. The Γ of the hypothetical non-magnetic metal is shown by dashed line (this line can also be traced experimentally, using data for the isoelectronic Mo(Nb/Re) and W(Ta/Re) systems which have no SDW). It is clear that the formation of the SDW decreases $\rho(\epsilon_F)$ by gapping away part of the Femi surface, and that the effect is quite dramatic for pure Cr. That charge carriers are missing, is also clear from the resistivity data (Fig. 7.11, left). Another remarkable feature is the appearance of a strong resistivity anisotropy in single-domain samples.

Our previous arguments would indicate that the transition to the SDW state is of second order. However, experiments show that chromium undergoes a weakly first-order transition at T_N=311K. Apparently, it is quite difficult to devise a microscopic model of Cr, which would predict a mildly discontinuous transition. In any case, the coupling to the lattice degrees of freedom is likely to be important. It is observed that the SDW is coupled to a strain wave which, in turn, causes a charge density wave. Higher-order harmonics of the SDW are also seen.

Solutions to the Problems

Problem 7.1 a) Using the steps which lead to (7.46), for non-interacting electrons

$$\chi^{(0)}(\mathbf{q}=0) = \int d\epsilon \left(-\frac{\partial f}{\partial \epsilon}\right) \rho(\epsilon). \tag{7.107}$$

We follow the standard procedure of the Sommerfeld expansion. Since $(-\partial f/\partial\epsilon)$ becomes vanishingly small beyond a distance of a few k_BTs on either side of the chemical potential μ, the integral can be extended to $\pm\infty$. Expanding ρ in a Taylor series about $\epsilon = \mu$

$$\begin{aligned} \frac{\chi}{2\mu_B^2} &\approx \int_{-\infty}^{\infty} d\epsilon \left(-\frac{\partial f}{\partial \epsilon}\right) [\rho(\mu) + \rho'(\mu)(\epsilon-\mu) + \frac{1}{2}\rho''(\mu)(\epsilon-\mu)^2] \\ &\rightsquigarrow \rho(\mu) + \frac{\pi^2}{12}\rho''(\mu)(k_BT)^2 \\ &\rightsquigarrow \rho(\epsilon_F) + \frac{\pi^2}{12}\left[\rho''(\epsilon_F) - 2\frac{(\rho'(\epsilon_F))^2}{\rho(\epsilon_F)}\right] \cdot (k_BT)^2 \end{aligned} \tag{7.108}$$

where, in the last line, we used the standard result for the temperature dependence of the chemical potential. It is apparent that, depending on the detailed form of the density of states, the coefficient of T^2 can be either positive, or negative.

b) Let the band lie in the interval $[-\frac{W}{2}, \frac{W}{2}]$, and the density of states satisfies $\rho(\epsilon) = \rho(-\epsilon)$. For half-filling, $\epsilon_F = 0$. It is easy to check that the chemical potential remains pinned at $\mu = 0$ for arbitrary temperatures. At high $T >> W$, expanding in terms of W/T gives

$$\chi = \frac{2\mu_B^2}{k_B T}\int_{-W/2}^{W/2} \frac{\rho(\epsilon)e^{\epsilon/k_B T}d\epsilon}{\left(e^{\epsilon/k_B T}+1\right)^2} \approx \frac{\mu_B^2}{2k_B T} + O(\frac{I_2}{T^3}) \qquad (7.109)$$

where I_2 is the second moment of the DOS. The leading term is Curie-like and corresponds to half of the sites having an unpaired electron. We were discussing (Sec. 4.2) that in a half-filled band, on the average, half of the sites are occupied by one electron each. At low T, the rapid hopping motion mixes the configurations and quenches most of the spins. At high T, the quantum-mechanical coherence between the configurations is destroyed by thermal fluctuations, and the spins "become visible". This is of mere academic interest for ordinary metals. However, for correlation-induced effective narrow bands of widths $W \leq k_B T_{\text{room}}$, the crossover from a low-temperature Pauli susceptibility to a high-temperature Curie susceptibility is an important effect.

Let us note that the assumption about a symmetrical DOS was made merely for convenience: we did not have to worry about the temperature dependence of μ. The result (7.109) holds for an arbitrary form of $\rho(\epsilon)$.

c) According to (7.108), the necessary condition for the spin susceptibility to increase as T is raised from $T = 0$ is that $\rho(\epsilon)$ must have a sufficiently large positive curvature at $\epsilon = \epsilon_F$. It is certainly satisfied if we place ϵ_F at the bottom of the central valley of the double-peak DOS shown in the left panel of Fig. 7.12. We choose

$$\rho(\epsilon) = \frac{15}{26}\left(\frac{1}{4} + \frac{7\epsilon^2}{16} - \frac{\epsilon^4}{8}\right) \qquad \text{for } |\epsilon| \leq 2 \qquad (7.110)$$

and assume that the band is half-filled, so that $\epsilon_F = 0$, and $\mu = 0$ at all T. The temperature dependent spin susceptibility (7.107) of non-interacting band electrons is shown (in units of $(g\mu_B)^2$) as the lower curve in the right-hand panel of Fig. 7.12.

The effect of switching on U is described by (7.44). We will learn from solving Problem 7.2 that the $T = 0$ Stoner instability sets in at $U \approx 4.23$, thus $U = 1.8$ is for this problem a modest interaction. Still, it leads to a much more pronounced maximum in $\chi(T)$ (Fig. 7.12, right, upper curve). The reason is that the enhancement factor also grows with increasing $\chi^{(0)}(T)$.

The temperature dependence of the measured susceptibility of Pd (and even more spectacularly $Y(Co_{.89}Al_{.11})_2$ [344]) shows a maximum (Fig. 7.7) which is reminiscent of the feature found by us. We should not jump to the conclusion that ϵ_F in Pd is sitting in a local minimum as in our model calculation. The necessary condition for the existence of a maximum is merely

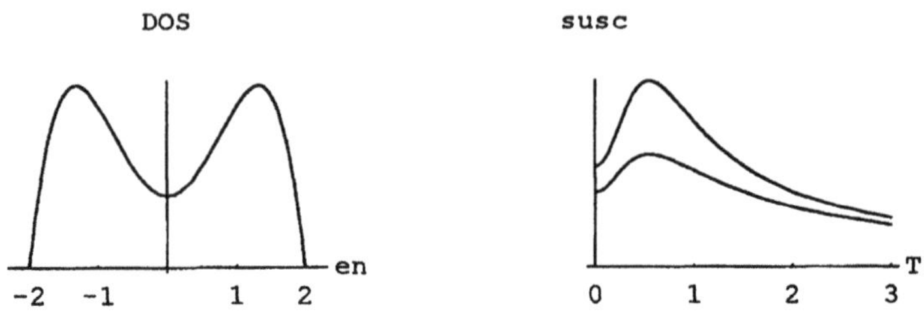

Figure 7.12: Assuming the model density of states (7.110) (left) gives rise to a maximum in the temperature dependence of the spin susceptibility (right, lower curve). Even a moderate interaction ($U = 1.8$) makes the maximum much more pronounced (upper curve).

$\rho''(\epsilon_F) > 2(\rho'(\epsilon_F))^2/\rho(\epsilon_F)$. It appears that in Pd, ϵ_F is sitting on the right-hand flank of a strong maximum of $\rho(\epsilon)$.

Problem 7.2 a) We want to find the condition for $E(m)$ having its minimum at some finite m. Let us first take $H = 0$. Taking the derivative of (7.21), and using

$$\frac{\partial \mu_\uparrow}{\partial m} = \frac{1}{\rho(\mu_\uparrow)} \quad \text{and} \quad \frac{\partial \mu_\downarrow}{\partial m} = -\frac{1}{\rho(\mu_\downarrow)}$$

which can be derived from (7.20), we get the finite-m Stoner criterion

$$U\bar{\rho}(\epsilon_F, m) = 1 \tag{7.111}$$

where $\bar{\rho}(\epsilon_F, m)$ is the DOS averaged over the suitably chosen environment of ϵ_F which accomodates $2m$ states:

$$\bar{\rho}(\epsilon_F, m) = \frac{\int_{\mu_\downarrow}^{\mu_\uparrow} \rho(\epsilon) d\epsilon}{\mu_\uparrow - \mu_\downarrow} = \frac{m}{\mu_\uparrow - \mu_\downarrow}. \tag{7.112}$$

It requires further investigation to decide whether the solution of (7.111) belongs to a local minimum, or a local maximum. Furthermore, it may happen that $\mathcal{E}(m)$ has its absolute minimum at the edge of the allowed domain of m, namely at $m = n$. If this is the case, (7.111) need not hold, and the minimum has to be found numerically.

b) Taking (7.110), the Stoner criterion (7.24) for the instability against an infinitesimal spin polarization is satisfied for $U = 104/15 \approx 6.93$. However, the condition for the appearance of finite-amplitude ferromagnetism is much less stringent. Plotting the energy (7.21) against m for various U (Fig. 7.13,

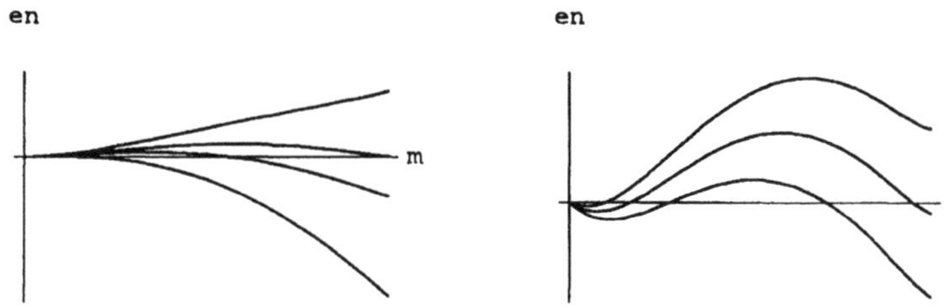

Figure 7.13: The m-dependence of the magnetic part of the energy density, assuming the DOS (7.110). Left: for $H = 0$, for $U = 3$, 4.23, 5, and 104/15 (curves from top down). The first-order transition to saturation magnetization sets in at $U = 4.23$. Right: for $U = 3.8$, in an external field $H = 0.15$, 0.225, and 0.3 (curves from top down). A metamagnetic transition occurs at $H \approx 0.225$.

left), we find that the absolute minimum belongs to the saturation polarization $m = 1/2$ for $U \geq 4.23$.

Finding that the $T = 0$ paramagnetic–ferromagnetic transition is first-order, is in itself not unphysical at all. However, it does not necessarily follow that m should immediately jump to its saturation value 1/2. Our (7.110) was meant to imitate merely a sharp structure around ϵ_F but not the entire band. Attaching tails to our $\rho(\epsilon)$, we should find an abrupt transition to an $m < 1/2$ state.

Choosing the marginally subthreshold value $U = 3.8$, we expect that the transition to the strongly magnetized case can be induced by a relatively small external field. Plotting the magnetic part of the energy density as a function of m for various fields H (Fig. 7.13, right), we see that the absolute minimum switches from $m = 0$ to $m = 1/2$ at $H \approx 0.225$. This is another example of the metamagnetic transition which apparently can have a number of different physical mechanisms (see also p. 536 and Sec. 3.8.3). Our calculation may be regarded as a naive model of the metamagnetic transition seen in the YCo_2 system. It has been surmised that for nearly ferromagnetic materials, there is a close relationship between the existence of a susceptibility maximum and the possibility of a metamagnetic transition. Our model DOS certainly exemplifies both.

Problem 7.3 The small parameter in the limit of very weak itinerant ferromagnetism is $U\rho(\epsilon_F) - 1 \ll 1$. Since in the temperature range of interest $T \ll T_F$, we can use the Sommerfeld expansion. Furthermore, because of $m \ll 1$, $\mu_\uparrow$ and $\mu_\downarrow$ deviate very little from ϵ_F, so the band can be characterized by ρ and its derivatives at ϵ_F.

a) Replacing (7.108) in (7.45), we get

$$\frac{\pi^2 k_B^2}{6} \cdot T_C^2 = \frac{U\rho(\epsilon_F) - 1}{\left(\frac{\rho'(\epsilon_F)}{\rho(\epsilon_F)}\right)^2 - \frac{\rho''(\epsilon_F)}{\rho(\epsilon_F)}}. \tag{7.113}$$

b) We have to minimize the free energy $\mathcal{F} = \mathcal{E}(T) - T\mathcal{S}$ the same way as we minimized the ground state energy in the solution of 7.2.a. Identifying $\mathcal{E}(T)$ from the finite-T version of (7.19), we write $\mathcal{F}$ as

$$\mathcal{F}(T) = \sum_\sigma \int_{-\infty}^{\infty} \epsilon\rho(\epsilon) f_\sigma(\epsilon) d\epsilon + U(n^2 - m^2) - g\mu_B Hm - \frac{(\pi k_B T)^2}{3} \sum_\sigma \rho(\mu_{\sigma 0}) \tag{7.114}$$

where f_σ is the Fermi distribution with the chemical potential μ_σ, satisfying

$$\int_{-\infty}^{\infty} \rho(\epsilon) f_\sigma(\epsilon) d\epsilon = n + m \cdot \eta(\sigma). \tag{7.115}$$

$\eta(\sigma) = \pm 1$ for $\sigma = \uparrow, \downarrow$ and $\mu_{\sigma 0}$ is the $T = 0$ value of μ_σ.

The term $-T\mathcal{S}$ was obtained from the following consideration: it is a basic assumption of the Stoner theory that the entropy arises solely from the electron–hole excitations, i.e., it is the entropy of band electrons. Using that the entropy is related to the specific heat via $C_V = T(\partial \mathcal{S}/\partial T)$, and the usual expression of the low-temperature specific heat of the spin-$\uparrow$ and spin-$\downarrow$ bands, the above form is obtained.

$\mathcal{F}$ depends on m both explicitely, and via μ_σ. Taking the derivative of both sides of (7.115) with respect m, using $(\partial f_\sigma/\partial \mu_\sigma) = -(\partial f_\sigma/\partial \epsilon)$, and the Sommerfeld expansion, we arrive at

$$\frac{\partial \mu_\sigma}{\partial m} = \eta(\sigma) \cdot \left[\rho(\mu_{\sigma 0}) - \frac{(\pi k_B T)^2}{6} \left(\frac{(\rho'(\mu_{\sigma 0}))^2}{\rho(\mu_{\sigma 0})} - \rho''(\mu_{\sigma 0})\right)\right]^{-1}. \tag{7.116}$$

Minimizing $\mathcal{F}$ with respect to m proceeds analogously to the derivation of (7.111), only we have to observe that μ_σ depends on T

$$\mu_\sigma = \mu_{\sigma 0} - \frac{(\pi k_B T)^2}{6} \cdot \frac{\rho'(\mu_{\sigma 0})}{\rho(\mu_{\sigma 0})}. \tag{7.117}$$

To order T^2, the extremum condition is found to be

$$\mu_{\uparrow 0} - \mu_{\downarrow 0} - \frac{(\pi k_B T)^2}{6} \left[\frac{\rho'(\mu_{\uparrow 0})}{\rho(\mu_{\uparrow 0})} - \frac{\rho'(\mu_{\downarrow 0})}{\rho(\mu_{\downarrow 0})}\right] = 2Um + g\mu_B H. \tag{7.118}$$

In what follows, we exploit that the magnetic order is weak even at $T = 0$, and expand to order m^2. The formulas are simpler if we abandon the general case and assume a half-filled band with the model DOS $\rho(\epsilon) = \rho_0 - a\epsilon^2$, with $a > 0$. Then

$$\rho(\mu_{\uparrow 0}) = \rho(\mu_{\downarrow 0}) \approx \rho_0 - a\frac{m^2}{\rho_0^2} \quad \text{and} \quad \rho'(\mu_{\uparrow 0}) = -\rho'(\mu_{\downarrow 0}) \approx -\frac{2am}{\rho_0}.$$

Let us first take the case $H = 0$. The spontaneous magnetization satisfies

$$\frac{1}{\rho_0}\left(1 + \frac{am^2}{\rho_0^3}\right) + \frac{(\pi k_B T)^2}{3} \cdot \frac{a}{\rho_0^2} = U. \tag{7.119}$$

At $T = 0$

$$m^2(0) = (U\rho_0 - 1) \cdot \frac{\rho_0^3}{a}. \tag{7.120}$$

T_C is obtained from requiring $m \to 0$

$$\frac{\pi^2 k_B^2}{3} \cdot T_C^2 = (U\rho_0 - 1) \cdot \frac{\rho_0}{a}. \tag{7.121}$$

(compare with this to (7.113) which we found by a different argument). Combining these with (7.119), we arrive at the celebrated result of the Stoner theory

$$\frac{m^2(T)}{m^2(0)} = 1 - \left(\frac{T}{T_C}\right)^2. \tag{7.122}$$

Allowing for the presence of $H \neq 0$, we get the Edwards–Wohlfarth [97] equation of the magnetic isotherms

$$\frac{m^2(T,H)}{m^2(0,0)} = 2\tilde{\chi}\frac{H}{m(T,H)} + 1 - \left(\frac{T}{T_C}\right)^2 \tag{7.123}$$

where $\tilde{\chi} = \rho_0 g \mu_B (U\rho_0 - 1)^{-1}$.

c) At $T > T_C$, and in a weak external field, we can omit the m^2 term in (7.123) to obtain the paramagnetic susceptibility

$$\chi(T) = \frac{2\tilde{\chi}}{\left(\frac{T}{T_C}\right)^2 - 1}. \tag{7.124}$$

The prediction of a quadratic T-dependence of the inverse susceptibility is a notable result of the Stoner theory.

Problem 7.4. We have to evaluate (7.42) at $T = 0$. With the free electron dispersion $\epsilon(k) = \hbar^2 k^2/2m$, this involves the integral $\int dk/(2k + q)$, leading to

$$\chi^{(0)}(q) = \frac{2m}{\hbar^2 q} \ln \frac{q + 2k_F}{|q - 2k_F|}. \tag{7.125}$$

Note that the apparent divergence at $q = 0$ is cancelled by the expansion of the ln. The only true divergence occurs at $q = \pm 2k_{\mathrm{F}}$, given by $\ln(q \pm 2k_{\mathrm{F}})$.

Solving the same problem with the tight binding dispersion $\epsilon(k) = -2t\cos k$, we are led to

$$\chi^{(0)}(q) = \frac{1}{t\sin(q/2)} \ln \frac{\tan(2k_{\mathrm{F}}+q)}{|\tan(2k_{\mathrm{F}}-q)|} . \tag{7.126}$$

Though the details of the formula differ from those of (7.125), the divergent behaviour is governed by the same $\ln(q \pm 2k_{\mathrm{F}})$.

Problem 7.5 One cannot use the Sommerfeld expansion because the zeroth-order contribution would diverge. Since the largest contribution is expected from a small neighbourhood of the band centre, it is acceptable to approximate the Fermi distribution by a formula which has the correct value, and first derivative, at $\epsilon = 0$:

$$f(\epsilon) \approx \begin{cases} 1 & \text{if } \epsilon < -2k_{\mathrm{B}}T \\ \dfrac{1}{2} - \dfrac{\epsilon}{4k_{\mathrm{B}}T} & \text{if } -2k_{\mathrm{B}}T < \epsilon < 2k_{\mathrm{B}}T \\ 0 & \text{if } \epsilon < -2k_{\mathrm{B}}T. \end{cases} \tag{7.127}$$

Using (4.31), the **q**=**0** response is described by (7.46)

$$\int \left(-\frac{\partial f}{\partial \epsilon}\right) \rho(\epsilon) d\epsilon \rightsquigarrow \frac{1}{4\pi^2 t k_{\mathrm{B}}T} \int_0^{2k_{\mathrm{B}}T} \ln\frac{t}{\epsilon} d\epsilon \rightsquigarrow \frac{1}{2\pi^2 t} \ln\frac{t}{2k_{\mathrm{B}}T} \tag{7.128}$$

thus for small T we find

$$\chi^{(0)}(\mathbf{q}=\mathbf{0}, T) \sim \frac{1}{t} \ln\frac{t}{2k_{\mathrm{B}}T}. \tag{7.129}$$

For **q**=**Q** the perfect nesting property (7.49) allows us to rewrite (7.42) as

$$\int \frac{f(\epsilon)-1/2}{-\epsilon} d\epsilon \rightsquigarrow \frac{1}{4\pi^2 t k_{\mathrm{B}}T} \int_0^{2k_{\mathrm{B}}T} \ln\frac{t}{\epsilon} d\epsilon + \frac{1}{\pi^2 t} \int_{2k_{\mathrm{B}}T}^{4t} \frac{1}{\epsilon} \ln\frac{t}{\epsilon} d\epsilon. \tag{7.130}$$

The largest contribution arises from the neighbourhood of the lower limit of the second integral

$$\chi^{(0)}(\mathbf{Q}, T) \sim \frac{1}{t} \left(\ln\frac{t}{2k_{\mathrm{B}}T}\right)^2 . \tag{7.131}$$

Thus at $T \to 0$, the **q**=**0** response diverges like $\ln T$, while **q**=**Q** response diverges as $\ln^2 T$. The antiferromagnetic instability is stronger.

Replacing the above result in (7.45), we get the following expression for the ordering (Néel) temperature in the weak coupling limit

$$T_{\rm N}(d=2) \sim t \cdot \exp\left[-\sqrt{\frac{t}{U}}\right]. \tag{7.132}$$

For hypercubic lattices in $d \neq 2$, $\rho(0)$ is finite. $\chi^{(0)}(\mathbf{Q})$ can be estimated as in (7.130), only the factor $\ln(t/\epsilon)$ has to be replaced by $\rho(0)$. The second term gives the leading contribution

$$\chi^{(0)}(\mathbf{Q},T) \sim \frac{1}{t}\ln\frac{t}{k_{\rm B}T}. \tag{7.133}$$

Hence

$$T_{\rm N}(d \neq 2) \sim t \cdot \exp\left(-\frac{t}{U}\right). \tag{7.134}$$

Naturally, we must not forget that (7.132) and (7.134) are mean field results. Mean field theory predicts finite $T_{\rm N}$ for any dimension, but we have to remember that the Mermin–Wagner theorem excludes the breaking of continuous symmetries at finite T for $d=1$ and $d=2$.

Problem 7.6 Using the mean field assumption $\langle n_{j\uparrow}\rangle = 1/2+(-1)^j m$, $\langle n_{j\downarrow}\rangle = 1/2-(-1)^j m$ for $j=1,2$ we are led to the mean field Hamiltonian

$$\mathcal{H}_{\rm MFA} = \mathcal{H}_\uparrow + \mathcal{H}_\downarrow + U - 2U(\frac{1}{4} - m^2) \tag{7.135}$$

where

$$\mathcal{H}_\sigma = -t(c_{1\sigma}^\dagger c_{2\sigma} + c_{2\sigma}^\dagger c_{1\sigma}) - Um\eta(\sigma)(\hat{n}_{1\sigma} - \hat{n}_{2\sigma}). \tag{7.136}$$

The spectrum is σ-independent: both $\mathcal{H}_\uparrow$ and $\mathcal{H}_\downarrow$ have the eigenvalues

$$\lambda^\pm = \pm\sqrt{t^2+U^2m^2}. \tag{7.137}$$

Proceeding as we did in Sec. 7.6, m can be obtained from minimizing

$$\mathcal{E} = 2Um^2 - 2\sqrt{t^2+U^2m^2} + \frac{U}{2} \tag{7.138}$$

which yields

$$m = \begin{cases} m = 0 & \text{if } U \leq 2t \\ m = \sqrt{\frac{1}{4} - \left(\frac{t}{U}\right)^2} & \text{if } U \geq 2t. \end{cases} \tag{7.139}$$

The simplicity of the problem makes it easy to follow the alternative route: writing down the self-consistency equation in the literal sense. In the basis

$|10\rangle = c_{1\uparrow}^{\dagger}|0\rangle$, $|01\rangle = c_{2\uparrow}^{\dagger}|0\rangle$ ($|0\rangle$ being the vacuum), the spin-$\uparrow$ ground state is (unnormalized)

$$|\phi_{\uparrow}\rangle = (mU + \sqrt{t^2 + U^2 m^2})|10\rangle + t|01\rangle. \tag{7.140}$$

Since in the two-electron ground state $|\phi_{\uparrow}\rangle$ holds one electron, self-consistency requires

$$m = \frac{\langle\phi_{\uparrow}|\hat{n}_{1\uparrow}|\phi_{\uparrow}\rangle}{\langle\phi_{\uparrow}|\phi_{\uparrow}\rangle} - \frac{1}{2} = \frac{mU}{2\sqrt{t^2 + U^2 m^2}}, \tag{7.141}$$

the same condition that derived by minimizing $\mathcal{E}$.

We learn that the mean field approximation gives symmetry-breaking ground states even for small systems. This is a spurious result[43]. What is more, we found here a $T = 0$ phase transition: the ground state is non-magnetic for weak coupling ($U \le 2t$), while it is 'antiferromagnetic' for sufficiently strong coupling ($U > 2t$). This is to be contrasted with the exact result that the ground state is always a singlet.

The quality of the MFA can be judged also by comparing the ground state energy derived here

$$\mathcal{E}_{\text{MFA}} = \begin{cases} -2t + \frac{U}{2} & \text{if } U \le 2t \\ -\frac{2t^2}{U} & \text{if } U \ge 2t \end{cases} \tag{7.142}$$

with the exact

$$\mathcal{E}_{\text{ex.}} = \frac{1}{2}(U - \sqrt{U^2 + 16t^2}) \tag{7.143}$$

It is easy to see that for intermediate U, $\mathcal{E}_{\text{ex.}}$ is significantly below $\mathcal{E}_{\text{MFA}}$. In the large-U limit $\mathcal{E}_{\text{MFA}} \approx -2t^2/U$, while $\mathcal{E}_{\text{ex.}} \approx -4t^2/U$. The energy scale $-t^2/U$ is right but MFA predicts a too small prefactor. We can argue that of the antiparallel spin correlations, MFA detects only the z–z correlations while in a true singlet, the x and y components also contribute.

7.7. a) We seek the $T = 0$ gap by expanding the ground state energy

$$E(\Delta) = U\left(\frac{1}{4} + \frac{\Delta^2}{U^2}\right) + \frac{1}{L}\sum^{\text{occ.}}\left(\epsilon_{\mathbf{k}} + \epsilon_{\mathbf{k+Q}} - \sqrt{(\epsilon_{\mathbf{k}} - \epsilon_{\mathbf{k+Q}})^2 + 4\Delta^2}\right) \tag{7.144}$$

in powers of Δ

$$E(\Delta) \approx E(0) + \frac{\Delta^2}{U}\left(1 - \frac{U}{U_{\text{cr}}}\right) + \frac{\Delta^4}{4}\sum^{\text{occ.}}\frac{1}{(\epsilon_{\mathbf{k+Q}} - \epsilon_{\mathbf{k}})^3} \tag{7.145}$$

[43] Except for ferromagnetism where the order parameter commutes with the Hamiltonian. We mentioned this earlier.

where $U_{\rm cr}$ is determined by the generalized Stoner criterion (7.45). Minimizing w.r.t. Δ gives $\Delta = 0$ if $U < U_{\rm cr}$, and

$$\Delta \approx AW^{3/2}\frac{\sqrt{U-U_{\rm cr}}}{\sqrt{UU_{\rm cr}}} \tag{7.146}$$

if $U > U_{\rm cr}$. The square root behaviour at the threshold is a characteristic mean field result. The dimensionless quantity A can be identified from the third-order term in (7.145).

(7.144) was written down for a half-filled, non-nested band. The result (7.146) is, however, quite general: it holds at arbitrary filling, assuming that $\chi^{(0)}(\mathbf{Q})$ is finite.

Keeping U fixed, we expect that the order parameter vanishes at a critical $\delta_{\rm cr}$ as $\Delta \sim \sqrt{\delta_{\rm cr}-\delta}$. The weak coupling result for large D is quoted in (10.86) (see p. 588).

7.8. For the sake of simplicity, we treat a one-dimensional system at $T=0$. Let us assume a circularly polarized SDW ground state with an arbitrary wavenumber $q = 2k_{\rm F}$

$$\langle \vec{S}_j \rangle = (\mathcal{S}\cos qj, \mathcal{S}\sin qj, 0)\,. \tag{7.147}$$

Recalling the mean field decoupling scheme (7.13), we find that the operator part of the mean field Hamiltonian is

$$\begin{aligned}\mathcal{H}_q &= -2U\sum_j \vec{S}_j\langle\vec{S}_j\rangle = -2U\mathcal{S}\sum_j [S_j^x\cos qj + S_j^y \sin qj] \qquad (7.148)\\ &= -U\mathcal{S}\sum_j [S_j^+ e^{-iqj} + S_j^- e^{iqj}] = -\frac{U\mathcal{S}}{2}\sum_p [c_{p+q\uparrow}^\dagger c_{p\downarrow} + c_{p-q\downarrow}^\dagger c_{p\uparrow}]\,.\end{aligned}$$

We envisage a weak SDW, so that gaps can open only in the immediate vicinity of Fermi points. An $\uparrow$-spin electron state at $k_{\rm F}$ gets mixed with a $\downarrow$-spin electron state at $-k_{\rm F}$, thus the $\uparrow$-spin spectrum is gapped at $k = k_{\rm F}$. However, an $\uparrow$-spin state at $-k_{\rm F}$ has only the $\downarrow$-spin state at $-3k_{\rm F}$ for a partner. Apart from the special case of half-filling[44], the band states at $-k_{\rm F}$ and $-3k_{\rm F}$ are not degenerate, thus there is no gap in the $\uparrow$-spin spectrum at $k = -k_{\rm F}$. Repeating the argument for $\downarrow$-spin states, we find that the $\downarrow$-spin spectrum is gapped at $k = -k_{\rm F}$ but not at $k = k_{\rm F}$ (Fig. 7.14, (b)).

We can also postulate a circularly polarized SDW with $q = -2k_{\rm F}$, which is turning in the opposite sense. Again, we find asymmetric spectra, only the roles of $k_{\rm F}$ and $-k_{\rm F}$ are interchanged (Fig. 7.14, (a)).

[44]We need not worry about this apparent loophole in the argument because at $q = \pi$, the SDW defined by (7.147) becomes linearly polarized.

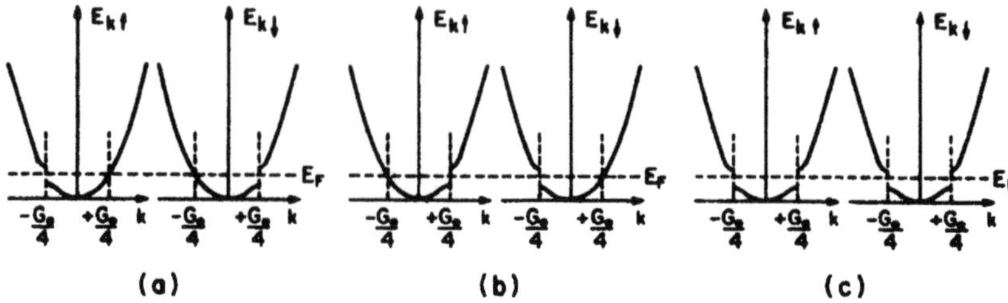

Figure 7.14: Circularly polarized SDWs open gaps at one of the Fermi points ((a) and (b)). A linearly polarized SDW opens gaps at both Fermi points, for both spin directions (after [103]).

Superposing the $q = k_F$ and $q = -k_F$ circularly polarized waves, we get the linearly polarized SDW $\langle \vec{S}_j \rangle = (2\mathcal{S} \cos qj, 0, 0)$. This has two essentially independent mechanisms for opening gaps at symmetrical positions (Fig. 7.14, (c)), thus (to order Δ) its stabilization energy is twice as large as for a circularly polarized SDW.

7.9. The mean field procedure of Sec. 7.6 was based on neglecting the correlation between $\uparrow$-spin and $\downarrow$-spin electrons; this corresponds to postulating a trial state which is factorized in the spin index

$$|\Psi\rangle = |\Psi_\uparrow\rangle|\Psi_\downarrow\rangle \,. \tag{7.149}$$

Furthermore, we found that an electron with wave vector $\mathbf{k}$ gets scattered into the state $\mathbf{k}+\mathbf{Q}$, but in no other state. This will be reflected by the wave function factorizing into two-element sums which describe the mixing of states with $\mathbf{k}$ and $\mathbf{k}+\mathbf{Q}$. Keeping the convention that if $\mathbf{k}$ is an element of the reduced Brillouin zone, then $c^\dagger_{\mathbf{k}\sigma}$ ($d^\dagger_{\mathbf{k}\sigma}$) is the creation operator of an electron with spin σ and wave vector $\mathbf{k}$ ($\mathbf{k}+\mathbf{Q}$), we seek the ground state in the form

$$|\Psi\rangle = \prod_{\mathbf{k}} \prod_{\sigma} \left[\cos\vartheta_{\mathbf{k}} c^\dagger_{\mathbf{k}\sigma} + \eta(\sigma)\sin\vartheta_{\mathbf{k}} d^\dagger_{\mathbf{k}\sigma}\right] |0\rangle \,. \tag{7.150}$$

$|\Psi\rangle$ is normalized. The $\vartheta_{\mathbf{k}}$ are independent variational parameters (we may assume $0 \le \vartheta_{\mathbf{k}} \le \pi/2$). We are going to see that inserting the spin-dependent factor $\eta(\sigma)$ (defined by $\eta(\uparrow) = 1$, $\eta(\downarrow) = -1$) introduces alternating spin order.

We have to calculate $\langle\Psi|\mathcal{H}|\Psi\rangle$, and minimize it with respect to the $L/2$ variational parameters. It is readily seen that

$$\langle\Psi|\mathcal{H}_{\rm band}|\Psi\rangle = 2{\sum_{\mathbf{k}}}' \epsilon_{\mathbf{k}} \cos 2\vartheta_{\mathbf{k}} \tag{7.151}$$

where $\sum'$ is taken over the reduced Brillouin zone.

As a prelude to evaluating the Hubbard term, let us determine the site occupation numbers (note that in the first line the sums extend over the whole Brillouin zone)

$$\langle\Psi|\hat{n}_{\mathbf{j}\sigma}|\Psi\rangle = \frac{1}{L}\sum_{\mathbf{k}_1}\sum_{\mathbf{k}_2} e^{i(\mathbf{k}_1-\mathbf{k}_2)\mathbf{j}}\langle\Psi|c^\dagger_{\mathbf{k}_1\sigma}c_{\mathbf{k}_2\sigma}|\Psi\rangle$$

$$= \frac{1}{L}{\sum_{\mathbf{k}_1}}'{\sum_{\mathbf{k}_2}}' e^{i(\mathbf{k}_1-\mathbf{k}_2)\mathbf{j}}\langle\Psi|[c^\dagger_{\mathbf{k}_1\sigma}+\eta(\sigma)e^{i\mathbf{Q}\mathbf{j}}d^\dagger_{\mathbf{k}_1\sigma}][c_{\mathbf{k}_2\sigma}+\eta(\sigma)e^{-i\mathbf{Q}\mathbf{j}}d_{\mathbf{k}_2\sigma}]|\Psi\rangle\,.$$

It is easy to check that there is no contribution from the $\mathbf{k}_1 \neq \mathbf{k}_2$ terms. The $\mathbf{k}_1 = \mathbf{k}_2$ contribution is of the form (7.51)

$$\langle\Psi|\hat{n}_{\mathbf{j}\sigma}|\Psi\rangle = \frac{1}{2} + \eta(\sigma)e^{i\mathbf{Q}\mathbf{j}}\cdot\frac{1}{L}{\sum_{\mathbf{k}}}'\sin 2\vartheta_{\mathbf{k}} = \frac{1}{2} + m\eta(\sigma)e^{i\mathbf{Q}\mathbf{j}}\,. \tag{7.152}$$

It is a consequence of the factorized form (7.149) that the Hubbard term decouples in the sense $\langle\hat{n}_{\mathbf{j}\uparrow}\hat{n}_{\mathbf{j}\downarrow}\rangle \rightsquigarrow \langle\hat{n}_{\mathbf{j}\uparrow}\rangle\langle\hat{n}_{\mathbf{j}\downarrow}\rangle$

$$\langle\Psi|\mathcal{H}_U|\Psi\rangle = LU\left[\frac{1}{4} - \left(\frac{1}{L}{\sum_{\mathbf{k}}}'\sin 2\vartheta_{\mathbf{k}}\right)^2\right]\,. \tag{7.153}$$

The total energy is the sum of (7.151) and (7.153). Minimizing with respect to $\vartheta_{\mathbf{k}}$ leads to

$$\epsilon_{\mathbf{k}}\tan 2\vartheta_{\mathbf{k}} = -Um\,. \tag{7.154}$$

Replacing $\sin = \tan/\sqrt{1+\tan^2}$ in the expression of m, we recover the gap equation (7.63)

$$1 = \frac{1}{L}{\sum_{\mathbf{k}}}'\frac{U}{\sqrt{\epsilon_{\mathbf{k}}^2 + U^2m^2}}\,. \tag{7.155}$$

Chapter 8

Ferromagnetism in Hubbard Models

We have repeatedly mentioned that studying the simplest version of the Hubbard model does not seem to permit a satisfactory understanding of itinerant ferromagnetism. In this chapter, we briefly review the arguments why we think that such is the case, and also the efforts to find particular variants, modifications, or extensions of the Hubbard model which can become ferromagnetic in a physically relevant range of the parameters.

8.1 Preliminary Overview

Let us recall the Hartree–Fock phase diagram of the cubic lattice Hubbard model (Fig. 8.1, left). Observe that now (in contrast to the representation shown in Fig. 4.11), the vertical range encompasses the entire semi-axis $0 \leq U/t \leq \infty$; this can be achieved by introducing an interaction strength variable like $U/(U+t)$. A very extended, and seemingly robust, ferromagnetic (FM) phase is seen. Ferromagnetism appears at all fillings $0 < n < 1$ if U/t is large enough. Sufficiently far from half-filling, where there is no intervening antiferromagnetic (AF) phase[1], the

[1] A more sophisticated Hartree–Fock calculation could allow incommensurate (IC) states which would appear in some intermediate range of n and/or U/t (such a phase is shown in the right panel of Fig. 8.1). The left and middle pictures of Fig. 8.1 were

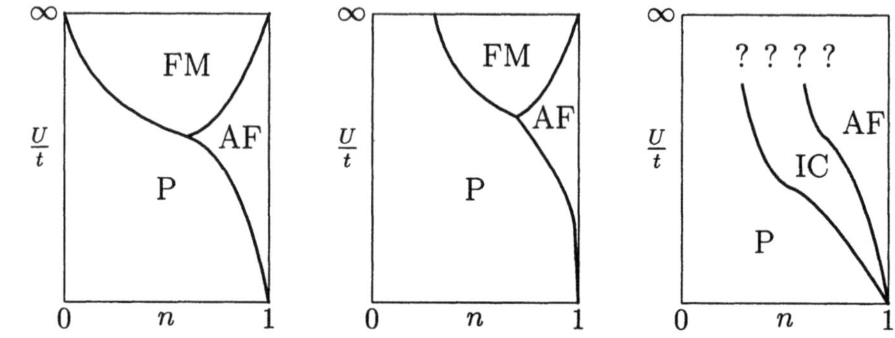

Figure 8.1: Schematic representation of different approximations for the $T = 0$ phase diagram of the cubic lattice Hubbard model in the n–U/t plane. Notice that the entire $0 \leq U/t \leq \infty$ domain has been compressed into a finite interval on the vertical axis. Only the $n \leq 1$ part of the mirror-symmetrical phase diagrams is shown. In the Hartree–Fock approximation (*left*) ferromagnetic (FM) order appears at all n if U/t is large enough [324]. *Middle*: An approximate treatment of correlations (e.g., [108]) predicts the shrinking of all ordered phases and the stabilization of the paramagnetic (P) state at low enough n for all U/t. *Right*: Some approximate treatments [383] show AF and incommensurate (IC) order, but no ferromagnetic phase at intermediate U/t. However, the nature of the large-U regime remains obscure: we cannot exclude FM at sufficiently high U/t.

P–FM phase boundary is given by the Stoner criterion (7.24). Notice that this implies that ferromagnetism survives into the low-density limit $n \ll 1$, even though it requires increasingly large values of U. For $D = 3$, $\rho(\epsilon) \sim \sqrt{\epsilon}$ near the bottom of a band[2]; it follows that $\epsilon_F \sim n^{2/3}$, and $\rho(\epsilon_F) \sim n^{1/3}$, thus U_{cr} diverges like $n^{-1/3}$.

Let us observe that the FM–AF boundary touches the $n = 1$ axis at $U = \infty$. This means that in the strong-coupling limit, the ground state should become ferromagnetic at an arbitrarily small deviation from half-filling. This may sound implausible because we know that strictly at $n = 1$, the ground state is antiferromagnetic. Let us, however, notice that at $U/t = \infty$, the coupling $J = 4t^2/U$ of the kinetic exchange vanishes[3],

---

drawn on the assumption that even in such an improved treatment, a substantial piece of the FM phase would survive.

[2]In these arguments, the bottom of the band is the energy zero.

[3]The equivalent t–J model becomes a pure t-model, describing the motion of holes

and the nature of the ground state is determined by optimizing the kinetic energy of the holes. It may well turn out that the unhindered motion of the holes requires the full polarization of the spins. Indeed this is the content of the celebrated *Nagaoka theorem* which states that for a large class of Hubbard models, the presence of a single hole in a half-filled band ($N = L - 1$) leads to a fully polarized ground state with $S = (L-1)/2$ (see Sec. 8.2).

The Nagaoka theorem[4] became known at about the same time as the mean field diagram [324], and it had been held that the former is, in some vague sense, a corroboration of the latter. It seemed reasonable to assume that a fully polarized ground state should be found also with two, three, etc. holes, or even with a finite density of holes, and thus the Nagaoka state could be continued into an itinerant ferromagnetic phase. Now we are more cautious: though we would still like to believe that such a continuation is, at least in a number of cases, possible, the details are far from being clear. We return to this subject shortly.

Naturally, it has always been admitted that the Hartree–Fock approximation overestimates the ordering tendencies and therefore in a better approximation, both the FM and AF regions would shrink (see Fig. 8.1, middle). Such a phase diagram can be obtained (for example) from a variational approach which includes correlation effects by Gutzwiller-projecting the Hartree–Fock states (see sections 8.4 and 10.6.4 for details). It is remarkable that ferromagnetism is completely eliminated from the low-n regime: below a lower critical density $n < n_{\rm cr}^{<}$, the ground state remains paramagnetic even at $U \to \infty$. It has to be emphasized that the theories we are referring to here are not exact, in the sense that we cannot be sure that the area in the phase diagram which bears the label FM, is indeed ferromagnetic. However, the claims about the *absence* of ferromagnetism in complementer regions are more dependable. Furthermore, Kanamori [201] brought a cogent argument to show that the Hubbard model cannot have low-density ferromagnetism in three-dimensional systems (Sec. 8.1.1).

Skipping the description of some further developments, let us just say that using more sophisticated approximations tends to shrink the

on a background of non-interacting spins (see Sec. 5.1.4).

[4]Discovered independently by Nagaoka [302] and Thouless [397].

FM area of the phase diagram, pushing it to higher values of U/t, and to a narrower range of n. We have no argument to exclude the existence of a FM phase for the cubic lattice Hubbard model but it appears that, even if it exists, it is such a fragile phenomenon that we should hesitate to invoke it as an explanation of ferromagnetism in any real system.

The fundamental difference between the status of ferro- and antiferromagnetism can be illuminated by the following argument: Both kinds of ordering are supposed to be present in the large-U limit. In this limit, the arising of AF is readily understood while there is no straightforward reason why FM should exist. We know the large-U effective Hamiltonian: for any band filling, it is the t–J model (5.36) in which the leading term of the spin–spin coupling is antiferromagnetic. At half-filling, only the spin–spin interaction remains, and AF order arises from effective AF interactions. Away from half-filling, the only possible origin of FM is the hopping term: the kinetic energy gain has to prevail over the spin–spin interaction. This is entirely possible but we would feel on safer ground if the origin of FM could be traced to an effective ferromagnetic spin–spin interaction. We do not have this option for simple bipartite Bravais lattices but for some carefully chosen close-packed lattice models, we do. In Sec. 8.3, we show on the example of a three-site toy model how ferromagnetism can arise from ring exchange processes.

The difficulty can be formulated also in the following way [432]: the on-site Hubbard interaction term is not sensitive to either the structure or the dimensionality of the lattice, thus if we are hunting for an ordering phenomenon which is obviously not a universal feature of the Hubbard model, we have to do so by fine-tuning the one-electron part of the Hamiltonian. Many of the results reviewed here rest on postulating a specific band structure: some of them plausible-looking, while some others disquietingly artificial.

The general lesson is that while there are only a few Hubbard models for which ferromagnetism can be excluded rigorously, well-developed "healthy" ferromagnetism is apparently reserved for non-bipartite, close-packed lattices. Several different lines of reasoning indicate that the triangular (or the fcc) lattice is much more inclined towards FM than the square (or the simple cubic) lattice. This holds for the simple, single-orbital Hubbard model. Incorporating the direct exchange would certainly help for any lattice structure, and it may (at least in certain regions) push forward the FM phase boundary quite substantially. Furthermore, extending the treatment to degenerate models, we can introduce Hund's coupling which may provide a vigorous, structure-insensitive mechanism of itinerant ferromagnetism. Apparently, there are several different routes to get FM in lattice fermion models; we do not know yet whether we will pick out one of these as *the* essential model of itinerant ferromagnetism, or the canonical model will rely on a combination of them.

8.1.1 The Low-Density Limit

It was pointed out by Kanamori [201] that the mean field criteria (7.24) and (7.45) overestimate the effect of a large U. To see why this must be the case, one can argue as follows: The ground state energy is determined by the competition of U with the kinetic energy which is measured by the bandwidth W. In order to keep the energy cost of doubly occupied sites as low as possible, the system sacrifices a certain fraction ΔE_{kin} of the kinetic energy. It is reasonable to claim that ΔE_{kin} gives a measure of the effective interaction strength U_{eff}. The biggest kinetic energy sacrifice is made when the electrons stop moving altogether; then $\Delta E_{\text{kin}} \sim W$. Thus the effective interaction cannot be greater than an energy of the order of the bandwidth: $U_{\text{eff}} \leq f(n) \cdot W$, where the form of $f(n)$ is determined by the band density of states. The solution of the two-electron problem [201, 264] makes it clear that the U_{eff} introduced here determines the energy splitting between high-spin and low-spin states, thus it is plausible that it should replace U in an improved version of the Stoner criterion:

$$U_{\text{eff}} \cdot \rho(\epsilon_{\text{F}}) = 1 \,. \tag{8.1}$$

Using the Kanamori criterion (8.1) instead of the Stoner criterion (7.24) makes a lot of difference when the latter would require that U be much bigger than W. Such is the case in the low-density limit of three-dimensional systems where $\rho(\epsilon_{\text{F}})$ vanishes like $n^{1/3}$. U_{eff} cannot diverge and it follows that ferromagnetism is suppressed if the band filling is smaller than some critical value $n_{\text{cr}}^{<}$. The detailed calculation shows that the threshold value is quite sizeable ($n_{\text{cr}}^{<} \sim 0.16-0.2$).

The same argument indicates that ferromagnetism has much better chances in $D = 2$. $\rho(\epsilon)$ approaches a constant value at the lower band edge, thus the Stoner criterion predicts $U_{\mathrm{cr}} \sim W$ as $n \to 0$. There is no qualitative argument which would exclude that Kanamori's criterion can be satisfied. In fact, a quite different approach indicates the existence of low-density ferromagnetism for certain two-dimensional Hubbard models [326, 170].

It is worth mentioning that even in $D = 3$, one easily finds examples of a large density of states at a band edge. The fcc tight binding band has a logarithmically diverging DOS at one of the band extrema, and the sign of t can be chosen so that it happens at the bottom of the band. It is not clear whether Kanamori's argument excludes low-density ferromagnetism in this situation. Actually, with the conventional sign of t, the DOS diverges at the top of the band, and that is also nice because it raises the possibility of FM at almost complete filling. It is tempting to think of the ferromagnetism of fcc Ni which has an almost completely filled $3d$ band. Naturally, we have to remember that in Ni, we have a fivefold degenerate $3d$ band hybridized with the $4s$ band. However, the canonical DOS [14] of the fcc $3d$ band still possesses an impressive peak near the upper edge. It is not important whether the peak corresponds to a diverging DOS, or it is merely large[5]; neither whether ϵ_{F} is truly at the peak, or just somewhere near. Let us remember that for sufficiently strong interaction the finite-moment version of the Stoner criterion (7.111) should give guidance and it requires only that the integrated DOS be large enough in an interval around ϵ_{F}. Such a qualitative argument is, of course, far too naive as an "explanation" of the magnetism of nickel. However, we understand why structures related to the fcc lattice are popular as high-dimensional model systems which may show robust ferromagnetism even within the confines of the simple, single-orbital Hubbard model.

It would be misleading to proceed in the same spirit and claim that $D = 1$ band structures, being characterized by a divergent $\rho(\epsilon) \sim 1/\sqrt{\epsilon}$, are particularly favourable for (low-density) ferromagnetism. The ground state of the one-dimensional Hubbard model is exactly known, and

[5]In fact, a recent study [416] shows that a large (but finite) peak near the band edge is more effective in producing FM than a logarithmic singularity at the edge.

it is never ferromagnetic. More precisely: low-spin and high-spin states (including the fully polarized state) become degenerate at $U = \infty$ but we would not want to call this ferromagnetism. We may visualize the situation as follows: an infinite U prevents an $\uparrow$-spin from switching places with a neighbouring $\downarrow$-spin, thus quite literally, there is no possibility for quantum-mechanical exchange. At $U = \infty$, exchange needs loops. Keeping this in mind, we do not feel surprised to find that a generalized Hubbard chain with first- *and* second-neighbour hopping has a sizeable domain of ferromagnetism in its ground state phase diagram. The model will be discussed a bit later.

8.2 Exactly Proven Cases of High-Spin Ground States

Though the mean field prediction of a broad domain of ferromagnetic phase away from half-filling in D-dimensional cubic lattices turned out to be either false or at least undependable, it would be wrong to say that the Hubbard model is never ferromagnetic. There are proven cases of high-spin ground states for some special choice of the parameters, or certain lattice structures, or in extended Hubbard models. Though at the moment they may seem to be curiosities of merely academic interest, we briefly review[6] them in the hope that one (or several) of them will turn out to hold the clue for understanding ferromagnetism in real systems. For the purposes of this Section, we introduce the following general form of the (non-degenerate) Hubbard model:

$$\mathcal{H} = -\sum_{\sigma}\sum_{x,y} t_{xy} c^{\dagger}_{x\sigma} c_{y\sigma} + \sum_{x} U_x \hat{n}_{x\uparrow} \hat{n}_{x\downarrow} \tag{8.2}$$

where x, y denote lattice sites, the $t_{xy} = t_{yx}$ are real hopping matrix elements and $U_x \geq 0$.

First, what shall we mean by ferromagnetism? For not more than half-filling ($N \leq L$), it is possible to align all electron spins, giving rise

[6]We will not attempt to state the theorems in their full rigour. For more details, see the authoritative review by Lieb [244]. In part of our reasoning, we follow a brief pedagogical review by Tasaki [393]. The short reviews [432, 105] discuss also generalized Hubbard models and other lattice fermion models.

to a total spin $\mathcal{S} = N/2$. This could be called saturated ferromagnetism, but we will want to admit also a partial polarization of the band, and call the system ferromagnetic if the total spin is an extensive quantity: $\mathcal{S} \propto L$. However, adopting this broad definition necessitates a careful distinction between ferrimagnetism and "real" ferromagnetism.

8.2.1 Lieb's Ferrimagnetism

Consider bipartite lattices which can be divided into two sublattices A and B in such a manner that the hopping $t_{xy} \neq 0$ only if x and y belong to different sublattices. The number of sites in the sublattices are denoted by $|A|$ and $|B|$; they may be unequal. The Hubbard U is the same everywhere: $U_x = U > 0$. Then *Lieb's theorem* [243] says that the ground state of the half-filled Hubbard model ($N = L$) is unique apart from a trivial $(2\mathcal{S} + 1)$-fold spin degeneracy, and it has the total spin $\mathcal{S} = ||A| - |B|| /2$.

The nature of the ground state becomes clear in the limit of large U where the Hubbard model can be transformed into a nearest-neighbour antiferromagnetic Heisenberg model. A spin at an A site will want to be antiparallel to the spin of its B-site neighbours. Since the number of A-sites differs from the number of B-sites, the fully *antiparallel* alignment of A- and B-spins leads to the total spin stated in Lieb's theorem[7]. Thus the microscopic origin is the same as for antiferromagnetism, and the reason for a non-zero total spin is merely geometrical: $|A| \neq |B|$. Hence, we will refer to this kind of ordering as "Lieb's ferrimagnetism", and hold it fundamentally different from "true" ferromagnetism which is associated with *parallel* nearest-neighbour correlations.

Note that for $|A| = |B|$, the theorem says that the ground state is a singlet. In the thermodynamic limit, this may be compatible with the arising of antiferromagnetic long-range order (see Sec. 6.3) but this is not necessarily so: the one-dimensional Hubbard model offers a counter-example.

---

[7]Taking all spins $\uparrow$ on A, and $\downarrow$ on B gives $\mathcal{S}^z = (|A| - |B|)/2$. Naturally, the ground state is not identical to this classical configuration but it is lying in the subspace generated by acting with the spin-flip terms of the Heisenberg Hamiltonian on this starting state.

You may find it perplexing that the theorem holds for arbitrarily small $U > 0$: it seems to suggest a discontinuity at $U = 0$ since it is conventional wisdom that the non-interacting band has a singlet ground state. The explanation is that $|A| \neq |B|$ implies that the singlet is only one of the $U = 0$ ground states and that there is a vast number of other ground states, with a total spin up to $\mathcal{S} = ||A| - |B||/2$. Instead of an abstract argument, think of our example of the CuO_2 lattice where $|A| = L/3$ and $|B| = 2L/3$. (Problem 4.4). The single-particle spectrum consists of three bands, including a dispersionless branch at $\epsilon = 0$. Half-filling of the lattice implies that this band is also half-filled, which means that the spins of $L/3$ electrons can be arbitrarily chosen. Thus the state with $\mathcal{S} = L/6$, $\mathcal{S}^z = L/6$ is also one of the ground states. Switching on an infinitesimal U selects the $\mathcal{S} = L/6$ state as the unique ground state. We may rationalize this result by invoking the exchange hole effect: the interaction energy is minimized for parallel spins and for a flat band, this is the only consideration[8] since there is no kinetic energy to gain. Generally, Lieb's ferrimagnetism has to do with a dispersionless part of the spectrum at $\epsilon = 0$; if the interaction competes with a vanishing kinetic energy, it necessarily wins.

It should be appreciated that though we could find qualitative arguments to illuminate the result both for small U and large U, it is very far from being obvious that the entire range $0 < U < \infty$ can be covered by a single exact statement. We also stress that it is only at large Us that we can ascribe a conventional ferrimagnetic nature (with antiparallel nearest-neighbour correlations) to the large-spin ground state. At intermediate U, we have no similar argument for an antiferromagnetic nearest-neighbour exchange, but Lieb's theorem continues to be valid. The intermediate-U ground state is likely to have an interesting structure which we will perhaps want to associate with the partially polarized state of certain itinerant ferromagnets.

[8] Our reasoning does not constitute a proof since it neglects the effect of interaction on electrons inhabiting the dispersive part of the spectrum. The rigorous proof uses completely different arguments [243].

8.2.2 Flat-Band Ferromagnetism

Motivated by the finding of Lieb's ferrimagnetism in systems for which the $U = 0$ spectrum shows a dispersionless band at the centre of the spectrum, we may ask what happens if a flat band lies at the bottom of the spectrum. The reasonable expectation is ferromagnetism, at least if N does not exceed the number of unit cells. A handwaving argument could be the following: in the $U = 0$ ground state, the electrons populate the dispersionless band, and the same energy is obtained for all possible total spins including the case of full polarization $\mathcal{S} = N/2$. Switching on U, the exchange hole effect will select the maximum-spin state as the (essentially) unique ground state.

To see a specific example, let us consider the railroad trestle lattice with Tasaki's parametrization (4.78), and $t' = 0$ (see Problem 4.3 on p. 167 and p. 190). Deleting the lines belonging to t' in Fig. 4.5, we get a "sawtooth lattice" which can be redrawn (see Fig. 8.2) as a linear chain with hopping v, and a second-neighbour hopping t for the sublattice of odd sites, while the even sites have a site energy ϵ.

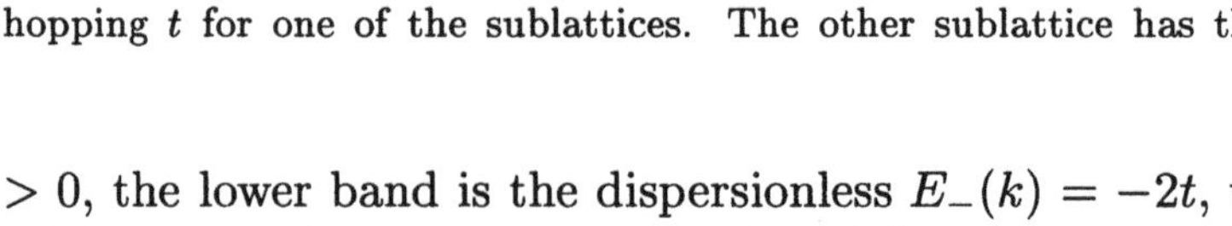

Figure 8.2: The linear chain with nearest-neighbour hopping v, and a second-neighbour hopping t for one of the sublattices. The other sublattice has the site energy ϵ.

For $t > 0$, the lower band is the dispersionless $E_-(k) = -2t$, which is separated by a gap from the upper band $E_+(k) = 2t\cos k + \lambda^2 t$. It can be proven that for $U > 0$, for $N = L$ electrons, the ground state is indeed a fully polarized ferromagnet[9] with $\mathcal{S} = N/2$.

Looking at Fig. 8.2 we might become worried because the previous finding seems to contradict a celebrated theorem by Lieb and Mattis

[9]In fact, it is an insulating ferromagnet: there is a gap even at $U = 0$ and it is likely to increase when U is switched on. However, even though the exact proof does not cover the case $N < L$, a large-U argument does [323], and there seems to be no doubt that metallic ferromagnetism will be found for a range of fillings.

[245], which states that for a wide class of one-dimensional models of interacting electrons, the ground state has the smallest possible spin ($\mathcal{S} = 0$ for an even, $\mathcal{S} = 1/2$ for an odd number of electrons). The apparent paradox is resolved by scrutinizing the conditions for the fulfillment of the theorem: it holds for strictly one-dimensional systems where one electron cannot "overtake" the other by going around on a different path. Narrowed down to Hubbard models, the theorem presupposes that there is only nearest-neighbour hopping. The presence of a second-neighbour hopping in our model introduces loops and means a deviation from strict one-dimensionality.

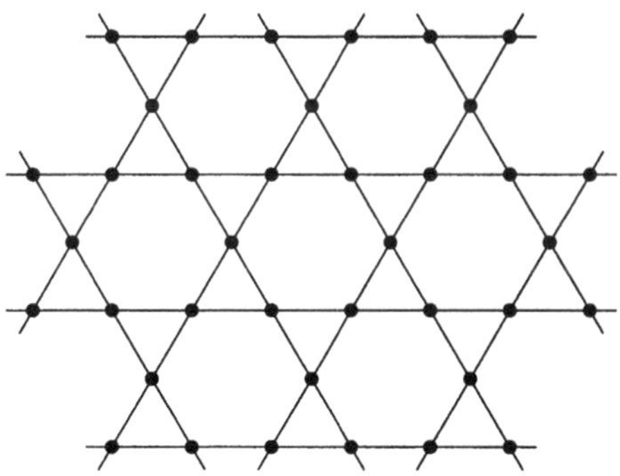

Figure 8.3: The kagome lattice. The unit cell contains three lattice sites.

Analogous results have been proven for a number of similar two- and three-dimensional models [274]. Perhaps the most interesting example is the kagome[10] lattice (Fig. 8.3). In its Mott-insulating state (at exact half-filling $N = L$, and a sufficiently large U), the kagome lattice provides a spectacular case of strongly frustrated antiferromagnetism (see Sec. 6.3.4). It is remarkable that in another range of the parameters, the same lattice shows flat-band ferromagnetism [273]. The one-electron spectrum possesses (in addition to two ordinary-looking bands) a flat band at $\epsilon = 2t$; if $t < 0$, then this is the bottom of the spectrum (Fig. 8.4). The first $L/3$ electrons ($L/3$ being the number of unit cells) can be put into the dispersionless band, and they prefer to have their

[10]Kagome is the Japanese word for basketwork.

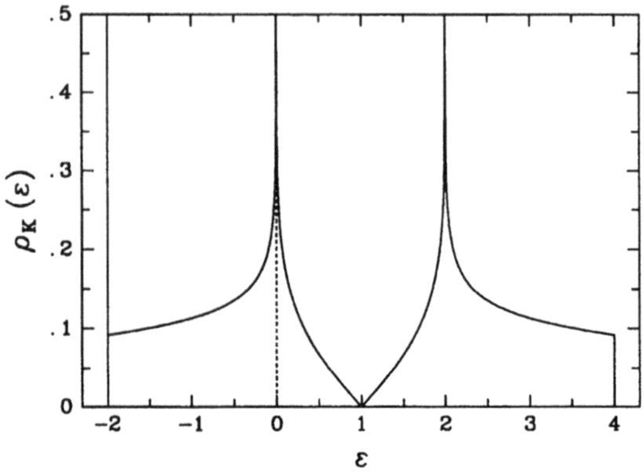

Figure 8.4: The tight binding density of states of the kagome lattice (ϵ in units of $|t$—). Observe the δ-function peak at $\epsilon = -2$, which belongs to the flat band at the bottom of the spectrum.

spins parallel for any $U > 0$. Later on, we will understand that it is no accident that frustration and ferromagnetism appear in the same system.

8.2.3 Nagaoka Ferromagnetism

We have already mentioned that ferromagnetic ordering may be driven by kinetic energy and that for the square lattice or the simple cubic lattice, this is the only possibility we can think of. To see how this might work, Nagaoka [302] studied the limit $U = \infty$ where the antiferromagnetic kinetic exhange $\sim t^2/U$ is suppressed. At exact half-filling $N = L$, the charge degrees of freedom are completely frozen. Removing a single electron, the resulting hole[11] moves on a background of spins. It turns out that under very general conditions, the "smoothness" of a completely spin-polarized background is optimal for the propagation of the hole, thus kinetic energy favours ferromagnetism. This is stated in

[11] "Holes" and "particles" are understood here as deviations from half-filling. In semiconductor physics, these expressions have a different meaning. – We prefer to state the theorem for a single hole, rather than for a single electron, which would be also usual. For bipartite lattices, electron–hole symmetry guarantess that the two cases are equivalent. For non-bipartite lattices, the single-hole and single-electron cases will have to be discussed separately.

the celebrated

Nagaoka theorem: *Let* $t_{xy} \leq 0$ *for any* $x \neq y$, $U_x = \infty$ *for all* x, *and let the number of electrons be* $N = L - 1$. *If the lattice satisfies a certain connectivity condition, then the ground state has the total spin* $S = N/2$, *and it is unique, apart from the trivial* $(N+1)$*-fold spin degeneracy.*

The notion of "connectivity" required here should become clear from the solution to Problem 8.1. Let it suffice that it requires the existence of loops [264]. Furthermore, the loops should pass through no more than four sites [390]. The theorem does not apply to the linear chain but it is valid for many standard two- and three-dimensional lattices (the square, triangular, and kagome lattices in $D = 2$, the simple cubic, bcc, fcc, and hcp lattices in $D = 3$, etc.). The formulation makes it clear that the number of lattice sites has to be finite. Thus when speaking of, say, the square lattice, we mean arbitrarily large but finite pieces of the square lattice.

It should not disturb us that the required sign of $t_{xy} < 0$ is just the opposite of that assumed in most discussions of the Hubbard model. For bipartite lattices, there is a canonical transformation connecting $t_{xy} > 0$ and $t_{xy} < 0$ models (see. p. 164). Combining it with the electron–hole transformation (4.45), it follows that the Nagaoka theorem is equally applicable to the cases of a single added electron ($N = L+1$), and a single hole ($N = L - 1$), for either sign of t. This holds, for example's sake, for the square, simple cubic, and bcc lattices. The honeycomb lattice is also bipartite, but the loops consist of six sites, so the connectivity condition is not satisfied. Numerical results for finite systems indicate that the Nagaoka state is not the only ground state: it becomes degenerate with lesser-spin states [168].

For non-bipartite lattices, the $t_{xy} > 0$ and $t_{xy} < 0$ cases are genuinely different. At the same time, different statements apply to the electron and hole cases. For instance, the triangular lattice with nearest-neighbour couplings $-t$ has a high-spin ground state for $N = L-1$ only if $t < 0$. Applying the transformation[12] $c^{\dagger}_{\mathbf{j}\sigma} \to c_{\mathbf{j}\sigma}$, $c_{\mathbf{j}\sigma} \to c^{\dagger}_{\mathbf{j}\sigma}$ changes the sign of t, and at the same time causes $N = L-1 \to N = L+1$, thus

[12] Observe that we did not include the sign factor (4.46). The transformation (4.45) is useful only for bipartite lattices.

with $t > 0$ we find a fully polarized ground state for $N = L + 1$. It is worth emphasizing that for non-bipartite lattices, the Nagaoka theorem makes *two* definitive statements: if the sign of t such that the $N = L+1$ $(N = L-1)$ system has a high-spin ground state, then it is also asserted that the spin of the $N = L - 1$ $(N = L + 1)$ system is less than the maximum value.

We do not prove the theorem but we will check its validity by studying small systems. The solution to Problems 8.1 and 8.2 should convince us that the stabilization of the high-spin state relies on the constructive interference between processes in which the hole reaches a given lattice point via different routes [264].

What we would really like is, of course, to understand an entire ferromagnetic phase by continuation from the Nagaoka point. We would like to see if the argument can be extended to finite U, and to finite hole density[13]. As to the first question: a continuity argument shows that if the ground state is fully polarized at $U = \infty$, then it stays so at sufficiently large finite Us. However, the threshold value of U diverges as $L \to \infty$ [381].

Extending the rigorous treatment to finite hole densities is an apparently hopeless enterprise. We might have envisaged treating the case of a few holes first. So what about two holes? If one hole can gain kinetic energy by propagating on a spin-polarized background, why could not two holes do the same? The suggestion sounded really plausible until one tried to check it and found counter-examples [386, 91, 101]. There are many proven cases of the *absence* of a high-spin state for two (or several) holes, for essentially all the lattices we might be interested in [381]. However, the proof often relies on specific assumptions about the parity of the number of sites along certain directions, and the existence of high-spin ground states is likely (or sometimes even proven) in situations not covered by the "absence of..." theorems. At the least, however, we can say that there seems to be no simple general statement guaranteeing the continuation of the single-hole Nagaoka state.

[13]The question is whether the Nagaoka theorem is "thermodynamically relevant". It is clear that in the thermodynamic limit $L \to \infty$ the ordering effect of a single hole (or any finite number of holes) must vanish. A single hole in a finite system represents the hole density $\delta = 1/L$.

Here we must be extremely cautious: the above findings do not constitute a proof of the non-existence of a ferromagnetic phase! That issue is still open. To get a feeling why even exact results can be misleading, envisage the following situation: say, we found that if along one of the principal axes the lattice contains an even number of sites, then the ground state is a singlet. Now, it is possible that the state in question is something like this: $\Uparrow\Uparrow \ldots \Uparrow\Uparrow\Downarrow\Downarrow \ldots \Downarrow\Downarrow$. We would naturally say: why, that is really a ferromagnet! We could fully polarize it with an infinitesimal external field! It is quite possible that the system is hiding its essentially ferromagnetic nature, just to accomodate a node of the wave function[14]. It is very difficult to infer the thermodynamic properties at finite hole density from the available few-hole results. The proven lack of Nagaoka ferromagnetism in two-hole cases may, or it may not, mean that the single-hole Nagaoka state is not connected to a ferromagnetic phase in the phase diagram. But at least it means that at present, we cannot hope to establish the existence of a ferromagnetic phase with the same rigour as it was achieved in proving the Nagaoka theorem. Great effort has been spent on delineating the boundary of the FM phase by devising instability criteria for the fully polarized state: these will be described later.

Problem 8.1. *Nagaoka Theorem: An Example.* Consider the $U \to \infty$ four-site Hubbard model in the shape of a square, with the hopping amplitude on the bond 12 differing from the other three:

$$\mathcal{H} = \sum_{\sigma}[t'(c_{1\sigma}^{\dagger}c_{2\sigma} + \text{H.c.}) + t(c_{2\sigma}^{\dagger}c_{3\sigma} + c_{3\sigma}^{\dagger}c_{4\sigma} + c_{4\sigma}^{\dagger}c_{1\sigma} + \text{H.c.})]\,. \tag{8.3}$$

Let the number of electrons be $N = L - 1 = 3$. What is the spin of the ground state for $t' = t$, and for $t' = -t$? Discuss the result!

8.2.4 Ferromagnetism in a Nearly Flat Band

We should be worried that all the previously described cases of rigorously established ferromagnetism belong to the singular case $U/|E_{\text{kin}}| = \infty$, either because $U/t = \infty$ (Nagaoka), or because the relevant band is flat. It is clear that only finite values of $U/|E_{\text{kin}}|$ are physically relevant. It is therefore of great interest that the existence of a ferromagnetic ground state could be proven for a non-singular system with finite

[14]Evidence for such a behaviour was found for the Hubbard ladder [407], and the t_1–t_2 Hubbard chain.

bandwidths and U [392]. The system can be envisaged either as a Hubbard chain with different second-neighbour hoppings for the even and odd sublattices (Fig. 8.5), or as the railroad trestle lattice shown in Fig. 4.5. We have seen in the solution to Problem 4.3 (see p. 190) that with the parametrization

$$t_1 = \lambda(t_2 - t_2') \qquad \epsilon_0 = (\lambda^2 - 2)(t_2 - t_2') \tag{8.4}$$

the band structure consists of two cosine bands. In particular, replacing $\lambda = \sqrt{2}$ in (8.4), and fixing $t_2' < 0$, the lower band has the width $4|t_2'|$, the upper band the width $4t_2$, and they are separated by a gap $2(t_2 - t_2')$. Tasaki [392] found that at exact quarter-filling ($N = L/2$, where L is the number of lattice sites) the ground state has the maximum value of the total spin if $U/|t_2'|$ is large enough, and at the same time $t_2/|t_2'|$ is large enough. Representative values are $t_2/|t_2'| \geq 4.5$ for $U/|t_2'| = 50$, and $t_2/|t_2'| \geq 2.6$ for $U/|t_2'| = 100$ [393]. The width of the occupied band being much smaller than U, we may speak about a nearly flat band. But both the kinetic and interaction energies are finite; certainly, there is nothing pathological about the model.

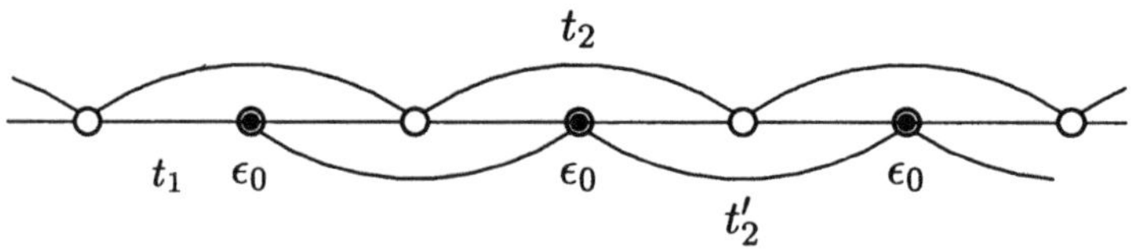

Figure 8.5: The linear chain with nearest-neighbour hopping t_1, and a second-neighbour hoppings t_2 and t_2' for the odd and even sublattices, resp. The black (even) sublattice has the on-site energy ϵ_0.

Tasaki's result is important because it assures us that there is a Hubbard model with non-singular parameters which is capable of becoming ferromagnetic in a finite domain of the parameter space. Unfortunately, the technique of the exact proof [392] does not give us any immediate insight into the mechanism of ferromagnetic ordering. For instance, we may suspect that postulating the parametrization (8.4) cannot be important for ferromagnetism, even though it must be instrumental in making the mathematics managable. We would like to identify a readily understandable physical mechanism which would hopefully work in a wider class of models.

8.3 The Ring Exchange Mechanism

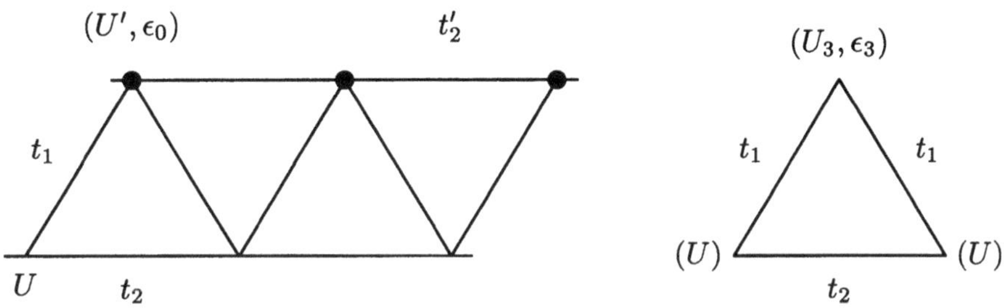

Figure 8.6: *Left*: This redrawing of the linear chain shown in Fig. 8.5 emphasizes the triangular plaquettes. We may ascribe an $U' \neq U$ to the even sublattice. *Right*: The three-site toy model.

Constructing his 3-site "toy model", Tasaki [393] called attention to the importance of triangular plaquettes (they are easier to recognize in Fig. 8.6 than in Fig. 8.5). Early numerical studies on small systems made it clear that triangular loops must, for some reason, be helpful in stabilizing ferromagnetism [386]. We will understand shortly that the geometrical shape in itself is not sufficient but it must be accompanied by the property of "frustration": the plaquette product Πt_{ij} must have a certain sign[15]. We find it useful to introduce a generalized version of the toy model in which site 3 is inequivalent to the other two sites, possessing an on-site energy ϵ_3, and also a different value of the Hubbard U:

$$\mathcal{H}_3 = t_2 \sum_\sigma (c_{1\sigma}^\dagger c_{2\sigma} + c_{2\sigma}^\dagger c_{1\sigma}) + \epsilon_3 \sum_\sigma \hat{n}_{3\sigma} + U_3 \hat{n}_{3\uparrow} \hat{n}_{3\downarrow} \tag{8.5}$$

[15]This gives the opportunity to clarify that the frustration of triangular plaquettes has a different meaning in the theory of antiferromagnetism than here. In the large-U limit, the toy model has three nearest-neighbour antiferromagnetic couplings, and is thus a frustrated antiferromagnet, whatever the signs of t_1 and t_2 are. The significance of the rule $\Pi t_{ij} > 0$ is most clearly seen when we set $t_1 = t_2$, $\epsilon_3 = 0$: then it follows that the single-electron ground state is degenerate. Thus we can loosely say that satisfying the sign rule generates a larger density of states in the lower half of the band, favouring ferromagnetism for less than half-filling.

$$+t_1 \sum_\sigma (c^\dagger_{1\sigma} c_{3\sigma} + c^\dagger_{3\sigma} c_{1\sigma} + c^\dagger_{2\sigma} c_{3\sigma} + c^\dagger_{3\sigma} c_{2\sigma}) + U \sum_{j=1}^{2} \hat{n}_{j\uparrow} \hat{n}_{j\downarrow} \,.$$

Relegating the details of the derivation to the solution to Problem 8.2 (see p. 488), we quote and discuss the results. Let us take $N = 2$ electrons. Gauge symmetry allows to fix $t_1 > 0$; then it follows from the Nagaoka theorem that the $U = U_3 = \infty$ ground state is ferromagnetic (namely, a triplet) if $t_2 > 0$. In the special case when $\mathcal{H}_3$ possesses the full symmetry of an equilateral triangle ($U = U_3$, $t_1 = t_2$, $\epsilon_3 = 0$), the single-electron gound state is degenerate, and the ground state is a triplet for any $U > 0$. We recognize that this is a case of flat-band ferromagnetism, and as such, not particularly interesting. Reducing the symmetry of the one-electron problem to a mere mirror symmetry, the $U = U_3 = 0$ two-electron ground state is unambiguously a singlet, while in the Nagaoka limit, it is a triplet, thus a transition to the ferromagnetic state must set in at a finite U. The results shown in Fig. 8.19 demonstrate that ferromagnetism is a non-perturbative, intermediate-coupling phenomenon, taking place in a wide range of the parameters $t_2/t_1 > 0$, and ϵ_3/t_1. It has been argued [393] that the mechanism of this toy version of the low-spin–high-spin transition is quite like to what happens in Tasaki's linear chain model[16].

Applying the ideas of [323] to the toy model, we gain valuable insight into the driving force of the transition by studying the limit where ϵ_3 is large compared to t_1 and t_2. Let furthermore $U \gg \epsilon_3$, and we may take $U_3 \to \infty$. Then we can expand the exhange coupling in terms of the small quantities $1/\epsilon_3$ and $1/U$.

The low-energy configurations have one electron at each of the base sites, and the effective Hamiltonian can be expressed by the spins $\mathbf{S}_1$ and $\mathbf{S}_2$. To get the exchange coupling, let us start from $|\uparrow\downarrow 0\rangle = c^\dagger_{1\uparrow} c^\dagger_{2\downarrow} |0\rangle$. The sequence of three hopping events shown in Fig. 8.7 effects the exchange of the two spins. The matrix elements are t_1, t_2, and t_1, and the two intermediate states have the excitation energy ϵ_3. The

[16]You may be worried about this point because we argued elsewhere that a small system cannot have a symmetry-breaking transition. However, the total spin as an order parameter has a special status, thus we can use the lessons drawn from studying the toy model with confidence. More about this point in Sec. 6.1.1.

third-order effective Hamiltonian describes these third-order processes

$$\mathcal{H}_{\text{eff}}^{(3)} = -\frac{4t_1^2 t_2}{\epsilon_3^2} \mathbf{S}_1 \cdot \mathbf{S}_2 = J_3 \, \mathbf{S}_1 \cdot \mathbf{S}_2 \,. \tag{8.6}$$

Obviously, the sign of t_2 controls the sign of J_3. In particular, the coupling is ferromagnetic if $t_2 > 0$.

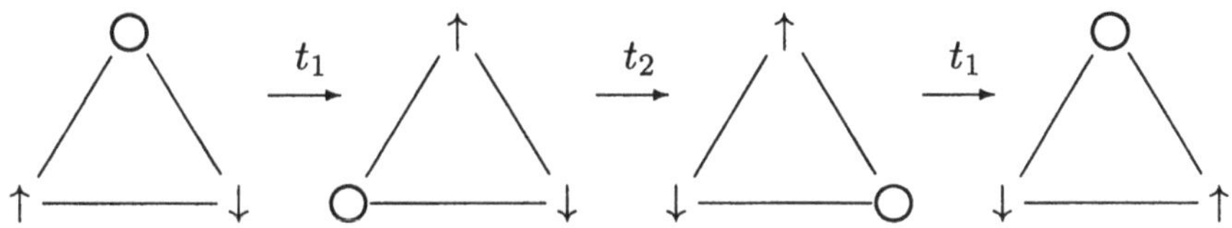

Figure 8.7: Ring exchange in the toy model proceeds via the top-site orbital which is kept empty by the large on-site energy ϵ_3. A sequence of three hopping processes exchanges the spins at the base of the triangle.

In our previous discussions of the exchange process (Sec. 2.2 and Sec. 5.1.6), we have met the problem of whether kinetic exchange is possible when the electrons are so strongly interacting that they can never be at the same place. The usual second-order contribution $\sim t^2/U$ is suppressed but it is still possible to exchange the spins by moving them in a loop. For this *ring exchange* process one needs empty orbitals where one electron can reside while the other is passing by. This is provided by the top site of the triangle which we kept empty by introducing a large ϵ_3.

If U is not kept so artificially big, there is also a contribution from the familiar second-order process

$$\mathcal{H}_{\text{eff}} = \mathcal{H}_{\text{eff}}^{(2)} + \mathcal{H}_{\text{eff}}^{(3)} = \left(\frac{4t_2^2}{U} - \frac{4t_1^2 t_2}{\epsilon_3^2} \right) \mathbf{S}_1 \cdot \mathbf{S}_2 \,. \tag{8.7}$$

The overall sign of the exchange coupling is decided by the competition between antiferromagnetic pair exchange, and ferromagnetic ring exchange.

The toy model can be extended into a quasi-one-dimensional lattice model: it is essentially Tasaki's model (Fig. 8.5), with the extra feature that $U \to \infty$ on the black (even) sublattice, but not on the odd sublattice. Assuming again a sufficiently large ϵ_0 on the even sublattice, and exact quarter-filling so that the odd sublattice can be filled with one electron per site, a systematic perturbation expansion [323] confirms that the effective spin–spin coupling is the same as in the toy model result (8.7)

$$\mathcal{H}_{\text{eff}} = \sum_{j=\text{odd}} \left(\frac{4t_2^2}{U} - \frac{4t_1^2 t_2}{\epsilon_0^2} \right) \mathbf{S}_j \cdot \mathbf{S}_{j+2} \, . \tag{8.8}$$

The system is ferromagnetic if U is large enough.

t_2' does not play a role at, or below, quarter-filling. Strictly at quarter-filling, the fully polarized system is an insulator because of a large single-particle gap, but it can be shown [323] that the system remains ferromagnetic even if we move away from quarter-filling.

The plaquette mechanism gives a very transparent reason for finding ferromagnetism in insulating and metallic cases alike, but at a price: U has to be so large that the leading (second-order) kinetic exchange is suppressed and the usually negligible higher-order processes become dominating. Besides, the presence of frustration has to be assumed in the sense that the plaquette product $\prod t_{ij}$ must have a definite sign. Finally, the possibility of the perturbation expansion relies on the seemingly ad hoc assumption of a large ϵ_0. But at least, it is a hopeful feature that the ferromagnetic order is understood to be stable in a finite domain of the parameter space, not only in singular limiting cases. Our basic hope is that ferromagnetism extends beyond the range of validity of the perturbational argument, and then perhaps the entire phase can be understood by continuation from the strong-coupling, large-ϵ_0 regime. There are reasons to think that indeed such is the case. Early numerical work [386] on small clusters used $\epsilon = 0$; nevertheless, it indicated that three-site exchange in triangular loops is instrumental in stabilizing strong ferromagnetism. Tasaki's [392] finding of a fully saturated ferromagnetic ground state requires a large U but it does not require a large ϵ_0. Penc et al. [323] studied an extended version of Tasaki's model by a variety of techniques and found that the extensive ferromagnetic phase is continuously connected to the ring-exchange regime. It is also interesting to note that this is our first argument which ascribes a distinguished role to quarter-filling. Though the reason here is almost trivial, it seems to have a more general validity: even higher-dimensional lattices built of triangular blocks seem to possess a particularly stable pocket of ferromagnetism around $n = 1/4$.

We can actually get an idea of the ϵ_0-dependence of the stabilization energy of the ferromagnetic state from (8.72) in the solution to Problem 8.2. The lesson is that the perturbative result (8.6) goes over smoothly into an expression of the order of the hopping energy. Since for medium or small ϵ_0 the electrons are no longer confined to the lower chain (Fig. 8.6, left), or to sites 1 and 2 in terms of the toy model, we cannot use an effective Heisenberg model. However, it should be still possible to write down an effective t–J model, where the qualitative behaviour of J is something like (8.72).

The "least decorated" version of the triangular strip (which is not covered by the arguments of [392] or [323]) was introduced by Müller–Hartmann [297]. It is obtained from the model shown in Fig. 8.6 (left) by setting $t_2 = t_2'$, $\epsilon_0 = 0$, and $U = U'$. Interpreting the model as the t_1–t_2–U Hubbard chain, we find another example of how the no-go theorem by Lieb and Mattis [245] can be circumvented. The point is that switching on the second-neighbour hopping, we have short loops which means an essential deviation from strict one-dimensionality. At the same time, the lattice ceases to be bipartite. A Nagaoka argument [265] shows that the single-hole ground state is fully polarized if t_2 is positive[17]. Let us remember that we met the tight-binding term in Problem 4.3.b where we found that the dispersion has two degenerate minima if $t_2 > -t_1/4$. Though the proven existence of a Nagaoka state shows that the model can become ferromagnetic also for $0 < t_2 < -t_1/4$, the possibility of a double-well dispersion seems to help[18] the FM phase to extend over an especially large region of the parameter space: it appears that (for sufficiently large U) the model can become ferromagnetic at all fillings $n < 1/2$, for a broad range of $t_2 > 0$. In particular, one finds

[17] We can gauge t_1 to be either positive or negative. Here we fix $t_1 < 0$.

[18] Though any naive application of the Stoner criterion should be avoided, it is certainly suggestive that the DOS diverges not only at the band edges but also somewhere inside the band (Fig. 4.13, right). Later, we will realize that the strongly asymmetrical shape of the DOS is more important than the presence of a divergency.

also low-density ferromagnetism continuously connected to the Nagaoka point [297, 326, 80, 79]. One does not need strong coupling to get ferromagnetism; in fact, the threshold value of U approaches 0 in the low-density limit [326, 80]. The t_1–t_2–U chain seems to provide an example of robust itinerant ferromagnetism in a single-orbital Hubbard model but we still do not quite understand how it works.

There has been no systematic attempt to extend the ring exchange argument to familiar higher-dimensional lattices with triangular plaquettes. Nevertheless, there are some hopeful developments: approximate theories which give an ever shrinking domain of ferromagnetism for bipartite lattices, often indicate impressively broad FM regions for non-bipartite close-packed lattices. These results will be reviewed in the next section.

Though lattices with triangular plaquettes are by no means uncommon, we may still feel uneasy that the assumption of triangles is a bit too restrictive and artificial. To allay such fears, let us make two remarks. First, one does not really need triangles: one can also construct a square plaquette with $\prod t_{ij}$ favourable for ferromagnetism, as we did in Problem 8.1. Second, the plaquette shape need not represent the real-space geometry: it is merely a graph showing how orbitals are connected by hopping. The Wannier states with on-site energy ϵ_0 can be also understood as crystal-field-split states centered at the same sites as the $\epsilon = 0$ Wannier orbitals. The loops can be created by inter-orbital hoppings. The ring exchange argument has good chances to become relevant for the Anderson lattice and other multiorbital models which we are going to discuss later.

Problem 8.2 *The three-site toy model of ferromagnetism* [393]. Consider the three-site Hubbard model illustrated in the right panel of Fig. 8.6. The Hamiltonian $\mathcal{H}_3$ is given in (8.5). Let the electron number be $N = 2$. We consider the system ferromagnetic if its ground state is a triplet, while finding a singlet ground state means that the system is non-magnetic. The interest of the model lies in the finding that in a wide parameter range, it has a triplet ground state and furthermore, the mechanism of this small-system ferromagnetism is very much the same as in Tasaki's linear chain model [392].

a) For a first orientation, determine the one-particle spectrum of $\mathcal{H}_3$. Does it give reason to expect a triplet ground state if U is switched on? In this respect, what is special about the $t_1 = t_2$, $\epsilon_3 = 0$ case?

How does the spectrum depend on the sign of t_1? Can you think of a deeper reason for what you find?

b) Now take two electrons. Let us consider first the case $U = U_3 = \infty$.

What is the condition for finding a triplet ground state?

c) Find the region of stability of the ferromagnetic ground state for finite $U = U_3$. Try to judge the effect of varying ϵ_3.

d) Derive the effective Hamiltonian.

8.4 Instability of the Nagaoka State

Up to now, we have been trying to locate the paramagnetic–ferromagnetic phase boundary. If we wish, we can delineate a further domain inside the ferromagnetic phase: the domain of saturated ferromagnetism. The saturation value of the total spin is $\mathcal{S}_{\text{sat}} = N/2$ for less than half filling, while it is $\mathcal{S}_{\text{sat}} = (2L - N)/2$ for more than half filling. The regime of fully polarized ground states is often called the Nagaoka state, or the Nagaoka phase. The name implies that we think that the Nagaoka theorem could be extended to a finite area in the phase diagram, even though we have a proof only for a singular point.

At the mean field level, it is easy to write down the condition for saturation magnetization: the exchange splitting (7.27) has to exceed the depth of the fully polarized Fermi sea

$$Un > \epsilon_{\text{F}} - \epsilon_{\text{b}} \tag{8.9}$$

where ϵ_{b} is the bottom of the band. This is a special case of the finite-m Stoner criterion (7.111). The threshold value of U required by (8.9) is larger than, but typically of the same order as, the critical U_{cr} predicted by the Stoner criterion (7.24). It is not unusual to find that the regime of Nagaoka state covers a sizeable portion of the ferromagnetic phase. It may even happen that there is nothing but saturated ferromagnetism; this depends on the form of the density of states. We have seen an example in Stoner theory (Problem 7.2) that the magnetization can jump straight to its saturation value. Naturally, we should not place too much trust in mean field theory. There is, however, evidence from a variety of techniques that the appearance of a fully polarized ground state need not be a rare phenomenon. We have seen that for flat-band models with the right value of n, the spin is saturated for any $U > 0$. For several non-pathological models, numerical experience indicates that whenever the ground state is ferromagnetic, it is fully saturated. For instance, preliminary studies suggested that the ground state of the t_1–t_2–U chain is either a singlet, or a saturated ferromagnet, even at moderate values of U [326]. A detailed analysis indicates that the U-driven non-magnetic to ferromagnetic transition is of first order[19] if the dispersion relation (4.80) has two minima, while it is continuous if there is only one minimum [79].

Apart from the very few cases covered by exact theorems, the Nagaoka phase is still merely hypothetical. Its existence cannot be proven by exact diagonalization for small systems, since we can never be certain what happens if we go over to larger systems. If we have bad luck, it may happen that upon approaching the thermodynamic limit, the conjectured Nagaoka phase shrinks to the single Nagaoka point whose existence is exactly proven. It might seem that we are doomed to lack dependable statements about the extent of the fully polarized domain.

Fortunately, there exists a generally applicable method to derive exact statements in the opposite sense: we can *disprove* the existence of a fully polarized ground state in part of the parameter space. This works by testing the *local stability* of the Nagaoka state: we examine whether the flipping of a spin leads to energy lowering. If it does, the true ground state cannot be fully polarized. We try to get an idea of the extent of the Nagaoka phase by pushing forward the boundaries of its guaranteed instability.

8.4.1 The Single-Spin-Flip State

In the fully polarized Nagaoka state $|N\rangle$ the $\uparrow$-spin Fermi sea is filled by N electrons up to the Fermi energy ϵ_F, while the $\downarrow$-spin Fermi sea is

[19] We may guess that this happens because the transition is not of the disorder–order kind, but rather it is a level crossing between states with different kinds of order. Invoking arguments from Sec. 6.3.2, we ascribe a spin gap to the $t_2/t_1 > 1/4$ non-magnetic state which is presumably related to the dimerized state of the J_1–J_2 Heisenberg model. Since we are not aware of any mechanism which would cause the spin gap to vanish at the same point where ferromagnetism appears, we conclude that the transition is discontinuous. The author is grateful to Stephane Daul for a discussion of this point.

empty. The energy density is simply

$$\frac{1}{L}E_{\rm Nag} = \int_{\epsilon_{\rm b}}^{\epsilon_{\rm F}} \epsilon\rho(\epsilon)d\epsilon\,. \tag{8.10}$$

The reason why the Nagaoka state may become unstable, strikes the eye: it appears that a substantial energy lowering could be achieved by transferring an electron from the top of the ↑-spin Fermi sea to the bottom of the ↓-spin band, because this is accompanied by the kinetic energy gain $\Delta E_0 = \epsilon_{\rm b} - \epsilon_{\rm F}$. The corresponding uncorrelated (Hartree–Fock) single-spin-flip state is

$$|\Phi\rangle = c^{\dagger}_{\mathbf{q}\downarrow} c_{\mathbf{k}_{\rm F}\uparrow}|N\rangle\,, \tag{8.11}$$

where the wavevector $\mathbf{q}$ belongs to the band minimum[20]. There is, however, a substantial energy penalty coming from the Hubbard term, because in an uncorrelated state, the ↓-electron cannot avoid the ↑-electrons, thus the spin-flip energy at the Hartree–Fock level is

$$\Delta E_{\rm HF} = \langle\Phi|\mathcal{H}|\Phi\rangle - E_{\rm Nag} = \epsilon_{\rm b} - \epsilon_F + Un\,, \tag{8.12}$$

wherewith we rederived (8.9). We can introduce the Hartree–Fock estimate for the Nagaoka threshold

$$U_{\rm HF}^{\rm Nag}(n) = \frac{\epsilon_F - \epsilon_{\rm b}}{n}\,. \tag{8.13}$$

The ground state is fully polarized if $U \geq U_{\rm HF}^{\rm Nag}$. Let us observe that in the independent electron approximation, the $U = \infty$ ground state is predicted to be fully polarized for all $N < L$.

However, $\Delta E_{\rm HF}$ should not be equated with the true energy difference

$$\Delta E = E_{\mathcal{S}=N/2-1} - E_{\rm Nag} \tag{8.14}$$

between the $\mathcal{S} = N/2$ and $\mathcal{S} = (N-2)/2$ ground states because, in reality, electron correlations play an important role, enabling the ↓-electron to get out of the way of the ↑-electrons, and thus reducing

[20]There may be several degenerate band minima and, in any case, the choice of the Fermi wavevector $\mathbf{k}_{\rm F}$ is not unambiguous. Here we proceed with the argument assuming that it is irrelevant which of the possible $\mathbf{k}_{\rm F}$ and $\mathbf{q}$ is chosen.

the interaction energy. We find that the domain where the Nagaoka state may still be stable, is much smaller than (8.9) would predict. In particular, there is a broad range of band filling in which the Nagaoka state is unstable even at $U = \infty$.

The reduction of the interaction energy is not the only effect of allowing for correlations. Both the $\uparrow$- and $\downarrow$-spin kinetic energies are modified (the bands narrowed) because the $\downarrow$-spin electron has to avoid the $\uparrow$-spin electrons. These correlation corrections are difficult to evaluate. In contrast to the Nagaoka state, the lowest-energy state of the $\mathcal{S} = (N-2)/2$ subspace is not exactly known, and we cannot determine the exact position of the spin-flip instability line which is defined by $\Delta E(U, n) = 0$. Nevertheless, we can set a *variational bound* on the $\mathcal{S} = (N-2)/2$ energy. The true energy gain cannot be smaller than the variational estimate and thus if the single-spin-flip process gains energy even variationally, then the Nagaoka state is certainly unstable.

What should a correlated state be like? Its simplest form is easily guessed at $U = \infty$, where it has to express that electrons are strictly prohibited to share the same site. The trial ground state is obtained by applying the projection operator

$$\hat{P}_{U=\infty} = \prod_{\mathbf{j}} (1 - \hat{n}_{\mathbf{j}\uparrow}\hat{n}_{\mathbf{j}\downarrow}) \tag{8.15}$$

to the Hartree–Fock state $|\Phi\rangle$

$$|\Psi_{\mathrm{SKA}}(U = \infty)\rangle = \hat{P}_{U=\infty}\, |\Phi\rangle = \prod_{\mathbf{j}} (1 - \hat{n}_{\mathbf{j}\uparrow}\hat{n}_{\mathbf{j}\downarrow})\, c^{\dagger}_{\mathbf{q}\downarrow} c_{\mathbf{k}_F\uparrow} |N\rangle\,. \tag{8.16}$$

The operator $\hat{P}_{U=\infty}$ projects out configurations with a doubly occupied site (in the present problem, there can be at most one doubly occupied site, since there is only one $\downarrow$-electron).

At finite U, double occupancy is no longer prohibited; it is merely undesirable, to a degree which depends on the value of U. The application of the "softened" projector

$$\hat{P} = \prod_{\mathbf{j}} [1 - (1-\eta)\hat{n}_{\mathbf{j}\uparrow}\hat{n}_{\mathbf{j}\downarrow}] \tag{8.17}$$

achieves that the amplitude of configurations with a doubly occupied site is reduced by a factor η with respect to the Hartree–Fock value.

This is easy to verify: if $\hat{P}$ finds a doubly occupied site, it acts like $\hat{P}|d\rangle = [1-(1-\eta)\hat{n}_\uparrow\hat{n}_\downarrow]|d\rangle = \eta|d\rangle$. On the other hand, $\hat{P}$ has no effect on singly occupied or empty sites. Thus the structure of the finite-U trial ground state

$$|\Psi_{\rm SKA}(\eta)\rangle = \hat{P}\,|\Phi\rangle = \prod_{\mathbf{j}}[1-(1-\eta)\hat{n}_{\mathbf{j}\uparrow}\hat{n}_{\mathbf{j}\downarrow}]\,c^\dagger_{\mathbf{q}\downarrow}c_{\mathbf{k}_F\uparrow}|N\rangle \tag{8.18}$$

is similar to that of the uncorrelated $|\Phi\rangle$, except for the systematic suppression of double occupation. The variational parameter η is obtained from demanding that the energy

$$E(\eta) = \frac{\langle\Psi_{\rm SKA}(\eta)|\mathcal{H}|\Psi_{\rm SKA}(\eta)\rangle}{\langle\Psi_{\rm SKA}(\eta)|\Psi_{\rm SKA}(\eta)\rangle} \tag{8.19}$$

be minimum. In Ch. 9 we will realize that what we are doing here, is the application of the Gutzwiller variational technique to the single-spin-flip problem; the operator $\hat{P}$ is the Gutzwiller projector. Ch. 9 will bring a more detailed argument for using such a variational approach. The presently considered trial state (8.16), and its finite-U version (8.18), were introduced by Shastry, Krishnamurty and Anderson (SKA) in [359]. In much of our discussion, we follow [152].

The finite-U state (8.18) is flexible: it can be optimized by the right choice of η. In contrast, it is clear that we have to set $\eta = 0$ in the limit $U = \infty$, thus (8.16) is a trial state without a variational parameter.

The energy $E(\eta)$, and thus the variational estimate for the energy change due to the single spin flip, can be evaluated exactly[21]. The derivation is given in Appendix B; here we quote the results, and comment on their physical meaning.

The general structure of the spin-flip energy (8.14) is

$$\begin{aligned}\Delta E(n,U) &= \Delta E_\uparrow + \Delta E_\downarrow + \Delta E_{\rm int} \\ &= (\mathcal{K}_\uparrow - E_{\rm Nag}) + \mathcal{K}_\downarrow + \Delta E_{\rm int}\ ,\end{aligned} \tag{8.20}$$

[21] We cannot speak about the exact determination of the spin-flip energy because the trial state (8.18) is not the exact single-spin-flip state. We merely mean that once the trial state is accepted, its consequences can be worked out in closed form without further approximation. This feature makes the SKA states almost unique among the Gutzwiller states. The small-$\mathcal{S}$ Gutzwiller states of the Hubbard model have many doubly occupied sites which leads to great technical difficulties (see Chs. 9, 10, and Appendix C).

i.e., it consists of the change in the kinetic energy of the $\uparrow$-electrons, the kinetic energy of the $\downarrow$-electron, and an interaction term.

Here we discuss mainly the $U = \infty$ case, because then the formulas are simpler. $\eta = 0$ guarantees that there is no interaction energy contribution. The kinetic energy of the $\downarrow$-spin electron is given in (B.9)

$$\mathcal{K}_\downarrow = \epsilon_b(1-n)\left\{1 - \left(\frac{E_{\rm Nag}}{L(1-n)zt}\right)^2\right\} . \tag{8.21}$$

Since the $U = 0$ result would be ϵ_b, the above equation describes that the band motion is slowed down due to electron correlation. The *correlation narrowing* factor appearing in (8.21) has a straightforward interpretation: for the motion of the $\downarrow$-electron on the background of the $\uparrow$-electrons, one needs *two* nearest neighbour holes in the $\uparrow$-electron sea: one, where the $\downarrow$-electron resides, and another where it can hop to. $(1-n)$ is the probability of finding a hole, and the factor in the curly brackets is the hole–hole correlation function of the $\uparrow$-electron sea (see Problem 9.2).

The $\uparrow$-spin kinetic energy term (B.15) describes, in addition to the removal of an electron from the Fermi level, the narrowing of the band

$$\Delta E_\uparrow = \mathcal{K}_\uparrow - E_{\rm Nag} = -\epsilon_{\rm F} - \frac{1}{1-n} \cdot \frac{E_{\rm Nag}}{L} . \tag{8.22}$$

Since the $U = \infty$ spin-flip energy $\Delta E = \Delta E_\uparrow + \Delta E_\downarrow$ consists of kinetic energy pieces only, it is interesting to compare it with the similarly dissected Hartree–Fock kinetic energy change $\Delta E_0 = \Delta E_{\uparrow 0} + \Delta E_{\downarrow 0}$, where $\Delta E_{\uparrow 0} = -\epsilon_{\rm F}$, and $\Delta E_{\downarrow 0} = \epsilon_b$. Envisaging a band centered at 0, it is clear that $\Delta E_{\downarrow 0} < 0$, and the same is true for $\Delta E_\downarrow$, only the energy gain is reduced by the narrowing of the $\downarrow$-spin band. As for $\Delta E_{\uparrow 0}$, it can be either positive, or negative; but in either case, $\Delta E_0 < 0$. On the other hand, $\Delta E_\uparrow$ given by (8.22) cannot be negative. Since the energy of a full band is 0, $-E_{\rm Nag}$ is the total energy of the $L - N$ states lying between $\epsilon_{\rm F}$, and the top of the band, thus $-E_{\rm Nag}/(L-N) > \epsilon_{\rm F}$. It is remarkable that there can be no kinetic energy gain associated with the $\uparrow$-spin electrons, even if we take away an electron from near the top of the band. We may say that hard-core collisions with the $\downarrow$-electron "stir up" the Fermi sea so much that the total increment of the kinetic energy of the $\uparrow$-electrons exceeds ϵ_F.

The vanishing of $\Delta E(n, U = \infty)$ gives the $U = \infty$ instability criterion

$$\mathcal{K}_\uparrow - E_{\text{Nag}} + \mathcal{K}_\downarrow = 0 \tag{8.23}$$

which is easily solved if one knows the band density of states. In the following, we discuss the square lattice. Using the DOS is given in (4.30), one finds that the Nagaoka state is stable only if $n > n_{\text{cr}}^{\text{SKA}} \approx 0.51$ [359]. Even a relatively simple variational theory sufficed to shrink the domain where saturated ferromagnetism may be possible from $0 < n < 1$ to $0.51 < n < 1$.

The finite-U version of the SKA theory works by minimizing the total energy (which consists of the terms (B.4), (B.8), and (B.15)) with respect to η. This procedure gives a simple formula for the boundary of the Nagaoka phase

$$U_{\text{cr}}(n) = U_{\text{HF}}^{\text{Nag}}(n) \left(1 + \frac{(1-n)(\epsilon_F - \epsilon_b)}{\Delta E(n, U = \infty)} \right) . \tag{8.24}$$

$U_{\text{cr}}(n)$ diverges as $n \to n_{\text{cr}}^{\text{SKA}} \approx 0.51$ from above, and tends to $\approx 18.2t$ as $n \to 1$. This latter feature shows that, along with its basic virtues, the SKA trial state must also have some defects because the correct result would be $\lim_{n\to 1} U_{\text{cr}}(n) = \infty$. To understand why this should be so, recall that the large-U $n \to 1$ Hubbard model is essentially an antiferromagnet with $J = 4t^2/U$, thus ferromagnetism can prevail only if $U \to \infty$. We can even propose the form for the functional dependence of $U_{\text{cr}}(n)$ in the limit $n \to 1$. We will show that $\Delta E \propto 1 - n$ for $1 - n \ll 1$ (see equation (8.30)). On the other hand, flipping a spin gains the exchange energy $-4zt^2/U$, thus U_{cr} diverges as

$$U_{\text{cr}} \propto t/(1-n) \qquad \text{as } n \to 1 . \tag{8.25}$$

Simple as this argument is, the SKA Ansatz "does not know about it", because it has no built-in magnetic correlations. Better variational trial states have to be (and have been) found. These can only push n_{cr} and $U_{\text{cr}}(n)$ upwards; the SKA predictions set exact bounds on the extent of the Nagaoka phase.

8.4.2 Improved Variational Methods: The Case of the Square Lattice

The SKA Ansatz is the simplest of a large family of single-spin-flip trial ground states. These were usually tested by applying them to the square lattice as the simplest $D > 1$ model which may become ferromagnetic. Numerical work on finite systems as large as 100×100 gave $n_{\rm cr} \approx 0.71$ [433]; comparing this to the SKA value $n_{\rm cr} \approx 0.51$ shows that there is much room (and need) for improvement. However, impressive as these results are, numerical data for finite systems cannot give strict bounds which would be valid in the thermodynamic limit, thus it is useful to trace the development of analytic approaches which are valid for arbitrarily large systems, and at the same time give a pictorial idea of the features of the wave function.

Improved bounds can be obtained from more flexible trial states whose nature has to be guessed [46, 154, 448]. For instance, one can argue that projecting out double occupation in (8.16) creates a drastic disturbance in the ↑-electron sea at the site of the ↓-electron, and this makes the ↑-electron kinetic energy unduly large. A better state can be obtained by letting the ↑-electrons jump away (with variable amplitudes) from the vicinity of the ↓-electron. The $U = \infty$, nearest-neighbour version of such a Basile–Elser-type state is [46]

$$|\Psi_{\rm BE}\rangle = \frac{1}{\sqrt{L}} \sum_{\mathbf{j}} e^{i\mathbf{q}\mathbf{j}} c_{\mathbf{j}\uparrow}[c^\dagger_{\mathbf{j}\uparrow} + \alpha \sum_{\mathbf{m}} c^\dagger_{\mathbf{m}\uparrow}] c^\dagger_{\mathbf{j}\downarrow} c_{k_F\uparrow}|N\rangle \tag{8.26}$$

where the $\mathbf{m}$ are nearest neighbours of $\mathbf{j}$, and α is a variational parameter. What we did is the following: removing the topmost ↑-electron from the Fermi sea, we created a ↓-spin electron at the band bottom with wavevector $\mathbf{q}$. Expanding the ↓-spin state in a Wannier basis, the electron is created at the site $\mathbf{j}$ with phase factor $e^{i\mathbf{q}\mathbf{j}}$. *If* there is also an ↑-spin electron at $\mathbf{j}$, we let it jump with amplitude α to one of the neighbouring sites; if not, we do nothing. Minimizing the ground state energy with respect to α, one finds that $\mathcal{K}_\uparrow$ is reduced in comparison to its SKA value, while $\mathcal{K}_\downarrow$ is left essentially unchanged. The procedure leads to a substantial increase of the critical density, to $n_{\rm cr} \approx 0.585$, which shows that the *local polarization around the flipped spin* is indeed an important effect. Allowing hopping to distances up to $\sqrt{13}$, using eight different variational parameters, gives only a marginal further improvement to $n_{\text{cr}} \approx 0.595$. On the other hand, the Basile–Elser states share the basic shortcoming of the SKA states, inasmuch as they predict a finite U_{cr} at $n \to 1$ [154].

A different approach (explored by Gebhard and Zotos [128]) seeks to improve upon $\mathcal{K}_\downarrow$. The idea is to dig a hole around the $\downarrow$-spin so that it can jump away. Let us recall that the SKA Ansatz (B.2)

$$|\Psi_{\text{SKA}}(U=\infty)\rangle = \frac{1}{\sqrt{L}} \sum_{\mathbf{j}} e^{i\mathbf{q}\mathbf{j}} \left[1 - \hat{n}_{\mathbf{j}\uparrow}\right] c^\dagger_{\mathbf{j}\downarrow} c_{\mathbf{k}_F\uparrow} |N\rangle \tag{8.27}$$

makes the creation of a $\downarrow$-electron at $\mathbf{j}$ dependent on the occupation of that site by an $\uparrow$-electron. Generalizing this idea, a nearest-neighbour Gebhard–Zotos-type trial state

$$|\Psi_{\text{GZ}}\rangle = \frac{1}{\sqrt{L}} \sum_{\mathbf{j}} e^{i\mathbf{q}\mathbf{j}} (1 - \hat{n}_{\mathbf{j}\uparrow})[1 + \beta \sum_{\mathbf{m}} (1 - \hat{n}_{\mathbf{m}\uparrow})] c^\dagger_{\mathbf{j}\downarrow} c_{\mathbf{k}_F\uparrow} |N\rangle \tag{8.28}$$

makes it dependent on the occupation of the neighbouring sites as well. Optimizing the value of β, one finds $n_{\text{cr}} \approx 0.601$.

Since (8.26) and (8.28) represent different strategies to minimizing the ground state energy, they can be combined. Hanisch and Müller-Hartmann carried out a systematic investigation of such possibilities [154]. A 28-parameter Ansatz recovered the numerically indicated $n_{\text{cr}} \approx 0.71$ within an analytic treatment. What is more, the finite-U treatment gives a phase boundary with the physically correct $\lim_{n\to 1} U_{\text{cr}} = \infty$ behaviour. The smallest $U_{\text{cr}} \approx 63.6t$ is found at $n \approx 0.86$.

A mighty effort [448] allowing for 1100 local correlation terms probably pushed this kind of approach to its limits, and found that the Nagaoka state is definitely unstable for $U < U_{\text{cr}} \approx 77.7t$ for any n, and for $n < n_{\text{cr}} \approx 0.749$ even at $U = \infty$. It should be added that finite-size extrapolations from exact diagonalization results suggest the even stricter $n_{\text{cr}} \approx 0.805$ [168].

So much about the local stability where one assumes a homogeneous Nagaoka state and studies its stability against the flipping of a single spin. It is, however, possible that a state is stable against all kinds of small perturbations, but still not the ground state, because it is unstable against a large-scale rearrangement of all the spins. To judge the *global stability*, one should be able to guess all the possible rearrangement patters, and that is asking for too much. However, there is a straightforward possibility suggested by the nearness of the insulating SDW state at $n = 1$. The transition from the Nagaoka state to the SDW state can only be of first order, and this raises the question whether a heterogeneous state caused by phase segregation may not be possible. Well-known arguments [430] show that it may: global instability carves out a slice from the Nagaoka domain near the $n = 1$ axis [448]. We return to the question of phase segregation in Ch. 10 (see p. 568).

Where does all this leave us: is there a finite Nagaoka domain in the $T = 0$ phase diagram of the square lattice Hubbard model, or not? It is difficult to tell. The ever-shrinking region indicated by consecutive approximations may lead us to believe that if we could improve the wave function ad infinitum, only the Nagaoka point would remain. On the other hand, the most systematic work [448] which took into account 1100 correlation terms influencing a 9×9 plaquette surrounding the flipped spin, still left a sizeable core region of Nagaoka phase, and it would be difficult to claim an insight into why correlations with more distant parts of the lattice should destroy the Nagaoka phase completely[22].

It is a quite different question whether the square lattice Hubbard model has a $T = 0$ ferromagnetic phase. Guided more by feeling than by reason, one may imagine that inside the ferromagnetic phase there should be also a Nagaoka phase, and that the area covered by the latter should give a fair idea of the extent of the former. However, this is not necessarily true. We may envisage that there in no finite area with full polarization, but a broad range of unsaturated ferromagnetism (like in iron). We have no comparable theoretical tools to test this possibility. It should be mentioned that a high temperature expansion study indicates that there may be ferromagnetism, and that it is never saturated [329].

[22]Available evidence suggests that the Nagaoka phase survives. The recently developed resolvent method [155] amounts to finding *the* optimal infinite-range Basile–Elser-type state, with infinitely many variational parameters. It turns out that the first five decimal places of the result come from correlations within the plaquette size studied in [154]. However, the non-Basile–Elser-type correlations are also important, so this result is still not definitive, even for the square lattice. Furthermore, it is found that correlations beyond a small range have more influence in other structures, e.g., the triangular lattice.

It is difficult to judge the power of high-temperature expansion to decide the nature of $T = 0$ ordering, but we have to keep an open mind.

Rather than trying to guess the eventual outcome, let us look back, and draw conclusions from our solid results. The Hartree–Fock criterion (8.13) would have made us believe that the Nagaoka phase covers most of the ground state phase diagram, extending over all $0 < n < 1$, and from $U \sim 4t$ upwards. Taking correlations into account has drastically reduced the area which may be still permitted for a Nagaoka state, changing its character from a supposedly robust, intermediate-coupling phenomenon to a marginally surviving strong-coupling regime. The square lattice Hubbard model may be ferromagnetic somewhere, but it certainly is not a model noted for its ferromagnetism.

8.4.3 Lattice Structure Dependence of Ferromagnetism in a Single-Band Model

Extending the studies to other lattices in $D = 2$, 3, and ∞, we learn that the shape and extent of the area in the $T = 0$ phase diagram, which is covered by the fully spin-polarized state, depends very much on the lattice structure [152]. Though the Nagaoka phase cannot be equated with the ferromagnetic phase (there are reasons to worry about systems with partially polarized ground states), still we can claim that systems with a large Nagaoka region are genuinely inclined towards ferromagnetism, while for systems with a small or vanishing Nagaoka phase, the very appearance of ferromagnetic order is in doubt.

Though we know that the SKA trial state does not perform so well as states with intersite correlations do, its simplicity allows to make a comparative study of a number of lattices[23]. At the level of the SKA states, the stability of the Nagaoka state hinges upon the band density of states. To see this, let us discuss the limiting forms of $\Delta E(n, U = \infty)$ for $n \to 0$, and for $n \to 1$. First, let us assume that the DOS is finite at both band edges: let it be ρ_{b} at the bottom of the band ϵ_{b}, and ρ_{t} at the top of the band ϵ_{t} (the assumption works in $D = 2$).

Near half-filling, we collect the terms of order $(1 - n)$. Replacing

[23]Using the resolvent method, very similar conclusions can be reached for the most general Basile–Elser states [155].

$\epsilon_{\rm F} \approx \epsilon_{\rm t} - (1-n)/\rho_{\rm t}$, and $E_{\rm Nag} \approx -(1-n)\epsilon_{\rm t}$ into (8.21), we find

$$\mathcal{K}_\downarrow \approx \epsilon_{\rm b}(1-n)\left\{1-\left(\frac{\epsilon_{\rm t}}{zt}\right)^2\right\} + O\left((1-n)^2\right) . \tag{8.29}$$

On the other hand, the change in the kinetic energy of the $\uparrow$-electrons is (see (8.22))

$$\Delta E_\uparrow \approx \frac{1-n}{\rho_{\rm t}} + O\left((1-n)^2\right) . \tag{8.30}$$

Let us reconsider the square lattice first. $\epsilon_{\rm t} = zt = 4t$, thus near half-filling, $\mathcal{K}_\downarrow$ is of order $(1-n)^2$, becoming utterly negligible compared to $\Delta E_\uparrow \approx (1-n)/\rho_{\rm t} > 0 \rightsquigarrow$ the Nagaoka state is stable. We consider this to be in essential agreement with the proven existence of the $N = L-1$ Nagaoka state[24].

What about low densities? It is easy to check that the total energy change is

$$\Delta E(n \ll 1, U=\infty) \approx -n\left(\frac{1}{\rho_{\rm b}} + 2\epsilon_{\rm b}\right) . \tag{8.31}$$

Replacing $\epsilon_{\rm b} = -4t$ and $\rho_{\rm b} = 1/(4\pi t)$ (see (4.62)) into (8.31), we find that the spin-flip is accompanied by energy gain $\rightsquigarrow$ the Nagaoka state is unstable. Let us observe that we did not find anything basically wrong with the idea of (saturated) low-density ferromagnetism in $D=2$; it is just that the square lattice with nearest-neighbour hopping happens to fail to satisfy the criterion, because $\rho_{\rm b}$ is too small. Switching on the second-neighbour hopping, it is possible to devise models whose ground state remains fully polarized at $n \to 0$ [326, 170, 155]. The low-density ferromagnetism of the t_1–t_2 Hubbard chain is an even simpler realization of basically the same idea [297, 326].

Next, let us consider the *triangular lattice*. Choosing the hopping integral in the conventional manner as $-t < 0$, the band energies lie

[24] The Nagaoka theorem is derived for a finite system, while here, we are discussing results which hold in the thermodynamic limit. All the same, whenever we have proof for a Nagaoka state, we find a remaining Nagaoka region even with the most sophisticated trial states. Conversely, if the case is not covered by Nagaoka's theorem, we usually find a variational argument for the instability. However, "islands of stability" (see later) around quarter-filling in certain lattices are exceptions from the latter statement.

in the interval $[\epsilon_b, \epsilon_t] = [-6t, 3t]$. The asymmetrical shape of the band (see Fig. 4.4, right) is a consequence of the non-bipartite nature of the lattice. In contrast to the case of the (bipartite) square lattice, we find now $\epsilon_t < zt$. Recalling (8.29), it is clear that near half-filling $\mathcal{K}_\downarrow \propto (1-n)$, i.e., it is of the same order of magnitude as $\Delta E_\uparrow$, and can outweigh it. Putting in the numerical factors, it turns out that the total spin-flip energy $\Delta E < 0$, i.e., the Nagaoka state is unstable for $1 - n \ll 1$. A detailed argument [153] shows that the Nagaoka state is completely eliminated below half-filling ($n < 1$).

However, it does not follow that the triangular lattice is incapable of ferromagnetism; far from it! Since the lattice is not bipartite, its phase diagram need not be symmetrical about the axis $n = 1$, and the physics of the $n > 1$ and $n < 1$ regimes can be quite different. In fact, the Nagaoka theorem assures that it must be so: since the $t > 0$ and $t < 0$ models are related by an electron–hole transformation, it must be true that adding a single electron to the half-filled band ($N = L + 1$), the ground state has the total spin $\mathcal{S} = (L-1)/2$. Using a variety of stability arguments [153, 152], one finds that at $U = \infty$, the Nagaoka state extends over the wide region $1 < n < 1.824$, while the smallest value of U_{cr} is $9.62t$, only slightly more than the bandwidth. Though the fulfillment of local stability criteria does not constitute a proof of the existence of a high-spin ground state, a comparison of this behaviour with that of the square lattice strongly suggests that the ferromagnetism of the triangular lattice is a robust, intermediate-coupling phenomenon. However, the Mermin–Wagner theorem restricts its presence to $T = 0$.

We briefly mention two other two-dimensional lattices. The *kagome lattice* (Fig. 8.3) gives an even more extreme example of phase diagram asymmetry. As discussed on p. 429, it is possible to choose the sign of t in such a manner that a flat band containing $L/3$ states is lying at the bottom of the spectrum (but not at the top). For any $U > 0$ we find flat band ferromagnetism (Nagaoka state) if $n \leq 1/3$. The flat band region is only a part of a huge Nagaoka phase which extends over all $0 < n < 1$ ($U_{cr} > 0$ for $n > 1/3$). In addition, there is a disjoint part of the Nagaoka phase, a small "island of stability" around $n = 3/2$ at large U. The physical nature of the "island" is quite unclear, since – in contrast to the Nagaoka "continent" at $n < 1$ – it is not continuously connected to a point where full polarization would be guaranteed by the Nagaoka theorem.

The *honeycomb lattice* is bipartite. Its interest lies in the fact that it is built up of six-site loops, thus we cannot invoke the Nagaoka theorem to ensure full polarization at $N = L - 1$. The application of local stability criteria shows that the part of the Nagaoka phase, which is pushing up against the $n = 1$ axis, is even smaller than for the square lattice. However, again we find an island of stability at quarter-filling.

Going over to *three- and higher-dimensional simple cubic lattices*, we have to reformulate the small-n and small-(1-n) expansions. Assuming the D-dimensional canonical behaviour $\rho(\epsilon) \sim (\epsilon - \epsilon_b)^{(D-2)/2}$ at the bottom of the band, for small n one finds (omitting prefactors) $\epsilon_F - \epsilon_b \sim n^{2/D}$, $E_{\text{Nag}} - \epsilon_b n \sim n^{(D+2)/D}$, $\Delta E_\uparrow + \epsilon_b \approx -n^{2/D} + O(n)$, $\mathcal{K}_\downarrow \approx \epsilon_b(1-n)$, thus the leading term of ΔE is $\sim -n^{2/D} < 0 \leadsto$ there is no Nagaoka ferromagnetism at low densities.

Similarly, let us assume that the DOS near the top of the band is $\rho(\epsilon) = c(\epsilon_t - \epsilon)^{(D-2)/D}$. Then, using $\epsilon_t = 2Dt$, we find

$$\mathcal{K}_\downarrow \approx -\frac{2D}{D+2}\left(\frac{D}{2c}\right)^{2/D}(1-n)^{(D+2)/D} \tag{8.32}$$

and

$$\Delta E_\uparrow \approx \frac{2}{D+2}\left(\frac{D(1-n)}{2c}\right)^{2/D} . \tag{8.33}$$

It is clear that $\Delta E_\uparrow$ dominates, leading to the conclusion that the Nagaoka state remains stable sufficiently close to half-filling in arbitrarily high (but finite) dimensions. However, the Nagaoka phase shrinks with increasing D, and vanishes at $D = \infty$. The reason for this is immediately seen: the $D = \infty$ hypercubic lattice has a gaussian DOS, thus the first $\downarrow$-electron can be put infinitely deep in energy. Replacing $\epsilon_b = -\infty$ in (8.13), we see that even the Hartree–Fock threshold diverges for any n, thus it is certain that the finite-U ground state cannot be fully polarized. It can be shown that the $T = 0$ magnetization remains unsaturated even at $U = \infty$ [108]. However, the gaussian-tail decrement of the spin polarization (see (10.54) on. p. 563) is so small that we may be wondering whether it can do any harm if there is a genuine tendency to ferromagnetism. The are conflicting pieces of evidence. A high-temperature series expansion carried to order T^{-10} indicated that there is no ferromagnetic order at any finite temperature [396]. However, it may have failed to detect low-temperature ordering at very strong coupling [308]. We return to this question in Sec. 10.6.5.

Returning to $D = 3$: the SKA Ansatz gives $n_{\rm cr} \approx 0.68$, and $U_{\rm cr}^{\rm min} \approx 12t$. Using more flexible trial states, the bounds can be improved to $n_{\rm cr} \approx 0.763$ and $U_{\rm cr}^{\rm min} \approx 48.9t$ [155], and in a subsequent study using 241 local operators, to $n_{\rm cr} \approx 0.815$ and $U_{\rm cr}^{\rm min} \approx 78.6t$ [447]. Even though computational difficulties preclude reaching the level of sophistication which can be achieved for the square lattice, we can safely conclude that the simple cubic lattice is *less* inclined to ferromagnetism than the square lattice. The tendency continues in higher dimensions: Müller-Hartmann [296] showed that the large-D expression is $n_{\rm cr} \approx 1 - 1/\sqrt{2D \log D}$. This casts some light on the remarkable fragility of ferromagnetic order in simple cubic lattices: we understand that it is basically a low-dimensional phenomenon[25]. Our idea of a robust order would be that it is stabilized by increasing the dimensionality. Seen in this light, it looks hopeful that the Nagaoka phase of the fcc lattice remains essentially unchanged as D is increased from 3 to ∞.

The fcc Lattice

The effects of "fine-tuning" the kinetic energy are seldom more spectacular (and pleasing, if you want to find ferromagnetism) than in the case of the fcc lattice. We have seen in Problem 4.2 (see p. 166 and p. 189) that with nearest-neighbour hopping t_1, the tight binding band behaves in a manner which is untypical for three-dimensional lattices: the DOS has a logarithmic divergency at one of the band edges. In what follows, we choose the sign of t_1 so that the singularity is at the bottom of the band. It is clear that our previous discussion of the $D = 3$ behaviour does not apply to this case. We should rather recall the general formulae (8.29)–(8.31). Replacing $\epsilon_{\rm b} = -4t_1$ and $1/\rho_{\rm b} \to 0$ in (8.31) gives that in the low-density limit $\Delta E > 0 \rightsquigarrow$ the Nagaoka state is stable. The statement remains true even if we use Basile–Elser states [155]. As

[25] Assuming that it exists at all. Let us not forget that our inability to disprove the existence of a Nagaoka phase does not prove that it exists.

for $n \to 1-0$, $\epsilon_t = zt$, thus $\mathcal{K}_\downarrow \sim (1-n)^2$, and ΔE is dominated by $\Delta E_\uparrow > 0$. All in all, at the SKA level and even in better approximations, the Nagaoka state is stable for all $0 < n < 1$. On the other hand, saturated ferromagnetism seems to be absent for $n > 1$.

Naturally, we should not forget that the SKA treatment gives a very optimistic forecast about the chances of ferromagnetism, and we should seek independent evidence if we insist on exploring the "fcc route to ferromagnetism". Since we are now speaking about a $D = 3$ system, nothing prevents the order (if any) to be present up to a Curie temperature T_C. Ulmke [416] performed a Quantum Monte Carlo calculation to find out about ordering at finite temperatures and found that T_C is either very small, or vanishing. This sounds quite bad but not all is lost yet: switching on the next-nearest-neighbour hopping t_2, we can explore a whole range of fcc models. For instance, we may recall (4.75) which says that for $t_2 = t_1/2$, the singularity is strengthened to a $1/\sqrt{\epsilon - \epsilon_b}$ behaviour. However, demanding $t_2 = t_1/2$ would be an artificial assumption and luckily, we do not need it at all. For the range $0 < t_2 < t_1/2$, the DOS does not diverge; it has merely a big peak near (but not exactly at) the bottom of the band. This is not unlike the canonical behaviour of the $3d$ band on the fcc lattice, based on realistic electronic structure calculations [14]. Ulmke found that choosing $t_2 = t_1/4$, and $U \approx 3W$ (W being the bandwidth), T_C is non-vanishing between $n \approx 0.15$ and $n \approx 0.85$, rising to the respectable value $T_C \sim 0.02$–$0.025W$ around, and somewhat above, quarter-filling (Fig. 8.8, left). The figure also shows that a narrow range of antiferromagnetism pushes the FM phase away from the immediate vicinity of half-filling. One of the reasons why the fcc lattice gives good chances to FM is just that it is a lattice for which antiferromagnetic interactions are frustrated, and this is reflected in the narrowness of the AFM phase.

Using a realistic $W \sim 2$eV, and choosing $n = 0.65$, one arrives at $T_C \approx 600$K, which is not bad at all for fcc Ni which has a $3d$-hole density $10 - n_{3d} \sim 0.6$! Perhaps too good, considering that we did not take into account the degeneracy of the $3d$ band, or Hund's rule coupling, to arrive at the result? Well, though one could argue that Ni has so few $3d$ holes that the degeneracy of the d-shell cannot play any significant role, we had better take the numerical coincidence as merely suggestive. It

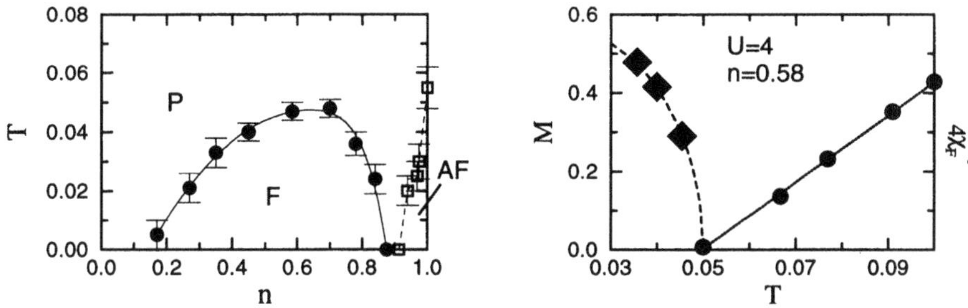

Figure 8.8: Results of QMC studies of the ferromagnetism of the fcc lattice, after [416]. *Left*: the phase diagram of the three-dimensional t_1–t_2 fcc Hubbard model in the band filling–temperature plane, for $U \approx 3W$. F denotes the ferromagnetic phase. *Right*: the magnetization (dotted line) and the inverse susceptibility (full line) of the $D = \infty$ fcc lattice.

is, however, an encouraging sign that according to the calculation, the $T = 0$ exchange splitting is predicted to be about 8 times as large as $k_B T_C$; the large value of this ratio is indeed a feature of the iron group ferromagnets, and it has been our worry that the Stoner theory has no explanation for it.

Remembering the lesson of the vanishing Nagaoka regime of the high-D simple cubic lattices, it is advisable to check if fcc ferromagnetism survives into $D = \infty$. We studied the DOS of the D-dimensional fcc lattice in Problem 4.7 (see p. 177 and p. 195) and found that strictly at $D = \infty$, varying the ratio $0 \leq t_2/t_1 \leq 1/2$ has no effect[26] on the DOS (4.96) which has a $1/\sqrt{\epsilon - \epsilon_b}$ singularity at $\epsilon_b = -\tilde{t}_1/\sqrt{2}$. Our previous simple arguments for $D = 3$ could be repeated here, leading us to expect a Nagaoka ground state for the entire $0 < n < 1$, but could it be trusted? It is nice to be assured that the answer is yes: Uhrig [413] has shown that in $D = \infty$, it is possible to calculate the single-spin-flip energy *exactly*, and the resulting phase boundary indeed reaches from $n = 0$ (where $U_{cr} \to 0$) to $n \to 1$ (where $U_{cr} \to \infty$). Note that (in contrast to $D = 3$) there is no narrow AFM range at $n = 1$: the degree of frus-

[26]Taking the limit $D \to \infty$ is a particularly delicate affair here, since for any finite D (and $t_2 < t_1/2$), the second term on the right-hand side of (4.94) causes the DOS to extend a "foot" down to $-\sqrt{2}\tilde{t}_1$. It is true that the spectral weight of the foot vanishes as $D \to \infty$, but putting there a single $\downarrow$-electron does not require a finite spectral weight in the thermodynamic limit, either! See [296, 413, 152] for a detailed discussion of this point.

tration increases with increasing D, and AFM is completely suppressed at $D = \infty$. Essentially the same conclusion is reached by a Quantum Monte Carlo study [416]. The infinite-dimensional fcc lattice shows a spectacular tendency towards ferromagnetism, and it is of much the same kind that we found for $D = 3$. Fcc ferromagnetism is apparently a very stable phenomenon. The finite-T behaviour (Fig. 8.8, right) is just what we welcome to see: the susceptibility follows the Curie–Weiss law down to T_C, and for $T < T_C$, the magnetization closely follows a Brillouin function, reaching 80% of the saturation value at $T \approx 0.7T_C$. It may well be true that the ground state magnetization is saturated.

Discussion

The appearance of robust ferromagnetism is not a generic feature of the single-band Hubbard model. However, it is possible to choose lattice structures and/or kinetic energy terms which favour ferromagnetism, and some of these assumptions are not unrealistic.

Bipartite lattices are primarily antiferromagnetic at (and presumably near) half-filling, and ferromagnetic order – if it exists at all – is a strong-coupling phenomenon confined to a rather limited range of band filling. In contrast, non-bipartite lattices have bands of asymmetrical shape, and this turns out to favour ferromagnetism. Formulating the statement for the case when ferromagnetism appears at $n < 1$, we found that increasing ρ_b/ρ_t, and decreasing $|\epsilon_b|/\epsilon_t$, lead to enlarging the Nagaoka regime. From among the three-dimensional structures, a class of fcc lattices has been identified as the most promising model for itinerant ferromagnetism[27].

It may be that bipartite lattices like the square, honeycomb, and simple cubic lattices, have their own, relatively weak mechanism for producing large-U ferromagnetism. Whatever this may be, it is becoming inefficient in the limit of large D.

In contrast, fcc-type ferromagnetism survives the limit $D \to \infty$ with

[27] It has to be mentioned that the idea that the fcc lattice favours ferromagnetism, is by no means new. All the three inventors of the Hubbard model [142, 174, 201] emphasized in their early works that the condition for ferromagnetism depends strongly on the form of the DOS. They knew that a maximum in the middle of a symmetrical band is no good, and inferred that one needs a big peak near one of the band edges.

undiminished intensity, maybe it becomes even stronger. We should also recall the large Nagaoka phases of the triangular and kagome lattices which we consider related to the fcc lattice[28]. What could be the common mechanism for all these lattices? We cannot do more than trying to guess. Let us observe that for all of these structures,

1. the density of states is asymmetrical, with a strong peak near one of the band edges, but not at the other.

Furthermore, because of the presence of triangular plaquettes,

2. antiferromagnetism is at least partially frustrated and thus destabilized, allowing a broader FM region.

3. Finally, the presence of triangular plaquettes, or short loops combined of jumps with different t's, raises the hope that ferromagnetism in the physically relevant parameter range is continuously connected to a part of the parameter space where the ring exchange argument becomes applicable.

It should be emphasized that the edges of "plaquettes" are defined by non-vanishing hoppings t, and not by the primary geometrical arrangement of the sites. A square lattice with first *and* second neighbour hoppings does have triangular plaquettes, and is thus not bipartite.

It is interesting that reason 1. which appeals merely to the form of the DOS, appears so forceful that it might be used alone. This transpires from a study [437] of a class of $D = \infty$ models with the tunable[29] DOS

$$\rho_a(\epsilon) = \frac{1 + \sqrt{1 - a^2}}{2\pi} \frac{\sqrt{4 - \epsilon^2}}{2 + a\epsilon}. \tag{8.34}$$

[28] According to our definition of D-dimensional half-hypercubic lattices, the "$D = 2$ fcc lattice" would be the square lattice, which is not very satisfactory. In contrast, the alternative definition of D-dimensional laminated lattices [413] gives the triangular and fcc lattices in $D = 2$ and $D = 3$, and a fairly ferromagnetic $D \to \infty$ limit which is, however, different from the one treated by us.

[29] In the dynamical mean field theory of $D = \infty$ Hubbard models, the lattice structure appears only through the non-interacting DOS which can be treated as an input parameter.

At $a = 0$ the DOS is semi-elliptic $\rightsquigarrow$ symmetrical, and the system shows no inclination to FM, unless U is very high. The extent of the FM phase is the largest for $a = 1$ where the DOS has a square root singularity at $\epsilon = -2$, imitating the $t_2 = t_1/2$ fcc model. However, one does not need to go to this extreme. The DOS develops a strong, but finite, peak near the lower band edge, as a exceeds 0.9 and this is all that one needs to observe a robust ferromagnetism setting in at intermediate values of U, and stable up to a Curie temperature of $\sim 0.02W$. It should be emphasized that, in spite of a superficial similarity, this is by far not the same as the Stoner criterion (7.24) which also requires a large DOS. A large peak of DOS satisfying the Stoner criterion could be at the center of the band, or it could be one of two peaks which are at symmetrical positions. These cases are, however, not favourable for ferromagnetism: one needs a peak in one half of the band but not in the other.

We can understand this in simple terms as follows: for an electron density n, the paramagnetic Fermi level is $\epsilon_F(n/2)$, and the average kinetic energy $\bar{\epsilon} \sim -t$. As we switch on a strong interaction, correlation will effectively narrow the band, and the kinetic energy gain is diminished. The question arises whether we are not doing better by fully polarizing the Fermi sea, filling it up to $\epsilon_F(n)$ with $\uparrow$-electrons. If the band is symmetrical, this involves filling band states lying much higher than $\epsilon_F(n/2)$, and the energy gain is doubtful. However, with much weight near the lower edge of the band, going to $\epsilon_F(n)$ implies filling states which are still fairly low-lying, and the formation of a ferromagnetic state is favourable. – All in all, if we want to name the "driving force" of ferromagnetism in the single-band Hubbard model, it is the kinetic energy gain, in the sense that it does not get renormalized for the fully polarized state, while it suffers a reduction in low-spins states, thus it *may* lead to FM order if the band shape and band filling are right. Technicalities aside, the reasoning is basically the same as for Nagaoka's theorem.

However, even if a ferromagnetic phase really exists, mean field theory is grossly misleading as to its extent and stability. Let us compare the (essentially exact) QMC results with the Hartree–Fock (HF) predictions for the band filling dependence of the Curie temperature T_C (Fig. 8.9, left). The band shape belongs to $a = 0.98$, particularly favourable for FM. The HF width of the FM area is almost three times as large as in QMC, and the maximum T_C exceeds its true value by a factor of $\sim$ 15. This latter feature can be analyzed in the plot of T_C vs U for $n = 0.4$ and $a = 0.98$ (Fig. 8.9, right). As shown by the inset, the HF T_C is unbounded, rising linearly with large U. It is easy to find, where this comes from: assuming that T_C is large, we replace the Curie susceptibility $\chi_0 = Cn(2-n)/2T$ in the Stoner criterion (the number of uncompensated $S = 1/2$ spins being $2{\cdot}n(2-n)/4$); this leads to $T_C^{HF} \propto U/n(2-n)$. In contrast, the QMC result saturates at $T_C \sim O(W)$. This is in accordance with Kanamori's idea that at a large (bare) U, the effective interaction strength saturates at $O(W)$.

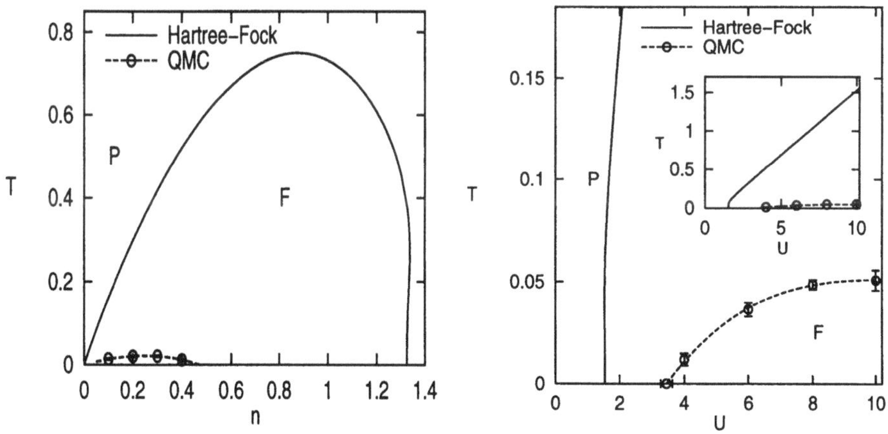

Figure 8.9: A comparison between Hartree–Fock and QMC results shows that mean field theory vastly overestimates the extent of the ordered phase. *Left*: the Curie temperature versus n, for $a = 0.98$ and $U = 4$. *Right*: the Curie temperature versus U, for $a = 0.98$ and $n = 0.4$. *Inset*: the Stoner criterion gives $T_C \propto U$ at large U, while the true T_C saturates (after [437]).

Hartree–Fock theory cannot be trusted in its predictions for ferromagnetism in the Hubbard model. We stress that this holds even for $D = \infty$ where all fluctuations resulting from finite dimensionality have ceased. The basic reason is that here FM cannot be ascribed to an effective intersite interaction but remains a truly itinerant (kinetic-energy-driven) feature. By the way, we have no reason for a similar distrust of mean field results for AFM which can be traced to antiferromagnetic intersite interactions $J \sim t^2/U$.

We do not attempt to review the results on the extended one-band Hubbard model. Let us, however, quote the large-U effective exchange coupling found in the solution to Problem 5.5 (see (5.108) on p. 254)

$$J_{\text{eff}} = \frac{4(t-X)^2}{U-V} - 2F\,. \tag{8.35}$$

J_{eff} tends to change sign as U is increased. Thus though the large Hubbard U in itself may not make the system ferromagnetic, it gives rise to a situation where even a quite weak direct exchange F is sufficient to turn the system into a ferromagnet (see [437] for a detailed study).

What is the upshot of all these arguments: did the study of the single-orbital Hubbard model enable us to understand itinerant ferromagnetism, or did it not? It is possible to take the point of view that, to some extent, it did. We may declare ourselves content with the available evidence that the fcc lattice shows robust ferromagnetism, and accept it as a semi-quantitative explanation for the properties of Ni. As for Co, its lattice has the hcp structure, and that must be just as inclined towards ferromagnetism as the fcc lattice[30]. This still leaves us with the ferromagnetism of Fe to account for. Though there is a widely used definition according to which iron is a weaker ferromagnet[31] than Co and Ni, and one could say that this is in agreement with the fact the bipartite bcc lattice shows only a weak tendency to ferromagnetism, we prefer to take a more balanced view. A Curie temperature $T_C = 1044$K does not suggest anything weak, and the value $p_{\text{par}}/p_{\text{sat}} \sim 1$ also shows that Fe is as far from being a weak itinerant ferromagnet as a system can be (see Table 7.1, p. 400). There is really no need to hurry and try to explain everything in terms of the single-orbital model, since allowing for the degeneracy of the $3d$ band, and the concomitant intraatomic exchange, adds further important features to our picture of itinerant ferromagnetism.

[30]It would take some arguing to show this. With nearest-neighbour hopping, the fcc and hcp lattices have exactly the same DOS, though the band structures are different. It is only at the SKA level, that a band is sufficiently characterized by its DOS; more complicated states are sensitive to the details of the band structure. Furthermore, it turned out that the relevant fcc model is a t_1–t_2 model, and the corresponding generalized hcp models have not been studied yet.

[31]In the sense that its $T = 0$ magnetization is decidedly smaller than the saturation value allowed by the band filling.

8.5 Effects of Degeneracy

We cannot even contemplate discussing degenerate bands in any detail, and will confine ourselves to the simplest arguments. For a first orientation, let us consider the Stoner criterion of ferromagnetism for a Hubbard model with $\mathcal{N}$-fold orbital degeneracy. The Hamiltonian

$$\begin{aligned}\mathcal{H} &= \sum_{\mathbf{k}}\sum_{\alpha=1}^{\mathcal{N}}\sum_{\sigma}\epsilon_{\mathbf{k}}\hat{n}_{\mathbf{k}\alpha\sigma} + U_0\sum_{\mathbf{j}\alpha}\hat{n}_{\mathbf{j}\alpha\uparrow}\hat{n}_{\mathbf{j}\alpha\downarrow} + U\sum_{\mathbf{j}}\sum_{\alpha\neq\alpha'}\sum_{\sigma}\hat{n}_{\mathbf{j}\alpha\sigma}\hat{n}_{\mathbf{j}\alpha'-\sigma} \\ &+(U-J_{\mathrm{H}})\sum_{\mathbf{j}}\sum_{\alpha\neq\alpha'}\sum_{\sigma}\hat{n}_{\mathbf{j}\alpha\sigma}\hat{n}_{\mathbf{j}\alpha'\sigma} \\ &-\frac{J_{\mathrm{H}}}{2}\sum_{\mathbf{j}}\sum_{\alpha\neq\alpha'}(S^{+}_{\mathbf{j}\alpha}S^{-}_{\mathbf{j}\alpha'}+S^{-}_{\mathbf{j}\alpha}S^{+}_{\mathbf{j}\alpha'})\,. \end{aligned} \tag{8.36}$$

is analogous to the doubly degenerate model (5.72)–(5.74), except that it is highly symmetrical in the orbital index α. We have, however, distinguished the intra-orbital U_0 from the inter-orbital U which are all taken equal. The intrasite (Hund's rule) exchange J_{H} stabilizes the local triplets. The uniform susceptibility in the Hartree approximation is

$$\chi = \frac{\chi_0}{1-\rho(\epsilon_{\mathrm{F}})(U_0+(\mathcal{N}-1)J_{\mathrm{H}})}\,. \tag{8.37}$$

We observe that J_{H} plays an important role and that increasing $\mathcal{N}$ reduces the critical value of the interaction parameters. In a qualitative sense, both observations are valid but it turns out that the Hartree–Fock criterion vastly overestimates the chances of ferromagnetism, just as it did for the single-orbital model. A different approach is needed to get a better estimate, and to understand the separate roles of mere degeneracy, and the Hund's rule J_{H}.

8.5.1 Double Exchange

Let us return to the $\mathcal{N}=2$ case. It was shown in Sec. 5.4 that strictly at $n=1$ (quarter filling), and assuming strong Coulomb interaction, we end up with a Mott insulator in which the kinetic exchange favours spin ferromagnetism *and* staggered orbital order. If we dope away from

$n = 1$, we have a metal and ask whether ferromagnetism survives into the metallic phase. The system can remain metallic even at $n = 1$ if the U's are not so large, and we may enquire whether it still has ferromagnetic or/and orbital order.

Let us take $N = 3$ electrons on $L = 2$ sites, which can be thought to represent the filling range $1 < n < 2$ in the framework of the "e_g molecule model" introduced in (5.72)–(5.74). In the process

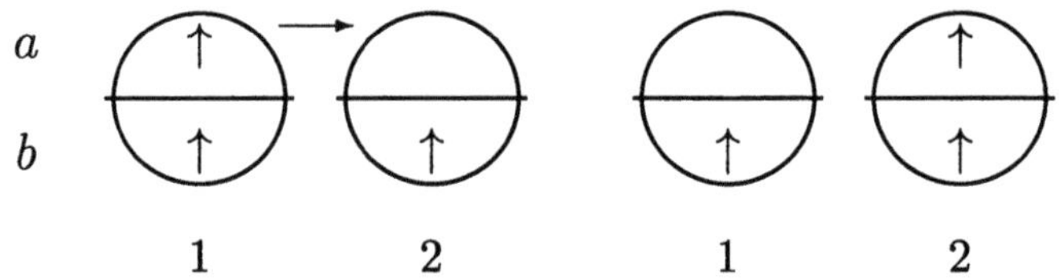

Hund's first rule is satisfied both before, and after, the hopping and the motion of the electron in the a orbitals in unhindered. If we start with antiparallel spins in the b orbitals, the following picture can be drawn

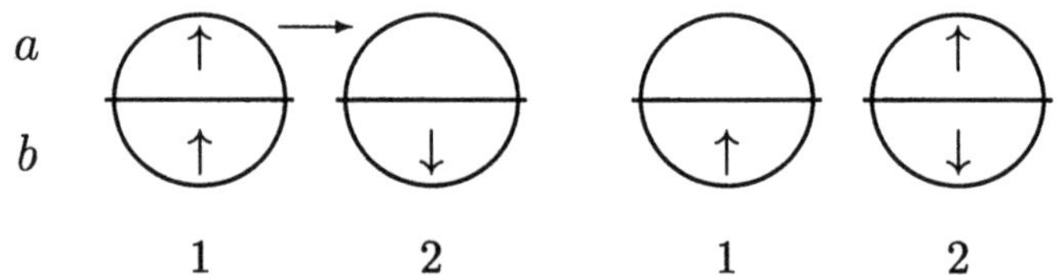

Here it is not unambiguous what the $S^z = 0$ state of site 2 at the end of the process means. It may stand for a singlet, which would mean that the hopping requires an energy investment of order J_{H}. It may also signify the $S^z = 0$ component of the local triplet, in which case Hund's rule is not violated but even then, it is apparent that the electron motion is not unhindered (it stirs up the spin state). All in all, we conclude that the kinetic (band) energy is optimized by making all the spins parallel. This is the essence of the *double exchange* mechanism of ferromagnetism [466, 26, 82].

To turn this idea into a theory, we have to describe how the motion of the electrons is influenced by the relative orientation of spins, and this is not straightforward at all. We had better deal separately with the case when t and $J_{\rm H}$ are more-or-less comparable (when a considerable fraction of local configurations violates Hund's rule), and the case of a strong $J_{\rm H}$.

Let us consider the latter case first because it is conceptually simpler. Here we wish to model systems like $La_{1-x}Sr_xMnO_3$ and $La_{1-x}Ca_xMnO_3$. In these materials, the valence states in question are $Mn^{4+} \rightsquigarrow t_{2g}^3$ and $Mn^{3+} \rightsquigarrow t_{2g}^3 e_g^1$. The crystal field splitting between the t_{2g} and e_g levels, and the fact that the t_{2g} lobes are not directed towards the nearest neighbour ions, help to accept the idea that it is only the "surplus" e_g electrons that can hop. Since the three parallel t_{2g} electrons form a kind of a ridig ion core, we might as well model them by an $S = 3/2$ core spin which is coupled ferromagnetically to an e_g electron which happens to visit the ion. We assume that t is the smallest energy parameter: $J_{\rm H} \gg t$. Then the low-energy subspace is spanned by configurations in which the ionic states satisfy Hund's rule[32] everywhere. If we further simplify the problem by neglecting the twofold degeneracy of the e_g orbital, then the inclusion of the e_g–e_g and e_g–t_{2g} Coulomb interactions is unnecessary because $J_{\rm H}/t \gg 1$ is sufficient to guarantee that charge fluctuations are kept to the unavoidable minimum[33], i.e., if n is lying between the integers n_0 and $n_0 + 1$, then there are $(n - n_0)L$ ions with $n_0 + 1$ electrons, and $(1 + n_0 - n)L$ ions with n_0 electrons ($n_0 = 3$ for doped $LaMnO_3$). The motion of the e_g electrons on the background of the t_{2g}^3 cores is described by the double exchange Hamiltonian

$$\mathcal{H}_{\rm dex} = -t \sum_{\mathbf{i},\mathbf{j}} \sum_{\sigma} (c_{\mathbf{i}\sigma}^\dagger c_{\mathbf{j}\sigma} + c_{\mathbf{j}\sigma}^\dagger c_{\mathbf{i}\sigma}) - J_{\rm H} \sum_{\mathbf{j}} \mathbf{s}_{\mathbf{j}} \cdot \mathbf{S}_{\mathbf{j}} \tag{8.38}$$

where $\mathbf{s}_{\mathbf{j}}$ is the e_g (henceforth conduction electron) spin at site $\mathbf{j}$

$$s_{\mathbf{j}}^z = \frac{1}{2}(c_{\mathbf{j}\uparrow}^\dagger c_{\mathbf{j}\uparrow} - c_{\mathbf{j}\downarrow}^\dagger c_{\mathbf{j}\downarrow}) \qquad s_{\mathbf{j}}^+ = c_{\mathbf{j}\uparrow}^\dagger c_{\mathbf{j}\downarrow} \qquad s_{\mathbf{j}}^- = c_{\mathbf{j}\downarrow}^\dagger c_{\mathbf{j}\uparrow}\,. \tag{8.39}$$

---

[32] In this section, always Hund's first rule is meant by Hund's rule.

[33] Having neglected the degeneracy of the e_g ("conduction electron") orbital, we would loose Hund's rule energy if we put two e_g electrons to the same site. Here $J_{\rm H} = \infty$ has the same effect as $U = \infty$: it keeps charge ("polarity") fluctuations to an unavoidable minimum. This "minimum polarity" picture is certainly not applicable to the transition metals but it should be quite acceptable for the "bad metals" obtained by doping certain transition metal compounds. – See p. 502 for the history of the minimum polarity concept.

In contrast to the starting (8.36), we now divide the electrons sharply into two classes: one kind is always localized, and can be treated as localized spins, and the other kind is itinerant, forming a conduction band.

Having neglected the degeneracy of the e_g orbitals, we actually renounced the possibility of recovering the ferromagnetic kinetic exchange at exact quarter filling (see Sec. 5.4). We did so in the hope that the double exchange processes give a stronger ($\propto t$) effective ferromagnetic coupling when the conduction electron concentration is $0 < n_{\rm c} < 1$. The simplified Hamilton operator (8.38) is often called the Kondo lattice model with ferromagnetic coupling[34].

It is simplest to treat the case when the electrons hop between classical ($S \to \infty$) spins because then spin-flip processes cease to operate. Let us furthermore assume that $J_{\rm H} \to \infty$ thus the electron spin has to be strictly aligned with the localized spin at every site. Let us assume that the spin S is pointing up at site **i**, while at site **j** it subtends an angle ϑ with the z-axis, lying in the x–z plane. If an electron jumps from **i** to **j**, it begins with pointing in the z-direction, and ends up with being rotated to a state with polar angles ϑ, and $\varphi = 0$

$$|\mathbf{j}, \vartheta, 0\rangle = \cos\frac{\vartheta}{2}|\uparrow\rangle_{\mathbf{j}} + \sin\frac{\vartheta}{2}|\downarrow\rangle_{\mathbf{j}}\,. \tag{8.40}$$

The amplitude of an $\mathbf{i} \to \mathbf{j}$ hopping event is (omitting the usual negative sign)

$$\tilde{t}_{\mathbf{j}\,\mathbf{i}} = t\langle \mathbf{j}, \vartheta, 0| \sum_{\sigma} c^{\dagger}_{\mathbf{j},\sigma} c_{\mathbf{i},\sigma} |\mathbf{i}, 0, 0\rangle = t\cos(\vartheta/2)\,. \tag{8.41}$$

The effective hopping amplitude is reduced if the core spins are not perfectly aligned. This result is sufficient for the discussion of coplanar structures of large spins.

Generally, the orientation of the localized spin at **j** is specified by the polar angles $\vartheta_{\mathbf{j}}$ and $\varphi_{\mathbf{j}}$, and the conduction electron state pointing in the same direction is

$$|\mathbf{j}, \vartheta_{\mathbf{j}}, \varphi_{\mathbf{j}}\rangle = \cos\frac{\vartheta_{\mathbf{j}}}{2}|\uparrow\rangle_{\mathbf{j}} + \sin\frac{\vartheta_{\mathbf{j}}}{2} e^{i\varphi_{\mathbf{j}}}|\uparrow\rangle_{\mathbf{j}}\,. \tag{8.42}$$

---

[34] The Kondo lattice with antiferromagnetic coupling will be introduced in Ch. 11.

Proceeding as in (8.41), we get the effective hopping amplitude [227, 299]

$$\begin{aligned}\tilde{t}_{\mathbf{j}\,\mathbf{i}} &= t\,\langle \mathbf{j}, \vartheta_{\mathbf{j}}, \varphi_{\mathbf{j}}| \sum_{\sigma} c^{\dagger}_{\mathbf{j},\sigma} c_{\mathbf{i},\sigma} |\mathbf{i}, \vartheta_{\mathbf{i}}, \varphi_{\mathbf{i}}\rangle \\ &= t\left\{\cos\frac{\vartheta_{\mathbf{j}}}{2}\cos\frac{\vartheta_{\mathbf{i}}}{2} + \sin\frac{\vartheta_{\mathbf{j}}}{2}\sin\frac{\vartheta_{\mathbf{i}}}{2}e^{i(\varphi_{\mathbf{i}}-\varphi_{\mathbf{j}})}\right\}\,. \end{aligned} \tag{8.43}$$

The maximum strength of $\tilde{t}$ is reached if the localized spins are parallel. Intraatomic exchange, and the induced double exchange, provide a rather robust (of $O(t)$) intersite ferromagnetic coupling from low to high dimensions [226].

Remember that the above conclusion was reached with the simplifying assumptions that the localized spins are classical, and Hund's $J_{\mathrm{H}} \to \infty$. We should enquire what happens if we relax either of these conditions.

If finite spins S are sitting at neighbouring sites, and a single electron hops between them, the total spin of the system can be $S_{\mathrm{tot}} = 2S + 1/2, 2S - 1/2, \ldots, 1/2$. We convince ourselves by considering simple cases in Problem 8.3 that the effective hopping amplitude is given by [26, 227, 299]

$$\tilde{t}(S_{\mathrm{tot}}) = t\frac{S_{\mathrm{tot}} + 1/2}{2S+1}(-1)^{2S - S_{\mathrm{tot}} + 1/2}\,. \tag{8.44}$$

It is interesting that $\tilde{t}$ changes sign if S_{tot} is changed by one step. This may have drastic consequences for an electron moving around a loop with various values of S_{tot} on the connecting edges [299]. $\tilde{t}$ does not depend on S^z_{tot} which reflects the fact that the total spin is conserved. It follows that the levels are $(2S_{\mathrm{tot}} + 1)$-fold degenerate.

It is apparent that the limit $S \to \infty$ is not "smooth" because $\tilde{t}$ has ever more sign reversals within the interval $-t < \tilde{t} \le t$. However, we can convince ourselves that (8.41) holds for the absolute value. Let us introduce the total spin of the site $\vec{S}_{\mathbf{j}}$, whereby it is understood that $\vec{S}_{\mathbf{j}}$ includes the spin of the conduction electron if it happens to sit at site $\mathbf{j}$, and reduces to the localized core spin $\mathbf{S}_{\mathbf{j}}$, if the electron is not there. Understanding that at one of the sites $|\vec{S}| = S + 1/2$, while at the other site $|\vec{S}| = S$, it is straightforward to derive that

$2\vec{S}_{\rm i}\cdot\vec{S}_{\rm j} = (\vec{S}_{\rm i} + \vec{S}_{\rm j})^2 - \vec{S}_{\rm i}^2 - \vec{S}_{\rm j}^2 = S_{\rm tot}(S_{\rm tot} + 1) - (2S^2 + 2S + 3/4)$. It follows that[35]

$$|\tilde{t}| = t\sqrt{\frac{2\vec{S}_{\rm i}\cdot\vec{S}_{\rm j}}{(2S+1)^2} + \frac{S+1}{2S+1}}\,. \tag{8.45}$$

In the limit $S \gg 1$, we may replace $\vec{S}_{\rm i}\cdot\vec{S}_{\rm j} \approx \mathbf{S}_{\rm i}\cdot\mathbf{S}_{\rm j} = S^2\cos\vartheta$, and recover (8.41).

Since $J_{\rm H} \to \infty$ projects onto the subspace of local high-spin states, it is an attractive possibility to express the hopping Hamiltonian in terms of the operators $\vec{S}_{\rm j}$ [299]. It follows from the even (odd) character of the eigenstates that whatever the on-site spin states are, they are interchanged by the hopping of the electron, thus the effective hopping term for the pair of sites **i** and **j** can be written as

$$\mathcal{H}_{\text{eff. hop.}} = -t\,\mathcal{P}_{\rm i,j}\, f(\vec{S}_{\rm i}\cdot\vec{S}_{\rm j}) \tag{8.46}$$

where $\mathcal{P}_{\rm i,j}$ permutes the states of the two sites, and f is an S-dependent polynomial of its argument. It is shown in the solution to Problem 8.3 (see p. 493) that the spectrum of the $S = 1/2$ case is reproduced by

$$\mathcal{H}^{S=1/2}_{\text{eff. hop.}} = -t\,\mathcal{P}_{\rm i,j}\left(\vec{S}_{\rm i}\cdot\vec{S}_{\rm j} + \frac{1}{2}\right)\,. \tag{8.47}$$

Indeed, this form of the double exchange term was suggested for describing the hole-doped Haldane system $Y_{2-x}Ca_xBaNiO_5$, in which Ni^{2+} ($S = 1$) and low-spin[36] Ni^{3+} ($S = 1/2$) sites coexist [77].

Discussing the above formula permits an insight into the role of the sign factor figuring in (8.44) [299]. Had we omitted the extra sign, we would have got $f' = (5 + 2\vec{S}_{\rm i}\cdot\vec{S}_{\rm j})/6$ instead of $f = (1 + 2\vec{S}_{\rm i}\cdot\vec{S}_{\rm j})/2$. This would have implied, for example's sake, the existence of the process (the

[35]The expression in [299] is recovered by replacing our S by $S-1/2$. It corresponds to the idea that the mobile entity is a hole, thus at one of the sites an electron is missing from the core.

[36]Actually, Y_2BaNiO_5 is a charge transfer insulator so holes go to O atoms, and one should envisage a chain of $S = 1$ Ni^{++} ions bridged by oxygens, some of which carry a $S = 1/2$ spin [322]). However, it should be possible to ascribe the 1/2 spins to Ni sites by constructing low-spin O–Ni–O complexes which are analogous to the Zhang–Rice singlets in CuO_2 planes (see p. 249).

notations are introduced in the solution to Problem 8.3, see p. 493)

$$\mathcal{P}f'|\Downarrow,\Uparrow\uparrow\rangle \rightsquigarrow \mathcal{P}\,4|\Downarrow,\Uparrow\uparrow\rangle \rightsquigarrow 4|\Uparrow\uparrow,\Downarrow\rangle. \tag{8.48}$$

This would be, however, unphysical because hopping can give only

$$|\Downarrow,\Uparrow\uparrow\rangle \rightsquigarrow |\Downarrow\uparrow,\Uparrow\rangle.$$

The z-component of a site spin can only change in steps of $1/2$ because the underlying process is the hopping of an electron. This is obeyed if we use the proper f because the amplitude of the process (8.48) becomes $1 + 2(\vec{S}_1)^z(\vec{S}_2)^z = 0$. Keeping the sign factor does not merely replace one numerical value with another: it also serves to filter out unphysical processes.

We have to include a biquadratic term in order to describe the case of $S = 1$ core spins

$$\mathcal{H}^{S=1}_{\text{eff. hop.}} = -\frac{t}{6}\,\mathcal{P}_{\mathrm{i,j}}\left[2(\vec{S}_{\mathrm{i}}\cdot\vec{S}_{\mathrm{j}})^2 + 3\vec{S}_{\mathrm{i}}\cdot\vec{S}_{\mathrm{j}} - 3\right]. \tag{8.49}$$

The effective Hamiltonian for $S = 3/2$ (which is relevant for doped $LaMnO_3$) is still more complicated, and includes a bicubic term [299, 337]. It might appear as if we have got an electronic mechanism for generating spin models with large biquadratic, and higher-order, couplings. We have to remember, however, that the status of the models treated here is very different from that of the Heisenberg models we met earlier. The couplings (8.47) and (8.49) act only between sites with unequal spins (different number of electrons), and describe the real motion of charge in a conductor[37]. In contrast, the terms of the bilinear–biquadratic model (5.142) (see p. 261) arise from virtual hopping processes in a Mott insulator.

If we want to relate (8.47) to a Hamiltonian we met earlier, we have to recall the $J = 0$ case of the t–J model, the pure t-model

$$\mathcal{H}_t = -t\sum_{\langle \mathrm{i,j}\rangle}\sum_{\sigma}[(1-\hat{n}_{\mathrm{i}-\sigma})c^\dagger_{\mathrm{i}\sigma}c_{\mathrm{j}\sigma}(1-\hat{n}_{\mathrm{j}-\sigma}) + \text{H.c.}]. \tag{8.50}$$

[37] It is a conductor in the sense of being doped away from the Mott insulator state. It is not clear, however, whether it is a "good metal" with sharp quasiparticle states.

Both (8.47) and (8.50) describe the hopping of a carrier between sites under the condition that the initial and final states are projected onto the ground state set of an infinite on-site interaction. However, in the t-model a spinless hole hops between non-degenerate orbitals which makes the Hamiltonian deceptively simple-looking (we learned in our discussion of ferromagnetism at $U \to \infty$ that the corresponding eigenvalue problem is not simple at all). In contrast, (8.47) and (8.49) describe the motion of a spinfull object between sites with two or three orbitals.

Does the double exchange model order, and in which sense? That the answer is anything but simple, can be guessed from experience with the comparatively simple t-model. Exact diagonalization results for small systems tend to be bewildering: whether the ground state has high, low, or intermediate spin, depends on the parity of the number of sites, on boundary conditions, etc. However, it can be argued that these embarassements are of a similiar nature as what we met in studying finite Nagaoka magnets (see p. 433): the appearance of a node of the wave function (a "soliton") in the middle of the system may mask its essentially ferromagnetic nature. There is accumulating evidence that $\mathcal{H}_{\text{dex}}$ has an extended and robust fully polarized phase in $D = 1$, 2, 3, and ∞ dimensions, for $S = 1/2$, $3/2$ and ∞ core spins, in a wide range of the conduction electron concentration n_c, and above some n_c-dependent threshold value of J_H/t [337, 78]. It appears that in most cases, the ground state is fully polarized.

Problem 8.3. Verify (8.44) for $S = 1/2$ and $S = 1$ by diagonalizing the double exchange model (8.38) on two sites. Express the effective Hamiltonian in terms of $\vec{S}_1$ and $\vec{S}_2$, where $\vec{S}_1$ ($\vec{S}_2$) is the total spin of site 1 (2).

8.5.2 Two-Band Model

The double exchange model (8.38), and particularly its $J_H \to \infty$ limit, is an extremely asymmetric two-band model: the mobile electrons are confined to a non-degenerate uncorrelated band, while electrons in the other band are completely Mott-localized. In the language of the model introduced in Sec. 5.4, we have taken $U_b/t_b \to \infty$, and $t_a \ll J_H$, so that the Hund coupling does the job of U_a as well, and U_{ab} plays no role. Clearly we are speaking about two very different kinds of electrons. Such a situation is often encountered in rare earth systems (Ch. 11), but it takes some exaggeration to take it literally for two kinds (e_g and t_{2g}) of $3d$ electrons. At any rate, the degeneracy of the e_g shell is not yet taken into account, though we have already convinced ourselves that in the right limit, it can give rise to spin ferromagnetism combined with orbital ordering.

The logical continuation of our previous lengthy study of the single-orbital Hubbard model is to take a twofold degenerate Hubbard model with $U_a = U_b$, and study what is brought by band degeneracy, and what is the additional effect of switching on the intraatomic exchange $J_{\rm H}$. The minimal interaction of two electrons in different orbitals at the same site is $U_{ab} - J_{\rm H}$, so now we cannot allow $J_{\rm H}$ to become larger than U_{ab}, otherwise an unphysical on-site attraction would result. Though we cannot yet speak about a thorough knowledge of the entire parameter space, representative cases have been considered [161], allowing to draw interesting conclusions about the conditions for ferromagnetism.

A twofold degenerate band with the symmetrical density of states $\rho(\epsilon) = \sqrt{4-\epsilon^2}$ was chosen, to accentuate that, in contrast to the single-orbital case, now we do not need a fine-tuned, strongly asymmetrical form of the DOS. Because of electron–hole symmetry, it is sufficient to map out the region up to half-filling ($n = 2$). At $n = 2$, the system is an antiferromagnetic Mott insulator. At exact quarter-filling, we know that the strong-coupling ($U_a > U_{ab} > J_{\rm H}$, $U_{ab} \gg t$) ground state is a ferromagnetic Mott insulator. The considerations of the previous subsection are not directly applicable but they allow to guess that there should be an itinerant mechanism for ferromagnetism operating at $1 < n < 2$. Assuming $D \to \infty$, dynamical mean field theory allows a nominally exact treatment. However, technical difficulties of the Quantum Monte Carlo method necessitated that the Hund coupling be replaced by its z-z term [161]. It remains to be seen whether the inclusion of the intrasite terms $S_a^+ S_b^- + S_a^- S_b^+$ may alter the outcome in any essential way.

A single Hubbard band with the same DOS is certainly not ferromagnetic up to $U = 30$, and possibly not until a much higher U is reached. Actually, the twofold degenerate band with the moderate interactions $U_a = U_b = 9$, $U_{ab} = 5$ (but $J_{\rm H} = 0$) is not FM, either. However, it must be near to being so, since the relatively small $J_{\rm H} = 1.6$ is sufficient to drive the system to ferromagnetism. Fig. 8.10 (left) shows the J_H-dependence of the Curie temperature. It is clear that J_H is the driving force of ferromagnetism (the plot was done for $n = 1.25$ which, as it turns out immediately, is the most favourable case for FM).

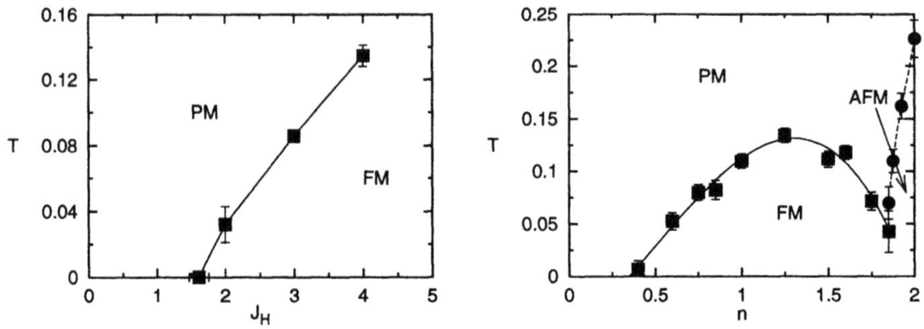

Figure 8.10: Ferromagnetism in the two-band Hubbard model at moderately strong interactions (after [161]). *Left*: Curie temperature vs J_H for $n = 1.25$, $U_a = 9$, and $U_{ab} = 5$. *Right*: magnetic phase diagram in the n–T plane for $U_a = 9$, $U_{ab} = 5$, and $J_H = 4$. Ferromagnetic ordering is driven by the Hund coupling, and extends over a wide range of band filling.

Fig. 8.10 (right) gives the phase diagram[38] in the $n - T$ plane, for the previous U's and $J_H = 4$. The ferromagnetic phase extends over the wide range $0.4 < n < 1.7$; the upper end belongs to a sudden transition to AFM near half-filling. The highest Curie temperature T_C is found somewhat above quarter-filling (at $n \approx 1.25$). In this example, J_H is probably a bit too high to be realistic, but the extension of the FM phase along the n-axis should be similar also for lower J_H's, only T_C would be scaled down. Remaining with the present values, the highest Curie temperature is about 0.1, i.e., for a bandwidth of 4eV it would be 1000K. Let it be remarked that in this model, too, the Hartree–Fock approximation overestimates T_C by more than a factor of 10 (we would get $T_C \approx 2$ with the parameters of the right panel of Fig. 8.10 [161]).

The reader may feel apprehensive because we have not made contact with our previous results: nothing particular seems to happen at quarter filling, and we have not mentioned orbital ordering (OO). Indeed, with

[38] In spite of the apparent similarity of this phase diagram to that seen in Fig. 8.8 (left), the underlying physics is completely different.

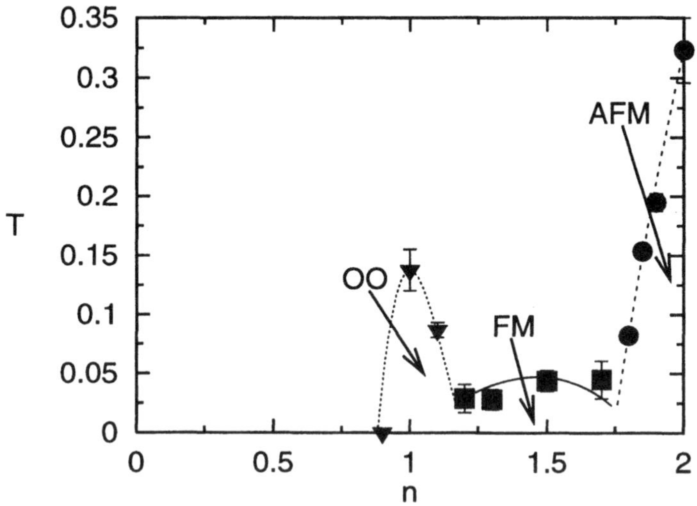

Figure 8.11: The phase diagram in the n–T plane for $U_a = 8$, $U_{ab} = 6$, and $J_{\rm H} = 2$ shows three ordered phases: AFM at and below half-filling, orbital ordering (OO) around quarter-filling, and FM in between. The system is a Mott insulator at $n = 1$ and $n = 2$, and a metal at other band fillings. (By courtesy of Karsten Held).

the large $J_{\rm H}/U_{ab} = 4/5$ adopted before, the high-temperature transition is from the paramagnet to a spatially uniform ferromagnet, and an OO region may be buried deep underneath at $n \approx 1$. In contrast, with the parameters $U_a = 8$, $U_{ab} = 6$, and $J_{\rm H} = 2$, the first transition of a nearly quarter-filled system is to a paramagnet with staggered orbital order (Fig. 8.11). With this smaller (and more realistic) value for $J_{\rm H}/U_{ab}$, the Curie temperature is modest in comparison with the Néel and OO transition temperatures. There is no doubt that the line of the FM phase boundary should cross into the OO region, perhaps with a change of slope, or a jump; these details are yet to be clarified. It seems to be a genuine feature that orbital order can preceed, and prepare the ground for, magnetic order which sets in at a lower temperature.

8.5.3 The $LaMnO_3$ System

The major theme of these notes is the study of strongly correlated metallic phases which are obtained by doping Mott insulators. Making the antiferromagetic Mott insulator V_2O_3 off-stoichiometric, we get a strange SDW phase (Sec. 5.5.1). Doping $LaTiO_3$ with Sr gives a heavy Fermi liquid (Sec. 10.5). Doping La_2CuO_4 with Sr gives a high-T_c superconductor; we cautiously avoid that one. Substituting 30% of La with Sr in $LaMnO_3$ gives a remarkable itinerant ferromagnet; we try our luck with that. Some of the groundwork was laid in Secs. 3.6 and 5.4.

The rich phase diagram of the system $La^{3+}_{1-x}A^{2+}_{x}Mn^{3+}_{1-x}Mn^{4+}_{x}O^{2-}_{3}$ (with A standing for Ca, Sr, Ba, or Pb) is the outcome of the interplay of a variety of effects: Jahn–Teller distortion, double exchange, orbital ordering, and different forms of kinetic (super)exchange. The experimentally determined phase diagram of $La_{1-x}Sr_xMnO_3$ [30] shown in Fig. 8.12 reveals more features than what we can endevour to interpret. Our main interest will be in low-temperature magnetic order, i.e., in the doping-induced sequence of transformations which lead from the antiferromagnetic insulator $LaMnO_3$ to the ferromagnetic metal $La_{0.7}Sr_{0.3}MnO_3$. To some degree, we are prepared to deal with this, thanks to our previous discussion of double exchange and two-band ferromagnetism. However, we observe that in the real world, the magnetic phase change involves the crossing of a line of structural phase transitions, and the sensitive interdependence of these two phenomena will become clear from the existence of the magnetostructural transitions we mention a bit later. A further great complication is the enigmatic nature of the high-temperature paramagnetic (PM) phase: it is much less conducting than it ought to be. We are accustomed to situations where the disappearance of magnetic order makes the system more metallic. Here, the resistivity increases by at least an order of magnitude when T is raised from $\sim$20K below the Curie temperature T_C of the ferromagnetic metal (FM) to slightly above T_C. Though the PM is not a clear-cut insulator with a well-developed gap, it is certainly not a good metal, either. It looks more like a paramagnetic lattice gas of carriers which move by thermally activated hops[39]. The ferromagnetic insulator (FI) phase is qualitatively similar. The carrier localization may be polaronic in nature (local Jahn–Teller effect), or a corollary of the strong correlation (double exchange) scenario, or a mixture of both. This may

[39]Photoemission experiments ascribe an activation energy $\sim$ 1.5eV to the hopping process Mn^{3+}+Mn^{4+} $\rightsquigarrow$ Mn^{4+}+Mn^{3+} in the paramagnetic phase [319].

well be the least-understood aspect of the phase diagram, and we do not consider it any further, except as a warning against hasty simplifications in dealing with other features.

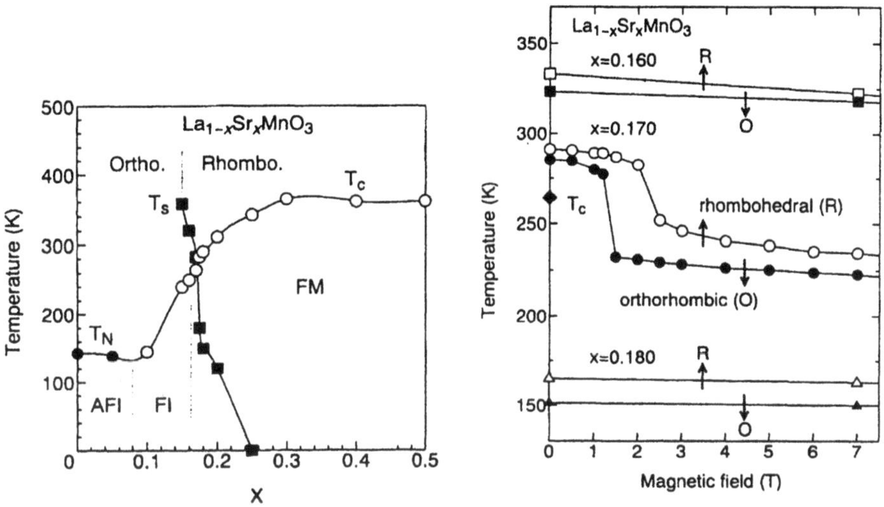

Figure 8.12: *Left*: The experimentally determined phase diagram of $La_{1-x}Sr_xMnO_3$ shows magnetic, structural, and metal–non-metal transitions. The phases are: antiferromagnetic insulator (AFI), ferromagnetic insulator (FI), ferromagnetic metal (FM), and a poorly characterized paramagnetic insulator above the magnetic ordering temperatures. The orthorombic–rhombohedral structural transition sets in at T_s. *Right*: the field dependence of T_s for $x = 0.16$, 0.17 and 0.18. We find a magnetostructural transition for $x = 0.17$: a field of several teslas switches the structure from orthorombic to rhombohedral if T is in the range 230–270 K. $\uparrow$ and $\downarrow$ indicate warming and cooling (after [30]).

In what follows, we try to discuss magnetic ordering phenomena with electronic models, and bring in structural aspects only when it is unavoidable. The prerequisite for describing the doped system should be a clear understanding of the pure compounds $LaMnO_3$ and $CaMnO_3$. Their magnetic structures are shown in Fig. 5.3. We feel safe about $CaMnO_3$: it is a Mott insulator consisting of $Mn^{4+} \rightsquigarrow t_{2g}^3$ ions whose $S = 3/2$ spins build a straightforward alternating antiferromagnet. As for $LaMnO_3$, its case is not straightforward at all, but its understanding from a variety of viewpoints is nearing a satisfactory stage thanks to an intense effort in the late 1990's [277, 368, 362, 113]. We cannot afford to go much beyond the considerations presented in Sec. 5.4.

For a start, we take the view that $LaMnO_3$ is a robust Mott insulator, i.e., its insulating nature is due to strong local correlations and it is not contingent upon such additional features as magnetic ordering or a Jahn–Teller distortion[40]. Furthermore, we assume spin-only moments, and neglect spin–orbit effects.

We learned in Sec. 5.4 that exploiting the two-fold degeneracy of the e_g orbital, we can derive a ferromagnetic kinetic exchange interaction which at the same time gives rise to staggered orbital order. Considering only this, $LaMnO_3$ might have assumed the orbital order $d_{x^2-y^2}/d_{3z^2-r^2}$ shown in Fig. 5.2 (left), and become a ferromagnet. This does not happen, but it almost does. $LaMnO_3$ is a foiled ferromagnet, consisting of ferromagnetic planes which are coupled antiferromagnetically (Fig. 5.3, right). The in-plane ferromagnetic coupling $J \approx -10\text{K}$ is somewhat stronger than the inter-plane antiferromagnetic $J' \approx 7\text{K}$ [277].

There are a number of factors to consider. First recall that $Mn^{3+} \rightsquigarrow t_{2g}^3 e_g^1$ is a Jahn–Teller ion in its $S = 2$ high-spin state (see Fig. 3.5, right). The above-mentioned orbital order $d_{x^2-y^2}/d_{3z^2-r^2}$ would be accompanied by the corresponding antiferrodistortive pattern of Jahn–Teller distortions. However, a cooperative Jahn–Teller effect has an energetics of its own; intercell elastic forces may prefer a different pattern of distortions, and thus enforce a different kind of orbital order $\rightsquigarrow$ a different magnetic order. It is also well to remember that perovskites are noted for their tendency to undergo structural phase transitions even if there is no Jahn–Teller effect or magnetism to speak about ($BaTiO_3$ and $SrTiO_3$ are well-known examples). Thus it is best to allow the lattice to play with the distortion pattern $\rightsquigarrow$ orbital order, and check what follows for magnetism[41]. Finally, we should not completely forget the antiferro-

---

[40]LSDA band structure calculations do not confirm the strongly localized moment picture of $LaMnO_3$ and ascribe its insulating nature to the tetragonal Jahn–Teller distortion. It is a usual dichotomy in solid state physics that density functional theory does not describe Mott insulating behaviour in any direct sense, but it offers an attractive explanation of many details in recompense (see Sec. 7.7.2). In our case, this includes the small ($\sim 0.3\mu_B$) orbital component of the ordered moment, and the weak canting of sublattice magnetizations due to spin–orbit interaction which gives an overall $\sim 0.1\mu_B$ uniform moment [368]. Our discussion neglects these fine features.

[41]We note, however, the claim that the magnetic order of $LaMnO_3$ can be under-

magnetic superexchange couplings between the t_{2g}^3 cores which make $CaMnO_3$ the antiferromagnet it is. It is a difficult many-parameter problem and the outcome seems to be that $LaMnO_3$ misses becoming a ferromagnet, but it is not wide off the mark [362]. The orbital order is nearly of the $d_{3x^2-r^2}/d_{3y^2-r^2}$ kind (Fig. 5.2, right). One should add to it a small uniform $d_{3z^2-r^2}$ component because it is helpful for the interplane antiferromagnetic coupling. All in all, the actual magnetic structure of $LaMnO_3$ is understood.

Given that $LaMnO_3$ is an antiferromagnet (AFM), and (for example's sake) $La_{0.7}Sr_{0.3}MnO_3$ is definitely a ferromagnetic metal (FM), we should ask how does the system get from one state to the other. The simplest possibility is to use a pure spin model of $La_{1-x}Sr_xMnO_3$, which contains the AFM t_{2g}–t_{2g} superexchange, and the double exchange term with the amplitude (8.41). It is understood that the latter is really the spin-orientation-dependent hopping energy of the e_g electrons. We envisage that the sublattice magnetizations gradually tilt towards each other as x increases. In other words, the [100] Mn planes remain uniformly polarized, but the angle subtended by the total spins of nearest-neighbour planes is an x-dependent ϑ. $\vartheta = \pi$ at $x = 0$, and $\vartheta = 0$ at some critical $x_{\rm cr}$. The expression of the mean field energy density (valid for infinite core spin[42] and $J_{\rm H}/t \to \infty$) is [82, 279]

$$E(\vartheta) = J\cos\vartheta - x[4t + 2t\cos(\vartheta/2)]\,. \tag{8.51}$$

Here we assumed that x is sufficiently small so that the band bottom energy can be ascribed to all e_g electrons. Minimizing $E(\vartheta)$ with respect to ϑ, we get

$$\cos(\vartheta/2) = \frac{tx}{2J}. \tag{8.52}$$

The notable features of the mean field solution are: a) the *canting* of the sublattice magnetizations at arbitrarily small x (at $x \ll 1$, $\vartheta \approx \pi - (tx/J)$), and b) the existence of a critical $x_{\rm cr} = 2J/t$. We predict that the transition from AFM to FM is gradual, with a *canted phase* for $0 < x < x_{\rm cr}$, and full ferromagnetic alignment for $x > x_{\rm cr}$ [82].

---

stood purely from superexchange, and that coupling to the lattice influences only the numerics [113]. There is continuing dispute concerning such details.

[42]The finite-S mean field theory is developed in [227].

Naturally, the canted phase also has a total moment which is, however, less than the saturation value[43]. The observed $x_{cr} \approx 0.1 - 0.2$ can be fitted with reasonable parameters.

This is an attractive scenario and there may be systems to which it is straightforwardly applicable. However, at the time of writing it is not clear whether it is realized in $La_{1-x}(Ca,Sr)_xMnO_3$, though it is not excluded that the FI phase (say, for $0.1 < x < 0.17$) shows features of the canted state; at least, it has an unsaturated moment. Let us, in any case, point it out that the theoretical case for canting is less compelling than it would seem from the previous argument. First, it was shown that dropping some of the simplifying assumptions on which the above derivation was based, a direct FM–AFM transition is found in a wide range of parameters [279]. Perhaps more significantly, we may question whether the restriction to homogeneous states is justified. Various studies indicate the existence of a wide range of phase segregation[44] into hole-rich FM and insulating AFM phases [337, 29, 78]. If this indeed happens, we have to address the extremely complicated question of the scale of the segregated regions, which may range from almost macroscopic to nm size (a small droplet of the FM phase may be called a ferromagnetic polaron, but it is also possible that one of the ionic species segregates into a network of charged domain walls, etc.).

Magnetostructural Transition

The dominant component of the Jahn–Teller distortion is tetragonal: the Mn–O distance along the major axis of the elongated MnO_6 octahedron is $\sim 13\%$ longer than in other directions. There are additional smaller low-symmetry distortions, and the low-temperature structure of $La_{1-x}Sr_xMnO_3$ is either orthorombic (O), or rhombohedral (R) (see Fig. 8.12, left). We do not characterize these structures in detail but it is useful to know that in the distorted perovskite phases, the MnO_6

[43]This should be distinguished from the small canting in undoped $LaMnO_3$ which we mentioned in footnote 40. That is a relativistic effect, occuring in an insulator. Here we mean a much stronger non-relativistic effect which presupposes itinerancy.

[44]This is a very general phenomenon which is likely to be seen whenever we expect a doped system to traverse the path from a Mott insulator to a uniform itinerant phase. We study related questions in more detail in Ch. 10 (see p. 568).

octahedra are tilted and rotated with respect to their neighbours and as a result, the Mn–O–Mn bonds are bent [30]. Bond bending reduces the hopping amplitudes, severely in the O phase, and less so in the R phase. Thus the rhombohedral phase is more itinerant. Since the only truly metallic phase of the system is the FM, we may also say that ferromagnetism is at least one of the factors[45] driving the O–R structural transition.

The balance between various possibilities for the lattice/electronic/ magnetic structure becomes particularly delicate when the Curie T_C and the structural transition temperature T_s are roughly equal, which happens at about $x = 0.17$. For $T < (T_C, T_s)$, we can envisage two free energy minima for the system: one of them belonging to an orthorombic, moderately ferromagnetic non-metal, while the other to a rhombohedral, strongly ferromagnetic metal. Going from one of the two states to the other requires a substantial shift of the atomic positions, thus it must be a first order transition. Though the two minima are far away in phase space, they are approximately degenerate and their small splitting can be fine-tuned by an external magnetic field which couples more strongly to the rhombohedral phase [30]. The effect is clearly seen in the right panel of Fig. 8.12 (middle). In zero field, the system goes through the ferromagnetic–paramagnetic transition at $T_C = 264$K within the O phase. However, at slightly higher x, it would have the R structure, and a higher T_C. The $x = 0.17$ sample is driven towards this state by the magnetic field. Keeping, say, $T = 270$K the structure is switched from orthorombic to rhombohedral at $H \approx 1.5$T: it is a *magnetostructural transition.* The results displayed in Fig. 8.12 (right) are summed up in the statement that T_s of $La_{0.83}Sr_{0.17}MnO_3$ is decreased from 280K to 220K by applying a field of merely 7 Tesla! In order to properly appreciate the enormity of the effect, we have to recall that the thermal equivalent of the free electron Zeeman splitting is 0.7K/T, thus the simple expectation would have been a shift of T_s by 5K. The intimate relationship between lattice structure and magnetism is seldom demonstrated so convincingly as in this case. – The data for $x = 0.16$ and $x = 0.18$ also indicate that the giant magnetostructural

[45]Certainly not the only one, since the O–R phase boundary continues into the paramagnetic phase.

effect is confined to the composition range where $T_C \sim T_s$. If T_C and T_s are sufficiently different then $H = 7$T cannot shift T_s by more than a couple of degrees.

We mention in passing another field-induced effect: namely, that the electrical resistivity ρ can be strongly affected by an external magnetic field. Manganite physics abounds with remarkable examples of this: we choose the behaviour $Pr_{0.5}Sr_{0.5}MnO_3$ [402] as an illustration (Fig. 8.13) because we will return to this system in a different context shortly. For the FM phase, it is generally found that if T is near T_C, then applying a field can strongly suppress spin disorder scattering, and thereby ρ is decreased. This *negative magnetoresistance effect* (which can be measured by $(\rho(H) - \rho(0))/\rho(H))$ tends to be quite strong in the manganites, which is indeed one of the reasons for the great interest in this class of materials [192]. Still stronger effects are seen when the field induces a sharp metal–insulator transition; since ρ may change by several orders of magnitude, the absolute value of the ratio defined above can easily attain 100000%, justifying the widely used phrase "colossal magnetoresistance effect".

Charge Ordering

Substituting La with a divalent cation in $LaMnO_3$, the resulting system can be envisaged as a mixture of Mn^{3+} and Mn^{4+} ions. Up to now, we have only considered on-site interactions, and the induced intersite spin–spin interactions, but certainly there is also a Coulomb repulsion between the excess e_g electrons of the Mn^{3+} sites. This tries to keep the Mn^{3+} sites away from each other. The effect should be most noticeable at the 50–50 composition, because then the e_g–e_g repulsion is minimized by forming a two-sublattice structure[46]. Whether this actually takes place, depends on the effective e_g bandwidth; the kinetic energy prefers that the e_g electrons form a uniform Fermi sea. We are

[46]Easily done since in an ideal perovskite, the sublattice of the Mn ions is cubic. It may get distorted in various ways but its bipartite nature remains recognizable. If the charge densities on the two sublattices differ only mildly, we speak of a small-amplitude, commensurate density wave. However, if the charge modulation is strong, approximating the alternation of Mn^{3+} and Mn^{4+} sites, then the system may be called a Wigner crystal.

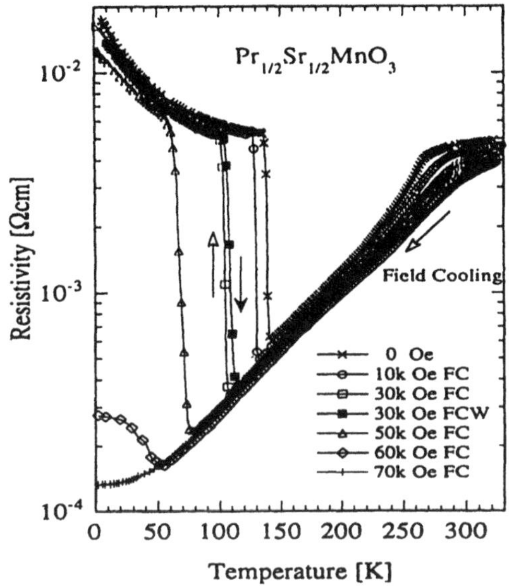

Figure 8.13: Negative magnetoresistivity in $Pr_{0.5}Sr_{0.5}MnO_3$ in fields up to 7T. ρ may get halved due to the suppression of spin disorder scattering for T near the Curie $T_C \approx 270K$. At 100–150K, the change in ρ is "colossal", corresponding to a field-induced insulator-to-metal transition (after [402]).

back to the problem of Wigner crystallization which we posed first in Ch. 1, Sec. 1.3.3. However, while the complicated structure of magnetite precludes any straightforward discussion of the charge ordering (CO), some doped manganites provide beautiful examples of Wigner electron crystals.

This may be the right occasion to clarify our standpoint on Wigner crystals, and to explain why neither of the sections bears the title "Wigner crystallization" (actually, the present one might have been named so). Originally, Wigner [440] predicted that an electron liquid moving on a uniform compensating charge background crystallizes if the electron density is sufficiently small, because the interelectronic repulsion eventually outweighs the kinetic energy. The critical value of the density was repeatedly revised, always to smaller values. We take the view that the Wigner crystal in this original sense is not realizable. True that related phenomena seem to be observed in doped semiconductors (p. 55), and in low-density Quantum Hall systems (p. 721), but these issues are hopelessly complicated because of disorder aspects. Wigner crystals formed on liquid He^3 and He^4 surfaces [364, 136] are very far from the original suggestion because their behaviour is presumably controlled by effective interactions mediated by the quantum fluid substrate. The case of electrons moving on a lattice is quite different. We think that the low-T phases of magnetite, and of certain manganites, are justly called Wigner crystals. Let us, however, note that in the latter cases, charge order tends to be accompanied with spin and/or orbital ordering, and thus the intersite interaction is not purely Coulombic but has also some superexchange components.

Our first example is not a simple perovskite, but a perovskite-type layer structure: $La_{1/2}Sr_{3/2}MnO_4 \rightsquigarrow La^{3+}_{1/2}Sr^{2+}_{3/2}Mn^{3+}_{1/2}Mn^{4+}_{1/2}O^{2-}_4$ [301]. The MnO_2 planes look the same as they would in the perovskite structure. The envisaged in-plane ordering pattern of the low-temperature phase is shown in Fig. 8.14. The system is antiferromagnetic Wigner crystal below the Néel temperature $T_N = 110K$, and it also has orbital ordering. These three orders are not independent but mutually contingent upon each other. The AF order can be envisaged as the spin-alternating sequence of ferromagnetic zig-zag chains. The zig-zag chains can be identified by following the lobes of the e_g orbitals of the Mn^{3+} sites to the nearest Mn^{4+} sites. It is along these paths that the ferromagnetic kinetic exchange can act. Perpendicularly to the lobes the t_{2g}–t_{2g} antiferromagnetic superexchange prevails. However, the relatively weak exchange interactions merely assist in stabilizing the charge order but do not constitute the primary driving force: at $T_N < T < T_{CO} = 217K$ the system has only charge and orbital order but no magnetic order. As far as it can be ascertained, charge and orbital order cannot be dissociated from each other which suggests that Wigner crystallization is driven by the sum of direct intersite Coulomb force, and a kinetic exchange piece which causes staggered orbital order.

It is interesting that this is probably the first instance when orbital order was seen directly by a sophisticated optical technique, and not merely inferred from its effect on spin order and lattice distortion. Let us note, however, that the resolution was not sufficient to differentiate between the $d_{3x^2-r^2}/d_{3y^2-r^2}$ pattern shown in Fig. 8.14, and the

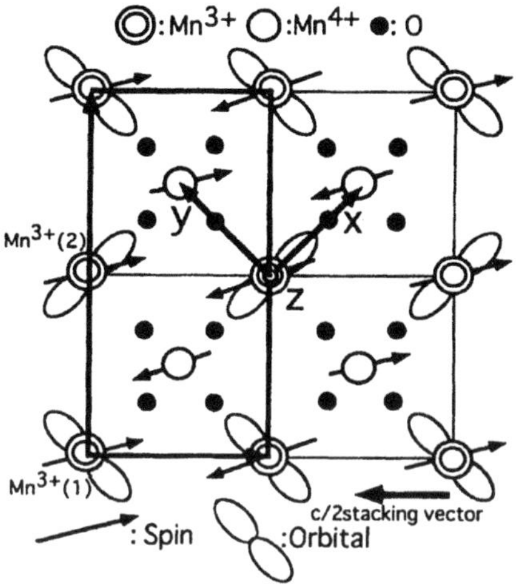

Figure 8.14: Schematic representation of simultaneous charge, spin and orbital ordering in the low-temperature phase of $La_{0.5}Sr_{1.5}MnO_4$. The two-dimensional unit cell is drawn in thick lines (after [301]).

$d_{x^2-z^2}/d_{z^2-y^2}$ pattern. For our purposes, it would not make a difference, if the latter were true.

The antiferromagnetic Wigner crystal optimizes the charge–charge repulsion energy, but looses much in kinetic energy, apart from perturbatively small pieces collected in the kinetic exchange term. On the other hand, away from a doping range centered on $x = 0.5$, the behaviour of doped manganites is governed by optimizing the kinetic energy which makes them metallic *and* ferromagnetic. Even at $x \approx 0.5$, the two states must be quite near.

The perovskite $Pr_{1/2}Sr_{1/2}MnO_3$ gives a nice example of the close competition between the metallic and insulating states [402]. The material is antiferromagnetic, charge-ordered (CO) insulator up to $T_V = 140$K, where it makes a first-order transition to a homogeneous ferromagnetic metal (FM). The latter is stable below the Curie temperature $T_C = 270$K. Both aspects of the transition are shown in Fig. 8.15.

The first-order transition at T_V is not merely the disappearance of one kind of order (CO-AFM), but also the simultaneous appearance of a completely different kind of order (FM), thus there must be two faraway, nearly degenerate minima in free energy, offering the possibility

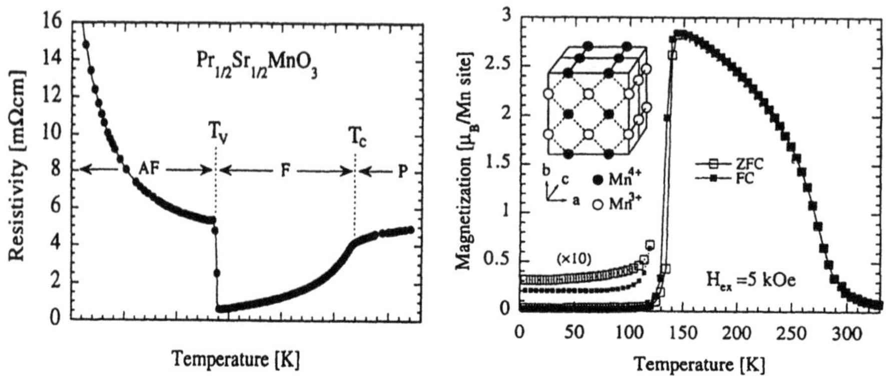

Figure 8.15: Close competition between metallic and insulating phases in $Pr_{1/2}Sr_{1/2}MnO_3$. The system is an AF insulator up to T_V where it makes a first-order transition to metallic ferromagnet. T_C is the Curie temperature. The inset shows the charge order at $T < T_V$ (after [402]).

of a field-induced transition. Choosing $T = 100$K (well below $T_v = 140$K), a field of $H \approx 2.7$T suffices to switch the system to a strongly polarized state with $\mu_{Mn} \approx 3.2\mu_B$ (Fig. 8.16, left). This is another striking example of metamagnetic behaviour, with a mechanism quite different from that of the other cases we have met.

Actually, a standard laboratory field of 7T suppresses the charge-ordered insulator even at $T = 0$. The T–H phase diagram (Fig. 8.16, right) shows that the Wigner crystal "melts" if we raise either the temperature or the external field. The $T = 0$ transition is certainly a "quantum melting" and one may ask why a magnetic field should seem to induce fluctuations when in other circumstances (say, in InSb, see p. 55) it drives towards localization. The reason is that in double-exchange systems the charge and spin degrees of freedom are coupled in a peculiar way, namely that polarizing the spins makes the charge flow. Seen in this light, even the $H = 0$ transition at T_V is not really thermal melting in the sense of merely disordering the Wigner crystal, but rather it is a transition between two quantum phases with different orders. The true disordering transition sets in at the much higher T_C.

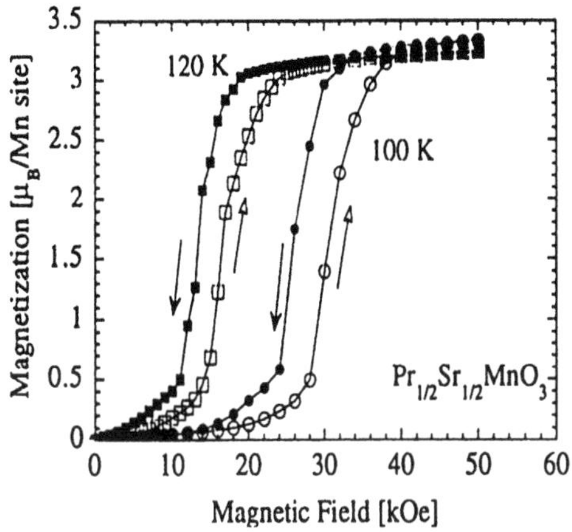

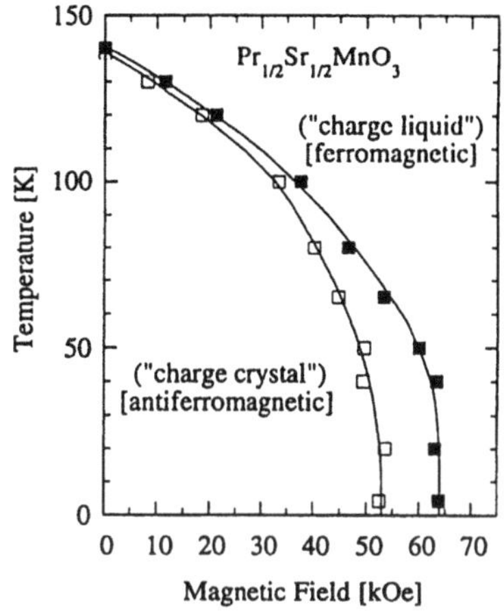

Figure 8.16: *Left*: Field-induced transition from an insulating antiferromagnet to a conducting ferromagnet in $Pr_{1/2}Sr_{1/2}MnO_3$. *Right*: Phase diagram for the Wigner crystal to charge liquid transition in the T–H plane (after [402]).

Solutions to the Problems

Problem 8.1. First, we solve the problem by exact diagonalization. The relationship to the Nagaoka theorem will be discussed afterwards.

The Hamiltonian is mirror symmetrical about the axis bisecting the edges 12 and 34, which allows to seek the eigenstates in the even and odd subspaces separately. The total spin is either $S = 3/2$ or $S = 1/2$. In the $S^z = 3/2$ subspace we choose the following basis states

$$\begin{aligned} \psi_1^{\pm} &= (1/\sqrt{2})(c_{1\uparrow}^{\dagger}c_{3\uparrow}^{\dagger}c_{4\uparrow}^{\dagger} \pm c_{2\uparrow}^{\dagger}c_{3\uparrow}^{\dagger}c_{4\uparrow}^{\dagger})|0\rangle \\ \psi_2^{\pm} &= (1/\sqrt{2})(c_{1\uparrow}^{\dagger}c_{2\uparrow}^{\dagger}c_{3\uparrow}^{\dagger} \pm c_{1\uparrow}^{\dagger}c_{2\uparrow}^{\dagger}c_{4\uparrow}^{\dagger})|0\rangle\ . \end{aligned} \tag{8.53}$$

The eigenvalues are (with the degeneracy given in parentheses)

$$\begin{aligned} &\text{for } t' = t \qquad -2t,\ \ 0\ (2),\ \ 2t \\ &\text{for } t' = -t \qquad -\sqrt{2}t\ (2),\ \ \sqrt{2}t\ (2)\ . \end{aligned} \tag{8.54}$$

Constructing an orthogonal basis of $S = 1/2$ states would be a cumbersome process. We rather prefer to work in the somewhat larger $S^z = 1/2$ subspace, and identify the $S = 1/2$ energy levels by omitting the states which are degenerate with the previously found $S = 3/2$ levels. The six even (odd) basis states can be chosen as

$$\begin{aligned}
\phi_1^\pm &= (1/\sqrt{2})(c_{1\downarrow}^\dagger c_{3\uparrow}^\dagger c_{4\uparrow}^\dagger \pm c_{2\downarrow}^\dagger c_{3\uparrow}^\dagger c_{4\uparrow}^\dagger)|0\rangle \\
\phi_2^\pm &= (1/\sqrt{2})(c_{1\downarrow}^\dagger c_{2\uparrow}^\dagger c_{4\uparrow}^\dagger \pm c_{1\uparrow}^\dagger c_{2\downarrow}^\dagger c_{3\uparrow}^\dagger)|0\rangle \\
\phi_3^\pm &= (1/\sqrt{2})(c_{1\downarrow}^\dagger c_{2\uparrow}^\dagger c_{3\uparrow}^\dagger \pm c_{1\uparrow}^\dagger c_{2\downarrow}^\dagger c_{4\uparrow}^\dagger)|0\rangle \\
\phi_4^\pm &= (1/\sqrt{2})(c_{2\uparrow}^\dagger c_{3\uparrow}^\dagger c_{4\downarrow}^\dagger \pm c_{1\uparrow}^\dagger c_{3\downarrow}^\dagger c_{4\uparrow}^\dagger)|0\rangle \\
\phi_5^\pm &= (1/\sqrt{2})(c_{1\uparrow}^\dagger c_{3\uparrow}^\dagger c_{4\downarrow}^\dagger \pm c_{2\uparrow}^\dagger c_{3\downarrow}^\dagger c_{4\uparrow}^\dagger)|0\rangle \\
\phi_6^\pm &= (1/\sqrt{2})(c_{1\uparrow}^\dagger c_{2\uparrow}^\dagger c_{4\downarrow}^\dagger \pm c_{1\uparrow}^\dagger c_{2\uparrow}^\dagger c_{3\downarrow}^\dagger)|0\rangle \,. \qquad (8.55)
\end{aligned}$$

The Hamiltonian matrix in the even (odd) subspace is found to be

$$\text{even:} \begin{pmatrix} t' & t & 0 & 0 & 0 & 0 \\ t & 0 & t & 0 & 0 & 0 \\ 0 & t & 0 & t & 0 & 0 \\ 0 & 0 & t & 0 & t' & 0 \\ 0 & 0 & 0 & t' & 0 & t \\ 0 & 0 & 0 & 0 & t & t \end{pmatrix} \qquad \text{odd:} \begin{pmatrix} -t' & t & 0 & 0 & 0 & 0 \\ t & 0 & t & 0 & 0 & 0 \\ 0 & t & 0 & t & 0 & 0 \\ 0 & 0 & t & 0 & t' & 0 \\ 0 & 0 & 0 & t' & 0 & t \\ 0 & 0 & 0 & 0 & t & -t \end{pmatrix} \qquad (8.56)$$

After eliminating the eigenvalues belonging to $S = 3/2$, we are left with the $S = 1/2$, $S^z = 1/2$ levels (degeneracies):

$$\begin{aligned}
&\text{for } t' = t \qquad \pm t\ (2),\ \pm\sqrt{3}t\ (2) \\
&\text{for } t' = -t \qquad \pm\sqrt{2-\sqrt{3}}t\ (2),\ \pm\sqrt{2+\sqrt{3}}t\ (2)\,. \qquad (8.57)
\end{aligned}$$

We may fix $t > 0$. We find that for $t' = t$ the ground state is a high-spin state ($S = 3/2$), with the energy $-2t$. For $t' = -t$ the ground state is a low-spin state ($S = 1/2$), and the ground state energy is $-\sqrt{2+\sqrt{3}}t \approx -1.93t$.

The problem is fully solved by finding the previous results, but we did not get an insight into why the sign of t'/t matters so much. The choice $t' = t$ gives an example of the Nagaoka theorem: all hopping amplitudes are positive, and indeed we found a fully polarized ground state. The significance of $\text{sg}(t') = \text{sg}(t)$ is that it allows an optimal choice of the signs in the ground state wave functions, so that the maximum energy gain results from each hopping process. Let us illustrate this by considering the action of $\mathcal{H}$ in the $S^z = 1/2$ subspace, starting with the configuration $|A\rangle = c_{2\uparrow}^\dagger c_{3\uparrow}^\dagger c_{4\downarrow}^\dagger|0\rangle$. The configurations reached by subsequent hopping events form the "superlattice" of three-electron configurations[47] shown in Fig. 8.17. Let us notice that the superlattice contains all the elements of the $N_\uparrow = 2$, $N_\downarrow = 1$ subspace. This is meant by "connectivity" when stating the conditions for the Nagaoka theorem: all configurations of a given $(N_\uparrow, N_\downarrow)$ subspace must be reachable from any initial configuration by a series of hoppings.

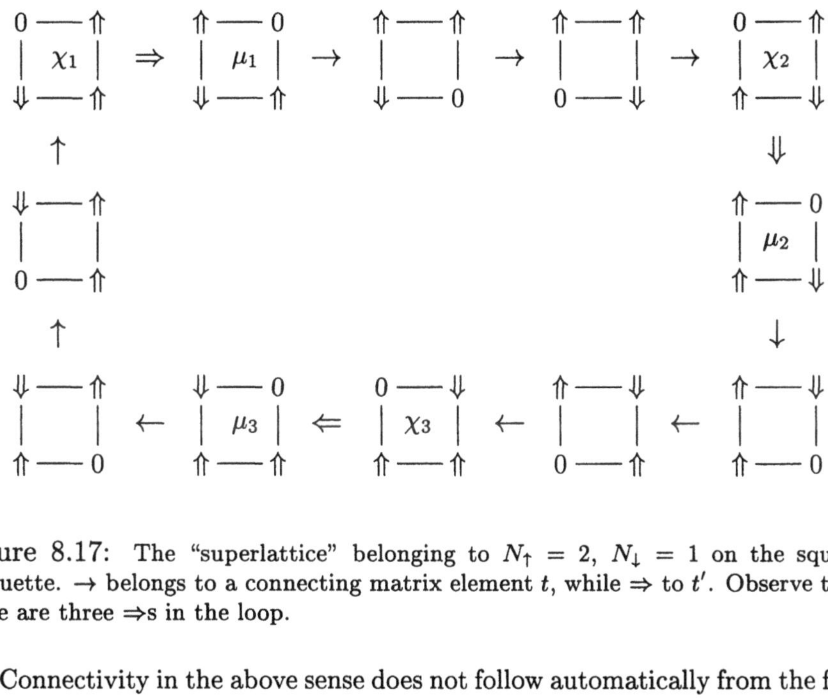

Figure 8.17: The "superlattice" belonging to $N_\uparrow = 2$, $N_\downarrow = 1$ on the square plaquette. $\rightarrow$ belongs to a connecting matrix element t, while $\Rightarrow$ to t'. Observe that there are three $\Rightarrow$s in the loop.

Connectivity in the above sense does not follow automatically from the fact that all sites are connected to at least one other site by hopping. For instance, for the $L = 5$ linear chain (with open boundary condition), the configuration $\uparrow -0- \uparrow - \downarrow - \downarrow$ is not connected to $\uparrow - \downarrow -0- \uparrow - \downarrow$. This is the reason why the Nagaoka theorem is not applicable to the linear chain.

Let us return to Fig. 8.17, and try to figure out the relative signs of the coefficients $c(\chi_1), \ldots$ of the various configurations. With $t' = t$, the configurations will be superposed with alternating signs, and fixing $c(\chi_1) > 0$, we find that $c(\chi_2)$ and $c(\chi_3)$ are positive, while $c(\mu_1)$, $c(\mu_2)$, and $c(\mu_3)$ are negative. Making a complete tour around the superlattice, we reconfirm that $c(\chi_1)$

[47]This usage of the term "superlattice" is confined to the literature on the Nagaoka theorem. We have shown the $N_\uparrow = 2$, $N_\downarrow = 1$ superlattice. The $N_\uparrow = 3$, $N_\downarrow = 0$ superlattice would be isomorphic with the original lattice. Generalizing the concept for larger systems is straightforward.

should be positive. The ferromagnetic ground state of the $t' = t$ case results from a constructive interference.

For $t' = -t$, we have to pay attention to the $\Rightarrow$s which show connections via t'. Configurations connected by $\Rightarrow$ should be superposed with the same sign. Fixing $c(\chi_1) > 0$, let us try to find the sign of $c(\chi_2)$. Going clockwise from χ_1 to χ_2, we get $c(\chi_2) < 0$. However, if we start counter-clockwise, we get $c(\chi_2) > 0$. Following different paths from χ_1 to χ_2 leads to a destructive interference.

There is a very easy way to check that the unambiguous requirement $\mathrm{sgn}(c(\chi_1)) = \mathrm{sgn}(c(\chi_2)) = \mathrm{sgn}(c(\chi_3))$ must be associated with a high-spin ground state. Then the ground state in the $S^z = 1/2$ subspace must be obtained by "stepping down" the $S^z = 3/2$ ground state. χ_1, χ_2, and χ_3 arise from turning one spin down in the parent configuration $|0\uparrow\uparrow\uparrow\rangle$ and therefore must have the same coefficient.

Problem 8.2.

a) The only geometrical symmetry of $\mathcal{H}_3$ is the C_2 axis through the lattice point 3. The one-particle eigenstates are (orbitally) non-degenerate, and can be classified as either even or odd under the π-rotation.

The odd combination $(|1\rangle - |2\rangle)/\sqrt{2}$ does not mix with $|3\rangle$ and is thus an eigenstate, with the eigenvalue

$$\lambda_o = -t_2\,. \tag{8.58}$$

t_1-type hopping mixes the even combination $(|1\rangle + |2\rangle)/\sqrt{2}$ with $|3\rangle$. The 2×2 eigenvalue problem gives the roots

$$\lambda_e^{\pm} = \frac{t_2 + \epsilon_3 \pm \sqrt{(\epsilon_3 - t_2)^2 + 8(t_1)^2}}{2}\,. \tag{8.59}$$

The ground state energy is either λ_o, or λ_e^-, depending on the size of $|t_1|/t_2$.

We notice that t_1 appears only through $(t_1)^2$, i.e., the sign of t_1 does not influence the spectrum. The reason is that changing the sign of t_1 is an invariance property of the Hamiltonian: it can be effected by the canonical transformation $c_{3\sigma} \to -c_{3\sigma}$, $c_{3\sigma}^\dagger \to -c_{3\sigma}^\dagger$ (the other operators remain unchanged). Henceforth we fix $t_1 > 0$.

At $U = 0$, the two-electron ground state is obtained by filling the lowest level with two electrons: this gives a singlet ground state which is separated by a finite gap from the lowest-lying triplet. Thus there is no reason to expect ferromagnetism in the weak-coupling limit. The special case $t_1 = t_2$, $\epsilon_3 = 0$ is a notable exception. For $t_2 > 0$, the one-particle ground state is twofold degenerate[48], thus the two-electron ground state can be chosen either as a

[48]This has a symmetry reason: $t_1 = t_2$, $\epsilon_3 = 0$ restores the cyclic symmetry

singlet, or a triplet. Switching on U lifts the degeneracy, and favours the triplet.

b) *Quick argument.* Both the singlets and the triplets are represented in the $S^z = 0$ subspace. For $U = U_3 = \infty$, there are only 6 allowed states. These we denote as $|\uparrow\downarrow 0\rangle = c^\dagger_{1\uparrow} c^\dagger_{2\downarrow}|0\rangle$, etc. As shown in Fig. 8.18, the hopping-induced transitions can be arranged in a ring[49]. The ground state wave function can be chosen as real, and the relative signs of the configurations are obtained from the requirement that each hopping process should lead to energy lowering. Thus if the connecting matrix element is negative, then the configurations are superposed with the same sign, while if it is positive, then we choose opposite signs.

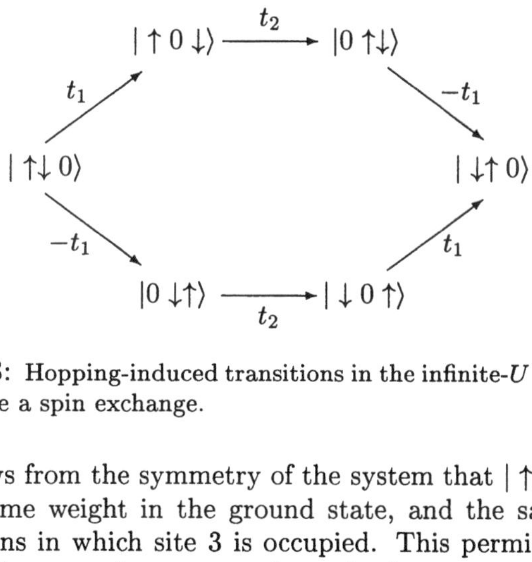

Figure 8.18: Hopping-induced transitions in the infinite-U limit. Three consecutive hoppings give a spin exchange.

It follows from the symmetry of the system that $|\uparrow\downarrow 0\rangle$ and $|\downarrow\uparrow 0\rangle$ should have the same weight in the ground state, and the same is true about the configurations in which site 3 is occupied. This permits us to conclude that for $t_2 < 0$, the ground state must have the form

$$|\Phi\rangle_{t_2<0} = \frac{1}{\sqrt{2}}(|\uparrow\downarrow 0\rangle - |\downarrow\uparrow 0\rangle) - \frac{\alpha}{2}(|0\uparrow\downarrow\rangle - |0\downarrow\uparrow\rangle + |\uparrow 0 \downarrow\rangle - |\downarrow 0 \uparrow\rangle). \quad (8.60)$$

The first term describes a singlet pair of spins connecting sites 1 and 2; similarly, the second term is composed of singlet pair states. *For $t_2 < 0$, the ground state is a singlet.*

---

of a equilateral triangle. The eigenstates can be chosen as Bloch states belonging to $k = 0, \pm 2\pi/3$. Time reversal invariance requires that the states belonging to $k = \pm 2\pi/3$ are degenerate. With the choice $t_2 > 0$, these are the low-lying states.

[49]Recalling Fig. 8.17 from the solution of Problem 8.1, we realize that this is the $N_\uparrow = 1$, $N_\downarrow = 1$ superlattice of the three-site plaquette.

If $t_2 > 0$, the ground state must look like

$$|\Phi\rangle_{t_2>0} = \frac{1}{\sqrt{2}}(|\uparrow\downarrow 0\rangle + |\downarrow\uparrow 0\rangle) + \frac{\beta}{2}(|0\uparrow\downarrow\rangle + |0\downarrow\uparrow\rangle - |\uparrow 0 \downarrow\rangle - |\downarrow 0 \uparrow\rangle). \quad (8.61)$$

Now the ↑-, and ↓-spins are arranged in symmetrical combinations. Thus we have found that *for $t_2 > 0$, the ground state is a triplet.* Note that this result is in agreement with the Nagaoka theorem.

It is remarkable that this argument (which we have taken over from [393]) did not require the solution of the eigenvalue problem: it used merely a "sign rule" for the ground state wave function[50]. What is more, the argument works for a general $\epsilon_3 \neq 0$, and it tells us that though the $U = 0$ case is non-magnetic, the $U = \infty$ model is ferromagnetic for $t_2 > 0$. There must be a phase boundary at some finite U. To learn its position, we have to solve the eigenvalue problem.

c) The paramagnetic–ferromagnetic phase boundary is determined by comparing the lowest-lying singlet, and triplet, eigenenergies.

Since in a triplet the two spins are parallel, the electrons are never at the same site. The triplet levels are always the same as at $U = 0$, thus the lowest-lying triplet level is

$$\mathcal{E}^{\text{tr.}}_{\text{min.}} = \lambda_o + \lambda_e^- = \frac{\epsilon_3 - t_2 - \sqrt{(\epsilon_3 - t_2)^2 + 8(t_1)^2}}{2}\,. \quad (8.62)$$

Next, we consider the eigenspace generated by acting with $\mathcal{H}_3$ on the singlet pair state

$$|A\rangle = \frac{1}{\sqrt{2}}(|\uparrow\downarrow 0\rangle - |\downarrow\uparrow 0\rangle) \quad (8.63)$$

which leads to

$$\mathcal{H}_3|A\rangle = t_1\sqrt{2}|B\rangle + 2t|C_1\rangle\,. \quad (8.64)$$

$|B\rangle$ is composed of intersite singlet pairs

$$|B\rangle = \frac{1}{2}(|0\uparrow\downarrow\rangle - |0\downarrow\uparrow\rangle + |\uparrow 0\downarrow\rangle - |\downarrow 0\uparrow\rangle) \quad (8.65)$$

while $|C_1\rangle$ involves double occupation at the base of the triangle

$$|C_1\rangle = \frac{1}{\sqrt{2}}(c^\dagger_{1\uparrow}c^\dagger_{1\downarrow} + c^\dagger_{2\uparrow}c^\dagger_{2\downarrow})|0\rangle\,. \quad (8.66)$$

[50] Our "sign rule" is intuitively clear. A formal argument could have been given by invoking the Perron–Frobenius theorems of matrix theory.

We landed in the singlet subspace which is even under C_2. The fourth basis function $|C_3\rangle = c^\dagger_{3\uparrow}c^\dagger_{3\downarrow}|0\rangle$ appears in

$$\begin{aligned}\mathcal{H}_3|B\rangle &= t_1\sqrt{2}|A\rangle + (t_2+\epsilon_3)|B\rangle + t_1\sqrt{2}|C_1\rangle + 2t_1|C_3\rangle \\ \mathcal{H}_3|C_1\rangle &= 2t_2|A\rangle + t_1\sqrt{2}|B\rangle + U|C_1\rangle \\ \mathcal{H}_3|C_3\rangle &= 2t_1|B\rangle + (2\epsilon_3 + U_3)|C_3\rangle. \end{aligned} \tag{8.67}$$

First, we reconsider the $U = U_3 = \infty$ case when the Hilbert space is restricted to $|A\rangle$ and $|B\rangle$. The lowermost singlet energy is

$$\mathcal{E}^{\text{s.}}_{\text{min.}} = \frac{\epsilon_3 + t_2 - \sqrt{(\epsilon_3+t_2)^2 + 8(t_1)^2}}{2}. \tag{8.68}$$

Though it is not quite obvious, it can be checked that $\mathcal{E}^{\text{tr.}}_{\text{min.}} < \mathcal{E}^{\text{s.}}_{\text{min.}}$ if $t_2 > 0$, which confirms our previous result.

For finite Hubbard interaction, deciding the nature of the ground state requires the numerical solution of the full 4×4 problem. Fig. 8.19 gives a sample of the results, showing that (for $t_1 \neq t_2$) the singlet–triplet transition occurs at a finite U. *The occurence of ferromagnetic ordering is a non-perturbative, intermediate-coupling phenomenon.*

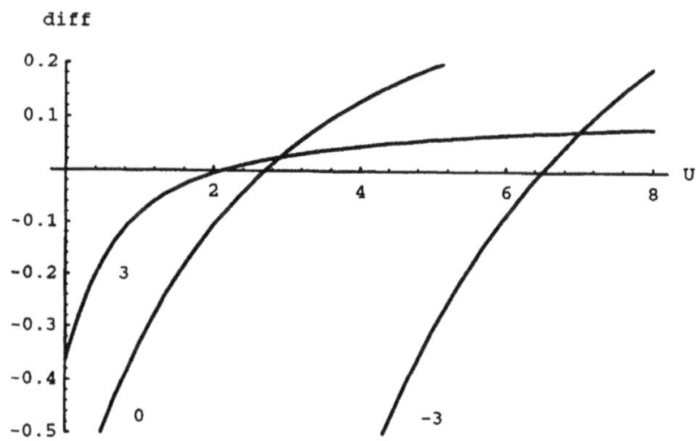

Figure 8.19: The energy difference $\mathcal{E}^{\text{s.}}_{\text{min.}} - \mathcal{E}^{\text{tr.}}_{\text{min.}}$ as a function of U (in units of t_1) for $\epsilon_3 = 3t_1$, 0 and $-3t_1$. We have chosen $U_3 = U$ and $t_2 = t_1/2$. Increasing ϵ_3 favours ferromagnetism.

We have performed the calculation for three different values of ϵ_3: $\epsilon_3 = 3t_1$, 0, and $-3t_1$. It strikes the eye that increasing ϵ_3 favours ferromagnetism.

d) To get an insight into the competition between the ferromagnetic and the non-magnetic ground state, let us envisage[51] the following situation: Let $U_3 = \infty$, U large but finite, and ϵ_3 smaller than U but still large compared to t and t_1. The singlet eigenvalue problem is restricted to the subspace spanned by $|A\rangle$, $|B\rangle$, and $|C_1\rangle$. Expanding in t_2 and t_1 we get

$$\mathcal{E}^{\text{s.}}_{\text{min.}} \approx -\frac{4t_2^2}{U} - \frac{2(t_1)^2}{\epsilon_3} + \frac{2t(t_1)^2}{\epsilon_3^2} \tag{8.69}$$

(we neglected a term of the order $t_2(t_1)^2/(U\epsilon_3)$ which would in any case drop out when writing down J). To the same order

$$\mathcal{E}^{\text{tr.}}_{\text{min.}} \approx -\frac{2(t_1)^2}{\epsilon_3} - \frac{2t_2(t_1)^2}{\epsilon_3^2}\,. \tag{8.70}$$

Under the assumptions made above, the electrons are essentially confined to the sites 1 and 2, and the effective Hamiltonian can be written as $J\mathbf{S}_1 \cdot \mathbf{S}_2$. The exchange coupling can be identified from the singlet–triplet splitting

$$J = \frac{4t_2^2}{U} - \frac{4(t_1)^2 t_2}{\epsilon_3^2}\,. \tag{8.71}$$

Thus a second-order and a third-order process are competing with each other. The second-order term is the usual antiferromagnetic coupling due to virtual hopping to a nearest-neighbour site. The third-order term describes a *ring exchange process*: as shown on the right-hand side of Fig. 8.18, three consecutive hoppings around the triangular plaquette effect the interchange of the two spins. The sign of this term depends on the sign of t_2 and thus we can control via the appropriate choice of t_2 whether the coupling is ferromagnetic or antiferromagnetic. The arising of ferromagnetism in Tasaki's toy model is now seen to result from the fact that choosing a large U suppresses the nearest-neighbour antiferromagnetic process sufficiently so that the usually negligible ring-exchange processes take over.

We may be worried about the behaviour at small ϵ_3 where (8.71) diverges even though we know form our previous results that the singlet–triplet splitting is perfectly well-behaved. For convenience, let us consider the limit $U \to \infty$. The difference of (8.68) and (8.62)

$$\Delta_{U\to\infty} = \mathcal{E}^{\text{tr.}}_{\text{min.}} - \mathcal{E}^{\text{s.}}_{\text{min.}} = -t_2 + \frac{1}{2}\left(\sqrt{(\epsilon_3+t_2)^2+8t_1^2} - \sqrt{(\epsilon_3-t_2)^2+8t_1^2}\right) \tag{8.72}$$

[51] Here we paraphrase the argument of [323] for the toy model.

is the energy stabilizing the high-spin state. The large-ϵ_3 expansion recovers (8.71). This is smoothly connected to the $\epsilon_3 = 0$ value $-t_2$. Though the latter still describes a singlet-triplet splitting, we do not want to call it an effective exchange parameter, because we do not have a spin Hamiltonian of the form $J\mathbf{S}_1 \cdot \mathbf{S}_2$ any more, into which we could insert it. The electrons visit all three lattice sites with equal amplitudes, and we cannot ascribe localized spins to certain lattice sites. At zero (or small) ϵ_3 the ferromagnetism becomes purely itinerant, being sustained by the kinetic energy, but this regime is continuously connected to the perturbative regime described by (8.71).

Actually, this is still only part of the story since Fig. 8.19 shows that the ferromagnetic state is not confined to the large-U regime but sets in at $U \sim (t_1, t_2)$. But this intermediate-U ferromagnetism is continuously connected to the large-U regime, and the results in Fig. 8.19 confirm that increasing ϵ_3 indeed helps ferromagnetism, thus we consider the insight gained from the effective-Hamiltonian-approach valuable.

Problem 8.3. Though the exact diagonalization for arbitrary $t/J_{\rm H}$ would be straightforward, we confine ourselves to the $J_{\rm H} \to \infty$ case where the character of the solution is transparent[52]. All the energies are measured from $-J_{\rm H}S/2$.

a) *Core spin* $S = 1/2$. Let us denote the core spin states by $\Uparrow$, $\Downarrow$, and the conduction electron states by $\uparrow$ and $\downarrow$. The $S_{\rm tot} = 3/2$, $S^z_{\rm tot} = 3/2$ subspace of the two-site system is spanned by $|\Uparrow\uparrow, \Uparrow\rangle$ and $|\Uparrow, \Uparrow\uparrow\rangle$. The even combination $(1/\sqrt{2})(|\Uparrow\uparrow, \Uparrow\rangle + |\Uparrow, \Uparrow\uparrow\rangle)$ has the energy $-t$, while the odd combination $(1/\sqrt{2})(|\Uparrow\uparrow, \Uparrow\rangle - |\Uparrow, \Uparrow\uparrow\rangle)$ the energy t. On a fully polarized background, the electron hops with the bare amplitude $\tilde{t}(S_{\rm tot} = 3/2) = t$.

Both $S_{\rm tot} = 3/2$ and $S_{\rm tot} = 1/2$ states are represented in the $S^z_{\rm tot} = 1/2$ subspace. Excluding local singlets, the even (odd) subspace is spanned by

$$(1/\sqrt{2})(|\Uparrow\uparrow, \Downarrow\rangle \pm |\Downarrow, \Uparrow\uparrow\rangle)$$

and

$$(1/2)\left[(|\Uparrow\downarrow, \Uparrow\rangle + |\Downarrow\uparrow, \Uparrow\rangle) \pm (|\Uparrow, \Uparrow\downarrow\rangle + |\Uparrow, \Downarrow\uparrow\rangle)\right] .$$

The eigenvalues in the even subspace are $-t$ and $t/2$, while in the odd subspace t and $-t/2$. $\pm t$ belong to $S_{\rm tot} = 3/2$, while the energies $\mp t/2$ to $S_{\rm tot} = 1/2$. Observe that the low-energy $S_{\rm tot} = 1/2$ state is *odd*, which we ascribe to $\tilde{t}(S_{\rm tot} = 1/2) = -t/2$.

All in all, the spectrum can be characterized in two steps: first, we take the spectrum of the renormalization factors $\tilde{t}/t$ (here 1 and -1/2), and then the spectrum of the permutation operator $(\pm\tilde{t})$. This is expressed by the general form (8.46).

---

[52]The author is grateful to K. Penc and K. Itai for valuable discussions.

Introducing the total spin operators of the sites $\vec{S}_1$ and $\vec{S}_2$, $2\vec{S}_1{\cdot}\vec{S}_2 = (\vec{S}_1+\vec{S}_2)^2 - \vec{S}_1^2 - \vec{S}_1^2 = S_{\text{tot}}(S_{\text{tot}}+1) - 2 - 3/4$ takes the values 1 ($S_{\text{tot}} = 3/2$) and -2 ($S_{\text{tot}} = 1/2$). It follows that the spectrum is correctly given by

$$\frac{\tilde{t}}{t} = \frac{1+2\vec{S}_1{\cdot}\vec{S}_2}{2}. \tag{8.73}$$

b) *Core spin* $S = 1$. If an electron is coupled to the core spin, only $S = 3/2$ states are allowed. The $S^z = 3/2$, $1/2$, and $-1/2$ states of the site are given by $|1\uparrow\rangle$, $(1/\sqrt{3})(|\sqrt{2}|0\uparrow\rangle + |1\downarrow\rangle)$, and $(1/\sqrt{3})(|\sqrt{2}|0\downarrow\rangle + |-1\downarrow\rangle)$, respectively. Taking now two sites with one electron, we find $S_{\text{tot}} = 5/2$, $3/2$, and $1/2$ states. The fully polarized state is described, as before, by the undiminished $\tilde{t}(S_{\text{tot}} = 5/2) = t$.

The basis states of the even (odd), $S^z_{\text{tot}} = 3/2$ subspace can be chosen as

$$(1/\sqrt{2})(|1\uparrow,0\rangle \pm |0,1\uparrow\rangle)$$

and

$$(1/\sqrt{6})\left[\sqrt{2}|0\uparrow,1\rangle + |1\downarrow,1\rangle \pm (\sqrt{2}|1,0\uparrow\rangle + |1,1\downarrow\rangle)\right].$$

The matrix of the hopping Hamiltonian is

$$\begin{pmatrix} 0 & -t/\sqrt{2} \\ -t/\sqrt{2} & -t/2 \end{pmatrix} \quad \text{(even)} \qquad \begin{pmatrix} 0 & t/\sqrt{2} \\ t/\sqrt{2} & t/2 \end{pmatrix} \quad \text{(odd)}.$$

The new set of eigenvalues $\mp 2t/3$ belongs to $S_{\text{tot}} = 3/2$ eigenstates. Again, we find a change of sign when S_{tot} is changed by unity: the low-energy $S_{\text{tot}} = 5/2$ state is even, but the low-energy $S_{\text{tot}} = 3/2$ state is odd. This can be ascribed to $\tilde{t}(S_{\text{tot}} = 3/2) = -2t/3$.

The calculation proceeds similarly in the $S^z_{\text{tot}} = 1/2$ subspace. In the local high-spin basis for even (odd) subspaces

$$(1/\sqrt{2})(|1\uparrow,-1\rangle \pm |-1,1\uparrow\rangle)$$

$$(1/\sqrt{6})\left[\sqrt{2}|0\uparrow,0\rangle + |1\downarrow,0\rangle \pm (\sqrt{2}|0,0\uparrow\rangle + |0,1\downarrow\rangle)\right].$$

$$(1/\sqrt{6})\left[\sqrt{2}|0\downarrow,1\rangle + |-1\uparrow,1\rangle \pm (\sqrt{2}|1,0\downarrow\rangle + |1,-1\uparrow\rangle)\right]$$

the hopping matrices are

$$\begin{pmatrix} 0 & 0 & \mp t/\sqrt{3} \\ 0 & \mp 2t/3 & \mp t\sqrt{2}/3 \\ \mp t/\sqrt{3} & \mp t\sqrt{2}/3 & 0 \end{pmatrix}.$$

The eigenvalues are $-t$, $-t/3$, and $2t/3$ (even), and $-2t/3$, $t/3$, and t (odd). $-t/3$ (even), and $t/3$ (odd) are new. We conclude that $\tilde{t}(S_{\rm tot} = 1/2) = t/3$; the magnitude is reduced, but there is no extra sign factor.

The regularities with increasing core spin are becoming apparent, and conform to the general rule stated in (8.44).

In deriving the effective spin Hamiltonian, we proceed as we did in the $S = 1/2$ case. Now $2\vec{S}_1 \cdot \vec{S}_2 = S_{\rm tot}(S_{\rm tot} + 1) - 23/4$ takes the values 3 ($S_{\rm tot} = 5/2$), -2 ($S_{\rm tot} = 3/2$), and -5 ($S_{\rm tot} = 1/2$). It is easy to check that

$$\frac{\tilde{t}}{t} = \frac{-3 + 3\vec{S}_1 \cdot \vec{S}_2 + 2(\vec{S}_1 \cdot \vec{S}_2)^2}{6}. \tag{8.74}$$

Chapter 9

The Gutzwiller Variational Method

We resume our study of the ground state of the Hubbard model described by the Hamiltonian

$$\begin{aligned}\mathcal{H} &= \mathcal{H}_{\text{band}} + \mathcal{H}_U \\ &= \sum_{\mathbf{k}\sigma} \epsilon(\mathbf{k}) \hat{n}_{\mathbf{k}\sigma} + U \sum_{\mathbf{j}} \hat{n}_{\mathbf{j}\uparrow} \hat{n}_{\mathbf{j}\downarrow}. \end{aligned} \tag{9.1}$$

In particular, we would like to follow the changes in the nature of the *ground state of the half-filled Hubbard model* as U is increased from 0 to very large values. We have seen convincing qualitative arguments which show that when U becomes comparable to the bandwidth W, a metal-to-insulator transition must occur. We have understood that this so-called Mott transition is driven by local charge correlations and that it must set in as we reach $U \sim W$. We wish to describe the behaviour of the system as the transition point is approached from either the metallic or the insulating side.

In Sec. 5.1.4, we learned about the effective Hamiltonian which governs the low-energy behaviour of the system at large U, i.e., deep in the insulating phase. The large-U description in terms of a suitable Heisenberg Hamiltonian is valid for any form of the tight-binding part of the Hubbard Hamiltonian, but it is expected to break down as U is reduced below $\sim W$, and it cannot tell us anything about the transition itself.

We know that if the band structure satisfies the perfect nesting condition (7.49) then the ground state is insulating for any $U > 0$. Since the non-interacting system is metallic, we may say that the metal–insulator transition is taking place at $U = 0$ but it is *not* a Mott transition; it is a consequence of a special feature of the band structure which causes an antiferromagnetic instability for arbitrarily small U. The insulating nature of the SDW ground state is due to magnetic long-range-order and it has nothing to do with a Mott transition. The true Mott phenomenon would be a local correlation effect which sets in at $U \sim W$. There is no doubt that as U is increased, the gap gradually assumes the features of a Mott–Hubbard gap, and the nature of the system changes from a weakly correlated SDW insulator to an antiferromagnetic Mott insulator[1]. However, there is no known singularity associated with this and apparently, the ground state evolves smoothly as U is increased from small to large values. For our purposes, its evolution is satisfactorily described by the SDW mean field theory. The discussion of the magnetically ordered insulating state is very important since in the real world, almost all Mott insulators are at the same time ordered magnets, but it hides from us the local correlation effects which we consider to be the true reason why Mott insulators exist.

In the present Chapter, we outline a description of the Mott transition as a local correlation effect. To achieve this, we will artificially constrain the ground state to be non-magnetic. The (hypothetical) "pure" Mott transition discussed here leads from a paramagnetic metal to a paramagnetic insulator. The starting point is to notice that the Hubbard U controls the structure of the ground state via n_d, the density of the doubly occupied sites. We have to find a way to calculate[2] n_d as a function of U. A popular approach is to use the Gutzwiller variational method: we devise a trial wave function in which n_d can be optimally chosen for a given value of U.

---

[1] We were arguing that these features must be present in the true solution of the problem, but they are missing from the simple mean-field SDW solution.

[2] In Section 5.1.5, we showed how this can be done for large U. Now we want a method which works for the range from small to intermediate U's.

9.1 Minimum Polarity Principle

We were repeatedly arguing that the uncorrelated Fermi sea, i.e., the paramagnetic band ground state

$$|\mathrm{FS}\rangle = \prod_{\mathbf{k}}^{\epsilon_{\mathbf{k}}<\epsilon_F} c^+_{\mathbf{k}\uparrow} c^+_{\mathbf{k}\downarrow}|0\rangle \tag{9.2}$$

has a lot of charge fluctuations in it, e.g., for a half-filled, nondegenerate band, on the average 1/4th of all lattice sites are doubly occupied, and 1/4th are empty. In other words, if we were making a "snapshot" of $|\mathrm{FS}\rangle$, to "see where the electrons are", typically we would find the situation[3] shown in the left-hand picture of Fig. 9.1:

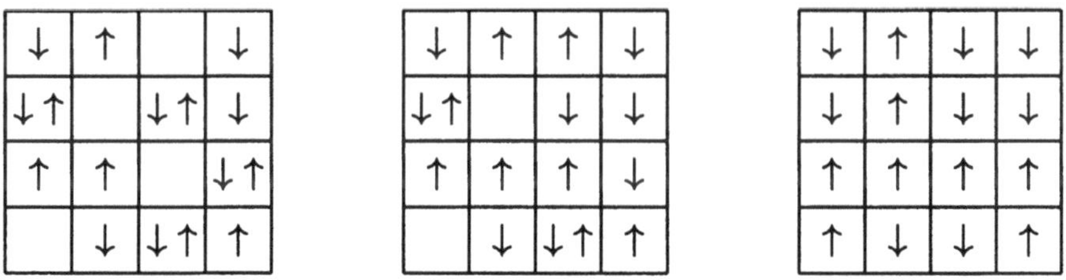

a. $U = 0$ b. intermediate U c. $U \to \infty$

Figure 9.1: The evolution of the ground state of the half-filled Hubbard model, illustrated by "snapshots" of the electron distribution. Increasing U suppresses charge fluctuations.

If we were using $|\mathrm{FS}\rangle$ as a trial ground state for the Hubbard model (9.1), the ground state energy density/site would be (cf. (4.9), and Fig. 4.3)

$$\frac{1}{L}\langle\mathrm{FS}|\mathcal{H}|\mathrm{FS}\rangle = \frac{1}{L}\langle\mathrm{FS}|\mathcal{H}_{\mathrm{band}}|\mathrm{FS}\rangle + \frac{1}{L}\langle\mathrm{FS}|\mathcal{H}_U|\mathrm{FS}\rangle = \bar{\epsilon} + \frac{U}{4} \tag{9.3}$$

where $\bar{\epsilon}$ is the average band energy per lattice site. Now we set the energy zero at the centre-of-mass of the entire band thus for any partial ($n < 2$) filling, $\bar{\epsilon} < 0$.

[3]In the snapshots in Fig. 9.1 the number of each species of lattice site was chosen to correspond to the average concentration. This is only for illustration: in any finite cluster taken from the lattice, the numbers may fluctuate.

According to the estimate (9.3), for $U > 4|\bar{\epsilon}|$ the ground state energy would become positive! It would be absurd to accept this conclusion since, for large U's, any of the neutral states $|\psi_{\mathrm{neut}}\rangle$ such as the one shown in the right-hand picture of Fig. 9.1, would be a better trial state, giving exactly 0 energy (the Wannier states have 0 band energy and there is no double occupation). It means that for sufficiently large Us, we must arrive at a Heitler–London type description of the solid. This was our naive argument for a Mott–transition in Sec. 4.2. Here we attempt to be less naive. Instead of letting n_d jump from its uncorrelated value 1/4 to zero, we allow for the obvious fact that, as soon as we switch on the Hubbard U, there must be a tendency to reduce the density of doubly occupied sites.

Let us return to Fig. 9.1.c. Note that such a fully neutral state is expected only at $U \to \infty$. At any finite U, hopping will inevitably mix in the charged states by the process

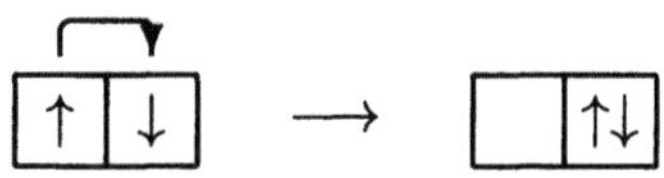

(9.4)

so the optimally chosen ground state must be something in between the extremes shown in Fig. 9.1.a and Fig. 9.1.c, with the concentration of doubly occupied sites

$$n_d = \frac{1}{L}\sum_{\mathbf{j}} \langle \hat{n}_{\mathbf{j}\uparrow}\hat{n}_{\mathbf{j}\downarrow}\rangle \tag{9.5}$$

lying at some intermediate value

$$0 < n_d < \frac{1}{4}. \tag{9.6}$$

As U is increased from 0 towards ∞, n_d must decrease from the uncorrelated value 1/4 to 0. We want to find out how n_d depends on U/W.

Remember that we found one possible answer to this question in our spin density wave mean field theory: according to eqn. (7.89), $n_d = 1/4 - m^2$ (m being the sublattice magnetization) which indeed performs a smooth interpolation between the aforementioned limits. The derivation relied on the assumption of magnetic long-range order, and neglected local correlation effects completely. Now we want an answer which is valid within the fully symmetrical, paramagnetic state, describing the suppression of n_d as a pure correlation effect.

At exact half-filling, simple counting gives

$$n_d = n_e \tag{9.7}$$

where n_e is the density of empty sites. (On the average, there is one electron per site. To put two electrons on a site, the second electron must have been taken from somewhere else, leaving behind an empty site.) – The densities of sites singly occupied by $\uparrow$-spin, or $\downarrow$-spin electrons will be denoted by $n_\uparrow$, and $n_\downarrow$, respectively. Obviously

$$n_\uparrow + n_\downarrow + n_e + n_d = 1. \tag{9.8}$$

Usually, we will assume a paramagnetic state for which $n_\uparrow = n_\downarrow$.

In equation (9.5), we defined n_d as a global average, taken over the whole lattice. In principle, $\langle \hat{n}_{\mathbf{j}\uparrow}\hat{n}_{\mathbf{j}\downarrow}\rangle$ could vary from site to site. However, here we are interested in a featureless paramagnetic state which has the full translational symmetry of the underlying lattice. Thus the relevant expectation values are site-independent, for instance $\langle \hat{n}_{\mathbf{j}\uparrow}\hat{n}_{\mathbf{j}\downarrow}\rangle = n_d$ for all $\mathbf{j}$, so the global average is the same as the local expectation value.

An alternative way to state the problem is to remember that (9.3) is nothing but the simplest[4] Hartree–Fock result for the ground state energy. The true ground state energy can be represented as

$$\frac{1}{L}\langle\Psi|\mathcal{H}|\Psi\rangle = \frac{1}{L}\langle \mathrm{FS}|\mathcal{H}|\mathrm{FS}\rangle + \mathcal{E}_{\mathrm{corr}} = \bar{\epsilon} + \frac{U}{4} + \mathcal{E}_{\mathrm{corr}} \tag{9.9}$$

where $|\Psi\rangle$ is the (normalized) exact ground state, and $\mathcal{E}_{\mathrm{corr}}$ is the so-called *correlation energy*. We have to find $\mathcal{E}_{\mathrm{corr}}$. Simple reasoning tells us that the correlation energy must be negative. By definition, $\mathcal{E}_{\mathrm{corr}}$ is the energy due to inter-particle correlations, i.e., due to the fact that electrons can avoid each other more effectively than it is predicted by Hartree–Fock theory. The Hartree–Fock approximation grossly overestimates the interaction energy, and this is corrected by $\mathcal{E}_{\mathrm{corr}}$.

[4] Derived on the assumption that there is no symmetry breaking, i.e., $\langle \hat{n}_{\mathbf{j}\uparrow}\rangle = \langle \hat{n}_{\mathbf{j}\downarrow}\rangle = 1/2$.

Our picture of the correlated, paramagnetic ground state is that "switching on" $U > 0$ does not change the overall appearance of the ground state, it just diminishes the density of doubly occupied sites n_d (and then, because of (9.7), also n_e). Fig. 9.1.b depicts an intermediate situation between the extremes of the uncorrelated Fermi sea, and full localization. For each U/W, n_d has to be chosen optimally. We achieve this by devising a trial ground state in which n_d is controlled by a variational parameter. The procedure is sometimes described as determining the optimum degree of "polarity fluctuations". To understand the phrase, we have to recall some of the early history of the subject.

9.1.1 Digression: The Early History

It was realized already in the 1930s that the independent-electron description may break down completely, as it does for the insulating oxides NiO, CoO and Fe_3O_4. Later it was recognized that it may be inadequate even for certain metals such as Ni. Here we follow an insightful and highly influential discussion given in 1953 by Van Vleck [423]. Van Vleck's arguments were critically re-examined and much elaborated by Herring [163].

Accepting[5] that the band filling in Ni can be given as $3d^{9.4}4s^{0.6}$, it is clear that a "snapshot" of metallic nickel should consist mainly of $3d^9$ and $3d^{10}$ sites (for the purposes of the present discussion, the whereabouts of the $4s$ electrons are immaterial). However, if – in the spirit of band theory – we neglect the correlations between electrons with different spin and orbital quantum numbers, we would predict a fair concentration of $3d^8$, $3d^7$, ... sites as well. Let us estimate the concentration of $3d^7$ sites! Consider an arbitrary lattice site, with ten $3d$ spin-orbitals. According to independent-particle theory, the probability of finding any of these spin-orbitals occupied is 0.94, *independently* of the occupation of the other spin-orbitals. Thus the probability of finding seven specified spin-orbitals occupied, and the other three empty, is $0.94^7 \times 0.06^3$. However, there are $C_7^{10} = 120$ ways to pick the seven

[5] The value $n_d \approx 9.4$ appears in many solid-state texts, e.g., in [209]. Since the present argument serves merely to illustrate an idea, we are not worried by the fact that some contemporary electronic structure calculations give a rather different value.

states from ten. Therefore, according to band theory, the concentration of $3d^7$ sites should be $0.94^7 \times 0.06^3 \times 120 \approx 1.68\%$. Generally, the probability of finding a site in the configuration $3d^n$ is

$$p_{ind.}(n) = 0.94^n \times 0.06^{10-n} \times C_n^{10}. \tag{9.10}$$

For the concentration of $3d^8$ sites, we find the appreciable value 9.9%.

We can argue, however, that this is a serious overestimate. Let us try to envisage how a $3d^8$ site arises from the band motion of electrons! Most of the sites are, even in the independent-particle picture, $3d^{10}$ and $3d^9$ (their concentrations being predicted to be 53.8 percent, and 34.4 percent, respectively). A typical snapshot of a small cluster would show just these two major species

d^{10}	d^{10}	d^9	d^9	d^{10}

(9.11)

There are two kinds of hopping events which involve the cell in the middle: either an electron jumps there from the left, according to the scheme

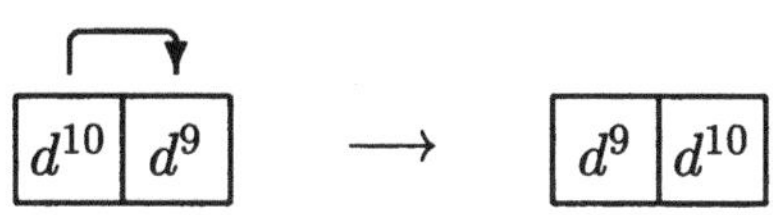

(9.12)

or an electron jumps away from the third cell to the right, leading to

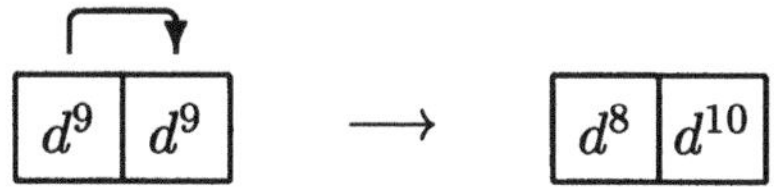

(9.13)

According to the independent-particle description, given the starting configuration in (9.11), the subsequent hopping events (9.12) and (9.13) are equally likely to occur. Consideration of energetics should, however, speak against (9.13), and give preference to (9.12). (9.12) just rearranges the d^9 and d^{10} sites, and does not change the ionic energies. In contrast, the process (9.13) is easily seen to cost intrasite Coulomb energy. For the sake of simplicity, let us assume that the interaction energy between any two electrons on the same site is the same Hubbard U_{dd}. Between 9 electrons there are $C_2^9 = 36$ pairwise interactions, so the on-site energy of a $3d^9$ site is

$$E(3d^9) = 9\epsilon_d + 36U_{dd} \tag{9.14}$$

where ϵ_d is the atomic d-level. Determining the energies of the $3d^8$ and $3d^{10}$ configurations similarly, the energy expenditure of the hopping event (9.13) is found to be

$$\begin{aligned}(E(3d^8) + E(3d^{10})) - (E(3d^9) + E(3d^9)) &= \\ (8\epsilon_d + 28U_{dd} + 10\epsilon_d + 45U_{dd}) - 2(9\epsilon_d + 36U_{dd}) &= U_{dd}\,.\end{aligned} \tag{9.15}$$

Creating also $3d^7$, etc. sites would require additional energy investments of U_{dd}. Hence, consideration of the electron–electron interaction shows that hopping processes leading to large fluctuations of local charge must be hampered by energy barriers, and the concentration of species like $3d^8$, $3d^7$, will be quite a bit lower than the independent-electron estimate (9.10). The mechanism responsible for this suppression of charge fluctuations is that *the hopping amplitudes must be dependent on the occupation numbers* of the d-orbitals. To include this feature, we must go *beyond the independent-particle picture*, and incorporate *local correlation effects*.

For very strong U_{dd} (i.e., for $U_{dd} \gg W$, the d-bandwidth), the average band filling $3d^{9.4}4s^{0.6}$ is realized by 40% of the sites being $3d^{10}$, and 60% $3d^9$. For non-integer filling, even $U_{dd} \to \infty$ leaves us with a residue, an unavoidable minumum, of charge fluctuations: in Van Vleck's classic work[6], this was referred to as the "model of minimum polarity". By "polarity" we should understand the deviation of the local charge from its mean value. We will often refer to them as *polarity fluctuations*: these are just the entities which can carry electric current.– Referring to Sec. 1.3, we now realize that by restricting the charge states of iron to Fe^{++} and Fe^{+++}, we were introducing a minimum polarity model of magnetite. Similarly, in Sec. 8.5.3 we used the minimum polarity picture for $La_{1-x}(Sr/Ca)_xMnO_3$.

[6]Actually, Van Vleck credits H. Hurwitz with having introduced the idea in his unpublished Thesis work in 1941.

Adopting the principle of minimum polarity is, of course, going to an extreme: for Ni, U_{dd} is likely to be of the same order as the $3d$ bandwidth, but not larger. We expect that the local configurations $3d^8$, etc. will be still present, only in smaller concentration than implied[7] by (9.10). In other words, instead of sticking to the principle of *minimum* polarity, we should find the *optimum* degree of polarity fluctuations. The need for this was acknowledged by Van Vleck; however, a formal execution of the program had to wait for another 12 years. It was the outstanding achievement of Gutzwiller [142, 143] to develop a formalism which turned Van Vleck's ideas into a calculational tool. This was accomplished by devising a variational wave function which *optimizes* the degree of polarity fluctuations.

9.2 The Variational Ground State

Let us consider the ground state of the half-filled Hubbard model as U is increased from 0 towards the critical value $U_{\rm cr} \sim W$, at which the Mott transition into an insulator occurs. At $U = 0$ we have an uncorrelated metal. Increasing U gradually suppresses the polarity fluctuations but the system is still metallic; it has what we call a *correlated metallic state.* It is of great interest to find out how the local correlations influence the observable features of a metal.

Our qualitative idea of the correlated metallic state is illustrated in Fig. 9.1.b: the ground state has the "same overall appearance" as the uncorrelated metallic ground state $|\mathrm{FS}\rangle$, only the polarity fluctuations are diminished, to a degree determined by U/W. Now we have to find the formal expression of this idea. It is achieved by using a variational trial wave function which was introduced by Gutzwiller [143]

$$|\Psi\rangle = \prod_{\mathbf{j}} [1 - (1-\eta)\hat{n}_{\mathbf{j}\uparrow}\hat{n}_{\mathbf{j}\downarrow}]|\mathrm{FS}\rangle \tag{9.16}$$

where $0 \le \eta \le 1$ is a variational parameter. It is easy to convince ourselves that $|\Psi\rangle$ does exactly what Fig. 9.1.b suggests: it starts with the

[7] We should note that (9.10) was deduced for a hypothetical paramagnetic ground state of nickel. For the actual ferromagnetic ground state, a similar reasoning leads to somewhat smaller polarity fluctuations.

Fermi sea $|\text{FS}\rangle$, and uniformly diminishes the concentration of doubly occupied sites. At $\eta = 1$, $|\Psi\rangle = |\text{FS}\rangle$, thus $\eta = 1$ must correspond to $U = 0$. For $U > 0$, we should find $\eta < 1$. The way the variational Ansatz (9.16) works is this: whenever a site $\mathbf{j}$ is found to be doubly occupied, the amplitude is multiplied by η; while if the site is not doubly occupied, the local Fermi sea structure remains unchanged. Configurations with a lot of doubly occupied sites accumulate a lot of $\eta < 1$ factors, and get suppressed. The resulting many-electron state will have a diminished $n_d < 1/4$ concentration of doubly occupied sites.

It is obvious that the same kind of correlated metallic trial ground state can be introduced for any band filling, not just half-filling. What is special about half-filling is that there a Mott transition is expexted to occur, while away from half-filling, the ground is conducting even at $U = \infty$. Thus the trial state (9.16) should have an even greater range of applicability than in the non-half-filled case. To avoid having to repeat the same steps again, henceforth we consider a general band filling: N electrons distributed over L lattice sites. Half-filling corresponds to $N = L$. We usually assume $N \leq L$ since for bipartite lattices, the physics at $N > L$ is simply related to that at $N < L$ (see Sec. 4.5). Because we restrict the discussion to non-magnetic states, there will be $N/2$ electrons with $\uparrow$-spin, and $N/2$ electrons with $\downarrow$-spin.

To find the optimum degree of polarity fluctuations, we have to calculate the ground state energy density

$$\frac{E(\eta)}{L} = \frac{1}{L} \cdot \frac{\langle\Psi|\mathcal{H}|\Psi\rangle}{\langle\Psi|\Psi\rangle} = \frac{1}{L} \cdot E_{\text{band}}(\eta) + U n_d(\eta) \tag{9.17}$$

and minimize it with respect to η. This we will do in several steps.

9.2.1 The Gutzwiller Trial State

To be able to count the doubly occupied sites, we have to express the ground state $|\Psi\rangle$ in real space. To gain experience, we start with the uncorrelated Fermi sea ground state $|\text{FS}\rangle$, by expressing the Bloch operators in terms of Wannier operators

$$|\text{FS}\rangle = \prod_{\mathbf{k}}^{\epsilon_{\mathbf{k}}<\epsilon_F} c_{\mathbf{k}\uparrow}^{+} \prod_{\mathbf{q}}^{\epsilon_{\mathbf{q}}<\epsilon_F} c_{\mathbf{q}\downarrow}^{+}|0\rangle \tag{9.18}$$

$$= L^{-N/2} \prod_{\mathbf{k}}^{\epsilon_{\mathbf{k}}<\epsilon_F} \left(\sum_{\mathbf{g}} e^{i\mathbf{k}\mathbf{g}} c_{\mathbf{g}\uparrow}^{+} \right) \prod_{\mathbf{q}}^{\epsilon_{\mathbf{q}}<\epsilon_F} \left(\sum_{\mathbf{h}} e^{i\mathbf{q}\mathbf{h}} c_{\mathbf{h}\downarrow}^{+} \right) |0\rangle .$$

In order to form an idea what the result should be looking like, let us consider the simplest example: the real-space expansion of a state with two Bloch electrons:

$$\begin{aligned} |\mathbf{k}_1\mathbf{k}_2\rangle &= c_{\mathbf{k}_1\uparrow}^{+} c_{\mathbf{k}_2\uparrow}^{+} |0\rangle \\ &= \left(\frac{1}{\sqrt{L}} \sum_{\mathbf{g}_1} e^{i\mathbf{k}_1\mathbf{g}_1} c_{\mathbf{g}_1\uparrow}^{+} \right) \left(\frac{1}{\sqrt{L}} \sum_{\mathbf{g}_2} e^{i\mathbf{k}_2\mathbf{g}_2} c_{\mathbf{g}_2\uparrow}^{+} \right) |0\rangle . \end{aligned} \tag{9.19}$$

Now, a particular localized configuration

$$|\mathbf{G}_1\mathbf{G}_2\rangle = c_{\mathbf{G}_1\uparrow}^{+} c_{\mathbf{G}_2\uparrow}^{+} |0\rangle \tag{9.20}$$

arises in the expansion (9.19) twice: once when $\mathbf{g}_1 = \mathbf{G}_1$ and $\mathbf{g}_2 = \mathbf{G}_2$, and once when $\mathbf{g}_1 = \mathbf{G}_2$ and $\mathbf{g}_2 = \mathbf{G}_1$

$$\begin{aligned} &\frac{1}{L} e^{i\mathbf{k}_1\mathbf{G}_1} c_{\mathbf{G}_1\uparrow}^{+} e^{i\mathbf{k}_2\mathbf{G}_2} c_{\mathbf{G}_2\uparrow}^{+} |0\rangle + \frac{1}{L} e^{i\mathbf{k}_1\mathbf{G}_2} c_{\mathbf{G}_2\uparrow}^{+} e^{i\mathbf{k}_2\mathbf{G}_1} c_{\mathbf{G}_1\uparrow}^{+} |0\rangle \\ &= \frac{1}{L} \left[e^{i\mathbf{k}_1\mathbf{G}_1} e^{i\mathbf{k}_2\mathbf{G}_2} - e^{i\mathbf{k}_1\mathbf{G}_2} e^{i\mathbf{k}_2\mathbf{G}_1} \right] c_{\mathbf{G}_1\uparrow}^{+} c_{\mathbf{G}_2\uparrow}^{+} |0\rangle \\ &= \frac{1}{L} \begin{vmatrix} e^{i\mathbf{k}_1\mathbf{G}_1} & e^{i\mathbf{k}_1\mathbf{G}_2} \\ e^{i\mathbf{k}_2\mathbf{G}_1} & e^{i\mathbf{k}_2\mathbf{G}_2} \end{vmatrix} c_{\mathbf{G}_1\uparrow}^{+} c_{\mathbf{G}_2\uparrow}^{+} |0\rangle . \end{aligned} \tag{9.21}$$

We can interpret this by saying that the event that the electrons arrived at the lattice sites $\mathbf{G}_1$ and $\mathbf{G}_2$, is resulting from the interference of two processes: first when the electron at $\mathbf{G}_1$ "came" from the Bloch state $|\mathbf{k}_1\rangle$, and the electron at $\mathbf{G}_2$ from the Bloch state $|\mathbf{k}_2\rangle$; and second, when the electron at $\mathbf{G}_1$ came from the Bloch state $|\mathbf{k}_2\rangle$, and the electron at $\mathbf{G}_2$ from the Bloch state $|\mathbf{k}_1\rangle$. The two terms in the first line of (9.21) describe these two processes. We have to pay due attention to the fact that according to (9.20), the localized configuration $|\mathbf{G}_1\mathbf{G}_2\rangle$ is defined by the Wannier operators acting in a definite sequence on the vacuum $|0\rangle$. (9.21) warns us that it is not necessarily true that the operators appear in the expansion (9.19) in this fixed order. We always have to commute back to the standard order, which tends to introduce factors of (-1) in the expansion. This causes the amplitudes to combine into a determinant of Bloch-factors.

The full expansion (9.19) of the itinerant two-electron state $|\mathbf{k}_1\mathbf{k}_2\rangle$ in terms of localized configurations can be written as

$$|\mathbf{k}_1\mathbf{k}_2\rangle = \frac{1}{L}\sum_{\{\mathbf{g}_1\mathbf{g}_2\}}\begin{vmatrix} e^{i\mathbf{k}_1\mathbf{g}_1} & e^{i\mathbf{k}_1\mathbf{g}_2} \\ e^{i\mathbf{k}_2\mathbf{g}_1} & e^{i\mathbf{k}_2\mathbf{g}_2}\end{vmatrix}|\mathbf{g}_1\mathbf{g}_2\rangle \tag{9.22}$$

where the summation is over all the C_2^L ways in which two sites can be chosen from L (C_2^L denotes a binomial).

The evaluation of (9.18) proceeds analogously, only now the summation extends over a vastly larger number of localized configurations. Let us begin with enumerating the localized configurations which form a complete, orthonormal set of N-electron states in the Wannier representation. Considering only non-magnetic ($S^z = 0$) states, there are $N/2$ electrons with spin $\uparrow$, and $N/2$ electrons with spin $\downarrow$. Let the $\uparrow$-spin electrons sit at lattice sites $\mathbf{g}_1$, $\mathbf{g}_2$, ... $\mathbf{g}_{N/2}$. There are $C_{N/2}^L$ possibilities of distributing the $\uparrow$-spin electrons over the lattice. We introduce the short-hand notation $\{\mathbf{g}_1, \mathbf{g}_2, \ldots \mathbf{g}_{N/2}\} \equiv \{\mathbf{g}_1^{N/2}\}$ for the set of sites with $\uparrow$-spin electrons. Similarly, the $\downarrow$-spin electrons can be placed in $C_{N/2}^L$ different ways at lattice sites $\{\mathbf{h}_1, \mathbf{h}_2, \ldots \mathbf{h}_{N/2}\} \equiv \{\mathbf{h}_1^{N/2}\}$. There are altogether $(C_{N/2}^L)^2$ localized basis states of the form

$$|\mathbf{g}_1, \ldots \mathbf{g}_{N/2}; \mathbf{h}_1, \ldots \mathbf{h}_{N/2}\rangle \equiv |\mathbf{g}_1^{N/2}; \mathbf{h}_1^{N/2}\rangle = \prod_{\mathbf{g}}^{\{\mathbf{g}_1^{N/2}\}} c_{\mathbf{g}\uparrow}^{+} \prod_{\mathbf{h}}^{\{\mathbf{h}_1^{N/2}\}} c_{\mathbf{h}\downarrow}^{+}|0\rangle\,. \tag{9.23}$$

The "snapshots" shown in Fig. 9.1 are illustrating various kinds of localized configurations. It strikes the eye that they differ in the amount of *charge fluctuation* present in them. It is easiest to discuss the half-filled case $N = L$, the case presented in the snapshots. The average number of electrons/site is 1; since a half-filled band arises naturally in a lattice of ions with nuclear charge $Z = +1$ (such as Na), we will say that on the average, a lattice site is neutral. Deviations[8] from local neutrality occur when a site is doubly occupied (charge $-e$), or empty (charge $+e$).

[8]These are the polarity fluctuations mentioned earlier.

Proceeding analogously to (9.20-9.22), $|\mathrm{FS}\rangle$ defined in (9.18) can be expanded in terms of the localized basis (9.23) as

$$
\begin{aligned}
|\mathrm{FS}\rangle \;=\; & L^{-N/2} \sum_{\{\mathbf{g}_1^{N/2}\}} \sum_{\{\mathbf{h}_1^{N/2}\}} \\
& \begin{vmatrix} e^{i\mathbf{k}_1\mathbf{g}_1} & \dots & e^{i\mathbf{k}_1\mathbf{g}_{N/2}} \\ \vdots & \ddots & \vdots \\ e^{i\mathbf{k}_{N/2}\mathbf{g}_1} & \dots & e^{i\mathbf{k}_{N/2}\mathbf{g}_{N/2}} \end{vmatrix} \cdot \begin{vmatrix} e^{i\mathbf{k}_1\mathbf{h}_1} & \dots & e^{i\mathbf{k}_1\mathbf{h}_{N/2}} \\ \vdots & \ddots & \vdots \\ e^{i\mathbf{k}_{N/2}\mathbf{h}_1} & \dots & e^{i\mathbf{k}_{N/2}\mathbf{h}_{N/2}} \end{vmatrix} \\
& |\mathbf{g}_1, \ldots \mathbf{g}_{N/2}; \mathbf{h}_1, \ldots \mathbf{h}_{N/2}\rangle . \qquad (9.24)
\end{aligned}
$$

Recalling the Hubbard model, we realize that the doubly occupied sites must be playing a significant role because each of them "costs" an on-site Coulomb energy U. Thus an important characteristic of the localized basis state in (9.23) is the number of common elements of the $\{\mathbf{g}_1^{N/2}\}$-, and $\{\mathbf{h}_1^{N/2}\}$-sets, i.e., the number of doubly occupied sites

$$
N_d(\mathbf{g}_1^{N/2}; \mathbf{h}_1^{N/2}) = \langle \mathbf{g}_1 \ldots \mathbf{g}_{N/2}; \mathbf{h}_1 \ldots \mathbf{h}_{N/2} | \sum_{\mathbf{j}} \hat{n}_{\mathbf{j}\uparrow} \hat{n}_{\mathbf{j}\downarrow} | \mathbf{g}_1 \ldots \mathbf{g}_{N/2}; \mathbf{h}_1 \ldots \mathbf{h}_{N/2} \rangle . \qquad (9.25)
$$

For the snapshots Fig. 9.1.a, c, and b, $N_d = 4$, 0, and 2, respectively.

The Gutzwiller expansion (9.24) of the non-interacting Fermi sea $|\mathrm{FS}\rangle$ encompasses *all* possible real-space distributions of the electrons, with N_d ranging[9] from 0 to $N/2$. There are configurations which you would rightly consider as highly implausible: for instance, when all electrons are crammed into one end of the crystal, where all sites are doubly occupied, while the rest of the lattice remains empty. Such configurations are, however, necessarily present in the exact expansion (9.24); the fact that they are indeed very improbable must be a property of the determinant factors figuring in that equation.

The determinant factors in (9.24) can be considered as the *coherence factors* of the Fermi sea. By "coherence" we usually mean a property of an ordered state; that is not the case here. The uncorrelated Fermi sea $|\mathrm{FS}\rangle$ is the clearest example of a non-ordered ground state. Still, it has

[9] Here we consider only $N \leq L$. For $N > L$, N_d could not be less than $N - L$.

a distinctive feature: a *sharp Fermi surface.* The existence of a sharp Fermi surface is a $\mathbf{k}$-space feature; it is not easily seen in the real-space expansion (9.24). The determinant factors must "know about" the existence of the Fermi surface! They are complicated, inherently complex quantities, expressing the *relative phases* of the localized configurations.

The expansion (9.24) highlights the basic difficulty of describing strong local correlations in itinerant systems. Itinerancy is a $\mathbf{k}$-space feature, easily expressed in the Bloch-representation. Local correlations are, on the other hand, best handled in the Wannier representation. To include both aspects, one has to be able to go to and fro between the two representations; for a many-electron system, this gets complicated.

The following argument should help to understand the relevance of the determinant coefficients in (9.24). The same determinants would appear in the Wannier representation of the ground state of a sea $|\mathrm{FS}\rangle_{\mathrm{sl}}$ of $N/2$ spinless fermions:

$$
\begin{aligned}
|\mathrm{FS}\rangle_{\mathrm{sl}} &= \prod_{\mathbf{k}}^{\epsilon_{\mathbf{k}}<\epsilon_F} c_{\mathbf{k}}^{+}|0\rangle = L^{-N/4} \prod_{\mathbf{k}}^{\epsilon_{\mathbf{k}}<\epsilon_F} \left(\sum_{\mathbf{g}} e^{i\mathbf{k}\mathbf{g}} c_{\mathbf{g}}^{+} \right) |0\rangle \qquad (9.26) \\
&= L^{-N/4} \sum_{\{\mathbf{g}_1^{N/2}\}} \begin{vmatrix} e^{i\mathbf{k}_1\mathbf{g}_1} & \ldots & e^{i\mathbf{k}_1\mathbf{g}_{N/2}} \\ \vdots & \ddots & \vdots \\ e^{i\mathbf{k}_{N/2}\mathbf{g}_1} & \ldots & e^{i\mathbf{k}_{N/2}\mathbf{g}_{N/2}} \end{vmatrix} |\mathbf{g}_1, \ldots \mathbf{g}_{N/2}\rangle .
\end{aligned}
$$

In the absence of orbital degeneracy, the Pauli principle prevents two electrons from sharing the same site, so there is no on-site interaction to speak about. Still, two fermions tend to keep away from each other: due to the antisymmetrization implicit in (9.26), an *exchange hole* surrounds each fermion (see Problem 9.1). The determinants in the above equation can therefore be regarded as exchange hole factors. They express the fact that a configuration in which the fermions are more-or-less evenly spaced, is much more probable than one in which the fermions are crowded together. Similarly, in (9.24) the same determinant factors describe the exchange hole effect for parallel-spin electrons. If you like, you can say that certain spatial correlations are present even in the non-interacting Fermi sea $|\mathrm{FS}\rangle$, but these are due to statistics, and therefore act only between particles with like quantum numbers (i.e., the same spin).

For an easy demonstration of this fact, consider the expansion (9.22). We introduce the correlation function which measures the probability of the two electrons sitting a lattice vector $\mathbf{G}$ away:

$$
\begin{aligned}
f(\mathbf{G}) &= \sum_{\mathbf{g}} \langle \mathbf{k}_1\mathbf{k}_2 | \hat{n}_{\mathbf{g}} \hat{n}_{\mathbf{g}+\mathbf{G}} | \mathbf{k}_1\mathbf{k}_2 \rangle \\
&= \frac{1}{L} \sum_{\mathbf{g}} \left| \begin{vmatrix} e^{i\mathbf{k}_1\mathbf{g}} & e^{i\mathbf{k}_1(\mathbf{g}+\mathbf{G})} \\ e^{i\mathbf{k}_2\mathbf{g}} & e^{i\mathbf{k}_2(\mathbf{g}+\mathbf{G})} \end{vmatrix} \right|^2 = \left| \begin{vmatrix} 1 & e^{i\mathbf{k}_1\mathbf{G}} \\ 1 & e^{i\mathbf{k}_2\mathbf{G}} \end{vmatrix} \right|^2 \\
&= 2[1 - \cos(\mathbf{k}_1 - \mathbf{k}_2)\mathbf{G}]
\end{aligned} \tag{9.27}
$$

(the occupation numbers are site-diagonal, therefore the expression contains the squared absolute values of the determinants). The density–density correlation has an oscillating character. For a more elaborate example, cf. Problem 9.1.

One has to be careful about the terminology: the non-interacting Fermi sea is, by definition, *uncorrelated* because the occupation numbers for antiparallel-spin electrons are independent. Such is the nature of the ground state of non-interacting electrons. As we have seen, this does *not* imply that all intersite correlations should vanish: far from it! There are appreciable spatial correlations for parallel-spin electrons. This is, however, not caused by any interaction thus in our terminology, it is not a correlation effect.

Now we are in the position to expand the trial state (9.16) in the orthonormal basis (9.23)

$$
\begin{aligned}
|\Psi\rangle = {} & L^{-N/2} \sum_{\{\mathbf{g}_1^{N/2}\}} \sum_{\{\mathbf{h}_1^{N/2}\}} \eta^{N_d(\mathbf{g}_1^{N/2};\mathbf{h}_1^{N/2})} \\
& \begin{vmatrix} e^{i\mathbf{k}_1\mathbf{g}_1} & \dots & e^{i\mathbf{k}_1\mathbf{g}_{N/2}} \\ \vdots & \ddots & \vdots \\ e^{i\mathbf{k}_{N/2}\mathbf{g}_1} & \dots & e^{i\mathbf{k}_{N/2}\mathbf{g}_{N/2}} \end{vmatrix} \cdot \begin{vmatrix} e^{i\mathbf{k}_1\mathbf{h}_1} & \dots & e^{i\mathbf{k}_1\mathbf{h}_{N/2}} \\ \vdots & \ddots & \vdots \\ e^{i\mathbf{k}_{N/2}\mathbf{h}_1} & \dots & e^{i\mathbf{k}_{N/2}\mathbf{h}_{N/2}} \end{vmatrix} \\
& |\mathbf{g}_1, \dots \mathbf{g}_{N/2}; \mathbf{h}_1, \dots \mathbf{h}_{N/2}\rangle .
\end{aligned} \tag{9.28}
$$

A comparison of (9.24) and (9.28) allows us to reformulate the reasoning behind the Gutzwiller Ansatz: $|\Psi\rangle$ is derived from $|\mathrm{FS}\rangle$ by systematically changing the coefficients in its real-space expansion. The determinant coefficients appearing in (9.24) are complex quantities, specified by their absolute value and phase. $|\Psi\rangle$ is derived from $|\mathrm{FS}\rangle$ by changing merely the absolute values of the coefficients, but leaving their phases intact. *The relative phases of the configurations are inherited from the non-interacting Fermi sea and this guarantees the continuing existence of a sharp Fermi surface*, which is the defining feature of a metallic state. The weights of the configurations are obtained from minimizing the ground state energy.

For a moment, let us confine our attention to the half-filled case. First, note that $|\Psi\rangle$ gives the correct ground state energy for both limiting cases: $U = 0$, and $U \to \infty$. For $U = 0$, choosing $\eta = 1$ we recover $|\mathrm{FS}\rangle$ as given by (9.24), the proper ground state. For $U \to \infty$, double occupation should be completely suppressed: this can be achieved by setting $\eta = 0$. This limit is a bit tricky because all the amplitudes in (9.28) seem to vanish. We should remember $0^0 = 1$ which says that only the configurations with $N_d = 0$ survive. The same is immediately seen from our starting form (9.16): it goes over into the fully Gutzwiller-projected Fermi sea

$$|\Psi\rangle_{\eta=0} = \prod_{\mathbf{j}} [1 - \hat{n}_{\mathbf{j}\uparrow}\hat{n}_{\mathbf{j}\downarrow}]|\mathrm{FS}\rangle \tag{9.29}$$

which (for $N = L$) consists only of neutral configurations.

It is a loose but plausible-sounding argument in favour of the Gutzwiller Ansatz that it interpolates (at least as far as the energy is concerned) between two correct limits. If the end-points are right what can go wrong in between?

Well, the main thing that can go wrong is that the two end-points belong to different phases (the metal and the Mott insulator), and we should not try to interpolate smoothly through a phase transition[10]. We will not be really surprised to find that the Gutzwiller method gives a reasonable description of only one of the phases: the metallic state. However, it is just the description of a correlated metal we are most interested in; for the Mott-insulator, the understanding gained

[10]A phase transition is always associated with a non-analyticity, so one cannot make a smooth continuation through a phase boundary.

from the large-U expansion (Section 5.1) is satisfactory. Besides, the Gutzwiller method gives valuable insight into what happens when the Mott-transition is approached from the metallic side.

Away from half-filling, there is no phase boundary to cross, and our variational description is valid for all values of U/W.

Problem 9.1 Determine the shape of the exchange hole in the ground state of a one-dimensional tight-binding band of spinless fermions!

There are N fermions, and L lattice sites, with cyclic boundary condition. You have to determine ${}_{sl.}\langle \mathrm{FS}|\hat{n}_0\hat{n}_g|\mathrm{FS}\rangle_{sl.}$ with $|\mathrm{FS}\rangle_{sl.}$ given by (9.26).

Discuss the result for a half-filled band!

Problem 9.2 Find the expression of the nearest-neighbour hole–hole correlation function g_{hh} in the ground state of spinless fermions on the d-dimensional cubic lattice. What is the effect of increasing the dimensionality?

Hint: Find the relationship between g_{hh} and the energy density.

9.2.2 The Cluster Method

The next step is to carry out the variational procedure, i.e., to evaluate the expectation value (9.17), and to minimize it with respect to η. Unfortunately, this is fairly difficult. A complete solution to the problem was obtained [271] only 22 years after its formulation by Gutzwiller. In the interim period, people resorted to various approximations, most of which were of the kind first used by Gutzwiller himself [143]. The eventual full solution showed that though the so-called *Gutzwiller approximation* becomes exact only in the limit $D = \infty$, with due caution it can be used also for finite D, even for low-dimensional systems.

Looking at the expansion (9.24) of $|\mathrm{FS}\rangle$, you may be worried that the presence of configurations with very large polarity fluctuations invalidates any simple picture of the metallic state. For instance, at half filling ($N = L$) the concentration of doubly occupied sites must be $n_d = \langle \hat{N}_d\rangle/L = 1/4$. Now we see that it results as an average over exponentially many (in fact, $(C^L_{L/2})^2 \sim 4^L/L$) configurations, with the individual values of N_d ranging from 0 to $L/2$. If the fluctuations about the mean value were large then a configuration with the average $\langle \hat{N}_d\rangle = L/4$ (such as shown in Fig. 9.1.a) could not be considered "typical" and we would be in deep trouble, having to take the intricate sums seriously. The situation is, luckily, not as bad as that. Just as in classical statistical mechanics, sums over the exponentially large number of states of a macroscopic system tend to be dominated by their largest term (see Appendix C) and the overwhelming majority of configurations looks very much like an average configuration. It is these "typical" configurations which dominate the behaviour of the system. The same argument applies also to the case of finite U, where in a typical configuration (illustrated in Fig. 9.1.b), the concentration of doubly occupied sites is $n_d(\eta_0)$ where η_0 is the optimum value of η. All this is yet very non-formal and vague but we can already see how the possibility of developing a mean-field-like language for local correlations in an itinerant quantum system emerges. One should be able to carve out a small cluster from the inside of the system, and subject it to an effective field from the environment. It can be achieved that the average behaviour of the cluster will be characteristic of the whole lattice. The "mean field" is, however, not like the fields familiar from models of ordering in statistical physics, but rather the typical environment in a fermionic reservoir. Among other things, fermions must be allowed to travel to and fro between the cluster and its environment, and this must be done phase-coherently so that a sharp Fermi surface survives.

Next, we describe a cluster method for the evaluation of expectation values, which puts the above idea into practice. The cluster method is one of the possible ways to arrive at the results of the Gutzwiller approximation; for a different approach, cf. Appendix C. In order not to burden the main text with too many technicalities, we will be satisfied with the most handwaving argument[11]. In this simplest form, our treatment is applicable to the half-filled case only.

Let us observe that $\mathcal{H}$ consists of an on-site term, and a nearest-neighbour pair term, so to evaluate $E(\eta)$, it is sufficient to know the average state of a pair of sites $\mathbf{i}$ and $\mathbf{j}$.

The essence of our effective medium treatment is to replace the true expectation values in (9.17) with values calculated with a cluster method. The simplification is obvious: the two-site cluster has only $4 \times 4 = 16$ states. However, we certainly do not mean to replace the lattice with an isolated pair of sites! The pair of sites is immersed in

[11]It can be regarded as a simplified version of the argument presented in the review [431] by Vollhardt.

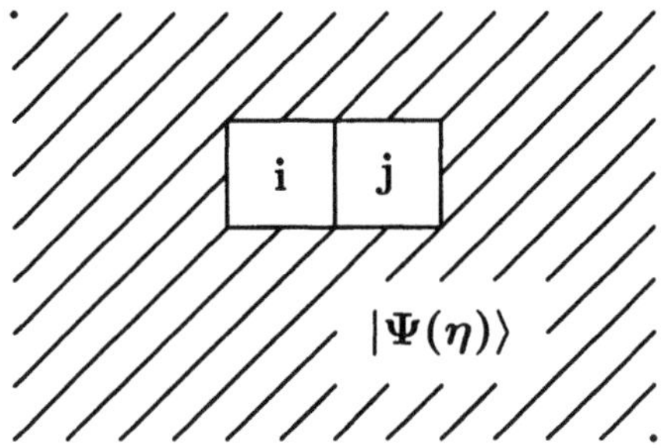

Figure 9.2: A pair of sites embedded in an effective medium characterized by the wavefunction $|\Psi(\eta)\rangle$.

the surrounding correlated Fermi sea (Fig. 9.2). We have to force the cluster to imitate, as closely as possible, the inside of a homogeneous macroscopic system.

We try to achieve this by assigning amplitudes to cluster configurations, which correspond to the recipe for constructing $|\Psi(\eta)\rangle$. According to (9.16), each time when a doubly occupied site appears, the amplitude of the configuration picks up a factor η. Constricting ourselves to the case of a half-filled band, (9.7) applies. In each configuration, the number of doubly occupied sites must be equal to the number of empty sites $N_d = N_e$. This means that in (9.16), the same variational amplitudes result if we associate a factor $\sqrt{\eta}$ with each doubly occupied, and also with each empty, site:

$$\eta^{N_d(\mathbf{g}_1^{N/2};\mathbf{h}_1^{N/2})} \equiv \eta^{(N_d(\mathbf{g}_1^{N/2};\mathbf{h}_1^{N/2})+N_e(\mathbf{g}_1^{N/2};\mathbf{h}_1^{N/2}))/2} . \tag{9.30}$$

Symmetrizing the prescription in e and d does not make any difference to the Gutzwiller Ansatz but simplifies the cluster procedure we are considering: it automatically ensures that exact half-filling will be kept.

Let us envisage the following decomposition of the N-electron wave function

$$\begin{aligned} |\Psi_N\rangle &= a(\uparrow\uparrow)|\uparrow\uparrow\rangle|\Psi_{N-2}^{(\uparrow\uparrow)}\rangle + a(\uparrow\downarrow)|\uparrow\downarrow\rangle|\Psi_{N-2}^{(\uparrow\downarrow)}\rangle + \ldots \\ &\quad + a(de)|de\rangle|\Psi_{N-2}^{(de)}\rangle + \ldots + a(d\uparrow)|d\uparrow\rangle|\Psi_{N-3}^{(d\uparrow)}\rangle + \ldots \end{aligned}$$

$$+ a(e\uparrow)|e\uparrow\rangle|\Psi_{N-1}^{(e\uparrow)}\rangle + a(\downarrow e)|\downarrow e\rangle|\Psi_{N-1}^{(\downarrow e)}\rangle + \ldots$$

$$+ a(dd)|dd\rangle|\Psi_{N-4}^{(dd)}\rangle + a(ee)|ee\rangle|\Psi_{N}^{(ee)}\rangle\,. \tag{9.31}$$

The amplitude of the event of the pair **ij** being in the state $\uparrow\uparrow$ is $a(\uparrow\uparrow)$; at the same time, the rest of the system accomodates $N-2$ electrons in the state $|\Psi_{N-2}^{(\uparrow\uparrow)}\rangle$. Since the two-site cluster is in communication with the other $N-2$ sites, the $(N-2)$-electron state depends on the cluster state: $|\Psi_{N-2}^{(\uparrow\uparrow)}\rangle$ differs from $|\Psi_{N-2}^{(\uparrow\downarrow)}\rangle$, which is, in turn, different from $|\Psi_{N-2}^{(de)}\rangle$, etc. Similarly, if both sites are doubly occupied (amplitude $a(dd)$), then the remaining $N-2$ sites are in the $(N-4)$-electron state $|\Psi_{N-4}^{(dd)}\rangle$, etc. We do not aim at finding the exact form of this decomposition; rather, we seek a simplified treatment which gives quick, and yet reasonable-looking, results for expectation values involving the sites **i** and **j**.

Let us think of the component states of the rest of the system: $|\Psi_{N-2}^{(\uparrow\uparrow)}\rangle$ etc., as normalized; then the norm of the total state is

$$\mathcal{N} = \sum_{\alpha\beta} |a(\alpha\beta)|^2 \tag{9.32}$$

where α, β run through the single-site configurations $\uparrow$, $\downarrow$, d, and e.

What should we think of the relative weights $|a(\uparrow\uparrow)|^2$, $|a(\uparrow\downarrow)|^2$, $|a(de)|^2$,...? Let us, for a moment, return to the uncorrelated Fermi sea. Picking out one site, this would be found in its four possible states $\uparrow$, $\downarrow$, d, and e with equal probabilities 1/4. Would it follow that a pair of sites is found with probability 1/16 in any of its sixteen possible configurations? Not at all: the two sites are not independent, their simultaneous occupation brings in the exchange hole effect which makes the probabilities unequal due to corrections which depend on the dimensionality and the structure of the lattice, the separation of the sites, etc. Since the exchange hole effect decays with increasing distance (see Problem 9.1), the assumption of independent sites would be justified only for sufficiently widely separated **i** and **j**, if the lattice is finite-dimensional. There is, however, another possibility for simplification: The fall-off of the exchange hole effect is the faster, the higher the dimensionality [332]. Therefore we can claim that assuming all free-electron coefficients

$a(\alpha\beta) = 1$ is justified in the limit of high-dimensional lattices[12]. It turns out that, for many purposes, three dimensions is high enough.

Switching on U, the amplitudes $a(\alpha\beta)$ pick up a variational factor $\sqrt{\eta}$ due to each d or e: thus $a(d\uparrow) = \sqrt{\eta}$, $a(de) = \eta$, while $a(\downarrow\uparrow)$ remains 1. Table 9.1 gives a list of the amplitudes of all 16 configurations.

Table 9.1: The 16 cluster configurations, and their amplitudes $a(\alpha\beta)$.

configuration	amplitude
ee, ed, de, dd	η
$e\uparrow$, $e\downarrow$, $d\uparrow$, $d\downarrow$	$\sqrt{\eta}$
$\uparrow e$, $\downarrow e$, $\uparrow d$, $\downarrow d$	$\sqrt{\eta}$
$\uparrow\uparrow$, $\downarrow\uparrow$, $\uparrow\downarrow$, $\downarrow\downarrow$	1

The total statistical weight for the cluster is

$$\mathcal{N} = 4(1+\eta)^2 . \tag{9.33}$$

$\mathcal{N}$ serves as a normalization factor in the calculation of expectation values. For instance, the density of doubly occupied sites, identified with the probability of finding the site $\mathbf{i}$ in the configuration d, turns out to be

$$n_d = \frac{\langle \Psi_N | \hat{n}_{\mathbf{i}\uparrow} \hat{n}_{\mathbf{i}\downarrow} | \Psi_N \rangle}{\mathcal{N}} = \frac{2\eta + 2\eta^2}{\mathcal{N}} = \frac{\eta}{2(1+\eta)} . \tag{9.34}$$

Note that this yields correct limiting values: $n_d = 0$ for $\eta = 0$, and $n_d = 1/4$ for $\eta = 1$.

We have to be a bit more circumspect in finding the cluster estimate for the kinetic energy $E_{\text{band}}(\eta)$. First, let us make it explicit that in (9.1), we consider the band arising from nearest-neighbour hopping.

[12] It must be mentioned that the cluster methods described in the literature [332, 312, 104] are much more sophisticated, inasmuch as they take into account the exchange hole effect within the cluster, and can be adapted to any finite dimension. Clusters with more than two sites give more satisfactory results [104].

Then, since all nearest-neighbour pairs, and also both spin directions, are equivalent, we can write

$$\frac{1}{L}\langle\sum_{\mathbf{k}\sigma}\epsilon(\mathbf{k})\hat{n}_{\mathbf{k}\sigma}\rangle = -2zt\langle c^+_{\mathrm{i}\uparrow}c_{\mathrm{j}\uparrow}\rangle \tag{9.35}$$

where z is the coordination number; on a D-dimensional cubic lattice, $z = 2D$. The evaluation of the band energy simplifies to finding the amplitude of an intra-cluster hopping process. For $\eta = 1$, we must recover the non-interacting value of the hopping amplitude $\bar{b}$, which is defined as

$$\begin{aligned}\bar{b} &= \frac{\langle\Psi_N|c^+_{\mathrm{i}\uparrow}c_{\mathrm{j}\uparrow}|\Psi_N\rangle_{\eta=1}}{16} = \frac{1}{16}(\langle\Psi^{(\uparrow e)}_{N-1}|\Psi^{(e\uparrow)}_{N-1}\rangle + \langle\Psi^{(\uparrow\downarrow)}_{N-1}|\Psi^{(ed)}_{N-1}\rangle \\ &+\langle\Psi^{(d\downarrow)}_{N-1}|\Psi^{(\downarrow d)}_{N-1}\rangle + \langle\Psi^{(de)}_{N-1}|\Psi^{(\downarrow\uparrow)}_{N-1}\rangle) = \frac{|\bar{\epsilon}|}{zt}\,.\end{aligned} \tag{9.36}$$

The full Fermi sea value of the hopping amplitude arises from letting the $\uparrow$-spin electrons hop, whatever the $\downarrow$-spin electrons are doing in the meantime. This includes processes like (9.4) which create doubly occupied sites. Decreasing η from 1, which suppresses double occupation, must therefore renormalize (in fact, reduce) the hopping amplitude

$$\langle c^+_{\mathrm{i}\uparrow}c_{\mathrm{j}\uparrow}\rangle_{\eta<1} = q(\eta)\cdot\langle c^+_{\mathrm{i}\uparrow}c_{\mathrm{j}\uparrow}\rangle_{\eta=1} = q(\eta)\cdot\bar{b}\,. \tag{9.37}$$

Our task is to find the *kinetic energy renormalization factor* $q(\eta)$.

Observe that four processes contribute to the amplitude in question:

1 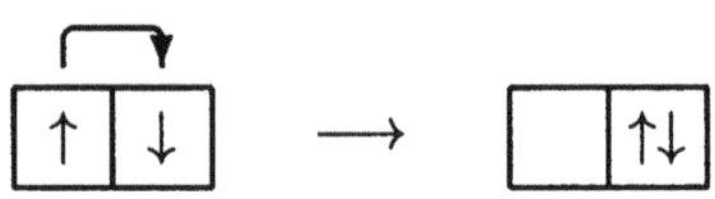 η (9.38)

$\sqrt{\eta}$ 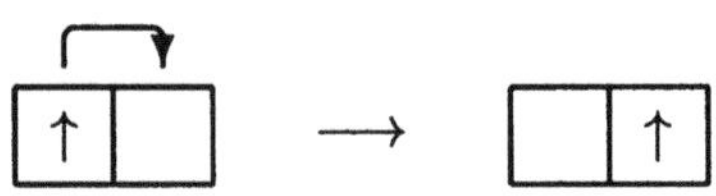 $\sqrt{\eta}$ (9.39)

$\sqrt{\eta}$ 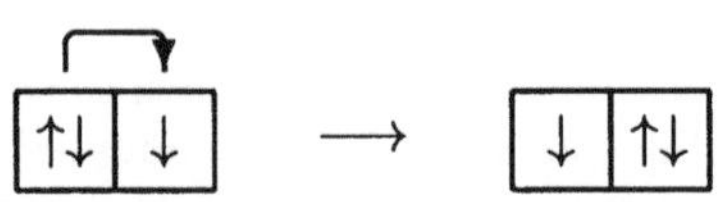 $\sqrt{\eta}$ (9.40)

and

η 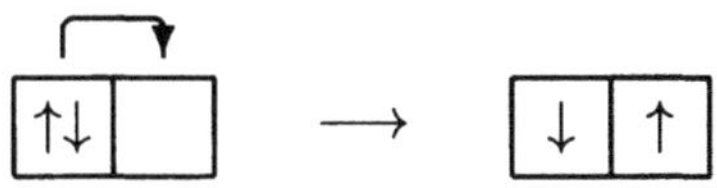 1 . (9.41)

We have indicated the amplitude factors acquired from the initial state on the left, and from the final state on the right. For all the four processes, the overall factor is η. Hence, a calculation analogous to that in (9.36) immediately yields

$$\begin{aligned}\langle c^+_{\mathrm{i}\uparrow} c_{\mathrm{j}\uparrow}\rangle_{\eta<1} &= \frac{\langle \Psi_N | c^+_{\mathrm{i}\uparrow} c_{\mathrm{j}\uparrow} | \Psi_N\rangle}{4(1+\eta)^2} = \frac{\eta}{4(1+\eta)^2}\big(\langle \Psi^{(\uparrow e)}_{N-1} | \Psi^{(e\uparrow)}_{N-1}\rangle \\ &\quad + \langle \Psi^{(\uparrow\downarrow)}_{N-1} | \Psi^{(ed)}_{N-1}\rangle + \langle \Psi^{(d\downarrow)}_{N-1} | \Psi^{(\downarrow d)}_{N-1}\rangle + \langle \Psi^{(de)}_{N-1} | \Psi^{(\downarrow\uparrow)}_{N-1}\rangle\big) \\ &= \frac{4\eta}{(1+\eta)^2}\cdot \bar{b}\,. \end{aligned} \tag{9.42}$$

In the last step, we replaced the factor describing the transition amplitudes in the fermion reservoir by its non-interacting value taken from (9.36). We could loosely argue that the suppression of the hopping amplitude due to the Hubbard U is a local effect, expressed by including η in the part of the wavefunction, which describes the cluster. The rest of the system provides the characteristic phase coherence of the Fermi liquid which is like that of a non-interacting Fermi sea.

We found that the renormalization factor of the hopping amplitude is

$$q(\eta) = \frac{\langle c^+_{\mathrm{i}\uparrow} c_{\mathrm{j}\uparrow}\rangle}{\langle c^+_{\mathrm{i}\uparrow} c_{\mathrm{j}\uparrow}\rangle_{\eta=1}} = \frac{4\eta}{(1+\eta)^2}\,. \tag{9.43}$$

Combining the results (9.34) and (9.43), the cluster method has given the following expression for the ground state energy density

$$E(\eta) = -2ztq(\eta)\langle c^\dagger_{\mathrm{i}\uparrow} c_{\mathrm{j}\uparrow}\rangle_{\eta=1} + Un_d = -\frac{4|\bar{\epsilon}|\eta}{(1+\eta)^2} + \frac{U\eta}{2(1+\eta)}\,. \tag{9.44}$$

9.3 Brinkman–Rice Transition

The remaining task is to minimize the energy expression (9.44) with respect to η, which yields[13]

$$\eta_0 = \begin{cases} \dfrac{1 - U/U_{\rm cr}}{1 + U/U_{\rm cr}} & \text{if } 0 \le U \le U_{\rm cr} \\ 0 & \text{if } U > U_{\rm cr} \end{cases} \tag{9.45}$$

where the critical value of U is

$$U_{\rm cr} = 8|\bar{\epsilon}| \; . \tag{9.46}$$

Replacing this in (9.34), we obtain

$$n_d = n_e = \begin{cases} \dfrac{1}{4}\left(1 - \dfrac{U}{U_{\rm cr}}\right) & \text{if } 0 \le U \le U_{\rm cr} \\ 0 & \text{if } U > U_{\rm cr} \; . \end{cases} \tag{9.47}$$

The salient feature of the solution is the prediction of a sharp metal–insulator transition at $T = 0$. In a half-filled band, the charge carriers are just the doubly occupied d, and the empty e sites. For $U < U_{\rm cr}$ the density of charge carriers $n_d + n_e$ is finite, so we have a metal. For $U > U_{\rm cr}$, $n_d + n_e$ vanishes, so the system must be an insulator. The Gutzwiller variational treatment gives a very simple scenario for a Mott transition taking place within the paramagnetic phase, by ascribing it to the gradual suppression of polarity fluctuations, i.e., of the density of charge carriers.

The fact that the Gutzwiller method allows us to arrive at the result (9.47), was pointed out by Brinkman and Rice [51]. Therefore we refer to this particular description of the Mott transition as the *Brinkman–Rice transition.*

The n_d versus U plot shows how the metallic nature is gradually suppressed. It must be true that at the same time, the system becomes

[13]Requiring $dE(\eta)/d\eta = 0$ would give just the first line of the expression. Formally, nothing seems to prevent us from using it also for $U > U_{\rm cr}$, where η would become negative. However, this would take us beyond the preset bounds of variation $0 \le \eta \le 1$. – There are many ways to check that the $\eta < 0$ solution has to be rejected as meaningless. For instance, n_d would become negative.

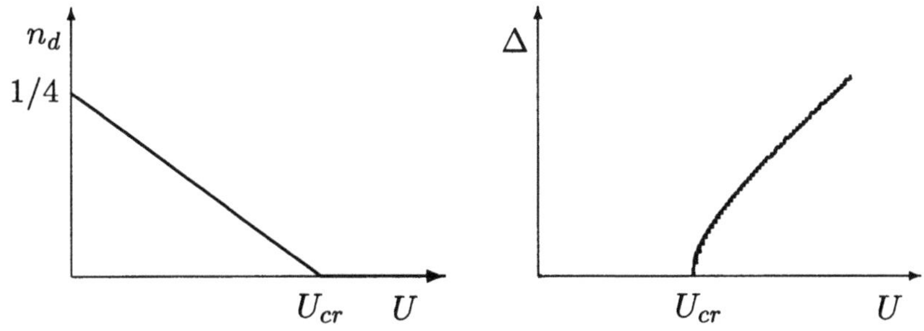

Figure 9.3: The two faces of the Brinkman–Rice transition. Left: the density of doubly occupied sites n_d vanishes as we approach U_{cr} from the metallic side. Right: The Mott–Hubbard gap Δ closes as we approach from the insulating side.

more magnetic. This can be measured by the square of the instantaneous magnetic moment (which is the same for all sites)

$$\langle (M_{\mathbf{j}}^z)^2 \rangle = \langle (\hat{n}_{\mathbf{j}\uparrow} - \hat{n}_{\mathbf{j}\downarrow})^2 \rangle = 1 - 2n_d \,. \tag{9.48}$$

As U is raised from 0 to $U_{\rm cr}$, $\langle (M_{\mathbf{j}}^z)^2 \rangle$ increases from 1/2 to 1 (see Fig. 9.4). It is useful to rewrite it as the sum of a Fermi sea term, and a correlation term [123]

$$\langle (M_{\mathbf{j}}^z)^2 \rangle = \langle (M_{\mathbf{j}}^z)^2 \rangle_{\rm FS} + \langle (M_{\mathbf{j}}^z)^2 \rangle_{\rm C} \,, \tag{9.49}$$

where $\langle (M_{\mathbf{j}}^z)^2 \rangle_{\rm FS} = n_\uparrow + n_\downarrow - 2n_\uparrow n_\downarrow \rightsquigarrow 1/2$, while $\langle (M_{\mathbf{j}}^z)^2 \rangle_{\rm C}$ is related to an on-site correlation function

$$\langle (M_{\mathbf{j}}^z)^2 \rangle_{\rm C} = -2g_{\uparrow\downarrow}(\mathbf{j},\mathbf{j}) = -2\langle (\hat{n}_{\mathbf{j}\uparrow} - \langle \hat{n}_{\mathbf{j}\uparrow} \rangle)(\hat{n}_{\mathbf{j}\downarrow} - \langle \hat{n}_{\mathbf{j}\downarrow} \rangle) \rangle \rightsquigarrow \frac{1}{2}(1 - 4n_d) \,. \tag{9.50}$$

We may be wondering at the large $U = 0$ moment: does it not make the Fermi sea appear more magnetic than it actually is? Should we expect a Curie susceptibility corresponding to $\langle (M_{\mathbf{j}}^z)^2 \rangle_{\rm FS}$? Well, no[14], because the instantaneous moments are very short-lived: they have a lifetime $\sim \hbar/W$. We may envisage that the fast fluctuations quench the moments before they would have time to respond to slow probes. The greatly enhanced susceptibility of strongly correlated metals arises not so much from the increase of $\langle (M_{\mathbf{j}}^z)^2 \rangle$ (which cannot be more than

[14] The Fermi sea moments begin to show up in the Curie susceptibility only at very high temperatures ($T \sim T_{\rm F}$). See the discussion of (7.109) on p. 407.

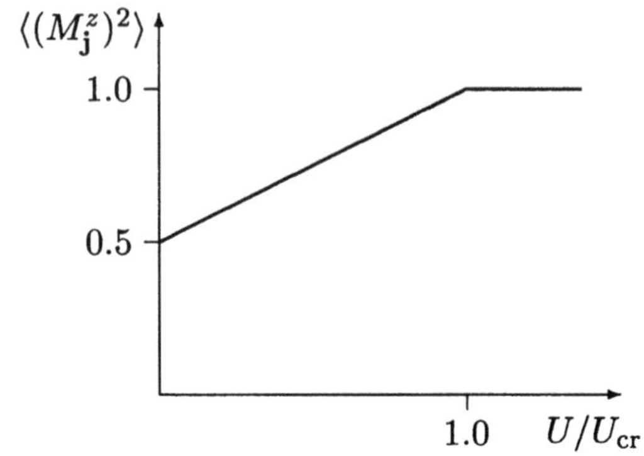

Figure 9.4: The U-dependence of the local magnetic moment defined in (9.48).

twofold) but rather from the reduced effective bandwidth, and the corresponding increase of the lifetime of the moments (see Sec. 10.2).

It is worth clarifying which is the particular feature of $E(\eta)$ which led to the prediction of a phase transition: it is that for small η, both terms of (9.44) are linear in η. Assuming that the transition is continuous, near the critical point $E(\eta)$ can be approximated as

$$E(\eta) \approx \left(\frac{U}{2} - 4|\bar{\epsilon}| \right) \eta \, . \tag{9.51}$$

The coefficient changes sign at $U_{cr} = 8|\bar{\epsilon}|$, thus $\eta = 0$ for $U > U_{cr}$, and $\eta \neq 0$ for $U < U_{cr}$.

The ground state energy density is

$$\frac{\langle \mathcal{H} \rangle}{L} = \begin{cases} -|\bar{\epsilon}| \left(1 - \dfrac{U}{U_{cr}} \right)^2 & \text{if } 0 \leq U \leq U_{cr} \\ 0 & \text{if } U > U_{cr} \; . \end{cases} \tag{9.52}$$

Comparing to (9.9)), we note the important finding of a large second-order contribution to the correlation energy

$$E_{corr}^{\text{Gutz.}} = -\frac{U^2}{64|\bar{\epsilon}|} \, . \tag{9.53}$$

It is gratifying that (9.53) represents 85% of the correct second-order result for the infinite-dimensional Hubbard model, which, in turn, is lying remarkably close to the result for the three-dimensional case [272].

The Brinkman–Rice transition does *not* affect the global symmetries of the system: on both sides of the transition, we find a translationally, and spin-rotationally invariant paramagnetic state. Order parameters in the conventional sense express the breaking of a global symmetry, so here we are somewhat at a loss, what order parameter to assign to the metal–insulator transition. The physical phenomenon we have in mind is that the zero-temperature static conductivity $\sigma(T=0)$ is finite for $U < U_{\rm cr}$, and zero for $U > U_{\rm cr}$, so why not use $\sigma(T=0)$ as an order parameter? Unfortunately, the electrical conductivity is rather difficult to calculate, therefore we are trying to identify other quantities which are relatively straightforward to evaluate and which we think to be associated with either the metallic, or the insulating, character of the system. The problem of the order parameter of the Mott transition is a delicate one and we will return to it several times in this text.

It is particularly difficult to find a good order parameter for the metallic paramagnet. At the present stage, the straightforward choice is n_d (Fig. 9.3, left) because that is the "controll parameter" of the variational treatment. However, in the next chapter we introduce another, more satisfying, order parameter (the height of the Fermi step) which is related to the Fermi-liquid character of the metallic state.

On the insulating side, we have a rather straightforward choice for an order parameter: the Mott–Hubbard gap. However, as long as we stick to the case of exact half-filling, it might seem that the Gutzwiller method is incapable to say something about the gap, for two reasons. First, the gap is an excitation energy, and the variational method is for the ground state. Second, the gap is certainly a function of U, while our characterization of the insulating state (namely, that $n_d = 0$) is perfectly featureless and U-independent. However, we will learn in Sec. 10.3 (see Problem 10.3) that the Gutzwiller method can be used to derive the Mott–Hubbard gap illustrated in Fig. 9.3 (right)

$$\Delta \sim \sqrt{U(U - U_{\rm cr})}\,. \tag{9.54}$$

At large U, the gap behaves as $\Delta \sim U - W$, which is in qualitative agreement with our previous picture of Hubbard subbands of width $\sim W$ lying at a distance U.

It should be understood that the present treatment is not meant to give us a complete description of the Mott transition. We should not be surprised: after all, it is based on an approximate handling of the simplest variational Ansatz which includes only on-site correlations. The variational method opens an easy way to such important results as the large enhancement of the effective mass in the pre-transition region (Sec. 10.2). It has also a number of obvious shortcomings. First, (9.47) predicts that $n_d = 0$ if U is larger than a critical value of the order of the bandwidth. However, it is clear that the process (9.4) guarantees that for any finite U, n_d remains finite. Recall that in Sec. 5.1.5, we derived the correct large-U expression for n_d, and found that it vanishes only asymptotically, according to $n_d \sim (t/U)^2$. To make it quite clear: we do accept that reducing the value of n_d is instrumental to driving the Mott transition. However, the transition is not accompanied by the complete suppression of n_d. In reality, the Mott transition should be taking place in a range of U where n_d is sufficiently small but finite[15].

A closely related erroneous feature of the present result is that the optimized ground state energy is identically zero in the insulating state even though we know (from Sec. 5.1.4) that, for sufficiently large U, it must have the characteristic scale of the Heisenberg coupling $\sim -t^2/U$.

The featurelessness of the insulating state is a failing of the present (simplest) Gutzwiller-type treatment. Starting from $U = 0$, we are reasonably happy about the scenario up to, and including, the metal–insulator transition but we find the characterization of the Mott insulator unsatisfactory. However, we contend that the basic virtue of the Gutzwiller treatment is that it does predict a Mott-type transition at all. We postpone worrying about the features we haven't got right, and go on to exploit what we got. In the next Chapter, we discuss the subject where the Gutzwiller method finds its most successful application: the description of heavy Fermi liquids.

[15]There is a rather artificial model (cf. Sec. 10.6.3) for which a complete description of the Mott transition is possible. Fig. 10.8 illustrates the relationship between three approaches: the exact solution of the model, the approximation-free evaluation of the Gutzwiller Ansatz, and the Gutzwiller approximation. It appears that the true behaviour of n_d is perfecty smooth as we cross the transition.

Besides, an extended Gutzwiller Ansatz gives an acceptable description of the Mott insulating state: as we are going to see in Sec. 10.6.4, a Gutzwiller-projected SDW state has a finite n_d, and a well-behaved energy density.

Remark. The relationship between the suppression of polarity fluctuations, and the insulating nature of the ground state, looks so obvious that we easily overlook that it is a single-orbital concept[16]. Having two Wannier orbitals (c and d) at each site can give rise to two hybridized bands. Assuming that the two bands are separated by a hybridization gap, the non-interacting $2L$-electron ground state belongs to filling the lower hybridized band completely

$$|\Phi\rangle = \prod_{\mathbf{k}}^{\mathrm{BZ}} \prod_{\sigma} (u_{\mathbf{k}} c_{\mathbf{k}\sigma}^{\dagger} + v_{\mathbf{k}} d_{\mathbf{k}\sigma}^{\dagger})|0\rangle \, . \tag{9.55}$$

The ground state is insulating, even though it is characterized by large polarity fluctuations. The density of $\uparrow$-spin c-electrons ($\downarrow$-spin d-electrons) is

$$n_{c\uparrow} = \sum_{\mathbf{k}} \frac{u_{\mathbf{k}}^2}{u_{\mathbf{k}}^2 + v_{\mathbf{k}}^2} \qquad\qquad n_{d\downarrow} = \sum_{\mathbf{k}} \frac{v_{\mathbf{k}}^2}{u_{\mathbf{k}}^2 + v_{\mathbf{k}}^2} \, .$$

Both $n_{c\uparrow}$ and $n_{d\downarrow}$ can be ~ 0.5. Since $\uparrow$- and $\downarrow$-occupations are independent, the probabilities of finding a $c\uparrow$ and a $d\downarrow$ electron at the same site, or neither of them, or only one of them, are all appreciable. It is not so straightforward to estimate the probability of finding two parallel-spin electrons at the same site, because it is influenced by the exchange hole effect, but it is not much less than $n_{c\uparrow}n_{d\uparrow}$. The ground state contains all configurations c^0d^0, c^1d^0, ..., c^2d^2.

Switching on the on-site Coulomb interactions modifies the weights of the various configurations (and renormalizes the hybridization gap), but for a wide range of parameters, it remains true that the system shows polarity fluctuations in spite of being a *mixed valent* insulator (see Ch. 11).

Solutions to the Problems

Problem 9.1 Expressing the occupation numbers in terms of Bloch-operators

$$\langle \hat{n}_0 \hat{n}_g \rangle = \frac{1}{L^2} \sum_{k_1} \sum_{k_2} \sum_{k_3} \sum_{k_4} e^{i(k_1-k_2)g} \langle c_{k_1}^{+} c_{k_2} c_{k_3}^{+} c_{k_4} \rangle . \tag{9.56}$$

One gets non-vanishing contributions either when $k_1 = k_2$, and $k_3 = k_4$, or when $k_1 = k_4$, and $k_2 = k_3$. In the first case

$$\frac{1}{L^2} \sum_{k_1} \sum_{k_3} \langle \hat{n}_{k_1} \hat{n}_{k_3} \rangle = \frac{1}{4\pi^2} \int_{-k_F}^{k_F} dk_1 \int_{-k_F}^{k_F} dk_3 = n^2$$

[16]Even within the framework of the Gutzwiller approximation. We have discussed that the physically correct description of a single-orbital Mott insulator has to include virtual charge fluctuations.

where $n = N/L$, and the Fermi wave vector is $k_F = \pi n$. In the second case, we find

$$\begin{aligned}
&\frac{1}{L^2}\sum_{k_1}\sum_{k_2} e^{i(k_1-k_2)g}\langle \hat{n}_{k_1}(1-\hat{n}_{k_2})\rangle \\
&= \frac{1}{4\pi^2}\int_{-k_F}^{k_F} e^{ik_1 g}dk_1\left(\int_{-\pi}^{-k_F} e^{-ik_2 g}dk_2 + \int_{k_F}^{\pi} e^{-ik_2 g}dk_2\right) \\
&= -\frac{1}{(\pi g)^2}\sin^2 \pi n g
\end{aligned}$$

so all in all

$$\langle \hat{n}_0 \hat{n}_g\rangle = n^2 - \frac{1}{(\pi g)^2}\sin^2 \pi n g. \tag{9.57}$$

For a half-filled band, $n = 1/2$, and the quantum-mechanical correction to the asymptotic (and, if you like to put it that way, classical) value 1/4 vanishes for even g, while it is always negative for odd g, falling off as $1/g^2$. Antisymmetrization of the wave function suppresses the probability of the simultaneous occupation of nearest-neighbour sites by about 40 per cent!

Remark: We could have invoked Wick's theorem to rewrite (9.56) as

$$\langle c_0^\dagger c_0 c_g^\dagger c_g\rangle = \begin{vmatrix} \langle c_0^\dagger c_0\rangle & \langle c_0^\dagger c_g\rangle \\ \langle c_g^\dagger c_0\rangle & \langle c_g^\dagger c_g\rangle \end{vmatrix} = n^2 - f_{g0}^2 \tag{9.58}$$

where

$$f_{g0} = \langle c_g^\dagger c_0\rangle = \frac{1}{\pi g}\sin n\pi g \tag{9.59}$$

is an element of the one-particle density matrix.

Problem 9.2. Let $|\Phi_0\rangle$ be the ground state of the spinless fermion sea (arbitrary filling). The probability of finding holes at the nearest-neighbour sites $\mathbf{g}$ and $\mathbf{h}$ is $(1-n)^2 g_{\mathrm{hh}}(\mathbf{gh})$, where $g_{\mathrm{hh}}(\mathbf{gh})$ is the hole–hole correlation function. Using Wick's theorem (see the *Remark* above)

$$\langle \Phi_0|(1-\hat{n}_{\mathbf{g}})(1-\hat{n}_{\mathbf{h}})|\Phi_0\rangle = 1 - 2n + \langle \Phi_0|\hat{n}_{\mathbf{g}}\hat{n}_{\mathbf{h}}|\Phi_0\rangle = (1-n)^2 - f_{\mathbf{gh}}^2\,. \tag{9.60}$$

For a hypercubic lattice with nearest-neighbour hopping $-t$, the kinetic energy density E_{kin} can be expressed as $E_{\mathrm{kin}} = -2Dtf_{\mathbf{gh}}$, thus we find

$$g_{\mathrm{hh}}(\mathbf{gh}) = 1 - \left(\frac{E_{\mathrm{kin}}}{2Dt(1-n)}\right)^2 . \tag{9.61}$$

The exchange hole effect is expressed by the correction to 1. Since a reasonable definition of high-dimensional lattices keeps E_{kin} finite, we conclude that the exchange hole gets weaker in high dimensions, and it vanishes as $D \to \infty$.

Chapter 10

The Correlated Metallic State

The variational method introduced in Chapter 9 is admittedly a very approximate way to describe the ground state of the Hubbard model. However, it offers us a nice pictorial way of thinking about local correlation effects. Besides, it allows an easy access to discovering two important features of the correlated metallic state: the reduction of the size of the step-like discontinuity in the band state occupation number, and the closely related correlation enhancement of the effective mass.

10.1 The Reduced Fermi Step

It is instructive to consider the renormalization factor of the kinetic (band) energy

$$q = \frac{\langle\Psi|c_{\mathbf{i}\sigma}^{+}c_{\mathbf{j}\sigma}|\Psi\rangle}{\langle\mathrm{FS}|c_{\mathbf{i}\sigma}^{+}c_{\mathbf{j}\sigma}|\mathrm{FS}\rangle}. \tag{10.1}$$

For a paramagnetic state, q does not depend on the spin. Our cluster method (cf. (9.37), (9.43), and (9.45)) gives

$$q = \begin{cases} 1 - \left(\dfrac{U}{U_{\mathrm{cr}}}\right)^2 & \text{if } 0 \le U \le U_{\mathrm{cr}} \\ 0 & \text{if } U > U_{\mathrm{cr}}. \end{cases} \tag{10.2}$$

q describes the gradual "slowing down" of the hopping motion with increasing electron–electron interaction.

q has a further interpretation which is even more enlightening. A momentary reflection convinces us that, in deriving the result (9.43) for the quantity defined in (9.37), we did not anywhere make use of the fact that $\mathbf{i}$ and $\mathbf{j}$ are nearest neighbours. The same renormalization factor should be obtained[1] for all analogous off-diagonal processes connecting more distant sites:

$$q = \frac{\langle\Psi|c^+_{\mathbf{i}\sigma}c_{\mathbf{j}\sigma}|\Psi\rangle}{\langle\mathrm{FS}|c^+_{\mathbf{i}\sigma}c_{\mathbf{j}\sigma}|\mathrm{FS}\rangle} \qquad \text{for arbitrary } \mathbf{i} \neq \mathbf{j}. \tag{10.3}$$

Just these matrix elements appear in the real-space expansion of the band state occupation numbers

$$\langle\Psi|\hat{n}_{\mathbf{k}\sigma}|\Psi\rangle = \langle\Psi|c^+_{\mathbf{k}\sigma}c_{\mathbf{k}\sigma}|\Psi\rangle = \frac{1}{L}\sum_{\mathbf{j}}\sum_{\mathbf{h}} e^{i\mathbf{k}(\mathbf{j}-\mathbf{h})}\langle\Psi|c^+_{\mathbf{j}\sigma}c_{\mathbf{h}\sigma}|\Psi\rangle. \tag{10.4}$$

Considering that the matrix element depends only on $\mathbf{j}-\mathbf{h}$, and using (10.3), an easy manipulation yields

$$\begin{aligned}
\langle\Psi|\hat{n}_{\mathbf{k}\sigma}|\Psi\rangle &= \frac{1}{2} + \sum_{\mathbf{j}\neq\mathbf{0}} e^{i\mathbf{k}\mathbf{j}}\langle\Psi|c^+_{\mathbf{j}\sigma}c_{\mathbf{0}\sigma}|\Psi\rangle \\
&= \frac{1}{2} + q\left[\sum_{\mathbf{j}} e^{i\mathbf{k}\mathbf{j}}\langle\mathrm{FS}|c^+_{\mathbf{j}\sigma}c_{\mathbf{0}\sigma}|\mathrm{FS}\rangle - \frac{1}{2}\right] \\
&= \frac{1}{2}(1-q) + q\langle\mathrm{FS}|\hat{n}_{\mathbf{k}\sigma}|\mathrm{FS}\rangle\,,
\end{aligned} \tag{10.5}$$

leading to the remarkable result

$$\langle\Psi|\hat{n}_{\mathbf{k}\sigma}|\Psi\rangle = \frac{1}{2}(1-q) + q\cdot\Theta(\epsilon_{\mathrm{F}}-\epsilon_{\mathbf{k}}) \tag{10.6}$$

where Θ is the step function.

The essential feature is that *the band state occupation number of an interacting Fermi system still shows a Fermi step*, but its height is reduced from 1 to q, and the upper half of the band gets partially filled

---

[1]This holds only in the simplest version of Gutzwiller approximation.

with electrons (Fig. 10.1). We can rationalize the latter finding by saying that increasing localization in real space must be accompanied by delocalization in the reciprocal space.

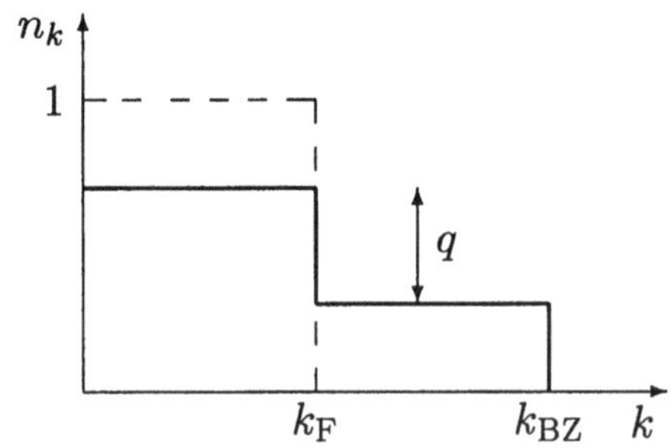

Figure 10.1: The band state occupation number n_k of an interacting Fermi system (thick lines) still shows a discontinuity at the Fermi wave vector k_F, but its height is reduced to $q < 1$. k_{BZ} is the Brillouin zone boundary. The Fermi step of the non-interacting system is shown with dashed lines.

It can be argued that the best[2] "order parameter" of the metallic, Fermi-liquid state is the height of the Fermi step. In our present approximation, the size of the Fermi step is given by (10.2). The behaviour is shown in Fig. 10.2. For non-interacting electrons, $q = 1$; in an interacting system, $q < 1$. As long as there is a finite Fermi step, however small, we still speak of a Fermi liquid. At the Brinkman–Rice transition $q \to 0$, and $\langle \hat{n}_k \rangle$ becomes flat. The finding $\langle \hat{n}_k \rangle = 1/2$ for all k if $U > U_{cr}$, is an extreme realization of the requirement that in an insulator, $\langle \hat{n}_k \rangle$ should be a smooth function of k.

We have seen two plots showing the behaviour on the metallic side of the Brinkman–Rice transition: the left panel of Fig. 9.3 and Fig. 10.2. At the level of the Gutzwiller approximation, these have the same status (actually, eqn. (C.19) shows that for $n = 1$, $q \sim n_d$ when n_d becomes small). However, if we look beyond the present simple treatment and recall what we understand about the Hubbard model from other

[2] Within the scope of our presentation. It has been pointed out [278] that it is more satisfactory to measure the metallic character by the sensitivity to changes in boundary conditions.

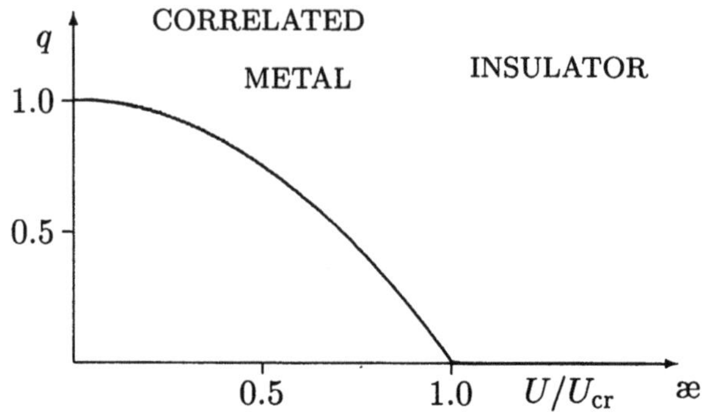

Figure 10.2: The height of the Fermi step as a function of $U/U_{\rm cr}$.

approaches, we can argue that while the n_d plot in Fig. 9.3 is certainly misleading, Fig. 10.2 should be basically right. We have pointed it out that the sharp transition seen in the n_d versus $U/U_{\rm cr}$ plot is an artefact of the approximation. On the other hand, it *is* true that the Mott insulator has no Fermi surface and thus can be characterized by $q = 0$. *If* there is a Mott transition at some finite $U_{\rm cr}$ then[3] $q > 0$ for $U < U_{\rm cr}$, and $q = 0$ for $U > U_{\rm cr}$. However, we will want to check whether the dependence of q on $U/U_{\rm cr}$ is like the one shown in Fig. 10.2.

10.2 Heavy Fermions: Half-Filled Band

The reduced Fermi step is a universal feature of interacting electron systems[4] but it is a theoretician's concept rather than an experimental observable. Compton scattering is supposed to yield the kind of information we need [257] but it measures the distribution over plane-wave states rather than over Bloch states, and identifying the signature of the Fermi step takes considerable effort even for light metals whose

[3]Actually, it may happen that the correlated metallic state has a Fermi surface but no finite Fermi step. The theory of the so-called Luttinger liquids demonstrates that the Fermi surface may be associated with an algebraic singularity, but no discontinuity, of the occupation number $n_{\mathbf{k}}$ [310].

[4]At least in $D = 3$, with the possible exception of a few exotic heavy fermion alloys.

conduction electron states are quite plane-wave-like [228]. For narrower bands which are better described in the tight binding framework, the measured Compton profile is dominated by the Fourier transform of the atomic wave functions, and it is difficult to make out the effect of the Fermi step. This is the reason why the lasting controversy about the (non) Fermi liquid nature of certain quasi-two-dimensional doped cuprates cannot be easily settled by measuring $n(\mathbf{k})$ [41].

A more immediate and indeed spectacularly useful connection with measurable properties is obtained by quoting a standard result of many-body theory: If the effective mass renormalization arises solely from the frequency dependence of the self-energy then the correlation enhancement of the effective mass m^* is predicted to be [90, 257]

$$\left(\frac{m^*}{m}\right)_{\text{corr.}} = \frac{1}{q} = \frac{1}{1-\left(\dfrac{U}{U_{\text{cr}}}\right)^2} \qquad \text{for } U < U_{\text{cr}}\,. \tag{10.7}$$

m^* is enhanced for $U > 0$, and diverges as $U \to U_{\text{cr}} - 0$. (With the disappearance of the Fermi surface, our definition of m^* breaks down for $U > U_{\text{cr}}$).

We were taking care to emphasize that we have discussed only one possible cause of the effective mass enhancement. In real physical systems, there are other possible causes such as the electron–phonon interaction, etc.

$$\left(\frac{m^*}{m}\right) = \left(\frac{m^*}{m}\right)_{\text{corr.}} \cdot \left(\frac{m^*}{m}\right)_{\text{el-phon}} \cdots \tag{10.8}$$

For simple metals, the enhancement factors on the right-hand side are close to 1, and it takes some trouble to disentangle them. For instance, for Na the experimental value is $m^*/m \approx 1.27$; and it is thought that $(m^*/m)_{\text{el-phon}} \approx 1.18$ and $(m^*/m)_{\text{corr.}} \approx 1.1$ [327]. However, our main interest lies in *strongly correlated systems* where the first factor is large and it can be claimed with some confidence that the main source of mass enhancement is the electron–electron interaction.

In solid state physics, several different definitions of the electron effective mass are used, so it should be made clear that here we are talking about the *thermal effective mass* which is directly measured by the coefficient γ of the *low-temperature electronic specific heat* $C_{el} = \gamma T$

$$\left(\frac{\gamma^*}{\gamma}\right) = \left(\frac{m^*}{m}\right)_{\text{corr.}} = \frac{1}{q}. \tag{10.9}$$

Strong electron correlation is manifested by a large enhancement of the low-temperature electronic specific heat. As we are going to discuss in more detail in Chapter 11, we can indeed find (particularly f-electron) metals with an anomalously large low-temperature specific heat. Alluding to the relationship between γ^* and m^*, these metals are called *heavy fermion systems.*

This enhancement of the specific heat coefficient γ can be looked upon as the enhancement of the (effective) density of states at the Fermi level. Using the simple rule that the density of states is inversely proportional to the bandwidth, we can also say [63] that an *effective narrow band* with width W^* has been created

$$\frac{W^*}{W} \sim \frac{\rho(\epsilon_F)}{\rho^*(\epsilon_F)} \sim q. \tag{10.10}$$

Finally, there must be a corresponding new temperature scale. The degeneracy temperature of the non-interacting system is $T_F \sim W$, usually much higher than room temperature. The arising of the effective narrow band is, however, a delicate quantum effect, having to do with carefully controlling the relative phases of the localized configurations in (9.16), and is thus an inherently low-temperature concept. An upper estimate for the temperature which is high enough to destroy the effective narrow band, is its degeneracy (or Fermi) temperature

$$\frac{T_F^*}{T_F} \sim \frac{W^*}{W} \sim q. \tag{10.11}$$

Let us envisage the temperature dependence of the entropy density S/L for a strongly correlated system not too far from half-filling. Raising T to T_F^* means that equipartition has been reached for the effective narrow band. However, narrowness implies that the suppression of double occupation is still in force, and thus the system consists mainly of neutral sites (spins). For U near U_{cr}, the non-degenerate Fermi system is a hot gas of uncompensated spins and a number of holes, with

$S/L \approx k_B \ln 2$. This can be thought of as an "intermediate high temperature limit" where the quantum coherence of the heavy quasiparticles has been lost but $k_B T$ is not large enough to compete with U. Since the effective spin–spin interactions are of $O(t^2/U)$, we infer that this temperature scale is $T_{\ln 2} \sim t^2/U k_B$. Much higher temperatures ($T_{\ln 4} \sim U/k_B$) are needed to saturate the entropy at the band value $k_B \ln 4$.

We can also argue backwards, and deduce the existence of heavy fermions from postulating that there are metallic systems with a spin entropy $\sim k_B \ln 2$ (or any other value of $O(1) \cdot k_B$) at T_F^*, which nonetheless do not order magnetically as the temperature is lowered. Then the entropy has to disappear in the Fermi liquid fashion [425]

$$S \propto \frac{k_B T}{T_F^*} \ln 2 \qquad \text{implying} \qquad \gamma^* \propto \frac{k_B \ln 2}{T_F^*}\,. \tag{10.12}$$

We certainly have a heavy fermion system if $T_F^* \sim$ 10–100K. The argument covers a large number of f-electron systems[5].

Remark. This is the right occasion to mention why we do not try to extend the Gutzwiller variational treatment to finite temperatures. Evaluating the partition function, and the entropy, would require the evaluation of a sum over all excited states. Since the interacting ground state is expressed as the Gutzwiller-projected Fermi sea ground state, it would seem natural to seek the excited states as Gutzwiller-projected $U = 0$ excitations. However, in the large-U limit we find a severe discrepancy: there are $\sim 4^L$ unprojected states, but only $\sim 2^L$ projected states [336]. There is no satisfactory way to overcome this counting problem. However, it is possible to devise low-temperature interpolation schemes which give reasonable results up to T_F^* [125, 356, 373].

In the theory of simple metals, $\rho(\epsilon_F)$ enters also the expression of the Pauli susceptibility [209], so we may be wondering whether it is enhanced like γ^*. Within the present approximate treatment, this is indeed the case: Brinkman and Rice [51] found that the dominating U-dependence of the spin susceptibility χ_s comes from the mass en-

[5]It should be remarked that, in point of fact, many f-electron systems undergo a magnetic ordering transition but the ordered moments are so small that the onset of magnetism cannot account for the disappearance of entropy.

hancement factor

$$\chi_s \sim \chi_P \cdot \frac{m^*}{m} \tag{10.13}$$

where χ_P is the non-interacting value of the Pauli susceptibility. Like γ^*, χ_s diverges as we approach the metal–insulator transition. The derivation of the above result is relegated to Problem 10.1.

The *Wilson ratio* measures the relative enhancements of the specific heat coefficient and the spin susceptibility, normalized to the free electron value

$$R_W = \left(\frac{\chi_s}{\gamma^*}\right) \cdot \left(\frac{3\mu_B^2}{\pi^2 k_B^2}\right)^{-1} . \tag{10.14}$$

According to the Brinkman–Rice theory, the Wilson ratio approaches a constant of the order of 1 at $U \to U_{cr}$. This shows that the large enhancements of the specific heat and the spin susceptibility have the same origin. The relevant instability of the system is the transition into a Mott-insulator. This should be contrasted with the behaviour seen at a ferromagnetic instability. There too, the spin susceptibility (7.23) was found to diverge; however, it was not accompanied by a diverging m^*.

Let us recall the U-dependence of the instantaneous magnetic moment (9.48) (see Fig. 9.4). The density of uncompensated spins is mildly increased but this cannot account for the large enhancement of the susceptibility. The latter is to be ascribed to the reduction of the bandwidth which increases the lifetime $\sim \hbar/W^*$ of the moments (longer-lived moments can be polarized more efficiently). Furthermore, $\langle (M_j^z)^2 \rangle$ appears in the numerator of a Curie-type susceptibility at relatively low temperatures $T \geq T^*$ (see (7.109) and the accompanying discussion on p. 407.). Large and strongly T-dependent Pauli-susceptibilities at $T < T_F^*$, and localized-moment behaviour at $T > T_F^*$, are wellknown characteristics of heavy fermion materials [375].

The reader should be reminded again that here we are speaking about the results of an approximate treatment and though some of the conclusions are confirmed by more elaborate methods, some others have to be corrected. Recent work on the infinite-dimensional Hubbard model indicates that though the uniform susceptibility may become large as we approach the metal–insulator transition, it does not diverge in proportion to m^*. On the other hand, the divergence of m^* is confirmed, and (10.14) gets replaced by the analogous relationship (10.74) in which the local susceptibility appears. More about this in Sec. 10.6.5.

Let us conclude the discussion of the correlated metallic state with a result about the dielectric properties. We recall the standard formula for the Thomas–Fermi screening length [209]

$$\lambda_{\mathrm{TF}}^2 = 3n^{1/3}/a_{\mathrm{H}} \tag{10.15}$$

where a_{H} is the Bohr radius. Now, it is our intuitive understanding that in the interacting case, it is only the polarity fluctuations that can carry current, or shift their position to contribute to screening, and thus n, the total electron density, should be replaced essentially by n_d, the density of doubly occupied sites. Such was indeed the finding of Rice and Brinkman [334]: near the metal–insulator phase boundary

$$(\lambda_{\mathrm{TF}}^*)^2 \approx \frac{U_{\mathrm{cr}}}{8\pi e^2}\left(1 - \frac{U}{U_{\mathrm{cr}}}\right)^{-1} . \tag{10.16}$$

As $U \to U_{\mathrm{cr}}$, the screening radius diverges, indicating that the system becomes less and less capable to screen external charges, i.e., less and less metallic. – This is an occasion to recall that in the Brinkman–Rice scenario, the metal–insulator transition is continuous; the results (vanishing n_d and q, diverging m^* and λ_{TF}) may remind us of those obtained in the usual mean-field description of second-order phase transitions. Remember, however, that here we are speaking about a ground state phase transition driven by a change in U/t, and not about a conventional phase transition driven by a change in temperature.

The Effect of a Strong Magnetic Field

The equilibrium properties of usual wide-band metals or wide-gap semiconductors change only very little if we place the sample in a magnetic field. The reason is that the energy of interaction with even the highest attainable laboratory fields is minute in comparison with the $\sim 1e$V energy scale of the electronic structure.

Strong correlation tends to introduce small energy scales. As shown by (10.10), heavy fermion metals have a renormalized bandwidth which vanishes as we approach the Brinkman–Rice transition. Later we are going to meet correlated insulators with extremely narrow gaps. The Zeeman energy in a field of, say, 10T may well turn out to be comparable with such small energies, and thus the field may have a drastic influence on the electronic structure. Strongly correlated systems often show an interesting and complicated behaviour in magnetic field.

Vollhardt [431] noticed that switching on the magnetic field has a similar effect as increasing U: it drives the system towards the fully localized $n_d = 0$ state (the field polarizes the electron spins, and parallel-spin electrons cannot go to the same lattice site). Thus in a finite field, a smaller U is sufficient to drive the system through a metal–insulator transition. If the system is near the Brinkman–Rice transition, then even small fields can have a drastic effect.

Let us see what happens when we approach the transition from the metallic side. The first effect of increasing the field is a departure from linearity in the magnetization curve. The lowest-order correction can be described by introducing the third-order nonlinear susceptibility

$$m \approx \chi_s H + \chi^{(3)} H^3 . \tag{10.17}$$

It is found that

$$\chi^{(3)} \propto \frac{m^*}{m} \cdot \chi_s^3 \tag{10.18}$$

(see the solution to Problem 10.1 for a derivation of this relationship). Upon approaching the Brinkman–Rice transition, $\chi^{(3)}$ diverges like

$$\chi^{(3)} \propto (U_{cr} - U)^{-4} , \tag{10.19}$$

so that it quickly becomes very large, signalling a substantial deviation from linearity. Furthermore, for U sufficiently near U_{cr}, $\chi^{(3)}$ is positive, i.e., the magnetization curve has a strong upward curvature. By going beyond a power expansion in H, we find that (at least in a range of U) the field can even cause a first-order transition: the system jumps from a metallic state with a finite n_d straight into the insulating state which has $n_d = 0$, and saturated magnetization. It is reassuring to find that this *metamagnetic transition* is not an artefact of the Gutzwiller approximation but a true feature of the Hubbard model: at very low temperatures, evidence of a sharp metamagnetic transition was found for the infinite-dimensional Hubbard model [231]. A simple demonstration of the metamagnetic behaviour is given in Problem 10.2.

Problem 10.1 Derive the relationship (10.13) for the spin susceptibility using the results of Appendix C. For the sake of simplicity, take a band with a constant density of states at half-filling. Discuss the result a) at small U, and b) as $U \to U_{\rm cr}$.

Expanding the magnetization to third order in powers of the magnetic field allows to deduce whether the magnetization curve has a downward, or an upward curvature. What do you find?

Problem 10.2 Extend the treatment developed in Problem 10.1 to fields of arbitrary strength, and calculate the magnetization curve.

10.3 Arbitrary Band Filling

Up to now, we were speaking about the correlated metallic state which is found in the pre-Mott-transition region. The unique feature of the half-filled case is that m^* diverges as U approaches a finite critical value $U_{\rm cr}$ from below. It is clear, however, that the metallic state becomes strongly correlated at large U for arbitrary band filling, even thogh there is no Mott transition for $n < 1$.

The formal description of the $n < 1$ case requires that we abandon the simple treatment of Sec. 9.2.2. Actually, the cluster method can be generalized for arbitrary filling [332, 312, 104]. An alternative is to use a different formulation of the Gutzwiller approximation (GA) which emphasizes that it is based on neglecting the intersite correlations which are caused by antisymmetrization (i.e., by the Fermi statistics). The formal expression of this is that the GA uses configurational averages: we pick out one or two sites, and average over the configurations of electrons at all other sites[6]. In order not to burden the main text with cumbersome formalism, we relegate to Appendix C the description of the averaging procedure, and quote here just the results.

The hopping amplitude renormalization factor (or the Fermi step) is given by (C.19)

$$q = \frac{2(n - 2n_d)}{n(2 - n)} \cdot [\sqrt{n_d} + \sqrt{1 - n + n_d}]^2 \tag{10.20}$$

[6]Doing the configurational average amounts to characterizing the "bath" which surrounds the pair of sites in Fig. 9.2. Thus this formal interpretation of the GA justifies the introduction of the simple cluster method we used previously.

where n_d is the density of doubly occupied sites. The dependence of n_d on n and U/t is obtained from the numerical solution of the optimization problem. However, we get a simple expression for $U \to \infty$ when (for $n < 1$) $n_d \to 0$

$$(q)_{U\to\infty} = \frac{2(1-n)}{2-n}. \tag{10.21}$$

It stands to reason that the result should be proportional to $1-n$, since this is the probability of finding a site empty. Let us consider the hopping of an $\uparrow$-spin electron. If the neighbouring site is also occupied by an $\uparrow$-spin electron, the electron cannot hop there because of the Pauli principle. If it is occupied by a $\downarrow$-spin electron, hopping is impossible because of the infinitely large U. Thus hopping can take place only onto an empty site. All methods for treating the strong correlation problem give the factor $1-n$. However, they may give different results for the constant of proportionality. Indeed, it is not obvious how to interpret the factor $1/(1-n/2)$ that we can read off from (10.21). Why do not we get simply $q = 1-n$?

One can argue that *some* factor should accompany $1-n$. We gain insight into this question by generalizing the variational treatment to the case of a Gutzwiller-projected spin-polarized Fermi sea. Let us allow that the $\uparrow$-spin and $\downarrow$-spin electron densities differ: $n_\uparrow \neq n_\downarrow$. The total electron density is $n = n_\uparrow + n_\downarrow$, and the state is homogeneous. The spin-dependent q_σ is given in (C.20). Taking the limit $U \to \infty$, and $\sigma = \uparrow$, we get

$$(q_\uparrow)_{U\to\infty} = \frac{1-n}{1-n_\uparrow}. \tag{10.22}$$

In the non-magnetic case $n_\uparrow = n_\downarrow = n/2$ and (10.21) is recovered. On the other hand, if we consider the fully spin-polarized case $n_\uparrow = n$, $q_\uparrow$ becomes 1. It is clear why: if all electrons have $\uparrow$-spins, then the Pauli principle (which has been taken into account by the antisymmetrical nature of the wave function of the Fermi sea) guarantees that two electrons never sit on the same site, and the Hubbard U does not play any role. Thus (10.21) and (10.22) can be thought to express that the hopping amplitude is reduced by the correlation effect according to the factor $(1-n)$, to the extent that double occupation is permitted by the Pauli principle.

For $n \neq 1$, the system remains a Fermi liquid for arbitrarily large U. Increasing U from zero, q decreases relatively fast until U reaches the order of magnitude of $U_{\rm cr}$, and afterwards gradually approaches the asymptotic value (10.21). The typical behaviour is illustrated for $n = 0.8$ in Fig. 10.3 (left). The calculation was done for a band with a constant density of states; for $n = 1$, the Brinkman–Rice transition sets in at $U_{\rm cr} = 2W$, where W is the bandwidth. Fig. 10.3 (right) shows the development of the Fermi step. The $U \to \infty$ limit is

$$n_{\mathbf{k}} = \frac{n^2}{2(2-n)} + \frac{2(1-n)}{2-n}\,\Theta(\epsilon_{\rm F} - \epsilon_{\mathbf{k}})\,. \tag{10.23}$$

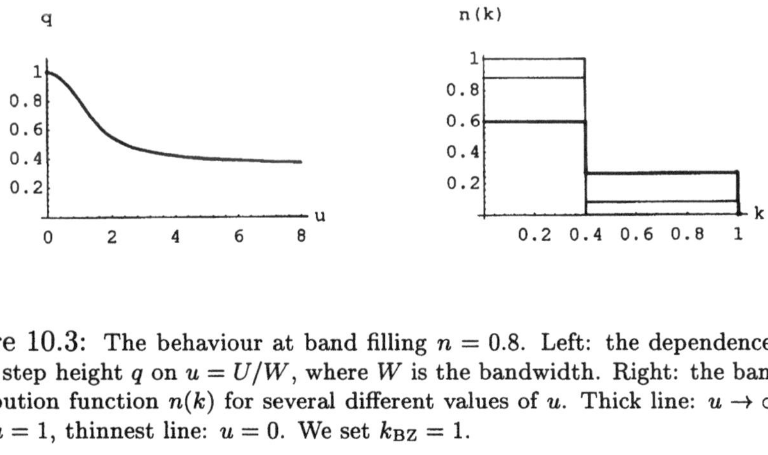

Figure 10.3: The behaviour at band filling $n = 0.8$. Left: the dependence of the Fermi step height q on $u = U/W$, where W is the bandwidth. Right: the band state distribution function $n(k)$ for several different values of u. Thick line: $u \to \infty$, thin line: $u = 1$, thinnest line: $u = 0$. We set $k_{\rm BZ} = 1$.

The relationship for the correlation enhancement of the effective mass

$$\frac{m^*}{m} = \frac{1}{q} \tag{10.24}$$

allows us to map out m^* in the entire n–U plane. The result for the case of a band with a constant density of states is shown in Fig. 10.4. The Mott-insulating region lies at $n = 1$, $U/W \geq 2$. The contour plot shows that m^* diverges if we approach the Mott-insulating state from any direction. If we set $n = 1$, (10.7) describes the divergence with increasing U. We can also fix a $U > U_{\rm cr} = 2W$, and change n

towards half-filling. At very large U, (10.21) shows[7] that m^* diverges as $m^* \sim |1-n|^{-1}$. It is easy to verify that the divergence is controlled by the factor $|1-n|^{-1}$ for arbitrary U (Problem 10.3).

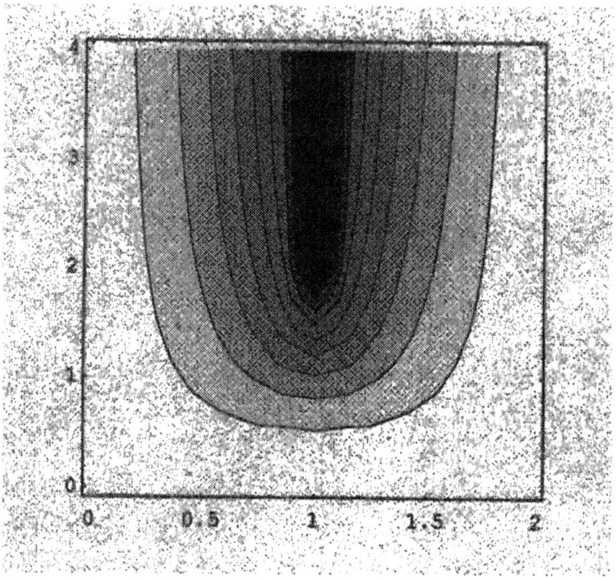

Figure 10.4: Contour plot of the correlation enhancement of the effective mass in the n–U/W plane. Darker shading indicates higher effective mass. The Brinkman–Rice transition occurs at $n = 1$, $U/W = 2$. The effective mass diverges if we approach the Mott insulating region either with increasing U, or with changing n towards half-filling.

An important corollary is that *we expect to see heavy electron behaviour in a system which is "doped away" from a Mott insulating state.* This is the result which is relevant for the vast majority of the experimentally found heavy fermion systems. It is difficult to realize a "fine-tuned" situation where the electron concentration is just right for a Mott insulator and at the same time, U is slightly below its critical value. However, it is relatively easy to find systems where U is large to begin with, and the concentration of the strongly correlated electrons does not deviate too much from an integral value. Remembering that

[7]For $n < 1$. For $n > 1$, we have to refer back to (10.20) and take into account that for $U \to \infty$, $n_d \to n - 1$.

the valence of an ion can often be defined by counting its strongly correlated (d-, or f-) electrons, we are led to expect that "nearly integral valent" systems tend to show heavy fermionic features. As we are going to discuss in Chapter 11, such is indeed the case.

Now we can return to the question posed in Sec. 9.3: how to calculate the Mott–Hubbard gap? Generally speaking, the variational approach is devised to determine the ground state. Furthermore, we are dealing here with a large system and we learned from Appendix C that in the Gutzwiller approximation, expectation values are obtained with a relative error $O(1/L)$. Thus an extensive quantity like the ground state energy is obtained with an error $O(1)$. Usually it does not matter, because what we are really interested in is the energy density which we get accurately. However, if we wanted to determine an excitation energy (of order $O(1)$) as the difference of two total energies (each of order $O(L)$), then we would be in trouble because the quantity we want would be of the same order of magnitude as the errors in the two energies! The point to notice is that the gap is a very special kind of excitation energy which can be derived also from the n-dependent ground state energy density. Namely, just as for usual semiconductors, the gap can be identified as the discontinuity of the chemical potential μ which, in turn, can be obtained from the ground state energy density as $\mu = (\partial\mathcal{E}/\partial n)$. Thus the gap is

$$\Delta = \left(\frac{\partial\mathcal{E}}{\partial n}\right)_{n\to 1+0} - \left(\frac{\partial\mathcal{E}}{\partial n}\right)_{n\to 1-0} . \tag{10.25}$$

The actual calculation is left to Problem 10.3 and it leads to (9.54).

The Gutzwiller projection suppresses charge fluctuations and increases the number of cells with an uncompensated spin. We understand that since hopping is tantamount to charge fluctuation, the hopping amplitude must be reduced from its Fermi sea value, as shown in (10.1). For the spin degrees of freedom we expect the reverse effect, namely, that Gutzwiller projection magnifies the spin correlations which are present even in the noncorrelated Fermi sea. The effect on the first-neighbour spin–spin correlation function can be expressed as

$$\langle\Psi(\eta)|\mathbf{S_i}\cdot\mathbf{S_j}|\Psi(\eta)\rangle = g_{SS}(\eta)\cdot\langle\mathrm{FS}|\mathbf{S_i}\cdot\mathbf{S_j}|\mathrm{FS}\rangle. \tag{10.26}$$

It has been found [467] that for $U \to \infty$

$$g_{SS}(\eta = 0) = \frac{4}{(2-n)^2}. \tag{10.27}$$

Problem 10.3 Use the result (10.20) to calculate the Mott–Hubbard gap as a function of U! Assume that the band has a constant density of states.

What is the behaviour of m^* as $n \to 1$ for $U > U_{\text{cr}}$?

Problem 10.4 Derive the spin correlation renormalization factor (10.27) using the method developed in Appendix C.

10.4 The Fermi Volume

Looking at the Figures 10.1 and 10.3, we observe that, for any $q > 0$, the Fermi step appears at the same k_{F} that we had at $U = 0$. The position of the discontinuity (or generally, any kind of non-analytic behaviour) of $n_{\mathbf{k}}$ defines a surface in **k**-space, namely, the Fermi surface. In the Gutzwiller approach we find that switching on the Coulomb interaction does not change the Fermi surface. The statement that, as long as interactions do not lead to a non-analyticity (i.e., a phase transition), *the Fermi volume remains unchanged*, is known as *Luttinger's theorem*. Here we cannot claim to have derived the theorem because the argument leading to (10.6) was anything but rigorous. However, we can at least say that the Gutzwiller variational method respects Luttinger's theorem.

Note that (10.6) implies more than what is required by Luttinger's theorem: namely that the precise shape of the Fermi surface remains unchanged as the interaction is turned on. We have reasons to believe that this is unreasonably restrictive. Luttinger's theorem would allow that switching on U causes a change of the shape of the Fermi surface, only the enclosed volume is preserved. There are indications that this is indeed what happens [298].

Let us assume that $U > U_{\text{cr}}$, and let us consider what happens to the Fermi volume if n is varied. For $n < 1$, the ground state is metallic, and the Fermi volume increases with increasing n. As $n \to 1$, the limiting value of the Fermi volume is half of the volume of the Brillouin zone. However, right at $n = 1$, we have an insulator which has no Fermi surface at all. We can also say that the Fermi volume of an insulator is zero. For $U > U_{cr}$, the Fermi volume behaves discontinuously at $n = 1$.

Let us note that in this respect, doping a Mott insulator is very different from doping an ordinary semiconductor. In either case, the Fermi volume of the insulator is zero. If we dope a semiconductor by adding a few electrons or holes, a small Fermi surface arises which encloses a volume corresponding to the number of added charge carriers. In this case, the smallness of the Fermi volume shows that we are near to an (ordinary) insulating state. In contrast, if we dope a Mott insulator, its Fermi volume immediately jumps to a large value: the Fermi volume is proportional to n (and not to $1-n$, what we would expect if the volume corresponded to the number of the doped-in carriers). The variational treament predicts that *the strongly correlated metal has a "large" (or Luttinger) Fermi surface.*

Again: the variational method is not rigorous so we cannot be certain that our finding is a true feature of the Hubbard model. It is, however, reassuring to know that there are important cases (one of them will be described in the next section) where experiments confirm that near to a Mott insulating state, the Fermi volume has its Luttinger value. Thus we can claim that the state with a large Fermi surface is at least one of the possible states of the system, and the Gutzwiller method can be usefully applied for the description of this state.

10.5 The $La_{1-x}Sr_xTiO_3$ System

The essential conclusion of the foregoing discussion is that a lightly doped Mott insulator should be a heavy Fermi liquid with a large Fermi surface. The behaviour of the $La_{1-x}Sr_xTiO_3$ system seems to fulfill our expectations [401].

The perovskite $LaTiO_3$ is an antiferromagnetic[8] Mott insulator with a small gap [400]. The ideal (cubic) coordination of the Ti ion was shown in Fig. 3.1. Measurements show that $LaTiO_3$ is not exactly cubic; it has a slight orthorombic distortion. The orbital momentum seems to

[8]From our point of view, it is an unessential complication that $LaTiO_3$ seems to be weakly ferromagnetic due to a small canting of the almost antiparallel sublattice magnetizations.

be quenched, thus to a good approximation, the magnetic properties of $LaTiO_3$ are ascribed to the $S = 1/2$ spins of the Ti^{3+} ions (see, however, footnote 13). We want to understand the $LaTiO_3$ system in terms of our previous results for the Hubbard model.

Establishing a connection between theoretical results for the non-degenerate Hubbard model, and the behaviour of a real compound with a lot of electrons in different bands, has to rely on a number of simplifying assuptions. Stoichiometric $LaTiO_3$ has one $3d$ electron per chemical unit and thus, loosely speaking, it can be thought of as realizing the $n = 1$ case for a Hubbard model. In truth, it should be a degenerate Hubbard model. Nominally, the $3d$-band would have fivefold orbital degeneracy but the large cubic field is expected to split it into t_{2g} and e_g subbands, with the threefold degenerate t_{2g} subband lying lower. We have not discussed the extension of the previous results to degenerate Hubbard models but we should not be overly worried about this. The integer filling $n = 1$ permits a Mott transition at arbitrary orbital degeneracy, and it is contended that at fillings $n \leq 1$ the behaviour looks much the same[9] as in the non-degenerate case [225].

It has been argued that substituting a fraction of the La atoms by Sr is a particularly neat way of doping a Mott insulator: according to the chemical formula $La_x^{3+}Sr_{1-x}^{2+}Ti_x^{3+}Ti_{1-x}^{4+}O_3^{2-}$ it produces the filling $n = x$ in our effective Hubbard model. Note that the interacting electrons move on the Ti sublattice, while the doping happened on the La sublattice. Thus the disorder potential of the randomly distributed Sr-ions should be relatively weak. When doping a Mott insulator, it is always a problem how to achieve non-integer filling without introducing a strong disorder potential which would change the physics drastically. In this respect, the recently discovered $La_xSr_{1-x}TiO_3$ system seems to be cleaner than the celebrated Mott–Hubbard system $V_{2-x}Cr_xO_3$ (cf. Sec. 5.5.1). In the latter, disorder is produced on the same sublattice where the interacting electrons reside.

$La_xSr_{1-x}TiO_3$ is apparently a Fermi liquid up to $x \simeq 0.95$ [401]. The inset in the left panel of Fig. 10.5 demonstrates that a metal–insulator

[9]However, the story may be quite different for $1 < n < 2$, as shown by our discussion of double exchange (Sec. 8.5).

transition must be taking place somewhere[10] between $x \simeq 0.95$ and x=1. As discernible from the left part of Fig. 10.5, the low-temperature metallic resistivity is dominated by a quadratic contribution

$$\rho(T) \approx \rho_0 + AT^2. \tag{10.28}$$

The T^2-term of the resistivity can be ascribed to electron–electron scattering [3]. It is thought to be a common characteristic of Fermi liquids, and it has been detected even in simple metals like aluminum and indium, where $A \approx 5 \cdot 10^{-13}\Omega\text{cm/K}^2$ (Al), and $5 \cdot 10^{-12}\Omega\text{cm/K}^2$ (In), respectively [122]. In weakly correlated systems, this term is much smaller than the contribution coming from electron–phonon scattering, and it takes careful measurements at helium temperatures to detect it. In contrast, $A \approx 10^{-9}\Omega\text{cm/K}^2$ for $La_{0.7}Sr_{0.3}TiO_3$, and the T^2-term is the dominant contribution to the resistivity up to at least 200K. Generally speaking, in strongly correlated systems A is found to be hugely enhanced, roughly following [197]

$$\frac{A^*}{A} \sim \left(\frac{m^*}{m}\right)^2. \tag{10.29}$$

The enhancement of A is one of the most spectacular correlation effects, and it is one of the hallmarks of heavy fermion metals (see also Fig. 11.13 (right) on p. 627).

We content ourselves with the most handwaving of arguments to see where a resistivity contribution $\sim AT^2$ may originate from. In the framework of the Landau theory of Fermi liquids[11], the simplest phase-space argument gives that in three-dimensional systems, the inverse electron lifetime $1/\tau \sim \rho(\epsilon_F)T^2 \rightsquigarrow m^*T^2$. It takes some arguing why the quasiparticle lifetime τ should be essentially the same as (or proportional to) the transport relaxation time τ_{tr} which appears in the standard formula for the conductivity $\sigma = ne^2\tau_{\text{tr}}/m \rightsquigarrow ne^2\tau_{\text{tr}}/m^*$. Let us note that if the Fermi liquid existed on a homogeneous background then because of $\mathbf{j} \propto \mathbf{v} \parallel \mathbf{k}$ the conservation of the total momentum

---

[10] Even a weak disorder should suffice to explain that the insulating regime extends over a finite range of x.

[11] For the present purposes, it is sufficient to refer to the short account given in Ch. 17 of [32]. A detailed description is given in [327].

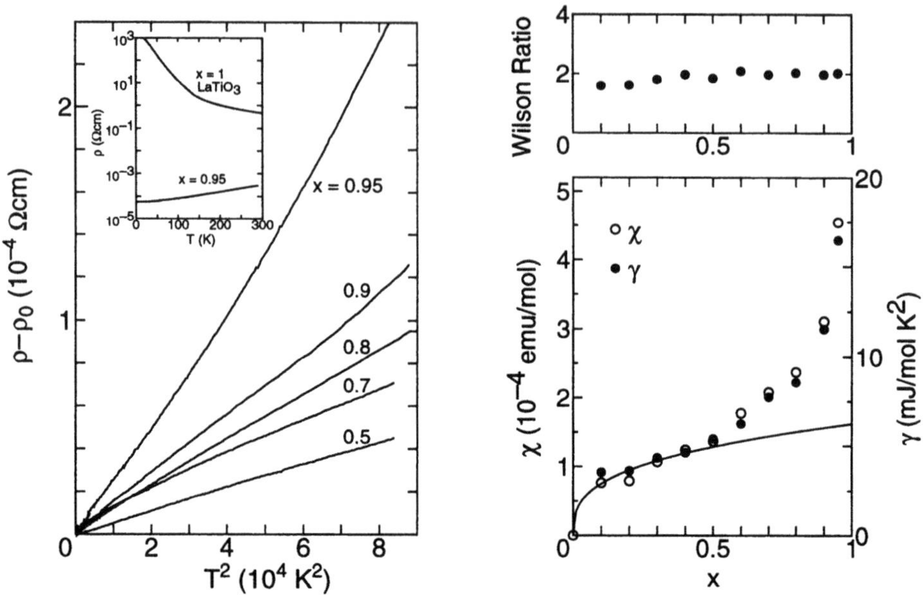

Figure 10.5: The behaviour of the $La_xSr_{1-x}TiO_3$ system, after [401]. *Left*: the resistivity (in units of Ωcm) as a function of T^2, for several values of x. The large quadratic resistivity characteristic of strongly correlated systems is clearly seen. Inset: a comparison of the T-dependence of the resistivities (logarithmic scale) of the nominally stoichiometric sample, and of the $x = 0.95$ sample shows that $LaTiO_3$ is an insulator while $La_{.95}Sr_{.05}TiO_3$ is a metal. – *Right, lower part*: Experimental results for the low-temperature susceptibility χ (empty circles) and the specific heat coefficient γ (filled circles) of $La_xSr_{1-x}TiO_3$, as a function of x. The continuous line gives an impression of the x-dependence of the one-electron density of states. *Right, upper part*: The Wilson ratio is approximately constant, showing that m^* and χ are similarly enhanced.

would be equivalent to the conservation of the current, and the conductivity would be infinite[12] even at finite temperatures. However, if the interacting fermions are moving on the background of a lattice, then we immediately find two reasons why the current should decay: the fact that **v** and **k** need not be parallel, and Umklapp scattering. Thus we expect that $\tau_{tr} \sim \tau$ is justified. Actually, in finite-dimensional systems the band filling has to exceed a threshold value for the Umklapp processes to become effective [414] but three dimensions is high enough and the

[12]This would be the so-called ideal conductivity of non-condensing electrons, not superconductivity!

condition is apparently satisfied for the band fillings shown in Fig. 10.5. Furthermore, n is fixed by Luttinger's theorem to be the same as for the non-interacting system, thus for the resistivity $\rho = 1/\sigma \sim (m^*/m)^2 T^2$ follows.

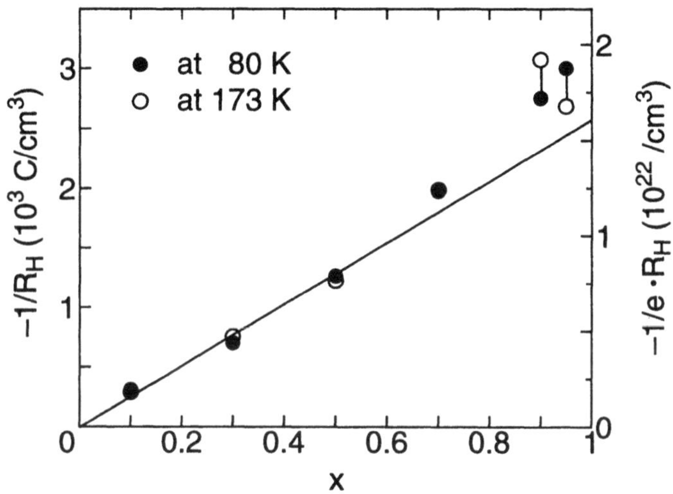

Figure 10.6: The inverse Hall coefficient as a function of x (after [401]). Filled and open circles show data taken at 80K, and 173K, resp. The solid line shows the behaviour calculated on the assumption that each La^{3+} site contributes one electron-like carrier. The observed behaviour indicates a Luttinger Fermi volume.

We show the x-dependence of the low-temperature specific heat coefficient γ, and of the enhanced Pauli susceptibility χ in the right-hand panel of Fig. 10.5. Some of the x-dependence is a one-electron effect, reflecting the x-dependence of the density of states. The continuous line in Fig. 10.5 indicates the result expected for non-interacting electrons. The correlation enhancement is identified as the ratio of the measured value and the estimated independent-electron result. The data show that both γ and χ are progressively enhanced as x approaches 1. Actually, as far as one is able to judge from data limited to $x \leq 0.95$, the observed behaviour is suggestive of a diverging enhancement of the kind $\gamma \sim \chi \sim (1-x)^{-1}$, just what we would guess from (10.21). The approximately x-independent Wilson ratio (upper part on the right) also shows that, within the range accessible to the measurement, the enhancements of the susceptibility and of the specific heat follow the same trend[13].

The resistivity plots show that with increasing x, the system becomes less and less metallic. In principle, this could arise from either a descreasing number of charge carriers, or from an increased scattering rate. The decisive experiment is the measurement of the Hall coefficient which, according to $R_H \sim 1/n|e|$, gives the concentration and character of the charge carriers. The data shown in Fig. 10.6 clearly prove that the charge carriers are electron-like and their concentration is $n \sim x$. This finding is in accordance with our reasoning (Sec. 10.4) that the strongly correlated metallic state has a large (Luttinger) Fermi surface. The number of charge carriers keeps on increasing right up to the metal–insulator transition where it suddenly drops to zero.

10.6 Discussion and Outlook

On the whole, we may be inclined to feel satisfied with the progress made in this chapter: we developed a pictorial way of thinking about correlation phenomena, obtained a number of nice results with relatively little effort, and these seem to be applicable to real physical systems. Remembering, however, that this was based on an approximate treatment of a simple-looking trial state for a model Hamiltonian, we should be worried about the following points:

1. Is the approximation good enough? Our results do not follow strictly from the Gutzwiller trial state (9.16) but were derived using the Gutzwiller approximation. What is the nature of the Gutzwiller approximation? Does it ever become exact? If it is not exact, did we merely make a small quantitative mistake, or is it possible that the true consequences of the Gutzwiller Ansatz are drastically different from our findings?

2. Is the trial state good enough? We were pointing it out that getting $n_d \to 0$ at the Brinkman–Rice transition must be wrong. Could

[13] This should be basically right but we might be troubled by the following observation: the approximately constant value of the Wilson ratio is ~ 2 down to $x \sim 0.1$ where the electron gas is dilute and we would rather expect that the ratio approaches the free-electron value ~ 1. It can be argued that the observed behaviour can be ascribed to a contribution from the hitherto neglected orbital magnetic moment [130].

we have done better by introducing a more complicated trial state? Can we get a meaningful description of the insulating state? And what about magnetic long-range order?

We do not raise here the even more fundamental question whether the (non-degenerate) Hubbard model is good enough. In the previous chapters, we had ample opportunity to convince ourselves that orbital degeneracy, and the extra terms included in the extended Hubbard model, can be important. However, the simplest Hubbard model should suffice for the discussion of the metal–insulator transition.

10.6.1 Gutzwiller Method: Exact Treatment

For decades, working with the Gutzwiller Ansatz involved approximations of unknown quality (such as the Gutzwiller approximation), or numerical evaluation for finite systems, or series expansions of limited range of validity. Since one of the main objectives was to describe such a subtle phenomenon as the metal–insulator transition, which (if it happens at all) is certain to be associated with non-analytic behaviour at some intermediate value of U, the lack of secure knowledge was most worrying. The matter was finally settled when Vollhardt and coworkers [271] devised a method for carrying out the Gutzwiller variational procedure exactly, for arbitrary interaction strength, and lattice dimensionality. Unfortunately, the solution is far too lengthy and complicated to be included here; we have to content ourselves with quoting some of the essential results.

Though the scheme works for any dimension, closed-form results can be obtained only for $D = 1$, and $D = \infty$. It turns out, however, that the character of the results is the same for all finite D. Thus it is enough to cite the $D = 1$ results to get an impression of the behaviour in finite dimensions.

The first step is to express the expectation values in terms of the variational parameter η; the second is to determine the optimum value of η. It is interesting that at half-filling ($n = 1$) the discontinuity of the band state occupation number $n_{k\sigma}$ is

$$q(\eta) = \frac{4\eta}{(1+\eta)^2} , \tag{10.30}$$

the same as eqn. (9.43) which we derived in our simple cluster approximation! Furthermore, (10.30) turns out to be valid for hypercubic lattices in arbitrary dimensions. However, such coincidences between the exact results, and those obtained in the Gutzwiller approximation are exceptional. Typically, the exact results are complicated, and depend on D. We cite the expression for the density of doubly occupied sites

$$n_d(\eta, n, D = 1) = \frac{\eta^2}{2(1-\eta^2)^2}\left\{-\ln\left[1 - n(1-\eta^2)\right] - n(1-\eta^2)\right\}. \tag{10.31}$$

Replacing $n = 1$, and assuming that η is small, we can write

$$n_d(\eta, n = 1, D = 1) \approx -\eta^2 \ln \eta. \tag{10.32}$$

The essential thing about the small-η behaviour is that it is not linear in η. We remind the reader of the discussion on page 522, according to which the Brinkman–Rice transition resulted from both terms of the energy expression being $\propto \eta$ for small η. Now we have no right to expect a Brinkman–Rice transition: n_d remains non-zero for arbitrarily large U/t. Using (10.30) and (10.32), the small-η energy expression can be written as

$$E_{D=1}(\eta) \approx -U\eta^2 \ln \eta - \beta t \eta \tag{10.33}$$

with some constant β. It is minimized by

$$\eta \approx -\frac{\beta t}{2U}\left(\ln \frac{\beta t}{2U}\right)^{-1} \tag{10.34}$$

from which

$$n_d \approx \frac{\beta^2 t^2}{4U^2}\left(\ln \frac{2U}{\beta t}\right)^{-1} \quad \text{and the Fermi step} \quad q \approx \frac{2\beta t}{U}\left(\ln \frac{2U}{\beta t}\right)^{-1}. \tag{10.35}$$

Observe that the Fermi step is finite for arbitrarily large U, thus the system is always metallic[14].

[14] We should keep in mind that here we are discussing the character of the variational solution and not the truthful picture of the Hubbard model. The exact solution for $D = 1$ shows the half-filled system to be insulating for all $U \neq 0$. The gap is associated with antiferromagnetic quasi-long-range order.

The detailed form of n_d for other dimensions is, perhaps, different from (10.31) but whatever it may be, it can be stated that the leading term in the small-η expansion is never linear in η, as long as the lattice dimensionality is finite. According to the exact treatment of the Gutzwiller variational problem, in finite-dimensional systems there is no Brinkman–Rice transition at all!

The nature of the Gutzwiller approximation became clarified by observing that it becomes exact for $D = \infty$. Thus all our previous results are found to be correct within the Gutzwiller variational description of the infinite-dimensional Hubbard model. It follows that it is true that at half-filling, a Brinkman–Rice transition is taking place in the manner described in Sec. 9.3.

The reason why the quality of the Gutzwiller approximation improves as we increase the dimensionality D, is the following: as we argue in Appendix C, the approximation is based on neglecting the intersite correlations caused by the antisymmetrical nature of the wave function, i.e., the exchange hole effect. The exchange hole is quite long-ranged in low dimensions, but it cuts off sharply in high dimensions and, roughly speaking, shrinks to the central site as $D \to \infty$. This justifies using the Gutzwiller approximation which is basically a single-site approximation.

At exact half-filling, the infinite-dimensional limit is an isolated case: there is a Brinkman–Rice transition for $1/D = 0$ but not for any $1/D \neq 0$ [420]. Therefore, just as far as the transition is concerned, we cannot make a meaningful $1/D$-expansion.

Returning to the case $D = 1$, the leading term of the ground state energy is found to be

$$\frac{\langle \mathcal{H} \rangle}{L} = -\left(\frac{4}{\pi}\right)^2 \frac{t^2}{U} \cdot \left(\ln \frac{2U}{\beta t}\right)^{-1} . \tag{10.36}$$

On the one hand, in finite dimensions the Gutzwiller wave function appears to do better than the Gutzwiller approximation would have led us to think: in contrast to (9.52), we find a nonzero energy density at large U. On the other hand, we are not happy to see that this energy density is ascribed to a metallic state: contrary to our hopes, we did not find the Mott transition. Besides, the above asymptotic result does not show the correct U/t-dependence: the correct energy density would be $\sim -t^2/U$, without the logarithmic factor.

We are in an embarrassing situation: we feel quite certain that there must be a metal–insulator transition at a finite U/t; in Sec. 9.3 we were happy to find it; and now we find ourselves compelled to abandon this appealing result. The Brinkman-Rice scenario developed in Sec. 9.3 is certainly a nice one but we could not justify applying it to any finite-dimensional system. Thus taken in itself, the Gutzwiller wave function does not enable us to predict a Mott transition. A bit later, we will discuss the reason why the construction of the trial state (9.16) should have made it clear from the outset that it cannot describe anything but a metallic state.

All this must sound very alarming: it might appear that all the work done from Chapter 9 onwards was a waste of effort. This would be quite a mistaken impression. We were dwelling long upon this particular failing of the Gutzwiller variational approach because the Mott transition is one of our central themes. It must be, however, emphasized that it remains valid that the success of the Gutzwiller–Brinkman–Rice approach lies in the description of the correlated metallic state. A comparison to the accurately known behaviour of the $D = \infty$ Hubbard model (cf. Sec. 10.6.5) shows that the results we found in Sec. 10.2 are basically right, and can be systematically improved by a $1/D$-expansion [124]. It turns out that the results of the Gutzwiller approximation are quite acceptable even in two dimensions, and can be said to be excellent for three-dimensional systems. For our purposes, $D = 3$ is already near to $D = \infty$. The only thing which is quite wrong about our previous results is the finding of the $n_d = 0$ insulator. However, even at $n = 1$, we can use the results of Sec. 9.3 as long as we do not get too near (measured on the energy scale of (10.36)) to the Brinkman–Rice critical point (9.46). A similar criterion holds if we approach the $n = 1$ line at large U. It is comforting to see that the picture of the heavy fermion metal, which we developed in Sec. 10.2, is applicable to real two- and three-dimensional systems up to fairly large values of the mass enhancement factor.

10.6.2 Metallic and Insulating States

The presence of the unwanted logarithmic factor in the asymptotic energy density (10.36) is a clear indication that the Gutzwiller trial state (9.16) fails to capture an essential feature of the large-U behaviour of the Hubbard model. We guess that the predicted ground state energy lowering due to polarity fluctuations is too small because the density of doubly occupied sites (10.35) is too strongly suppressed. The reason for this was understood [202] as follows: starting from a neutral background, the hopping process (9.38) creates a nearest-neighbour pair of d and e sites. It is preferred that in the next step, the system returns to a low-energy configuration via the process (9.41). Thus for large U we should see an "effective attraction" between d and e sites: most of the d sites are found next to an e site. The Gutzwiller wave function does not incorporate this feature: the weight factor η^{N_d} counts merely the total number of the d sites, without paying attention to the spatial distribution of the sites. For this reason, the creation of d sites looks energetically more unfavourable than it actually is.

The Gutzwiller trial state (9.16) has the general structure of a projection operator applied to a reference state

$$|\Psi\rangle = \hat{P}_{\mathrm{G}}|\Psi_0\rangle \tag{10.37}$$

where the Gutzwiller projector

$$\hat{P}_{\mathrm{G}} = \prod_{\mathbf{j}}[1-(1-\eta)\hat{n}_{\mathbf{j}\uparrow}\hat{n}_{\mathbf{j}\downarrow}] \tag{10.38}$$

contains the variational parameter and $|\Psi_0\rangle = |\mathrm{FS}\rangle$, the non-interacting Fermi sea. If we want to include the possibility of an effective e-d attraction, it is straightforward to augment the Gutzwiller Ansatz in the following way

$$|\Psi'\rangle = \hat{P}_{de}\cdot\hat{P}_{G}|\mathrm{FS}\rangle \tag{10.39}$$

where the new intersite projector

$$\hat{P}_{de} = \prod_{\langle\mathbf{i},\mathbf{j}\rangle}\{1-(1-\lambda)[(1-\hat{n}_{\mathbf{i}\uparrow})(1-\hat{n}_{\mathbf{i}\downarrow})\hat{n}_{\mathbf{j}\uparrow}\hat{n}_{\mathbf{j}\downarrow}+(1-\hat{n}_{\mathbf{j}\uparrow})(1-\hat{n}_{\mathbf{j}\downarrow})\hat{n}_{\mathbf{i}\uparrow}\hat{n}_{\mathbf{i}\downarrow}]\} \tag{10.40}$$

brings a factor λ if the nearest-neighbour pair $\langle \mathbf{i}, \mathbf{j} \rangle$ consists of a d and an e site, and a factor of 1 otherwise. $\lambda \geq 1$ is a new variational parameter which controls the degree of the d–e pair binding. Approximate analytic [110], and variational Monte–Carlo [457], studies with (10.39) suggest that the inclusion of intersite correlations brings essential improvements, especially in the large-U regime: for $n = 1$, the ground state energy density is of the order $-t^2/U$.

One might have hoped that once we have managed to get the energy scale $-t^2/U$ which is characteristic of Mott insulators, the trial ground state itself can be identified as an insulating state. As it turns out, this is not the case. Fig. 10.7 shows typical results for the band state occupation number of the one-dimensional Hubbard model. It has a number of features which indicate a substantial improvement over the piecewise constant behaviour[15] shown Fig. 10.1. First, $n(\mathbf{k})$ is a monotonically decreasing function of the band energy $\epsilon(\mathbf{k})$. We are inclined to think this reasonable[16]: the occupation of higher-lying band states looks energetically unfavourable. Second, the leading term in the approximate result [110]

$$n(\mathbf{k}) \approx \frac{1}{2} - \alpha \frac{\epsilon(\mathbf{k})}{U} + \frac{\beta t^2}{U^2} \cdot \Theta(\epsilon_{\rm F} - \epsilon(\mathbf{k})) \tag{10.41}$$

is basically right, though the coefficient α is not the same as in the exact expansion (5.100). Note, however, the last term: the generalized Ansatz (10.39) describes a metal because $n(\mathbf{k})$ has a finite Fermi step. It appears that by introducing nearest-neighbour d-e correlations, we managed to get a qualitatively correct behaviour to order t/U, but features from order $(t/U)^2$ upwards are still associated with a metallic character.

We could try to remedy the situation by including farther-neighbour, or many-site, correlations which improve the behaviour in order $(t/U)^2$,

[15] That was obtained in the Gutzwiller approximation and is strictly correct only for $D = \infty$. Accurate results for finite dimensions [271] reveal a puzzling feature of the Gutzwiller state: $n(k)$ is found to be a piecewise *increasing* function.

[16] Reasonable as it seems, it is not always true. Examples of a markedly non-monotonic behaviour have been found in the case of the $U = \infty$ one-dimensional Hubbard model [310]. Remember that it is the sum of the band and interaction energies that has to be minimized and apparently, it may happen that the electrons can best avoid each other in real space if they preferentially occupy a higher-lying part of the band. All the same, we think that such cases are exceptional. For three-dimensional systems, perturbation theory predicts a monotonically falling $n(\mathbf{k})$ [257].

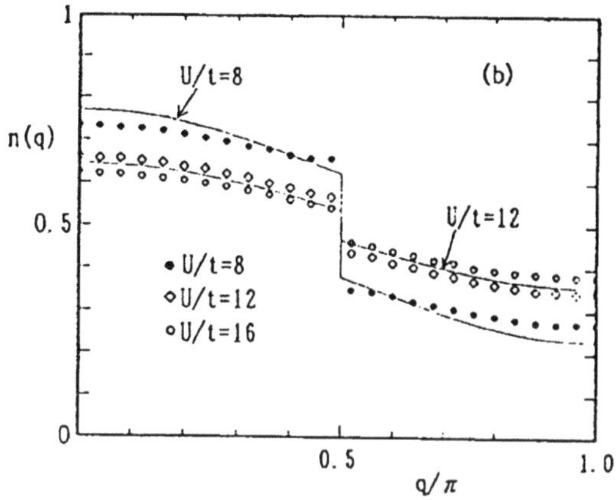

Figure 10.7: Combined variational Monte Carlo (circles) [457] and approximate analytic (lines) [110] results for the band state occupation number of the one-dimensional Hubbard model with the trial state (10.39) (after [457]).

or at subsequent levels of accuracy. The undoubtedly increasing sophistication of the wave function is dearly bought: the calculations are soon becoming quite complicated. However, what we achieve is still merely a better characterization of the correlated Fermi liquid state; we never see a Mott transition at finite U. It has been argued [278] that any finite-order generalization of the Gutzwiller state $|\Psi\rangle = \hat{P}|\mathrm{FS}\rangle$, however complicated the projector $\hat{P}$ is, is bound to be metallic. The basic reason why the Gutzwiller-type trial states fail to describe the Mott-insulator, is the same why they are so successful for the metallic phase. The phase relationships between the localized configurations are inherited from the Fermi sea reference state, and guarantee the survival of a sharp Fermi surface. The only case when the Gutzwiller-projected Fermi sea is an insulator is the limit $\eta = 0$ where charge fluctuations are completely eliminated. However, $\eta = 0$ is really justified only if $U = \infty$. If we like we can say that in the Gutzwiller treatment of finite-dimensional systems, the Mott transition is pushed up to $U = \infty$.

Thus we have come to the conclusion that the wave function of the Mott insulator cannot be sought as a projected Fermi sea; quite on the contrary, it has to be radically different from a projected Fermi sea. Still, we can attempt to seek it by a strategy which is analogous to what we followed in constructing (9.16). Let us reformulate the Gutzwiller wavefunction as

$$|\Psi_{\rm G}\rangle = \exp\left\{\frac{\mathcal{H}_U}{U}\cdot\ln\eta\right\}\cdot|\Psi_{U=0}\rangle. \tag{10.42}$$

Thus the idea was to decompose the Fermi sea into localized configurations, and weight these according to the interaction energy. Baeriswyl [34] suggested that the large-U insulating state can be sought by starting from the opposite limit: take the ground state which belongs to infinite interaction, decompose it into k-space configurations, and weight these according to the band energy:

$$|\Psi_{\rm B}\rangle = \exp\left\{\frac{\mathcal{H}_{\rm band}}{2t}\cdot\zeta\right\}\cdot|\Psi_{U=\infty}\rangle \tag{10.43}$$

where ζ is a variational parameter.

There is an obvious difficulty with constructing the Baeriswyl state: while $|\Psi_{U=0}\rangle$ is immediately known, $|\Psi_{U=\infty}\rangle$ is not. To date, explicit use of (10.43) has been confined to models where the fully Gutzwiller projected Fermi sea (9.29) can be used as the $U\to\infty$ ground state: it is a surprisingly good approximation in the case of the $D=1$ Hubbard model [137], and it is exact for the $1/r$ Hubbard chain. Since the latter model provides particularly valuable insight into the workings of the variational method, we review the pertinent results in some detail.

Problem 10.5 Introducing the Baeriswyl state (10.43) permits a simple derivation of the results of the large-U expansion. Work out the $1/U$-corrections to the ground state energy.

10.6.3 Digression: The $1/r$ Hubbard Chain

We should be troubled by the following consideration: We are seeking an accurate variational ground state of the Hubbard model at $n=1$ for arbitrary U, with the eventual aim of understanding the Mott transition. However, if we really succeeded in this pursuit, we should not see a Mott transition at all! We have been studying the Hubbard model with nearest-neighbour hopping on hypercubic lattices; these are characterized by the perfect nesting property (7.49), and we are quite certain that the ground state is insulating for any $U > 0$. A metal–insulator transition occurs at $U = 0$ but we would not like to call this a Mott transition because it is not a strong correlation phenomenon: it is an SDW instability. We can, to some extent, get around this difficulty by constraining the ground state to be paramagnetic but it would be nicer to work with a model which *really* has a Mott transition at finite U.

One possibility would be to make our previous models more complicated by including some second-neighbour hopping: this would destroy the perfect nesting. In any case, such amendments would be necessary to make the band term more realistic-looking. Another possibility is to study the nearest-neighbour hopping model on non-bipartite lattices, such as the triangular lattice. However, there is little secure knowledge about the behaviour of these models. It is clearly desirable to have at least one model, however artificial it may be, where the existence of a Mott transition at finite U is rigorously known. This could be used as a testing ground to decide whether, and in which sense, the variational approach is capable of detecting the Mott transition. Such a model was constructed by Gebhard and Ruckenstein [126]. Consider L sites arranged equidistantly on the perimeter of a circle with radius L. The hopping is long-range: electrons can hop between any two sites

$$\mathcal{H} = \sum_{\substack{j,m=1 \\ j \neq m}}^{L} \sum_{\sigma} (t_{jm} c^{\dagger}_{\mathbf{j}\sigma} c_{\mathbf{m}\sigma} + \text{H.c.}) + U \sum_{\mathbf{j}} \hat{n}_{\mathbf{j}\uparrow} \hat{n}_{\mathbf{j}\downarrow} \tag{10.44}$$

with the distance-dependent hopping

$$t_{jm} = \frac{it(-1)^{j-m}}{d_{jm}} = t^*_{mj} \tag{10.45}$$

being inversely proportional to the chord distance between j and m

$$d_{jm} = \frac{L}{\pi} \sin \frac{\pi(j-m)}{L}. \tag{10.46}$$

Rigorous arguments show that the Mott transition occurs when U reaches the bandwidth: $U_{cr} = W = 2\pi t$ [126].

Let us discuss the ground state of the half-filled ($n = 1$) model. The Gutzwiller trial state is the familiar (10.42), where $|\Psi_{U=0}\rangle$, the ground state of the long-range hopping term, is a rather curious object (see Problem 10.6).

Somewhat miraculously, it turns out that the fully Gutzwiller-projected Fermi sea is the exact ground state of the large-U effective Hamiltonian which is now the $1/r^2$ Heisenberg chain [149, 358]

$$\mathcal{H} = \sum_{\substack{j,m=1 \\ j \neq m}}^{L} \frac{4t^2}{d_{jm}^2 U} \mathbf{S}_j \cdot \mathbf{S}_m \,. \tag{10.47}$$

Thus the Gutzwiller wave function is exact at $U = 0$, and also at $U \to \infty$, but nowhere in between. In accordance with our previous discussion, the Gutzwiller treatment predicts that the Mott transition is taking place at $U = \infty$. The best that can be said about $|\Psi_{\rm G}\rangle$ is that it gives an acceptable approximation for $U < W$. What we wish for is a description of the insulating ($U > W$) state at the same level of accuracy. This is achieved [96] by constructing Baeriswyl states which is now a practicable proposition, thanks to the exact knowledge of $|\Psi_{U=\infty}\rangle = |\Psi_{\rm G}(\eta = 0)\rangle$. It turns out that the Gutzwiller and Baeriswyl wave functions are complementary: the former can be used in the metallic region, while the latter for the Mott insulator. Improvements of the Gutzwiller trial state give a kind of a small-U expansion, while systematic improvements of the Baeriswyl trial state are similar to a large-U expansion. The Mott transition occurs between two states whose character is completely different and neither of them can be meaningfully continued into the range of validity of the other.

Fig. 10.8 shows results for the density of doubly occupied sites n_d (in the figure, the notation is d). For $U/W < 1$ the Gutzwiller, while for $U/W > 1$ the Baeriswyl estimate is used. The approximate variational result would be that the Mott transition is a first-order one, since n_d has a jump at $U/W = 1$. But according to the exact solution, the Mott transition is continuous, with n_d behaving perfectly smoothly across the transition. Subsequent refinements of the Gutzwiller, and Baeriswyl, states could reduce the discontinuity but would not eliminate it. Note that the figure also shows the result (9.47) which was obtained in the Gutzwiller approximation.

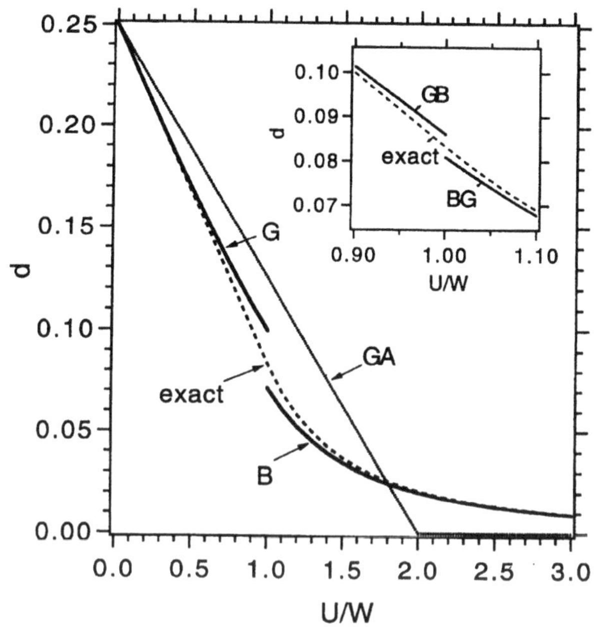

Figure 10.8: The U/W-dependence of the density of the doubly occupied sites for the half-filled $1/r$ Hubbard chain, after [96]. Thick lines show the results from the approximation-free evaluation of the Gutzwiller (G) and Baeriswyl (B) trial states; the thin line shows the Gutzwiller approximation. The dotted line gives the exact solution. The inset shows a magnification of the neighbourhood of the Mott transition.

The reason why we are reluctant to delve deeper into the solution of the $1/r$ model to learn more about the Mott transition, is that it has some pathological features just as far as electrical conduction is concerned. As the reader can convince himself by solving Problem 10.6, the noninteracting ground state carries a large current and we conjecture that even the nominally insulating ground state is current-carrying. The reason why we are bold enough to call it a Mott insulator is that at $U = W$, a gap opens in the single-particle excitation spectrum.

The studies of the $1/r$ Hubbard chain give a particularly convincing illustration of the wellknown warning: the accuracy of the predicted ground state energy is *not* a reliable criterion of the quality of a variational wave function. As discussed also in the previous section, improved

Gutzwiller states may turn out to give quite good ground state energies even at large Us but they are qualitatively wrong for $U > W$ because they characterize a Mott insulator as a metal. Similarly, a sophisticated Baeriswyl state may give a good ground state energy for $U < W$ but it would ascribe insulating character to a state which is, in truth, metallic. The model discussed here is a lucky case: it turns out that if we use wave functions of the same level of sophistication then Gutzwiller states have a lower-lying ground state energy for $U < W$, while Baeriswyl states for $U > W$. Thus the predicted phase boundary is accurate even though the characterization of the individual phases is not. This feature is, however, probably unique to the model (10.44) and we cannot count on such luck again. We should, in particular, keep in mind that, when seeking a phase boundary, it is dangerous to use trial states of differing quality for the two phases: it can result in a large error in the predicted critical value of the coupling constant.

Before closing this section, we should comment upon the insight provided by the finding that $|\Psi_G(\eta = 0)\rangle$ is the exact ground state of (10.47). For the commonly studied Hubbard models, the fully projected Gutzwiller state cannot be the ground state of the corresponding Heisenberg models, because in realistic two- and three-dimensional structures, the Heisenberg model has a Néel-ordered ground state[17], while $|\Psi_G(\eta = 0)\rangle$ is clearly a *disordered quantum paramagnet*, also called a *spin liquid*. Now, the $1/r^2$ Heisenberg model describes an antiferromagnetic coupling between any two sites, thus it is highly frustrated. It is the combined effect of quantum fluctuations (which are strongest for $S = 1/2$; see Sec. 6.2.2), and frustration, which prevents the model from having an ordered ground state and makes the fully Gutzwiller projected Fermi sea a viable ground state.

Problem 10.6 Discuss the ground state of the $1/r$ hopping model specified in (10.44)–(10.46). It is sufficient to consider the limit $L \to \infty$. – Try to make a guess about the ground state current at large U.

[17]With the possible exception of the kagome lattice antiferromagnet. See Sec. 6.3.

10.6.4 Gutzwiller States with Magnetic Order

Up to now, we have focussed our attention to local correlation effects, and disregarded tendencies to magnetic ordering. However, we might as well admit that it is in the nature of the Hubbard model on hypercubic lattices that it wants to develop antiferromagnetic order at half-filling, and it may want to order in a different fashion at other band fillings. Can the variational technique help us to devise a better description of magnetic ordering?

It certainly can, by combining features of the effective field treatment given in Chapter 7, and of the local correlation picture developed in this Chapter. We discuss the simplest magnetic instabilities of the infinite-dimensional hypercubic lattice. Since the Gutzwiller approximation becomes exact in $D = \infty$ [272], the calculations are quite elementary [108]. It is sufficient to discuss the case $n \leq 1$.

Ferromagnetism

First, let us investigate whether the paramagnetic Fermi sea is unstable against the onset of uniform spin polarization $m = n_\uparrow - n_\downarrow \neq 0$. The trial state is a Gutzwiller-projected spin-polarized Fermi sea

$$|\Psi_{\rm F}\rangle = \hat{P}_{\rm G}|{\rm FS}\uparrow\rangle|{\rm FS}\downarrow\rangle = \prod_{\mathbf{j}}[1-(1-\eta)\hat{n}_{\mathbf{j}\uparrow}\hat{n}_{\mathbf{j}\downarrow}] \prod_{\mathbf{k}}^{\rm FS\uparrow} c^\dagger_{\mathbf{k}\uparrow} \prod_{\mathbf{k}}^{\rm FS\downarrow} c^\dagger_{\mathbf{k}\downarrow}|0\rangle \quad (10.48)$$

Recalling the form (4.56) of the Gaussian density of states, the Fermi energy $\epsilon_{{\rm F}\sigma}$ of spin-σ electrons is determined by

$$n_\sigma = \frac{1}{t\sqrt{2\pi}} \int_{-\infty}^{\epsilon_{{\rm F}\sigma}} \exp\left(-\frac{\epsilon^2}{2t^2}\right) d\epsilon = \frac{1}{2}\left[1 + {\rm erf}\left(\frac{\epsilon_{{\rm F}\sigma}}{t\sqrt{2}}\right)\right] \quad (10.49)$$

where erf denotes the error function [4]. The corresponding average band energy is

$$\bar{\epsilon}_\sigma = -\frac{t}{\sqrt{2\pi}} \exp\left(-\frac{\epsilon^2_{{\rm F}\sigma}}{2t^2}\right) = -t^2\rho(\epsilon_{{\rm F}\sigma}) \,. \quad (10.50)$$

We have to minimize the energy density

$$E_{\rm F}(m) = \sum_\sigma q_\sigma \bar{\epsilon}_\sigma + U n_d \quad (10.51)$$

with respect to m (or $n_\uparrow$) and n_d. The σ-dependent q-factors given in (C.20) depend on n_d, and the optimization problem has to be solved numerically. Accepting that the paramagnetic–ferromagnetic transition is continuous, the phase boundary is determined by

$$\left.\frac{\partial^2 E_{\mathrm{F}}(m)}{\partial m^2}\right|_{m=0} = 0\,, \tag{10.52}$$

where it is understood that η (or n_d) has its optimized value. The result is shown in Fig. 10.9 (left). It is worth pointing out that the problem becomes particularly simple in the limit $U \to \infty$ where (see (C.22))

$$E_{\mathrm{F}}(m) = \sum_\sigma \frac{1-n}{1-n_\sigma}\,\bar{\epsilon}_\sigma \tag{10.53}$$

which yields the critical band filling $n_{\mathrm{cr}} = 0.5819$. The paramagnetic state remains stable even at $U \to \infty$ if $n < 0.5819$. The F domain is much reduced in comparison with the Hartree–Fock prediction, according to which the ground state is ferromagnetic at $U \to \infty$ for $0 < n < 1$.

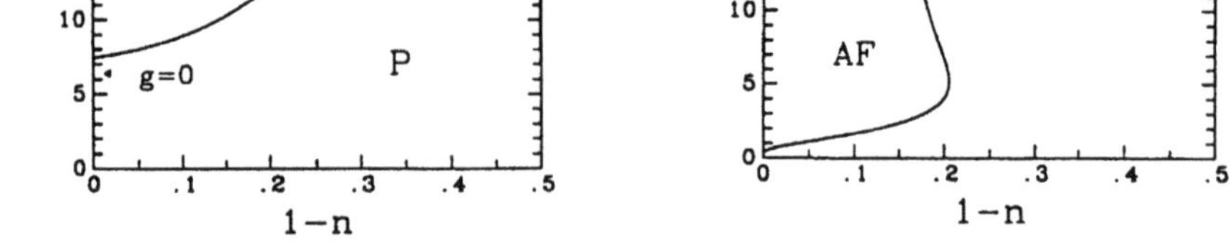

Figure 10.9: The ferromagnetic–paramagnetic (F–P, left) and the antiferromagnetic–paramagnetic (AF–P, right), phase boundary in the ground state of the infinite-dimensional Hubbard model, after [108]. Note that the band filling is measured from $n = 1$, the vertical axis belongs to half-filling, and n decreases to the right. The energy unit is t. The $U \to \infty$ position of the phase boundary is shown by an arrow.

Taking the limit $n \to 1$, we would find that the half-filled band becomes ferromagnetic at $U_{\mathrm{F}}/t \approx 7.35$. However, this prediction is inconsequential because at $n = 1$, SDW ordering takes precedence.

It is easy to accept that the magnetization remains unsaturated ($m < n$) even at $U = \infty$. The Nagaoka state is unstable because the bottom of the empty spin-$\downarrow$ band is at $\epsilon_b = -\infty$. However, the Gaussian tail is very thin, and the minority band quickly gets filled up to energies where the interaction blocks further electron transfer. Near half-filling, one finds [108] that the deviation of the magnetization

$$m_{U=\infty} \approx n - \exp\left\{\frac{1}{4(1-n)^2 \ln(1-n)}\right\} \tag{10.54}$$

from its saturation value is exceedingly small, e.g., for $n = 0.95$ we get $n - m \approx 3{\cdot}10^{-15}$.

Finally, let us discuss the form of $E_{\rm FM}$ in the limit $U \to \infty$, $1-n \ll 1$. Here $n_\downarrow \ll n_\uparrow$, and the requirement

$$1 - n \approx 1 - n_\uparrow = \int_{\epsilon_{\rm F\uparrow}}^{\infty} \rho(\epsilon) d\epsilon \approx \frac{t^2 \rho(\epsilon_{\rm F\uparrow})}{\epsilon_{\rm F\uparrow}} \left(1 - \frac{t^2}{\epsilon_{\rm F\uparrow}^2}\right)$$

is (to lowest order) satisfied by

$$\epsilon_{\rm F\uparrow} \approx t\sqrt{2|\ln(1-n)|}\,.$$

$E_{\rm FM}$ is dominated by the contribution of $\uparrow$-spin electrons, $q_\uparrow \approx 1$, thus

$$E_{\rm FM} \approx -(1-n)\epsilon_{\rm F\uparrow} \approx -t(1-n)\sqrt{2|\ln(1-n)|}\,. \tag{10.55}$$

The logarithmic factor which results from having reached the Gaussian tail of the DOS, helps in stabilizing the ferromagnetic state (no such factor appears in the corresponding expressions for the P and AF phases).

Antiferromagnetism

The mean field theory of spin density waves is based on the independent-electron state (see the solution to Problem 7.9 on p. 416)

$$|\Phi_{\rm SDW}\rangle = |\Phi_\uparrow\rangle|\Phi_\downarrow\rangle = \prod_{\mathbf{k}}\prod_{\sigma} \left[\cos\vartheta_{\mathbf{k}} c_{\mathbf{k}\sigma}^\dagger + \eta(\sigma)\sin\vartheta_{\mathbf{k}} d_{\mathbf{k}\sigma}^\dagger\right]|0\rangle\,. \tag{10.56}$$

Imposing the condition (7.154), $|\Phi_{\rm SDW}\rangle$ is optimized for the degree of two-sublattice magnetic order, but it is still uncorrelated, as shown by the very fact that it factorizes according to the spin index. We can build a correlated SDW state by acting with the Gutzwiller projector on a general two-sublattice reference state

$$|\Psi_{\rm SDW}\rangle = \prod_{\mathbf{j}}[1-(1-\eta)\hat{n}_{\mathbf{j}\uparrow}\hat{n}_{\mathbf{j}\downarrow}]\cdot\prod_{\mathbf{k}}\prod_{\sigma}[u_{\mathbf{k}\sigma}c^{\dagger}_{\mathbf{k}\sigma}+v_{\mathbf{k}\sigma}c^{\dagger}_{\mathbf{k}+\mathbf{Q}\sigma}]|0\rangle. \quad (10.57)$$

The structure of the reference state is the same as of $|\Phi_{\rm SDW}\rangle$, but we do not demand that the amplitudes $u_{\mathbf{k}}$ be given by (7.154). It is to be expected that the degree of magnetic order is influenced by the presence of the Gutzwiller projector. We were arguing that part of the energetical reason for the emergence of antiferromagnetic order is that in a mean-field state, the only way for opposite-spin electrons to avoid each other is to go to different sublattices. Since allowing for local correlations provides an alternative way to keep the Coulomb energy down, we expect a somewhat reduced tendency to magnetic ordering. (10.57) has to be simultaneously optimized with respect to the Gutzwiller parameter η, and the SDW parameters $u_{\mathbf{k}\sigma}/v_{\mathbf{k}\sigma}$.

The $((L/2)+1)$-parameter optimization problem is far more difficult than the single-parameter problem (P phase), or the two-parameter problem (F phase) we had to solve before. Still, adapting the Gutzwiller approximation to the present case allows finding the optimal form of $u_{\mathbf{k}\sigma}/v_{\mathbf{k}\sigma}$, and leaves us with a two-parameter problem which (apart from limiting cases) has to be solved numerically.

A ground state of the form (10.57) can be sought at arbitrary band filling n. Let us note that it reduces to the paramagnetic Gutzwiller state if $v_{\mathbf{k}} \equiv 0$, thus it permits finding the boundary of the continuous antiferromagnetic–paramagnetic transition from a condition which is analogous to (10.52)

$$\left.\frac{\partial^2 E_{\rm AF}(m_{\rm sl})}{\partial m_{\rm sl}^2}\right|_{m_{\rm sl}=0} = 0\,. \quad (10.58)$$

$E_{\rm AF}$ is the optimized energy density and $m_{\rm sl}$ is the sublattice magnetization.

Like in the Hartree–Fock theory, it is useful to construct the reference state $|\Psi_{\rm SDW}\rangle$ in an explicitely normalized form, expressing $u_{\mathbf{k}}$ and $v_{\mathbf{k}}$ with the variables[18] $0 \le \theta_{\mathbf{k}} \le 1$

$$\begin{aligned} u_{\mathbf{k}\sigma} &= \frac{1}{\sqrt{2}}[1+\sqrt{1-\theta_{\mathbf{k}}^2}]^{1/2} && (10.59) \\ v_{\mathbf{k}\sigma} &= -\frac{\eta(\sigma)}{\sqrt{2}}\,\mathrm{sgn}(\epsilon_{\mathbf{k}})[1-\sqrt{1-\theta_{\mathbf{k}}^2}]^{1/2}\,. && (10.60) \end{aligned}$$

It can be shown [272, 270] that the structure of the energy expression is quite similar to (9.44) and (10.51)

$$E_{\rm AF} = q\bar{E}_{\rm HF} + Un_d \qquad (10.61)$$

($\bar{E}_{\rm HF}$ is like (7.151)) but q has become complicated

$$q = \frac{4(n-2n_d)\sqrt{n_d(n_d+1-n)} + 2(2n_d+1-n)\sqrt{(n-2n_d)^2-m_{\rm sl}^2}}{\sqrt{(n^2-m_{\rm sl}^2)((2-n)^2-m_{\rm sl}^2)}}\,. \qquad (10.62)$$

Note that (10.62) reduces to (10.20) for $m_{\rm sl}=0$.

It turns out that the **k**-dependence of the optimized $\theta_{\mathbf{k}}$ arises from that of $\epsilon_{\mathbf{k}}$

$$\theta_{\mathbf{k}} = \frac{A\epsilon_{\mathbf{k}}^2 + \Delta\sqrt{\epsilon_{\mathbf{k}}^2+\Delta^2}}{B\epsilon_{\mathbf{k}}^2+\Delta^2}. \qquad (10.63)$$

In the Hartree–Fock theory $A=0$ and $B=1$. Here, A and B still depend on η in a complicated manner. However, the relationship between the sublattice magnetization $m_{\rm sl}$, and the gap parameter Δ turns out to be the same (7.63) as in our previous SDW theory:

$$m_{\rm sl} = \frac{2}{L}\sum_{\mathbf{k}}\frac{\Delta}{\sqrt{\epsilon_{\mathbf{k}}^2+\Delta^2}}. \qquad (10.64)$$

The energy (10.61) remains to be minimized with respect to η (or n_d) and $m_{\rm sl}$ (or Δ).

[18] The present parametrization is an alternative to that quoted in (10.56).

The AF–P phase boundary is found from (10.58). The result of a numerical evaluation is shown in Fig. 10.9 (right panel). Only the regime $0.5 \leq n \leq 1$ is shown (the full phase diagram should be symmetrical about the $n = 1$ axis, and the range $n < 0.5$ would be covered by the featureless paramagnetic phase). It follows[19] from the perfect nesting property that at half-filling, the AF-P boundary reaches down to $U = 0$. Let us observe that the antiferromagnetic phase is confined to the relatively narrow range $0.883 \leq n \leq 1.117$. This should be contrasted with the Hartree–Fock prediction according to which the P phase becomes unstable against the onset of antiferromagnetism at arbitrary n if U is large enough. Again we find that mean-field theories may grossly overestimate the tendency to form states with long-range order. The essential reason is that in the mean field approximation, the system is prevented from developing short-range order, and it tries to lower its energy by building up too much long-range order. This can be followed in detail by analyzing the results for the large-U limit.

We consider the large-U limit at exact half-filling ($n = 1$). Analytic results can be derived by expanding in powers of $t/U = 1/u$ [108, 268]. The Hartree–Fock solution is obtained by fixing $\eta = 1$, and optimizing with respect to $\theta_{\mathbf{k}}$, with the results

$$n_d^{(\mathrm{HF})} \approx \frac{1}{u^2} - \frac{6}{u^4} \qquad m_{\mathrm{sl}}^{(\mathrm{HF})} \approx 1 - \frac{2}{u^2} + \frac{10}{u^4} \qquad E_{\mathrm{AF}}^{(\mathrm{HF})} \approx -\frac{1}{u} + \frac{2}{u^3}\,. \tag{10.65}$$

In contrast, the present variational approach leads to

$$n_d \approx \frac{1}{u^2} - \frac{3}{u^4} \qquad m_{\mathrm{sl}} \approx 1 - \frac{2}{u^2} - \frac{2}{u^4} \qquad E_{\mathrm{AF}} \approx -\frac{1}{u} + \frac{1}{u^3}\,. \tag{10.66}$$

It is interesting to observe that though the central idea of the Gutzwiller method is an explicit suppression of double occupation, we found that n_d is *larger* than in the Hartree–Fock calculation! The energy price of the increased n_d is paid for by the reduction of the sublattice magnetization which lets the electrons propagate more freely from site to site, and thus gives a lower band energy contribution. The outcome is a decreased

[19]The instability at arbitrarily small U was a mean-field prediction. However, just at $U \to 0$ correlation effects should become negligible, thus we are not surprised that this mean-field result remains valid.

total energy. Let us, however, note that the whole business is a next-to-leading-order effect.

It should be emphasized that n_d remains finite as long as U is finite. The SDW–Gutzwiller (SDWG) variational method avoids the pitfall of the paramagnetic solution, and does not predict a suppression of polarity fluctuations at a critical U. Nevertheless, the system is insulating[20] for all $0 < U < \infty$. We can proceed as in (10.25), and calculate the gap $G_{\rm SDWG}$ as the discontinuity of the chemical potential. Exploiting the electron–hole symmetry (4.48), we find

$$G_{\rm SDWG} = \left.\frac{\partial E_{\rm AF}}{\partial n}\right|_{n=1+0} - \left.\frac{\partial E_{\rm AF}}{\partial n}\right|_{n=1-0} = U - 2\left.\frac{\partial E_{\rm AF}}{\partial n}\right|_{n=1-0} . \quad (10.67)$$

Let us hasten to add that the half-gap is not given by the gap parameter Δ which appears in the formal solution (10.63). We quote the formulas for limiting behaviour. At small U

$$G_{\rm SDWG} \approx 2\Delta\left(1 - \left(\frac{U}{U_{\rm BR}}\right)^2\right) \quad (10.68)$$

where $U_{\rm BR} = 8\bar{\epsilon}(n = 1)$ is the Brinkman–Rice value for the Gaussian band[21]. (10.68) bears an appealing similarity to (9.52) in the sense that both equations say that characteristic itinerant energies (and the nearly-free-electron-like SDW gap is such a quantity) are scaled down by q. However, the validity of this relationship is confined to $U \ll U_{\rm BR}$. At large U, the gap

$$G_{\rm SDWG} \approx U + O\left(\frac{t^2}{U}\right) \quad (10.69)$$

is essentially the Mott–Hubbard gap.

In a substantial part of the n–U plane, the fully symmetrical paramagnetic state is predicted to be unstable against both ferro-, and antiferromagnetism, so we may ask, which kind of ordering do we really

[20] As a result of magnetic cell doubling, there are two orbitals per unit cell. Let us recall the *Remark* on p. 525 where it was pointed out that an insulating ground state is compatible with the presence of polarity fluctuations in the multi-orbital case.

[21] Remember that we find the Brinkman–Rice transition preempted by the onset of a SDW at $U > 0$. $U_{\rm BR}$ plays a purely formal role.

expect? This can be decided by finding out whether $|\Psi_F\rangle$ or $|\Psi_{AF}\rangle$ gives the lower ground state energy. The F–AF phase boundary is defined by $E_F(U,n) = E_{AF}(U,n)$. Since the AF, and F, order is well-developed on the corresponding side of the boundary, this is a line of first-order transitions. The resulting complete phase diagram is shown in Fig. 10.10. It represents a substantial improvement over the Hartree–Fock phase diagram we met[22] in Fig. 4.11, and in Fig. 8.1. We see that the inclusion of local correlations has led to a substantial shrinking of the two ordered phases. Hartree–Fock theory predicts instabilities also for $n < 0.58$ (or $n > 1.42$); these are now seen to be spurious. Let us, in particular, observe that ferromagnetism is eliminated from the low-filling regime even at $U = \infty$.

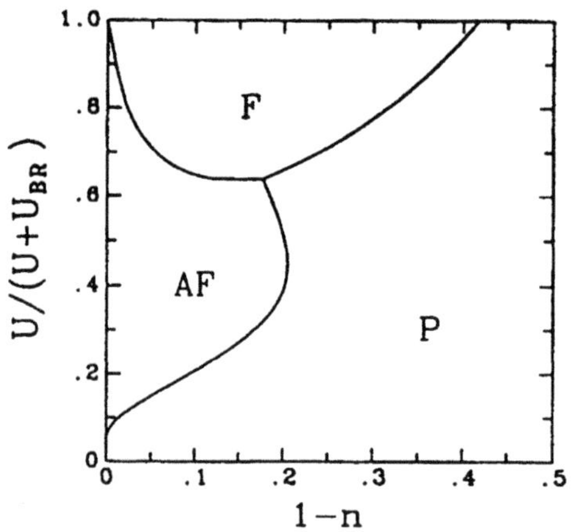

Figure 10.10: The phase diagram for the homogeneous phases, showing the AF, F, and P phases (after [108]).

Phase Segregation versus Stripe Phases

We would be right to suspect that, even where it gives a lower energy than the paramagnetic state, two-sublattice antiferromagnetism is not

[22]The figures cited here were meant to show the Hartree–Fock phase diagram for finite-D cubic lattices, while here we discussed $D = \infty$. However, the shrinking of ordered phases due to allowing for local correlations is similar in all dimensions.

necessarily the optimal ordering pattern away from half-filling. In seeking better ground states, we can follow two different strategies. One of them is to postulate a more complicated trial state (say, a spiral wave with variable $\mathbf{Q}$) for a homogeneous phase. A different strategy is to utilize what we have already got (the optimized energies $E_{\rm F}(U,n)$, $E_{\rm AF}(U,n)$, and the paramagnetic $E_{\rm P}(U,n)$), and check whether the system signals its displeasure with this restricted set of choices by segregating into two phases. Following the latter approach, we find that phase separation is predicted in a substantial part of the phase diagram.

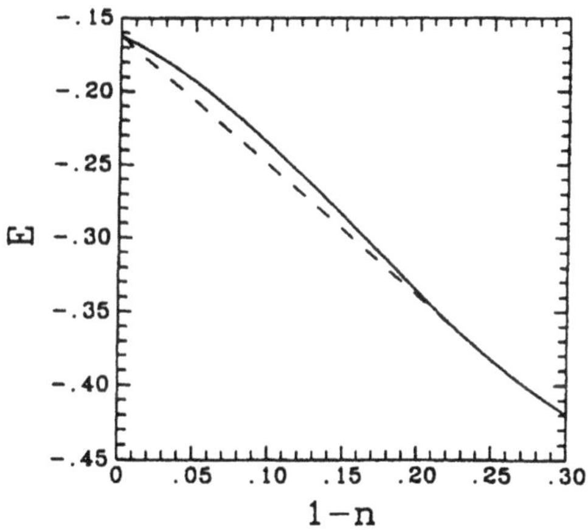

Figure 10.11: The combined ground state energy of the AF and P states for $U/t = 6$, after [108]. The tangent drawn from the $n = 1$ point is shown. The fact that the curve is concave from below indicates that phase segregation must occur.

The reason transpires from the following observation: let us fix $U/t = 6$ where the ground state is either AF or P, and plot the ground state energy as a function of $1 - n$ (Fig. 10.11). Let us notice that the curve is concave from below. More precisely, it is concave between the points $n = 1$ and n_0, where n_0 belongs to the point where the tangent drawn from $n = 1$ touches the curve. Paraphrasing the usual argument for phase separation, we can argue that at nominal (average) densities $n_0 < n < 1$, the system cannot remain homogeneous, but instead phase-segregates into domains with local densities $n = n_0$ and $n = 1$. Namely, a system with overall density n can be realized either as a homogeneous system, or as a heterogeneous system with a volume fraction $(n-n_0)/(1-n_0)$ with density 1, and a fraction $(1-n)/(1-n_0)$ with density n_0. It is apparent from Fig. 10.11 that the overall energy density of the heterogeneous system

$$E_{\text{heter.}}(n) = \frac{n-n_0}{1-n_0}E(n=1) + \frac{1-n}{1-n_0}E(n=n_0) , \qquad (10.70)$$

being the equation of the tangent, lies deeper than the energy density $E(n)$ of the corresponding homogeneous system. Hence we predict that for sufficiently small values of U, and away from half-filling, the homogeneous AF (metallic) state does not exist: instead, we should find a *heterogeneous system consisting of antiferromagnetic insulating, and paramagnetic metallic domains.* – It has to be added that in Fig. 10.11, the tangent touches the curve inside the P regime. Hence the phase-separated AF–P regime is wider than the homogeneous AF regime would have been.

For larger values of U, we similarly find that the $n < 1$ homogeneous AF state is unstable against decomposition into insulating antiferromagnetic, and metallic ferromagnetic regimes. The situation can be summed up by saying that "the antiferromagnet gets all the spins while all the holes go into the (non-antiferromagnetic) metal". We can rationalize this finding as follows: The antiferromagnet lives from spin–spin interactions thus it is energetically unfavourable to dilute it with holes. The way to gain (kinetic) energy from the presence of holes is to make them as itinerant as possible, and this can be achieved by letting them move on a paramagnetic (or ferromagnetic) background.

The revised phase diagram shown in Fig. 10.12 features both homogeneous and heterogeneous phases. A homogeneous antiferromagnetic state occurs only at exact half-filling. In the regions lying between the $n = 1$ axis and the respective phase boundaries[23], AF domains segregate from P domains ($U < 12.4t$), or FM domains ($U > 12.4t$). Furthermore, we find pure FM and PM phases.

Though the calculation we have described was carried out for the $D = \infty$ case [108], there is little doubt that the result would be similar for finite dimensions, including the $D = 3$ simple cubic lattice, and

[23]Note that the boundaries of the two-phase regions differ from those of the corresponding pure phases in Fig. 10.10.

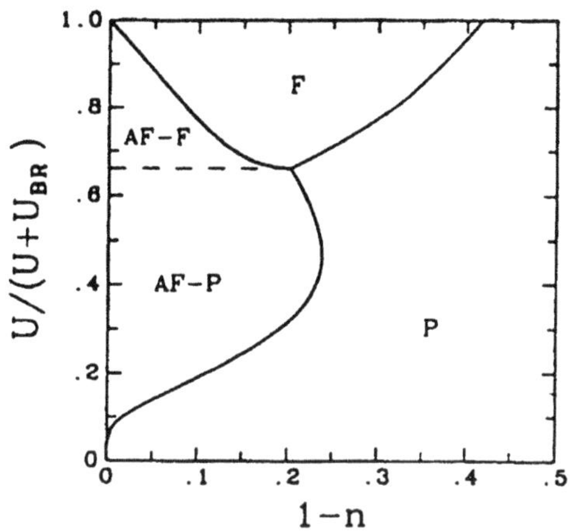

Figure 10.12: The complete phase diagram (after [108]) contains pure ferromagnetic (F) and paramagnetic (P) regions, in addition to the AF–F and AF–P two-phase regions. The homogeneous antiferromagnetic (insulating) state is confined to the axis $n = 1$.

even the square lattice. Also the t–J model shows phase separation in a wide range of parameters in $D = 2$, and $D = 1$. It has long been realized [430] that in lattice models with short-range electron–electron interactions, it is unlikely that one could find a dilute gas of holes on an antiferromagnetic background; instead, the holes tend to segregate into a hole-rich metallic phase, leaving behind an AF insulator.

This raises the question, what should we think about the finding of phase segregation: is it merely an artefact of a not sufficiently sophisticated theory, or does it point towards a genuine physical phenomenon? We cannot give a conclusive answer but it seems that the tendency to segregation is (at least, in some cases) genuine.

One of the options would be to argue that the system became unstable because we did not find the correct homogeneous state. Leaving aside the question whether phase segregation has something to do with an electronic mechanism of superconductivity, let us restrict ourselves to states with magnetic order. To "interpolate" between ferromagnetism and two-sublattice antiferromagnetism, we may consider ferrimagnetism [315], or spiral spin density waves. Generalizing the previous Gutzwiller treatment to spiral SDWs, it has been found [213] that the the region of phase segregation shrinks, but it remains appreciable. At very large U's, we find a surviving pocket of the previously described AF–F segregation. At smaller U's, AF–P segregation is ruled out but it becomes replaced by AF–spiral phase separation[24].

Resigning ourselves to the outcome that in these model calculations, phases seem to want to segregate, we have to deal with a basic objection which recalls that in reality, the Coulomb force is long-ranged. Creating large domains with different values of the electron (or hole) concentration would mean the accumulation of macroscopic amounts of charge, and this would certainly be penalized by a horrendous Coulomb energy. If we still insist on a separation of phases, we have to make the domain structure as fine-grained as possible because this keeps the deviation from local neutrality at minimum. In $D = 2$, we may envisage a structure like that shown in Fig. 10.13, in which the holes are condensed into thin domain walls. A model calculation of the competition between a "segregation force" and a long-range repulsive force shows that the formation of such *striped phases* is a natural outcome [248]. We expect that at low doping, the wavevector of charge order is $q_{\text{charge}} \sim |1 - n|$. It has been argued that it is advantageous to form antiphase domains of spin order, thus $2q_{\text{spin}} = q_{\text{charge}}$ [463], as illustrated in the figure.

We have drawn the domain walls as consisting purely of holes. These empty regions cannot conduct, and since the inside of the domains is also insulating, we expect that the system as a whole is insulating, in spite of its being doped away from half-filling. This would be a disorder-free mechanism for producing an insulator in a finite range of n. However, this is only one of the possibilities. If we take it seriously that the domain walls must be thin slices of the (hole-rich, non-magnetic) minority phase[25], then we have to allow that they have a small but finite concentration of electrons, and are thus quasi-one-dimensional

---

[24]This is the variational result. Second-order perturbation theory shows that in the weak coupling limit, the segregating phases are the AF insulator and the PM metal [419].

[25]The reader may ask how could we afford not to take this seriously, since the argument leading to the stripes relied on the concept of phase separation. The alternative would be to claim that postulating the kind of long-period order shown in Fig. 10.13 obviates phase segregation, and therefore we do not have to make reference to the results of that scenario. We leave the question open.

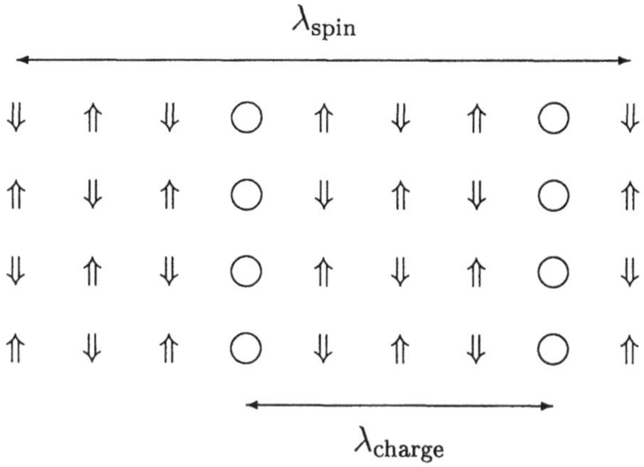

Figure 10.13: Schematic representation of the striped phase. The holes accumulate at domain walls which separate antiphase domains of antiferromagnetic order. The wavelength of spin order is twice as large as that of charge order.

conductors in an insulating matrix. The outcome should be a gapless system with esoteric electrical properties. The consequences of this scenario are as yet little explored.

Our discussion, and the accompanying Fig. 10.13, tacitly assumed that we are in the large-U limit, where the AF domains can be represented by localized spins. It is an interesting corollary that the lines of holes provide a new, quantum-mechanical, kind of antiferromagnetic domain wall, presumably with high mobility, and little (if any) hysteresis. Since the overall hole concentration is fixed, this implies an unusual domain structure in which the total length of the walls is an *intrinsic* characteristic.

It should be emphasized that stripe structures can appear also in the (suitably extended) weak-coupling Hubbard model. It is interesting that for $\delta = 1 - n \ll 1$ and $U \ll t$, a long-period Hartree–Fock solution (essentially the same as in Fig. 10.13) was found for the $D = 2$ pure Hubbard model [353]. If we could consider this result as definitive, we may say that the striped state is not "heterogeneous on a mesoscopic scale", but rather homogeneous with a large repeat distance, and then its driving force would be the Hubbard U. However, a systematic perturbation theory[26] for small U/t and large D shows [419] that fluctuations are important even in the weak-coupling limit, and that incommensurate Hartree–Fock solutions are overlaid by a heterogeneous mixture of pure AFM insulator and PM metal. It is not immediately clear whether these $D \geq 3$ results are in conflict with the abovementioned $D = 2$ study but if they are, we had better comclude that for all U/t, the ground state of short-range models is segregated (with indefinite spatial structure), and the striped phase arises because of the long-range components of the Coulomb interaction.

On the experimental side, it is gratifying to learn that striped states have been observed in quasi-two-dimensional nickelates and cuprates, and that $q_{\text{charge}} \sim |1 - n|$ is approximately satisfied [404]. While these findings are often discussed in connection with the electronic mechanism described above, it should not be forgotten that oxygen ions are also mobile, and can shift their position to reduce the overall Coulomb energy. The possibility that the electronic superstructure is accompanied (or driven) by a modulation of chemical composition, is completely beyond the scope of our arguments.

10.6.5 The Infinite-Dimensional Hubbard Model

When trying to infer the behaviour of three-dimensional systems from the results derived in Sections 9.3 and 10.2, two sources of error have to be kept in mind: first, the Gutzwiller approximation is valid only for $D = \infty$, and second, even in $D = \infty$, the Gutzwiller trial state is certain to differ from the true ground state. It is of great interest that a nominally exact description of the behaviour of the infinite-dimensional Hubbard model has become possible (see [130, 123] for reviews). Thanks to this development, we can judge how much of the true physics of the Hubbard model is captured by the Gutzwiller–Brinkman–Rice approach.

We introduced the infinite-dimensional hypercubic Hubbard model in Sec. 4.6. We understood that in order to get a non-trivial model in which the kinetic and interaction energy terms can compete, the hopping matrix element has to be scaled as $t \sim 1/\sqrt{D}$. It turns out

[26]We review the results in Sec. 10.6.5.

that the propagator of the non-interacting system

$$G^0_{\mathbf{ij},\sigma} \sim \frac{1}{\sqrt{D}} \qquad (\mathbf{i} \neq \mathbf{j}) \tag{10.71}$$

scales accordingly [295, 270]. Since sums over products of $G^0_{\mathbf{ij},\sigma}$ appear in consecutive orders of perturbation theory, it is easy to accept that such contributions become negligible in the limit $D \to \infty$, unless the site indices coincide. The resulting real-space collapse of perturbation-theoretical contributions is the underlying reason why the infinite-dimensional Hubbard model poses a simpler problem than its finite-dimensional counterparts [272]. It is intuitively clear that the simplification arises because the lattice model becomes equivalent to an effective single-site model. It remains, however, a difficult problem to solve. The effective on-site model in question can be understood as an Anderson impurity model, where the impurity is subject to a fluctuating effective field. Owing to the complicated frequency-dependence of the effective field one should, in principle, impose an unlimited number of self-consistency conditions. All methods of solution amount to keeping a finite (and not necessarily large) number of conditions. For certain quantities the results may be regarded as numerically exact, while for small energy scales (like the effective bandwidth which is expected to vanish near the Mott transition) the character of the solution is the subject of continuing debate.

The reason why the Hubbard model remains non-trivial even in the limit of infinite dimensionality, can be traced to the on-site character of the interaction. For models with intersite interactions between localized degrees of freedom[27], the limit $D \to \infty$ brings great simplification because fluctuations of order $1/\sqrt{D}$ become negligible, and thus the usual mean field approximation becomes exact. In contrast, here we are dealing with itinerant degrees of freedom, and the purely on-site interaction $U\hat{n}_{\mathbf{j}\uparrow}\hat{n}_{\mathbf{j}\downarrow}$ apparently does not care about the number of neighbours. For this reason, the dynamics of the Hubbard model remains complicated even in $D = \infty$. Still, there is a subtler sense, embodied in the effective impurity model, in which the limit $D \to \infty$ permits a simpler, mean-

[27] The spin models of magnetic ordering are an important example. See Ch. 6.

field-like treatment. In view of the fluctuating character of the mean field, one often calls it a dynamical mean field theory (DMFT)[28].

We are primarily interested in the results for $T = 0$ because those are directly comparable to our previous variational results. Herein lies some difficulty because most non-perturbative works were done with the quantum Monte Carlo (QMC) method which is confined to finite temperatures, and the extrapolation to $T = 0$ is not without uncertainties. Still, reasonable guesses can be made.

Another difficulty is associated with the lattice and/or band structure. The nearest-neighbour hopping model on the hypercubic lattice retains its perfect nesting property up to $D = \infty$, and thus if we take the model quite seriously, then the Mott transition which would occur at finite U is preempted (and overlaid) by the transition into the antiferromagnetic insulating state at $U = +0$. Keeping the solution paramagnetic "by hand" is certainly an option but it is more convincing to postulate a different band structure and/or band density of states which does not imply a magnetic instability at small U. The $D = \infty$ fcc (half-hypercubic) lattice would be an ideal choice but its Mott transition has not been studied yet. However, the semicircular density of states

$$\rho(\epsilon) = \frac{1}{\pi W}\sqrt{1 - \left(\frac{2\epsilon}{W}\right)^2} \tag{10.72}$$

is a popular choice [130]. (10.72) is non-vanishing in a finite interval[29] and bounded, thus hopefully it allows to draw conclusions about the generic behaviour. Besides, it happens to be the DOS for a completely connected graph with random matrix elements t_{ij}. Realizing that in the corresponding Heisenberg model all bonds are equally antiferromagnetic, we can readily accept that the large-U limit corresponds to a fully frustrated magnet which does not wish to order at all. Thus we are

---

[28]DMFT results can be very different from those of usual (static) mean field theory. See Fig. 8.9 on p. 461 for an interesting example.

[29]The gaussian tails of the DOS for many $D = \infty$ structures may give rise to confusing results. If we envisage the Hubbard subbands as reduced-scale copies of the original band, we should fear that their tails overlap in the purported Mott–Hubbard gap, which would make the identification of the Mott transition a delicate affair.

justified in looking for a Mott transition from a paramagnetic metal to a paramagnetic insulator, without having to restrain any of the natural inclinations of the model[30].

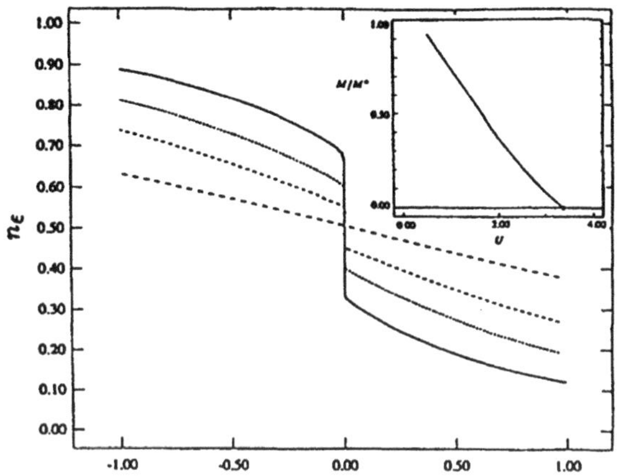

Figure 10.14: The band state occupation number n_k for the half-filled $D = \infty$ Hubbard model, for $2U/W$=2, 2.5, 3, and 4, after [470]. The semicircular density of states (10.72) was assumed. The inset shows the vanishing of the inverse effective mass at the Mott transition.

The Correlated Metal

Let us consider first the half-filled ($n = 1$) case. Increasing the interaction from small values, the development is remarkably similar to that predicted by the Brinkman–Rice scenario [470]. Up to a critical value $U_{\rm cr} \approx 1.68W \approx 8\bar{\epsilon}$, the system is a Fermi liquid, with the band state distribution function (Fig. 10.14) not unlike that shown in Fig. 10.7. The Fermi step is related to the effective mass which follows

$$\frac{1}{q} = \frac{m^*}{m} \propto \left[1 - \left(\frac{U}{U_{\rm cr}}\right)^2\right]^{-1} \tag{10.73}$$

---

[30]Unfortunately, this comfort is bought at a price: the effective Heisenberg model is essentially $J(\mathbf{S}_{\rm tot})^2$, thus all singlets are ground states, and the insulating phase has a finite zero point entropy. It remains a matter of concern whether a part of the observed behaviour is ascribable to this artificial feature.

just as we found in (10.7).

(10.73) predicts a perfectly Brinkman–Rice-type divergence – assuming that the metallic solution can be continued up to $U \to U_{\rm cr}$. Let us, however, consider a later QMC result (Fig. 10.15). For most of the regime $U < U_{\rm cr}$ the solution seems to be well-described by (10.73). However, in an interval about $U = U_{\rm cr}$, m/m^* follows different branches, according to whether U is decreased or increased. This kind of hysteresis indicates a first-order transition.

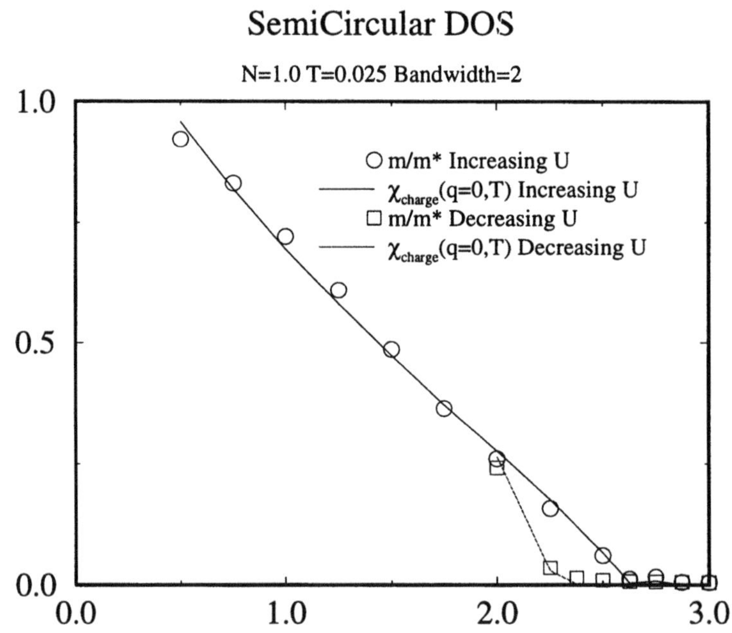

Figure 10.15: QMC results for the U/W-dependence of m/m^* (data points) and the charge compressibility (continuous line) for the DOS (10.72). The presence of hysteresis indicates a weakly first-order transition. (Unpublished results, by courtesy of M. Jarrell).

The question of the order of the Mott transition is truly delicate one [130, 123] and we would not like to pass judgement on the issue here. Equation (10.73) describes how the transition point is approached from the metallic side, and it points to an inherent instability: there is a

$U_{\rm cr}$ where m^* would diverge, thus we could not continue the metallic solution to $U > U_{\rm cr}$. The Brinkman–Rice scenario for the MIT predicts a critical-point transition: m^* diverges as $U \to U_{\rm cr} - 0$, and the gap vanishes continuously as $U \to U_{\rm cr} + 0$. A continuous transition (in fact, a third-order transition) was also the outcome of the exact solution of the $1/r$-model. However, we seem to have no convincing argument to exclude that the Mott transition is generically first-order. Let us recall our discussion in Sec. 10.6.2: we understood that the insulating ground state is not a distorted (projected) metallic state but an entirely different state[31]. Thus the ($T = 0$) Mott transition is basically a level-crossing transition, where the energies of two completely different many-electron states become equal: $E_{\rm M} = E_{\rm PI}$ (M: metal, PI: paramagnetic insulator). There is a metallic ground state which, taken in itself, would become unstable at an $U \to U_{\rm cr}^{\rm M} - 0$ where $m^* \to \infty$. There is also a localized-moment paramagnetic insulating ground state which may also become inherently unstable at some $U \to U_{\rm cr}^{\rm PI} + 0$ because it cannot sustain the gap any more. If we find a mechanism to impose $U_{\rm cr}^{\rm M} = U_{\rm cr}^{\rm PI}$, the transition is continuous. Otherwise, the transition point need not coincide with the inherent instability of either phase but is determined by $E_{\rm M}(U_{\rm tr}) = E_{\rm PI}(U_{\rm tr})$.

The idea of level-crossing can be extended to finite temperatures where the corresponding free energies cross. Actually, this is the only way to have a phase transition without symmetry breaking. Let us remember that at $T \neq 0$ even a gapped system has a finite conductivity, thus the contrast between a metal and an insulator becomes blurred, and the transition has to be defined in thermodynamic terms.

Let us notice that Fig 10.16 shows a finite-temperature phase transition. Recalling that (in Sec. 5.5) we presented a loose argument showing that the spin entropy term gives an added reason for the MIT to be of first order at finite T, we may be wondering whether the discontinuous behaviour at $T = 0.025W$ would not be compatible with a second-order transition at $T = 0$. The ground state phase transition is not directly accessible to the QMC method, and it is an open question whether the

[31] Here we are speaking about a non-magnetic Mott insulator. If the metal-insulator transition coincides with a magnetic ordering transition, the SDW picture may provide a valid unified description of metallic and insulating states.

discontinuity shrinks to zero as $T \to 0$. If it does, the metal–insulator phase boundary is a line of first-order transitions with a usual critical end point at some $T_{\rm cr}$, and a quantum critical point as lower end point at $T = 0$.

Though m^*/m may not diverge, it can become quite large. The correlation-narrowed band should have a correspondingly low effective Fermi temperature $T_{\rm F}^* \sim qT_{\rm F} \sim qW$. Evidence for this is seen in the T-dependence of the entropy density S/L: for $U/W = 1$ (still inside the metallic phase), S/L reaches $k_{\rm B}\ln 2$ at $T \approx 0.04W$ [470]. At $T \sim T_{\rm F}^*$ (which can be quite low), the system behaves like a hot gas of uncompensated spins. This confirms our view that the large effective DOS detected by the enhanced γ^* is coming mainly from the spin degrees of freedom.

The development of localized moments can be followed by measuring $\langle (M_{\rm j}^z)^2\rangle = \langle (\hat{n}_{{\rm j}\uparrow} - \hat{n}_{{\rm j}\downarrow})^2\rangle$ with the Quantum Monte Carlo method. Representative results are shown in Fig. 10.16. The overall U/W-dependence is quite similar to what we have seen in Fig. 9.4, showing that the Gutzwiller approximation can give a fair idea of the true behaviour. Let us, however, observe two features which are different. First, $\langle (M_{\rm j}^z)^2\rangle$ approaches 1 gradually in the limit of large U's. We infer that the calculation recovers the correct asymptotic behaviour $1 - \langle (M_{\rm j}^z)^2\rangle \sim n_d \sim (t/U)^2$ which we derived in (5.44). Second, like in the m^*/m plot, there is a hysteresis in the transition region.

In contrast to the Brinkman–Rice prediction, the enhanced Pauli susceptibility does not diverge, though it is enhanced. What diverges instead, is the so-called local susceptibility $\chi_{\rm loc} \sim \int d\mathbf{q}\chi(\mathbf{q})$. We can try to understand this in very simple terms[32] as follows: the approach to the Mott–insulating phase is certainly accompanied by the arising of uncompensated spins, or local moments. The large, and eventually diverging, contribution of the low-lying spin excitations to the total density of states is evident in m^*. However, it is not only local moments that are generated, but also spin-spin interactions $J \propto t^2/U$. In hypercubic lattices, the spin–spin interaction causes antiferromagnetic ordering, i.e., $\chi(\mathbf{Q})$ is expected to diverge, while $\chi(\mathbf{q} = 0)$ is kept finite,

[32] Our argument is best suited for the case when the Mott–insulator is approached by letting $n \to 1$, at fixed large U. See (10.79) and the accompanying discussion.

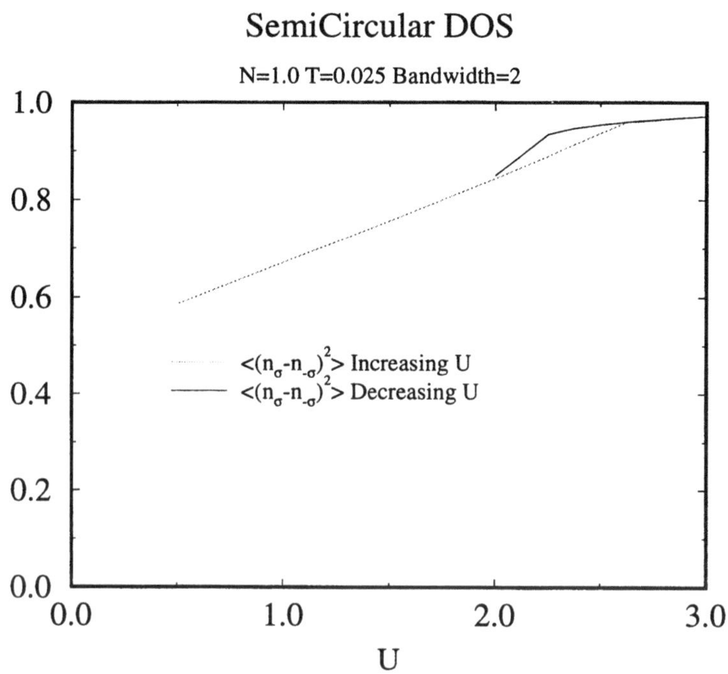

Figure 10.16: QMC results for the square of the instantaneous moment $\langle (M_{\mathrm{j}}^{z})^2 \rangle$. The presence of hysteresis indicates a weakly first-order transition. (Unpublished results, courtesy of M. Jarrell).

and approaches the AF Heisenberg value $\sim 1/J$. For other lattice or band structures, the generalized susceptibility may diverge at a different $\mathbf{q}$, while for fully frustrated lattices, we may expect that the enhancement $\chi(\mathbf{q})$ is roughly independent of $\mathbf{q}$. In either case, we feel entitled to expect an enhancement in χ_{loc}. In the simplest Brinkman–Rice theory, we apparently ascribed all the enhancement to $\chi(\mathbf{q}=0)$.

Having realized that it is really χ_{loc} which has to diverge (or at least, become large) at the same time when $m^* \sim \gamma$ does, we may enquire whether these quantities increase at the same rate. This is indeed the case: it was found [283] that approaching the transition from the metallic side, the generalized Wilson ratio

$$R = \frac{\chi_{\mathrm{loc}}/\chi_{\mathrm{loc}}^{\mathrm{free}}}{\gamma/\gamma^{\mathrm{free}}} \approx 2.8 \tag{10.74}$$

tends to a constant value. According to the understanding based on the study of the $D = \infty$ limit, (10.74) should replace our earlier (10.14).

All the above conclusions were drawn from studying the narrow band of low-lying excitations which appears as a sharp resonance centered about the chemical potential. The width $\sim qW$ of the resonance defines the scale of small energies in the problem, and it is in this regime where parallels with the Brinkman–Rice scenario can be found. However, in contrast to the variational treatment, the DMFT does not miss the high energy scale: roughly at the energies $\pm U/2$, strong bands are seen, which describe charge excitations. These are the Hubbard subbands which we introduced in Sec. 4.7. Loosely speaking, we can say that the reason why we still cannot speak about a definitive analytic description of the immediate neighbourhood of the Mott transition is that such results as we have, had to rely on a clear separation of low- and high-energy scales, and this cannot be taken for granted.

Coming away from half-filling, the system – when non-ordered – is found to be a Fermi liquid at all interaction strengths. For a fixed $U > U_{\rm cr}$, the effective mass appears to diverge as half-filling is approached [189]. We have already commented upon the T^2-behaviour of the low-temperature resistivity in Sec. 10.5.

Phase Diagram

The combination of QMC with other techniques holds the eventual promise of mapping out the finite-temperature phase diagram of various infinite-dimensional Hubbard models, but more work has to be done before a generally accepted picture could emerge. Let us first consider the status of the large-U effective models because these should be relevant for establishing connection with many finite-dimensional results. At half-filling, the connection with the antiferromagnetic $D = \infty$ Heisenberg model is unproblematic: $J \propto t^2/U \sim 1/D$ gives finite ordering energy densities, because the number of neighbours is[33] $\propto D$. On the other hand, away from half-filling we have a difficulty with constructing a well-behaved t–J model, because $J/t \propto (1/D)(1/\sqrt{D})^{-1} \propto 1/\sqrt{D}$.

[33] Actually, this does not hold for tightly packed lattices. To cite a counter-example, the fcc lattice has $z \propto D(D-1)$, and needs a special consideration.

If one insists on interpreting the behaviour of the Hubbard model in terms of an effective Hamiltonian, one has to study a t–J model with "parametrically small" values[34] of J/t [328].

A t–J model with correctly scaled coefficients is (dropping the density–density interaction)

$$\mathcal{H}_{t-J} = -\frac{t}{\sqrt{2D}} \sum_{\langle \mathbf{i},\mathbf{j} \rangle} \sum_{\sigma} (X_{\mathbf{i}}^{\sigma\leftarrow 0} X_{\mathbf{j}}^{0\leftarrow\sigma} + X_{\mathbf{j}}^{\sigma\leftarrow 0} X_{\mathbf{i}}^{0\leftarrow\sigma}) + \frac{J}{2D} \sum_{\langle \mathbf{i},\mathbf{j} \rangle} \mathbf{S}_{\mathbf{i}} \cdot \mathbf{S}_{\mathbf{j}} \tag{10.75}$$

where we used the Hubbard operators (5.21) to express the kinetic energy. The projected hopping term still keeps the on-site dynamics, and the problem remains difficult even at $J = 0$ where we recover the $U = \infty$ Hubbard model. It is, however, a welcome simplification that the limit $D \to \infty$ allows to replace the exchange term with its static Hartree-decoupled form

$$\frac{J}{2D} \sum_{\langle \mathbf{i},\mathbf{j} \rangle} \mathbf{S}_{\mathbf{i}} \cdot \mathbf{S}_{\mathbf{j}} = \sum_{\mathbf{q}} J_{\mathbf{q}} \mathbf{S}_{\mathbf{q}} \cdot \mathbf{S}_{-\mathbf{q}} \rightsquigarrow \sum_{\mathbf{q}} J_{\mathbf{q}} \mathbf{S}_{\mathbf{q}} \cdot \langle \mathbf{S}_{-\mathbf{q}} \rangle \tag{10.76}$$

where the Fourier component $J_{\mathbf{q}}$ is

$$J_{\mathbf{q}} = \frac{J}{D} \sum_{j=1}^{D} \cos(q_j a) \tag{10.77}$$

It follows that in its paramagnetic phase, the t–J model becomes identical to the $U = \infty$ Hubbard model.

Noticing that the decoupling (10.76) is analogous to the decoupling of the Hubbard term in (7.37), and that a mean field treatment is now strictly justifiable, we can make a derivation parallel to that in Sec. 7.5 to arrive at the **q**-dependent susceptibility of the t–J model [328]

$$\chi_{t-J}(n, \mathbf{q}) = \frac{\chi_{\mathrm{H}}(U = \infty, n, \mathbf{q})}{1 + J_{\mathbf{q}} \chi_{\mathrm{H}}(U = \infty, n, \mathbf{q})} \,. \tag{10.78}$$

This expression has the same structure as the RPA result (7.44). Notice, however, that the role of the free-electron susceptibility has been taken

[34]It is a different question that one can postulate a large-D t–J model with a different parameter range [59].

over by $\chi_{\rm H}(U=\infty,\delta,\mathbf{q})$ (the susceptibility of the $U=\infty$ Hubbard model).

Strictly at $n=1$ the $U=\infty$ model is a set of decoupled spins, and its susceptibility is the $\mathbf{q}$-independent $\chi(U=\infty)=C/T$ with the appropriate Curie constant. It follows that the susceptibility of the paramagnetic phase of the t–J model satisfies

$$\lim_{n\to 1}\chi_{t-J}(n,\mathbf{q},T)=\frac{C/T}{1+J_{\mathbf{q}}C/T}\,. \tag{10.79}$$

At low temperatures, the uniform susceptibility approaches $\chi\sim 1/J$ as $n\to 1$. Thinking of reasonable t/J ratios, the enhancement with respect to the Pauli value $\chi\sim 1/t$ is large, but certainly not diverging. On the other hand, the $\mathbf{q}=\mathbf{Q}$ suscepibility can diverge at sufficiently low T and we expect antiferromagnetism in a finite interval of n centered on $n=1$.

The same idea can be exploited to map out the large-U part of the n–U phase diagram of the Hubbard model [308]. (10.78) says that the uniform susceptibility

$$\chi(U,\mathbf{q}=0,n)=\frac{\chi(U=\infty,\mathbf{q}=0,n)}{1+(4t^2/U)\chi(U=\infty,\mathbf{q},n)} \tag{10.80}$$

is suppressed as we lower U. At the same time, the staggered susceptibility

$$\chi(U,\mathbf{q}=\mathbf{Q},n)=\frac{\chi(U=\infty,\mathbf{q}=\mathbf{Q},n)}{1-(4t^2/U)\chi(U=\infty,\mathbf{q}=\mathbf{Q},n)} \tag{10.81}$$

is enhanced. In the intermediate-coupling regime, the surviving order is antiferromagnetism (or SDW). Whether we have ferromagnetism in the strong coupling limit, depends on finding $\chi(U=\infty,\mathbf{q}=0,n)\to\infty$ in a range of n. Our variational study (Sec. 10.6.4) indicated that there is such a range but it would be comforting to have independent evidence from DMFT. QMC cannot be applied at $U=\infty$, but the question can be investigated with the so-called non-crossing approximation (NCA was developed originally for the Anderson impurity problem). The result is that the $U=\infty$ model is ferromagnetic for $n_{\rm cr}<n<1$, where $n_{\rm cr}$ is estimated to lie between 0.6 and 0.7. The Curie temperature is maximum (about $T_C \sim 0.022t$) at $n \sim 0.92$. The $T = 0$ moment is large and it may be near the Nagaoka limit[35].

The strong-coupling phase boundaries can be traced by determining where the susceptibilities given in (10.80) and (10.81) diverge. The result is shown by the curves A (boundary of the FM phase) and B (part of the boundary of the AFM phase) in Fig. 10.17. The lower part of the AFM phase is bounded by curve C, the result of a small-U calculation [188]. B and C do not match properly since they are derived in approximations which are valid in different limits; nevertheless, the figure gives a good impression of the extent of the AFM phase. Though the boundaries are derived by extrapolation from finite-T results, we should not hesitate to accept Fig. 10.17 as one preliminary version of the ground state phase diagram. A different version, based on QMC which is an intermediate-coupling scheme, would show AFM and incommensurate SDW phases, but no FM [118, 383]. It remains to be seen what an all-encompassing phase diagram will look like.

In its gross features, the phase diagram shown in Fig. 10.17 is not unlike the variationally derived Fig. 10.10. Let us, however, point to the differences. Now we find a regime where the PM phase protrudes in between the FM and AFM phases. This has become possible because the ferromagnetic area, though horizontally almost as broad as in the Gutzwiller theory, has been pushed to higher values of U. Ferromagnetism seems to be a marginally surviving strong-coupling phenomenon whose driving force is the (projected) kinetic energy. This should be contrasted with the robustness of itinerant ferromagnetism in the fcc lattice (see Sec. 8.4 and, in particular, Fig. 8.8).

Our discussion of the variational results showed that away from half-filling, the homogeneous AFM phase is likely to be unstable against phase segregation. DMFT has not yet addressed this question at strong and intermediate couplings, but there are results for the weak coupling limit where segregation occurs in an extremely small vicinity of $n = 1$. As a prelude, let us discuss exact half-filling.

It follows from the perfect nesting property that at $n = 1$ for all

[35]In sections 8.4 and 10.6.4, we learned about some cogent arguments why the Nagaoka limit cannot be reached. Extrapolation from the finite-T results of [308] does not allow detecting a small deviation of the order of magnitude shown in (10.54).

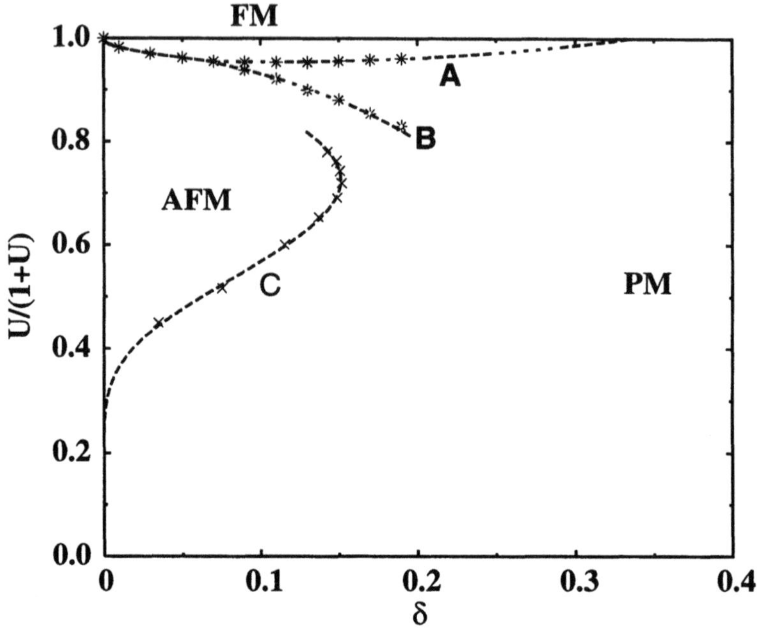

Figure 10.17: The ground state phase diagram of the $D = \infty$ hypercubic lattice, based on a combination of large-U and small-U results (after [308]). Notice the broad range of strong-coupling ferromagnetism.

$U > 0$, the system is an antiferromagnetic (or **q**=**Q** SDW) insulator below a Néel temperature T_N. The phase diagram in the U/t-T plane is shown in Fig. 10.18 (it is not an accurate plot, but merely a sketchy graphics which is meant to convey an impression of the results obtained in [188].) Though there is no analytical treatment covering intermediate U/t's, we can cite results for $U/t \ll 1$ and $U/t \gg 1$ [418]. Furthermore, the weak-coupling results include $1/D$-corrections which are supposed to describe even $D = 3$ fairly well.

The large-U limit can be understood as the $D = \infty$ antiferromagnetic Heisenberg model with $J = 4t^2/U$, thus $T_N \propto t^2/U$. Since the effective Hamiltonian contains intersite interactions, the Curie–Weiss theory of ordering becomes exact.

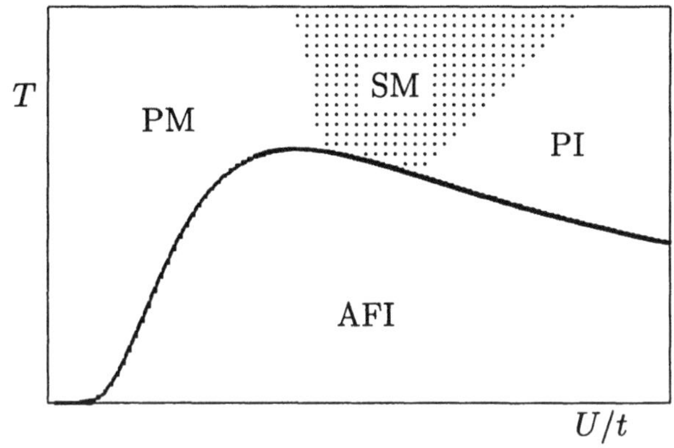

Figure 10.18: Schematic representation of the phase diagram of the infinite-dimensional half-filled Hubbard model in the U/t–T plane, based on [188]. The low-temperature phase is an antiferromagnetic insulator (AFI) for any $U > 0$. For moderate Us, the high-temperature phase is a correlated paramagnetic metal (PM), while at high Us, a transition into a paramagnetic Mott insulator (PI) is observed. The semi-metallic (SM) crossover region (indicated by dots) between the PM and PI rapidly broadens with increasing T.

As for the weak coupling limit: collecting contributions up to order $1/D$ in an expansion about $D = \infty$ allows an essentially complete treatment which describes the non-analytic energy scale of the SDW as well as perturbative corrections up to second order in $u = U/t$ [418, 419]. Fluctuations originate from the on-site interaction thus they have no reason to vanish at $D = \infty$; on the contrary, the character of the solution is basically the same in $D = \infty$ and $D = 3$, and (at $T = 0$) presumably even in $D = 2$. The calculation confirms that though ordering is weakened by correlations, it is not eliminated at any $U \neq 0$. Formally, the results look like renormalized Hartree formulas. However, the D-dependent renormalization factors do not approach 1 as $U \to 0$ but retain distinct values of $O(1)$. This finding highlights the intrinsically fluctuating nature of order[36] derived from an on-site interaction: it would be wrong to think of the mean field value of the order parameter

[36]Meaning order with a non-conserved order parameter. Ferromagnetism is an exception.

as the correct "leading order" solution.

We cite a few results. All relevant quantities can be related to the $T = 0$ Hartree value[37] of the exponentially small SDW gap of the half-filled band

$$\Delta_0^{\rm H} \sim 2\sqrt{2}t \exp\left(-\frac{1}{U\rho_D(0)} - \frac{\gamma}{2} + \frac{1}{4D}\right) \tag{10.82}$$

where ρ_D is the D-dimensional DOS, and $\gamma \approx 0.577$ is Euler's constant. The inclusion of terms of order u^2 renormalizes the $T = 0$ gap by the factor

$$r = \exp\left\{-\ln(\sqrt{2}+1)\cdot\sqrt{2}\left(1+\frac{3}{32D}\right) - \frac{1}{16D}\right\}. \tag{10.83}$$

The same holds for the SDW Néel temperature $T_{\rm N} = rT_{\rm N}^{\rm H}$, while the temperature dependence of the gap follows

$$\Delta(T) = r\Delta^{\rm H}(T/r)\,. \tag{10.84}$$

The functional dependence of $\Delta/k_{\rm B}T$ on $T/T_{\rm N}$ is the same as in the Hartree theory.

It is of great interest that the treatment can be extended to band fillings $n \neq 1$ because it supplies a well-founded argument for phase segregation. Let us again begin with reviewing the Hartree results (here we stick to $T = 0$). As a homogeneous ground state, the two-sublattice SDW state is stable up to an exponentially small critical value of doping

$$\delta_{\rm cr}^{\rm H} = \rho_D(0)\Delta^{\rm H} \tag{10.85}$$

(where $\delta = 1 - n$), with the sublattice magnetization following

$$m_{\rm sl}^{\rm H} = \frac{2\Delta^{\rm H}}{U}\sqrt{1-\frac{\delta}{\delta_{\rm cr}}}\,. \tag{10.86}$$

Plotting the energy $E_{\rm AF}$ against δ, one finds a curve which is concave from below (as in Fig. 10.11), leading to the result that the ground state is segregated into an insulating antiferromagnet and a paramagnetic metallic phase if $0 < \delta < \sqrt{2}\delta_{\rm cr}$. Allowing for incommensurate states

[37] We should arrive at this result by replacing (4.55) in (7.66).

does *not* change the conclusion. – So much about the Hartree behaviour. It is found that analogous statements hold in second-order perturbation theory, only $\delta_{\rm cr}$ and m become renormalized [419]

$$\delta_{\rm cr} = r\delta_{\rm cr}^{\rm H} \qquad m_{\rm sl}(\delta, T=0) = r\cdot m_{\rm sl}^{\rm H}(\delta/r, T=0)\,. \tag{10.87}$$

The shape of the curve $E_{\rm AF}(\delta)$ remaining (apart from a change in scale) the same as in the mean field calculation, it follows that the ground state is phase segregated below the critical concentration $\delta_{\rm cr} = \sqrt{2}r\delta_{\rm cr}^{\rm H}$. It has been shown that incommensurate states do not come into play, and that the conclusion holds also for an extended Hubbard model with a not too strong nearest neighbour interaction. All in all, a variety of evidence suggests that phase segregation into a heterogeneous mixture of the AF insulator and the PM metal is a generic feature of lattice fermion models near half-filling. In the sense of the theory outlined above, the statement is supposed to hold in sufficiently high dimensions, but we are fairly certain that $D = 3$ is high enough. It is less clear whether the suppression of incommensurate states remains valid beyond the weak coupling regime. Alternative approaches (such as QMC) predict that the domain of two-sublattice antiferromagnetism is flanked by regions with incommensurate order [118, 383], and this may well be the valid picture at intermediate couplings.

Finally, let us discuss another feature of the $n = 1$ phase diagram shown in Fig. 10.18. In the weak-coupling limit, $T_{\rm N}$ is also the critical temperature for a metal–insulator transition. On the other hand, there is no doubt that for sufficiently high values of U/t, the high-temperature phase is a paramagnetic insulator (PI) with a Mott–Hubbard gap. It is customary to assume (as illustrated in Fig. 7.6) that at sufficiently low temperatures, the PM–PI transition is of first order. However, Fig. 10.18 displays no sharp PM–PI phase boundary. Instead, the thermal filling-in of the gap region leads to a rapid broadening of a semimetallic (SM) crossover regime (indicated by dotting). This should not be taken as necessarily negating the possibility of first-order transitions. It is also possible that above $T_{\rm N}$, we are in the supercritical regime, and the line of first-order transitions is buried under the AFI phase.

Solutions to the Problems

Problem 10.1 Let the band have the constant density of states $\rho(\epsilon) = 1/W$ in the interval $\epsilon \in [-W/2, W/2]$. In response to switching on the magnetic field H, the band becomes polarized with spin density $m = (n_\uparrow - n_\downarrow)/2$. At half-filling $n_\uparrow + n_\downarrow = 1$, the kinetic energy renormalization factor (C.20) turns out to be spin-independent

$$q_\uparrow = q_\downarrow = \frac{n_d}{\frac{1}{4} - m^2} \cdot [1 - 2n_d + \sqrt{(1-2n_d)^2 - 4m^2}]. \tag{10.88}$$

The energy density (C.21) has to be augmented by the Zeeman energy term

$$E(n_d, m) = -Wn_d[1 - 2n_d + \sqrt{(1-2n_d)^2 - 4m^2}] + Un_d - Hm. \tag{10.89}$$

If the field is weak, we can expand to second order in m. The part of the energy which depends on m

$$\frac{2m^2 W n_d}{1 - 2n_d} - mH$$

is minimized by

$$m = \frac{H(1-2n_d)}{4Wn_d}. \tag{10.90}$$

Replacing this in (10.89), the resulting expression can be minimized with respect to n_d, which yields

$$n_d \approx n_d^0 - \frac{H^2}{(8Wn_d^0)^2}. \tag{10.91}$$

As expected, polarizing the spins diminishes the concentration of d sites.

In (10.90), we can replace n_d by the $H = 0$ value n_d^0, which we quote from (9.47). The susceptibility is

$$\chi_s = \frac{\mu_B^2}{W} \cdot \frac{1 - 2n_d^0}{n_d^0} = \chi_P \cdot \frac{U_{cr} + U}{U_{cr} - U} = \chi_P \cdot \frac{m^*}{m} \cdot \left(1 + \frac{U}{U_{cr}}\right)^2. \tag{10.92}$$

This simple form is a special case of the general result [51]

$$\chi_s = \chi_P \cdot \frac{m^*}{m} \cdot \left\{1 - \frac{U_{cr}\rho(\epsilon_F)}{2}\left[1 - \frac{1}{4(1-2n_d^0)^2}\right]\right\}^{-1}. \tag{10.93}$$

(10.92) is recovered by setting $U_{cr}\rho(\epsilon_F) = 2$. The limiting value of the Wilson ratio depends on the density of states but in any case, χ_s is predicted to diverge as $U \to U_{cr}$, in essentially the same manner as m^*/m.

Expanding to first order in U, (10.92) coincides with the Hartree–Fock (or RPA) result $\chi_s = \chi_P \cdot (1 - U\rho(\epsilon_F))^{-1}$ (see (7.23)). However, the agreement ceases to hold at higher values of U. The Hartree–Fock instability against ferromagnetism at $U = W$ is now seen to be spurious, and it is replaced by the Brinkman–Rice transition at $U = 2W$. It is a general lesson that taking correlation effects into account shows that long range order is less likely to occur than mean field theories would make us believe.

The calculation of the third-order susceptibility proceeds similarly, only we have to expand E to fourth order in m, and m to third order in H. Omitting the details, one arrives at

$$\chi^{(3)} = \frac{4U^2}{(U_{cr} - U)^4} \propto \frac{m^*}{m} \cdot \chi_s^3 \tag{10.94}$$

which confirms (10.18). (the result differs in some details from those in [431] and [231], which were derived using a semicircular density of states). The magnetization curve deviates upwards from linearity. We can understand this by referring to (10.91): the field diminishes the concentration of the nonmagnetic d sites, and therefore at higher fields the system is more polarizable.

Problem 10.2 We proceed as with Problem 10.1, only now we cannot use expansion in powers of m, or H. Minimizing the m-dependent part of (10.89) with respect to m, we get

$$m = \frac{H(1 - 2n_d)}{\sqrt{4n_d^2 + H^2}} \tag{10.95}$$

which we replace back in (10.89), and seek the minimum with respect to n_d numerically. Typical results are shown in the upper half of Fig. 10.19. The essential observation is that, for large enough U, and at high enough fields, the energy plotted against n_d (called d in the figure) has *two minima*: one at a finite n_d (which evolved smoothly from the low-field regime), and another at $n_d = 0$. At some critical field, the absolute minimum switches from the finite-n_d solution to the $n_d = 0$ solution. The optimum values of n_d and the magnetization (bottom left and right plots in the figure) were found by tracing the field dependence of the absolute minimum. At the field where the two minima become degenerate, a first-order metal–insulator transition is taking place. At the same time, there is also a sharp metamagnetic transition, with the magnetization jumping to its saturation value.

Our considerations were based on the assumption of spatially uniform ground states. In cases where a first-order transition can take place, attention has to be paid to the possibility of heterogeneous solutions (phase segregation). We refer to [231] for further details.

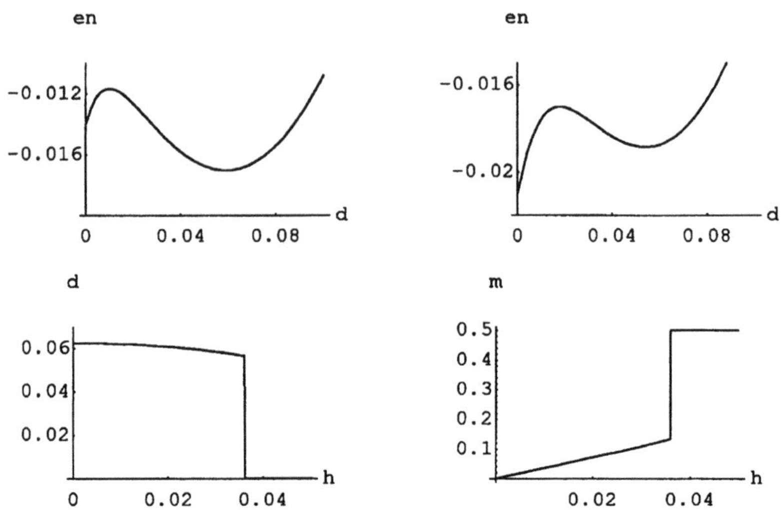

Figure 10.19: *Top*: At $U = 0.75U_{\text{cr}}$, the ground state energy (10.89) shows two minima as a function of the density d of doubly occupied sites. On the left ($H/W = 0.028$), the absolute minimum is at finite d, while on the right ($H/W = 0.042$) the absolute minimum is at $d = 0$. *Bottom*: the field dependence of d (left) indicates a first-order metal–insulator transition simultaneously with the sharp metamagnetic transition (right).

Problem 10.3 Using (10.20), the variational ground state energy density for a band $\epsilon \in [-W/2, W/2]$ with the DOS $\rho(\epsilon) = 1/W$ can be written as

$$\mathcal{E} = -W\left(\frac{n}{2} - n_d\right)[\sqrt{n_d} + \sqrt{1 - n + n_d}]^2 + Un_d. \tag{10.96}$$

We are interested in the behaviour when $|n - 1| \ll 1$, and $U > U_{\text{cr}}$. Let us first take $n < 1$. Since at $n = 1$ the state is localized, both n_d and $1 - n$ must be small. We make the Ansatz

$$n_d = \alpha(1 - n). \tag{10.97}$$

This does not constrain our solution: it turns out that for finite U, α is also finite, as long as $U > U_{\text{cr}}$. Keeping only the dominant terms (of order $1 - n$), the expression

$$\mathcal{E} \approx (1 - n) \cdot \left\{U\alpha - \frac{W}{2}[\sqrt{\alpha} + \sqrt{1 + \alpha}]^2\right\} \tag{10.98}$$

has to be minimized with respect to α, which yields

$$\alpha = \frac{1}{2}\left[-1 + \frac{U-W}{\sqrt{U(U-2W)}}\right]. \tag{10.99}$$

Remember that here $U_{\rm cr} = 2W$. Replacing this back into the energy, we get

$$\mathcal{E}_- = \frac{1-n}{2}[-U + \sqrt{U(U-U_{\rm cr})}]. \tag{10.100}$$

wherefrom the chemical potential on the $n < 1$ side is

$$\mu_- = \left(\frac{\partial \mathcal{E}_-}{\partial n}\right)_{n\to 1-0} = \frac{1}{2}[U - \sqrt{U(U-U_{\rm cr})}]. \tag{10.101}$$

For the more-than-half-filled case, the calculation proceeds analogously by assuming that for $n - 1 \ll 1$, $n_d \approx \beta(n-1)$, and leads to

$$\mu_+ = \frac{1}{2}[U + \sqrt{U(U-U_{\rm cr})}]. \tag{10.102}$$

The gap is determined from (10.25)

$$\Delta = \mu_+ - \mu_- = \sqrt{U(U-U_{\rm cr})} \tag{10.103}$$

whereby we proved (9.54).

Replacing our results in (10.20), we get

$$q \sim 2(1-n)\sqrt{\frac{U}{U-U_{\rm cr}}}. \tag{10.104}$$

For $U > U_{\rm cr}$, m^* diverges like $|1-n|^{-1}$. As $U \to \infty$, (10.21) is recovered. q seems to blow up as we approach $U_{\rm cr}$ from above. It should be kept in mind that (10.104) was derived assuming that $1 - n \ll (U - U_{\rm cr})/W$.

Problem 10.4 Taking it for granted that the Gutzwiller approximation respects rotational invariance, it is sufficient to calculate

$$\langle S_{\mathbf{i}}^+ S_{\mathbf{j}}^- \rangle = -\langle c_{\mathbf{i}\uparrow}^\dagger c_{\mathbf{j}\uparrow} c_{\mathbf{j}\downarrow}^\dagger c_{\mathbf{i}\downarrow} \rangle. \tag{10.105}$$

We immediately find

$$\langle FS|S_{\mathbf{i}}^+ S_{\mathbf{j}}^-|FS\rangle = -\left(\frac{1}{L}\sum_{\mathbf{k}}^{(FS)} e^{-i\mathbf{k}(\mathbf{i}-\mathbf{j})}\right)^2. \tag{10.106}$$

The limit $U \to \infty$ is described by the fully Gutzwiller-projected state (9.29). Analogously to (C.13), applying (C.16), and remembering that now $\{\mathbf{g}_1^{N/2}\} \cap \{\mathbf{h}_1^{N/2}\} = \emptyset$, we can write

$$\langle\Psi(\eta=0)|S_{\mathbf{i}}^+ S_{\mathbf{j}}^-|\Psi(\eta=0)\rangle = -L^N \sum_{\{\mathbf{g}_1^{N/2}\}}^{\mathbf{i}\notin,\mathbf{j}\in} \sum_{\{\mathbf{h}_1^{N/2}\}}^{\mathbf{j}\notin,\mathbf{i}\in}$$

$$\cdot \begin{vmatrix} e^{-i\mathbf{k}_1\mathbf{g}_1'} & \cdots & e^{-i\mathbf{k}_1\mathbf{g}_{N/2}'} \\ \vdots & \ddots & \vdots \\ e^{-i\mathbf{k}_{N/2}\mathbf{g}_1'} & \cdots & e^{-i\mathbf{k}_{N/2}\mathbf{g}_{N/2}'} \end{vmatrix} \cdot \begin{vmatrix} e^{i\mathbf{k}_1\mathbf{g}_1} & \cdots & e^{i\mathbf{k}_1\mathbf{g}_{N/2}} \\ \vdots & \ddots & \vdots \\ e^{i\mathbf{k}_{N/2}\mathbf{g}_1} & \cdots & e^{i\mathbf{k}_{N/2}\mathbf{g}_{N/2}} \end{vmatrix}$$

$$\cdot \begin{vmatrix} e^{-i\mathbf{k}_1\mathbf{h}_1'} & \cdots & e^{-i\mathbf{k}_1\mathbf{h}_{N/2}'} \\ \vdots & \ddots & \vdots \\ e^{-i\mathbf{k}_{N/2}\mathbf{h}_1'} & \cdots & e^{-i\mathbf{k}_{N/2}\mathbf{h}_{N/2}'} \end{vmatrix} \cdot \begin{vmatrix} e^{i\mathbf{k}_1\mathbf{h}_1} & \cdots & e^{i\mathbf{k}_1\mathbf{h}_{N/2}} \\ \vdots & \ddots & \vdots \\ e^{i\mathbf{k}_{N/2}\mathbf{h}_1} & \cdots & e^{i\mathbf{k}_{N/2}\mathbf{h}_{N/2}} \end{vmatrix}$$

$$= -\left(\frac{1}{L}\sum_{\mathbf{k}}^{(FS)} e^{-i\mathbf{k}(\mathbf{i}-\mathbf{j})}\right)^2 \cdot \frac{1}{(C_{N/2-1}^{L-2})^2} \sum_{\{\mathbf{g}_1^{N/2}\}}^{\mathbf{i}\notin,\mathbf{j}\in} \sum_{\{\mathbf{h}_1^{N/2}\}}^{\mathbf{j}\notin,\mathbf{i}\in} 1$$

The configuration sum in the last line gives $C_{N/2-1}^{L-2} C_{N/2-1}^{L-N/2-1}$. Recalling (C.4), we find that the renormalization factor of the spin correlation function is

$$\begin{aligned} g_{SS} &= \frac{\langle\Psi(\eta=0)|S_{\mathbf{i}}^+ S_{\mathbf{j}}^-|\Psi(\eta=0)\rangle}{\langle\Psi(\eta=0)|\Psi(\eta=0)\rangle} \cdot \frac{1}{\langle FS|S_{\mathbf{i}}^+ S_{\mathbf{j}}^-|FS\rangle} \\ &= \frac{L(L-1)}{(L-N/2)^2} \rightsquigarrow \frac{4}{(2-n)^2} \end{aligned} \tag{10.107}$$

whereby (10.27) is derived.

Problem 10.5 Our derivation closely follows [34]. To simplify the notations, let us write the Hubbard Hamiltonian (with nearest neighbour hopping) as

$$\mathcal{H} = -t\hat{T} + U\hat{D} \tag{10.108}$$

where the dimensionless operators $\hat{T}$ and $\hat{D}$ are readily identified. The definition of the Baeriswyl state

$$|\Psi_{\mathrm{B}}\rangle = \exp\left\{-\frac{\zeta}{2}\cdot\hat{T}\right\}\cdot|\Psi_\infty\rangle \tag{10.109}$$

presupposes the knowledge of the $U = \infty$ ground state $|\Psi_\infty\rangle$. We do not know it in detail but for the present purposes, it is enough to know that it contains only neutral sites, thus only even powers of $\hat{T}$ have a non-vanishing expectation value.

We expand to the lowest relevant order in ζ. The expectation value of the kinetic energy is

$$\langle \Psi_{\mathrm{B}}| - t\hat{T}|\Psi_{\mathrm{B}}\rangle \approx \zeta \cdot \langle \Psi_\infty|\hat{T}^2|\Psi_\infty\rangle \tag{10.110}$$

while for the interaction term we find

$$\begin{aligned}\langle \Psi_{\mathrm{B}}|U\hat{D}|\Psi_{\mathrm{B}}\rangle &\approx U\langle\Psi_\infty|\left(1-\frac{\zeta}{2}\hat{T}+\frac{\zeta^2}{8}\hat{T}^2\right)\hat{D}\left(1-\frac{\zeta}{2}\hat{T}+\frac{\zeta^2}{8}\hat{T}^2\right)|\Psi_\infty\rangle \\ &\rightsquigarrow \frac{\zeta^2}{4}\cdot\langle\Psi_\infty|\hat{T}\hat{D}\hat{T}|\Psi_\infty\rangle = \frac{\zeta^2}{4}\cdot\langle\Psi_\infty|\hat{T}^2|\Psi_\infty\rangle. \end{aligned}\tag{10.111}$$

To lowest order

$$\frac{\langle \Psi_{\mathrm{B}}|\mathcal{H}|\Psi_{\mathrm{B}}\rangle}{\langle \Psi_{\mathrm{B}}|\Psi_{\mathrm{B}}\rangle} = \left[\zeta t + \frac{\zeta^2}{4}U\right]\cdot\frac{\langle\Psi_\infty|\hat{T}^2|\Psi_\infty\rangle}{\cdot\langle\Psi_\infty|\Psi_\infty\rangle}. \tag{10.112}$$

Minimizing with respect to ζ gives $\zeta = -2t/U$. Using the arguments we learned in Sec. 5.1.4, the leading correction to the ground state energy density can be written as

$$\mathcal{E}_{gr.st.} \approx -\frac{t^2}{LU}\cdot\frac{\langle\Psi_\infty|\hat{T}^2|\Psi_\infty\rangle}{\cdot\langle\Psi_\infty|\Psi_\infty\rangle} = \frac{4t^2}{U}\cdot\frac{\langle\Psi_\infty|\left(\mathbf{S_i}\cdot\mathbf{S_j}-\frac{1}{4}\right)|\Psi_\infty\rangle}{\cdot\langle\Psi_\infty|\Psi_\infty\rangle}. \tag{10.113}$$

We recover the result that the antiferromagnetic Heisenberg model is the low-energy effective Hamiltonian in the large-U limit of the half-filled Hubbard model.

Problem 10.6 In the limit $L \to \infty$, the hopping Hamiltonian (10.44)-(10.46) can be rewritten as

$$\mathcal{H} = it\sum_j\sum_\sigma\sum_{m>0}\frac{(-1)^{m+1}}{m}[c^\dagger_{j+m,\sigma}c_{j,\sigma} - c^\dagger_{j-m,\sigma}c_{j,\sigma}]. \tag{10.114}$$

It follows[38] that the band dispersion is

$$\epsilon(k) = 2t\sum_{m>0}\frac{(-1)^{m+1}}{m}\cdot\sin mk = tk \tag{10.115}$$

[38] See, e.g., (509) in: L.B.W. Jolley: *Summation of Series*, 2nd ed., Dover Publ., Inc. (New York, 1961).

where $k \in (-\pi, \pi)$. For a half-filled band, the negative-energy states with $k < 0$ are occupied. Since all Bloch-states have the same velocity $v = (1/\hbar) \cdot (\mathrm{d}\epsilon/\mathrm{d}k) = t/\hbar$, the non-interacting ground state carries a large current. In fact, *all* N-electron states carry the *same* current which is thus not influenced at all by the Mott transition!

This unusual state of affairs is related to the fact that the band energy is an *odd* function of k: $\epsilon(-k) = -\epsilon(k)$. Remember that for the usual hopping Hamiltonians, the evenness $\epsilon(-k) = \epsilon(k)$ is guaranteed by time reversal invariance. We realize now that the innocuous-looking (10.114) was constructed to break time reversal invariance; in fact, $\mathcal{H}$ is self-adjoint but not real: $\mathcal{H}^* \neq \mathcal{H}$.

Chapter 11

Mixed Valence and Heavy Fermions

Placed into a solid environment, rare earth atoms donate two or three electrons to wide bands, and are thus left with $4f$ valence electrons only. The valence of a rare earth ion in a solid is defined as the number of electrons donated to the $5d$–$6s$ type conduction bands. Rare earth ions are usually trivalent. Notable exceptions are Eu, and Yb, which tend to be divalent, and Ce which may loose more than three electrons, becoming tetravalent.

An alternative way to characterize the valence state of an ion is to count its f-electrons. Though the energies of the $5d$-, $6s$-, and $4f$-like states tend to overlap, it turns out that the number of the strongly interacting $4f$ electrons which remain at the ion, is often a sharply defined integer. For instance, Sm^{3+} is described as a $4f^5$ ion, and Yb^{2+} has the closed-shell configuration $4f^{14}$.

In many systems, including most elemental rare earths, all $4f$ sites have the same integral valence. In addition, there are also systems in which the valence can have any of two neighbouring integral values; these are the so-called *mixed-valent systems.* Quite often, the local valence states are mixed by either the quantum-mechanical motion of the electrons, or thermally; this gives rise to a non-integral, *intermediate valence.* Both integral and intermediate valence arise in a natural manner as a result of strong correlations.

11.1 Lanthanides and Actinides

In the periodic table, there are two series of elements with incomplete f-shells. In the lanthanide series of rare earths[1], the $4f$-shell is gradually filled up starting with cerium, and ending with lutetium. Lanthanum with its empty $4f$-shell is a useful reference material, and we will discuss it along with the rare earths.

The $5f$-shell is partially filled in the actinides. For the purposes of solid state physics, the actinide series is rather incomplete: from the transurans, only neptunium, plutonium, and americium are available in quantities larger than a few grams, and towards the end of the series, we may have only a small number of short-lived atoms[2].

The f-electron count in solids is often different from that in atoms. We have mentioned (see footnote 15 on p. 45) that it is debatable whether the cerium atom has two or one f-electrons. For solid state physics, this controversy hardly matters because Ce tends to donate at least three electrons to wide bands, thus the ionic configuration is either Ce^{3+}: $[Xe]4f^1$, or Ce^{4+}: $[Xe]4f^0$. Similarly, the praseodymium atom ($[Xe]4f^3 6s^2$) usually looses one f-electron and is left in the ionic configuration $4f^2$ which we discussed at length in Sec. 3.8. For most of the rare earths, the $5d$-$6s$ shell of the isolated atom harbours only two electrons, but the ionic state in a solid tends to be trivalent[3]. The

[1]A note about terminology: Lanthanides are the elements 58 through 71 (Ce through Lu). For many solid state physicists, "rare earths" would be La and the lanthanides, and this is how we will understand the term. Let us, however, note that The International Union for Pure and Applied Chemistry recommended (1968) that the name "rare earth" should be applied to the lanthanides, and also to Sc, Y, and La. The reasons are, apart from the great similarity in the chemical behaviour of these elements, historical: the name "rare earth" was first given to yttrium oxide which, at the time of its discovery in 1794, was thought to be an element.

[2]Whether available in larger quantities or not, experiments on the actinides are confined to a few specially equipped laboratories, because the substances are usually radioactive, toxic, and unpleasantly reactive. From the pre-uran elements protactinium, though naturally occurring in certain ores, is quite scarce and avaliable only at the gram scale. Even more scarce (mg scale) is actinium which is produced by isotope irradiation in high-flux reactors. Let it be mentioned here that the artificially produced radioactive rare earth element promethium (Pm) is also very intractable, and little discussed in solid state texts.

[3]Exceptions occur in the middle of the series, where Hund's rule energy can be

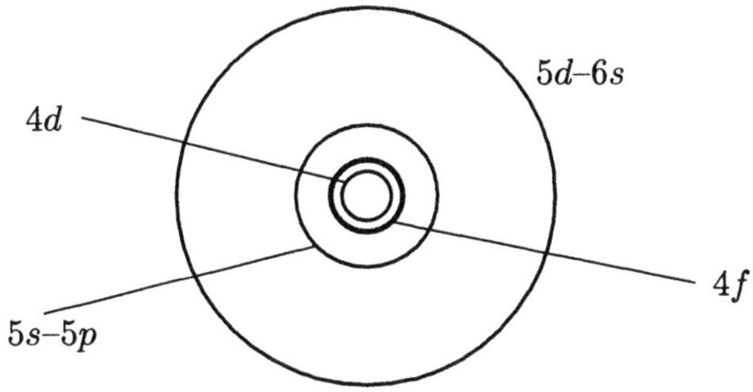

Figure 11.1: Schematic representation of the atomic orbitals of a rare earth atom (after [134]). The circles give the approximate position of the charge density maxima. The 5s-5p orbit indicates the size of the xenon core. The lattice constant is about ten times as large as the diameter of the 4f shell.

explanation of this finding is that the binding energy of the solid is enhanced by donating a third electron to the wide 5d-6s bands.

The unique and interesting features of rare earth systems originate from the strong intrasite correlations between 4f electrons, which seem to be essentially unaffected by the surrounding crystal. The underlying reason is the extraordinary compactness of the 4f shell. As shown in Fig. 11.1, the 4f shell is lying well inside the xenon core!

Since the core electrons outside the 4f orbital shield the fields arising from neighbouring atoms very effectively, the physics of rare earth systems is dominated by the *strong intra-ionic correlations* of the f-electrons. The behaviour of a lanthanide ion in a crystal is pretty much the same as it would be for free ions. At least, this holds for relatively high temperatures (room temperature is, in this respect, a high temperature). At sufficiently low temperatures, the effects of a *weak crystal field* may become apparent. As we discussed in Ch. 3, the weakness of the crystal field means that all three of Hund's rules apply; but in spite of this fact, magnetism may turn out to be largely quenched for temperatures lower than the crystal field splittings.

---

gained by aligning the maximum number of spins. Thus in order to keep the half-filled shell configuration $4f^7$, europium tends to give up only two electrons, while gadolinium has three 5d-6s electrons even as a neutral atom.

The effective magnetic moments deduced from room-temperature susceptibility data (see Fig. 2.1 on p. 49) give a good idea of the quality of the single-free-ion approximation. We see that (apart from two cases in the middle of the series) the predictions based on Hund's rules are in good agreement with the observations. This tells us that, for most of the rare earth ions, the spin–orbit splitting is larger than the thermal energy $k_B T$ in the interesting temperature range. For instance, for Pr^{3+}, the energy difference between the ground state $J = 4$ term, and the excited $J = 5$ is $\Delta \approx 3150$K. Thus it is a good approximation to confine our attention to the (9-fold degenerate) Hund's rule ground state, and neglect the effect of higher-lying levels. We have also understood the reason for the exceptional behaviour of Sm^{3+} and Eu^{3+}: for these ions, the multiplet splitting $\Delta \sim 400 - 500$K is quite low, and the first excited level has to be taken into account (see p. 49 and p. 72).

The $5f$-series of the actinides is also unique in its own way, but at the same time rather different from the rare earths [52]. Speaking of the elements, the beginning of the series, the light actinides, are reminiscent of d-metals[4] while the heavy actinides are similar to the rare earths. Plutonium and americium lie in the transition region.

11.2 The Kinds of Valence: Integral, Mixed, and Nearly Integral

Why should we want to discuss valence? For the majority of solids, the question about the valence states of the individual atoms/ions is neither too interesting, nor particularly meaningful. Assume that we are trying to determine valence by counting the electrons in the atomic-like states which we equate with the Wannier orbitals[5]. In band insulators, bands are either full, or empty. The filling of all Bloch states being equivalent to the filling of all Wannier states, all sites are in the same valence

[4]A historical note: in earlier times, it was thought that the light actinides belong to the $6d$ series of transition metals, and the filling-up of the $5f$ shell begins only in the middle of the series. Now we know otherwise: all the actinides are $5f$ metals.

[5]This works smoothly if we describe bands in the tight binding scheme. Band theory uses different basis states for wide bands. In any case, the question of valence is interesting only for narrow-band systems.

state. On the other hand, for wide-band metals it is clear from the discussion in Sec. 4.2, and Sec. 9.1, that all sites fluctuate between *all* the valence states permitted by the shell degeneracy[6]. In this sense, band theory implies the largest possible degree of valence fluctuation. Furthermore, the fluctuation is *fast*: the lifetime of any particular local valence state is $\tau \sim \hbar/W$, where W is the bandwidth. The shortness of τ implies that the local configurations get averaged over on the time scale of most measurements; that's why the local approach would be useless for wide-band metals.

We have also understood that the concept of valence becomes relevant in strongly correlated systems. The Gutzwiller approach is based on the variational control of valence fluctuations (which we called polarity fluctuations in Ch. 9). When the system (usually a transition metal compound) becomes a Mott insulator, polarity fluctuations are suppressed, and the valence becomes sharply defined[7], the partial filling of the d-shell notwithstanding.

f-electron systems are remarkable because we find very strong restrictions on the valence state of the ions, even in metals. This is particularly true for rare earth materials, but there are similar findings for some actinide systems, too. To see how this comes about, we consider the crudest model of a rare earth system, and use it for qualitative arguments.

The fact that the $4f$-shell is very compact has two consequences: first, there is a strong on-site, Hubbard-like interaction between electrons sitting in the same shell; and second, there is hardly any overlap, or hopping matrix element, between adjacent f-shells. Thus the subsystem of $4f$-electrons can be modelled by the atomic limit of a degenerate Hubbard model. In addition, there are also electrons moving in wide ($5d$–$6s$-like) bands. To model this situation, let us consider a set of $\mathcal{N}$-fold degenerate Hubbard atoms (which we call $4f$ ions) embedded in

[6]Metallic Na was argued to contain Na^+, Na^-, and Na^0; and independent-electron theory would have it that metallic Ni has everything, from $3d^{10}$ to $3d^0$.

[7]We understand that this statement is compatible with the existence of the residual virtual fluctuations which give rise to the Heisenberg coupling.

a conduction electron sea:

$$\mathcal{H}^{(0)} = \mathcal{H}_{\text{band}} + \mathcal{H}_f = \sum_{\mathbf{k}\sigma} \epsilon(\mathbf{k}) c^{\dagger}_{\mathbf{k}\sigma} c_{\mathbf{k}\sigma} + U \sum_{\mathbf{j}} \sum_{\alpha<\beta}^{\mathcal{N}} f^{\dagger}_{\mathbf{j}\alpha} f_{\mathbf{j}\alpha} f^{\dagger}_{\mathbf{j}\beta} f_{\mathbf{j}\beta} \,. \quad (11.1)$$

The degeneracy $\mathcal{N}$ counts orbital and spin degeneracies together, thus $\mathcal{N} = 14$ for f-electrons; and we assume that the interaction of any two electrons on the same site is given by the same Hubbard U. Superficially, (11.1) may look like the degenerate Hubbard model but note the important difference: in (8.36) the same electrons move in the band *and* interact locally while here, the two terms contain different fermion operators. (8.36) is a one-band model, while (11.1) is a two-band model. As yet, we have not included a hybridization term; the only way in which the two sets of electrons communicate with each other is by being in contact with the same particle number reservoir which has the chemical potential μ.

The spectrum of the Hubbard sites consists of $\mathcal{N}$ equidistant levels with spacing U (Fig. 11.2, middle), the ionic energy levels being ϵ_f, $\epsilon_f + U$, $\epsilon_f + 2U$, etc. There is one Hubbard atom per unit cell; the total concentration of electrons (per unit cell) is $n = n_c + n_f$, where n_c is the concentration of conduction electrons, and n_f is that of the f-electrons.

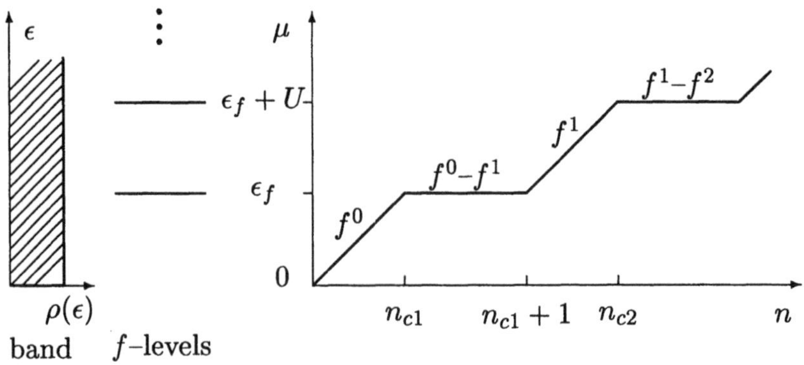

Figure 11.2: A wide band (left, schematic) overlapping with the Hubbard spectrum of strongly correlated states (middle) gives rise to an alternation of integral valent and mixed valent regimes (right). Increasing the total electron concentration n, the chemical potential μ passes through a series of plateaus in which μ is pinned to one of the f-levels.

Let us discuss the relationship between the overall electron concentration n, the chemical potential μ, and the $4f$ count n_f. The situation is depicted in Fig. 11.2. Assuming that ϵ_f is lying above the bottom of the band, increasing n from zero, μ rises gradually while the band is filling up, and the $4f$-shells remain empty, i.e., the ionic configuration is $4f^0$. However, when μ reaches ϵ_f at some $n = n_{c1}$, the electrons begin to populate the $4f$ shells. In the range $n_{c1} < n < n_{c1} + 1$, *the chemical potential is pinned at* ϵ_f, and a fraction of the f-sites is in the valence state $4f^1$, while the remaining sites are empty. This is the case of $4f^0$–$4f^1$ *valence mixing.* To put it differently, *the average valence is intermediate*: $0 < n_f = n - n_{c1} < 1$. We see, however, that the term "intermediate" has to be interpreted carefully: because of the strong Coulomb interaction between $4f$-electrons, the non-integral value of n_f arises from the mixture of the two neighbouring valence states $4f^0$ and $4f^1$ – and nothing else. This should be contrasted with the situation for uncorrelated electrons, where the contribution of all valence states is significant. Mixed valent systems are unique in the sense that the allowed values of the valence are restricted to just two values. The phenomenon of mixed valence offers us the most clear-cut realization of the principle of minimum polarity.

Raising μ further, the filling of band states begins again, and every site is in the $4f^1$ state. This integral valent situation is apparently stable over a wide range of band filling. Then, after the electron concentration has reached n_{c2}, further added electrons go into the f-shells, turning some of the sites into $4f^2$. In the regime of $4f^1$–$4f^2$ valence mixing, μ remains pinned to $\epsilon_f + U$ until every f-shell has taken up a second electron. The alternation of integral valent and mixed valent (generally $4f^n$–$4f^{n+1}$) behaviour continues like this until the f-shells get completely filled with $\mathcal{N} = 14$ electrons each.

Allowing for a weak quantum-mechanical mixing (hybridization) between f and c states, the sharp corners of the μ–n plot shown in Fig. 11.2 become rounded, but plateaus with approximately constant μ are still recognizable. However, the bondary between integer and mixed valence regimes is no longer sharply defined but becomes replaced by a transition regime of nearly integral valence. We resume this discussion after a short *Digression* about valence states in a system with negative-U

Hubbard centers.

11.2.1 Digression: Valence Skipping

If electrons are added to a $U > 0$ system, the valence states follow each other in an orderly fashion: $f^0 \to f^1 \to f^2 \to \ldots$. However, the story becomes entirely different if we assume that the Hubbard U is negative: $U = -|U|$. Let us reconsider the reasoning which gave us Fig. 11.2. Though putting the first electron to a site[8] costs the energy ϵ_f, putting a second electron to the same site costs merely $\epsilon_f - |U|$. Hence we conclude that as soon as the chemical potential reaches the value $\mu = \epsilon_f - |U|/2$, we start creating f^2 sites. In a range of the overall electron concentration n, we find $f^0 - f^2$ *valence mixing*. At the upper end of this range, the filling-up of the band continues and all sites remain f^2. The valence f^1 is *skipped*.

Though the assumption of negative-U Hubbard centres may sound unfounded, it is interesting to observe that there are about fifteen elements which rather consistently skip certain valence states [426]. For instance, Sb and Bi are tri- or pentavalent but tetravalence is essentially missing. Pt likes to be Pt^{2+} or Pt^{4+} but apparently shuns Pt^{3+}. Though such observations are often conveniently interpreted with the empirical rule of preference for (half)filled shells[9], it is tempting to try to give an explanation in terms of an effective $U < 0$. It would not be simple to justify that the effective U is negative: clearly, it should result from the full treatment of a problem in which the shell electrons in question interact with the lattice displacements of the surrounding complex, and with other electronic subsystems. However, we may accept it that experiments tell us that such a justification should be possible.

To take an example, the average valence state of Bi in the perovskite $BaBiO_3$ is Bi^{4+}, i.e., $6s^1$ (apart from closed shells). Based on the picture of a half-filled $6s$ band, we expect a paramagnetic metal but we find a diamagnetic insulator. What are the $6s$ electrons doing in $BaBiO_3$?

Let us model the $6s$ band with a half-filled negative-U Hubbard model and assume that $|U|$ is sufficiently large. The Bi sublattice of $BaBiO_3$ is bipartite, thus we may use the transformation (4.86) (see p. 193) to establish the correspondence with the half-filled $-U > 0$ Hubbard model. The single-site basis states are mapped (apart from an alternating sign factor) according to

$$|\uparrow\rangle \to |d\rangle \qquad |\downarrow\rangle \to |0\rangle \qquad |0\rangle \to |\downarrow\rangle \qquad |d\rangle \to |\uparrow\rangle \tag{11.2}$$

[8]Since it is just a formal argument, we do not bother to the change notation, and go on to speak about "local f-shells". As it turns out, the physical situation arises for s-, p-, or d-electrons.

[9]For example, Bi^{5+} is $[Xe]4f^{14}5d^{10}$, while Bi^{3+} is $[Xe]4f^{14}5d^{10}6s^2$.

Thus the $\uparrow - \downarrow$ alternation of the antiferromagnetic ground state of the positive-U model is mapped onto an alternating $6s^2 - 6s^0$ charge order of the negative-U model. We conclude that the ground state of $BaBiO_3$ is a *charge density wave.* Since it amounts to an alternation of Bi^{3+} and Bi^{5+} valence states, we may also speak about valence disproportionation. The argument implies that (at sufficiently large $|U|$) the charge order is stabilized by a coupling $\sim t^2/|U|$.

The negative-U concept has a nice corollary. Let us notice that the binding into singlet pairs (d-sites) is energetically favoured whether the band is half-filled or not. However, away from half-filling, the singlet pairs are free to move. This seems to be the recipe for a superconductor with a small coherence length. And indeed, doped $BaBiO_3$ (notably, $Ba_{1-x}K_xBiO_3$ [263]) is a representative of one of the classes of high-T_c superconductors.

11.2.2 Valence Mixing in Rare Earth Compounds

For the rest of this chapter, we return to the study of positive-U systems whose zeroth order Hamiltonian is (11.1). In the mixed valent regimes, the ground state of $\mathcal{H}^{(0)}$ is macroscopically degenerate. Taking the example of $4f^0$–$4f^1$ mixing, there are $C^L_{Ln_f}$ ways to distribute the $4f^1$ sites over the lattice, and the ground state entropy is further increased by the fact that each $4f^1$ site has a degenerate ionic ground state.

The most significant omission in writing down (11.1) is that it neglects any quantum-mechanical mixing between the f-electrons and the conduction electrons. In reality, the two sorts of electron states are *hybridized.* The complete Hamiltonian should include at least the one-electron hybridization processes[10] which turn an f-electron into a conduction electron, or vice versa

$$\mathcal{H} = \mathcal{H}^{(0)} + \mathcal{H}_{\text{hyb}} = \mathcal{H}^{(0)} + \sum_{\mathbf{k}\sigma} \sum_{\mathbf{j}\alpha} [V_{\mathbf{k}\sigma\mathbf{j}\alpha} c^{\dagger}_{\mathbf{k}\sigma} f_{\mathbf{j}\alpha} + \text{H.c.}] \,. \qquad (11.3)$$

To see the effect of switching on $\mathcal{H}_{\text{hyb}}$, let us consider the two consecutive hybridization processes shown in Fig. 11.3. We envisage a

[10] The notation $V_{\mathbf{k}\sigma\mathbf{j}\alpha}$ for the hybridization amplitude is very slipshod. Properly written, it should make it clear that hybridization satisfies the requirements of symmetry. Furthermore, some off-diagonal Coulomb processes also hybridize. However, we do not need the detailed form here.

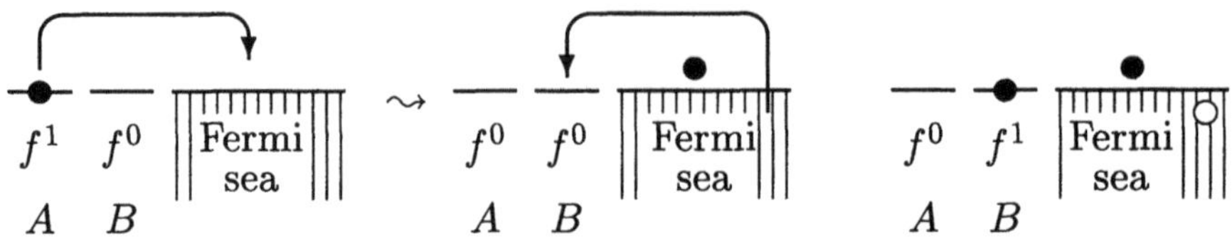

Figure 11.3: Two consecutive hybridization processes allow the f-electron to propagate from site to site (discussion in the text).

system well inside the regime of f^0–f^1 valence mixing where both f^0 and f^1 are abundant, and the band is filled up to $\mu = \epsilon_f$. Consider two sites (not necessarily nearest neighbours) so that initially A has an f-electron, and B does not. In the first step, the f-electron is "hybridized away" from A into a band state above μ. In the second step, a Fermi sea electron is hybridized onto B. If it is the "same" electron that we excited in the first step then in end-effect, the Fermi sea is undisturbed, and an f-electron propagated from A to B. Thus hybridization mixes the ground states of $\mathcal{H}^{(0)}$, and thereby lifts the ground state degeneracy. However, the second step may involve (as illustrated in Fig. 11.3) taking an arbitrary conduction electron from the Fermi sea, and place it onto B. Thus in addition to letting the f-electrons propagate, hybridization redistributes the band electrons in the vicinity of the (zeroth-order) chemical potential μ.

As it transpires from the above discussion, the distribution of the f^0 and f^1 sites over the lattice is not static but dynamic: any particular site *fluctuates* between the two valence states[11]. This is the justification for speaking about *fluctuating valence.* At the same time, the intermediate (non-integer) value of n_f becomes identical with the quantum-mechanical expectation value $\langle \hat{n}_{\mathbf{j}}^f \rangle$ for any site $\mathbf{j}$. It is in this sense that one speaks about *homogeneous mixed valence*[12]. The terms

[11] In this Chapter, we emphasize the role of hybridization because it is important for f-electron systems. It is, however, clear that band motion can play a similar role: the metallic state of the large-U Hubbard model at non-integral fillings is also intermediate valent. Our discussion of double exchange was based on the assumption that doped $LaMnO_3$ is a Mn^{3+}–Mn^{4+} valence fluctuator (Sec. 8.5).

[12] The phenomenon of homogeneous mixed valence is widespread in rare earth systems. There are, however, also cases of *heterogeneous mixed valence*, where the two valencies form a superlattice (Wigner crystal). This is realized for $4f$-electrons

"mixed", "intermediate", and "fluctuating" are used more-or-less interchangeably, depending on which aspect of the phenomenon one wishes to emphasize.

Hybridization matrix elements are usually quite small in comparison to either U, or the bandwidth W, thus hybridization can be treated as a weak perturbation. However, what we need is degenerate perturbation theory: the lowest-order effect of switching on hybridization is to mix the vast number of the zeroth-order ground states. This is a difficult problem which can be compared to that of the large-U Hubbard model.

A simple argument (presented a little later) shows that the process discussed above has an essential influence within a hybridization width of order $\sim V^2\rho(\epsilon_f)$ around ϵ_f (V^2 is a suitable average over $|V_{\mathbf{k}\sigma\mathbf{j}\alpha}|^2$, and ρ is the band density of states). $V^2\rho(\epsilon_f)$ can be identified as the broadening of the f-level, or – in another context – as the hybridization-induced effective f-bandwidth. It sets a temperature scale, too: at low temperatures ($V^2\rho > k_BT$), valence fluctuation is quantum-mechanical, while for $T > V^2\rho/k_B$, it is mainly thermal.

There is also an associated time scale: $\tau_{\text{fluc}}^{-1} \sim V^2\rho/\hbar$. For measuring times longer than τ_{fluc}, the state of a lattice site gets averaged over f^0 and f^1, and the valence is seen to be intermediate. On the other hand, an observation over a timespan shorter than τ_{fluc} yields a snapshot of the mixed valent state: an array of sites which appear randomly frozen either into the configuration f^1, or f^0. The interesting point is that for parameter values characteristic of rare earth systems, $\tau_{\text{fluc}} \approx 10^{-12}s$ falls between the timescales of two widely used techniques: Mössbauer effect (ME) with its 10^{-8}–10^{-10} s, and X-ray photoelectron spectroscopy (XPS) with 10^{-15}–10^{-18} s [436]. It is a favourite experimental check on mixed valent character that XPS should see a mixture of two ionic species, while Mössbauer effect finds an average charge state which is intermediate between the ionic values.

Typical results are shown in Figs. 11.4 and 11.5 for SmB_6 which has $4f^5$–$4f^6$ valence mixing. The isomer shift is almost temperature-independent, and corresponds to the average valence state which is approximately $Sm^{2.6+}$. XPS sees two sets of lines, and can be fitted with

in Eu_3S_4, and for $3d$-electrons in $Pr_{1/2}Sr_{1/2}MnO_3$, in $La_{1/2}Sr_{3/2}MnO_4$ (Sec. 8.5), and also in magnetite (Chapter 1).

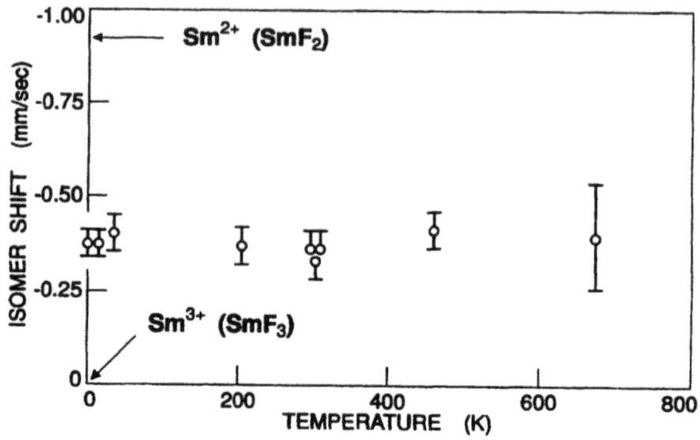

Figure 11.4: Mössbauer effect isomer shift measurements on SmB_6 indicate that the Sm ions are in an approximately temperature-independent intermediate valence state $Sm^{2.6+}$ (after [66], reproduced in [436]).

the assumption that $\sim$ 60% of the sites are Sm^{3+}, and $\sim$ 40% Sm^{2+}. The comb-like structures in the lower part of Fig. 11.5 show the multiplet splittings of the respective configurations. Though both configurations correspond to an energy range with several peaks, we can read off an overall $4f^5$–$4f^6$ splitting which allows to estimate that for Sm, the Hubbard $U_{4f} \approx 7$eV. The largeness of U_{4f} shows that it is not unwarranted to think of mixed valence as a strong correlation phenomenon.

The slowest measurements are, of course, the static ones, for instance the measurement of the lattice constants. Fig. 11.6 shows the metallic radii of the rare earth (RE) elements, along with those of barium and lanthanum. Most of the points (including La) are lying on a gently falling line which belongs to trivalent ions. The fact that the atomic volume of the trivalent substances decreases slowly with increasing atomic number, is known as the lanthanide contraction. It can be ascribed to the fact that though the electron content of the $4f$-shell increases, it cannot completely screen the increase in the nuclear charge and therefore the outer electrons get sucked in.

We notice that europium and ytterbium have outstandingly large values of the metallic radius; values which are similar to that found for divalent barium. We conclude that Eu and Yb are divalent. In either case, the deviation from the standard trivalent behaviour is associated

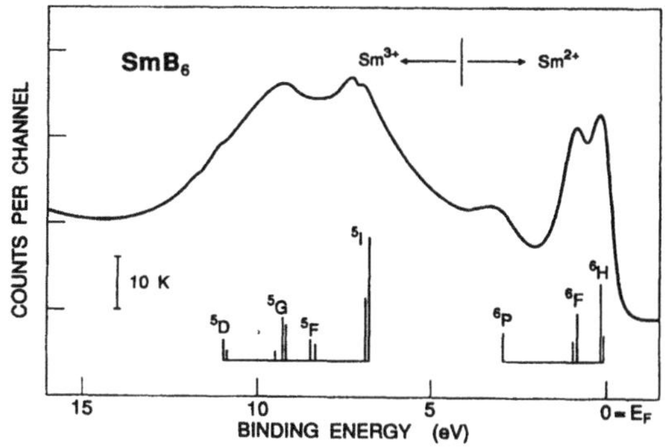

Figure 11.5: XPS data on intermediate valent SmB_6 can be fitted by assuming a 40%–60% mixture of Sm^{2+} and Sm^{3+} valence states (after [55], reproduced in [436]). The $4f^5$–$4f^6$ Hubbard splitting is $U_{4f} \approx 7$eV.

with the drive towards particularly stable $4f$ shell configurations. We have mentioned that Eu prefers divalency in order to have an exactly half-filled f-shell; similarly, Yb^{2+} ions have a completely filled $4f$ shell.

We find also a value which is remarkably small: one of the points for Ce. Cerium has two modifications: γ-Ce looks trivalent and behaves like the majority of rare earths but α-Ce appears to be strongly intermediate valent, say like $Ce^{+3.6}$, and has a much smaller atomic volume.

An important general lesson is that the valence state of the constituent ions has a large effect on the lattice constant. The relationship is so marked that (after establishing suitable benchmarks) it can be used to determine the valence state from lattice constant measurements. Naturally, it is best to take an isostructural series of certain compounds: e.g., the RE hexaboride series, or the family of RE monosulphides.

A further hybridization effect is that the chemical potential suffers a shift of the order $\sim V^2\rho(\mu)$, and becomes slightly n-dependent. Thus the plateaus shown in Fig. 11.2 would be no longer perfectly flat but gently rising. Still, the picture of the mixed valence regimes remains essentially valid: there are wide ranges of n where the chemical potential is pinned within a distance $\sim V^2\rho(\epsilon_f)$ to one of the f-levels, and the main effect of changing n is the corresponding change in the f^n–f^{n+1} mixing ratio.

The model argument based on Fig. 11.2 may lead us to believe that

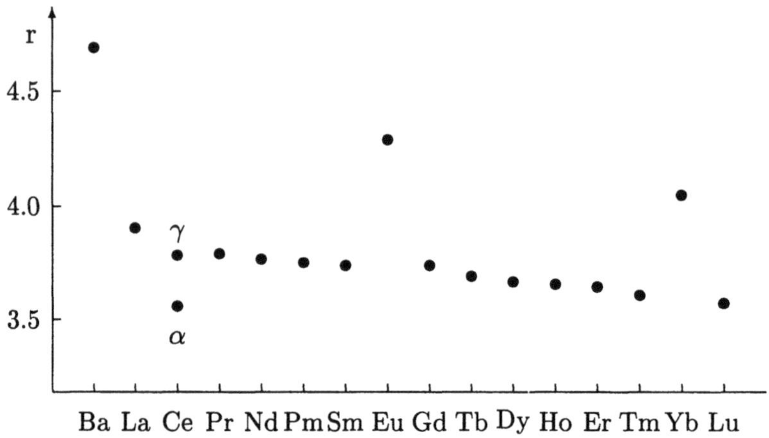

Figure 11.6: The metallic radii (in atomic units) as a function of the atomic number, from Ba through the rare earth series. Note the exceptionally large values for the divalent elements, and the small radius of α-Ce which is nearer to tetra- than to trivalency.

valence mixing between any two neighbouring integers can occur as we pass along the rare earth series. However, in reality, requirements of the stability of the solid restrict the range over which the chemical potential, or the valence of any particular rare earth ion, can be varied. Furthermore, the positions of the f-levels should be corrected by Hund's rule energies, which is an important effect if we are near one of the stable configurations of the f-shell. It would be very hard to work this out in detail, thus we just accept it as a matter of experimental fact that not all rare earths are capable of mixed valent behaviour, only those at the beginning (Ce and perhaps[13] Pr), in the very middle (Eu, Sm), or at the end (Tm, Yb), of the series. Even these ions are not always mixed valent (for instance, in elemental Eu and Sm they are not), only in the environment provided by particular substances [424, 237, 134, 436].

[13] Pr is usually trivalent. There are a few compounds like Na_2PrF_6, PrF_4, and PrO_2 (or at least Pr_6O_{11}) in which Pr is stable as Pr^{4+}. This suggests that homogeneous valence mixing may also occur, but the author is not aware of any case in which non-integer valency of Pr would be as clearcut as that of Sm is in SmB_6.

11.2.3 Renormalized Hybridized Bands

We can get rather far in visualizing the behaviour of mixed valence systems by assuming that it arises from the crossing of a wide conduction band and a dispersionless f-band, and that the strong Hubbard U merely renormalizes the parameters, but it does not change the qualitative aspects of the behaviour. For an elementary discussion, it is sufficient to consider the non-degenerate periodic Anderson model (PAM)

$$\begin{aligned}\mathcal{H}_{\text{PAM}} &= \sum_{\mathbf{k},\sigma} \epsilon_{\mathbf{k}} c^{\dagger}_{\mathbf{k}\sigma} c_{\mathbf{k}\sigma} + \sum_{\mathbf{k},\sigma} \epsilon^{f}_{\mathbf{k}} f^{\dagger}_{\mathbf{k}\sigma} f_{\mathbf{k}\sigma} - \sum_{\mathbf{k},\sigma} v_{\mathbf{k}} \left(c^{\dagger}_{\mathbf{k}\sigma} f_{\mathbf{k}\sigma} + f^{\dagger}_{\mathbf{k}\sigma} c_{\mathbf{k}\sigma} \right) \\ &+ U_f \sum_{\mathbf{j}} \hat{n}^{f}_{\mathbf{j}\uparrow} \hat{n}^{f}_{\mathbf{j}\downarrow} \end{aligned} \tag{11.4}$$

where $\hat{n}^{f}_{\mathbf{j}\sigma} = f^{\dagger}_{\mathbf{j}\sigma} f_{\mathbf{j}\sigma}$. The first term describes the wide conduction band of width W, the second the narrow f-band, and the third term their hybridization. In what follows, we neglect direct f–f hopping and assume that the f-band is dispersionless $\epsilon^{f}_{\mathbf{k}} \equiv \epsilon_f$. The last term describes the strong on-site interaction of f-electrons; all other Coulomb matrix elements are omitted. It would be typical to take $|v_{\mathbf{k}}| \sim 0.1\text{eV}$, W of the order of several eV, and U_f still larger. It transpires from our previous qualitative discussion that v (the average $v_{\mathbf{k}}$) is important in spite of its numerical smallness: it gives a finite width $W_f \sim v^2/W$ to the f-level. – The PAM can hold up to $n = 4$ electrons (plainly speaking, two f-electrons, and two c-electrons) per lattice site.

The non-interacting part of the PAM is diagonalized by a pair of hybridized bands (Fig. 11.7)

$$\lambda^{\pm}_{\mathbf{k}} = \frac{1}{2} \left(\epsilon_{\mathbf{k}} + \epsilon_f \pm \sqrt{(\epsilon_{\mathbf{k}} - \epsilon_f)^2 + 4v^2_{\mathbf{k}}} \right) . \tag{11.5}$$

Assuming $v \ll W$, the branch $\lambda^{-}_{\mathbf{k}}$ has f-character if $\epsilon_f - \lambda^{-}_{\mathbf{k}}$ is of the order of v^2/W (which holds if $\epsilon_{\mathbf{k}} > \epsilon_f$), and c-character outside this range (roughly speaking, for $\epsilon_{\mathbf{k}} < \epsilon_f$). The salient feature of the band structure is the hybridization gap opening around ϵ_f. More precisely, the gapping is complete if $v_{\mathbf{k}}$ does not vanish anywhere (e.g., for the often assumed $\mathbf{k}$-independent hybridization[14] $v_{\mathbf{k}} \equiv v$ which we use in

[14]This is equivalent to saying that we have only on-site hybridization. At the first

Fig. 11.7). If $v_{\mathbf{k}}$ vanishes at certain points of the Brillouin zone, the gapping may be incomplete but in any case, we expect a strong dip in the DOS at ϵ_f.

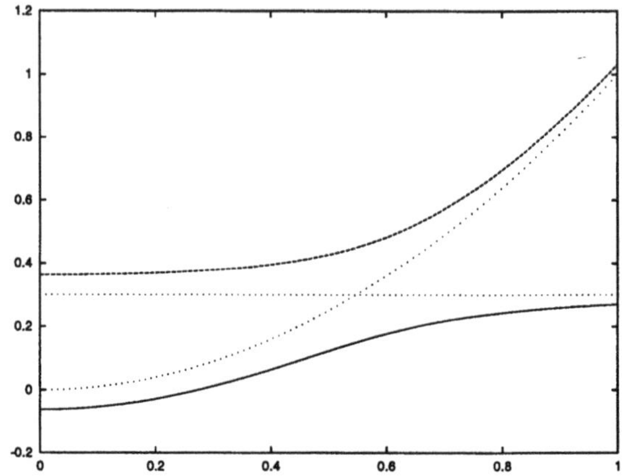

Figure 11.7: **The hybridization of the conduction band $\epsilon_{\mathbf{k}} = k^2$, and a dispersionless f-band at $\epsilon_f = 0.3$ (thin dotted lines) with $v = 0.15$ gives rise to two hybridized bands (bold lines). Horizontal axis: k (in units of π/a), vertical axis: energy.**

In Fig. 11.7, the hybridization gap is *indirect*: the minimum of $\lambda_{\mathbf{k}}^+$ is associated with the conduction band minimum at the zone center $((\lambda_{\mathbf{k}}^+)_{\min} \approx \epsilon_f + v^2/(W - \epsilon_f))$, while the maximum of $\lambda_{\mathbf{k}}^-$ is at the zone boundary $((\lambda_{\mathbf{k}}^-)_{\max} \approx \epsilon_f - v^2/\epsilon_f)$. Assuming that v is rather smaller than W, we find that the hybridization gap

$$\Delta(U_f = 0) \approx v^2 \left(\frac{1}{W - \epsilon_f} + \frac{1}{\epsilon_f} \right) \tag{11.6}$$

is an order of v/W smaller than v.

With $n \leq 2$ electrons per site, the $U_f = 0$ ground state (the non-interacting Fermi Sea) is obtained by filling up the lower hybridized

---

sight, it may appear that on-site hybridization is forbidden if the RE atom sits at a site with central symmetry, because $4f$-states are odd, while typical conduction band states have $5d$ or $6s$ (i.e., even) character. However, one should envisage that it is possible to construct odd linear combinations of the, say, $6s$ states of the anions surrounding the central RE site, and thus we may speak about on-site hybridization in a model in which only the RE sites are regarded as lattice sites.

band

$$|\mathrm{FS}\rangle = \prod_{\mathbf{k}\in\mathrm{FS}} \prod_{\sigma} \left(c^\dagger_{\mathbf{k}\sigma} + a_0(\mathbf{k}) f^\dagger_{\mathbf{k}\sigma} \right) |0\rangle \tag{11.7}$$

where we include the **k** vectors for which $\lambda^-_{\mathbf{k}} \leq \mu$. The number of **k** in the Fermi volume is $N/2$ (the number of electrons per spin direction). The $U_f = 0$ mixing amplitudes are

$$a_0(\mathbf{k}) = \frac{2v_{\mathbf{k}}}{\epsilon_{\mathbf{k}} - \epsilon_f + \sqrt{(\epsilon_f - \epsilon_{\mathbf{k}})^2 + 4v^2_{\mathbf{k}}}}. \tag{11.8}$$

Now let us consider the effect of switching on a large U_f. If we wish to discuss the case of f^0–f^1 valence mixing only, we may let $U_f \to \infty$, assuming that ϵ_f lies within the c-band, or not far below. Replacing the expansion of $f^\dagger_{\mathbf{k}\sigma}$ in terms of the atomic f states in (11.7), we can easily convince ourselves that the uncorrelated hybridized-band Fermi sea $|\mathrm{FS}\rangle$ contains a substantial concentration of f^2 sites. However, the f^2 configuration is energetically forbidden by $U_f \to \infty$. Our first idea might be to act with the Gutzwiller projector $\hat{P}_f = \prod_{\mathbf{j}} \left(1 - \hat{n}^f_{\mathbf{j}\uparrow}\hat{n}^f_{\mathbf{j}\downarrow}\right)$ on $|\mathrm{FS}\rangle$ [335, 427]. However, we might suspect that the spatial redistribution of the f-electrons changes also the hybridization amplitudes, and therefore a more flexible variational trial state is called for, in which $\hat{P}_f$ acts on a freely adjustable hybridized band state [107]

$$|\Psi\rangle = \prod_{\mathbf{j}} \left(1 - \hat{n}^f_{\mathbf{j}\uparrow}\hat{n}^f_{\mathbf{j}\downarrow}\right) \cdot \prod_{\mathbf{k}} \prod_{\sigma} \left(c^\dagger_{\mathbf{k}\sigma} + a(\mathbf{k}) f^\dagger_{\mathbf{k}\sigma} \right) |0\rangle. \tag{11.9}$$

Here the $a(\mathbf{k})$ are independent variational parameters. $|\Psi\rangle$ has the general structure shown in (10.37), but we have to remember that here we have to do with a two-band model, and the projector $\hat{P}_f$ acts only on one of the fermion species. We also note the great formal similarity of (11.9) to the Gutzwiller-projected SDW state[15] (10.57).

Expanding $f^\dagger_{\mathbf{k}\sigma}$ in terms of the atomic f states in (11.9), we get a direct-space expansion of $|\Psi\rangle$, which is akin to (9.28) (see [107]). It

[15] We mentioned in Sec. 7.6 that the SDW effective field acts like a spin-dependent hybridization amplitude mixing the **k** and (**k**+**Q**) states. The many-parameter SDW state (10.57) allows the readjustment of the mixing amplitudes upon switching on local correlations.

becomes explicit that every f-shell is in a linear superposition of f^0 and f^1 configurations, and that a Bloch-like phase coherence connects the intrashell valence fluctuations. $|\Psi\rangle$ was devised to fit our phenomenologically developed ideas about the nature of a homogeneously mixed valent ground state. Though it incorporates the effects of a strong interaction, it does not break any symmetries just as the Gutzwiller state (9.16) preserved the symmetries of the single-band Fermi sea.

The Gutzwiller approximation can be extended to the many-parameter trial state (11.9) [107]. We do not describe this in any detail because the conclusions can be made plausible by recalling analogous results for the single-orbital Hubbard model (see Chaps. 9 and 10). The mixed valent system can be envisaged as a strongly correlated effective f-band immersed into an uncorrelated conduction band. If there were a direct f–f hopping then, analogously to (10.1) and (10.21), its amplitude should be suppressed by the Q-factor

$$Q = \frac{1-n_f}{1-(n_f/2)}. \tag{11.10}$$

However, in the PAM (11.4), inter-f-shell propagation has to be a two-step process: first, an f-electron gets converted into a conduction electron (with matrix element $v_{\mathbf{k}}$), and then it can hybridize into another f-shell (another $v_{\mathbf{k}}$), assuming that it finds the shell empty (Fig. 11.3). Thus in the present case Q multiplies $v_{\mathbf{k}}^2$, or in other words, $U_f \to \infty$ gives rise to the *effective hybridization*

$$\tilde{v}_{\mathbf{k}} = v_{\mathbf{k}}\sqrt{\frac{1-n_f}{1-(n_f/2)}}. \tag{11.11}$$

The optimum mixing amplitude

$$a(\mathbf{k}) = \frac{2v_{\mathbf{k}}}{\epsilon_{\mathbf{k}} - \tilde{\epsilon}_f + \sqrt{(\tilde{\epsilon}_f - \epsilon_{\mathbf{k}})^2 + 4Qv_{\mathbf{k}}^2}} \tag{11.12}$$

has the same form as $a_0(\mathbf{k})$, but with renormalized parameters [107]. In addition to $\tilde{v}^2$ appearing in the denominator (but not in the numerator), ϵ_f gets replaced by the *effective f-level* $\tilde{\epsilon}_f$ which satisfies

$$\tilde{\epsilon}_f = \epsilon_f + \frac{1}{1-(n_f/2)}\frac{1}{L}\sum_{\mathbf{k}} \frac{v_{\mathbf{k}}^2}{\sqrt{(\epsilon_{\mathbf{k}} - \tilde{\epsilon}_f)^2 + 4v_{\mathbf{k}}^2 Q}}. \tag{11.13}$$

The region $\epsilon_{\mathbf{k}} \approx \epsilon_f$ gives a contribution $\sim -(v^2/W)\ln(Qv^2/W^2)$ to the $\mathbf{k}$-sum. As long as we are in the bona fide mixed valent regime (n_f being sufficiently far from 1 and 0), $\ln Q$ remains finite, and $\tilde{\epsilon}_f \approx \epsilon_f$. We see, however, that $\tilde{\epsilon}_f$ should suffer a substantial shift in the nearly integral valent regime which we discuss in the next subsection.

The variational solution (11.9)–(11.13) describes the independent-particle behaviour of dressed quasiparticles, and in a loose sense, it can be called a mean field approximation. However, it is *not* based on a mean field decoupling of the on-site interaction. Rather, the hybridization amplitude $\langle f^{\dagger}_{\mathbf{j}\sigma} c_{\mathbf{k}\sigma}\rangle$ should be thought of as a mean field[16]. Like in the case of the Hubbard model, the Gutzwiller approximation becomes exact in the limit of infinite spatial dimensionality [125].

Though (11.9) was written down to imply partial occupation of the lower hybridized band, the renormalized band picture can be shown to be complete with a gap and an upper hybridized band. However, the total electron content of the two hybridized bands has to satisfy $n \leq 3$, meaning at most 2 c- and 1 f-electron per site. $U_f = \infty$ prevents adding more electrons, since it would have to mean the presence of a second f-electron at some of the sites. Generally, the validity of the renormalized band scheme is confined to the low-energy subspace: similarly, the Gutzwiller treatment of the Hubbard model was not meant to describe the upper Hubbard subband.

The relationship between the bare and renormalized hybridized band ground states is clearly discernible in the f-state occupation number $n^f_{\mathbf{k}\sigma} = \langle f^{\dagger}_{\mathbf{k}\sigma} f_{\mathbf{k}\sigma}\rangle$. A simple illustration (one dimension, linear dispersion) for a mixed valent metal with $n = 1.4$, $n_f \approx 0.7$ is shown in Fig. 11.8. For $U_f = 0$, the lower hybridized band has strong f-character in the region[17] where $\epsilon_k > \epsilon_f$ (continuous line). Switching on $U_f = \infty$

[16]In the cluster treatment for the Hubbard model (Sec. 9.2.2), we treated the hopping amplitude $\langle c^{\dagger}_{\mathbf{j}\sigma} c_{\mathbf{i}\sigma}\rangle$ as a primary quantity which cannot be decoupled into a product of single-site amplitudes. Similarly, Gutzwiller approximation for the PAM [107] can be thought to be based on selecting a pair of f- and c-orbitals, and immersing it in a fermionic bath.

[17]The band has almost pure c-character in the region $\epsilon_k < \epsilon_f$, while c- and f amplitudes are equal at $\epsilon_{\mathbf{k}} = \epsilon_f$ which is the centre of a transition region of width $\sim v^2/W$. $n^c_{\mathbf{k}\sigma} = \langle c^{\dagger}_{\mathbf{k}\sigma} c_{\mathbf{k}\sigma}\rangle$ is not essentially modified by switching on $U_f \to \infty$ except that the crossover region becomes narrower. In contrast to the case of f-electrons,

compels the f-electrons to keep away from each other in direct space, and therefore they become redistributed in **k**-space, according to

$$\langle\Psi|f^{\dagger}_{\mathbf{k}\sigma}f_{\mathbf{k}\sigma}|\Psi\rangle = \begin{cases} (1-Q)\dfrac{n_f}{2} + \dfrac{Qa^2(\mathbf{k})}{Q^{-1}+a^2(\mathbf{k})} & \text{if} \quad \mathbf{k}\in \mathrm{FS} \\ (1-Q)\dfrac{n_f}{2} & \text{if} \quad \mathbf{k}\notin \mathrm{FS}\,. \end{cases} \tag{11.14}$$

Substantial f-electron occupation appears in the region which had been predominantly c-electron-like, and also outside the Fermi surface (curve plotted with squares in Fig. 11.8). The discontinuity at $k = k_{\mathrm{F}}$ is much reduced. On the whole, the behaviour is reminiscent of what we had seen in Fig. 10.1. In particular, the **k**-independent term is analogous to that seen in (10.6) and (10.23).

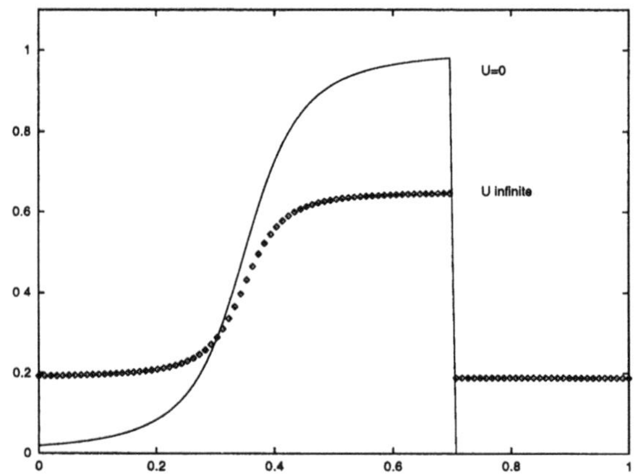

Figure 11.8: The f-electron occupation number as a function of k (in units of π/a), for $U_f = 0$ (line), and for $U_f = \infty$ (points). A strong U_f pushes a part of the f-electrons into the high-energy part of the Brillouin zone, and reduces the the discontinuity at $k_{\mathrm{F}} = 0.7$ ($\epsilon_k = k$, $\epsilon_f = 0.35$, $v = 0.05$).

Let us emphasize that in spite of the f-electron occupation becoming non-vanishing in the entire Brillouin zone, $n^f_{\mathbf{k}\sigma}$ (and of course also $n^c_{\mathbf{k}\sigma}$ as well as the total $n_{\mathbf{k}\sigma}$) still has a discontinuity at the same k_{F} as in the non-interacting case. Sticking to the family of states represented

no c-electrons appear outside the Fermi surface.

by the effective hybridization Ansatz (11.9), *Luttinger's theorem* holds in the strongly correlated regime. Naturally, the theorem might break down because of some symmetry breaking transition but, as a matter of fact, a large number of intermediate valent f-electron systems are non-magnetic and also otherwise non-ordering[18].

We learned in Sec. 10.2 that reducing the Fermi step[19] is related to the enhancement of the effective mass. In the present case, an analogous relationship [360]

$$\left(\frac{m^*}{m}\right)_{\rm corr} = \frac{1 - \Delta n_c - \Delta n_f}{\Delta n_f} \propto \frac{n_f}{2(1-n_f)} \tag{11.15}$$

holds, where Δn_c and Δn_f are the U_f-induced changes of the respective Fermi discontinuities. We do not predict any particularly large mass enhancement in the truly intermediate valent regime[20] (just as we did not for the Hubbard model well away from half filling).

A quantity for which even a moderate renormalization matters, is the hybridization gap

$$\Delta(U_f = \infty) = \frac{1}{L}\left.\frac{\partial\langle\mathcal{H}\rangle}{\partial n}\right|_{n=2+0} - \frac{1}{L}\left.\frac{\partial\langle\mathcal{H}\rangle}{\partial n}\right|_{n=2-0} \approx v^2 Q\left(\frac{1}{\epsilon_f} + \frac{1}{W-\epsilon_f}\right). \tag{11.16}$$

The non-interacting system is a semiconductor with $n = 2$ electrons per site, and Luttinger's theorem permits the strongly correlated ground state to remain insulating, with a renormalized hybridization gap separating the lower and upper effective hybridized bands. However, before

---

[18] A notable exception is antiferromagnetic TmSe (average valence state $\mathrm{Tm}^{+2.75}$). It seems to be a useful rule that a mixed valent system can be magnetic if it fluctuates between two magnetic moment-bearing configurations [424] (for Tm, these are $4f^{12}$ and $4f^{13}$). In contrast, Sm fluctuates between $4f^5$ and $4f^6$ (of which the latter has $J = 0$, see Problem 2.6), and Yb between $4f^{13}$ and $4f^{14}$. Magnetism tends to be quenched if one of the valence states is a singlet.

[19] In the present two-band model, the $U_f = 0$ Fermi step is unity for the total $n_{\mathbf{k}\sigma} = n^c_{\mathbf{k}\sigma} + n^f_{\mathbf{k}\sigma} = \langle c^\dagger_{\mathbf{k}\sigma} c_{\mathbf{k}\sigma}\rangle + \langle f^\dagger_{\mathbf{k}\sigma} f_{\mathbf{k}\sigma}\rangle$. f- and c-electrons have their respective discontinuities which are affected by U_f quite differently. In particular, $\Delta n_c \propto Q(v/W)^2$ can be neglected in (11.15).

[20] Meaning that the correlation enhancement of the band mass is not large. However, if μ lies in the flat part of a hybridized band, then the band mass can be substantial to begin with. $m^* \sim 3-6$ for Sm-based intermediate valent compounds.

taking this too literally, we should remember that we considered the orbitally non-degenerate PAM, and we might worry about whether in a real system a gap can be opened for all the relevant angular momentum channels, or there is hybridization only in some channels, while the others give rise to flat bands in the middle of the gap. It seems to be difficult to decide this by a general argument but the experimental fact is that mixed valent semiconductors (with an even number of electrons per unit cell) exist, and their behaviour is – roughly speaking – in accordance with our simple argument [436].

We cite the example of YbB_{12} in which the Yb^{2+} and Yb^{3+} valence states are mixed. With 52 valence electrons per formula unit, Luttinger's theorem allows the system to be an insulator, and that is what it appears to be[21]. The temperature dependence of the resistivity can be fitted with an Arrhenius law (Fig. 11.9), which allows the deduction of the activation energy $\sim$ 5.2meV. In the physics of ordinary semiconductors, such a small activation energy would be associated with an impurity level, but we have no qualms about ascribing it to intrinsic behaviour in the present case. Even for non-interacting electrons, assuming $v = O(0.1)\,\text{eV}$, and W and ϵ_f of $O(1)\,\text{eV}$, (11.6) readily gives a gap 5–20meV. Strong interactions cause this value to be divided by 2–5 (see (11.16)), yielding values typical for mixed valent semiconductors.

Gaps of order 1–10 meV get washed out at temperatures of order of 100K, or maybe below. First of all, this means that a huge number of carriers can be produced thermally. As for YbB_{12}, Hall effect measurements show that the carrier concentration is $7{\cdot}10^{-6}$ per formula unit at $T = 1.7\text{K}$, while it is 120000 times as high at room temperature [436]. Furthermore, thermal agitation destroys the intersite quantum coherence (sustained by an energy gain $\sim Qv^2/W$) which gave rise to the hybridization gap in the first place, and the system can be thought of as an assembly of independent f-shells, immersed into a light electron gas of c-electrons. As T is increased from 0 to T_{room}, we pass from a

[21]We have chosen YbB_{12} because SmB_6 which for decades had been thought of as the canonical mixed valent semiconductor, shows bewildering behaviour. Recent studies on good quality samples indicate that though the resistivity follows activated behaviour with an apparent gap of $\sim$ 4meV at intermediate T, it saturates at low T, as if the system were metallic with a very low concentration of carriers. If this is indeed intrinsic behaviour, it is one for which we can offer no explanation.

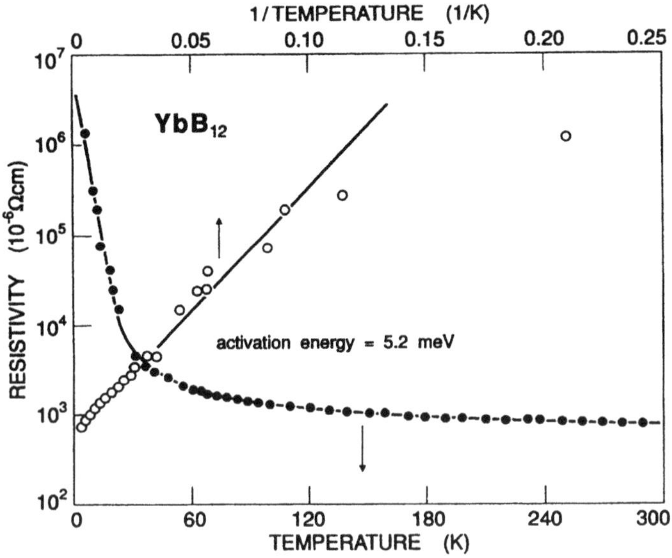

Figure 11.9: Resistivity of YbB_{12} versus temperature (upper scale: inverse temperature). The activation energy $\sim$ 5.2meV is argued to be the renormalized hybridization gap (after [291], reproduced in [436]).

quantum mechanical superposition of different valence states (intermediate valence) to a random mixture of $4f^n$ and $4f^{n+1}$ ionic species.

11.2.4 Nearly Integral Valence: Heavy Fermions

Weak hybridization smoothes out the piecewise linear μ–n plot seen in Fig. 11.2, but leaves its basic message unchanged. μ is pinned to ϵ_f (or $\epsilon_f + U_f$, etc.) within a width $\sim Qv^2\rho(\epsilon_f)$ as long as n_f is sufficiently intermediate between two integer values. The previous discussion holds for this homogeneously mixed valent regime. If n is increased further, μ becomes detached from the previous f-level, and climbes up in the band, whereby $\mu - \epsilon_f \sim O(W)$. If v were zero, this would mean strictly integral valence. However, even if μ is lying, say, well above ϵ_f, the process which converts an f-electron into a conduction electron at the top of the Fermi sea (Fig. 11.10), still has a non-negligible amplitude. As a result, though most of the sites have one f-electron, there will be a fraction of sites with empty f-orbitals. We call this state *nearly integral valent*, to distinguish it from the bona fide mixed valent behaviour which we associate with the regime well inside the plateaus. Clearly, the borderline between the mixed valent, and the nearly integral valent, regimes is not sharply defined but rather blurred. We might say that if the deviation of the valence n_f from an integral value is larger than $\sim (V\rho(\mu))^2$ then it is intermediate valence, and if not, then it is nearly integral valence.

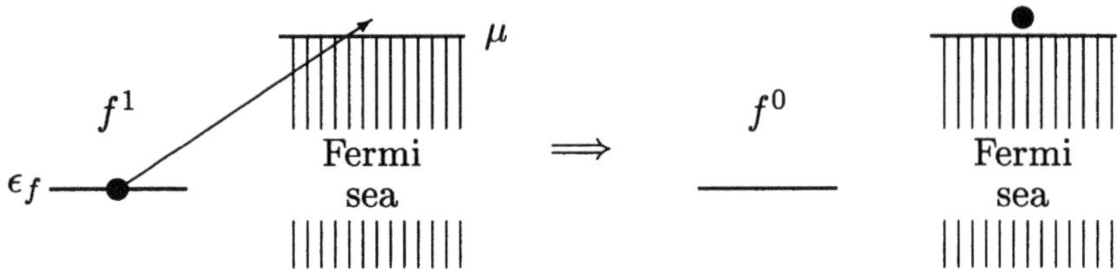

Figure 11.10: Even if μ is lying some distance above the f-level, hybridization can change an f-electron into a conduction electron at the top of the Fermi sea, giving rise to a small admixture of f^0 character.

The small admixture of f^0 character can be estimated by recalling our remark accompanying (11.13), which said that

$$\tilde{\epsilon}_f - \epsilon_f \approx -(v^2/W)\ln(Qv^2/W^2).$$

By definition, $\tilde{\epsilon}_f$ is the energy at which the valence fluctuation process $f^1 \leftrightarrow f^0$ is possible, thus it must be true that $\tilde{\epsilon}_f \approx \mu$. On the other hand, valence mixing is very weak in the limit we are discussing, thus it must hold that $\tilde{\epsilon}_f \approx \epsilon_{\mathrm{F}}^0$, the $v = 0$ value of the Fermi energy. It follows that $1 - n_f$ is exponentially small. We quote the results obtained for a constant c-band DOS [107]

$$n_f = 1 - \frac{\epsilon_{\mathrm{F}}^0 W}{4v^2}\exp\left\{-\frac{W(\epsilon_{\mathrm{F}}^0 - \epsilon_f)}{4v^2}\right\}. \qquad (11.17)$$

The small deviation from integral valence is driven by an exponentially small mixing energy[22]

$$\frac{\langle\mathcal{H}\rangle}{L} = \frac{(\epsilon_{\rm F}^0)^2}{W} + \epsilon_f - \epsilon_{\rm F}^0 \exp\left\{-\frac{W(\epsilon_{\rm F}^0 - \epsilon_f)}{4v^2}\right\} = \frac{\langle\mathcal{H}\rangle_{v=0}}{L} - \frac{4v^2}{W}(1 - n_f). \tag{11.18}$$

Actually, the above expression holds for $n \leq 2$, and a different relationship corresponds to $2 < n \leq 3$. The hybridization gap defined by (11.16) still survives

$$\Delta \propto W \exp\left\{-\frac{W(\epsilon_{\rm F}^0 - \epsilon_f)}{4v^2}\right\}. \tag{11.19}$$

In accordance with Luttinger's theorem, we predict that the $n = 2$ system is a nearly integral valent insulator with an exponentially small gap. *Kondo insulators* with gaps of order 1–10 K are thought to realize this; we return to them a bit later.

The picture of renormalized hybridized bands is applicable both in the mixed valent and the nearly integral valent regimes, but with quite different parameters. In the mixed valence regime, the renormalized bands look pretty much like the bare bands (Fig. 11.7), and the gap opens (roughly speaking) at the f-level. For nearly integral valence, we find an extremely narrow double-peak resonance pinned to the chemical potential. The total weight of the two peaks is enough to hold one electron per site[23].

Recalling (11.15), we find

$$\frac{m^*}{m} \propto \frac{n_f}{1 - n_f} \propto \frac{4v^2}{\epsilon_{\rm F}^0 W} \exp\left\{\frac{W(\epsilon_{\rm F}^0 - \epsilon_f)}{4v^2}\right\} \gg 1\,. \tag{11.20}$$

We are led to conclude that *nearly integral valent metals are heavy Fermi liquids* with exceptionally large values of the mass enhancement.

[22]It is a moot point whether the relevant energy scale, and the concomitant $1 - n_f$, are indeed exponentially small. The fact that this follows from a variational calculation, does not constitute a proof of the analytic form! However, for our present purposes it is sufficient to know that it is a scale weaker than the leading order v^2.

[23]The pair of narrow peaks is a further example of occupation-number dependent features of the DOS, which are characteristic of strongly correlated systems. The narrow feature is there only if it is actually occupied by almost 1 f-electron per site. The shape of the DOS depends on n, and it has to be determined self-consistently.

Let us emphasize that we do not need a fine-tuning of the parameters of the Hamiltonian to arrive at this conclusion, in spite of the narrowness of the range of n_f. $1 - n_f \ll 1$ arises in a natural manner for a wide range of $v/(\epsilon_{\rm F}^0 - \epsilon_f)$. Indeed, our worry should be the opposite: it might seem that all f-electron systems which are not mixed valent, should be heavy Fermi liquids. This is certainly not so: most of the elemental rare earths are not mixed valent, nevertheless they have quite ordinary values of m^*. As it turns out, they have to be characterized as strictly integral valent systems (in the same sense that we call a Mott insulator strictly integral valent). We are going to argue that there is a competition between a small-m^* and a large-m^* phase, and thus in principle, both kinds of behaviour can be explained.

In any case, there is a large class of compounds for which our previous conclusions are apparently valid. Many f-electron systems, most often Ce, Np, U, and Yb compounds, are heavy fermion systems with m^*/m ranging up to 200, and beyond [120, 135, 317, 375].

We outlined the phenomenology of the single-band heavy fermion state in Secs. 10.2, 10.5, and 10.6.5; now this has to be slightly modified because of the presence of *two* narrow bands. A large m^*/m is tantamount to a large density of states $\rho^*(\mu) \sim (m^*/m)\rho(\mu)$ which can be interpreted as evidence of a small effective bandwidth $W^* \sim (m/m^*)W$. W^* is the common order of magnitude of the widths of the lower and upper peaks of DOS, and the gap in between, but we may also think of it as the total width of the double-peak structure which we identify with the f-part of the renormalized hybridized bands. In contrast to the Hubbard model case where the narrow band could be treated as isolated, here the mass-enhanced part of the spectrum sits on the background of a wide band[24]. The overall width of the structure corresponds to a very low effective degeneracy temperature $T^* \propto (1 - n_f)W^2/vk_{\rm B}$. In heavy fermion materials $T^* \approx 10 - 100{\rm K}$, i.e., it is orders of magnitude smaller than the Fermi temperatures of simple metals. We illustrate[25] the consequences of such an extraordinarily small degeneracy temperature on the example of the long-studied heavy electron systems $CeAl_3$

[24]This is true even if the bare f-level is below the conduction band.

[25]It is understood that comparison to measurements has to remain qualitative, since we have not discussed the effects of orbital degeneracy and crystal fields.

and $CeCu_6$.

By the time T reaches T^*, the fine structure of the DOS gets completely washed over, and we are left with $\sim L$ essentially independent f^1-sites, surrounded by a light electron gas which sees the Ce ions as scattering centers. The measured T-dependence of the susceptibility of $CeAl_3$ (Fig. 11.11, right) can be fitted for $T > 35$K with a Curie–Weiss formula with $\mu_{\text{eff}} \approx 2.6\mu_B$ which is essentially the magnetic moment of free Ce^{3+} ions. On the other hand, χ is non-monotonic at low T, and eventually converges to a finite value as $T \to 0$. $\chi(T = 0)$ is two orders of magnitude larger than values typical for simple metals; we interpret it as an enhanced Pauli susceptibility: $\chi(T = 0) = (g\mu_B)^2 \rho^*(\mu) \propto (g\mu_B)^2 (m^*/m) W^{-1}$.

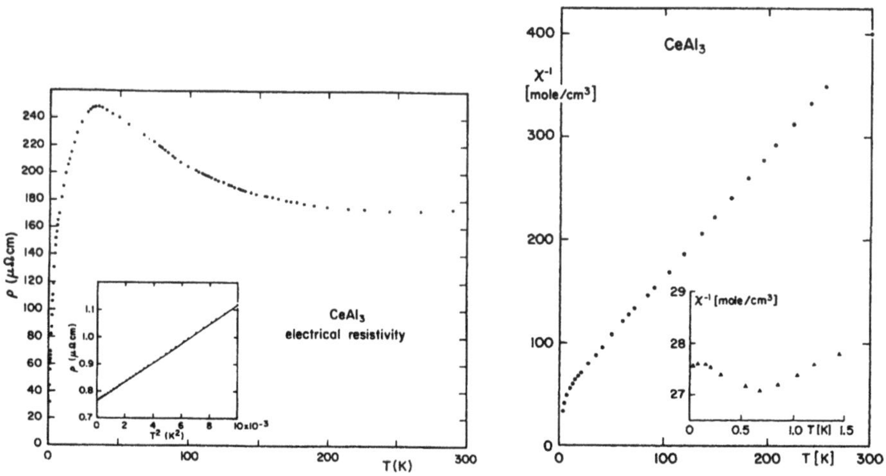

Figure 11.11: Measurements on $CeAl_3$ reveal the formation of a narrow effective band with a width of the order of 10K at low temperatures, and independent-ion behaviour at $T > 50$K (after [317]). *Left*: the T-dependence of the resistivity ρ. Inset: $\rho = \rho_0 + A^* T^2$ is well-obeyed for $T < 0.1$K. *Right*: the T-dependence of the inverse susceptibility χ^{-1} is well-fitted by a Curie–Weiss expression at high T, while at low T (inset), χ approaches a finite value which we identify as the enhanced Pauli susceptibility.

What must be happening as T is lowered from 100K to 1K, is that the f-electrons change their character completely: they cease to be independent scattering centers, and become charge carriers. It also follows that at low T, f-electrons merge into the Fermi sea which consequently becomes a heavy Fermi sea. The procedure can be followed in the T-dependence of the resistivity ρ (Fig. 11.11, left). For $T > 35\text{K}$, ρ is fairly large, and increases with decreasing temperature. This can be explained as the result of Kondo scattering on the spins of independent Ce^{3+} ions (see Sec. 11.3.1). At $T \sim 35\text{K}$ ρ bends over, and begins to drop sharply, decreasing by two orders of magnitude before saturating. We associate this with the development of intersite phase coherence, and the formation of the narrow effective bands by the f-electrons. The $T < 0.1\text{K}$ behaviour (inset) is well described by the quadratic law $\rho = \rho_0 + A^*T^2$. We met similar behaviour in $La_{1-x}Sr_xTiO_3$ (see Sec.10.5), and interpreted it as a low-temperature Fermi liquid feature: it is a consequence of strong electron–electron scattering in the vicinity of the Fermi surface. In the present context, it is worth emphasizing that accepting this explanation of the T^2 law implies that the strongly interacting f-electrons belong to the Fermi sea.

T must be reduced by two orders of magnitude from the resistivity maximum ($T \sim 35\text{K}$) to reach the Landau regime ($T < 0.1\text{K}$) where $\rho - \rho_0 \propto T^2$ holds. For $35\text{K} > T > 0.1\text{K}$, there is certainly some kind of inter-f-shell phase coherence but the electron system is still too hot to allow the use of Fermi liquid theory. In a strict sense, we can speak about heavy electrons only in the Landau regime $T < 0.1\text{K}$ where it is also true that the electronic specific heat obeys $C = \gamma^* T$, with $\gamma^* = (m^*/m)\gamma$. In fact, an anomalously large γ^* is the experimental criterion for having a heavy fermion system. The low temperature specific heat of $CeCu_6$ (Fig. 11.12) is linear in T, allowing to deduce $\gamma^* = 1550\text{mJ/mole K}^2$. For $CeAl_3$, one finds $\gamma^* \approx 1600\text{mJ/mole K}^2$ [317]. In order to appreciate how huge these values are, let us recall that for many metallic elements, γ is about a factor of 10^{-3} smaller (Cu: $\gamma = 0.695\text{mJ/mole K}^2$, K: $\gamma = 2.08\text{mJ/mole K}^2$, Ni: $\gamma = 7.02\text{mJ/mole K}^2$).

The heaviest Fermi liquids (some Ce, U, and Yb compounds) are characterized by $\gamma^* \sim O(1)\text{J/mole K}^2$, but there are many more moderately heavy[26] systems with $\gamma^* \sim O(100)\text{mJ/mole K}^2$ [135, 164, 317].

[26] It would be mistaken to look only at this single number: $\gamma^* \sim O(100)\text{mJ/mole K}^2$ is not an unambiguous signature of heavy fermion character! For instance, $NdGa_2$ has $\gamma^* = 150\text{mJ/mole K}^2$, and at the same time, it is known to be an integral valent system with the ionic value of the Nd moment [28]. In the conceptual framework

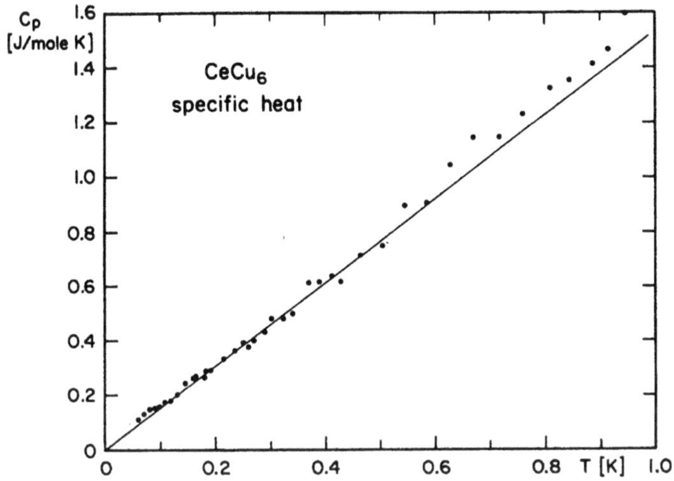

Figure 11.12: The low-temperature specific heat of $CeCu_6$ (after [317]).

URu_2Si_2 which we discussed at some length (though not from the present point of view) in Ch. 3, barely passes as a "light heavy fermion" system with $\gamma^* \approx 50\text{mJ/mole K}^2$.

It should be noted that the experimental definition of the mass enhancement $m^*/m = \gamma^*/\gamma_{\text{free}}$ (where γ_{free} is calculated for the total concentration n of valence electrons, counting also the f-electrons) uses the free electron m as unit, i.e., it does not distinguish band structure and correlation (and possibly still other) effects. The direct overlap between f-shells being small, the band structure of rare earth[27] systems can be envisaged as arising from the hybridization of a dispersionless band with wide bands (Fig. 11.7), and this gives rise to a sizeable enhancement over the free electron mass even if $U_f = 0$. If we see a really big enhancement, say $m^*/m \sim 200$, we may imagine that a factor ~10–12

adopted by us, heavy fermion behaviour is associated with non-integral valence, a tendency to overall spin compensation, Luttinger Fermi volume, and a close enough fit to the Kadowaki–Woods plot (Fig. 11.13). $NdGa_2$ fails to satisfy any of these criteria; its largish γ^* must have a physical origin which is quite different from what we are discussing here.

[27]The $5f$-shells of actinides are less compact and therefore it is not obvious that the argument applies there. Nevertheless, quite a few actinide systems show heavy fermion behaviour.

is due to band structure, and a factor ~15–20 is due to correlation. The latter may be ascribed to $1 - n_f \sim 0.02 - 0.04$ which is, in principle, measurable but we should not be overly optimistic about spectroscopies being always able to tell a nearly integral valent system from a truly integral valent one.

It is clear that heavy fermionic behaviour is driven by strong interactions, and doing a systematic derivation for any particular quantity, say m^* or $\chi(T = 0)$, is very difficult. However, we are helped by the empirical fact that various anomalously large quantities are connected by simple relationships. For instance, for the coefficient A^* of the T^2-dependence of the low-T resistivity, the relationship (familiar from Sec. 10.5 as (10.29))

$$\frac{A^*}{A} \propto \frac{(\gamma^*)^2}{\gamma^2} \propto \left(\frac{m^*}{m}\right)^2 \tag{11.21}$$

holds through several orders of magnitude [197], as shown in the so-called Kadowaki–Woods plot (Fig. 11.13, right).

Another important point is that γ^* and $\chi(T = 0)$ are enhanced in roughly the same measure, suggesting that the source of the susceptibility enhancement is essentially the same as that of the mass enhancement, and that for a first orientation, it is useful to ascribe them to a large $\rho^*(\epsilon_{\rm F})$. Actually, this seems to be more unambiguous here than in the case of the one-band model where we were forced to discuss whether the enhancement of the uniform, or of the local, susceptibility scales with m^*/m (see the discussion accompanying (10.74)). Fig. 11.13 (left) shows that for a large number of systems, ranging from 'light' heavy fermions to the heaviest ones, the Wilson ratio

$$R = \frac{\pi^2 k_{\rm B}^2}{J(J+1)(g\mu_{\rm B})^2} \frac{\chi(T=0)}{\gamma^*} \tag{11.22}$$

is not far from unity.

We close this subsection by briefly addressing two questions: first, whether the only possible explanation of heavy fermion behaviour is the one we have sketched, and second, whether only f-electrons can be heavy.

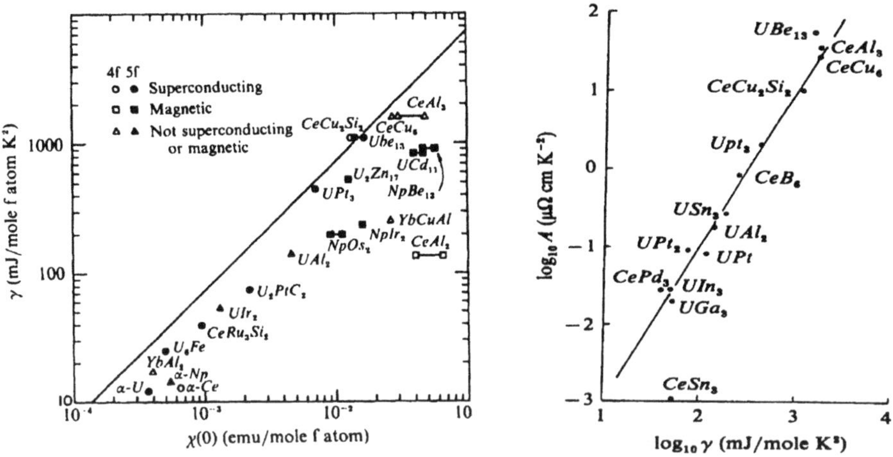

Figure 11.13: Simple empirical relationships between large enhancement factors prove the consistency of the idea of a heavy Fermi liquid. *Left*: The specific heat coefficient γ and the $T = 0$ susceptibility are enhanced in about equal measure (after [164]). *Right*: the Kadowaki–Woods plot. The double logarithmic plot of A versus γ (experimental values) shows that (11.21) holds over several decades of m^*/m (after [197]).

As to the first question: our basic scenario is that (apart from the obvious nearly-flat-band aspect) an f-electron is heavy because of the presence of the other f-electrons. This is embodied in (11.15): $(m^*/m)_{\text{corr}} \propto (1-n_f)^{-1}$ is *not* large, unless the system is nearly integral valent. In particular, we would not speak about heavy fermions for a small concentration of f-electrons. Now, this is only because we narrowed our scope, and considered only the f–f Hubbard term in (11.4). Allowing for f–c interactions, we would also find that the effective f-band is narrowed because of the formation of an electronic polaron (see, e.g., [464]). The more usual phononic polaron formation should not be forgotten, either. Such phenomena may prove to be important in certain f- and d-electron systems. Nevertheless, we believe that an essential part of heavy fermion behaviour is connected with nearly integral valence, at least in Ce-based systems.

Finally, what about heavy d-electrons? Our basic argument $m^*/m \propto (1-n)^{-1}$ is as applicable to a Hubbard model near half-filling, as to the PAM. Indeed, such was our explanation for the mass enhancement observed in the $La_{1-x}Sr_xTiO_3$ system, and we would not hesitate to read off $m^*/m \approx 3$ from Fig. 10.5. There may be other d-electron systems which show a larger mass enhancement but the fact remains that as a rule, transition metal compounds do not reach even the level of moderate heaviness by the standard of f-electron systems. We may think of some arguments why this should be the case. In terms of Hubbard-like models, $1-n \ll 1$ requires a fine-tuning of the composition which may be difficult to achieve. Also, if the system is on the verge of a Mott transition, localization due to disorder is likely to set in.

11.3 Kondo Lattice

Our picture of the nearly integral valent regime with $n_f \approx 1$ is depicted in Fig. 11.10: for $v = 0$, the decoupled system would consist of an array of f^1 sites, and a conduction band (light) Fermi sea with $\epsilon_F^0 > \epsilon_f$. Switching on a weak hybridization, it gives rise to a small admixture of f^0 character. The situation is akin to that found for the strongly interacting half-filled Hubbard band which consists almost exclusively of c^1 sites, only the hopping matrix elements connecting to the upper Hubbard subband give a small fluctuating c^2–c^0 component (see Sec. 5.1). We have seen how this permits description in terms of an effective spin model (the Heisenberg model), in which spin–spin interaction is driven by virtual charge fluctuations.

Similarly, in the present case f^1 sites are bearing uncompensated spins[28], thus it should be possible to construct an effective Hamiltonian describing an array of localized f-spins, interacting with a c-electron Fermi sea.

To guess what this Hamiltonian may look like, let us begin to collect its pieces. An $\uparrow$-spin f-electron is taken by hybridization from site $\mathbf{j}$ to

[28]Trying to translate our model results into the language of realistic situations, we may think of trivalent Ce^{3+} with a doubly degenerate Γ_7 ground state where the two states are not literally $S = \pm 1/2$ states, but the time-reversed $J = 5/2$ states given in (3.131). – If a U-based system is nearly tetravalent, it consists of $5f^2$ shells which may have the Γ_3 doublet ground state. Since Γ_3 is a pure quadrupolar doublet, the effective interaction should be a quadrupole–quadrupole coupling [73].

the unoccupied band state $\mathbf{k}_1$, and its place is taken by a $\downarrow$-spin electron coming from the state $\mathbf{k}_2$

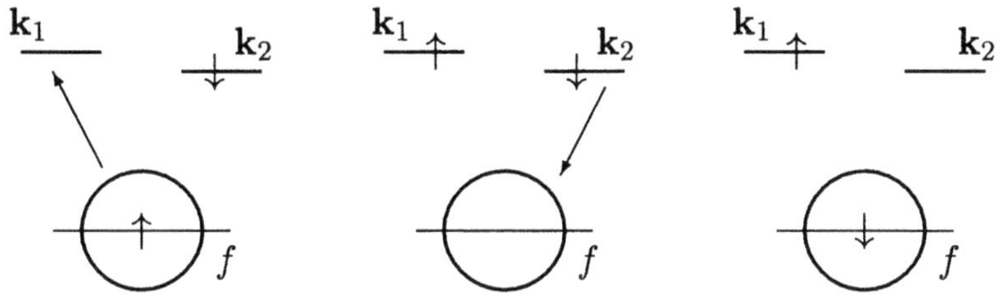

This perturbative process, in which f^0 appears as a higher-lying intermediate state, is described by

$$-\frac{1}{L}\frac{v_{\mathbf{k}_1}e^{-i\mathbf{k}_1\mathbf{j}}\,v^*_{\mathbf{k}_2}e^{i\mathbf{k}_2\mathbf{j}}}{\epsilon_{\mathbf{k}_1}-\epsilon_f}\,f^\dagger_{\mathbf{j}\downarrow}c_{\mathbf{k}_2\downarrow}c^\dagger_{\mathbf{k}_1\uparrow}f_{\mathbf{j}\uparrow}.$$

Allowing for a large but finite U_f, a similar process can be taking place via an intermediate f^2 state

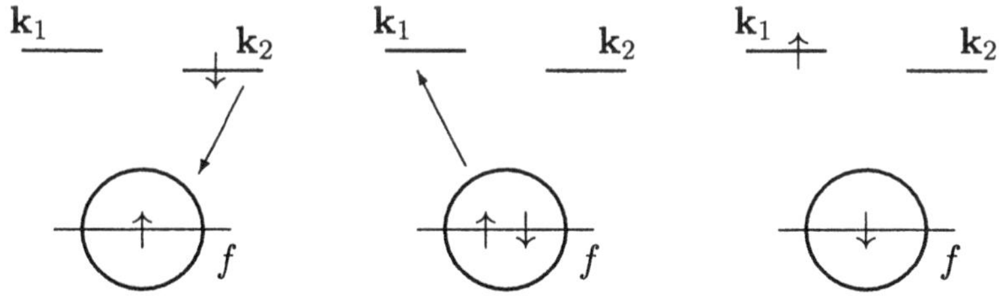

belonging to

$$-\frac{1}{L}\frac{v^*_{\mathbf{k}_2}e^{i\mathbf{k}_2\mathbf{j}}\,v_{\mathbf{k}_1}e^{-i\mathbf{k}_1\mathbf{j}}}{\epsilon_f+U_f-\epsilon_{\mathbf{k}_2}}\,c^\dagger_{\mathbf{k}_1\uparrow}f_{\mathbf{j}\uparrow}f^\dagger_{\mathbf{j}\downarrow}c_{\mathbf{k}_2\downarrow}.$$

Such contributions can be collected into

$$\frac{1}{L}\sum_{\mathbf{k}_1}\sum_{\mathbf{k}_2}\sum_{\mathbf{j}} v_{\mathbf{k}_1}e^{-i\mathbf{k}_1\mathbf{j}}\,v^*_{\mathbf{k}_2}e^{i\mathbf{k}_2\mathbf{j}}\left(\frac{1}{\epsilon_{\mathbf{k}_1}-\epsilon_f}-\frac{1}{\epsilon_{\mathbf{k}_2}-\epsilon_f-U_f}\right)S^-_{\mathbf{j}}\,c^\dagger_{\mathbf{k}_1\uparrow}c_{\mathbf{k}_2\downarrow} \tag{11.23}$$

describing that a conduction electron suffers a spin-flip scattering turning its spin up, while the f-spin gets turned down.

Since the underlying Anderson lattice Hamiltonian (11.4) is spin-rotationally invariant, we can be sure that the effective Hamiltonian we are contemplating here is invariant, too. Thus collecting all second-order contributions of the kind (11.23), we must find the isotropic c–f coupling

$$\mathcal{H}_{c-f} = \frac{1}{L}\sum_{\mathbf{k}_1}\sum_{\mathbf{k}_2}\sum_{\mathbf{j}} v_{\mathbf{k}_1} e^{-i\mathbf{k}_1\mathbf{j}} v^*_{\mathbf{k}_2} e^{i\mathbf{k}_2\mathbf{j}} \left(\frac{1}{\epsilon_{\mathbf{k}_1} - \epsilon_f} - \frac{1}{\epsilon_{\mathbf{k}_2} - \epsilon_f - U_f}\right)$$
$$\cdot \left\{ S^-_{\mathbf{j}} c^\dagger_{\mathbf{k}_1\uparrow} c_{\mathbf{k}_2\downarrow} + S^+_{\mathbf{j}} c^\dagger_{\mathbf{k}_1\downarrow} c_{\mathbf{k}_2\uparrow} + S^z_{\mathbf{j}} (c^\dagger_{\mathbf{k}_1\uparrow} c_{\mathbf{k}_2\uparrow} - c^\dagger_{\mathbf{k}_1\downarrow} c_{\mathbf{k}_2\downarrow}) \right\} \quad (11.24)$$

The coupling is fairly complicated if $v_{\mathbf{k}}$ is truly $\mathbf{k}$-dependent. A more transparent form is obtained if we adopt our usual simplifying assumption $v_{\mathbf{k}} \equiv v$ (real), and furthermore assume that ϵ_f ($\epsilon_f + U_f$) is sufficiently low (high) lying to replace the energy denominators by constants, because then the f-spin couples only to the local c-spin density[29]. The customary simplest form of the effective Hamiltonian contains the c-band kinetic energy, and the spin–spin interaction shown above

$$\mathcal{H}_{\mathrm{KL}} = \sum_{\mathbf{k}\sigma} \epsilon_{\mathbf{k}} c^\dagger_{\mathbf{k}\sigma} c_{\mathbf{k}\sigma} + \frac{J}{2} \sum_{\mathbf{j}} \sum_{\sigma\sigma'} \mathbf{S}_{\mathbf{j}} \cdot \left(c^\dagger_{\mathbf{j}\sigma} \vec{\sigma}_{\sigma\sigma'} c_{\mathbf{j}\sigma'} \right) \quad (11.25)$$

where $\vec{\sigma}$ is the vector of the Pauli matrices, and the *Kondo coupling*

$$J = v^2 \sum_{\mathbf{k}} \left(\frac{\Theta(\epsilon_{\mathbf{k}} - \mu)}{\epsilon_{\mathbf{k}} - \epsilon_f} + \frac{\Theta(\mu - \epsilon_{\mathbf{k}})}{\epsilon_f + U_f - \epsilon_{\mathbf{k}}} \right) \quad (11.26)$$

is *antiferromagnetic*. (11.25) is the simplest version of the *Kondo lattice model*[30].

Note that though both f^0 and f^2 appear as intermediate states, they need not play symmetrical roles. We may let $U_f \to \infty$ to get an f^0–f^1 model, and still find a sizeable J. Similarly, we could place the f-level at $\epsilon_f \to -\infty$, and insist that only $f^1 \to f^2$ transitions give a finite contribution. However, the symmetrical Kondo lattice model in which $\epsilon_f + U_f/2$ is exactly at the centre of a half-filled band, offers certain simplifications[31], and is therefore often postulated in the literature.

---

[29] Expanding $1/(-\epsilon_f + \epsilon_k)$ in powers of ϵ_k, and replacing $\epsilon_k = -2t\cos k$, we find additional intersite c–f exchange terms [249].

[30] The celebrated Kondo impurity problem will be discussed in Sec. 11.3.1.

[31] Namely, there is no need for a self-consistent determination of μ, which is a difficult step in the theory of asymmetric Anderson and Kondo lattices.

The systematic derivation of the Kondo lattice model (KLM) from the PAM uses the Schrieffer–Wolff transformation[32] which is analogous to the canonical transformation which led from the large-U Hubbard model to the t–J model (see Sec. 5.1). In (5.11), we rotated to a basis which (in lowest order) cut off the communication between the Hubbard subbands, allowing to confine attention to the low-energy subspace. Similarly, in the present problem, we rotate to a basis which cuts off one-electron hybridization processes, thus in lowest order, the f-electrons appear as localized moments. The derivation is sketched in Appendix D. It gives, in addition to the isotropic Kondo coupling (11.24), also the potential scattering term (D.12), the induced f–f hopping (D.13), and an additive constant (D.14). However, these extra terms are usually omitted and the discussion is based on (11.25).

What are the possible phases of the Kondo lattice model? $\mathcal{H}_{\rm KL}$, as defined in (11.25), consists of two terms: the c-electron kinetic energy, and the sum of the intracell Kondo interactions. For $J = 0$, the ground state is 2^L-fold degenerate: the c-electron Fermi sea surrounds L localized $S = 1/2$ spins (the f-electrons) which are free to point either up or down. What happens if we switch on the antiferromagnetic Kondo coupling? At the first glance, the model is unlikely to lead to magnetic ordering, since the interaction favours the formation of local c–f singlets. This is indeed one of the possible scenarios: the creation of the c–f singlets binds the f-spins to the c-electron Fermi sea which was a singlet to begin with; the outcome is a "large" $c + f$ Fermi sea which we will identify with the non-magnetic heavy fermion state. However, looking closer, we discover that the KLM does not lack the means to give magnetic order. In the attempt to create a local singlet, an f-spin has to polarize the surrounding Fermi sea; the next f-spin feels this disturbance, interprets it as an effective field, and lines up. Thus the c Fermi sea may serve either as a singlet glue to absorb the f-electrons into a joint big Fermi sea, or as a transmitter of f–f interactions, giving rise to magnetic order.

We will try to sketch a rudimentary phase diagram after some preliminaries. Both the idea that there is a phase which is basically the

[32] Originally devised as a transformation between the corresponding impurity models [352]. For the lattice case, see also [249] and the Appendix of [361].

superposition of f–c singlets, and the great physical and historical interest of the single impurity problem, demand that we briefly review the results about the single-site Kondo problem (Sec. 11.3.1). Afterwards, we discuss the indirect f–f exchange which is mediated by the conduction electrons (Sec. 11.3.2). This prepares the ground for a cursory discussion of the competition of magnetic and non-magnetic phases in the KLM, which is still an open problem. However, a little detour to the subject of the magnetism of rare earth elements (Sec. 11.4) makes it clear that the physics of real f-electron substances requires going beyond the simple KLM.

11.3.1 Kondo Impurity

Usually we confine our interest to lattice problems, but the remarkable similarity of the $T > T^*$ behaviour of periodic Kondo systems to that expected from dilute alloys makes a brief discussion of the impurity problem indispensable. Let us consider a single f-shell immersed into a Fermi sea, say, a single Ce atom in bulk $LaAl_2$. A Mn or Fe impurity in Cu or Al presents the same problem in d-electron version. This is the renowned magnetic impurity problem [139, 164] whose treatment can be based either on the single-impurity Anderson model [19], or – in the appropriate limit – on the single impurity Kondo Hamiltonian [223]. We start with the latter

$$\mathcal{H}_{\mathrm{imp}} = \sum_{\mathbf{k}\sigma} \epsilon_{\mathbf{k}} \hat{n}_{\mathbf{k}\sigma} + \frac{J}{2L} \sum_{\mathbf{k},\mathbf{q}} \left[S_0^+ c_{\mathbf{k}\downarrow}^\dagger c_{\mathbf{q}\uparrow} + S_0^- c_{\mathbf{k}\uparrow}^\dagger c_{\mathbf{q}\downarrow} + S_0^z (c_{\mathbf{k}\uparrow}^\dagger c_{\mathbf{q}\uparrow} - c_{\mathbf{k}\downarrow}^\dagger c_{\mathbf{q}\downarrow}) \right] \tag{11.27}$$

The derivation in Appendix D makes it plausible that the Kondo term is essentially an independent-impurity result, thus we may think of J as given by (11.26).

Let us try to anticipate the character of the solution. At $T = 0$, the AF Kondo interaction binds the f-spin to the electron sea, presumably forming an overall singlet, with a binding energy $E_0(J)$. At high temperatures ($T \gg E_0(J)/k_{\mathrm{B}}$), the f-spin is essentially free, contributing an almost full-size Curie term to the susceptibility, thus it is proper to speak about a magnetic impurity. Conduction electrons scattering off the f-shell may undergo spin-flip, as well as ordinary potential, scattering. As T is lowered, in a wide crossover regime around $T = E_0(J)/k_{\rm B}$, spin flip scattering events should bear witness to the gradual build-up of a spin compensating cloud around the impurity, thus scattering is expected to become more intense with decreasing T. At low temperatures, spin compensation is essentially complete, the localized spin is part of an overall c–f singlet, thus the impurity has become effectively non-magnetic. Nothing remains of the $T \to 0$ divergencies ordinarily associated with an uncompensated spin. Electrons coming from afar see a big singlet object, and suffer potential scattering with a large cross section. These expectations are born out by a large body of theoretical work which culminated in the exact solution for many physical quantities for both the Anderson, and the Kondo, version of the impurity problem.

We follow [459] in constructing the simplest variational ground state of (11.27). Let N be the number of conduction electrons, and let us assume that the total number $N+1$ of all the electrons (the c electrons and the single f electron) is even. If we neglected the presence of the f orbital, we could use $N+1$ electrons to form an unpolarized Fermi sea ground state $|{\rm FS}\rangle_{N+1}$, in which all band states are filled up to the Fermi energy $\epsilon_{\rm F}$. However, the f level is supposed to be lying well below $\epsilon_{\rm F}$, possibly below the bottom of the band, thus we have to let an electron "fall" from a higher-lying band state to the f level. At the first sight, this is best accomplished by taking an electron from the Fermi level, and convert it into an f electron. We may compose a singlet, ascribing symmetrical roles to $\uparrow$ and $\downarrow$ electrons

$$|\phi\rangle = \left(f_\uparrow^\dagger c_{{\bf k}_{\rm F}\uparrow} + f_\downarrow^\dagger c_{{\bf k}_{\rm F}\downarrow}\right)|{\rm FS}\rangle_{N+1} \equiv \left(f_\uparrow^\dagger c_{{\bf k}_{\rm F}\downarrow}^\dagger - f_\downarrow^\dagger c_{{\bf k}_{\rm F}\uparrow}^\dagger\right)|{\rm FS}\rangle_{N-1}$$

where $|{\rm FS}\rangle_{N-1}$ is the $(N-1)$-electron Fermi sea. We added to $|{\rm FS}\rangle_{N-1}$ (which is a singlet) a f–c singlet at the top of the Fermi sea. This would bring, in addition to the big diagonal energy lowering $\epsilon_f - \epsilon_{\rm F}$, also a spin flip energy contribution $\propto -(J/2L)$. However, this vanishes for a large system. In order to have a finite energy gain from the Kondo coupling, we have to "stir up" the Fermi sea to a finite energy depth, allowing c–f singlets involving $\epsilon_{\bf k} < \epsilon_{\rm F}$ band states. This is accomplished by

$$|\Phi_0\rangle = \sum_{\bf k} a({\bf k}) \left(f_\uparrow^\dagger c_{{\bf k}\uparrow} + f_\downarrow^\dagger c_{{\bf k}\downarrow}\right)|{\rm FS}\rangle_{N+1}\,, \tag{11.28}$$

the simplest member of a class of variational trial states introduced by Yosida [459]. The $a(\mathbf{k})$ are variational parameters. It is apparent that while the construction converts the orbital character of one of the electrons from c to f, it does not change the overall spin structure, thus $|\Phi_0\rangle$ is a singlet in the same sense as $|\mathrm{FS}\rangle_{N+1}$ is.

Replacing (11.28) in the Schrödinger equation $\mathcal{H}_{\mathrm{imp}}|\Phi_0\rangle = E|\Phi_0\rangle$, we get[33]

$$Ea(\mathbf{k}) = (E_{\mathrm{FS}} + \epsilon_f - \epsilon_{\mathbf{k}})a(\mathbf{k}) - \frac{3J}{4L}\sum_{\mathbf{q}} a(\mathbf{q}) \,, \tag{11.29}$$

where E_{FS} is the (purely band) energy of $|\mathrm{FS}\rangle$. Summing both sides over $\mathbf{k}$, $\sum_{\mathbf{q}} a(\mathbf{q})$ drops out and we find that E has to satisfy

$$1 = -\frac{3J}{4L}\sum_{\mathbf{k}} \frac{1}{E - E_{\mathrm{FS}} + \epsilon_{\mathbf{k}} - \epsilon_f} . \tag{11.30}$$

Converting the sum to an integral, and assuming that it is dominated by the contribution from states near the Fermi level, we may replace $\rho(\epsilon) \approx \rho(\epsilon_{\mathrm{F}}) = \rho_0$. Performing the integral

$$-\frac{4}{3J\rho_0} = \ln\frac{|E - E_{\mathrm{FS}} + \epsilon_{\mathrm{F}} - \epsilon_f|}{|E - E_{\mathrm{FS}} - \epsilon_f|} \approx \ln\frac{|\Delta E|}{\epsilon_{\mathrm{F}}} \,, \tag{11.31}$$

we assumed that the absolute value of the Kondo energy lowering $\Delta E = E - (E_{\mathrm{FS}} + \epsilon_f - \epsilon_{\mathrm{F}}) < 0$ is small in comparison with ϵ_{F}. We find that the singlet formation is accompanied by the energy gain[34]

$$\Delta E = -\frac{\epsilon_{\mathrm{F}}}{\exp\left\{\dfrac{4}{3J\rho_0}\right\} - 1} . \tag{11.32}$$

The accustomed form of the Kondo energy emerges in the weak coupling limit $J\rho_0 \ll 1$

$$\Delta E \approx -\epsilon_{\mathrm{F}}\, e^{-\dfrac{4}{3J\rho_0}} . \tag{11.33}$$

[33] ϵ_f is an additive constant which could be dropped. We kept it in order to emphasize the way the trial state is constructed.

[34] The zeroth order energy change $\epsilon_f - \epsilon_{\mathrm{F}}$ would be the same for the triplet $(f_\uparrow^\dagger c_{\mathbf{k}_{\mathrm{F}}\uparrow} - f_\downarrow^\dagger c_{\mathbf{k}_{\mathrm{F}}\downarrow})|\mathrm{FS}\rangle$. However, extending the Yosida argument to the triplet state, we find that it leads to less further energy gain from the Kondo terms, thus the ground state is a singlet.

Improved trial states can be obtained by dressing the basic singlet (11.28) with electron–hole excitations

$$|\Phi_1\rangle = |\Phi_0\rangle + \sum_{\mathbf{k}} \sum_{\mathbf{q}_1,\mathbf{q}_2} a(\mathbf{k},\mathbf{q}_1,\mathbf{q}_2) \sum_{\sigma'} c^{\dagger}_{\mathbf{q}_1,\sigma'} c_{\mathbf{q}_2\sigma'} \left(f^{\dagger}_{\uparrow} c_{\mathbf{k}\uparrow} + f^{\dagger}_{\downarrow} c_{\mathbf{k}\downarrow} \right) |\mathrm{FS}\rangle. \tag{11.34}$$

This does not change the character of the solution, but the numerical coefficient in the exponent is reduced from 4/3 to 1.22 [459]. Adding an arbitrary number of electron–hole pairs, the exponential factor of the Kondo energy becomes $\exp(-1/J\rho_0)$ [461]. The same exponent is readily found by using a similar variational method in the Kondo limit of the equivalent impurity Anderson model [428], and it is confirmed by the exact solution. It is useful to introduce the equivalent temperature scale, the *Kondo temperature*

$$T_{\mathrm{K}} = \frac{W}{k_{\mathrm{B}}} e^{-1/J\rho_0} \tag{11.35}$$

(in the prefactor, we replaced ϵ_{F} by the bandwidth W, which is of the same order of magnitude). We do not derive finite-T relationships, but we cite a couple of relevant results (see [164] and references therein).

Though the formation of the singlet ground state is accompanied by a non-analytic energy gain, and thus it is a non-perturbative process, we cannot speak about a phase transition. The single-impurity system is "small" even though it is coupled to an infinitely large fermionic bath, and cannot show cooperative behaviour. In particular, T_{K} should not be thought of as a transition temperature. Its significance becomes more clear, if we cite the result for the impurity contribution to the high-temperature susceptibility

$$\chi_{\mathrm{imp}}(T) \approx \frac{(g\mu_{\mathrm{B}})^2}{4k_{\mathrm{B}}T} \left\{ 1 - \frac{J\rho_0}{1 + J\rho_0 \ln(T/W)} \right\} = \frac{(g\mu_{\mathrm{B}})^2}{4k_{\mathrm{B}}T} \left\{ 1 - \frac{1}{\ln(T/T_{\mathrm{K}})} \right\}. \tag{11.36}$$

At $T \gg T_{\mathrm{K}}$, there is a logarithmic correction to the susceptibility. It is negative, showing that the conduction electron spins tend to align antiparallel to the f spin. The so-called logarithmic approximation which gives the above result, breaks down at $T = T_{\mathrm{K}}$, but this is not associated with any true divergence. Instead, T_{K} should be thought of as the center of a broad crossover regime where the behaviour changes gradually from the high-temperature localized moment behaviour to the low-temperature Fermi liquid behaviour. At $T \ll T_{\rm K}$, it becomes apparent that the f-electron has become (in the sense of (11.28) and its generalizations) part of the Fermi sea[35] which, consequently, has acquired a "heavy" component. We can isolate the impurity contribution to various quantities, which often show the same power law behaviour as the host conduction electron sea, however, with the scale of T set by $T_{\rm K}$ rather than $T_{\rm F}$. For instance, the impurity contributes

$$\gamma_{\rm imp} = \frac{\pi^2 k_{\rm B}}{6T_{\rm K}} \tag{11.37}$$

to the specific heat coefficient. The low-temperature susceptibility follows

$$\chi_{\rm imp}(T) = \chi_{\rm imp}(0)\left\{1 - a_\chi \left(\frac{T}{T_{\rm K}}\right)^2 + O\left(\frac{T}{T_{\rm K}}\right)^4\right\}. \tag{11.38}$$

The $T = 0$ susceptibility

$$\chi_{\rm imp}(0) = \frac{w(g\mu_{\rm B})^2}{4k_{\rm B}T_{\rm K}} \tag{11.39}$$

(where $w = 0.4128$ is the Wilson number) is finite, signalling that the impurity spin is fully compensated (the ground state is a singlet). However, the contribution to the Pauli susceptibility is $\sim T_{\rm F}/T_{\rm K}$ times larger than the conduction electron susceptibility per unit cell. The enhancements of γ and χ are of the same order: it can be shown that the Wilson ratio

$$R_{\rm W} = \frac{\chi_{\rm imp}(0)}{\gamma_{\rm imp}} \cdot \frac{\gamma_{\rm con}}{\chi_{\rm con}(0)} = 2\,. \tag{11.40}$$

[35] This phrase has to interpreted with due caution. We do not mean that the f-electron has become itinerant; the construction of (11.28) makes it clear that the f-orbital is always occupied by exactly one electron, thus it remained a localized spin, as it should be. However, the Fermi sea got swollen by one extra state so that the Fermi volume corresponds to the total number of c and f electrons taken together. Furthermore, c-electrons participate in singlet building with the f-electron in the same way as if it occupied an additional c-orbital.

Large as the enhancement factors may be, the contributions to the measured susceptibility and specific heat would be imperceptible, if we had literally a single impurity only. However, what we really have in mind, is the case of *dilute magnetic alloys* with a finite concentration c_{imp} of moment-bearing impurities, for which we expect a specific heat contribution

$$\Delta C = c_{\text{imp}} \gamma_{\text{imp}} T \,, \tag{11.41}$$

and a susceptibility contribution

$$\Delta \chi = c_{\text{imp}} \chi_{\text{imp}}(T). \tag{11.42}$$

Another interesting quantity is the electrical resistivity R

$$R(T) = R_{\text{phon}}(T) + c_{\text{imp}} R_{\text{res}} + c_{\text{imp}} R_{\text{imp}}(T) \tag{11.43}$$

where the first term is the usual phonon contribution, the second term gives the residual resistivity due to potential scattering, and the third term is associated with the building up of the Kondo singlet. At high T, we find the characteristic signature of the Kondo effect, a logarithmically increasing resistivity component

$$R_{\text{imp}}(T) = -R_1 \ln \frac{k_{\text{B}} T}{W}. \tag{11.44}$$

Assuming that this result holds in the temperature range where $R_{\text{phon}}(T) \sim T^5$, we predict that the resistivity passes through a minimum at $T_{\text{min}} \propto c_{\text{imp}}^{1/5}$. Such a resistivity minimum was observed in certain Au samples as early as 1934 [83], but it was only much later that it was ascribed to the unintended presence of magnetic impurities, probably Fe. The explanation was given 30 years later by Kondo [223]. Let us mention that in the original sense, (11.44) and the resulting resistivity minimum were meant by "Kondo effect". Nowadays, however, the term Kondo effect covers essentially everything which is associated with the gradual development of the Kondo spin compensation[36], either for

[36]It should be mentioned that all the remarkable phenomena described in this subsection originate from the antiferromagnetic nature of the c–f coupling. Assuming a ferromagnetic coupling in the model (11.27), we would not find the Kondo effect.

the model (11.27), or one of its numerous generalizations. Hewson's monography [164] covers much of the subject.

Like in the case of other physical quantities, the diverging character of (11.44) signals a merely technical difficulty. The exact $R_{\rm imp}(T)$ saturates at a large but finite value R_0 as $T \to 0$, and the low-T behaviour is described by

$$R_{\rm imp}(T) \approx R_0 \left\{ 1 - a_R \left(\frac{T}{T_{\rm K}} \right)^2 \right\} . \tag{11.45}$$

Let us note that the same $T_{\rm K}$ figures in the high-temperature and low-temperature relationships. The Kondo impurity problem has a single energy scale.

Naturally, (11.41), (11.42), and (11.43) are valid only if $c_{\rm imp}$ is small enough so that impurity–impurity interactions can be neglected. Well, how small should $c_{\rm imp}$ be? This is an unexpectedly difficult question, to which we can offer no conclusive answer.

Our basic idea is that the single-impurity trial state (11.28) describes a Kondo compensation cloud of diameter $\xi_{\rm K}$, which contains an f–c singlet. An electron looking on from farther than $\xi_{\rm K}$ sees a big singlet object which acts like a potential scatterer. We can use an uncertainty principle argument to derive $\xi_{\rm K}$. Assuming that we envisage $|\Phi_0\rangle$ as a wave packet with a momentum distribution width $\Delta p = |\Delta E|/v_{\rm F}$,

$$\xi_{\rm K} = \frac{\hbar}{\Delta p} \propto \frac{\epsilon_{\rm F}}{|\Delta E|} \cdot \frac{\hbar}{p_{\rm F}} \propto a e^{1/J\rho_0} , \tag{11.46}$$

where a is the lattice parameter. The Kondo cloud is exponentially large. It may seem reasonable to consider an alloy dilute if the Kondo clouds of two impurities separated by an average distance $c_{\rm imp}^{-1/3} a$ do not overlap. Taking $T_{\rm K} = 10{\rm K}$ (which would not be unusual), and $W = 10000{\rm K}$, we get $\xi_{\rm K} = 1000a$. According to the previous criterion, an alloy would be dilute if $c_{\rm imp} < 10^{-9}$. Since such levels of sample purity cannot be achieved, we might fear that the results (11.37)–(11.45) have no practical application. Fortunately, this is far from being the case. The Kondo effect is robust, observable in a wide range of systems, including rather concentrated alloys with strongly overlapping Kondo clouds. Indeed, in $\text{Ce}_x\text{La}_{1-x}$-type alloy systems, there seems to exist a continuity for the $T \geq T_K$ behaviour from the dilute limit through concentrated alloys to periodic heavy fermion systems [316].

Does this mean that the argument yielding (11.46) was flimsy, and ξ_K is a false concept? That would be unjust. Careful investigation confirms that the large Kondo cloud is a reality [372]. However, the situation is quite complicated. At finite T, the behaviour has to be described by the simultaneous use of two length scales: ξ_K, and the thermal length scale ξ_T. Furthermore, one has to be careful in interpreting measurements by local probes like NMR, because they are sensitive to the details of the internal structure of the Kondo singlet. Part of the story is that the Kondo exchange polarizes the electron sea, and the resulting f–c spin correlations have an oscillating $\propto \cos k_F r/r^2$ character. A little later, we will see that this gives rise to the RKKY interaction. The formation of the Kondo cloud imposes a T-dependent envelope upon the oscillating correlations; at low T, the remaining length scale is ξ_K. We refer to [372] and [45] for details.

The genuinity of the large Kondo cloud, juxtaposed with the survival of Kondo features in concentrated alloys, is paradoxical. It suggests that there should exist a theoretical framework in which typical large-T manifestations of the Kondo effect follow from single-impurity Kondo compensation for $c_{imp} < (a/\xi_K)^3$, and they are not only not destroyed, but conceivably even amplified by impurity–impurity interactions for $c_{imp} > (a/\xi_K)^3$. The behaviour is governed by a single energy scale T_K (which should not be much different from (11.35)), and scales with c_{imp}. The results of resistivity measurements on $Ce_xLa_{1-x}Cu_6$, for which a T_K of 3.5–5K was suggested [316], may serve as illustration (Fig. 11.14). The behaviour is unified and smooth[37] for $T > 10$K. However, substantial differences are seen at low T. Low-x samples follow by and large the independent-impurity Kondo behaviour: the resistivity increases monotonically with lowering T, saturating at $T = 0$ as described by (11.45). On the other hand, concentrated samples, and most clearly pure $CeCu_6$, show a resistivity maximum, and after a steep decrease of ρ, follow the Landau regime law $\rho \approx \rho_0 + A^*T^2$. In these systems, we may speak about the emergence of low temperature phase coherence between the

[37]The humpy character of some low-x curves is ascribed to crystal field excitations which we cannot discuss here.

Kondo clouds, or the formation of a heavy fermion band. Systems with a low temperature heavy fermion regime have at least two[38] characteristic temperatures: T_K, and the coherence temperature T^* which we introduced in Sec. 11.2.4.

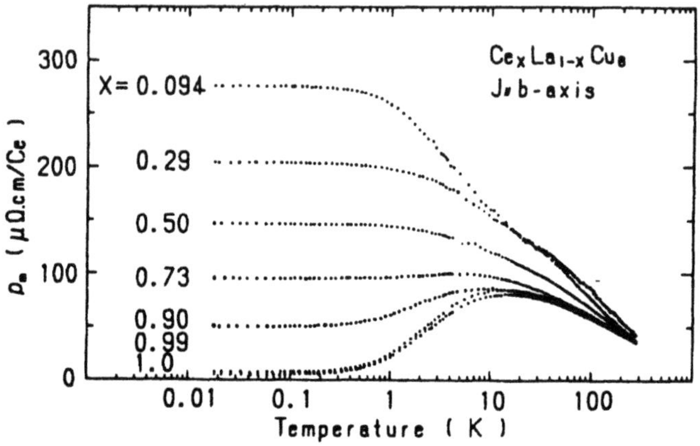

Figure 11.14: The f-shell-related resistivity of $Ce_xLa_{1-x}Cu_6$ samples (defined as $\rho_m = (\rho(Ce_xLa_{1-x}Cu_6) - \rho(LaCu_6))/x$), as a function of $\log T$, for a variety of x (after [316]).

It is a great outstanding problem to work out the theory of Kondo alloys for arbitrary c_{imp}. It should certainly provide an insight into the nature of the heavy fermion state, which at the moment we are still lacking: after all, if we could extend (11.37) and (11.41) to $c_{imp} = 1$, we would readily predict a large electronic specific heat which would often be quite near to the actual value. Unfortunately, we have no justification for proceeding in this manner. Great effort has been spent on the two-impurity problem, revealing the bewildering complexities of the competition between Kondo singlet formation, and RKKY magnetism. While the lessons drawn from such studies are of great value, it is also clear that we cannot hope to proceed in the same spirit to bridge the

[38] There may be more than two. If we postulate that the overall width of the narrow structure at the chemical potential is $\sim T^*$, then we should go to a lower temperature to resolve the narrow gap separating the effective hybridized bands. It may require a still lower temperature to observe Fermi liquid behaviour, whereby different experimental probes give slightly different characteristic temperatures [316].

gap between $c_{\rm imp} \ll 1$ and $c_{\rm imp} = 1$. In particular, looking at $c_{\rm imp} \ll 1$ models does not permit to realize the importance of the question of the Fermi volume. Therefore, after a brief discussion of the RKKY interaction, we return to the study of the periodic Kondo model.

11.3.2 Indirect Exchange

The local f–c exchange (in our case, the Kondo coupling (11.24)) polarizes the surrounding Fermi sea which carries this information to other f-sites, mediating an effective f-spin–f-spin interaction [210, 191, 264, 466]. To find its pair term, let us pick out two f-shells, situated at $\mathbf{R}_1$ and $\mathbf{R}_2$, which bear the spins $\mathbf{S}_1$ and $\mathbf{S}_2$, respectively. Considering the two-step process in which a spin-flip scattering at site $\mathbf{R}_1$ is followed by a spin-flip scattering at site $\mathbf{R}_2$

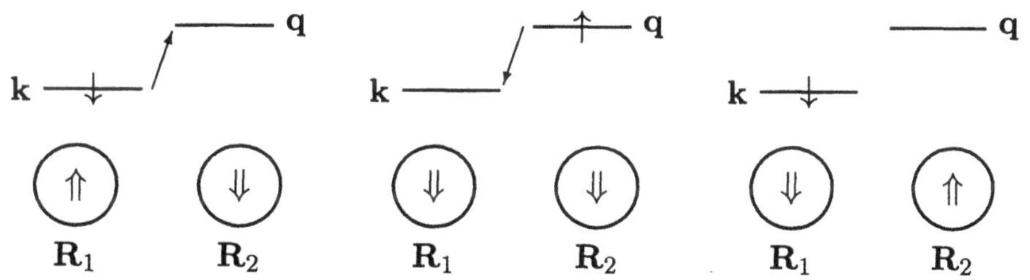

gives a contribution

$$\left(\frac{J}{L}\right)^2 e^{i(\mathbf{k}-\mathbf{q})(\mathbf{R}_1-\mathbf{R}_2)} \frac{n_{\mathbf{p}\downarrow}(1-n_{\mathbf{q}\uparrow})}{\epsilon_{\mathbf{q}}-\epsilon_{\mathbf{k}}} S_1^- S_2^+ . \tag{11.47}$$

Collecting all terms $\propto J^2/W$ is conveniently done by performing a canonical transformation on the Kondo lattice model. For the present purposes, it is sufficient to include the terms which couple the two selected f-sites to the Fermi sea

$$\mathcal{H} = \mathcal{H}_0 + \mathcal{H}_1 ,$$

where

$$\mathcal{H}_1 = \frac{J}{2L} \sum_{j=1}^{2} e^{i(\mathbf{k}-\mathbf{q})\mathbf{R}_j}\, \mathbf{S}_{\mathbf{j}} \cdot \sum_{\alpha,\beta} c_{\mathbf{q}\alpha}^\dagger \vec{\sigma}_{\alpha\beta} c_{\mathbf{k}\beta} . \tag{11.48}$$

If we transform out the f–c interactions, the transformed Hamiltonian

$$\mathcal{H}_{\text{eff}} = e^{i\mathcal{S}}\mathcal{H}e^{-i\mathcal{S}} \approx \mathcal{H} + [i\mathcal{S},\mathcal{H}] + \frac{1}{2}[i\mathcal{S},[i\mathcal{S},\mathcal{H}]] \tag{11.49}$$

will contain the effective f–f interaction we are interested in. Since we are familiar with the method from Sec. 5.1.2, and from Appendix D, we can keep the presentation short.

It is required that

$$\mathcal{H}_1 + [i\mathcal{S},\mathcal{H}_0] = 0 \tag{11.50}$$

which leads to

$$\mathcal{H}_{\text{eff}} = \mathcal{H}_0 + \frac{1}{2}[i\mathcal{S},\mathcal{H}_1]\,. \tag{11.51}$$

It can be checked that (11.50) is satisfied by

$$i\mathcal{S} = \frac{J}{2L}\sum_{\mathbf{j}=1}^{2}\sum_{\mathbf{k},\mathbf{q}}\frac{e^{i(\mathbf{k}-\mathbf{q})\mathbf{R}_\mathbf{j}}}{\epsilon_\mathbf{k}-\epsilon_\mathbf{q}}\left[S_\mathbf{j}^{+}c_{\mathbf{k}\downarrow}^{\dagger}c_{\mathbf{q}\uparrow} + S_\mathbf{j}^{-}c_{\mathbf{k}\uparrow}^{\dagger}c_{\mathbf{q}\downarrow} + S_\mathbf{j}^{z}(c_{\mathbf{k}\uparrow}^{\dagger}c_{\mathbf{q}\uparrow} - c_{\mathbf{k}\downarrow}^{\dagger}c_{\mathbf{q}\downarrow})\right] \tag{11.52}$$

where ill-defined terms with $\epsilon_\mathbf{k} = \epsilon_\mathbf{q}$ are omitted from the second sum.

When evaluating the commutator in (11.51), intersite contributions are obtained by taking $\mathbf{R}_1$ terms from $i\mathcal{S}$ and $\mathbf{R}_2$ terms from $\mathcal{H}_1$, or vice versa[39]. For instance,

$$\left[S_1^{-}c_{\mathbf{k}\uparrow}^{\dagger}c_{\mathbf{q}\downarrow}, S_2^{+}c_{\mathbf{q}\downarrow}^{\dagger}c_{\mathbf{k}\uparrow}\right] = S_1^{-}S_2^{+}(\hat{n}_{\mathbf{k}\uparrow} - \hat{n}_{\mathbf{q}\downarrow})\,,$$

corresponding to (11.47). Collecting all relevant pieces, the interaction is seen to be isotropic[40] $\propto \mathbf{S}_1\cdot\mathbf{S}_2$. Finally, we assume that the Fermi sea remains unpolarized and replace $\langle\hat{n}_{\mathbf{k}\sigma}\rangle$ by $f_\mathbf{k}$, the free electron Fermi occupation number.

The same kind of interaction can be derived for an arbitrary pair of sites, thus we can return to the case of the periodic lattice and write the effective Hamiltonian as the sum of the conduction band kinetic energy and the *indirect exchange interaction*

$$\mathcal{H}_{\text{RKKY}} = \sum_{\mathbf{k}\sigma}\epsilon_\mathbf{k}\hat{n}_{\mathbf{k}\sigma} + \frac{1}{2}\sum_{i\neq j}\mathcal{I}(\mathbf{R}_i - \mathbf{R}_j)\,\mathbf{S}_i\cdot\mathbf{S}_j \tag{11.53}$$

[39] Single-site contributions give an additive constant $\propto (J^2/W)S(S+1)$ per f-site.

[40] Because we started with the isotropic PAM (11.4). In rare earth systems, the f-electrons are subject to crystal fields, and the indirect f–f exchange interaction is, in fact, anisotropic.

where the coupling strength of the so-called Ruderman–Kittel–Kasuya–Yosida (RKKY) interaction[41] is

$$\mathcal{I}(\mathbf{R}_i - \mathbf{R}_j) = -\left(\frac{J}{L}\right)^2 \sum_{\mathbf{k},\mathbf{q}} \cos(\mathbf{k}-\mathbf{q})(\mathbf{R_i}-\mathbf{R_j}) \frac{f_{\mathbf{k}} - f_{\mathbf{q}}}{\epsilon_{\mathbf{q}} - \epsilon_{\mathbf{k}}} . \tag{11.54}$$

It is enlightening to rewrite this formula, introducing $\mathbf{p} = \mathbf{q} - \mathbf{k}$, and recalling the wave vector dependent free electron susceptibility (7.42)

$$\mathcal{I}(\mathbf{R}_i - \mathbf{R}_j) = -\frac{J^2}{L} \sum_{\mathbf{p}} \chi_0(\mathbf{p}) \cos \mathbf{p}(\mathbf{R_i} - \mathbf{R_j}) . \tag{11.55}$$

This result can be interpreted in the following manner: the local f–c exchange at $\mathbf{R}_1$ acts like the magnetic field

$$\mathbf{H}(\mathbf{r}) = J\mathbf{S}_1 \delta(\mathbf{r} - \mathbf{R}_1) = \frac{J}{L} \sum_{\mathbf{p}} e^{i\mathbf{p}(\mathbf{r}-\mathbf{R}_1)} \mathbf{S}_1 , \tag{11.56}$$

causing the conduction electron spin polarization

$$\mathbf{s}(\mathbf{R}_2) = -\frac{J}{L} \sum_{\mathbf{p}} \chi_0(\mathbf{p}) e^{i\mathbf{p}(\mathbf{R}_2 - \mathbf{R}_1)} \mathbf{S}_1 \tag{11.57}$$

at site $\mathbf{R}_2$. The Kondo interaction of $\mathbf{S}_2$ with this induced spin polarization amounts to the RKKY coupling (11.54).

The character of the RKKY interaction can be gleaned from evaluating (11.54) for the free electron dispersion $e_{\mathbf{k}} = \hbar^2 k^2/2m$. Rewriting the cosine as a sum of exponentials, using $\mathbf{R} = \mathbf{R}_1 - \mathbf{R}_2$, and denoting by ϑ the angle subtended by $\mathbf{q}$ and $\mathbf{R}$, we have integrals like

$$\int d\mathbf{q} \frac{e^{i\mathbf{q}\mathbf{R}}}{q^2 - k^2} = 2\pi \int_0^\infty q^2 dq \int_{-\pi}^{\pi} d\vartheta \sin\vartheta \frac{e^{iqR\cos\vartheta}}{q^2 - k^2}$$

$$= \frac{4\pi}{R} \mathcal{P} \int_0^\infty dq\, q \frac{\sin qR}{q^2 - k^2} = \frac{2\pi}{R} \mathcal{P} \int_{-\infty}^{\infty} dq\, q \frac{\sin qR}{q^2 - k^2}$$

[41]Ruderman and Kittel [342] solved the analogous problem of conduction-electron-mediated interaction between nuclear magnetic moments in metals. In this case, the role of Kondo coupling is taken over by the hyperfine contact interaction. Kasuya [203] and Yosida [458] studied the problem of a pair of magnetic impurity atoms in a simple metal.

where $\mathcal{P}$ stands for principal value (we omitted the prefactor $(a_0/2\pi)^3$ where a_0^3 is the unit cell volume). Extending the q-integral to $(-\infty,\infty)$ is convenient for using the method of residues [210]. For e^{iqR} (e^{-iqR}), the integration contour can be closed by an infinite semi-circle in the upper (lower) half-plane. For e^{iqR}, the pole at $q = k$ ($q = -k$) gives the residuum $(1/2)e^{ikR}$ ($(1/2)e^{-ikR}$). The same contribution comes from e^{-iqR}, thus all in all, integration over $\mathbf{q}$ gives

$$\int d\mathbf{q} \ldots = \frac{2\pi^2}{R} \cos kR.$$

We proceed with the polar angle part of the $\mathbf{k}$-integral similarly, finding the factor $k \sin kR$. The $T = 0$ behaviour is given by integrating from $k = 0$ to $k = k_F$. Using

$$8 \int x \sin x \cos x \, dx = \sin 2x - 2x \cos 2x \,,$$

we arrive at

$$\mathcal{I}(\mathbf{R}_i - \mathbf{R}_j) = -\frac{J^2}{\pi^3} \left(\frac{k_F a_0}{2} \right)^6 \frac{\sin 2k_F R_{ij} - 2k_F R_{ij} \cos 2k_F R_{ij}}{(2k_F R_{ij})^4} \tag{11.58}$$

where $R_{ij} = |\mathbf{R}_i - \mathbf{R}_j|$. The essential conclusion is that the RKKY interaction oscillates and it is long-ranged in the sense that it decays according to a power law[42]. The period of the oscillation is determined by the Fermi surface diameter. Let us note that the coupling is ferromagnetic within the first nodal surface of $\mathcal{I}$, which is a sphere with the radius $4.49/2k_F \approx 0.73 n^{-1/3} a_0$ (for a simple cubic lattice).

It is clear that the above free electron result gives only qualitative guidance; the conduction bands of rare earth elements have strong $5d$ character, and the anisotropy of the Fermi surface should be reflected in the true form of the RKKY coupling.

11.3.3 Kondo Singlet versus RKKY Magnetism

Periodic f-electron systems which we hope can all be modelled by some suitably chosen Anderson or Kondo lattice model, show extremely varied

[42] However, if we postulate that the electrons have a finite mean free path l, then the expression (11.58) gets multiplied with the exponentially decaying factor $e^{-R_{ij}/l}$ [264].

behaviour [135, 317, 375]. Some seem to remain non-ordering heavy Fermi liquids down to the lowest accessible temperatures. We discussed $CeAl_3$ and $CeCu_6$ in this sense in Sec. 11.2.4, and could add to the list $CeRu_2Si_2$, $CeNi_2Ge_2$, and many other substances. It should, however, be emphasized that found to be "non-ordering" is an uncertain status, subject to frequent experimental revisions. In particular, the sensitive technique of muon spin relaxation is apt to detect static magnetism of $\sim 10^{-3}\mu_B$ moments in systems which are, by other criteria, non-magnetic. Since we are not in the position to make any useful comment on this phenomenon, we leave it out of further consideration.

There are also clearcut cases of heavy fermion magnetism, exemplified by the antiferromagnet U_2Zn_{17}. The specific heat coefficient is quite large ($\gamma \approx 0.55 J/\text{mole U K}^2$) above the Néel temperature $T_N = 9.7$K. The sharp specific heat anomaly (Fig. 11.15) reminds us of spin density wave transitions (Sec. 7.6.2). However, SDW gapping is clearly incomplete and it leaves a moderately heavy ($\gamma \sim O(0.1 J/\text{mole U K}^2)$) Fermi liquid at $T \to 0$. The ordered moment $0.8\mu_B$ is much smaller than the paramagnetic moment $3.3\mu_B$, and this would also fit into the SDW framework, but let us note that it is a size which may also be explained by crystal field theory. – There are a few Kondo lattice compounds which order ferromagnetically: for instance, CePdSb does so at $T_C = 18$K [70]. Its low-T ordered moment of $0.9\mu_B$ is about a third of the free-ion value.

It is a further challenge to understand the localized moment f-electron magnets, for instance heavy rare earth elements like Gd. Also these are periodic f–c systems, so we should be wondering what makes them so different from both $CeAl_3$ and U_2Zn_{17}: they have very ordinary specific heat coefficients, and show no evidence of Kondo-like features in their disordered phase. We return to them in some detail in Sec. 11.4.

For the purposes of the present discussion, let us simplify the rich variety of Kondo lattice behaviour to these three categories[43]: 1) non-magnetic heavy Fermi liquids; 2) heavy fermion magnets with reduced

---

[43]It should not be left unmentioned that there is a further, completely different class of f-electron systems which are enjoying great attention: non-magnetic non-Fermi-liquids [435]. These are, however, beyond the scope of this text. – Besides, heavy fermion superconductivity can either compete, or coexist, with magnetism.

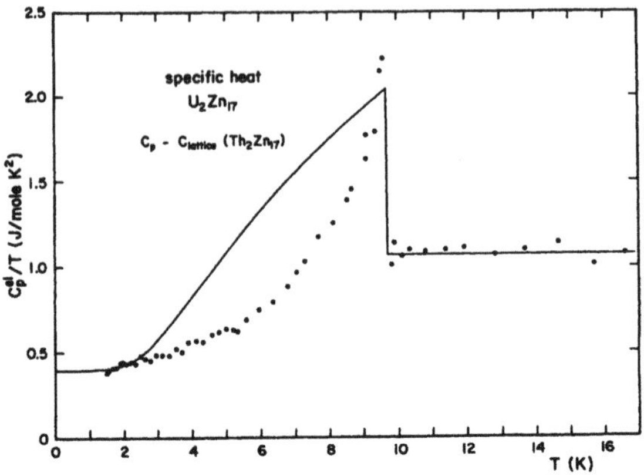

Figure 11.15: The $5f$ contribution to the specific heat coefficient of antiferromagnetic U_2Zn_{17} was obtained by subtracting the data for Th_2Zn_{17} (dots: experimental data, line: a SDW theory fit; after [317]).

ordered moments; and 3) the magnetism of localized f-electron moments coexisting with a light Fermi sea of conduction electrons. In principle, we would like to understand these three kinds of behaviour as three possible phases (or ground states) of the same Kondo lattice Hamiltonian. It had better be said at once that though we can claim some insight into how these phases may arise, much remains to be done.

Doniach Phase Diagram

Our discussion will be based on the simplest form (11.25) of the Kondo lattice model (KLM). The same Kondo interaction can drive two distinct phenomena which seem to be mutually exclusive: the building of an overall singlet (Sec. 11.3.1), and RKKY magnetism (Sec. 11.3.2). Let us assess the relative strength of these tendencies by comparing their energy (or characteristic temperature) scales. The strength of singlet binding may be measured by the Kondo temperature[44] T_K (11.35), and in a similar spirit, for the RKKY interaction (11.54) we introduce

[44] Actually, it would be better to use $k_B T_K = W/(\exp(1/J\rho)-1)$ which is analogous to (11.32). This coincides with (11.35) in the weak coupling limit, and it has the correct order of magnitude $k_B T_K \propto J$ for $J\rho \gg 1$.

$T_{\rm RKKY} = J^2/Wk_{\rm B}$. The left panel of Fig. 11.16 shows $T_{\rm RKKY}$ and $T_{\rm K}$ as a function of $J\rho(\epsilon_{\rm F})$ (in the figure which is taken from [70], the notation is $JN(E_{\rm F})$). It is easy to check that the two curves cross at $J\rho \sim 1$. This suggests the following scenario:

For small J/W, $T_{\rm RKKY}$ is much larger than $T_{\rm K}$, from which we conclude that the magnetic ordering at $T_{\rm RKKY}$ pre-empts Kondo singlet formation, and the system goes directly from a high-T paramagnetic state (which is an ordinary Fermi liquid since "heaviness" would arise only at $T < T_{\rm K}$), to an RKKY magnetic phase with well-localized moments. On the other hand, if $T_{\rm K} > T_{\rm RKKY}$, the localized moments get compensated by conduction electron spins before they could order, and the system remains non-magnetic. Upon lowering the temperature, it assumes the character of a heavy Fermi liquid, more-or-less according to the formulae (11.41)–(11.45). The magnetic ordering temperature $T_{\rm tr}$ follows $T_{\rm RKKY}$ while $J/W \ll 1$, but deviates downwards with increasing J/W, and drops to zero in the region where $T_{\rm K}$ overtakes $T_{\rm RKKY}$. If $T_{\rm tr}$ goes to 0 continuously, the point at which $T_{\rm tr} \to 0$, corresponds to a quantum critical point between the Kondo singlet, and RKKY magnetic, ground states.

Figure 11.16: Doniach's phase diagram. *Left*: the relative stability of the Kondo singlet and RKKY phases is judged by plotting the J-dependence of the characteristic temperatures (thin lines). The magnetic–non-magnetic phase boundary is shown in bold line, with the conjectured locations for the ferromagnetic–paramagnetic transition of CePdSb and CeAg. *Right*: the pressure dependence of the Curie temperature of CePdSb (after [70]).

The phase boundary drawn in bold line Fig. 11.16 (left) is not merely a graphic expression of the above-described expectations: there is a simple derivation based on Doniach's Kondo necklace model [89]. The necklace[45] model reduces the Kondo lattice problem to its barest minimum: an array of localized spins $\vec{S}$ (the f-spins) tend to be compensated by the spins $\vec{\tau}$ of itinerant fermions

$$\mathcal{H}_{\rm KNL} = J\sum_j \vec{S}_j\cdot\vec{\tau}_j - w_\tau \sum_{i,j}(\tau_i^x\tau_j^x + \tau_i^y\tau_j^y). \tag{11.59}$$

Using the x–y term for the intersite (nearest neighbour) $\vec{\tau}$–$\vec{\tau}$ coupling is meant to emphasize that the coupling originates from itinerant fermions. The charge degrees of freedom are omitted, assuming that they are not of primary importance for the process of spin compensation.

We wish to describe the competition of a singlet phase (which in the present model is composed of local $\vec{S}$–$\vec{\tau}$ singlets) with magnetic order. The latter is treated in the mean field approximation. We assume $w_\tau > 0$, and consequently a ferromagnetic order of the $\vec{\tau}$'s (the same result would be obtained assuming $w_\tau < 0$ and two-sublattice antiferromagnetism). The local Kondo interaction induces an equal polarization of the $\vec{S}$-spins in the opposite direction. Mean field decoupling yields the single-site Hamiltonian

$$\mathcal{H}_{{\rm KNL},j} = J\vec{S}_j\cdot\vec{\tau}_j - w_\tau z\langle\tau^x\rangle\tau_j^x\,, \tag{11.60}$$

where z is the coordination number. Notice that the Kondo term is left untouched, thus the singlet state is described correctly. In the ground state

$$\langle S^x\rangle_{T=0} = -\langle\tau^x\rangle_{T=0} = \sqrt{\frac{1}{4} - \left(\frac{J}{zw_\tau}\right)^2}\,. \tag{11.61}$$

The order parameter decreases with increasing J, and vanishes at the critical Kondo coupling strength $J_{\rm cr} = zw_\tau/2$. It appears that the partial Kondo compensation of ordered moments is at least a possibility.

---

[45]The model was originally put forward in a one-dimensional version, and was justly called "necklace". Our presentation assumes $D = 3$.

Standard mean field procedure gives the following self-consistency equation for the critical temperature $T_{\rm tr} = 1/\beta_{\rm cr} k_{\rm B}$

$$\frac{2J}{zw_{\tau}} = \frac{J\beta_{\rm cr} + e^{\beta_{\rm cr}J} - 1}{e^{\beta_{\rm cr}J} + 3}. \tag{11.62}$$

$T_{\rm tr}$ decreases monotonically from $zw_{\tau}/2k_{\rm B}$ at $J = 0$ to 0 as $J \to J_{\rm cr}$. At this point, however, it should be remembered that the intersite "magnetic" coupling w_{τ} should be really the RKKY coupling $w_{\tau} \to J^2/W$. It follows that $T_{\rm tr}$ increases like J^2/W at small J/W, goes through a maximum, and vanishes at $J \to J_{\rm cr}$, i.e., we obtained a phase boundary like that in Fig. 11.16.

It is not difficult to find debatable points in the above argument, and we will raise a few of them shortly, but first let us emphasize the virtues of the resulting scheme. Doniach's phase diagram, and its various adaptations, have proved remarkably valuable in systematizing and interpreting a great body of experimental observations. As an example, let us consider the pressure dependence of the Curie temperature of the Kondo lattice ferromagnet CePdSb (Fig. 11.16, right). Increasing pressure increases the hybridization, and thereby J, thus shifts the system to the right in Doniach's plot. Assuming that we started on the left hand slope of the hump of the phase boundary, we predict that with increasing P, $T_{\rm C}$ passes though a maximum, dropping at a fast rate on the other side. The experimental findings are in accordance with this scenario. Let us, however, remark, that the plot does not bring conclusive evidence about $T_{\rm C}$ approaching 0 continuously at a critical pressure; this may happen by a sudden drop. In any case, we are assured that no ferromagnetism is detected for $P > 15$GPa.

In spite of its great phenomenological success, the theoretical framework which is embodied in the Doniach phase diagram, leaves us unsure about a number of points[46]. In one way or another, these are all associated with the central question whether it is right to picture the collective singlet formation ("lattice Kondo effect") as a simple addition of single-site Kondo effects:

---

[46]We are not worried about the derivation by mean field theory, which – for the reader's convenience – was sketched in the simplest possible way. More sophisticated approaches lead to similar results.

- The idea that for $T_K > T_{RKKY}$, spin compensation "comes first" and so prevents ordering, does not really correspond to what we learned about Kondo screening. The Kondo cloud compensates the localized spin at its center only when integrated out to a large[47] distance ξ (11.46). Many other f-spins are well inside the cloud and see at least a reduced moment at its center, with which they could interact.

- At any rate, the magnetic-to-singlet phase transition is predicted to happen in the unphysical range $J = O(W)$. Both the Kondo and the RKKY interactions were derived by perturbation theory, and their use is questionable unless J is sufficiently small compared to W.

- The Kondo and RKKY interactions are really not two distinct terms in a Hamiltonian (as implied in the Kondo necklace model $\mathcal{H}_{KNL}$), but rather two manifestation of the same Kondo interaction in the Kondo lattice Hamiltonian[48] (11.25).

- Keeping the spin degrees of freedom only does not permit to raise the question about the Fermi volume. However, we are going to argue that the Kondo and RKKY phases belong to different Fermi volumes.

The variational treatment of the Kondo lattice model does, in some measure, deal with these issues. The idea is to construct trial states which correspond to our expectations about a heavy Fermi liquid, or a localized f-moment magnet, and draw the phase diagram by comparing the respective ground state energies. It should be added that though our introduction of the Hamiltonian (11.25) implied that the physically

[47] One may counter that the phase boundary lies at $J/W = O(1)$, where $\xi \sim a$, so spin compensation is already intracell. However, this contention brings us to the next item in the list.

[48] Or, going one step back, two manifestations of the c–f hybridization and f–f Coulomb terms in the PAM (11.4). However, physical reality is more complicated, and there is also the direct f–c exchange term which can give rise to RKKY magnetism, but can not favour overall singlet formation in the same way that the AF Kondo coupling does. We return to this question in Sec. 11.4.

relevant range is $J < W$, the variational phase diagram is studied for arbitrary coupling strength. The justification is twofold: first, the nature of some of the phases becomes transparent at large J/W. Second, similar-looking effective Hamiltonians where $J > W$ is physically acceptable, may be obtained in many-band situations.

Variational Approach

Let us start with the overall singlet state. One way to visualize this is to build a Kondo singlet around each f-spin, and then superpose them. Yosida's simplest wave function (11.28) for an isolated Kondo impurity located at $\mathbf{R}_0$ can be rewritten as

$$|\Phi_0\rangle = \left(f_{0\uparrow}^{\dagger}\tilde{c}_{0\uparrow} + f_{0\downarrow}^{\dagger}\tilde{c}_{0\downarrow}\right)|\mathrm{FS}\rangle_{N+1}\,, \tag{11.63}$$

where

$$\tilde{c}_{0\sigma} = \sum_{\mathbf{k}} a(\mathbf{k})\frac{e^{-i\mathbf{k}\mathbf{R}_0}}{\sqrt{L}}c_{\mathbf{k}\sigma} \tag{11.64}$$

annihilates[49] a conduction electron in the form of a cloud centered on $\mathbf{R}_0$. The variational parameters $a(\mathbf{k})$ specify the shape of the spin compensation cloud.

Now let us generalize (11.63) to the case of a periodic array of f-shells. Let there be N conduction electrons; the number of f-electrons is L. Assuming that $N \leq L$, the c-band could, in principle, hold all the $N + L$ electrons in a fictitious pure c ground state $|\mathrm{FS}\rangle_{N+L}$ which is a Fermi sea obtained by filling up the $N + L$ lowest-lying band states. Assuming that N and L are even, $|\mathrm{FS}\rangle_{N+L}$ is a singlet. Naturally, $|\mathrm{FS}\rangle_{N+L}$ is very far from being the ground state of the KLM (11.25): it populates the high-energy part of the band quite unreasonably. We have to let L electrons "fall" from c-states into f-states, and with a large probability, this will affect the upper L states of $|\mathrm{FS}\rangle_{N+L}$. We

---

[49] At the first sight, one may find it disturbing that we do not speak about the *creation* of a conduction electron with the same spatial distribution. However, what we do is equivalent to that. Say, $\tilde{c}_{0\uparrow}$ digs out an $\uparrow$-spin hole from a non-magnetic Fermi sea, and therefore leaves a $\downarrow$-spin conduction electron with the same wave function; this will be paired off with the $\uparrow$-spin f-electron we have just created.

seek the singlet ground state by repeating the Yosida construction at every lattice site

$$|\Psi\rangle = \prod_j \left(f^\dagger_{j\uparrow}\tilde{c}_{j\uparrow} + f^\dagger_{j\downarrow}\tilde{c}_{j\downarrow} \right) |\mathrm{FS}\rangle_{N+L}\,, \tag{11.65}$$

where $\tilde{c}_{j\sigma}$ is given by an expansion analogous to (11.64)

$$\tilde{c}_{j\sigma} = \sum_{\mathbf{k}} a(\mathbf{k}) \frac{e^{-i\mathbf{k}\mathbf{R}_j}}{\sqrt{L}} c_{\mathbf{k}\sigma}\,. \tag{11.66}$$

The variational parameters $a(\mathbf{k})$ are to be determined from optimizing the expectation value of (11.25) with (11.65). Naturally, we have no reason to expect that the solution would be the same as for the impurity problem.

Let us note some general features of $|\Psi\rangle$. It is a singlet with the same spin structure as the hypothetical "large" c-electron Fermi sea $|\mathrm{FS}\rangle_{N+L}$; we merely changed the orbital character of L electrons from c to f. It is also explicit that the expectation value of any of the f-spins is 0. Though it is a strong correlation problem we did not, as yet, need any projection operators to ensure that every site is occupied by exactly one f-electron. All the intricacies of the ground state are associated with the c-electrons which, among themselves, are non-interacting. That the problem is nonetheless very difficult, is due to the fact that the same Bloch operators $c_{\mathbf{k}\sigma}$ appear in the expansion of all the $\tilde{c}_{j\sigma}$; the Kondo clouds are strongly non-orthogonal. So to speak, a c-electron near $\mathbf{R}_j$ participates not only in the compensation of the f-spin at that particular site, but also in the compensation of a very large number of other f-spins, and therefore its state is subject to a large number of conflicting demands. Formally, the strong overlapping of the Kondo clouds is signalled by the fact that the $\tilde{c}_{j\sigma}$ operators do not even approximately anticommute

$$\left[\tilde{c}_{j\sigma}, \tilde{c}^\dagger_{m\sigma}\right] = \frac{1}{L}\sum_{\mathbf{k}} |a(\mathbf{k})|^2 e^{i\mathbf{k}(\mathbf{R}_m - \mathbf{R}_j)}\,. \tag{11.67}$$

It can be shown [361] that (11.65) is identical to a projected hybridized band state

$$|\Psi\rangle = \hat{P} \prod_\sigma \prod_{\mathbf{k}} \left(1 + a(\mathbf{k}) f^\dagger_{\mathbf{k}\sigma} c_{\mathbf{k}\sigma}\right) |\mathrm{FS}\rangle_{N+L}\,, \tag{11.68}$$

where the projector

$$\hat{P} = \prod_j \left[\hat{n}_{j\uparrow}(1 - \hat{n}_{j\downarrow}) + \hat{n}_{j\downarrow}(1 - \hat{n}_{j\uparrow})\right] \tag{11.69}$$

guarantees that there is strictly one f-electron at each site, and thus $|\psi\rangle$ is a legitimate wave function of the KLM which describes localized f-spins. (11.68) is analogous to the Fermi liquid trial state (11.9) which we introduced for the Anderson lattice model (11.4), except that the projector is now defined to exclude not only f^2 but also f^0. The technique of Gutzwiller expansion enables us to tackle the problem of non-orthogonality of the Kondo clouds [107]. Combined with the Gutzwiller approximation[50], a formal solution can be derived [111, 361, 109]. The optimal mixing amplitude looks similar to (11.12)

$$a(\mathbf{k}) = \frac{\lambda^2 \tilde{J}}{\tilde{\epsilon}_f - \epsilon_{\mathbf{k}} + \sqrt{(\tilde{\epsilon}_f - \epsilon_{\mathbf{k}})^2 + (\lambda \tilde{J})^2}}, \tag{11.70}$$

where the renormalized Kondo coupling $\tilde{J}$ plays the role of effective hybridization. We may be wondering where the finite hybridization amplitude may be coming from, since real f-electron motion is prohibited. The underlying mechanism is that the f–c spin-flip process can be pictured as put together of two simultaneous hybridization events. In the Gutzwiller approximation, this corresponds to the decoupling scheme

$$\langle S_f^+ S_c^- \rangle \to -\langle f_\uparrow^\dagger c_\uparrow \rangle \langle c_\downarrow^\dagger f_\downarrow \rangle. \tag{11.71}$$

Assuming that the band has a constant density of states $\rho_0 = 1/W$, allows to derive a closed-form expression valid for arbitrary J/W [109]. The ground state energy density of the singlet phase is

$$\frac{1}{L} E_{\text{sing}} = W \frac{n^2}{4} - \frac{n}{2} \frac{W}{\exp\left\{\frac{2W}{3J}\right\} - 1}. \tag{11.72}$$

[50] Like in the case of the Hubbard model and the PAM, this is expected to become exact at $D \to \infty$. However, comparison with variational Monte Carlo results shows that the approximation is acceptable even in $D = 1$ [361].

The first term is the $J = 0$ energy of the conduction electron Fermi sea (the "small" Fermi sea), and the second term gives the energy lowering associated with the singlet formation. This should be compared to the single impurity result (11.32). The functional form is the same, but the exponent is now half of the single-impurity value. This appears most clearly in the weak coupling form of the energy lowering density (analogous to (11.33))

$$\Delta E = \frac{1}{L}\left(E_{\text{sing}} - E_{\text{sing}}(J=0)\right) = -\frac{Wn}{2} e^{-\frac{2W}{3J}} . \tag{11.73}$$

Thus we have found that the singlet binding energy density of the periodic Kondo model is in an exponential sense larger than the Kondo energy for the impurity[51].

Let us make a remark about the strong-coupling limit of (11.72)

$$E_{\text{sing}} = E_{\text{sing}}(J=0) - N\frac{3J}{4} \tag{11.74}$$

This is easy to interpret: when the band energy becomes negligible, each electron pairs off with an f-spin. Since $N \leq L$, the number of on-site singlets is limited by N.

Returning to the weak coupling limit: in spite of the "lattice coherence enhancement", the energy (11.73) is still exponentially smaller than W (or ϵ_{F}), and this poses a riddle: how can all the f-spins be compensated? First, let us understand why this is a question. For the Kondo impurity, the answer would be clear: the conduction electrons do the spin compensation. Actually, only a small fraction of the conduction electrons. The derivation in Sec. 11.3.1 made it clear that the disturbance of the Fermi sea is confined to a thin shell of energy width

[51]One should not hastily reformulate this into the statement that the "lattice Kondo temperature" is correspondingly larger than the impurity T_{K}, because this would conflict with the standard usage of the term. As we mentioned on p. 640, heavy fermion systems have several characteristic temperatures. Their T_{K} is defined for the high-T (incoherent) regime, and it may not be very different from the single-ion T_{K}. The coherence temperature T^* is lower, but it is associated with the excitation spectrum, not the ground state energy. The stronger singlet binding of the lattice may not appear as a characteristic temperature, but it has a strong influence on the position of the magnetic–singlet phase boundary which we will discuss shortly.

$\propto k_B T_K$; from this, we estimate that only NT_K/T_F electrons participate in the Kondo cloud and the rest (which are deeper in the Fermi sea) are unperturbed. Using the same argument for the lattice, we must conclude that only $N \exp(-2W/3J) \ll N$ conduction electrons participate in spin compensation. They are too few to compensate more than a tiny fraction of the f-spins. This is the essence of the "exhaustion argument" against the possibility of finding an overall singlet ground state in the weak coupling limit [307].

The solution to the riddle is (in principle) simple: the f-electron spins largely compensate each other thus there is very little left for the c-electrons to do. This should not be wholly unexpected. The nature of the f-spin system, as described by (11.68), is the following: we have enlarged the c-electron Fermi sea by a thick **k**-space shell which contains N states. Then we acted with the projector (11.69) to turn these into a set of localized spins[52]. Let us recall (from Sec. 10.6.3) that fully Gutzwiller-projecting a half-filled band is a good recipe for making a quantum paramagnet, i.e., a non-ordered singlet with antiparallel spin correlations. The same principle is at work here, only the projected **k**-space region does not belong to the lower half of a band, but to a middle region. This is corroborated by variational Monte Carlo results which show that the Ansatz (11.68) implies strong antiparallel f–f spin correlations [361].

At this point, it is already clear that the nature of the lattice Yosida state (11.65) (or the equivalent (11.68)) is quite different from that of the single-impurity Yosida state (11.63). Though both involve some kind of spin compensation, it is only in a vague sense that we can speak about a lattice (or collective) Kondo effect for the KLM. It is well to remember that while the exponential form of the single ion Kondo energy is known to be exact, the similar-looking lattice result (11.73) may be an artefact of the variational procedure; we simply do not know whether E_{sing} really has an exponential term, or not.

Whatever the exact formula for the ground state energy may be, we trust that it is a solid result that the Kondo lattice model can have a

---

[52] Strictly speaking, this applies only at $J = 0$. At small J/W, there is a weak mixing described by (11.70), and a small part of the f-electrons is derived from **k**-states within the small Fermi sea.

non-symmetry-breaking, singlet ground state. For such a symmetrical state, Luttinger's theorem should hold, i.e., the Fermi volume should be the same as for the non-interacting system. Tracing back the origin of the KLM, the non-interacting model is not the c-band term in (11.25), but the $U_f = 0$ hybridized-band model (11.4) which has a "large" Fermi surface comprising both c- and f-electrons. In Sec. 11.2.3, we postulated that the non-magnetic heavy fermion state of the large-U_f Anderson lattice model can be understood by continuation from the $U_f = 0$ limit; the interaction redistributes the electrons over the Brillouin zone, but preserves the position of the $\mathbf{k}$-space discontinuity (Fig. 11.8). The mass enhancement factor is related to the reduced Fermi step (11.15). Since the KLM is equivalent to the large-U_f, nearly integral valent regime of the PAM, we have to be able to find the same kind of ground state for the Kondo lattice. $|\Psi\rangle$ as defined in (11.68) fills the bill: it is explicit that for any $J \neq 0$, the c-electron occupation number $\langle \hat{n}_c(\mathbf{k})\rangle$ is non-vanishing within the large Fermi surface. At the surface, there is a discontinuity, since $\langle \hat{n}_c(\mathbf{k})\rangle = 0$ outside[53]. Characteristic results are shown in Fig. 11.17 for $n = 3/4$ [361]. The occupation number changes steeply but continuously in the region where the Fermi step of the c Fermi sea "would be", but the actual discontinuity is the tiny step at the Luttinger Fermi surface (see inset with magnification).

One may feel uneasy about the concept of the large Fermi surface because the Kondo lattice model defines the f-electrons as localized spins, and therefore "they cannot be in the Fermi sea". Certainly, they are not. We are speaking about the occupation number of the c-electrons only. By developing the long thin tail of $\langle \hat{n}_c(\mathbf{k})\rangle$ which reaches up to high band energies (but the presence of holes deep below μ is just as significant), the system enlarges the phase space where exchange energy can be gained from processes like (11.71). The formation of the large Fermi sea is permitted because the exchange energy gain outweighs the kinetic energy loss.

It is interesting to observe that for $N = L$ conduction electrons (a

[53]The discontinuity of $\langle \hat{n}_c(\mathbf{k})\rangle$ is a physical result, but the vanishing of $\langle \hat{n}_c(\mathbf{k})\rangle$ outside the large Fermi surface is not: it is merely a limitation of the Ansatz which is the lattice version of the simplest Yosida state. Adding electron–hole pairs as in (11.34) would improve the energy, but it seems very difficult to carry through.

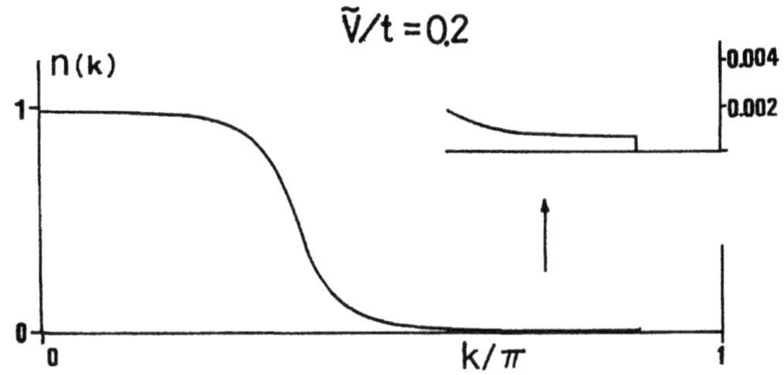

Figure 11.17: Band electron occupation number in the singlet ground state (11.68) of the one-dimensional Kondo lattice (after [361]). There is a small discontinuity at the Luttinger Fermi surface (inset). $\tilde{v}$ is the same as $\lambda\tilde{J}$ in (11.70).

half-filled c-band), the $U_f = 0$ reference state is the completely filled lower hybridized band (see Fig. 11.7), and thus Luttinger's theorem requires (or at least permits) that the system be a non-magnetic insulator. There is a fair number of lanthanide and actinide systems which are thought to belong to this category of *Kondo insulators*, for instance $Ce_3Bi_4Pt_3$, TmTe, or UNiSn [6]. Their gaps range from several K to several 1000K. One cannot really distinguish them from the mixed valent semiconductors we discussed earlier (see p. 619), but a compound with an intrinsic gap of $O(1-10)$K would be called Kondo, rather than mixed valent, insulator. The smallness of the gap is related to the small energy scale (11.73). The variational result for the charge gap is [109]

$$\Delta_{\text{gap}} = \frac{W}{\exp\left\{\dfrac{2W}{3J}\right\} - 1}. \tag{11.75}$$

This is exponentially small in the weak coupling limit, and becomes $3J/2$ for strong coupling. This is easy to interpret: to get an electron moving, we have to break up two local singlets: one, where the electron started from, and one, where it has landed. There is also a spin gap which is, in general, smaller[54] than the charge gap.

[54]This is easy to check in the large-J limit, where the spin gap is given by the local singlet–triplet excitation energy J. In the weak coupling limit, the disparity between the spin and charge gaps may be greater, especially in $D = 1$ [410].

Let us close the discussion of the large Fermi volume with two remarks which we cannot elaborate further. On the theoretical side: for the one-dimensional Kondo lattice it is exactly proven [451] that the Fermi volume satisfies Luttinger's theorem. This is reassuring but, as we will remark a bit later, the $D = 1$ model is so particular that it is difficult to draw inferences for higher-dimensional systems. On the experimental side, the shape and size of the Fermi surface are observable. De Haas–Van Alphen experiments tend to show that heavy fermion metals have large Fermi surfaces. And as for the non-magnetic Kondo insulators, their very existence proves that Luttinger's theorem applies: if it did not, these systems should be metals.

However, we do not mean to say that the Luttinger Fermi volume would be the inevitable consequence of having the Kondo lattice model (11.25); merely that it is one of the physically viable ground states. The situation somewhat resembles that found for the Hubbard model which can have either a metallic ground state with the Luttinger Fermi volume, or a Mott insulating state without a Fermi surface. Similarly, we can envisage that the f-electrons undergo a Mott localization of their own, and do not contribute to the Fermi volume which consequently belongs to the number of c-electrons. The "small Fermi sea" of conduction electrons mediates the RKKY interaction of the localized f-spins. The energetical argument presented on p. 647 still holds: in the weak coupling regime, the RKKY energy gain $\sim -J^2/W$ is superior to (11.73) which results from singlet formation. In the variational context, we have to construct a magnetic trial state and compare its energy to (11.72). The optimum magnetic order depends on the band filling n, but the general principle can be demonstrated on the example of the simple ferromagnetic trial state

$$|\Phi_{\rm FM}\rangle = \prod_{\mathbf{k}}^{N_\uparrow} c^\dagger_{\mathbf{k}\uparrow} \prod_{\mathbf{k}}^{N_\downarrow} c^\dagger_{\mathbf{k}\downarrow} \prod_{\mathbf{j}}^{L} f^\dagger_{j\downarrow}|0\rangle. \tag{11.76}$$

All the f-spins point $\downarrow$, and the band is polarized in the opposite sense. The total Fermi volume of the $\uparrow$ and $\downarrow$ Fermi seas corresponds to $N_\uparrow + N_\downarrow = N$, the total number of c-electrons. The energy consists of diagonal terms only, and it is readily minimized to give[55]

$$\frac{1}{L}E_{\text{FM}} = W\frac{n^2}{4} - \frac{J^2}{16W}. \tag{11.77}$$

Comparing this to (11.72), we find that the singlet heavy fermion state is unstable against the onset of RKKY ferromagnetism at arbitrary band filling, if J is small enough. However, instability against antiferromagnetism (which can be described by a suitably adapted SDW trial state [109]) takes precedence near $n = 1$. The combined phase diagram is shown in Fig. 11.18.

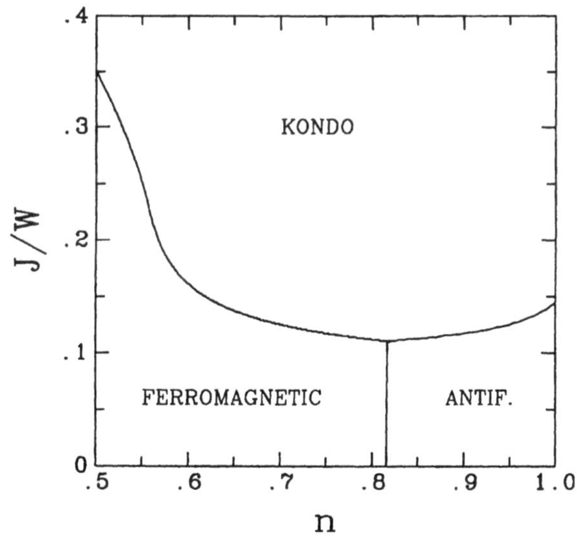

Figure 11.18: Ground state phase diagram of the Kondo lattice with a constant DOS [109]. The small-J magnetic phases are RKKY magnets with a small Fermi sea. The collective (Kondo) singlet state (11.65) has the Luttinger Fermi volume.

In a rough sense, the variational phase diagram corresponds to the $T = 0$ line in the Doniach phase diagram (Fig. 11.16, right): for any particular n, there is a magnetic-to-singlet phase transition at a critical J_{cr}. However, here we are speaking about a first-order transition:

[55]This expression holds until the minority band becomes empty. For larger J's, the energy is linear in J.

the magnetic phase transition coincides with the change of the Fermi volume which is necessarily discontinuous. The phase boundary is determined by the crossing of the energy levels (11.72) and (11.77) (or the analogous expression corresponding to a different magnetic structure). It is a virtue of the variational approach that, by detecting the lattice enhancement of the singlet energy, it predicts a transition in the range $J_{\rm cr} \sim O(0.1W)$ rather than $O(W)$. This allows both magnetic and non-magnetic states at physically acceptable values of J/W.

Let us note that we did not assume that there is a distinct RKKY interaction; rather, we deduced the RKKY energy scale from the Kondo interaction. At least in our simple treatment of lattice systems, the Kondo and RKKY scales belong to different sides of a phase boundary. Singlet building requires blowing up the Fermi surface to the Luttinger size, while RKKY interaction presupposes the existence of a small Fermi sea. This also transpires from the derivation in Sec. 11.3.2: the canonical transformation was done to cut off the f-electrons from the c-electrons, and the oscillating nature of the interaction (11.58) arises from the sharp small Fermi surface. Naturally, our arguments are far from proving anything, thus it is reassuring to see that there are systems which possess these two states as distinct thermodynamic phases. The α–γ transition of cerium can be qualitatively understood as an itinerant-to-localized transition of $4f$ electrons [193].

Additional Remarks

Since powerful numerical methods are available for studying one-dimensional systems, the ground state phase diagram of the Kondo chain is fairly accurately known [410, 411]. However, $D = 1$ is very special. Applying the formula (11.55) for the RKKY interaction, we realize that it contains the logarithmically diverging susceptibility of linear electron systems (see Problem 7.4, p. 411). Thus the RKKY interaction cannot be treated perturbatively, which may suggest that the drive towards magnetic ordering is strong; however, for one-dimensional systems quantum fluctuations foil all kinds of magnetic order except ferromagnetism whose special status we are familiar with. The phase diagram of the Kondo chain displays an extended FM region dominating at large J, a low-J Luttinger liquid phase, and (at $n = 1$), a non-magnetic Kondo insulator [411]. $D = 1$ systems do not manifest the Kondo–RKKY competition in the sense that higher-dimensional systems do.

The extent of the FM phase is much larger in $D = 1$ than what we expect in $D > 1$, but the appearance of a high-J pocket of ferromagnetism is not unexpected. At $J = \infty$, assuming $N < L$, sites are either occupied by a $c^1 f^1$ singlet, or by an unpaired f electron. In this limit, the problem can be reformulated as a $U = \infty$ Hubbard model in which the $c^1 f^1$ sites are the "empty sites", and the f^1 sites the "fermions" [230]. Notice that the fermion number is $N_f = L - N$. Since the $U = \infty$ Hubbard model is expected to become FM in the Nagaoka regime near half-filling (see Secs. 8.4 and 10.6.5), we expect that the $J = \infty$ Kondo lattice becomes likewise ferromagnetic at low n. It is likely that for bipartite lattices, the Nagaoka ferromagnetism of the KLM is a weak effect, and the large-J FM phase (which has the Luttinger Fermi volume like the Kondo state) is disconnected from the low-J RKKY ferromagnet[56].

Returning to low-J RKKY magnetism, let us notice that our crude variational description includes some moment compensation: in (11.76), band electron spins are polarized antiparallel to the f-spins. This would give up to 10–30 per cent reduction of the ordered moment on the magnetic side of the phase boundary. A more sophisticated treatment which works with placing a Kondo impurity in an effective field, yields similar results [27]. This theoretical framework may provide an acceptable description of magnets like $CeAl_2$ where the moment is visibly reduced, but still sizeable.

We should be, however, aware that there is a surprising number of heavy fermion magnets with "micromoments" in the range of 0.01–0.001μ_B. We do not offer any explanation why the ordering of such small moments should be a regular occurence. However, there may be cases where the understanding of micromoment magnetism does not require novel principles. For U compunds, the assumption of the $5f^2$

[56]The large-J FM region does not appear in Fig. 11.18, because for a constant DOS, the Gutzwiller approximation gives exactly the same energy for the Nagaoka state as for the $J = \infty$ Kondo state. However, for the square lattice, two distinct FM phases are found [109].

configuration permits that the crystal field ground state is a singlet. In this case, the ordered moment is exchange induced, and we have seen that it can be arbitrarily small (Sec. 3.8.4). Alternatively, it is possible that the band structure nearly fulfills a nesting condition, and then a small-amplitude spin density wave becomes a natural option (Sec. 7.6). Indeed, this is the basis of one of the models suggested for the explanation of the mysterious behaviour of URu_2Si_2 [365, 314].

11.4 Rare Earth Magnetism

There are several reasons why the simplest textbook variant of the RKKY interaction cannot be employed directly to explain the magnetic properties of rare earth (RE) elements [191]. One has to allow for the degeneracy of the f-shell, consider crystal field effects, use a realistic band structure for the $5d$ and $6s$ electrons, and a proper c–f interaction.

Let us, however, state at the beginning the one essential feature which even the simplest RKKY model (11.53) would get right: with the exception of α-Ce and Yb, the RE elements are localized moment magnets. The f-electrons can be thought of as Mott-localized, and they do not contribute to the Fermi volume. The shape, and diameter, of the Fermi surface is well accounted for by band structure calculations which assign the $4f$ electrons to the ionic cores. According to the *standard model* of rare earth magnetism, RE elements are integral valent systems.

As yet, we have considered only the strongest Coulomb matrix element U_f. However, it can be argued that the direct (ferromagnetic) exchange I_{f-c}^{dir} between f and conduction electrons is also important; in fact, for the majority of RE elements it is held to be more important than the antiferromagnetic Kondo coupling J we discussed before. For the moment, let us consider only direct exchange.

Assuming that we use an orthonormal basis, the arguments of Sec. 2.2 apply to the exchange interaction $-I_{f-c}^{\text{dir}}\mathbf{s}_1^f\cdot\mathbf{s}_2^c$ between an f-electron spin $\mathbf{s}_1^f$ at site $\mathbf{R}=0$ and a conduction electron spin $\mathbf{s}_2^c$. The relevant exchange integral is (see (2.48))

$$I_{f-c}^{\mathrm{dir}}(n',\mathbf{k}';n,\mathbf{k}) = \int d\mathbf{r}_1 \int d\mathbf{r}_2 \psi^*_{n'\mathbf{k}'}(\mathbf{r}_1)\phi^*_{4f}(\mathbf{r}_2)\frac{e^2}{|\mathbf{r}_1-\mathbf{r}_2|}\psi_{n\mathbf{k}}(\mathbf{r}_2)\phi_{4f}(\mathbf{r}_1) \tag{11.78}$$

where n, n' are band indices. If there are several f-electrons in the shell, the exchange interaction $-I_{f-c}^{\mathrm{dir}}\sum_\nu \mathbf{s}_\nu^f\cdot\mathbf{s}_2^c = -I_{f-c}^{\mathrm{dir}}\mathbf{S}\cdot\mathbf{s}_2^c$ acts on the total f-spin $\mathbf{S}$. Let us, however, recall that within the ground state multiplet the only conserved vector is $\mathbf{J}$, thus we have to express $\mathbf{S}$ with $\mathbf{J}$. Recalling (see p. 41) that within a given $(2J+1)$-dimensional subspace, $\mathbf{L}+2\mathbf{S} = g\mathbf{J}$, where g is the Landé factor, and subtracting from this equation $\mathbf{L}+\mathbf{S}=\mathbf{J}$, we get

$$\mathbf{S} = (g-1)\mathbf{J}\,. \tag{11.79}$$

The full expression of the c–f direct exchange interaction is analogous to (11.24)

$$\mathcal{H}_{c-f}^{\mathrm{dir}} = -(g-1)\sum_j\sum_{n,n'}\sum_{\mathbf{k},\mathbf{k}'} I_{f-c}^{\mathrm{dir}}(n',\mathbf{k}';n,\mathbf{k})\,e^{i\mathbf{R}_j(\mathbf{k}-\mathbf{k}')}$$

$$\cdot\left\{J_{\mathbf{j}}^- c^\dagger_{n'\mathbf{k}'\uparrow}c_{n\mathbf{k}\downarrow} + J_{\mathbf{j}}^+ c^\dagger_{n'\mathbf{k}'\downarrow}c_{n\mathbf{k}\uparrow} + J_{\mathbf{j}}^z(c^\dagger_{n'\mathbf{k}'\uparrow}c_{n\mathbf{k}\uparrow} - c^\dagger_{n'\mathbf{k}'\downarrow}c_{n\mathbf{k}\downarrow})\right\} \tag{11.80}$$

Like in (11.56), the f-shell at $\mathbf{R}_i$ can be considered to give rise to an effective field $\mathbf{H}_i \propto (g-1)\mathbf{J}_i$ acting on the conduction electrons; however, its source is now not point-like, but extended in space. A simple approximate form is obtained by dropping $n'\neq n$ terms, replacing the wave functions ψ by plane waves, and assuming that the Coulomb interaction is so strongly screened that it can be replaced by $\delta(\mathbf{r}_1-\mathbf{r}_2)$. Then we find

$$I_{f-c}^{\mathrm{dir}}(n,\mathbf{k}';n,\mathbf{k}) \approx \int d\mathbf{r}|\phi_{4f}(\mathbf{r})|^2 e^{i(\mathbf{k}'-\mathbf{k})\mathbf{r}} = f_{4f}(\mathbf{k}'-\mathbf{k})\,, \tag{11.81}$$

the form factor of the $4f$ electron density. The electron gas reacts to this Fourier component of $\mathbf{H}_i$ with the susceptibility $\chi_0(\mathbf{k}'-\mathbf{k})$, and the resulting band polarization couples to another f-shell spin at $\mathbf{R}_j$. It follows that the RKKY Hamiltonian (11.53) can be rewritten as

$$\mathcal{H}_{\mathrm{RKKY}} = \sum_{\mathbf{k}\sigma}\epsilon_{\mathbf{k}}\hat{n}_{\mathbf{k}\sigma} + \frac{1}{2}(g-1)^2\sum_{i\neq j}\mathcal{I}(\mathbf{R}_i-\mathbf{R}_j)\,\mathbf{J}_i\cdot\mathbf{J}_j\,. \tag{11.82}$$

Because of the long range of the interaction, the magnetic moment at any site is subject to a large number of aligning fields, pointing in different directions. The equilibrium structure arises from a compromise and as a result, it is often non-collinear or/and modulated. Furthermore, Nd, Sm, Tb, Dy, Ho, and Tm have two successive ordering transitions, and Er has three. There are examples of linearly polarized (Er, Tm), and circularly polarized (or helix, in Tb and Dy) magnetization waves, as well as more complicated structures (Ho, Er). Gd is a robust ferromagnet with $T_C = 293$K; the low-temperature phases of Dy and and Tb are also ferromagnetic. γ-Ce, Nd, Sm, and Eu are antiferromagnets.

Whatever the details of the ordering pattern may be, mean field theory tells us that the transition temperature $T_{\rm tr}$ should be proportional to the *de Gennes factor*

$$\lambda_{\rm dG} = (g-1)^2 J(J+1). \tag{11.83}$$

A joint plot of $\lambda_{\rm dG}$ for the trivalent RE ions, and of the highest transition temperature of the corresponding elements (Fig. 11.19), shows that the de Gennes factor captures most of the atomic number dependence of $T_{\rm tr}$, particularly for the heavy rare earths. There are, however, discrepancies. Most noticeable is that for Eu which is actually not tri- but divalent[57].

It is also clear that the anomalously low $T_N = 0.05$K for Pr is not compatible with de Gennes scaling: indeed, the nature of this extremely low-temperature magnetic ordering is completely different from that in other rare earth elements [191]. Generally speaking, lighter rare earths have larger $4f$ orbital diameters, and therefore feel more intense crystal fields. It turns out that for Pr, the crystal field is strong enough to prevent the kind of induced-moment magnetism we discussed for $PrRu_2Si_2$ (Sec. 3.8.4). However, at very low T, a combination of electron and nuclear magnetism manages to produce an ordered phase.

A few discrepancies notwithstanding, we should be impressed to see how well the de Gennes scaling works; it shows that the integrated ef-

[57]One may retort that then Eu should show the same behaviour as trivalent Gd, which is patently untrue. However, we should keep in mind that Eu and Gd have different lattice structures, and (as we saw in Fig. 11.6) very different lattice constants, thus the contrast in their magnetic properties is not surprising.

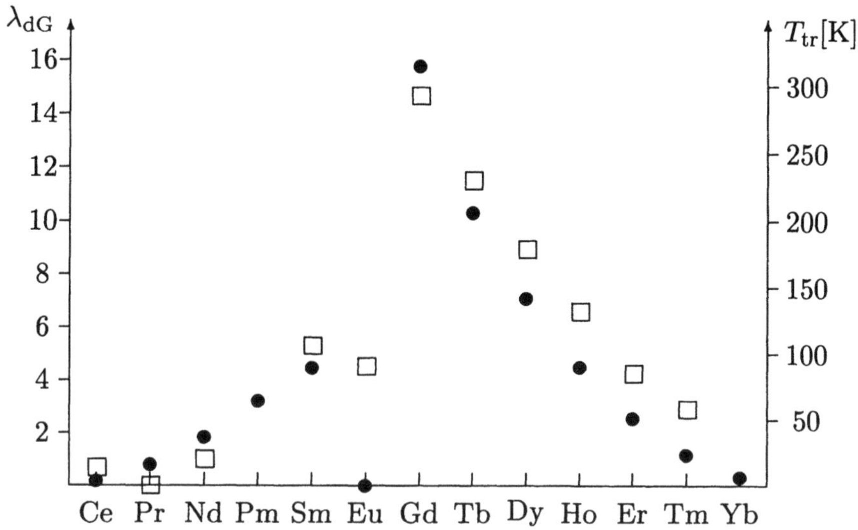

Figure 11.19: De Gennes factors for the trivalent RE ions (black circles, left-hand scale), and highest transition temperatures of the RE elements (squares, right-hand scale), based on data from [191].

fect of a lot of complicated hybridization and direct exchange matrix elements is almost constant through the RE series. It is a favourite game to work out plots like Fig. 11.19 for various families of RE compounds, because a large deviation from the de Gennes prediction gives a hint that the particular substance has different physics than the rest. For instance, learning that $GdRh_3B_2$ orders at $T_C = 93K$, we might conclude that $CeRh_3B_2$ becomes magnetic at $T_{tr} \approx 1K$. Instead, it is a ferromagnet with $T_C = 117K$ [71]. We argue that the fundamental difference between the two systems is that $GdRh_3B_2$ is a strictly integral valent RKKY magnet with a small Fermi surface, while $CeRh_3B_2$ is a nearly integral valent system ($n_f \approx 0.95$) with a large Fermi surface. The two systems are on the opposite sides of a phase boundary, and there is no reason to expect any straightforward relationship between their properties.

We have not yet mentioned the effect of the c–f coupling on the conduction electrons, though this can lead to observable effects. To take the simplest case, let us assume that the f-moments are ordered ferromagnetically in the z direction. The mean field decoupling of (11.80)

$$\mathcal{H}_{\mathrm{MF}} = -(g-1)\langle J^z\rangle \sum_j \sum_{n,n'} \sum_{\mathbf{k}} I_{f-c}^{\mathrm{dir}}(n',\mathbf{k};n,\mathbf{k})(c_{n'\mathbf{k}\uparrow}^{\dagger}c_{n\mathbf{k}\uparrow} - c_{n'\mathbf{k}\downarrow}^{\dagger}c_{n\mathbf{k}\downarrow}) \tag{11.84}$$

describes exchange-split bands, indicating that the c–f interaction causes a uniform band polarization which is (for $g > 1$) parallel to the f-polarization. To cite an example, the f-shell moment of Gd^{3+} is $7\mu_B$, while the observed saturation moment is $7.63\mu_B$. $0.63\mu_B$/atom is contributed by the conduction electrons.

Finally, let us discuss the likely nature of the net f–c coupling. It is clear that the appearance of a hybridization-induced antiferromagnetic Kondo coupling, whose simplified form is shown in (11.26), is a universal feature; but so is that of the direct c–f exchange (11.80). The total c–f interaction can be schematically written as

$$\mathcal{H}_{f-c} = \mathcal{H}_{f-c}^{\mathrm{Kondo}} + \mathcal{H}_{f-c}^{\mathrm{dir}} = \mathcal{H}_{f-c}^{\mathrm{Kondo}} - (g-1)I_{f-c}^{\mathrm{dir}} \sum_j \mathbf{J}_j \cdot \mathbf{s}(\mathbf{R}_j). \tag{11.85}$$

We did not specify $\mathcal{H}_{f-c}^{\mathrm{Kondo}}$ which should be derived from generalizing the hybridization term to the case of degenerate f-shell and conduction band states, and using this as the starting point of a Schrieffer–Wolff transformation [69, 72]. Whatever the detailed form, we do not doubt that the Kondo term is basically antiferromagnetic. As for $\mathcal{H}_{f-c}^{\mathrm{dir}}$, though direct exchange is always ferromagnetic as a spin–spin coupling, the sign of the corresponding J–s coupling $-(g-1)I_{f-c}^{\mathrm{dir}}$ depends on g. In the first half of the RE series (for instance, for Ce), $g < 1$, and thus direct exchange acts to couple the conduction electron spins antiparallel to the f-electron $\mathbf{J}$. It might seem that for light rare earths, the effect of direct exchange is indistinguishable from that of the hybridization-induced Kondo coupling, but it is not so. Without going into details, let us state that the form of $\mathcal{H}_{f-c}^{\mathrm{Kondo}}$ depends on the symmetry of the lattice and it is, in general, different from that of $\mathcal{H}_{f-c}^{\mathrm{dir}}$. Both terms of $\mathcal{H}_{f-c}$ drive towards some kind of moment compensation, but the nature of the ground state may depend on the ratio of the coupling constants [321]. Electronic structure calculations on several Ce compounds give evidence that both hybridization and direct exchange are important [285].

For the heavy rare earths (from Gd through Tm) $g > 1$, and the net f–c exchange which is roughly

$$I_{f-c}^{\text{eff}} = \frac{v^2 U_f}{(\mu - \epsilon_f)(\epsilon_f + U_f - \mu)} - (g-1) I_{f-c}^{\text{dir}} \tag{11.86}$$

may be either ferro- or antiferromagnetic. The sign of I_{f-c}^{eff} is immaterial for the RKKY interaction which contains $(I_{f-c}^{\text{eff}})^2$, but it matters very much for the overall nature of the system. Systems with $I_{f-c}^{\text{eff}} > 0$ are AFM coupling Kondo lattices in which the formation of a heavy fermionic phase with a large Fermi surface is at least an option. Even if the system forms an RKKY ordered phase, above its transition temperature it may still behave as a dense Kondo system showing a resistivity increment $\Delta\rho \propto \ln(T/T_K)$. On the other hand, if $I_{f-c}^{\text{eff}} < 0$, then the system is destined to become an RKKY magnet with a small Fermi sea, and a very ordinary effective mass, and it does not show any of the anomalies which are customarily associated with the name "Kondo". In other words, though periodic Kondo systems are capable of RKKY behaviour, there should exist a class of RKKY magnets which do not have an underlying Kondo physics[58].

Looking at (11.86), we see that the hybridization term is large whenever one of the f-levels is near to the Fermi level. It is then likely that those RE ions will form Kondo lattices, which are known to show valence instability, at least in certain environments. The list includes Ce, Sm, Eu, Tm, and Yb. For the latter two, it is an additional argument that their $(g-1)$ is quite small, which suppresses the second term of (11.86). Ce^{3+} and Yb^{3+} are well-known Kondo impurities, and Sm impurities have also been reported to show Kondo effect [164]. However, elemental Yb is divalent and non-magnetic.

As far as non-valence-fluctuating rare earths (i.e., the majority of the elements) are concerned, we follow [169] in taking the view that the very fact that de Gennes scaling is reasonably well obeyed, indicates that the direct exchange term dominates. It should then follow that in the second half of the series, the net f–c interaction is ferromagnetic.

[58]The impurity Hamiltonian (11.27) with a FM coupling does not lead to a Kondo effect: there is no $\ln T$ term in the high-T resistivity, etc.

This is well attested for Gd, where the sizeable conduction electron spin polarization is parallel to the $4f$ spins.

Chapter 12

Quantum Hall Effect

The discovery of the Quantum Hall Effect (QHE) [434] opened the way to a vast uncharted domain of solid state physics, whose mere existence was unsuspected theretofore. Indeed, while even the most recent developments described in other chapters can be connected to antecedents decades before (not seldom, to the beginnings of solid state physics), QHE came unanticipated. The area now comprises a variety of quantization phenomena in low-dimensional semiconductor devices.

Though QHE physics is rich and variegated and it could fill (in fact, it does fill) books of its own, it is also largely disconnected from the rest of magnetism. This necessitates that we limit ourselves to a few aspects which are easily accessible to the student. The quantum-mechanical treatment of the motion of an electron in a uniform magnetic field provides an illuminating exercise from which we can understand the nature of the eigenstates in the absence of time reversal invariance, and this we discuss in detail. From the theory of the fractional Hall quantization, we describe only the approach which relies on the explicit construction of many-electron states, because the idea of a correlated (Jastrow-type) ground state is a recurring theme in the text (see Ch. 9). However, we do not attempt to review the theory of electron localization in a random potential (Anderson localization) though we will have to cite some of its results in our arguments for the existence of Hall plateaus.

12.1 Introduction

First, let us recall the simplest classical description of the Hall effect. The schematic arrangement is shown in Fig. 12.1. A conductor is placed in the external magnetic field $\mathbf{H} = (0, 0, H)$ and we try to drive a current in the x direction. A charge carrier with a velocity v_x will be deflected in the y direction by the Lorentz force. This leads to charge accumulation at the y-boundaries of the system. The accumulated charge exerts a Coulomb force on the electrons moving in the bulk. A stationary state is reached when the Lorentz force is exactly compensated by the electrostatic force $\mathbf{E} = (0, E, 0)$

$$\frac{v_x e H}{c} - eE = 0\,, \tag{12.1}$$

so that a uniform propagation in the x direction with the *drift velocity*

$$v_{\text{drift}} = \frac{cE}{H} \tag{12.2}$$

becomes possible.

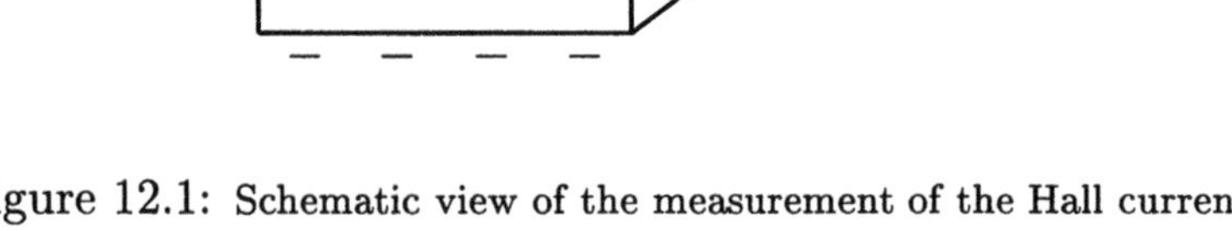

Figure 12.1: Schematic view of the measurement of the Hall current.

If the concentration of charge carriers is n, a Hall current $j_x = -nev_{\text{drift}}$ will flow and the classical result for the Hall conductivity is found to be[1]

$$\sigma_{xy} = -\sigma_{yx} = \frac{j_x}{E_y} = -\frac{nec}{H}. \tag{12.3}$$

[1]In our discussion, we will call the positive quantities σ_{yx} and ρ_{xy} the Hall conductivity, and resistivity, respectively. The usual definition of the Hall conductivity $\sigma_{xy} = -nec/H$ gives a negative number for electron-like carriers.

We derived this for free carriers but the same result is obtained in Drude theory [32, 187]. We consider a two-dimensional system lying in the x–y plane, with the magnetic field pointing in the z direction. During a lifetime τ, the electrons accelerate freely under the action of the electrical field $\mathbf{E}$ and the Lorentz force; then they suffer a collision, and the process begins again. The typical velocity is

$$\mathbf{v} = \frac{\mathbf{F}}{m} \cdot \tau = -\frac{e\tau}{m}(\mathbf{E} + \frac{1}{c}\mathbf{v} \times \mathbf{H}). \tag{12.4}$$

Since the current density is $\mathbf{j} = -ne\mathbf{v}$, the above equation can be rearranged so that we can read off the relevant components of the conductivity matrix from

$$\mathbf{j} = \sigma_0 \mathbf{E} - \frac{e\tau}{mc}\mathbf{j}\times\mathbf{H} \tag{12.5}$$

where $\sigma_0 = ne^2\tau/m$ is the zero-field Drude conductivity. One finds

$$\sigma_{xx} = \frac{\sigma_0}{1 + (\omega_c\tau)^2} \qquad \sigma_{yx} = \frac{\omega_c\tau\sigma_0}{1 + (\omega_c\tau)^2} = \omega_c\tau\sigma_{xx} \tag{12.6}$$

where $\omega_c = eH/mc$ is the so-called cyclotron frequency. The resistivity matrix is the inverse of the conductivity matrix

$$\rho_{xx} = \frac{\sigma_{xx}}{\sigma_{xx}^2 + \sigma_{yx}^2} \qquad \rho_{xy} = \frac{\sigma_{yx}}{\sigma_{xx}^2 + \sigma_{yx}^2}. \tag{12.7}$$

The parameter $\omega_c\tau$ controls the nature of the conduction process. In the weak-field limit $\omega_c\tau \ll 1$, $\sigma_{xx} \approx \sigma_0$ diverges if $\tau \to \infty$. In the strong-field limit $\omega_c\tau \gg 1$, $\sigma_{xx} \approx \sigma_0/(\omega_c\tau)^2 \propto 1/\tau$ vanishes if $\tau \to \infty$: one needs collisions (friction) to get longitudinal electrical current. (We can envisage the situation also classically: without collisions, the combined action of E_x and H_z may cause the electrons to move in the y direction, but motion along the x axis is oscillatory, carrying no net current. Collisions interrupt the oscillatory motion, enabling the electron to drift in the x direction). At the same time, $\sigma_{yx} \approx nec/H$ as in (12.3). The combination of $\omega_c\tau \gg 1$ with a large τ (a nearly ideal sample in a high magnetic field) may be called the friction model [187] which gives useful simple results. The longitudinal resistivity

$$\rho_{xx} \approx \frac{\sigma_{xx}}{\sigma_{yx}^2} = \frac{1}{\sigma_0} \tag{12.8}$$

is field-independent (absence of magnetoresistance). The Hall resistivity

$$\rho_{xy} = -\rho_{yx} = \frac{1}{\sigma_{yx}} = \frac{H}{nec} \tag{12.9}$$

is not influenced by scattering events, and depends linearly on H.

Fig. 12.2 shows the measured field dependence of the Hall resistivity of a relatively low-mobility GaAs–AlGaAs heterostructure sample which is an experimental realization of a two-dimensional system of nearly free electrons. For weak fields (say, up to half a Tesla), the predictions (12.8), and particularly (12.9), are fulfilled. From $H \approx 1$T upwards, however, spectacular deviations from the predicted linearity are found: ρ_{xy} develops *plateaus*. There are wide ranges of the magnetic field where ρ_{xy} is, to an excellent approximation, constant. In fact, this is the prevalent behaviour, interrupted every now and then with narrow intervals of the magnetic field, where the Hall resistivity increases steeply.

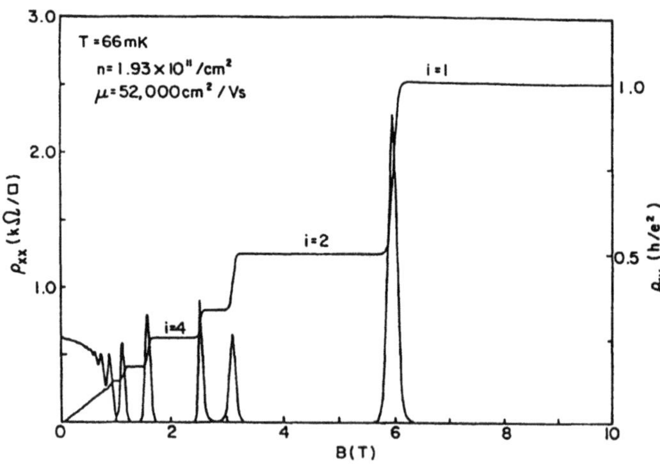

Figure 12.2: The longitudinal and transverse components of the resistivity as a function of the magnetic field, for a low-mobility GaAs–AlGaAs sample. ρ_{xy} has plateaus in regimes where ρ_{xx} vanishes. Integer plateaus up to $i = 6$ are clearly identifiable (after [408]).

The behaviour of ρ_{xx} is no less striking. In the plateau regimes, the longitudinal resistivity seems to vanish, while it has narrow spikes in the regimes where ρ_{xy} is changing.

The above-described phenomena (or, in a narrower sense, just the emergence of plateaus in the field dependence of ρ_{xy}) have become known as the Quantum Hall Effect (QHE). It was discovered by measurements on Si MOSFET structures by Klaus von Klitzing in 1980 [434], and later found also in GaAs-based devices. The plateau values of ρ_{xy} were identified as

$$\rho_{xy} = \frac{h}{i \cdot e^2} \qquad i = 1, 2, \ldots \tag{12.10}$$

The highest plateau value we see in Fig. 12.2 is $i = 6$ but in another measurement [54], even $i = 12$ was clearly visible[2]. It must be remarked, though, that from $i = 6$ upwards only the even-integer plateaus are recognizable; this is an indication of the importance of the spin degree of freedom (we return to this problem at the end of the chapter).

The accuracy of quantization is extremely high, at least 10^{-8}. Let us remark that in three dimensions, measuring the resistivity with this accuracy would require a comparably accurate characterization of the sample geometry[3]. However, we do not have to bother about this in the case under study, because in two dimensions, the Hall resistance $R_{xy} = -U_y/I_x = -L_y E_y / j_x L_y = \rho_{xy}$ is the same as the Hall resistivity[4].

The vanishing of ρ_{xx} in the plateau regions is concluded from the fact that there is no measurable voltage drop in the x direction (¡ 10^{-14}V). In this sense only, the system is an ideal conductor but this does *not* mean in infinite σ_{xx}. Quite on the contrary, it follows from (12.7) that if $\rho_{xy} \neq 0$ then σ_{xx} and ρ_{xx} vanish together.

It is useful to restate the effect in terms of the Hall conductivity σ_{yx}. The classical result (12.3) is that σ_{yx} is a linear function of the

---

[2]By now, this is a standard achievement in laboratory exercises for students. There is no doubt that many more Quantum Hall states could be identified, especially by counting the spikes of ρ_{xx}. However, there seems to be no point to struggling further in this direction because it is believed that no new physics would appear in the higher Hall plateaus.

[3]The linear dimensions of a cm-sized sample should be known with the accuracy of an atomic diameter!

[4]For the sake of simplicity, we assumed that the current density is constant. However, the conclusion holds also for a realistic non-uniform current distribution.

electron density n. The quantized behaviour (see also Fig. 12.5 on p. 691) consists in the appearance of plateaus at integer multiples of e^2/h which we can recognize as the "atomic unit" of conductivity:

$$\sigma_{yx} = \frac{i \cdot e^2}{h} \qquad i = 1, 2, \ldots \tag{12.11}$$

This is referred to as the Integer Quantum Hall Effect (IQHE).

Attaching the adjective "integer" to QHE became necessary when, a few years later [409], measurements on high-quality GaAs–AlGaAs heterostructures revealed a novel effect: the *fractional quantization* of the Hall conductivity. Namely, it was found that σ_{yx} can have plateaus (in addition to the integer ones mentioned before) at

$$\sigma_{yx} = \frac{p}{q} \cdot \frac{e^2}{h} \tag{12.12}$$

where p/q is a simple fraction with an *odd denominator*[5]. This is the Fractional Quantum Hall Effect (FQHE). There is a number of selected values (1/3, 2/5, 4/7, 5/11, 2/3, to mention a just a few) where the FQHE is clearly seen. No plateau is observed at some other, equally simple, fractions. In order to observe the behaviour shown in Fig. 12.3, we need a sample for which the electron mobility is an order of magnitude higher than in the sample used for obtaining Fig. 12.2. A typical value of the surface electron density is $3{\times}10^{11}/\text{cm}^2$. Let us emphasize that both the IQHE and the FQHE are low temperature phenomena, to be observed at not higher than helium temperatures, and preferably far below ($T < 0.01$K). At higher T, the plot of σ_{yx} gradually approaches the classical (12.3). There is no identifiable critical temperature.

We should appreciate the beauty, and also the perplexing nature, of the Quantum Hall Effect. It is one of the few solid state effects which permit the measurement of fundamental constants with an accuracy which rivals that of measurements in atomic physics. In the IQHE, the quantization of σ_{yx} to units of e^2/h seems to have an accuracy of at least 10^{-8}, with no detectable dependence on sample quality, size effects, etc. Apparently, the quantization phenomenon is of fundamental

[5] At least, for fractions less than 1. There seems to be a plateau at 5/2 [98].

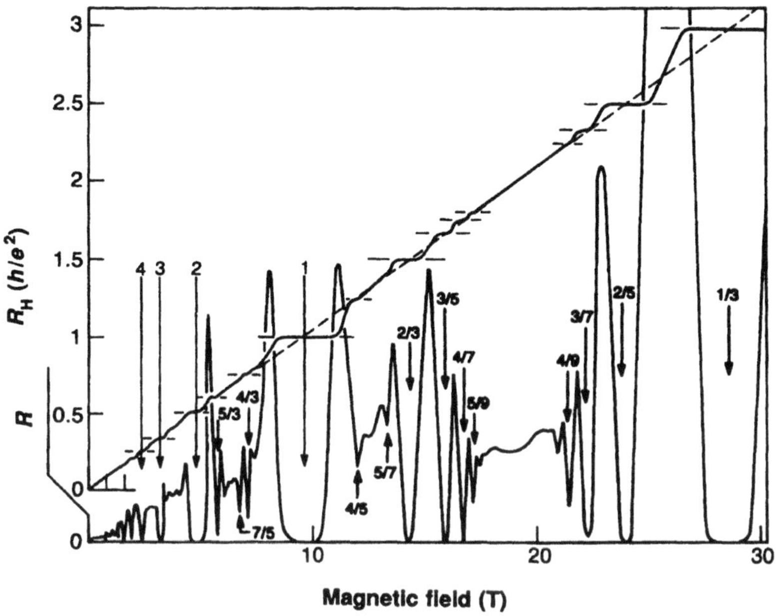

Figure 12.3: The longitudinal and transverse components of the resistivity as a function of the magnetic field, for a high-mobility GaAs–AlGaAs sample (after [98]). A large number of fractional QHE plateaus are identified. The dashed line shows the classical result (12.9).

nature, so it is all the more remarkable that its discovery came entirely unanticipated. Indeed, its understanding is still not complete.

In the present Chapter, we cannot hope to cover more than the most elementary aspects of QHE theory. We will usually restrict our attention to $T = 0$. It turns out that the σ_{yx} plateaus can be discussed in terms of equilibrium concepts (the vanishing of ρ_{xx} shows that there is no dissipative current in the corresponding parameter range). We will not attempt to develop a true transport theory which might describe finite temperatures and/or the regions in between the plateaus.

Our argument exploits in an essential way that QHE is an effect in a *two-dimensional electron system.* Without going into details, let us accept that the Si MOSFET devices, and interfaces in GaAs heterostructures, provide experimental realizations of a strictly two-dimensional electron system (see [209] for an introduction). The way this works is basically the following: let the surface be the x–y plane. By the application of a voltage, or some other means, it is possible to create a sharp potential well at the surface (or interface). The potential well does not hinder the motion of the electrons in the x–y plane, but it prevents their escape in the z-direction. More precisely, the z-motion is quantized into widely spaced levels, while the motion in the x–y plane is essentially free-electron-like. It is possible to create circumstances in which only the lowest quantized level is occupied by electrons. Then we have a truly two-dimensional electron system, where the degree of freedom associated with the z-coordinate is playing no role[6].

Given that we can restrict our attention to two-dimensional systems, the following simple argument gives us a feeling of where the quantized value e^2/h may be coming from:

Let the plane be the x–y plane, and the magnetic field H applied in the z-direction. In classical mechanics, the Lorentz force provides the centripetal force which keeps the electrons moving in circular cyclotron orbits. The angular velocity ω is determined by

$$\frac{evH}{c} = \frac{er\omega H}{c} = mr\omega^2. \tag{12.13}$$

Thus

$$\omega_c = \frac{eH}{mc} \tag{12.14}$$

the so-called cyclotron frequency, is independent of the radius r. Faster electrons move in larger orbits: $v = r\omega$. There is no restriction on r.

In quantum mechanics, only a discrete set of cyclotron orbits is permitted. Using quasi-classical quantization, the radius r_0 of the smallest orbit should belong to the angular momentum $\hbar$

$$r_0 = \sqrt{\frac{\hbar}{m\omega_c}} = \sqrt{\frac{\hbar c}{eH}}. \tag{12.15}$$

What is the area A of the smallest cyclotron orbit? Not πr_0^2, because the radius r_0 belongs to the maximum of the wave function; roughly

[6] It is quite realistic to assume that the separation of the lowest two levels is about 100K. At temperatures around $T \sim 1$K, thermal excitation to the next surface level is negligible.

speaking, the full wave function should cover an area twice as large; so we come upon $A = 2\pi r_0^2$. Now we can identify a presumably important value of the electron density (per unit area): it belongs to the case when the plane is fully covered with cyclotron orbits of the smallest size:

$$n_0 = \frac{1}{2\pi r_0^2} = \frac{eH}{hc}. \tag{12.16}$$

Replacing this in the result (12.3), we are led to surmise that the special value of the Hall conductivity

$$\sigma_{yx} = \frac{n_0 ec}{H} = \frac{e^2}{h} \tag{12.17}$$

must be playing an important role, and that it is associated with a magic value of the electron density. A proper quantum-mechanical treatment confirms that such is indeed the case.

12.2 Motion in a Magnetic Field

We have seen in Sec. 2.1 that the Hamiltonian describing the orbital motion of an otherwise free electron in a magnetic field is

$$\mathcal{H} = \frac{1}{2m}\left(\hat{\mathbf{p}} + \frac{e}{c}\mathbf{A}\right)^2 . \tag{12.18}$$

In classical mechanics, the velocity is obtained from

$$\mathbf{v} = \frac{\partial \mathcal{H}}{\partial \mathbf{p}} = \frac{1}{m}\left(\mathbf{p} + \frac{e}{c}\mathbf{A}\right) . \tag{12.19}$$

It would be usual to call $\mathbf{p}_{\text{kin}} = m\mathbf{v}$ the kinetic momentum. In quantum mechanics, the velocity operator is defined analogously

$$\hat{\mathbf{v}} = \dot{\mathbf{r}} = \frac{i}{\hbar}[\mathcal{H}, \mathbf{r}] = \frac{1}{m}\left(\hat{\mathbf{p}} + \frac{e}{c}\mathbf{A}\right) . \tag{12.20}$$

If we wish, we can rewrite the Hamiltonian as

$$\mathcal{H} = \frac{m}{2}\hat{\mathbf{v}}^2 = \frac{m}{2}\left(\hat{v}_x^2 + \hat{v}_y^2 + \hat{v}_z^2\right) \tag{12.21}$$

which looks exactly the same that we had in the absence of the magnetic field! Why do we consider the present eigenvalue problem as much more difficult to solve?

The reason is that the components of the velocity operator no longer commute

$$[\hat{v}_x, \hat{v}_y] = \frac{e}{m^2c}([\hat{p}_x, A_y] - [\hat{p}_y, A_x]) = -\frac{i\hbar e}{m^2c}H_z \,. \tag{12.22}$$

The above equation and the two similar ones obtained by cyclic permutation of x, y and z can be elegantly summarized in

$$\hat{\mathbf{v}} \times \hat{\mathbf{v}} = -\frac{i\hbar e}{m^2c}\mathbf{H} \,. \tag{12.23}$$

Though it may look puzzling that "the vector (operator) $\hat{\mathbf{v}}$ cannot be parallel to itself", in quantum mechanics the situation arises naturally, and it leads to important consequences. We are reminded of the components of the angular momentum whose commutation relations can be summarized like $\hat{\mathbf{L}} \times \hat{\mathbf{L}} \sim \hat{\mathbf{L}}$.

12.2.1 Landau Levels

Motivated by the Hall geometry shown in Fig. 12.1, let us consider the quantum-mechanical eigenvalue problem of the two-dimensional motion of an (otherwise free) electron moving in crossed magnetic and electric fields $\mathbf{H} = (0, 0, H)$ and $\mathbf{E} = (0, E, 0)$. The electron is confined to the (x, y) plane. Here we choose to describe the magnetic field in the so-called *Landau gauge* with the vector potential $\mathbf{A} = (-Hy, 0, 0)$. Including the spin degree of freedom, the Hamiltonian (12.18) becomes

$$\mathcal{H} = \frac{1}{2m}\left(\hat{p}_x - \frac{e}{c}Hy\right)^2 + eEy + \frac{\hat{p}_y^2}{2m} + \hat{S}^z g\mu_B H. \tag{12.24}$$

Here we have taken care to denote operators by $\hat{p}_x, \ldots$ because the hatless quantities $p_x, \ldots$ will stand for eigenvalues (or expectation values, whichever may be the case).

The system has the linear dimensions L_x and L_y. We prescribe the periodic boundary condition for the x-direction and assume free boundaries at $y = \pm L_y/2$.

The problem is S^z-diagonal so we can forget about the spin variable until the end when we reinsert the Zeeman term in the energy expression.

Since $\mathcal{H}$ does not depend on x, we can seek the orbital dependence of the eigenstates in the factorized form

$$\psi(x, y) = \exp\left(i\frac{p_x}{\hbar}x\right)\chi(y) \tag{12.25}$$

where $\chi(y)$ satisfies

$$\left[\frac{1}{2m}\left(p_x - \frac{mEc}{H} - \frac{eH}{c}y\right)^2 + \frac{\hat{p}_y^2}{2m}\right]\chi(y) = \epsilon\chi(y), \tag{12.26}$$

where ϵ includes a p_x-dependent energy shift which we write out explicitely in (12.34). (12.26) is the eigenvalue equation for a displaced harmonic ocillator with the frequency

$$\omega_c = \frac{eH}{mc}. \tag{12.27}$$

ω_c is the same as the classical cyclotron frequency (12.14). The motion in a uniform magnetic field is one of the few problems where quantum mechanics gives results closely resembling those of classical mechanics.

The eigenvalues of (12.26) are

$$\epsilon = \left(n + \frac{1}{2}\right)\hbar\omega_c \qquad n = 0, 1, 2, \ldots \tag{12.28}$$

corresponding to the eigenstates

$$\chi_n(y) = H_n\left(\frac{y - y_0}{l}\right)\exp\left(-\frac{(y - y_0)^2}{2l^2}\right) \tag{12.29}$$

where the H_n are Hermite polynomials. As it can be read off from (12.26), the centre of the oscillator function is displaced to

$$y_0 = -\frac{mEc^2}{eH^2} + \frac{cp_x}{eH}. \tag{12.30}$$

Note the intimate connection between y_0 and p_x.

The natural length scale of the problem (which we used to write down (12.29)) turns out to be the cyclotron radius l

$$l = \sqrt{\frac{c\hbar}{eH}}. \tag{12.31}$$

This is easy to check. Let us rewrite the Hamilton operator appearing in (12.26), by bringing to the front the natural energy unit $\hbar\omega_c$ (for simplicity's sake, here we set $p_x = E = 0$)

$$\hbar\frac{eH}{mc}\left[\frac{1}{2}\frac{eH}{\hbar c}y^2 - \frac{1}{2}\frac{\hbar c}{eH}\frac{\partial^2}{\partial y^2}\right] \rightsquigarrow \hbar\omega_c\left[\frac{1}{2}\tilde{y}^2 - \frac{1}{2}\frac{\partial^2}{\partial \tilde{y}^2}\right]. \tag{12.32}$$

The second step required the introduction of the dimensionless

$$\tilde{y} = y/\sqrt{\frac{\hbar c}{eH}}. \tag{12.33}$$

Returning now to the eigenvalue problem of (12.24): the energy of the eigenstate with quantum numbers n, p_x, and S^z is

$$\lambda(n, p_x, S^z) = \left(n + \frac{1}{2}\right)\hbar\omega_c + \frac{Ec}{H}p_x - \frac{mE^2c^2}{2H^2} + g\mu_B H S^z. \tag{12.34}$$

Let us first discuss the solution in the absence of the electric field $E = 0$. (The problem of the motion of an electron in a uniform magnetic field was solved by L.D. Landau in 1930 [232].) The energy expression becomes much simpler

$$\lambda(n, p_x, S^z) = \left(n + \frac{1}{2}\right)\hbar\omega_c + g\mu_B H S^z \tag{12.35}$$

inasmuch as it does not depend on the quantum number p_x. This indicates that each of the energy levels (12.35) has a degeneracy: in fact, a huge degeneracy. The levels will be referred to as the *Landau levels*.

There are two equidistant sequences of Landau levels: one for the $\uparrow$-spin, and one for the $\downarrow$-spin electrons. Their relative displacement is given by the Zeeman splitting $g\mu_B H$. We notice that if we choose the nominal free-electron value $g = 2$, the Landau level $n+1$ for $\downarrow$-spin electrons becomes degenerate with the Landau level n for $\uparrow$-spin electrons. This finding is spurious. Quite apart from the fact that g is not strictly 2 even for an electron in vacuum, solid state effects can change both $\hbar\omega_c$ and g quite substantially. We are mainly interested in GaAs-based heterostructures, and may take the parameters of bulk GaAs as a guide. The cyclotron energy $\hbar\omega_c = eH/m^*c$ is increased more than tenfold over the free electron value because the band effective mass $m^* = 0.07m$ is quite small. At the same time, the Zeeman splitting is strongly reduced, since $g \approx 0.44$. The disparity between $\hbar\omega_c$ and $g\mu_B H$ is even greater in the heterostructures. When we speak about a Landau level, it will always be for a given spin direction.

The eigenstates corresponding to (12.34) are (omitting the common spin factor of the wave functions)

$$\psi_{n,p_x}(x,y) = \exp\left(i\frac{p_x}{\hbar}x\right) H_n\left(\frac{y-y_0}{l}\right) \exp\left(-\frac{(y-y_0)^2}{2l^2}\right). \qquad (12.36)$$

Such a state can be visualized as a stripe of width $\sim l$, centered at y_0, and extending uniformly over the whole length of the system in the x-direction (Fig. 12.4). Where should we draw the stripe representing the next Landau state? We could not distinguish it from the state we have shown. The spacing l^2/L_x between consecutive eigenvalues of y_0 is much smaller than l, so the centres of $L_x/l \gg 1$ other Landau states lie within a stripe. The x-dependent exponential factor guarantees the orthogonality of Landau-states, even though they may occupy almost the same region of the plane.

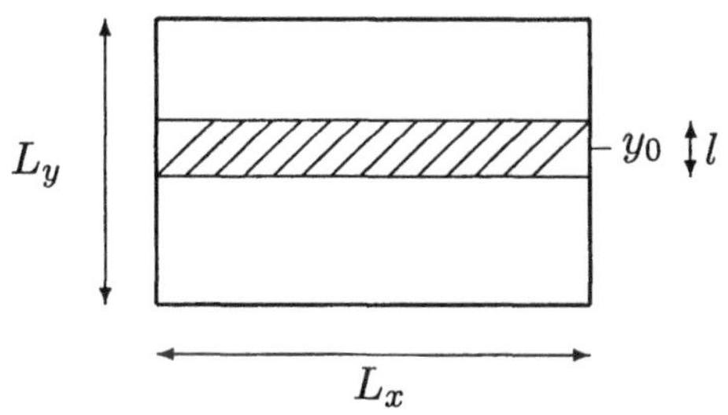

Figure 12.4: Most of the weight of a Landau state is lying within a belt of width $\sim l$ centered about $y = y_0$.

Now we determine the degeneracy of a Landau level. Imposing the periodic boundary condition $\psi(x + L_x) = \psi(x)$ gives

$$p_x = \frac{2\pi\hbar}{L_x} q \tag{12.37}$$

where q is an integer. The degeneracy of a Landau level is given by the number of the permissible values of p_x. According to (12.30) p_x is intimately related to y_0. We accept the following criterion[7]: those states are considered to belong to the system whose centre y_0 lies in the interval $(-L_y/2, L_y/2)$. Then it follows that the number of the allowed values of q is

$$\mathcal{N} = \frac{L_x L_y}{\frac{2\pi\hbar c}{eH}} = \frac{\mathcal{A}}{2\pi l^2}, \tag{12.38}$$

where $\mathcal{A}$ is the area of the system. Apparently, a Landau state "needs" an area $2\pi l^2$ which we ascribe to the smallest cyclotron orbit. A useful way of rewriting (12.38) is obtained if we recognize that

$$\phi_0 = \frac{hc}{e} \tag{12.39}$$

is the flux quantum[8]:

$$\mathcal{N} = \frac{\mathcal{A}H}{\phi_0} = \frac{\phi_{\text{tot}}}{\phi_0}. \tag{12.40}$$

We can say that each Landau state "carries" one flux quantum and thus the number of different states in a Landau level is just what is required to account for the entire flux ϕ_{tot} through the system.

Let us observe that the degeneracy is the same for all Landau levels. (12.38) was obtained from (12.30) in which n does not appear. The width of the stripes shown in Fig. 12.4 increases with n (roughly as $l\sqrt{n}$) but the spacing of the y_0 values does not.

---

[7]It may be called the Landau counting. An apparently dubious contribution is coming from the edges, where we count some states which are only partly within the system. In any case, the edge contribution becomes negligible in the thermodynamic limit $L_x L_y / l^2 \to \infty$.

[8]The flux quantum used in the theory of superconductivity is $hc/2e$ because the relevant current carriers are Cooper pairs. Note that though introducing ϕ_0 allows an attractive reformulation of some of our results, we do not imply that ϕ_{tot} should be quantized. However, it is convenient to assume that ϕ_{tot}/ϕ_0 is an integer.

12.2.2 Algebraic Approach

Identifying the Landau level $(n+1/2)\hbar\omega_c$ as comprised of states (12.36) relied on using the Landau gauge. However, the energy levels and their degeneracies must be gauge invariant. For these quantities, we should be able to find a derivation which does not depend on choosing the Landau (or any other) gauge. This is given by the algebraic method which uses nothing but the gauge-independent commutation relations. Writing the Hamiltonian as

$$\mathcal{H} = \frac{m}{2}(\hat{v}_x^2 + \hat{v}_y^2) \tag{12.41}$$

and recalling (12.22), we are reminded of the energy expression of the one-dimensional harmonic oscillator, which is the sum of the squares of two non-commuting operators. Remembering the standard algebraic method of diagonalizing the oscillator problem, we are led to introduce the boson operators

$$a^\dagger = \frac{lm}{\hbar\sqrt{2}}(\hat{v}_x + i\hat{v}_y) \qquad a = \frac{lm}{\hbar\sqrt{2}}(\hat{v}_x - i\hat{v}_y)\,. \tag{12.42}$$

It is easy to check that $[a, a^\dagger] = 1$, and that the Hamiltonian can be expressed as

$$\mathcal{H} = \hbar\omega_c\left(a^\dagger a + \frac{1}{2}\right)\,. \tag{12.43}$$

Adding the Zeeman energy, we recover the Landau levels (12.35).

We are still left with the problem of the degeneracies. The two-dimensional motion of the electron can be characterized by two quantum numbers thus it must, in some sense, be expressible in terms of *two* harmonic oscillators, or two kinds of bosons. We have found one of the oscillators (the a-bosons), which controls the energy. The other oscillator must be associated with displacing the cyclotron orbit, which has no influence on the energy.

Classically, the electron moves in a circle about the center of the cyclotron orbit

$$x = x_0 + r\cos\omega t \qquad y = y_0 + r\sin\omega t\,. \tag{12.44}$$

The corresponding velocities are

$$v_x = -r\omega \sin \omega t \qquad v_y = r\omega \cos \omega t\,. \tag{12.45}$$

The center-of-orbit coordinates can be expressed as

$$x_0 = x - \frac{v_y}{\omega} \qquad y_0 = x + \frac{v_x}{\omega}\,. \tag{12.46}$$

We introduce the corresponding operators

$$\hat{x}_0 = x - \frac{\hat{v}_y}{\omega} \qquad \hat{y}_0 = x + \frac{\hat{v}_x}{\omega}\,. \tag{12.47}$$

It can be checked that both $\hat{x}_0$ and $\hat{y}_0$ are conserved quantities

$$[\hat{x}_0, \mathcal{H}] = [\hat{y}_0, \mathcal{H}] = 0\,. \tag{12.48}$$

However, it is not possible to specify both coordinates simultaneously because the correponding operators do not commute[9]

$$[\hat{x}_0, \hat{y}_0] = il^2\,. \tag{12.49}$$

This, too, can be reformulated as a canonical Bose commutation relation, by introducing the boson operators

$$b^\dagger = \frac{1}{l\sqrt{2}}(\hat{x}_0 + i\hat{y}_0) \qquad b = \frac{1}{l\sqrt{2}}(\hat{x}_0 - i\hat{y}_0)\,. \tag{12.50}$$

It can be checked that $[b, b^\dagger] = 1$ and furthermore $[a, b] = [a, b^\dagger] = 0$, so that we really have two independent bosons. Creating b quanta does not change the energy, it merely transforms one eigenstate into another within the same Landau level. The mth state of the nth Landau level is

$$|m, n\rangle = \frac{1}{\sqrt{m!}}\frac{1}{\sqrt{n!}}(a^\dagger)^n (b^\dagger)^m |0\rangle\,. \tag{12.51}$$

[9]In our Landau gauge calculation, it became manifest that y_0 is a good quantum number because it was expressible with the constant of motion p_x (see (12.30)). However, we found no room for a second quantum number x_0. This is explained by $[\hat{x}_0, \hat{y}_0] \neq 0$. Only one quantum number of the energy eigenstates can be associated with the position of the center of orbit. It depends on the gauge whether this is conveniently chosen as y_0, x_0, or a combination of both.

The large degeneracy of the Landau levels follows from the fact that m can be macroscopically large. We should not feel troubled because there is no straightforward way to arrive at (12.38). Landau counting and its equivalents rely on choosing a boundary condition which is convenient for the particular choice of gauge. However, the outcome is always the degeneracy $\mathcal{N} = \phi_{\text{tot}}/\phi_0$ which measures the total area of the system but is insensitive to the form of the boundary.

12.2.3 Symmetrical Gauge

A particularly useful representation of the above general results is found by introducing the symmetrical gauge

$$\mathbf{A} = \left(-\frac{Hy}{2}, \frac{Hx}{2}, 0\right) \tag{12.52}$$

whereupon the Hamiltonian becomes

$$\mathcal{H} = \frac{1}{2m}\left(\frac{\hbar}{i}\frac{\partial}{\partial x} - \frac{eH}{2c}y\right)^2 + \frac{1}{2m}\left(\frac{\hbar}{i}\frac{\partial}{\partial y} + \frac{eH}{2c}x\right)^2 . \tag{12.53}$$

The manifest axial symmetry[10] of the problem makes it advantageous to introduce the complex coordinates [184]

$$z = x + iy \qquad z^* = x - iy , \tag{12.54}$$

whereby z and z^* are treated as independent variables. Using $\hbar\omega_c$ (l) as the unit of energy (length), the Hamilton operator appears as

$$\mathcal{H} = \frac{1}{2}\left[-4\frac{\partial}{\partial z} \cdot \frac{\partial}{\partial z^*} + \frac{1}{4}zz^* - z\frac{\partial}{\partial z} + z^*\frac{\partial}{\partial z^*}\right] , \tag{12.55}$$

while (12.42) and (12.50) turn into

$$a^\dagger = \frac{1}{\sqrt{2}}\left(-2\frac{\partial}{\partial z} + \frac{1}{2}z^*\right) \qquad a = \frac{1}{\sqrt{2}}\left(2\frac{\partial}{\partial z^*} + \frac{1}{2}z\right) \tag{12.56}$$

[10] Writing $\mathbf{A}$ in cylindrical coordinates, the only non-vanishing component is the tangential $A_\varphi = H\rho/2$. See [232] for an alternative (first quantization) treatment of this eigenvalue problem.

and

$$b^\dagger = \frac{1}{\sqrt{2}}\left(-2\frac{\partial}{\partial z^*} + \frac{1}{2}z\right) \qquad b = \frac{1}{\sqrt{2}}\left(2\frac{\partial}{\partial z} + \frac{1}{2}z^*\right) \tag{12.57}$$

(remember that $(\partial/\partial z)^\dagger = -\partial/\partial z^*$). The eigenstates are generated according to (12.51). Let us first find the vacuum state. It is easily verified that $a|0,0\rangle = b|0,0\rangle = 0$ for

$$|0,0\rangle = \frac{1}{\sqrt{2\pi}}\exp\left(-\frac{zz^*}{4}\right) \rightsquigarrow \frac{1}{\sqrt{2\pi l^2}}\exp\left(-\frac{zz^*}{4l^2}\right). \tag{12.58}$$

(In the last step, we show how to reattribute to z its proper dimension l). We may regard $|0,0\rangle$ as the state for which the center of the cyclotron orbit is as localized as possible. Namely, (12.49) can be associated with the Heisenberg uncertainty relation

$$\sqrt{(\Delta x_0)(\Delta y_0)} \geq l\,. \tag{12.59}$$

If we insist on specifying, say, y_0 with complete precision then we have to make Δx_0 divergent (or at least as large as the sample). This was the case with the Landau eigenstates which are y_0-eigenstates. If we want to know Δx_0 and Δy_0 with equal accuracy then the most we can achieve is to know that the center of orbit is somewhere within a disk of radius $\sim l$.

The next state of the lowest level is

$$|1,0\rangle = b^\dagger|0,0\rangle = \frac{1}{\sqrt{4\pi}}\left(-2\frac{\partial}{\partial z^*} + \frac{z}{2}\right)\exp\left(-\frac{zz^*}{4}\right) = \frac{z}{\sqrt{4\pi}}\exp\left(-\frac{|z|^2}{4}\right). \tag{12.60}$$

Generally

$$|m,0\rangle = \frac{z^m}{(2\pi 2^m m!)^{1/2}}\exp\left(-\frac{zz^*}{4}\right). \tag{12.61}$$

Introducing $z = \rho\exp(i\varphi)$, the solutions $|m,0\rangle \propto \rho^m \exp(im\varphi)$ are seen to be eigenstates of the angular momentum component $L^z = (1/i)(\partial/\partial\varphi)$ with eigenvalue m. This follows from the axial symmetry of the problem and holds for all Landau levels. However, electrons in the lowest Landau level can go around only in one direction (counterclockwise): all the L^z eigenvalues are non-negative. The non-equivalence of the clockwise and counter-clockwise senses of rotation (in other words, the absence of time reversal invariance) is caused by the magnetic field. As far as the charge distribution is concerned, the eigenstates can be envisaged as concentric rings of width $\sim l$. To determine the radius of the ring, let us use $z = \sqrt{2}(a + b^\dagger)$ to find

$$\langle m, n|zz^*|m, n\rangle = 2\langle(a + b^\dagger)(a^\dagger + b)\rangle \rightsquigarrow 2\langle aa^\dagger + b^\dagger b\rangle = 2(n + m + 1)\,. \tag{12.62}$$

For $n = 0$, the radius of the mth Landau orbital is $\sqrt{2(m+1)}l$. How large can m be? There is no inherent restriction; the perimeter of the system can be defined by the outermost orbital which is filled. For N electrons, we prepare the state with a filled Landau level by occupying all orbitals out to the radius $\sqrt{2N}l$. This gives a system area $2\pi Nl^2$, thus we recover the result that each Landau state occupies an area $2\pi l^2$. In the limit $N \to \infty$, the total electron density

$$\rho(z) = \frac{1}{2\pi l^2} \lim_{N\to\infty} \sum_{m=0}^{N} \frac{|z|^{2m}}{(2l^2)^m m!} \exp\left(-\frac{|z|^2}{2l^2}\right) = \frac{1}{2\pi l^2} \tag{12.63}$$

is uniform (in the formula z has the dimension of length, and we reinserted l in the proper places). The charge density of a large system is uniform inside the disk of radius $\sqrt{2N}l$, and it falls off smoothly at the edge within a transition region of width $\sim l$.

If we want to see clockwise rotation, we have to go to higher Landau levels. The innermost orbital of the second ($n = 1$) Landau level is

$$|0, 1\rangle = a^\dagger|0, 0\rangle = \frac{1}{\sqrt{4\pi}}\left(-2\frac{\partial}{\partial z} + \frac{z^*}{2}\right)\exp\left(-\frac{zz^*}{4}\right) = \frac{z^*}{\sqrt{4\pi}}\exp\left(-\frac{|z|^2}{4}\right). \tag{12.64}$$

This is an $L^z = -1$ state. The general expression of the $n = 1$ eigenstates is

$$|m, 1\rangle = \frac{1}{\sqrt{4\pi 2^m m!}}(-2mz^{m-1} + z^m z^*)\exp\left(-\frac{|z|^2}{4}\right). \tag{12.65}$$

z^* appears in each state. Similarly, z^{*2} appears in each $n = 2$ state, etc. The range of the L^z eigenvalues in the $(n + 1)$th Landau level is $L^z \geq -n$. One might say that the magnetic field picks a preferred sense of rotation, and it takes increasingly higher energies to make the electron rotate faster in the opposite sense.

It is educative to compare the energy eigenstates $|1,0\rangle$ and $|0,1\rangle$. $|1,0\rangle$ is the complex conjugate of $|0,1\rangle$, but it belongs to a different eigenenergy. Experience with the standard problems of quantum mechanics makes us familiar with the situation when the complex conjugate of an eigenstate is also an eigenstate, with the same eigenenergy. For instance, in the hydrogen atom problem, the conjugate of the p-state $f(r)e^{i\varphi}$ with $L^z = 1$ is the p-state $f(r)e^{-i\varphi}$ with $L^z = -1$. In the absence of a magnetic field, the two states are degenerate, and thus it was arbitrary to use complex wavefunctions; we might as well have used the real $f(r)\cos\varphi$ and $f(r)\sin\varphi$. In contrast, we might say that the eigenstate $ze^{-|z|^2/4}$ is essentially complex: it is not degenerate with its complex conjugate, and therefore we are not able to replace it with a real wavefunction. As we discuss in Sec. 3.7, this is a consequence of the lack of time reversal invariance which manifests itself in the fact that though the Hamiltonian is self-adjoint ($\mathcal{H}^\dagger = \mathcal{H}$), it is not a real operator: $\mathcal{H}^* \neq \mathcal{H}$.

12.3 Integer Quantum Hall Effect

Let us try to find out whether, and in which sense, the solution of the Landau eigenvalue problem can help us to understand the Quantum Hall Effect. To find the ground state, we start populating the Landau levels from bottom upwards (not forgetting to consider the levels for both spin directions). For any usual kind of system in weak fields, a large number of Landau levels would be populated. However, in the semiconductor devices of interest to us, it is possible to reach the quantum limit where only several lowest-lying levels are populated, or even the *extreme quantum limit* where only the lowest Landau level is partially occupied. Typical electron densities in the semiconductor devices are $\sim 10^{11}\text{cm}^{-2}$ which corresponds to an average interparticle separation $\sim 3{\cdot}10^{-6}\text{cm}$. The cyclotron radius being $l \approx 2.6{\cdot}10^{-6}/\sqrt{H}$ (with H in units of Tesla), in magnetic fields of 1–10T the cyclotron radius gets smaller than the interparticle distance and the (extreme) quantum limit is reached. The spectacular features arising from the Landau quantization of the energy levels become visible only if the thermal smearing of the level occupation is negligible. This condition is met for measurements done at $T < 1\mathrm{K}$.

For systems in the quantum limit, it is useful to express the surface density of electrons n_{surf} in units of the (Landau level) *filling factor* ν, defined by

$$n_{\text{surf}} = \frac{H}{\phi_0}\nu \qquad\qquad \nu = n_{\text{surf}} 2\pi l^2 \,. \tag{12.66}$$

For $\nu = 1$, the lowest Landau level is completely filled (for one spin direction) and other states are empty.

We have solved the problem of the two-dimensional motion in crossed magnetic and electric fields. The Hamiltonian is (12.24), and the spectrum is given by (12.34). The eigenstates are still of the same form (12.36) as in the $E = 0$ case, only all states are uniformly shifted in the y-direction according to (12.30). Recalling that the velocity operator is (12.20), we find that the expectation of value of v_x in a Landau state is

$$\begin{aligned} \langle \hat{v}_x \rangle &= \frac{1}{m}\left(\langle \hat{p}_x \rangle - \frac{eH}{c}\langle y \rangle\right) = \frac{1}{m}\left(p_x - \frac{eH}{c} y_0\right) \\ &= \frac{1}{m}\left(p_x - \frac{eH}{c}\left(-\frac{mEc^2}{eH^2} + \frac{cp_x}{eH}\right)\right) = \frac{cE}{H} = v_{\text{drift}} \end{aligned} \tag{12.67}$$

i.e., equal to the classical drift velocity (12.2), for all the states in all the Landau levels[11]. Thus all occupied states give the same contribution $-ev_{\text{drift}}$ to the Hall current j_x and the Hall conductivity is

$$\sigma_{yx} = -\frac{j_x}{E_y} = \frac{nec}{H}. \tag{12.68}$$

The quantum-mechanical treatment of a free electron in magnetic field reconfirmed the classical result (12.3). We will find that the inclusion of disorder (or electron–electron interaction *and* disorder) is needed to understand the emergence of plateaus.

However, our results help us to restate the problem in a way which makes it clear in which direction the solution should be sought. It is

[11] It is interesting to observe that, in spite of their appearance of waves running in the x-direction, the $E = 0$ Landau states do not carry a current. $\langle v_x \rangle$ arises solely from the E-induced shift of y_0.

useful to reexpress (12.3) in terms of the filling factor

$$\sigma_{yx} = \frac{e^2}{h}\nu \,. \tag{12.69}$$

e^2/h is the "atomic unit" of conductivity. If the number of electrons is just right to fill one Landau level, the free electron value of the Hall conductivity is e^2/h. This is fine. The problem is that in IQHE systems, σ_{yx} gets stuck at its quantized value, and does not change when a fraction of electrons is removed from a filled Landau level, or added to it. We must find a mechanism by which a number of electrons can be prevented from contributing to the Hall current. This is provided by the Anderson localization.

Let us allow for the inevitable imperfection of the samples, and consider electrons moving in a disorder potential. This causes the hitherto sharp Landau levels to broaden into peaks of finite width (Fig. 12.5, bottom). Let us, however, assume that the disorder is sufficiently weak in comparison to $\hbar\omega_c$, so that the peaks remain distinct, and do not merge into a structureless band of states. In the absence of a magnetic field, it would be a standard result of localization theory that all states are localized (albeit perhaps only weakly), and the system is an insulator. However, switching on a magnetic field has a very non-trivial effect: in the center ϵ_n of each Landau band, the localization length diverges as $\xi \sim |\epsilon - \epsilon_n|^{-7/3}$ (the non-analyticity of this behaviour is enough to show that the disorder potential cannot be handled by perturbation theory). The diverging ξ exceeds the sample dimension in a narrow interval of energies around each ϵ_n. Our picture is that most of the states in between the peaks are localized, and only a small fraction of states at the peak centers are capable of carrying a dc current.

Accepting this, we are in the position to explain how plateaus can emerge. While the chemical potential is moving across the broad regimes of localized states (the so-called mobility gaps), adding or withdrawing electrons does not change the number of current carriers, and σ_{yx} remains unaltered. Transitions from one plateau to the next are associated with sweeping through the narrow range of extended states. It is in these regimes that dissipative current can flow as evidenced by the peaks of σ_{xx} (Fig. 12.5, top).

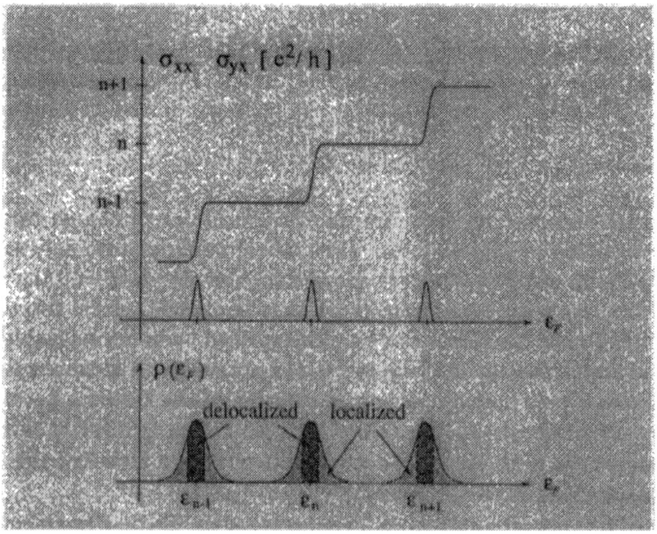

Figure 12.5: *Top*: schematic representation of the density (or equivalently, chemical potential) dependence of σ_{yx} (in units of e^2/h) and σ_{xx} in the IQHE. *Bottom*: The DOS of disorder-broadened Landau levels. Extended states (darker shading) are possible in the high-DOS regions at the level centers ϵ_n, while states is the tails are localized (after [187]). σ_{yx} remains at a plateau value while the chemical potential sweeps through a mobility gap.

Here we face the next problem. It seems alright to see plateaus in σ_{yx} vs n (or ρ_{xy} vs H) plots, but why are the plateau values the same as for filled Landau levels of free electrons? Conventional wisdom would tell us that the Hall coefficient measures the number of current-carrying electrons. Somehow, IQHE systems find a way to compensate for the apparent loss of electrons which settled down in localized states. Whatever the explanation is, it must be very general, free of approximations (as far as we know, the quantization (12.11) is exact), and independent of the details of the disorder potential, etc. We offer two different arguments. The high-field argument explains the quantized values at $H \to \infty$, but at finite H it leaves us with doubts as to how accurate the quantization can be. The so-called gauge argument [234] is quite general, and it reformulates the problem by requiring that in order to have a Hall plateau, the chemical potential must be lying in a gap, or at least in a mobility gap.

12.3.1 High-Field Argument

Let us discuss the motion of a Landau electron in the presence of a non-uniform electric field. We use the Landau gauge, and add to the Hamiltonian (12.24) the potential energy term $-e\varphi(x,y)$. The total potential is the sum of the potential of the externally applied electric field and a static disorder potential arising from charged impurities, etc.

$$\varphi(x,y) = \varphi_{\text{ext}}(x,y) + \varphi_{\text{dis}}(x,y) \,. \tag{12.70}$$

In the absence of an electric field, the center-of-orbit coordinates (12.47) are constants of motion. Switching on φ, the orbit center begins to drift

$$\dot{\hat{x}}_0 = \frac{i}{\hbar}[\mathcal{H}, x - \frac{c\hat{p}_y}{eH}] \rightsquigarrow -\frac{i}{\hbar}[-\frac{c}{eH}\frac{\hbar}{i}\frac{\partial}{\partial y}, -e\varphi(x,y)] = -\frac{c}{H}\frac{\partial\varphi}{\partial y} \,. \tag{12.71}$$

If we consider only the homogeneous external field

$$\dot{\hat{x}}_0 \rightsquigarrow -\frac{i}{\hbar}[\hat{x}_0, -e\varphi_{\text{ext}}] = \frac{c}{eH}[\frac{\partial}{\partial y}, eEy] = \frac{cE}{H} = v_{\text{drift}} \,. \tag{12.72}$$

Now we interpret (12.67) in the following manner: the Hall current is due to the slow drift of the centre of the cyclotron orbit. The fast cyclotron motion of the electron about the centre does not give rise to a macroscopic current.

It is straightforward to generalize the above result to the case when the scalar potential is an arbitrary function of the coordinates

$$\dot{\hat{x}}_0 \rightsquigarrow -\frac{i}{\hbar}[\frac{c\hat{p}_y}{eH}, e\varphi(x,y)] = \frac{cE}{H} - \frac{c}{H}\cdot\frac{\partial\varphi_{\text{dis}}}{\partial y} \,. \tag{12.73}$$

In the above equation, the potential is taken at the instantaneous position of the electron. Clearly, it would be very difficult to describe the motion of the electron if the electric field were changing substantially within a cyclotron radius.

Let us, however, note that because of $l^2 \propto 1/H$, a great conceptual simplification occurs in the *high field limit* [180, 187]. $\hat{x}_0$ and $\hat{y}_0$ approximately commute and (12.59) permits to localize the centre of the orbit with increasing accuracy. The cyclotron motion of the electron about the centre is taking place within a range $\sim l$, thus we can speak about a sharply defined classical trajectory when $H \to \infty$. This is the trajectory along which the cyclotron orbit center is drifting under the influence of the electric fields. In the high-field limit, the variation of the electric field over a distance l becomes negligible, thus the scalar potential can be regarded as a function of the classical variables x_0 and y_0, and the equation of motion (12.73) can be rewritten as

$$\dot{x}_0 = \frac{cE}{H} - \frac{c}{H} \cdot \frac{\partial \varphi_{\text{dis}}}{\partial y_0} . \tag{12.74}$$

Similarly

$$\dot{y}_0 = \frac{c}{H} \cdot \frac{\partial \varphi_{\text{dis}}}{\partial x_0} . \tag{12.75}$$

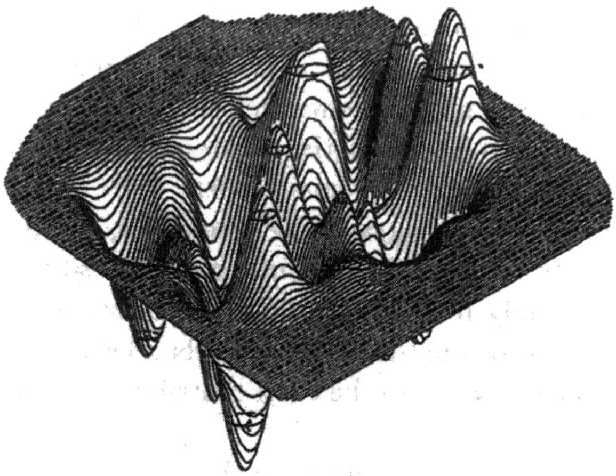

Figure 12.6: Mountains and valleys of a smooth random potential with average $\langle \varphi_{\text{dis}} \rangle = 0$. The quasi-classical orbits follow equipotential lines (after [187]).

Let us follow the change of the potential "seen" by the electron as it is drifting along in the random potential (let us, for the moment, forget about φ_{ext})

$$\frac{\mathrm{d}\varphi}{\mathrm{d}t} \rightsquigarrow \frac{\partial \varphi}{\partial x_0}\dot{x}_0 + \frac{\partial \varphi}{\partial y_0}\dot{y}_0 = \frac{c}{H}\left(-\frac{\partial \varphi}{\partial x_0} \cdot \frac{\partial \varphi}{\partial y_0} + \frac{\partial \varphi}{\partial y_0} \cdot \frac{\partial \varphi}{\partial x_0}\right) = 0. \tag{12.76}$$

The quasi-classical motion proceeds along equipotential lines. Taking the limit $H \to \infty$ is an acceptable approximation[12] only if the actual random potential is *smooth*, i.e., it changes little on the scale of the true value of l. Let us try to visualize the equipotential lines of $\varphi_{\rm dis}$. It is reasonable to assume that the random potential fluctuates about the mean value zero $\langle\varphi_{\rm dis}\rangle = 0$. It is easy to see that the equipotential line correponding to some fixed negative value of $\varphi_{\rm dis}$ must be a closed curve. Namely, starting from a point where we have this value, and going in any specific direction, it will be increasingly improbable that $\varphi_{\rm dis}$ remains negative (otherwise it could not have the mean value zero). Hence such equipotential lines enclose finite valleys of the disorder potential. Similarly, equipotential lines corresponding to a fixed positive value of $\varphi_{\rm dis}$ enclose finite potential mountains. In the limit of a macroscopic sample, the only equipotential lines reaching across the sample (percolation paths) belong to $\varphi_{\rm dis} = 0$.

Let us again switch on the weak external field acting in the y-direction. It stands to reason that most equipotential lines enclosing potential maxima and minima remain closed, and it is only in a narrow range about $\varphi = 0$ (i.e., about the centre of the Landau level), that equipotential lines become, or remain, open.

It is clear that electrons travelling around closed equipotential lines cannot contribute to the Hall current. A disorder-broadened Landau level contributes only if it is more-than-half-filled, so that the narrow range of extended states in the centre of the band is fully occupied. Once this filling is reached, we have the total contribution of the Landau level to σ_{yx}.

Since our problem is to understand the plateau values of σ_{yx}, let us assume that n Landau levels are completely filled. This way, we do not have to worry about exactly where the range of extended states is lying; we sum over all the states of n Landau bands.

In the limit $H \to \infty$, x_0 and $p_x = \hbar y_0/l^2$ become a pair of classical canonical variables, thus taking the trace over the states of a Landau

[12]It is quite possible that one can realize two different sorts if disorder: the smooth random potential arising from long range forces, and the sharply varying potential of short-range scatterers. In certain respects, these two classes of systems have different physics.

band is replaced by the phase space integral divided by the phase volume unit h. To see how this works let us re-evaluate the number of states per unit area

$$\mathcal{N} = \frac{1}{L_x L_y}\frac{1}{h}\int dx_0 \int dp_x = \frac{1}{L_x L_y}\frac{1}{2\pi l^2}\int dx_0 \int dy_0 = \frac{1}{2\pi l^2} \quad (12.77)$$

as it has to be. We proceed similarly when evaluating the Hall current density of i filled Landau bands

$$\begin{aligned} j_x &= \frac{1}{2\pi l^2}\sum_{\text{levels}}^{i}\int dx_0 \int dy_0 \, \frac{-e\dot{x}_0}{L_x L_y} \quad (12.78) \\ &= \frac{ne^2 H}{ch}\cdot\frac{1}{L_y}\int dy_0 \, \frac{c}{H}\frac{\partial\varphi}{\partial y_0} = \frac{ne^2}{h}\frac{\varphi(L_y)-\varphi(0)}{L_y} = -\frac{ie^2}{h}\cdot E_y . \end{aligned}$$

Hence we conclude that the plateau value of the Hall conductivity corresponding to i filled Landau levels must be

$$\sigma_{yx} = \frac{ie^2}{h} . \quad (12.79)$$

The high-field argument makes it plausible that plateaus occur at the integer quantized values which coincide with the free electron value for the same number of filled Landau levels, in spite of the fact that a large number of states became localized by disorder.

Though the result is impressive, we cannot be content. The limit $H \to \infty$ made the electrons move classically, and we would certainly like to see the QHE explained by a fully quantum-mechanical treatment. Allowing l to be finite, our argument could not exclude that corrections of order l/L_x, l/L_y appear. Now, this is what does *not* happen. Finite-size corrections of this sort could easily be as large as 10^{-4}. However, experiments tell us that the Hall conductivity is quantized with an accuracy of 10^{-8}, or better. This suggests that there must exist a more general and profound argument to justify (12.79).

12.3.2 The Gauge Argument

In the Landau gauge, we use periodic boundary conditions in the x-direction, and open boundary conditions at the edges $y = 0$ and $y = L_y$.

This makes it natural to think of the system as wrapped on a cylinder of height L_y. Now the x-direction belongs to walking around the cylinder, and the quantizing field is piercing the cylinder normal to its surface. The Hall current flowing around the cylinder must be giving rise to an auxiliary magnetic flux ϕ_{aux} along the cylinder axis. We may think of this flux as confined within a thin solenoid along the axis of the cylinder, so that the system surely never feels the extra magnetic field. It does, however, sense the auxiliary vector potential $\mathbf{A}_{\text{aux}} = (a_x, 0, 0)$ associated with ϕ_{aux}. Integrating along a line encircling the cylinder

$$\oint \mathbf{A}_{\text{aux}} \cdot \mathbf{dl} = a_x L_x = \phi_{\text{aux}}$$

wherefrom $a_x = \phi_{\text{aux}}/L_x$. Going over from the plane geometry to the cylinder geometry necessitated the inclusion of a new term of the vector potential which becomes

$$\mathbf{A} = (-Hy, 0, 0) + (\phi_{\text{aux}}/L_x, 0, 0). \tag{12.80}$$

The first term is giving rise, as before, to the quantizing magnetic field $(0, 0, H)$. The second belongs to a flux passing wholly outside the sample so we expect that we should be able to control it by a gauge transformation. Let us consider an increment $\delta\phi_{\text{aux}}$ of ϕ_{aux}, associated with a change in the vector potential $\delta\mathbf{A}_{\text{aux}} = \nabla\Omega$, where $\Omega = x \cdot \delta\phi_{\text{aux}}/L_x$. Performing $\mathbf{A} \to \mathbf{A} + \nabla\Omega$, the wave function ψ gets transformed to

$$\psi' = \psi \exp\left(\frac{ie\delta\phi_{\text{aux}}}{\hbar c} \cdot \frac{x}{L_x}\right). \tag{12.81}$$

But the cylinder geometry requires that ψ', too, satisfies the periodic boundary condition $\psi'(x + L_x) = \psi'(x)$. It follows that

$$\delta\phi_{\text{aux}} = 2\pi n \cdot \frac{\hbar c}{e} = n\phi_0. \tag{12.82}$$

We may put it in words like this: the physical state of the system is not altered if we change ϕ_{aux} with an integer number of flux quanta, because the increment in ϕ_{aux} can be "gauged away", while the boundary conditions remain respected.

Actually, this is not quite true because there is a subtle change accompanying the gauge transformation. Let us consider free electrons, and a filled Landau level. The gauge transformation (for $n = 1$) of a Landau state with momentum p_x shifts p_x according to

$$p_x \rightsquigarrow p_x + \frac{e\phi_0}{cL_x} \tag{12.83}$$

or equivalently, the center-of-orbit coordinate y_0 is shifted by

$$y_0 \rightsquigarrow y_0 + \frac{\phi_0}{HL_x}. \tag{12.84}$$

It is easy to check that the increment of y_0 is just the y-spacing between the centres of adjacent Landau states. Hence the filled Landau level is transformed into itself: each electron is transferred to the next Landau state. There are only two places where something strange happens: at one edge of the system, a state becomes empty; and at the other edge, an electron seems to get pushed out of the sample. In fact, what must happen is that the electron gets transferred (say, via a connecting wire, or along an edge channel[13]) to the other edge.

To summarize: for a filled Landau level of free electrons, the gauge transformation which adds a flux quantum ϕ_0 to ϕ_{aux}, is leaving the state of the system unaltered, except for the transfer of a single electron from one edge to the other.

The gauge argument [234] relies on the *assumption* that the same is true whenever there is a plateau in the Hall conductivity: adding ϕ_0 to ϕ_{aux}, the only change in the state of the system is the transfer of a number of electrons from one edge to the other.

First, let us see what the argument is leading to. We will try to justify the starting assumption afterwards.

We know from classical electrodynamics that the vector potential enters the energy density in the form $(1/c)\mathbf{j} \cdot \mathbf{A}$. The energy increment

[13]Though the bulk of the system is often gapped due to kinetic energy quantization (Landau levels), or due to electron–electron interaction (see later), we always find gapless excitations at the perimeter of the system. This is quite easy to accept for the symmetrical gauge where the edge of the system is where we stopped putting in more electrons.

due to the uniform increment δA_x of the vector potential is

$$\delta W = \frac{1}{c}\int\int dx dy j_x \cdot \delta A_x = \frac{L_x L_y}{c} j_x \cdot \delta A_x. \tag{12.85}$$

According to our assumption, the change in the energy of the system is due to the transfer of an integer number r of electrons via the potential drop U_y (the Hall voltage)

$$\delta W = -reU_y = -reE_y L_y. \tag{12.86}$$

The increment of $\mathbf{A}$ which corresponds to adding ϕ_0 to $\phi_{\rm aux}$ is

$$\delta A_x = \frac{\phi_0}{L_x} = \frac{ch}{eL_x}. \tag{12.87}$$

Putting these in (12.85), the Hall conductivity is found to be

$$\sigma_{yx} = -\frac{j_x}{E_y} = r \cdot \frac{e^2}{h}. \tag{12.88}$$

Apparently we have proven that, once the starting assumption is granted, the Hall conductivity is quantized in integer multiples of e^2/h. This is a very powerful result and we may feel a little surprised that its derivation (which we admittedly just sketched) can be done so simply. Here two remarks can be made. On the one hand, since the quantization is extremely accurate, and independent of experimental details, it is expected that it should follow from very general principles. On the other hand, the crucial point is to find a justification for the simple-sounding starting assumption which, in retrospect, is seen to be essentially equivalent to saying that there is a QHE.

First let us take an example when the assumption is not valid. If we take a partially filled Landau level, and free electrons[14], the gauge transformation (12.84) will reshuffle the electrons in the bulk of the sample. The energy change can be ascribed to the infinitesimal displacement

[14] "Free" means: non-interacting, and not subject to either a periodic, or a disordered, potential in the bulk. An edge potential is needed to confine the electrons to the sample in the y-direction, but its effect can be incorporated in the argument.

of a macroscopic number of electrons (rather than to a macroscopic displacement of a few electrons)

$$\delta W = -n \cdot \frac{\phi_0}{HL_x} \cdot eE_y. \tag{12.89}$$

Equating this with (12.85), we recover the free-electron result $\sigma_{yx} = nec/H$, as we should.

If we want to apply the gauge argument to a partially filled Landau level, and derive a plateau value for σ_{yx}, we have to exclude the possibility that energy is imparted to electrons in the bulk. The best guarantee is to assume that there is a *finite gap for bulk excitations*[15].

Landau quantization for free electrons provides such gaps of size $\hbar\omega_c$, but only between a completely filled, and the subsequent completely empty, level. This case we understand, anyway. Disorder broadens the Landau levels into bands, presumably with overlapping (approximately Gaussian) tails, thus the gaps are no longer really there, and the Fermi level is lying in a region with a non-vanishing density of states. The high-field argument tells us that σ_{yx} remains at its quantized value but to get the same result from the gauge argument, we have to assume that "a mobility gap is as good as a real gap". Thus we just assume that if ϵ_F is lying in a regime of localized states (which it does for most fillings) then the gauge argument still works, and integer quantization is predicted. Any detailed justification of this assumption should rely on the theory of Anderson localization in two-dimensional systems subjected to a strong magnetic field, a very difficult problem which is still not completely solved.

---

[15]It must be warned against a possible misunderstanding. Emphasizing the importance of the gap may sound as if we were ready to ascribe QHE to conventional insulators. We certainly do not mean that! (unless in the trivial sense $\sigma_{yx} \equiv 0$). The gap we are talking about must be above a mobile electron liquid ground state (a sort of a condensate). For instance, the gapped $\nu = 1$ Landau level is not pinned to the underlying substrate and it cannot become pinned by weak disorder just because it is gapfull. – This raises the question whether a movable CDW/SDW (a sliding wave, see [138]) gives rise to QHE with $\sigma_{yx} \neq 0$. The issue is delicate but the upshot of the argument is that it does not; however, Landau bands split off the main density wave condensate may do so [450].

12.3.3 Periodic Potential

Taken in itself, the gauge argument cannot give the integer r in the quantization rule (12.88). However, our previous considerations make it plausible that in a weak random potential, the plateau $\sigma_{yx} = e^2/h$ is centered on $\nu = 1$, etc. Perhaps it looks a trifle too obvious: could it be any other way? Indeed it could. It is educative to study a problem where the relationship between $\sigma_{yx}/(e^2/h)$ and the level filling is less evident [398, 117].

Let us consider tight binding electrons moving on a square lattice in the presence of a perpendicular magnetic field H [156, 171]. At $H = 0$ the band dispersion is

$$\epsilon(k_x, k_y) = -2t(\cos(k_x a) + \cos(k_y a)). \tag{12.90}$$

The lattice constant a is the characteristic length of the periodic potential. If we switch on the magnetic field, it brings its own length scale: the cyclotron radius l. The character of the solution must depend on l/a. To look at it from a different angle, let us compare characteristic frequencies [171]. We associate the frequency $\omega_{\text{kin}} = (1/\hbar)(\hbar^2\pi^2/2ma^2)$ with the motion through the lattice. Building the ratio $\omega_{\text{c}}/\omega_{\text{kin}}$, we come upon the important parameter

$$\alpha = \frac{ea^2H}{hc} = \frac{a^2H}{\phi_0}, \tag{12.91}$$

the number of flux quanta per unit cell.

It is usual to treat the motion of band electrons in a magnetic field by using the so-called Peierls substitution. It consists in replacing

$$\mathbf{k} \rightarrow \frac{1}{\hbar}\left(\frac{\hbar}{i}\nabla + \frac{e}{c}\mathbf{A}\right) \tag{12.92}$$

in the $H = 0$ dispersion relation (in our case (12.90)). This permits us to seek the eigenstates by expanding in the Wannier basis $\Psi = \sum_{g,h} w(g,h)\varphi(g,h)$, where $\varphi(g,h)$ is the Wannier state for the lattice point (ga, ha). Using the Landau gauge $\mathbf{A} = (0, Hx, 0)$, and performing the substitution, we get the equation

$$-t[w(g+1,h) + w(g-1,h) + \exp(-iega^2H/\hbar c)w(g,h+1)$$
$$+ \exp(iega^2H/\hbar c)w(g,h-1)] = 2\epsilon w(g,h) \tag{12.93}$$

where ϵ is the eigenvalue we seek. The exponential factors on the left-hand site conform to the general rule that the effect of a magnetic field on the transition amplitude from $\mathbf{R}_1$ to $\mathbf{R}_2$ (or the hopping matrix element) is

$$t(\mathbf{R}_1, \mathbf{R}_2) \to t(\mathbf{R}_1, \mathbf{R}_2) \exp\left\{ -\frac{2\pi}{\phi_0} \int_{\mathbf{R}_1}^{\mathbf{R}_2} \mathbf{A} \cdot d\mathbf{l} \right\}. \tag{12.94}$$

Since y does not appear in the coefficients in (12.93), the eigenstate can be sought as $w(g, h) = e^{i\zeta y} a(g)$. Having done this replacement, we arrive at a one-dimensional difference equation (the so-called Harper equation [156]) which contains the flux parameter α

$$a(g+1) + a(g-1) + 2\cos(2\pi g\alpha - \zeta) a(g) = \epsilon a(g). \tag{12.95}$$

Seeking bounded eigenstates, we have to require that the cosine has a finite period, say of length Q, i.e.,

$$2\pi\alpha(g + Q) - \zeta = 2\pi\alpha g - \zeta + 2\pi P \tag{12.96}$$

with some integer P. It follows that the flux per unit cell

$$\alpha = \frac{P}{Q} \tag{12.97}$$

must be a rational number. It is curious to find a physical problem where the rationality of a parameter is playing a role.

Let us discuss the α-dependence of the spectrum [171]. First, let us note that $\alpha \to \alpha + 1$ amounts to a shift by an integer period, and as such, not detectable; thus the spectrum is periodic in α, and it is enough to study the range $0 \leq \alpha \leq 1$. The results of extensive numerical work assuming rational α have given rise to the remarkable shape shown in Fig. 12.7, which became known as the Hofstadter butterfly. The butterfly has many features which could be discussed at length; for instance, its approximate self-similarity (meaning that one can pick small details from the picture which, after magnification, look like one of the larger parts, and this can be continued ad infinitum). However, from our present point view, the interesting thing is that the spectrum shows some well-developed *gaps*.

Figure 12.7: The Hofstadter butterfly represents the spectrum of a square lattice tight binding band in a perpendicular magnetic field (after [171]). The horizontal axis is the energy ranging from $-4t$ to $4t$. The vertical axis is the parameter α defined in (12.91) ($0 \leq \alpha \leq 1$).

At $H = 0$, the spectrum is the horizontal line at the bottom ($-4t \leq \epsilon \leq 4t$). At small H, the band apparently splits into a large number of levels. Noteworthy things happen at simple fractions. Let us take $\alpha = 1/3$. As shown in the picture, the band is split into three subbands separated by gaps of the order of t. Similarly, at $\alpha = 1/5$ (and also at $\alpha = 2/5$) we find five subbands. Generally, for $\alpha = P/Q$ there are Q subbands if Q is odd[16]. It can be shown that the subbands have equal weight. Q crystal cells constitute the magnetic supercell which is pierced by the flux $P\phi_0$, i.e., it corresponds to P Landau levels. Putting $1, 2, \ldots, Q$ electrons in a magnetic supercell, ν goes through the sequence P/Q, $2P/Q$, ..., P. At each of these fillings, the Fermi level is lying in a gap, and the gauge argument requires that $\sigma_{yx}/(e^2/h)$ be an integer even though ν may be an arbitrarily small fraction (modulo

[16]The situation is less clear-cut if Q is even; for instance, it appears that there are only three subbands at $\alpha = 1/4$. However, for even Q the central subband is degenerate.

1). This is confirmed by a detailed argument [398]. For example, for $\alpha = 11/7$ there are seven subbands, and $\sigma_{yx}/(e^2/h)$ goes through the plateau values $2, 4, -1, 1, -4, -2, 0$ [117]. σ_{yx} may jump wildly even upon a slight change of ν. Interestingly, some of the plateau values are negative, indicating that the Bragg reflection from the crystal potential causes the electrons to behave as hole-like carriers. Sign reversal in the sequence of IQHE states is thought to be a characteristic lattice effect.

At the time of the discovery of the butterfly [171], the experimental realization of α of the order of 0.1 seemed impossible. However, rapid progress in manufacturing heterostructures with a superlattice potential has yielded samples with interesting magnetotransport phenomena which have been the subject of intense study [325]. Furthermore, we call attention to the finding that $(TMTSF)_2PF_6$ under pressure goes through a series of Integer Quantum Hall states, and manifests the phenomenon of sign reversal [36].

12.4 Fractional Quantum Hall Effect

We have got the apparently impeccable gauge argument to show that the Hall conductivity is quantized in integer multiples of e^2/h. But stated in this way, the result is simply not true: we know that there exist also fractional quantum Hall plateaus, the most robust being at

$$\sigma_{yx} = \frac{1}{3} \cdot \frac{e^2}{h}. \tag{12.98}$$

What has gone wrong? Should we discard the gauge argument? That would be a cowardly thing to do; if we really believe in the profundity of the argument, we should stick to it through thick and thin, and see what it leads to, however strange the outcome may seem. So, once again: when ϕ_0 is added to ϕ_{aux}, a particle is transferred from one edge to the other. Since it cannot be an electron (that would lead to predicting integer quantization), let us assume that it is an as yet unknown quasiparticle of charge e^*. Repeating the previous derivation, e has to be replaced by e^* in (12.86), but not in (12.87), because that came from the fundamental electromagnetic interaction. Instead of (12.88), we find now

$$\sigma_{yx} = r \cdot \frac{ee^*}{h}. \tag{12.99}$$

Assuming that one quasiparticle got transferred, we are led to predict the existence of *quasiparticles with fractional charge*

$$e^* = \frac{e}{3} \tag{12.100}$$

in the many-electron state which gives rise to the 1/3-plateau. This provoking conclusion proved to be extraordinarily fruitful in understanding the many-body correlations which we hold responsible for the fractional quantum Hall effect [236].

12.4.1 Correlated Many-Electron States

Here we will attempt to understand FQHE by explicitely constructing the ground state wave function of a 1/3-filled (and later, a 2/5-filled) Landau level. First, let us assume that the external field is sufficiently large to spin-polarize the system completely, so that we may drop the spin index. The full Hamiltonian includes, in addition to the kinetic energy, the electron–electron interaction and the disorder potential

$$\mathcal{H} = \sum_{n,m} c^\dagger_{n,m} c_{n,m} \left(n + \frac{1}{2}\right) \hbar\omega_c + \mathcal{H}_{\text{int}} + \mathcal{H}_{\text{dis}} \,. \tag{12.101}$$

Let us assume that the typical scale of the interaction $e^2/\epsilon_0 l \propto \sqrt{H}$ is much smaller[17] than $\hbar\omega_c \propto H$ (ϵ_0 is the dielectric constant; for GaAs $\epsilon_0 = 13$). This allows us to seek the ground state within the lowest Landau level. Postponing the consideration of $\mathcal{H}_{\text{dis}}$, the one-electron energy is the same for all N-electron states, and we have only the Coulomb interaction energy to minimize.

[17]In point of fact, this is not true: in a typical experimental situation for heterostructures, $\hbar\omega_c$ would be 150K, while the interaction strength would be about 100K [253]. We hope, however, that the understanding achieved for the $H \to \infty$ limit remains valuable at realistic parameter values.

We use the symmetrical gauge. The completely filled lowest Landau level of N electrons is the Slater determinant

$$\begin{aligned}\Psi_1 &= \frac{1}{\sqrt{N!}}\begin{vmatrix} 1 & 1 & \dots & 1 \\ z_1 & z_2 & \dots & z_N \\ \vdots & \vdots & \ddots & \vdots \\ z_1^{N-1} & z_2^{N-1} & \dots & z_N^{N-1}\end{vmatrix} \cdot \exp\left\{-\frac{1}{4}\sum_j |z_j|^2\right\} \\ &= \frac{1}{\sqrt{N!}}\prod_{i<j}(z_i - z_j)\exp\left\{-\frac{1}{4}\sum_j |z_j|^2\right\} \\ &= \mathcal{N}^{-1}\mathcal{P}(z_1,\dots,z_N)\,\mathcal{G}(z_1,\dots,z_N)\,. \end{aligned} \tag{12.102}$$

The factorization follows because we are dealing with a Vandermonde determinant. It is generally true that many-electron wave functions for the lowest Landau level can be written as the product of a homogeneous polynom $\mathcal{P}$ of z, and a Gaussian factor $\mathcal{G}$ (and an uninteresting normalization factor). The homogeneity of $\mathcal{P}$ follows from the conservation of L^z. Since $\mathcal{G}$ is the same for all N-electron states, we may drop it.

The filling ν can be read off from the polynomial factor. Since the largest power of, say, z_1 is $(N-1)$, the outermost occupied orbital is the Nth Landau orbital, and $\nu = N/N = 1$. Similarly, if the largest power of z_1 is $(M-1)$, then $\nu = N/M$.

Let us consider the $\nu = 1/3$ ground state. A naive first try would be to fill every third Landau orbital

$$\begin{aligned}\Phi_{1/3} &= \frac{1}{\sqrt{N!}}\begin{vmatrix} 1 & 1 & \dots & 1 \\ z_1^3 & z_2^3 & \dots & z_N^3 \\ \vdots & \vdots & \ddots & \vdots \\ z_1^{3(N-1)} & z_2^{3(N-1)} & \dots & z_N^{3(N-1)}\end{vmatrix} \cdot \exp\left\{-\frac{1}{4}\sum_j |z_j|^2\right\} \\ &= \frac{1}{\sqrt{N!}}\prod_{i<j}(z_i^3 - z_j^3)\exp\left\{-\frac{1}{4}\sum_j |z_j|^2\right\}. \end{aligned} \tag{12.103}$$

This would be, however, unsatisfactory. The dominant contribution from $\mathcal{H}_{\text{int}}$ arises when electrons approach one another closely, and we have to strive to keep the electrons away from each other, as strictly as possible. In the filled level Ψ_1 there is no escape from close approaches which are governed by $(z_i - z_j)$, a linearly vanishing factor. Now, our $\Phi_{1/3}$ is doing no better because

$$z_i^3 - z_j^3 = (z_i - z_j)(z_1^2 + z_1 z_2 + z_2^3)$$

vanishes also linearly. It is clear that keeping the same degree of the polynom, a much better state could be obtained by requiring that the close approach of i and j is governed by the factor $(z_i - z_j)^3$ which vanishes much faster. Thus we are led to propose that the ground state of the 1/3-filled Landau level is the Laughlin state [236]

$$\Psi_{1/3} = \prod_{i<j} (z_i - z_j)^3 \mathcal{G} \,. \tag{12.104}$$

Since the largest power of z_1 is $3(N-1)$, the filling is indeed $\nu = N/3(N-1) \approx 1/3$ (for a large system).

Numerical evidence and analytic arguments show that $\Psi_{1/3}$ is very near to the true ground state. It is lucky that it is so because, had one missed making such an insightful guess, it would have been very difficult to find the ground state by a systematic approach. In terms of the one-electron basis (12.61), the N-electron ground state can be expressed as

$$|\Psi\rangle = \sum_{m_1,\ldots,m_N} a(m_1,\ldots,m_N) c_{m_1}^\dagger \ldots c_{m_N}^\dagger |0\rangle \tag{12.105}$$

and we are faced with a formidable eigenvalue problem of dimensionality C_N^{3N}, where the basis states are much more difficult to handle than the usual plane wave (or Bloch) states. The only constraint is the conservation of L^z which requires $\sum_{j=1}^N m_j = 3(N-1)N/2$. In contrast to $\Phi_{1/3}$ (which is one of the N-electron basis states), the Laughlin state is not a single-determinant state; in fact, it is a linear superposition of a very large number of Slater determinants. This is usually the case with optimally correlated many-electron states.

One way to grasp the uniqueness of the Laughlin state is this: all allowed N-electron states are homogeneous polynoms of degree $3(N-1)$, thus taken as functions of any of the coordinates z_j, they have $3(N-1)$

zeroes. For instance, taking $\Phi_{1/3}$ as a polynom of z_1, it has $N-1$ zeroes at $z_2, \ldots, z_N$, and $2(N-1)$ zeroes elsewhere; the latter are wasted from the point of view of keeping the electrons away from one another. The Laughlin state is unique inasmuch as it has all its zeroes at the positions of the electrons. If the interactions were of short range, one could do no better, and $\Psi_{1/3}$ would be the exact ground state. The fact that the actual interaction is of long range, should explain why $\Psi_{1/3}$ differs slightly from the true ground state.

Let us observe that $\Psi_{1/3}$ depends only on the differences of coordinates but not on the absolute positions. The Laughlin state describes a *homogeneous quantum fluid*[18]. For reasons which we discuss shortly, we may speak about an *incompressible quantum fluid.* – Using the expression "quantum fluid" may seem to suggest a degree of familiarity and a relationship to the long-studied helium quantum fluids, or BCS superconductors. Well, the relationship is extremely remote and every aspect of the FQHE is quite unfamiliar; the incompressible quantum fluids we are talking about constitute a family of startlingly novel phases of condensed matter.

What about the other magic values of ν: can we construct the corresponding ground states? The obvious recipe is to try $\mathcal{P} = \prod_{i<j}(z_i - z_j)^r$ with some other integer r. However, we have to ensure that $\mathcal{P}$ is a fully antisymmetrized many-fermion state[19]; this requires that $r = 2m+1$, an odd integer. The Laughlin states belong to $\nu = 1/3,\ 1/5,\ 1/7, \ldots$ Of these, only $\Psi_{1/3}$, and possibly $\Psi_{1/5}$, are useful as FQHE ground states; fillings around $1/7$ and below, are thought to belong to a (translational invariance breaking) Wigner crystal phase.

Fractionally Charged Quasiparticles

The fact that a Laughlin state can be constructed at $\nu = 1/3$, but not at nearby values of ν, shows that the 1/3-filled Landau level has

[18] It is a droplet of quantum fluid, since it has a circular edge (defined, in reality, by a confining potential). The statement about the uniform density holds well inside the droplet; within a distance $\sim l$ about the edge, the density falls to essentially zero. The behaviour at the edge is fundamentally different from the bulk behaviour, and it is important for the QHE.

[19] $\mathcal{G}$ is symmetric, so $\mathcal{P}$ must change sign if we interchange z_i and z_j.

a special property. Experiments tell us that it is the arising of the $\sigma_{yx} = e^2/3h$ quantum Hall plateau. However, the plateau has a (quite substantial) width in ν, thus it must be true that in a finite interval of ν, the behaviour is controlled by the presence of the 1/3 Laughlin state.

For ν differing slightly from 1/3, we will use a kind of a two-fluid picture: we consider the system as made up of the $\nu = 1/3$ quantum fluid, and a number of quasiparticles. We discuss quasielectrons and quasiholes from different points of view.

In the Laughlin state $\Psi_{1/3}$, the electrons keep their mutual distance because the relative angular momentum of any two particles is 3. This state is unique thus it is clear that, upon adding another electron to the bulk of the system[20], a relative angular momentum 1, and a closer approach of two electrons, must arise. The concomitant increase of the energy is a small fraction[21] of e^2/l. Thus the Laughlin ground state for N electrons is separated by a gap from the lowest bulk excitation with $N+1$ electrons. In contrast, there would be no such gap if we added an electron at the perimeter of the system where the average inter-electronic distance $l\sqrt{3}$ could be kept, giving rise to $\Psi_{1/3}$ with $N+1$ electrons.

Adding a single electron to a large system increases the density by an infinitesimal amount; the same could be achieved by slightly reducing the system area. Since an infinitesimal compression requires a finite energy, we may call the system incompressible[22]. The Laughlin states are *incompressible quantum fluids.*

We will be a bit more careful in discussing the *quasihole.* $\Psi_{1/3}$ is

---

[20]Here it is implied that the system is still fully spin-polarized, so that an electron with spin $\uparrow$ is added to the next Landau level. Since, in point of actual fact, the Zeeman splitting is much smaller than either $\hbar\omega_c$ or $e^2/\epsilon_0 l$, a lower-energy excitation could be constructed by adding a spin-$\downarrow$ electron to the lowest Landau level. We return to this point later.

[21]The numbers are $\epsilon_{qe} = 0.085e^2/l$ for quasielectrons and $\epsilon_{qh} = 0.0287e^2/l$ for quasiholes [254]. The important quantity is the gap $\epsilon_{qe} + \epsilon_{qh} = 0.114e^2/l$.

[22]Formally, the compressibility κ is defined by $\kappa^{-1} = -V(\partial P/\partial V) = V(\partial^2 E/\partial V^2)$, where V is the volume, P the pressure, and E the energy. For a system whose energy depends on N only through the density: $E = N\epsilon(n)$, where $n = N/A$ (A being the area of the system), $\kappa^{-1} = n^2 d\mu/dn$ follows, where $\mu = \partial E/\partial N$ is the chemical potential. The gap amounts to a discontinuity of the chemical potential, thus κ vanishes.

constructed to distribute N electrons over an area which is permeated by a total flux $3N\phi_0$. The precisely controlled correspondence between electrons and flux quanta makes it a natural idea that the Laughlin state can be excited also by the addition of a flux quantum. How to create an extra flux quantum, say, at the origin? We have to make sure that there is an area $2\pi l^2$ which is "not used" by any of the electrons. This is achieved by introducing the one-quasihole state

$$\Psi_{1/3}^{1\text{qh}} = \prod_j z_j \prod_{i<j} (z_i - z_j)^3 \,\mathcal{G}\,. \tag{12.106}$$

Indeed, the lowest power of any z_j is 1 (in contrast to the case of $\Psi_{1/3}$ where it was 0), thus the innermost Landau state is left unoccupied. Similarly, we may add a second and a third hole at the center of the system. The three-quasihole state is

$$\Psi_{1/3}^{3\text{qh}} = \prod_j z_j^3 \prod_{i<j} (z_i - z_j)^3 \,\mathcal{G}\,. \tag{12.107}$$

But this would be the $(N+1)$-electron Laughlin wave function, assuming that the $(N+1)$th electron is sitting at the origin! Consequently, it must be true that a single quasihole amounts to the absence of one-third of an electron. The quasi-hole charge is $e^*_{\text{qh}} = e/3$. Though we did not present the analogous argument for the quasielectron, we may infer that $e^*_{\text{qe}} = -e/3$.

The existence of *fractionally charged quasiparticles* is just the ingredient which we needed to extend the gauge argument to the FQHE (see (12.100)). Now we can draw up the scenario for the 1/3 plateau as follows: Strictly at $\nu = 1/3$, the system is an incompressible quantum liquid. Since the decay of a weak dc current would need the existence of low-lying excitations, and there are none, the system can move past obstacles (impurities, etc.) without dissipation, and it has the same Hall conductivity as a free-electron system would have $(e^2/3h)$. Changing ν slightly, we add quasielectrons (quasiholes) which become localized in the disorder potential. The quasiparticles make no contribution to the bulk current. However, fractional charge may be transported along the edge and, in the sense of (12.99), it gives rise to $\sigma_{yx} = e^2/3h$.

The quasihole is not a simple, local object. Writing down the wave-function of a quasihole situated at ζ

$$\Psi_{1/3}^{1\mathrm{qh}}(\zeta) = \prod_j (z_j - \zeta) \prod_{i<j} (z_i - z_j)^3 \, \mathcal{G} \tag{12.108}$$

we find that it depends, in addition to the variable ζ, also on all the electron coordinates as parameters. Nevertheless, most of the excess charge e^* is localized within a circle of radius $\sim l$.

Let us invoke a Gedankenexperiment to lend more credence to the idea of fractional charge [252]. We take the idea of adding a flux quantum literally: let us imagine that the plane is pierced by an infinitely thin solenoid at ζ, and drive a flux $\phi(t)$ through it, whereby $\phi(t)$ increases from 0 until it reaches ϕ_0. According to Faraday's law of induction, the time dependent flux induces an azimuthal electric field. At a distance R from ζ

$$E_\varphi = \frac{1}{c}\frac{d\phi}{dt}\frac{1}{2\pi R}. \tag{12.109}$$

In the incompressible Hall liquid state, σ_{xx} vanishes, and the current flows perpedicularly to the electric field, i.e., in radial direction

$$j_\rho = \sigma_{yx} E_\varphi = \frac{e^2}{3h}\frac{1}{c}\frac{d\phi}{dt}\frac{1}{2\pi R}. \tag{12.110}$$

The total charge which has been transferred towards the central point ζ is

$$Q = 2\pi R \int j_\rho dt = \frac{e^2}{3h}\frac{1}{c}\int \frac{d\phi}{dt} dt = \frac{e^2}{3h}\frac{1}{c}\phi_0 = \frac{e}{3}, \tag{12.111}$$

which reconfirms the contention that fractionally quantized Hall plateaus must be associated with fractionally charged quasiparticles.

Thinking of the solenoid as bounded by infinite potential walls, the field inside it is not directly visible to the electrons, and the whole operation may be regarded as a gauge transformation [232, 355]. However, for $0 < \phi < \phi_0$ it would drastically influence the states whose orbits surround ζ, because the change of phase acquired by passing around a closed loop $(2\pi/\phi_0)\oint \mathbf{A}\cdot d\mathbf{l}$ would not be compatible with the single-valuedness of the wavefunction. However, when ϕ reaches ϕ_0, the additional phase becomes 2π, which is alright[23]. The fact that each electron going around the flux tube gets 2π added to its phase, can be described as the result of multiplying the wavefunction by $\prod_j(z_j - \zeta)$. This is what we wrote down in (12.108).

It should be emphasized that "adding the flux ϕ_0 to the system" does not mean that we increased the magnetic field! The flux got increased at the expense of increasing the system size (by $2\pi l^2$), while we kept the field H strictly constant.

Is the fractional charge $e^* = e/3$ real? Up to this point, it may appear as a rather artificial theoretical construct introduced with the single aim of justifying $\sigma_{yx} = e^2/3h$. We should feel more convinced about its existence if $e/3$ turned up in other measurements as well. It is thus reassuring to find that $e^* = e/3$ has been found in an experimental study of the noise associated with inter-edge tunneling [347].

What are the quasiholes? Fermions, bosons, or something else? Well, statistics determines the change of phase of a two-particle wave function $\psi(x_1, x_2)$ when the particles are interchanged: there is a change of sign for fermions, but no change for bosons. These two possibilities can be summarized by introducing the statistical angle Θ, whereby $\psi(x_2, x_1) = e^{i\Theta}\psi(x_1, x_2)$. $\Theta = 2\pi$ for bosons, and $\Theta = \pi$ for fermions.

To find out about the Θ for quasiholes, we have to construct a two-quasihole wavefunction. In order to be able to effect the interchange of the quasiholes, we place them at different locations ζ_1 and ζ_2. Looking at (12.108), our first guess might be $\prod_j(z_j - \zeta_1)(z_j - \zeta_2)\mathcal{G}$, but on second thoughts, we feel inclined to modify it. The factor $\prod_j(z_j - \zeta_1)$ arose from the introduction of a flux quantum at ζ_1. Recalling the Gedankenexperiment, we may imagine that the field E_φ spinned up each electron, giving them an additional quantum of angular momentum with respect to the center ζ_1. Then, we repeat the process with another flux quantum at ζ_2. However, by that time the first quasihole is present in the system, so it should be spinned up, too! However, it has only a charge $e/3$, therefore it can be dragged around only by one-third of the force which

[23]At this point, the solenoid is imperceptible to the electrons, and can be removed. At the end of the Gedankenexperiment, we are left with the quasihole.

applies to the electrons, and it acquires one-third of the corresponding change of phase. We are led to infer the two-hole state [117]

$$\Psi^{2qh}_{1/3}(\zeta_1,\zeta_2) = (\zeta_2-\zeta_1)^{1/3}\prod_j (z_j-\zeta_1)(z_j-\zeta_2)\prod_{i<j}(z_i-z_j)^3\,\mathcal{G}\,. \quad (12.112)$$

(Actually, $\mathcal{G}$ should be modified as well, but we do not bother about that). Upon the interchange $\zeta_1 \leftrightarrow \zeta_2$, Ψ gets multiplied by $(-1)^{1/3} = e^{i\pi/3}$, the statistical angle is $\Theta = \pi/3$. Quasiholes can be characterized by *fractional statistics*. Though this an attractive outcome, it is not quite as compelling as the conclusion about the fractional charge. It is a peculiarity of the two-dimensional world that statistics is an ambiguous concept[24].

Jain States

We are still left with the problem of accounting for the FQHE at $\nu = 2/5$. The Laughlin states $\Psi_{1/(2n+1)}$ exhaust the store of correlated wave functions which can be formulated simply in terms of states from the lowest Landau level. However, we are well advised to stick to the Jastrow strategy[25]: we seek correlated states by applying a "correlator" to a simple reference state which is usually a single-determinant state. In fact, this was Laughlin's reasoning [236]: we start from the $\nu = 1$ state (12.102), and then enlarge the spacing between the electrons by introducing the correlator

$$\hat{C}_2 = \prod_{i<j}(z_i - z_j)^2\,. \quad (12.113)$$

(12.104) can be rewritten as $\Psi_{1/3} = \hat{C}_2\Psi_1$. – There are a number of ways to guess the form of the correlator. Early numerical work [235] made it clear that there is a relationship between the quantization of the relative distance and the relative angular momentum, and $\hat{C}_2$ achieves blowing up the system size to threefold. Alternatively, one can argue that since $\hat{C}_2$ multiplies a fermion wave function to yield another fermionic state, it

[24]However, the transformation from one statistics to another introduces long-range ("statistical") forces between the particles. See [208] for a popular introduction.

[25]The same strategy was applied in introducing the Gutzwiller trial state (9.29), and its generalizations (Sec. 10.6.2).

must look like a many-boson state built up as a product of even powers of $(z_i - z_j)$. $\hat{C}_2$ is the lowest-order homogeneous polynom of this kind. Naturally, the same procedure can be applied repeatedly but it leads to states which we already know: $\hat{C}_2\Psi_{1/3} = \Psi_{1/5}$, etc.

It was Jain's idea to use states with more than one filled Landau level as starting states [183]. Let us consider $\hat{C}_2\Psi_2$, where Ψ_2 consists of the filled levels $n = 0$ and $n = 1$. Having N electrons, we fill the first $N/2$ orbitals (12.61), and also the first $N/2$ orbitals (12.65). For large N, the largest power of any z_j is $N/2$. Applying $\hat{C}_2$, that power becomes $M = (N/2) + 2N = 5N/2$, and the filling factor is $\nu = N/M = 2/5$. Thus the prescription for constructing the homogeneous (fully spin-polarized) quantum liquid we associate with the $\nu = 2/5$ plateau is

$$\Psi_{2/5} = \hat{C}_2\Psi_2 \,. \tag{12.114}$$

Generally, starting from a state with n filled Landau bands

$$\Psi_{n/(2n+1)} = \hat{C}_2\Psi_n \,. \tag{12.115}$$

Thereby, we generated the fractions

$$\frac{n}{2n+1} = \frac{1}{3}, \frac{2}{5}, \frac{3}{7}, \ldots \tag{12.116}$$

which is the main sequence of the experimentally observed plateau values. In Fig. 12.3, we can follow the sequence up to 4/9, but an anomaly has been detected even at 9/19. Furthermore, it can be argued that weaker components of the electron–electron interaction necessitate the consideration of low-filling states in which the electrons are kept even farther apart. These can be generated by the application of the fourth-order correlator $\hat{C}_4 = \prod_{i<j}(z_i - z_j)^4$ which gives rise to the fractions

$$\frac{n}{4n+1} = \frac{1}{5}, \frac{2}{9}, \frac{3}{13}, \ldots \,. \tag{12.117}$$

$\nu > 1/2$ plateaus can be obtained by assuming an electron–hole symmetry within a single Landau level (it becomes exact at $H \to \infty$). Applying the transformation $\nu \to 1-\nu$ to (12.116), we get (with $n \geq 2$) the sequence

$$\frac{n}{2n-1} = \frac{2}{3}, \frac{3}{5}, \frac{4}{7}, \ldots \tag{12.118}$$

which is also prominent in Fig. 12.3. Let us note, however, that the presence of a fraction in either of the above series is not necessarily evidence that the description in terms of a fully spin-polarized state is correct. Fractions like 3/5 or 2/3 may be better handled by extending the method to states with incomplete spin polarization [446]. Even for our prime example 2/5, the maximum-spin state competes with the spin-singlet devised by Halperin [151]

$$\Psi = \prod_{i<j}(z_i^{\uparrow} - z_j^{\uparrow})^3 \prod_{k<l}(z_k^{\downarrow} - z_l^{\downarrow})^3 \prod_{i,k}(z_i^{\uparrow} - z_k^{\downarrow})^2 \, \mathcal{G}^{\uparrow}\mathcal{G}^{\downarrow} . \tag{12.119}$$

The construction is transparent: the correlation between separate 1/3 Laughlin states of $\uparrow$ and $\downarrow$ electrons is governed by a quadratic factor. A simple counting of powers confirms $\nu = 2/5$. The decision between (12.114) and (12.119) is not simple at all: we are going to see that in addition to the density–density interaction, exchange is also important. It may well happen that a transition between the low-spin and high-spin state can be induced by changing the g-factor [240].

Jain's construction has the pleasing aspect that it connects the FQHE to the IQHE. Indeed, looking at Fig. 12.3, we do not see any qualitative difference between integer and fractional plateaus: some are wider, others are narrower, but that is all. In the next section, we will learn about a more profound argument why the $\nu = 1$ plateau is in the same class as the $\nu = 1/3$ plateau. Here let it suffice that we map starting states whose incompressibility is manifest because it follows from kinetic energy quantization, onto fractionally occupied states which are consequently also incompressible. It turns out that the mapping can be extended also to small deviations from the magic fillings: the well-understood single-electron excitations of a filled Landau level map onto the quasi-electron excitations of the 1/3 state [183, 87]. The cyclotron energy gets scaled down to the size of the correlation gaps, but the gapfullness is preserved.

There is, however, a feature we should worry about: Jain's trial states are not really states of the lowest Landau level! Remember (12.65) which contained those z^*s which indicate that we are in an $n > 0$ level. After applying $\hat{C}_2$, these z^*s are still there, so we have an admixture of states from the $n = 1$ level. The formal device of getting rid of this is to project $\hat{C}_2\Psi_2$ onto the lowest Landau level, and strictly speaking, this step should always be included. However, we might fear that if this brings a major restructuring of the state, then the carefully designed correlations become lost. Fortunately, this is not the case: $\Psi_{2/5}$ as written down in (12.114) is almost completely within the lowest Landau level! An elementary argument as to how this can come about, may be the following: let us assume that in the expansion of $\Psi_{2/5}$, z_j appears in the form $z_j^* z_j^m$. This reminds us of the mth orbital of the $n = 1$ Landau level, but it is not quite that: combining (12.61) and (12.65), we find

$$z^* z^m = |m, 1\rangle + \sqrt{m}|m - 1, 0\rangle.$$

Since after "blowing up" the system with $\hat{C}_2$, the typical power is very large ($m \sim N$), the relevant one-electron states are almost entirely within the $n = 0$ level. A numerical study of the many-electron states of small systems [405] corroborates this conclusion.

It may look disquieting to refer to higher Landau levels if we want to construct a state within the lowest level, but it would be unjustified to think so. After all, also a fully Gutzwiller-projected state is quite unlike the parent Fermi sea. It is after the application of the correlators that the state has to be physically acceptable, not before.

It is worth recalling that Landau levels are indeed mixed in the physically relevant range of parameters. However, this is not the motivation for introducing the Jain states, and conversely, the small admixture from higher levels implied by the construction of these trial states is not meant to describe the true level mixing.

12.5 Discussion and Outlook

In the previous chapters, we acquired a certain feeling about the likely effects of electron–electron interactions. We understood that if the ratio $\langle\mathcal{H}_{\text{int}}\rangle/\langle\mathcal{H}^0_{\text{kin}}\rangle$ (where $\langle\mathcal{H}^0_{\text{kin}}\rangle$ is a suitable measure of the one-electron energy) exceeds a threshold of $O(1)$, and if the band filling is right (what this means, depends on the range and form of the interaction), then the system settles into a strongly correlated localized state such as a Mott insulator or a Wigner crystal. At the first sight, the problem of a 1/3-filled Landau level in the high-field limit looks a particularly easy one: the kinetic energy is "frozen out" by the magnetic field, and there remains only the repulsive Coulomb interaction which, one might have thought, is minimized by localizing the electrons onto the sites of a (triangular) Wigner crystal. In fact, prior to the discovery of the quantum Hall effect, it had been assumed that this would happen and one was looking forward to finding the first clear example of a Wigner crystal. Instead, there came the QHE. The idea of a Wigner crystal could immediately be disposed of on phenomenological grounds: an electron crystal would get pinned on impurities. To explain the QHE, one needs a gapfull electron liquid which can move past the obstacles.

The apparent contradiction with our previous conclusion (namely that $\langle \mathcal{H}_{\text{int}} \rangle / \langle \mathcal{H}^0_{\text{kin}} \rangle \to \infty$ should lead to collective electron localization) is resolved by realizing that the interaction is not purely diagonal in the Landau basis. In fact, just because using states from the lowest Landau level we cannot localize better than with an accuracy $\sim l$, the Coulomb force has strong non-diagonal components. So to speak, the interaction provides both kinetic and potential energy terms[26], and for a broad range of fractional filling, the kinetic energy pieces predominate. The outcome is an incompressible quantum fluid which is strongly correlated in the sense of (12.104), but it does not break translational invariance.

Before embarking upon a more detailed discussion of two specific issues (the Quantum Hall phase diagram, and the role of the spin degree of freedom), let us briefly address the question of dimensionality. We have hitherto assumed that the system is strictly two-dimensional ($2D$). Is this essential, or can one observe QHE-related phenomena also in systems of weakly coupled planes? We should be inclined to say yes. Let us recall our discussion of systems of weakly coupled spin chains (see p. 299*ff*): we learned that while gapless states are unstable against infinitesimal additional couplings, gapfull states (such as the Haldane phase) have a certain stability. QHE is associated with gaps, thus it may survive switching on the inter-plane coupling. There are, in fact, at least two experimental realizations. Double-layer devices are

[26]Let us remember our previous discussion of the extended Hubbard model in Sec. 5.3, where we found that the interaction gives rise to assisted, and two-electron, hopping terms.

manufactured to contain two weakly coupled $2D$ electron systems and show interesting QHE phenomena [253]. Another relevant class of systems can be envisaged as $3D$ arrays of chains which are weakly coupled into planes which are, in turn, coupled negligibly. For instance, the quasi-$1D$ conductor $(TMTSF)_2PF_6$ which we cited as an SDW-bearing system (p. 374), becomes quasi-$2D$ under pressure and shows a sequence of Quantum Hall states [36]. The interplay of SDW and Landau gaps raises many interesting issues (see the short review [450]).

Phase Diagram

We have met a bewildering variety of quantum fluids and we are still far from having finished with them. When and why do we find them, what is their relationship to each other? It would be helpful to arrange them in a phase diagram.

Let us begin with a question. Let us assume that $\nu = 0.6$. Which Hall plateau do we expect? If the system is sufficiently disordered so that only IQHE is seen, $\sigma_{yx} = e^2/h$. However, in a better sample, 0.6 should belong to the "domain of attraction" of the 2/3 plateau. In a still better sample, it is dead at the center of the 3/5 plateau. Clearly, we need a variable $\mathcal{D}$ which characterizes the degree of disorder. It would be difficult to say precisely[27] what $\mathcal{D}$ is but it should express that the $\sigma_{yx} = e^2/h$ plateau is the most robust of all, that $\sigma_{yx} = e^2/3h$ is more robust than either $\sigma_{yx} = 2e^2/h$ or $\sigma_{yx} = 2e^2/5h$, etc. Such a phase diagram is shown in Fig. 12.8. The horizontal axis is $H \propto \nu^{-1}$.

The diagram is not an arbitrary graphical arrangement of the known phases; its construction is based on the idea of *corresponding states* [185, 211]. In fact, we already know this idea: we used (12.115) to establish a correspondence between the integer plateaus 1, 2, and 3 and the fractional plateaus 1/3, 2/5, and 3/7. Now we extend this procedure

[27] This is as yet a conceptual, rather than a real, phase diagram! Changing H (or ν) in a real sample, we do not move on a horizontal line but follow a complicated path because $\mathcal{D}$ has to be readjusted in a suitable manner [211]. $\mathcal{D}$ is often suggestively denoted as $\rho_{xx}^{(0)}$ but it cannot be understood as the actual ρ_{xx} because that would not be a finite number at $T = 0$.

to arbitrary filling: correlating the starting state Ψ_ν with $\hat{C}_2$ yields

$$\Psi_{\bar{\nu}} = \hat{C}_2 \Psi_\nu \,, \tag{12.120}$$

where

$$\bar{\nu} = \frac{\nu}{2\nu + 1} \,. \tag{12.121}$$

Previously, we mapped isolated points (the midpoints of plateaus); now we map whole intervals, i.e., one incompressible quantum fluid phase (Hall liquid) into another. The procedure generates an infinite series of phases on both sides of $\nu = 1/2$, $1/4$, ..., which Fig. 12.8 cannot show.

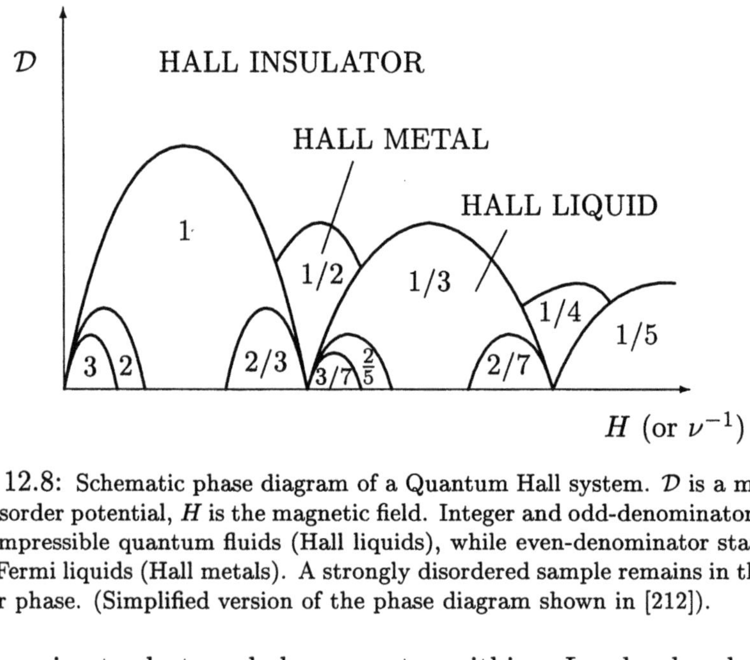

Figure 12.8: Schematic phase diagram of a Quantum Hall system. $\mathcal{D}$ is a measure of the disorder potential, H is the magnetic field. Integer and odd-denominator states are incompressible quantum fluids (Hall liquids), while even-denominator states are gapless Fermi liquids (Hall metals). A strongly disordered sample remains in the Hall insulator phase. (Simplified version of the phase diagram shown in [212]).

Approximate electron–hole symmetry within a Landau band allows to deduce the existence of a 2/3 Hall liquid. We know from experiments that it is by far not as robust as the 1/3 liquid, and this is reflected in our drawing. Once the 2/3 phase is given, it can be used as the starting point of (12.121), giving rise to the 2/7 Hall liquid, etc.

The finite-$\mathcal{D}$ phase boundaries are rather arbitrarily drawn, but we can make some definite statements about the limit[28] $\mathcal{D} \to 0$. First

[28]Strictly at $\mathcal{D} = 0$ (in the absence of any disorder) even an interacting electron

of all, let us emphasize that we are speaking about critical values of ν where we switch from one plateau to another. In Figs. 12.2 and 12.3, we saw intervals in between the plateaus, where ρ_{yx} is changing continuously. This is a finite-T effect, and there is evidence that the width of the transition regions vanishes according to a power law: $\Delta H \sim T^x$. This allows to raise the question: what is the critical filling factor $\nu_{\rm cr}^{1/3\to 2/5}$? A naive guess would be the arithmetic mean $((1/3)+(2/5))/2 \approx 0.3667$. However, the idea of corresponding states leads to another prediction. Namely, the transition should be the image of the IQHE transition $1 \to 2$, which surely happens at $\nu = 3/2$, so (12.121) gives $(3/2)(2(3/2)+1)=3/8=0.375$. The two predictions are pretty close; can experiment decide? Yes, it can: the measured value is 0.3745 ± 0.0012, vindicating our present scheme [132].

Theoretically, a sample with infinitesimal disorder should go through an infinite squence of incompressible fluids as ν is raised from 1/3 to 1/2 (the same happens coming from $\nu = 2/3$ via the electron–hole analogs). The ground state in the point of accumulation $\nu = 1/2$ should be an exotic beast and so it turns out: it is a *compressible* quantum fluid, very much like a Fermi liquid with a Fermi surface (see [377] for a brief review). We do not attempt to explain this in any detail. Nevertheless, just to get a glimpse of how a Fermi liquid may arise, we have to mention that it is a very fruitful idea to interpret the sequence of Jain states (12.116) as states of filled Landau bands of *composite fermions*[29]. Attaching the correlation factor $(z_i - z_j)^2$ to a wave function means that when the electron i passes along a loop around electron j, it acquires an extra phase $2 \cdot 2\pi$. We have argued that a phase change 2π can be induced by a thin flux tube carrying ϕ_0, thus here we say that C_2 acts as if two flux tubes had been attached to each electron. Thus the $\nu = 1/3$ state of electrons can be looked at as the $\nu' = 1$ state of (1 electron + 2 flux tube) composites, and the correlation gap can be identified as the cyclotron energy of the composite fermions (CF). The

---

system would slide freely and we would not see a QHE at all. However, the incompressible states are still there, even though they are not accompanied by plateaus. In any case, the plateaus appear at $\mathcal{D} \to 0$ but the infinitesimal disorder can be neglected in the discussion of the ground state.

[29]This was an essential motivation for introducing the states (12.115) [183].

CFs see only one ϕ_0 per particle as an external field because the rest of the flux piercing the sample has become part of the particles.

We carry over the idea to states with $\nu = n/(2n+1)$, which are pictured to contain n Landau bands filled with CFs. From the original flux $H \cdot 2\pi l^2/\nu = \phi_0/\nu$ per electron, $2\phi_0$ is defined into the CFs. The rest of the flux $(1/\nu - 2)\phi_0 = \phi_0/n$ corresponds to a residual external field H_{res} which governs the behaviour of the CFs. At the $\nu = n/(2n+1)$ plateau, CFs see a residual field $H_{\text{res}} \propto H/n$. The same holds for their cyclotron energies, and consequently, the correlation gap of the FQHE states should vanish as $n \to \infty$. This is indeed observed[30] [93]. At $\nu = 1/2$ $(n \to \infty)$, the composite fermions see no external field at all, and form a Fermi liquid. This is the origin of the "Hall metal" phase which we show in Fig. 12.8. The transformation (12.120)–(12.121) generates another Hall metal at 1/4, etc. This is also confirmed by experiment.

At the first sight, the (integer-charged) composite fermions appear even more artificial than the fractionally charged quasiparticles[31]. However, experiments show that they are amazingly real [377]. If the effective external field is really that weak, one should observe ballistic transport in a dilute gas of composite fermions; and indeed one does.

It is worth emphasizing that, in spite of its phenomelogical similarity to a free-electron metal, the $\nu = 1/2$ Fermi liquid is an even more exotic state of matter than the incompressible quantum fluid states are. For the latter, the Laughlin and Jain wave functions give a pretty good idea about the structure of the ground state. We do not have a comparable insight into the nature of the Hall metal.

At sufficiently large levels of disorder, the system is a Hall insulator characterized by the absence of Hall plateaus, and an activated σ_{xx}. Far to the right of the phase diagram (beyond 1/5), one expects a Wigner crystal ground state at $\mathcal{D} = 0$. It is believed that even a small disorder is enough to break up the Wigner crystal into a random array of crys-

---

[30] Experimentally, we have a sample with a given electron density, and the external field is swept in an interval containing $H_{\nu=1/2}$.

[31] These are not conflicting concepts! We use the CFs to construct the ground state at ideal fillings $\nu = n/(2n+1)$. The excitations above these ground states carry the fractional charge $\pm e/(2n+1)$.

tallites, so that it merges smoothly with the disordered Hall insulator which we find at larger ν and high $\mathcal{D}$.

Finally, let us remember that we have confined the discussion to fully spin-polarized states. A complete phase diagram should incorporate partially polarized ground states as well.

Spin Effects

All our previous discussion relied on using the high-field limit where both $(e^2/\epsilon_0 l)/\hbar\omega_c$ and $(e^2/\epsilon_0 l)/g\mu_B H$ are very small. We have to clarify whether, in reality, inter-Landau-level or inter-Zeeman-level mixings are important enough to change our picture of QHE systems drastically.

We have already quoted (on p. 681) the relevant parameters for bulk GaAs, which show that band structure effects enhance $\hbar\omega_c$ by at least a factor of ten, and at the same time suppress $g\mu_B H$ substantially. We may feel confident that inter-Landau-level mixing can be well described by the first few terms of a systematic $1/\sqrt{H}$ perturbation theory [370]. On the other hand, it is not true that $(e^2/\epsilon_0 l)/g\mu_B H$ is a small parameter. Quite on the contrary; it is held that in a typical high-field experiment, $\hbar\omega_c \approx 150\text{K}$, $e^2/\epsilon_0 l \approx 100\text{K}$, while the (bare) Zeeman splitting is $g\mu_B H \approx 2\text{K}$ [253]. Let us remark that g is smaller in heterostructures than in bulk GaAs, and it may even be tuned[32] through zero by choosing a suitable composition and applying pressure [267, 240].

It appears that spin-flip is a very low-lying excitation, and we would expect that $g\mu_B H \approx 2\text{K}$ shows up in transport measurements as a correspondingly small activation energy of σ_{xx}. However, this is not the case: measured activation energies are about ten times as large [417]. The observation can be interpreted as a huge renormalization of g which must be an effect of electron–electron interactions: it is an exchange enhancement. It follows that the observed essentially complete spin polarization of the $\nu = 1/3$ FQHE state, and even of the $\nu = 1$ IQHE state, is sustained mainly by interactions. We may say that these are *Quantum*

[32]The ratio $g\mu_B H/\hbar\omega_c$ can also be tuned by tilting the magnetic field: the component perpendicular to the plane defines ω_c while the total field appears in the Zeeman splitting. Though it sounds easy, the method has difficulties of its own.

Hall ferromagnets (QHFM) with a saturated spontaneous moment and the external field (which we previously credited with driving the system to full spin alignment) plays a merely subsidiary role by picking the direction of spin polarization. It was a startling development to find that at least some of the major QHE plateaus are associated with itinerant ferromagnetism[33].

It need not be true that all incompressible quantum fluid states are at the same time fully polarized itinerant ferromagnets, but this is thought to be the case at least for $\nu = 1$ and $\nu = 1/3$. It is, in any case, interesting to find a common mechanism behind an integer and a fractional plateau, which lends further justification to the view that IQHE and FQHE should be understood in a unified framework. Much work has been done on the $\nu = 1$ QHFM for which we cite a few results.

It is found that the exchange hole effect manages to keep the system fully polarized for all (bare) $g > 0$ and consequently, we may think of the ground state in terms of the simple (essentially spinless fermion) picture we used in the previous sections. However, the nature of the excitations changes dramatically as g is decreased. For large enough g, the lowest quasielectron excitation corresponds to putting an electron into the next Landau level; it is a simple localized $S^z = 1/2$ object. For lower g, it is becoming more and more favourable to compose the excitation from reversed-spin states. At the same time, the system is approaching the limit of an isotropic two-dimensional ferromagnet which can support topologically non-trivial vortex-like excitations called skyrmions[34]. In the original sense, skyrmions are spatially unbounded, infinite-spin excitations of an isotropic model. However, in the QHE context, one may speak about skyrmions also for sufficiently small $g > 0$, apparently including the values which are realized in GaAs heterostructures [370].

Let us visualize skyrmions. Assume that the external field is $\uparrow$, so that in the $\nu = 1$ ground state, all spins are $\downarrow$. Now create a single

[33]It should not be forgotten that we have in mind essentially isotropic ferromagnets so that at $T > 0$, the Mermin–Wagner theorem forbids spontaneous ordering. However, $k_B T$ being apparently much smaller than the exchange interaction, the system is almost ferromagnetic, and the external field takes care of the surviving long-wavelength fluctuations. The temperature dependence of the spin polarization is a difficult issue.

[34]Named after T.H.R. Skyrme who studied a related field theoretical model [367].

excitation. For independent particles, it would be a single upturned spin which we may place at the centre. Allowing for interactions, the added particle becomes a "charged spin-texture excitation" [114]. In the centre the spin is $\uparrow$, far out at the perimeter it is $\downarrow$, while at some intermediate distance it is lying in the x–y plane, turning around in a vortex-like fashion. This spin pattern must arise from an effective field oriented like $xS^x + yS^y \rightsquigarrow zS^+ + z^*S^-$. Thus when the spin is turned up (down), the orbital momentum is decreased (increased) by one step. A Hartree–Fock state with this property is

$$\psi_{\text{Skyr}} = \prod_m (u_m c^\dagger_{m,\uparrow} + v_m c^\dagger_{m+1,\downarrow})|0\rangle\,. \tag{12.122}$$

The free electron limit is recovered by $u_m = 0$ for $m \geq 1$. Generally, we expect that u_m/v_m decreases continuously as we move out from the center. – Postulating a vortex rotating in the opposite direction ($xS^x - yS^y \rightsquigarrow zS^- + z^*S^+$), we can create an anti-skyrmion

$$\psi_{\text{a.-Skyr}} = \prod_m (-u_m c^\dagger_{m+1,\uparrow} + v_m c^\dagger_{m,\downarrow}) c^\dagger_{0,\uparrow}|0\rangle\,. \tag{12.123}$$

Using (12.122) and (12.123) as variational states, it turns out that for a system with the heterostructure parameters, a skyrmion (anti-skyrmion) with charge $-e$ (e) contains 3–4 overturned spins [114].

Skyrmions are common to all $D = 2$ ferromagnets; however, they are usually high-energy excitations. QHFMs are exceptional in two respects: the skyrmions carry charge, and the lowest-lying charge-carrying excitation is the creation of a skyrmion–anti-skyrmion pair [370].

Skyrmions can appear even at $T = 0$. Taking away (adding) an electron from (to) the filled Landau level amounts to creating an anti-skyrmion (skyrmion). This scenario implies that the system looses its spin polarization rapidly if we change ν from 1. For non-interacting electrons, full polarization is possible if $\nu \leq 1$, while the degree of polarization is $(2-\nu)/\nu$ if $\nu > 1$. Switching on the interactions, let us assume that the removal of a charge $-e$ is accompanied by turning over A spins; then the ground state polarization is $(1-(1-\nu)A)/\nu$ for $\nu < 1$. Similarly, if adding a charge $-e$ goes with turning over S spins, the polarization is $(1 - S(\nu - 1))/\nu$. The spin polarization can be determined

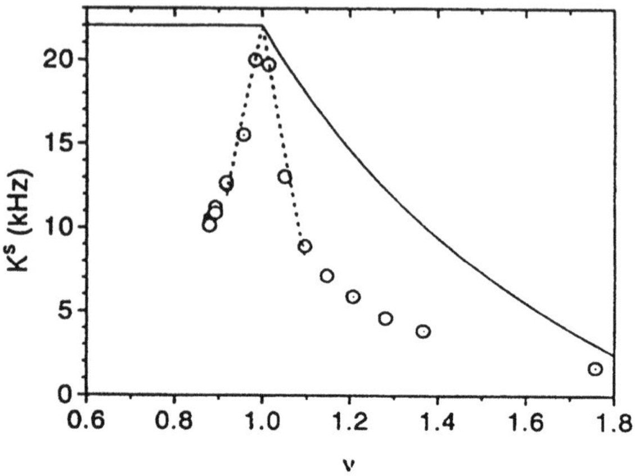

Figure 12.9: The Knight shift data on a GaAs quantum well device are normalized to give 1 for the free electron polarization at $\nu \leq 1$. The measured values differ markedly from the free electron prediction (dashed line), and indicate that the skyrmion contains 3.6 overturned spins (after [44]).

in an NMR experiment by measuring the Knight shift. Fig. 12.9 shows the results of such a measurement for a GaAs device [44]. It is apparent that depolarization is fast on both sides of $\nu = 1$; a fit with the assumption $A = S$ gives $A \approx 3.6$. Ascribing an effective spin $3.6/2$ to the excitations is an impressive proof of the validity of the skyrmion idea.

Appendix A

Hydrogen Atom

A.1 Hydrogenic Wave Functions

We give a list of the special functions appearing in the solution (2.2) of the hydrogen atom problem. Introducing the notation $\alpha = (Z/a_0)^{3/2}$, and using $\tilde{r} = Zr/a_0$ as variable, the radial wave functions are

$$R_{10}(\tilde{r}) = 2\alpha e^{-\tilde{r}} \tag{A.1}$$

$$R_{20}(\tilde{r}) = \frac{\alpha}{\sqrt{2}}\left(1 - \frac{\tilde{r}}{2}\right) e^{-\tilde{r}/2} \tag{A.2}$$

$$R_{21}(\tilde{r}) = \frac{\alpha}{2\sqrt{6}}\,\tilde{r}e^{-\tilde{r}/2} \tag{A.3}$$

$$R_{30}(\tilde{r}) = \frac{2\alpha}{3\sqrt{3}}\left(1 - \frac{2\tilde{r}}{3} + \frac{2\tilde{r}^2}{27}\right) e^{-\tilde{r}/3} \tag{A.4}$$

$$R_{31}(\tilde{r}) = \frac{4\sqrt{2}\alpha}{27\sqrt{3}}\,\tilde{r}\left(1 - \frac{\tilde{r}}{6}\right) e^{-\tilde{r}/3} \tag{A.5}$$

$$R_{32}(\tilde{r}) = \frac{2\sqrt{2}\alpha}{81\sqrt{15}}\,\tilde{r}^2 e^{-\tilde{r}/3}. \tag{A.6}$$

The angular dependence is described by the spherical harmonics Y_l^m. For the isotropic s-states $Y_0^0 = 1/\sqrt{4\pi}$, just a constant. For the p-states

$$Y_1^{\pm 1} = \mp\frac{1}{2}\sqrt{\frac{3}{2\pi}}\sin\vartheta e^{\pm i\varphi} \tag{A.7}$$

$$Y_1^0 = \frac{1}{2}\sqrt{\frac{3}{\pi}}\cos\vartheta. \tag{A.8}$$

For the d-states

$$Y_2^{\pm 2} = \frac{1}{8}\sqrt{\frac{30}{\pi}}\sin^2\vartheta e^{\pm 2i\varphi} \tag{A.9}$$

$$Y_2^{\pm 1} = \mp\frac{1}{4}\sqrt{\frac{30}{\pi}}\sin\vartheta\cos\vartheta e^{\pm i\varphi} \tag{A.10}$$

$$Y_2^0 = \frac{1}{4}\sqrt{\frac{5}{\pi}}(3\cos^2\vartheta - 1). \tag{A.11}$$

Let us note that multiplying and dividing by r^2, the same functions can be expressed as

$$Y_2^{\pm 2} = \frac{1}{8}\sqrt{\frac{30}{\pi}}\frac{(x \pm iy)^2}{r^2} \tag{A.12}$$

$$Y_2^{\pm 1} = \mp\frac{1}{4}\sqrt{\frac{30}{\pi}}\frac{z(x \pm iy)}{r^2} \tag{A.13}$$

$$Y_2^0 = \frac{1}{4}\sqrt{\frac{5}{\pi}}\frac{3z^2 - r^2}{r^2}. \tag{A.14}$$

Taking, for example, the $m = 0$ component of a $3d$ state

$$\psi_{320} = R_{32}Y_{20} \sim (3z^2 - r^2)e^{-Zr/3a_0} \tag{A.15}$$

we find that it is $(3z^2 - r^2)$-like.

The $l = 3$ spherical harmonics are

$$Y_3^{\pm 3} = \mp 15\sqrt{\frac{35}{\pi}}\sin^3\vartheta e^{\pm 3i\varphi} \tag{A.16}$$

$$Y_3^{\pm 2} = \frac{1}{8}\sqrt{\frac{210}{\pi}}\sin^2\vartheta\cos\vartheta e^{\pm 2i\varphi} \tag{A.17}$$

$$Y_3^{\pm 1} = \mp\frac{1}{8}\sqrt{\frac{21}{\pi}}\sin\vartheta(5\cos^2\vartheta - 1)e^{\pm i\varphi} \tag{A.18}$$

$$Y_3^0 = \frac{1}{4}\sqrt{\frac{7}{\pi}}(5\cos^3\vartheta - 3\cos\vartheta)\,, \tag{A.19}$$

while for $l = 4$

$$Y_4^{\pm 4} = \frac{3}{32}\sqrt{\frac{70}{\pi}}\,\frac{(x \pm iy)^4}{r^4} \tag{A.20}$$

$$Y_4^{\pm 3} = \mp\frac{3}{8}\sqrt{\frac{35}{\pi}}\,\frac{z(x \pm iy)^3}{r^4} \tag{A.21}$$

$$Y_4^{\pm 2} = \frac{3}{16}\sqrt{\frac{10}{\pi}}\,\frac{(7z^2 - r^2)(x \pm iy)^2}{r^4} \tag{A.22}$$

$$Y_4^{\pm 1} = \mp\frac{3}{16}\sqrt{\frac{20}{\pi}}\,\frac{(7z^3 - 3r^2 z)(x \pm iy)}{r^4} \tag{A.23}$$

$$Y_4^{0} = \frac{3}{16\sqrt{\pi}}\,\frac{35z^4 - 30z^2r^2 + 3r^4}{r^4}. \tag{A.24}$$

A.2 J-Eigenstates

We use the notation $|l, m\rangle = Y_l^m$ for the orbital angular momentum states, and $|\uparrow\rangle$, $|\downarrow\rangle$ for the spin. The action of the angular momentum raising (lowering) operators $L^{\pm} = L^x \pm iL^y$ is described by

$$L^{\pm}|l, m\rangle = \sqrt{(l \pm m + 1)(l \mp m)}|l, m \pm 1\rangle\ . \tag{A.25}$$

Introducing the spin–orbit coupling requires finding the eigenstates of the total angular momentum $\mathbf{J}^2$, and its z-component J^z. In fact, $\mathbf{J}^2$, J^z, $\mathbf{L}^2$, and $\mathbf{S}^2$ commute with one another and the standard basis is $|J, J^z, L, S\rangle$. In our case, $S = 1/2$ can be omitted and we use the notation $|J = l \pm 1/2, J = j^z, L = l, S = 1/2\rangle = |l \pm 1/2, j^z, l\rangle$. Obviously, $|l+1/2, l+1/2, l\rangle = |l, l\rangle|\uparrow\rangle$. To find $|l+1/2, l-1/2, l\rangle$, we apply $J^- = L^- + S^-$ to $|l+1/2, l+1/2, l\rangle$

$$\begin{aligned} |l + 1/2, l - 1/2, l\rangle &= (L^- + S^-)|l, l\rangle|\uparrow\rangle \\ &\rightsquigarrow \sqrt{\frac{2l}{2l+1}}|l, l-1\rangle|\uparrow\rangle + \sqrt{\frac{1}{2l+1}}|l, l\rangle|\downarrow\rangle \end{aligned} \tag{A.26}$$

where in the last step a normalizing factor was inserted.

$|l-1/2, l-1/2, l\rangle$ must be a linear combination of the same two component states that we see in $|l+1/2, l-1/2, l\rangle$, and it must be orthogonal to it, thus

$$|l-1/2, l-1/2, l\rangle = -\sqrt{\frac{1}{2l+1}}|l, l-1\rangle|\uparrow\rangle + \sqrt{\frac{2l}{2l+1}}|l, l\rangle|\downarrow\rangle. \quad \text{(A.27)}$$

Further states can be calculated similarly. It can be proven by full induction that

$$|l \pm 1/2, j^z, l\rangle = \alpha_{\pm}|l, j^z - 1/2\rangle|\uparrow\rangle + \beta_{\pm}|l, j^z + 1/2\rangle|\downarrow\rangle \quad \text{(A.28)}$$

where

$$\alpha_{\pm} = \pm\sqrt{\frac{l \pm j^z + 1/2}{2l+1}} \quad \text{and} \quad \beta_{\pm} = \sqrt{\frac{l \mp j^z + 1/2}{2l+1}}. \quad \text{(A.29)}$$

Appendix B

Single-Spin-Flip Ansatz

Let us start with the N-electron $\uparrow$-spin Nagaoka state $|N\rangle$, and examine whether it is stable against local disturbances, such as the turning over of a single spin. Shastry, Krishnamurty and Anderson [359] suggested using the following single-spin-flip variational trial state

$$|\Psi\rangle = \hat{P} c^{\dagger}_{\mathbf{q}\downarrow} c_{\mathbf{k}_F\uparrow} |N\rangle \, . \tag{B.1}$$

An electron is taken from the Fermi level, and placed at the bottom of the $\downarrow$-spin band with wave vector $\mathbf{q}$. The resulting state may contain a doubly occupied site. The Gutzwiller projector $\hat{P}$ serves to optimize the amplitude of double occupation.

The case of a single $\downarrow$-spin electron in an $\uparrow$-spin Fermi sea is special because it allows the exact evaluation of expectation values in a relatively straightforward manner[1]. Expanding $c^{\dagger}_{\mathbf{q}\downarrow}$ in Wannier basis, we find

$$\begin{aligned} |\Psi\rangle &= \prod_{\mathbf{h}} [1 - (1-\eta)\hat{n}_{\mathbf{h}\uparrow}\hat{n}_{\mathbf{h}\downarrow}] \cdot \frac{1}{\sqrt{L}} \sum_{\mathbf{g}} e^{i\mathbf{q}\mathbf{g}} c^{\dagger}_{\mathbf{g}\downarrow} |N-1\rangle \\ &= \frac{1}{\sqrt{L}} \sum_{\mathbf{g}} e^{i\mathbf{q}\mathbf{g}} [1 - (1-\eta)\hat{n}_{\mathbf{g}\uparrow}\hat{n}_{\mathbf{g}\downarrow}] \, c^{\dagger}_{\mathbf{g}\downarrow} |N-1\rangle \end{aligned}$$

[1]Remember that we resorted to using the Gutzwiller approximation when treating the case of a finite ratio $N_{\uparrow}/N_{\downarrow}$. An exact treatment of this more general problem is, in principle, also possible [271] but far too complicated for our purposes.

$$= \frac{1}{\sqrt{L}} \sum_{\mathbf{g}} e^{i\mathbf{q}\mathbf{g}} c^{\dagger}_{\mathbf{g}\downarrow} \left[1 - (1-\eta)\hat{n}_{\mathbf{g}\uparrow}\right] |N-1\rangle \tag{B.2}$$

where $|N-1\rangle$ is the lowest-energy state of $(N-1)$ $\uparrow$-spin electrons. The norm of $|\Psi\rangle$ is

$$\begin{aligned} \langle\Psi|\Psi\rangle &= \frac{1}{L}\sum_{\mathbf{g}} \langle N-1|[1-2(1-\eta)\hat{n}_{\mathbf{g}\uparrow} + (1-\eta)^2\hat{n}^2_{\mathbf{g}\uparrow}]|N-1\rangle \\ &= 1-(1-\eta^2)n_{\uparrow} \end{aligned} \tag{B.3}$$

It is straightforward to show that

$$\langle\Psi|\mathcal{H}_U|\Psi\rangle = U\eta^2 n_{\uparrow} \,. \tag{B.4}$$

To get the kinetic energy of the $\downarrow$-spin electron

$$\mathcal{K}_{\downarrow} = -t \sum_{\langle \mathbf{j},\mathbf{h}\rangle} \frac{\langle\Psi|c^{\dagger}_{\mathbf{j}\downarrow}c_{\mathbf{h}\downarrow}|\Psi\rangle}{\langle\Psi|\Psi\rangle} \tag{B.5}$$

we have to calculate the hopping amplitude

$$\begin{aligned} \langle\Psi|c^{\dagger}_{\mathbf{j}\downarrow}c_{\mathbf{h}\downarrow}|\Psi\rangle &= \frac{1}{L}\sum_{\mathbf{g}_1}\sum_{\mathbf{g}_2} e^{i\mathbf{q}(\mathbf{g}_2-\mathbf{g}_1)}\langle N-1|[1-(1-\eta)\hat{n}_{\mathbf{g}_1\uparrow}] \\ &\quad c_{\mathbf{g}_1\downarrow}c^{\dagger}_{\mathbf{j}\downarrow}c_{\mathbf{h}\downarrow}c^{\dagger}_{\mathbf{g}_2\downarrow}[1-(1-\eta)\hat{n}_{\mathbf{g}_2\uparrow}]|N-1\rangle \\ &= \frac{1}{L}e^{i\mathbf{q}(\mathbf{h}-\mathbf{j})}\{[1-2(1-\eta)n_{\uparrow}] \\ &\quad +(1-\eta)^2\langle N-1|\hat{n}_{\mathbf{j}\uparrow}\hat{n}_{\mathbf{h}\uparrow}|N-1\rangle\} \,. \end{aligned} \tag{B.6}$$

Using Wick's theorem

$$\langle N-1|\hat{n}_{\mathbf{j}\uparrow}\hat{n}_{\mathbf{h}\uparrow}|N-1\rangle = n^2_{\uparrow} - f^2_{\mathbf{jh}} \tag{B.7}$$

where $f_{\mathbf{jh}} = \langle N-1|c^{\dagger}_{\mathbf{j}\uparrow}c_{\mathbf{h}\uparrow}|N-1\rangle$ is an element of the density matrix, which is related to the (kinetic) energy density of the Nagaoka state via $f_{\mathbf{jh}} = E_{\text{Nag}}/(-zt)$ (for a macroscopic system, the energy density is the same for N as for $(N-1)$ particles). All in all, we find

$$\mathcal{K}_{\downarrow} = \frac{\epsilon(\mathbf{q})}{\langle\Psi|\Psi\rangle}\left\{[1-(1-\eta)n_{\uparrow}]^2 - (1-\eta)^2\left(\frac{E_{\text{Nag}}}{zt}\right)^2\right\} \tag{B.8}$$

where $\langle\Psi|\Psi\rangle$ is given by (B.3). For $U \to \infty$ ($\eta \to 0$), we get the result

$$\mathcal{K}_\downarrow = \epsilon(\mathbf{q})(1 - n_\uparrow)\left\{1 - \left(\frac{E_{\text{Nag}}}{Lzt(1-n_\uparrow)}\right)^2\right\} \tag{B.9}$$

which is easy to interpret. In order to move, the $\downarrow$-spin electron has to find a neighbouring site empty, which has the probability $1 - n_\uparrow$. However, the site of the $\downarrow$-spin electron is also a hole site, as far as the $\uparrow$-spin electron sea is concerned. The factor in curly brackets is the nearest-neighbour hole–hole correlation function of the $\uparrow$-spin electron sea (see Problem 9.2).

To calculate the kinetic energy of the $\uparrow$-spin electrons

$$\mathcal{K}_\uparrow = -t \sum_{\langle \mathbf{j},\mathbf{h}\rangle} \frac{\langle\Psi|c^\dagger_{\mathbf{j}\uparrow}c_{\mathbf{h}\uparrow}|\Psi\rangle}{\langle\Psi|\Psi\rangle} \tag{B.10}$$

we need

$$\begin{aligned}\langle\Psi|c^\dagger_{\mathbf{j}\uparrow}c_{\mathbf{h}\uparrow}|\Psi\rangle &= \frac{1}{L}\sum_{\mathbf{g}_1}\sum_{\mathbf{g}_2} e^{i\mathbf{q}(\mathbf{g}_2-\mathbf{g}_1)}\langle N-1|[1-(1-\eta)\hat{n}_{\mathbf{g}_1\uparrow}] \\ &\quad c_{\mathbf{g}_1\downarrow}c^\dagger_{\mathbf{j}\uparrow}c_{\mathbf{h}\uparrow}c^\dagger_{\mathbf{g}_2\downarrow}[1-(1-\eta)\hat{n}_{\mathbf{g}_2\uparrow}]|N-1\rangle \qquad \text{(B.11)} \\ &= \frac{1}{L}\sum_{\mathbf{g}}\langle N-1|[1-(1-\eta)\hat{n}_{\mathbf{g}\uparrow}]c^\dagger_{\mathbf{j}\uparrow}c_{\mathbf{h}\uparrow}[1-(1-\eta)\hat{n}_{\mathbf{g}\uparrow}]|N-1\rangle\end{aligned}$$

Let us commute $c^\dagger_{\mathbf{j}\uparrow}c_{\mathbf{h}\uparrow}$ to the left

$$\begin{aligned}[1-(1-\eta)\hat{n}_{\mathbf{g}\uparrow}]c^\dagger_{\mathbf{j}\uparrow}c_{\mathbf{h}\uparrow} &= c^\dagger_{\mathbf{j}\uparrow}c_{\mathbf{h}\uparrow} - (1-\eta)(\delta_{\mathbf{gj}}-\delta_{\mathbf{gh}})c^\dagger_{\mathbf{j}\uparrow}c_{\mathbf{h}\uparrow} \\ &\quad -(1-\eta)c^\dagger_{\mathbf{j}\uparrow}c_{\mathbf{h}\uparrow}c^\dagger_{\mathbf{g}\uparrow}c_{\mathbf{g}\uparrow}\end{aligned}$$

Multiplying this by $[1-(1-\eta)\hat{n}_{\mathbf{g}\uparrow}]$ from the right, we get

$$\begin{aligned}&[1-(1-\eta)\hat{n}_{\mathbf{g}\uparrow}]c^\dagger_{\mathbf{j}\uparrow}c_{\mathbf{h}\uparrow}[1-(1-\eta)\hat{n}_{\mathbf{g}\uparrow}] \\ =\ &[1-(1-\eta)(\delta_{\mathbf{gj}}-\delta_{\mathbf{gh}})]\,c^\dagger_{\mathbf{j}\uparrow}c_{\mathbf{h}\uparrow} + (\eta^2-1)c^\dagger_{\mathbf{j}\uparrow}c_{\mathbf{h}\uparrow}c^\dagger_{\mathbf{g}\uparrow}c_{\mathbf{g}\uparrow} \\ &+(1-\eta)^2(\delta_{\mathbf{gj}}-\delta_{\mathbf{gh}})c^\dagger_{\mathbf{j}\uparrow}c_{\mathbf{h}\uparrow}c^\dagger_{\mathbf{g}\uparrow}c_{\mathbf{g}\uparrow}\,.\end{aligned}$$

Performing $(1/L)\sum_{\mathbf{g}}$, we arrive at

$$\langle\Psi|c^{\dagger}_{\mathbf{j}\uparrow}c_{\mathbf{h}\uparrow}|\Psi\rangle = \left\{1-\frac{(1-\eta)^2}{L}\right\}\langle N-1|c^{\dagger}_{\mathbf{j}\uparrow}c_{\mathbf{h}\uparrow}|N-1\rangle + \frac{\eta^2-1}{L}\sum_{\mathbf{g}}\langle N-1|c^{\dagger}_{\mathbf{j}\uparrow}c_{\mathbf{h}\uparrow}c^{\dagger}_{\mathbf{g}\uparrow}c_{\mathbf{g}\uparrow}|N-1\rangle\,.$$

We use Wick's theorem to write

$$\begin{aligned} &\langle N-1|c^{\dagger}_{\mathbf{j}\uparrow}c_{\mathbf{h}\uparrow}c^{\dagger}_{\mathbf{g}\uparrow}c_{\mathbf{g}\uparrow}|N-1\rangle \\ = \;& \delta_{\mathbf{hg}}\langle N-1|c^{\dagger}_{\mathbf{j}\uparrow}c_{\mathbf{h}\uparrow}|N-1\rangle + \langle N-1|c^{\dagger}_{\mathbf{j}\uparrow}c^{\dagger}_{\mathbf{g}\uparrow}c_{\mathbf{g}\uparrow}c_{\mathbf{h}\uparrow}|N-1\rangle \\ = \;& \delta_{\mathbf{hg}}\langle N-1|c^{\dagger}_{\mathbf{j}\uparrow}c_{\mathbf{h}\uparrow}|N-1\rangle \\ &+\langle N-1|c^{\dagger}_{\mathbf{j}\uparrow}c_{\mathbf{h}\uparrow}|N-1\rangle\langle N-1|c^{\dagger}_{\mathbf{g}\uparrow}c_{\mathbf{g}\uparrow}|N-1\rangle \\ &-\langle N-1|c^{\dagger}_{\mathbf{j}\uparrow}c_{\mathbf{g}\uparrow}|N-1\rangle\langle N-1|c^{\dagger}_{\mathbf{g}\uparrow}c_{\mathbf{h}\uparrow}|N-1\rangle\,. \end{aligned} \tag{B.12}$$

It is to check that

$$\sum_{\mathbf{g}}\langle N-1|c^{\dagger}_{\mathbf{j}\uparrow}c_{\mathbf{g}\uparrow}|N-1\rangle\langle N-1|c^{\dagger}_{\mathbf{g}\uparrow}c_{\mathbf{h}\uparrow}|N-1\rangle = \langle N-1|c^{\dagger}_{\mathbf{j}\uparrow}c_{\mathbf{h}\uparrow}|N-1\rangle \tag{B.13}$$

thus upon summing over **g**, the first and third terms in (B.12) cancel, and we are left with

$$\langle\Psi|c^{\dagger}_{\mathbf{j}\uparrow}c_{\mathbf{h}\uparrow}|\Psi\rangle = \left\{1+\frac{N-1}{L}(\eta^2-1)-\frac{(1-\eta)^2}{L}\right\}\langle N-1|c^{\dagger}_{\mathbf{j}\uparrow}c_{\mathbf{h}\uparrow}|N-1\rangle\,. \tag{B.14}$$

The sum of the first two terms in the curly brackets is $\langle\Psi|\Psi\rangle$ (observe that the $n_\uparrow$ figuring in (B.3) is $(N-1)/L$, and at this stage, we still have to keep terms of order $1/L$).

The final result

$$\mathcal{K}_\uparrow - E_{\mathrm{Nag}} = -\epsilon_{\mathrm{F}} - \frac{(1-\eta)^2}{1-(1-\eta^2)n_\uparrow}\cdot\frac{E_{\mathrm{Nag}}}{L} \tag{B.15}$$

expresses the spin-flip-induced change in the $\uparrow$-spin kinetic energy.

Appendix C

Gutzwiller Approximation

There are several different ways in which the Gutzwiller approximation (GA) can be introduced. Of course, the most satisfactory approach would be to follow the exact treatment of the Gutzwiller variational problem [271], and realize that its results coincide with those of the GA in the limit when the lattice dimensionality D becomes infinite. Unfortunately, this lies far beyond the scope of this elementary treatise. Instead, we seek to demonstrate how the GA is done, and give some argument why it is expected to work better in higher dimensions.

The expansion (9.28) of the Gutzwiller trial state is difficult to work with because it contains an exponentially large ($\sim e^L$) number of large ($\sim L \times L$) determinants. The GA is introduced to simplify the task by treating the determinants in an averaged-over fashion. More precisely, the relevant expressions (which are composed of the determinants) will be replaced by their *configuration averages.*

Let us quote from (9.26) the expansion of the spinless fermion Fermi sea state for $N/2$ particles on L sites

$$|\mathrm{FS}\rangle_{\mathrm{sl}} = L^{-N/4} \sum_{\{\mathbf{g}_1^{N/2}\}} \begin{vmatrix} e^{i\mathbf{k}_1\mathbf{g}_1} & \dots & e^{i\mathbf{k}_1\mathbf{g}_{N/2}} \\ \vdots & \ddots & \vdots \\ e^{i\mathbf{k}_{N/2}\mathbf{g}_1} & \dots & e^{i\mathbf{k}_{N/2}\mathbf{g}_{N/2}} \end{vmatrix} |\mathbf{g}_1, \dots \mathbf{g}_{N/2}\rangle \,. \quad \text{(C.1)}$$

On the right-hand-side, the sum is over all possible local configurations.

The norm of $|\mathrm{FS}\rangle_{\mathrm{sl}}$ contains the sum of the absolute squares of the determinants

$$1 = L^{-N/2} \sum_{\{\mathbf{g}_1^{N/2}\}} \left\| \begin{array}{ccc} e^{i\mathbf{k}_1\mathbf{g}_1} & \dots & e^{i\mathbf{k}_1\mathbf{g}_{N/2}} \\ \vdots & \ddots & \vdots \\ e^{i\mathbf{k}_{N/2}\mathbf{g}_1} & \dots & e^{i\mathbf{k}_{N/2}\mathbf{g}_{N/2}} \end{array} \right\|^2 . \qquad \text{(C.2)}$$

The sum contains $C^L_{N/2}$ terms. We do not know the determinants one by one but we can use the above identity to determine the configuration average of their absolute squares

$$\left\| \begin{array}{ccc} e^{i\mathbf{k}_1\mathbf{g}_1} & \dots & e^{i\mathbf{k}_1\mathbf{g}_{N/2}} \\ \vdots & \ddots & \vdots \\ e^{i\mathbf{k}_{N/2}\mathbf{g}_1} & \dots & e^{i\mathbf{k}_{N/2}\mathbf{g}_{N/2}} \end{array} \right\|^2_{\text{ave.}} = \frac{L^{N/2}}{C^L_{N/2}} . \qquad \text{(C.3)}$$

Replacing the |determinant|^{2}s by the above average is a typical step in the Gutzwiller approximation[1].

In Sect. 9.2, we were pointing it out that the determinants describe the exchange hole effect which gives a larger weight to configurations in which the fermions are more-or-less evenly spaced, than to those in which the fermions are crowded together (Problem 9.1). The Gutzwiller approximation amounts to neglecting the exchange hole effect, i.e., to neglecting the intersite correlations caused by statistics. Later, we are going to present arguments which show that such an approximation is likely to work better for higher lattice dimensionalities.

Now we can write down the norm of the Gutzwiller trial state (9.28) in the GA (we assume $N \leq L$)

$$\langle \Psi | \Psi \rangle = L^{-N} \sum_{\{\mathbf{g}_1^{N/2}\}} \sum_{\{\mathbf{h}_1^{N/2}\}} \eta^{2N_d(\mathbf{g}_1^{N/2};\, \mathbf{h}_1^{N/2})}$$

$$\cdot \left\| \begin{array}{ccc} e^{i\mathbf{k}_1\mathbf{g}_1} & \dots & e^{i\mathbf{k}_1\mathbf{g}_{N/2}} \\ \vdots & \ddots & \vdots \\ e^{i\mathbf{k}_{N/2}\mathbf{g}_1} & \dots & e^{i\mathbf{k}_{N/2}\mathbf{g}_{N/2}} \end{array} \right\|^2 \cdot \left\| \begin{array}{ccc} e^{i\mathbf{k}_1\mathbf{h}_1} & \dots & e^{i\mathbf{k}_1\mathbf{h}_{N/2}} \\ \vdots & \ddots & \vdots \\ e^{i\mathbf{k}_{N/2}\mathbf{h}_1} & \dots & e^{i\mathbf{k}_{N/2}\mathbf{h}_{N/2}} \end{array} \right\|^2$$

[1] This particular formulation of the GA is presented in [107].

$$\rightsquigarrow \quad L^{-N} \sum_{\{\mathbf{g}_1^{N/2}\}} \sum_{\{\mathbf{h}_1^{N/2}\}} \eta^{2N_d(\mathbf{g}_1^{N/2};\mathbf{h}_1^{N/2})} \left(\frac{L^{N/2}}{C^L_{N/2}}\right)^2$$

$$= \quad \sum_{N_d=0}^{N/2} \eta^{2N_d} \; \frac{C^{N/2}_{N_d} C^{L-N/2}_{N/2-N_d}}{C^L_{N/2}} \,. \tag{C.4}$$

Arriving at the last line required a bit of combinatorics. We can argue like this: there are $C^L_{N/2}$ ways to pick out the lattice sites where the ↑-spin electrons are sitting. Of these, N_d harbour also a ↓-spin electron; these can be chosen in $C^{N/2}_{N_d}$ ways. The remaining $N/2 - N_d$ ↓-spin electrons have to be distributed over the $L - N/2$ sites which have no ↑-spin electrons; this can be done in $C^{L-N/2}_{N/2-N_d}$ ways.

The expectation value of the density of doubly occupied sites is

$$n_d = \frac{1}{L} \cdot \frac{\sum_{N_d=0}^{N/2} \eta^{2N_d} \;\; N_d C^{N/2}_{N_d} C^{L-N/2}_{N/2-N_d}}{\sum_{N_d=0}^{N/2} \eta^{2N_d} \;\; C^{N/2}_{N_d} C^{L-N/2}_{N/2-N_d}} \,. \tag{C.5}$$

Recalling Stirling's formula it becomes clear that the general structure of the N_d-sums is

$$\sum_{N_d} f(N_d) e^{Lg(N_d)}$$

where $g(N_d)$ is of $\mathcal{O}(1)$. The sums can be approximated by integrals

$$\sum_{N_d} f(N_d) e^{Lg(N_d)} \rightsquigarrow \int_0^{1/2} f(x) e^{Lg(x)} dx \tag{C.6}$$

where the integration variable is $x = N_d/L$ (we reserve the notation n_d for the expectation value of x). The error made by this replacement has the relative order of magnitude $1/L$, and is thus negligible for a large system.

The point is that the integrand contains an exponential with a very large ($\sim \mathcal{O}(L)$) exponent. If the function $g(x)$ has its maximum within the domain of integration, at some $x_0 \in (0, 1/2)$, then the maximum will be hugely amplified for $e^{Lg(x)}$, and the integral will be dominated by the contribution from a small neighbourhood of x_0. We can expand g about x_0

$$g(x) \approx g(x_0) + \frac{1}{2} g''(x_0)(x - x_0)^2 \tag{C.7}$$

where $g''(x_0) < 0$. Introducing the new integration variable $x - x_0$, the limits of integration can be taken as $\pm\infty$ (the error is of $\mathcal{O}(e^{-L})$ and thus utterly negligible). Expanding also $f(x)$ about x_0 and realizing that there is no contribution from the first-order term, we find

$$\int_0^{1/2} f(x) e^{Lg(x)} dx = \sqrt{\frac{2\pi}{L|g''(x_0)|}} f(x_0) e^{Lg(x_0)} \cdot \left(1 + \mathcal{O}\left(\frac{f''(x_0)}{L}\right)\right) . \tag{C.8}$$

Once again, we can neglect the correction term because it has the relative order of magnitude $1/L$.

An expectation value of an operator $\hat{A}$ is usually expressed as the ratio of two sums containing the same exponent

$$\frac{\langle\Psi|\hat{A}|\Psi\rangle}{\langle\Psi|\Psi\rangle} \rightsquigarrow \frac{\sum_{N_d} f_1(N_d) e^{Lg(N_d)}}{\sum_{N_d} f_2(N_d) e^{Lg(N_d)}} \approx \frac{f_1(x_0)}{f_2(x_0)} . \tag{C.9}$$

For the purpose of calculating expectation values, sums over an exponentially large number of configurations are dominated by the term corresponding to the largest exponent. It follows that it is useful to bring the sum in the numerator to such a form that its terms appear as the product of the corresponding term in the norm $\langle\Psi|\Psi\rangle$, times an extra factor. Then the expectation value will be this extra factor, taken at x_0. Note that the expectation value in (C.5) appears in such a form so we can immediately deduce $n_d = x_0$, where x_0 is the place where the function $g(x)$

$$g(x) = (2x - n)\ln\left(\frac{n}{2} - x\right) - x\ln x - (1 - n + x)\ln(1 - n + x) + x\ln\eta^2 \tag{C.10}$$

identifiable from the last line of (C.4), reaches its maximum. Hence n_d satisfies

$$\left(\frac{n}{2} - n_d\right)^2 \eta^2 = n_d(1 - n + n_d) . \tag{C.11}$$

We could easily write down the cumbersome expression for n_d but we do not really need it. Instead, let us note particular cases: for $n=1$, (9.34) is recovered. For $U \to \infty$ we expect $\eta \to 0$; then either $n_d = 0$ (band not more than half-filled), or $n_d = n-1$ (band more than half-filled).

We do not repeat the derivation for a spin-polarized (spatially uniform) state where $n_\uparrow \neq n_\downarrow$. The result turns out to be the straightforward generalization of (C.11)

$$(n_\uparrow - n_d)(n_\downarrow - n_d)\eta^2 = n_d(1-n+n_d) \,. \tag{C.12}$$

A bit more care is needed to evaluate the hopping amplitude

$$\langle\Psi|c^\dagger_{\mathbf{i}\uparrow}c_{\mathbf{j}\uparrow}|\Psi\rangle = L^{-N} \sum_{\{\mathbf{g}_1^{N/2}\}}^{\mathbf{i}\notin,\mathbf{j}\in} \sum_{\{\mathbf{h}_1^{N/2}\}} \frac{L^N}{C^L_{N/2}}\, \eta^{N'_d+N_d} \tag{C.13}$$

$$\cdot \begin{vmatrix} e^{-i\mathbf{k}_1\mathbf{g}'_1} & \dots & e^{-i\mathbf{k}_1\mathbf{g}'_{N/2}} \\ \vdots & \ddots & \vdots \\ e^{-i\mathbf{k}_{N/2}\mathbf{g}'_1} & \dots & e^{-i\mathbf{k}_{N/2}\mathbf{g}'_{N/2}} \end{vmatrix} \cdot \begin{vmatrix} e^{i\mathbf{k}_1\mathbf{g}_1} & \dots & e^{i\mathbf{k}_1\mathbf{g}_{N/2}} \\ \vdots & \ddots & \vdots \\ e^{i\mathbf{k}_{N/2}\mathbf{g}_1} & \dots & e^{i\mathbf{k}_{N/2}\mathbf{g}_{N/2}} \end{vmatrix} \,.$$

Note the restriction on the set of sites with $\uparrow$-spin electrons. By definition, the set "on the right" (of the matrix element) must contain $\mathbf{j}$ (because we take an electron from there), and must not contain $\mathbf{i}$ (because the electron has to hop there). After the hopping event, the set of sites with $\uparrow$-spin electrons is $\{\mathbf{g}'_1, \dots \mathbf{g}'_{N/2}\}$ which differs from $\{\mathbf{g}_1^{N/2}\}$ in that the site $\mathbf{j}$ has been replaced by $\mathbf{i}$. From the right, we have a determinant corresponding to the $\{\mathbf{g}_1^{N/2}\}$-set, and from the left, a complex conjugate determinant corresponding to the $\{\mathbf{g}'^{N/2}_1\}$-set. Though we can use (C.3) when a $|\text{determinant}|^2$ appears (and we did use it in writing down (C.13), for the quantity corresponding to the $\downarrow$-spin electrons), we cannot use it for the product of determinants in (C.13), which is an inherently complex quantity. To approximate this, a new "off-diagonal" averaging rule is needed.

Let us note that the same determinant product appears in the expression of the hopping amplitude for the spinless Fermi sea

$${}_{\rm sl}\langle \mathrm{FS}|c^\dagger_{\mathbf{i}}c_{\mathbf{j}}|\mathrm{FS}\rangle_{\rm sl} = \tag{C.14}$$

$$L^{-N/2}\sum_{\{\mathbf{g}_1^{N/2}\}}^{\mathbf{i}\notin,\mathbf{j}\in}\begin{vmatrix} e^{-i\mathbf{k}_1\mathbf{g}'_1} & \dots & e^{-i\mathbf{k}_1\mathbf{g}'_{N/2}} \\ \vdots & \ddots & \vdots \\ e^{-i\mathbf{k}_{N/2}\mathbf{g}'_1} & \dots & e^{-i\mathbf{k}_{N/2}\mathbf{g}'_{N/2}} \end{vmatrix} \cdot \begin{vmatrix} e^{i\mathbf{k}_1\mathbf{g}_1} & \dots & e^{i\mathbf{k}_1\mathbf{g}_{N/2}} \\ \vdots & \ddots & \vdots \\ e^{i\mathbf{k}_{N/2}\mathbf{g}_1} & \dots & e^{i\mathbf{k}_{N/2}\mathbf{g}_{N/2}} \end{vmatrix} .$$

At the same time, the amplitude is an element of the one-particle density matrix, and it is given by a simple **k**-sum

$$_{\rm sl}\langle \mathrm{FS}|c^\dagger_{\mathbf{i}}c_{\mathbf{j}}|\mathrm{FS}\rangle_{\rm sl} = \frac{1}{L}\sum_{\mathbf{k}_1,\mathbf{k}_2} e^{i(\mathbf{k}_1\mathbf{i}-\mathbf{k}_2\mathbf{j})}\,{}_{\rm sl}\langle \mathrm{FS}|c^\dagger_{\mathbf{k}_1}c_{\mathbf{k}_2}|\mathrm{FS}\rangle_{\rm sl} = \frac{1}{L}\sum_{\mathbf{k}}^{(\mathrm{FS})} e^{-i\mathbf{k}(\mathbf{i}-\mathbf{j})} . \tag{C.15}$$

Since the $\{\mathbf{g}_1^{N/2}\}$-sum in (C.14) runs over $C^{L-2}_{N/2-1}$ sets, we can introduce the following configuration average for the relevant determinant product

$$\left\{\begin{vmatrix} e^{-i\mathbf{k}_1\mathbf{g}'_1} & \dots & e^{-i\mathbf{k}_1\mathbf{g}'_{N/2}} \\ \vdots & \ddots & \vdots \\ e^{-i\mathbf{k}_{N/2}\mathbf{g}'_1} & \dots & e^{-i\mathbf{k}_{N/2}\mathbf{g}'_{N/2}} \end{vmatrix} \cdot \begin{vmatrix} e^{i\mathbf{k}_1\mathbf{g}_1} & \dots & e^{i\mathbf{k}_1\mathbf{g}_{N/2}} \\ \vdots & \ddots & \vdots \\ e^{i\mathbf{k}_{N/2}\mathbf{g}_1} & \dots & e^{i\mathbf{k}_{N/2}\mathbf{g}_{N/2}} \end{vmatrix}\right\}_{\rm ave.}$$

$$= \frac{L^{N/2-1}}{C^{L-2}_{N/2-1}}\sum_{\mathbf{k}}^{(\mathrm{FS})} e^{-i\mathbf{k}(\mathbf{i}-\mathbf{j})} . \tag{C.16}$$

Now (C.16) can be replaced in (C.13). The remaining configuration sum is complicated by the fact that the hopping event may change N_d. For instance, if there is no $\downarrow$-spin electron at **j**, but there is one at **i**, then the number of doubly occupied sites increased by one during the hopping. To admit such possibilities, we denoted by N_d the number of common elements of the $\{\mathbf{g}_1^{N/2}\}$- and $\{\mathbf{h}_1^{N/2}\}$-sets, and with N'_d that of the $\{\mathbf{g'}_1^{N/2}\}$- and $\{\mathbf{h}_1^{N/2}\}$-sets. We find

$$\begin{aligned}\langle\Psi|c^\dagger_{\mathbf{i}\uparrow}c_{\mathbf{j}\uparrow}|\Psi\rangle &= \frac{1}{L}\sum_{\mathbf{k}}^{(\mathrm{FS})} e^{-i\mathbf{k}(\mathbf{i}-\mathbf{j})}\sum_{N_d}\eta^{2N_d}\,\frac{C^{N/2}_{N_d}C^{L-N/2}_{N/2-N_d}}{C^L_{N/2}}\cdot\frac{4}{n(2-n)} \\ &\quad\cdot\left\{\eta\left(\frac{n}{2}-x\right)^2+\frac{1}{\eta}(1-n+x)x\right. \\ &\quad\left. x\left(\frac{n}{2}-x\right)+(1-n+x)\left(\frac{n}{2}-x\right)\right\} .\end{aligned} \tag{C.17}$$

where again $x = N_d/L$ and we have taken care to factor the terms in the way described in the remark following (C.9).

It would be long to derive (C.17) in detail but we show how to get the first term in the square brackets. $N'_d = N_d + 1$ implies $\mathbf{j} \notin \{\mathbf{h}_1^{N/2}\}$ and $\mathbf{i} \in \{\mathbf{h}_1^{N/2}\}$. Remembering also $\mathbf{i} \notin \{\mathbf{g}_1^{N/2}\}$ and $\mathbf{j} \in \{\mathbf{g}_1^{N/2}\}$, there are $C_{N/2-1}^{L-2}$ ways to pick the $\uparrow$-sites; $C_{N_d}^{N/2-1}$ ways to choose the doubly occupied sites; and $C_{N/2-N_d-1}^{L-N/2-1}$ ways to choose the sites for the unpaired $\downarrow$-spin electrons (since one of them has to be at $\mathbf{i}$). $C_{N/2-1}^{L-2}$ gets cancelled by the same factor in the denominators (cf. (C.16)), and bringing the remaining binomials to the standard form, the result (C.17) is found.

Applying (C.9), the hopping amplitude can be expressed as the non-interacting value, times the renormalization factor q

$$\frac{\langle\Psi|c_{\mathbf{i}\uparrow}^{\dagger}c_{\mathbf{j}\uparrow}|\Psi\rangle}{\langle\Psi|\Psi\rangle} = q \cdot \frac{1}{L}\sum_{\mathbf{k}}^{(\mathrm{FS})} e^{-i\mathbf{k}(\mathbf{i}-\mathbf{j})} \,. \tag{C.18}$$

Expressing η from (C.11), q can be brought to a more elegant form

$$\begin{aligned} q &= \frac{4}{n(2-n)} \cdot \left\{ \eta \left(\frac{n}{2} - n_d\right)^2 + \frac{1}{\eta}(1-n+n_d)n_d \right. \\ &\quad \left. + n_d\left(\frac{n}{2} - n_d\right) + (1-n+n_d)\left(\frac{n}{2} - n_d\right)\right\} \\ &= \frac{2(n-2n_d)}{n(2-n)} \cdot [\sqrt{n_d} + \sqrt{1-n+n_d}]^2 \,. \end{aligned} \tag{C.19}$$

(C.19) is a standard result of the Gutzwiller approximation.

Again, it would be straightforward to generalize the result to the spin-polarized case where q becomes spin-dependent (σ=$\uparrow$ or $\downarrow$, $n = n_\uparrow + n_\downarrow$). We quote the result from [431]

$$q_\sigma = \frac{1}{n_\sigma(1-n_\sigma)} \cdot [\sqrt{(n_\sigma - n_d)(1-n+n_d)} + \sqrt{n_d(n_{-\sigma} - n_d)}]^2 \,. \tag{C.20}$$

The σ-dependent renormalization factor appears in the energy expression for a Gutzwiller-projected spin-polarized Fermi sea

$$E = q_\uparrow \bar{\epsilon}_\uparrow + q_\downarrow \bar{\epsilon}_\downarrow + U n_d \,. \tag{C.21}$$

In the unpolarized case, (C.21) reduces to (9.44).

Let us note that (C.20) becomes very simple in the limit $U \to \infty$

$$q_\sigma = \frac{1-n}{1-n_\sigma}\,. \tag{C.22}$$

Appendix D

Schrieffer–Wolff Transformation

The Schrieffer–Wolff transformation was introduced [352] for deriving the Kondo impurity model [223] from the impurity Anderson model [19]. The derivation of the KLM from the PAM proceeds through the same steps, and gives essentially the same results, except for additional f–f hopping terms [361].

We proceed as we did in Sec. 5.1.2. However, the present problem differs from that of the single-band Hubbard model, since only one set of fermions (the f-electrons) is strongly correlated. For most of the calculation, it will be useful to represent f-electrons by Hubbard operators (see Sec. 5.1.3)

$$f^{\dagger}_{\mathbf{j}\sigma} = X^{\sigma\leftarrow 0}_{\mathbf{j}} + \eta(\sigma) X^{d\leftarrow -\sigma}_{\mathbf{j}} \tag{D.1}$$

$$f_{\mathbf{j}\sigma} = X^{0\leftarrow\sigma}_{\mathbf{j}} + \eta(\sigma) X^{-\sigma\leftarrow d}_{\mathbf{j}} \tag{D.2}$$

where $\eta(\sigma) = 1$ for $\sigma =\uparrow$, and $\eta(\sigma) = -1$ for $\sigma =\downarrow$. c-electrons are described by their usual Bloch state creation and annihilation operators.

Our aim is to transform to a basis in which the f-electrons are (in lowest order) decoupled from the band electrons. Correspondingly, we decompose the PAM (11.4) as

$$\mathcal{H}_{\mathrm{PAM}} = \mathcal{H}_0 + \mathcal{H}_{\mathrm{hyb}}\,, \tag{D.3}$$

where

$$\mathcal{H}_0 = \sum_{\mathbf{k},\sigma} \epsilon_{\mathbf{k}} c^\dagger_{\mathbf{k}\sigma} c_{\mathbf{k}\sigma} + \epsilon_f \sum_{\mathbf{j},\sigma} X_{\mathbf{j}}^{\sigma\leftarrow\sigma} + (2\epsilon_f + U_f) \sum_{\mathbf{j}} X_{\mathbf{j}}^{d\leftarrow d} \tag{D.4}$$

and

$$\begin{aligned}\mathcal{H}_{\mathrm{hyb}} &= -\frac{1}{\sqrt{L}} \sum_{\mathbf{j},\mathbf{k},\sigma} \left\{ v_{\mathbf{k}} e^{i\mathbf{k}\mathbf{j}} \left(X_{\mathbf{j}}^{\sigma\leftarrow 0} + \eta(\sigma) X_{\mathbf{j}}^{d\leftarrow -\sigma} \right) c_{\mathbf{k}\sigma} \right. \\ &\quad \left. + v^*_{\mathbf{k}} e^{-i\mathbf{k}\mathbf{j}} c^\dagger_{\mathbf{k}\sigma} \left(X_{\mathbf{j}}^{0\leftarrow\sigma} + \eta(\sigma) X_{\mathbf{j}}^{-\sigma\leftarrow d} \right) \right\} . \end{aligned} \tag{D.5}$$

The transformed Hamiltonian is

$$\begin{aligned}\mathcal{H}_{\mathrm{eff}} &= e^{i\mathcal{S}} \mathcal{H} e^{-i\mathcal{S}} \\ &\approx \mathcal{H}_0 + \mathcal{H}_{\mathrm{hyb}} + [i\mathcal{S}, \mathcal{H}_0] + [i\mathcal{S}, \mathcal{H}_{\mathrm{hyb}}] + \frac{1}{2} [i\mathcal{S}, [i\mathcal{S}, \mathcal{H}_0]] . \end{aligned} \tag{D.6}$$

We want to find an $\mathcal{S}$ which satisfies

$$\mathcal{H}_{\mathrm{hyb}} + [i\mathcal{S}, \mathcal{H}_0] = 0 , \tag{D.7}$$

whereupon the effective Hamiltonian becomes

$$\mathcal{H}_{\mathrm{eff}} = \mathcal{H}_0 + \frac{1}{2} [i\mathcal{S}, \mathcal{H}_{\mathrm{hyb}}] . \tag{D.8}$$

In principle, $\mathcal{H}_0$ still contains its f-terms but they got reduced to the role of additive constants, and are usually dropped.

Experience with the derivation of the t–J model suggests that $\mathcal{S}$ may closely resemble what we want to transform out, namely $\mathcal{H}_{\mathrm{hyb}}$. Let us try commuting pieces of $\mathcal{H}_{\mathrm{hyb}}$ with $\mathcal{H}_0$. For instance,

$$\left[X_{\mathbf{j}}^{\sigma\leftarrow 0} c_{\mathbf{k}\sigma}, \mathcal{H}_0 \right] = (\epsilon_{\mathbf{k}} - \epsilon_f) X_{\mathbf{j}}^{\sigma\leftarrow 0} c_{\mathbf{k}\sigma} ,$$

and

$$\left[X_{\mathbf{j}}^{d\leftarrow\downarrow} c_{\mathbf{k},\uparrow}, \mathcal{H}_0 \right] = (\epsilon_{\mathbf{k}} - \epsilon_f - U_f) X_{\mathbf{j}}^{d\leftarrow\downarrow} c_{\mathbf{k},\uparrow} ,$$

etc. Eventually, we arrive at

$$\begin{aligned}\mathcal{S} &= -\frac{i}{\sqrt{L}} \sum_{\mathbf{j},\mathbf{k},\sigma} \left\{ \frac{v_{\mathbf{k}} e^{i\mathbf{k}\mathbf{j}}}{\epsilon_{\mathbf{k}} - \epsilon_f} X_{\mathbf{j}}^{\sigma\leftarrow 0} c_{\mathbf{k}\sigma} + \frac{\eta(\sigma) v_{\mathbf{k}} e^{i\mathbf{k}\mathbf{j}}}{\epsilon_{\mathbf{k}} - \epsilon_f - U_f} X_{\mathbf{j}}^{d\leftarrow -\sigma} c_{\mathbf{k}\sigma} \right. \\ &\quad \left. - \frac{v^*_{\mathbf{k}} e^{-i\mathbf{k}\mathbf{j}}}{\epsilon_{\mathbf{k}} - \epsilon_f} c^\dagger_{\mathbf{k}\sigma} X_{\mathbf{j}}^{0\leftarrow\sigma} - \frac{\eta(\sigma) v^*_{\mathbf{k}} e^{-i\mathbf{k}\mathbf{j}}}{\epsilon_{\mathbf{k}} - \epsilon_f - U_f} c^\dagger_{\mathbf{k}\sigma} X_{\mathbf{j}}^{-\sigma\leftarrow d} \right\} . \end{aligned} \tag{D.9}$$

$\mathcal{H}_{\text{eff}}$ consists of many terms, and it is useful to divide them into classes. The most interesting pieces describe the scattering of a conduction electron on an f-site. To learn about such processes, let us collect $\mathbf{k} \to \mathbf{q}$ contributions which contain the factor $v_{\mathbf{k}} v_{\mathbf{q}}^* e^{i(\mathbf{k}-\mathbf{q})\mathbf{j}}$, and also $(\epsilon_{\mathbf{k}} - \epsilon_f)^{-1}$ or $(\epsilon_{\mathbf{q}} - \epsilon_f)^{-1}$. Here we assume $\mathbf{k} \neq \mathbf{q}$.

$$- \left[\frac{v_{\mathbf{k}} e^{i\mathbf{kj}}}{\epsilon_{\mathbf{k}} - \epsilon_f} \sum_{\sigma} X_{\mathbf{j}}^{\sigma \leftarrow 0} c_{\mathbf{k}\sigma} \, , \, v_{\mathbf{q}}^* e^{-i\mathbf{qj}} \sum_{\sigma'} c_{\mathbf{q}\sigma'}^{\dagger} \left(X_{\mathbf{j}}^{0 \leftarrow \sigma'} + \eta(\sigma') X_{\mathbf{j}}^{-\sigma' \leftarrow d} \right) \right]$$

$$+ \left[\frac{v_{\mathbf{q}}^* e^{-i\mathbf{qj}}}{\epsilon_{\mathbf{q}} - \epsilon_f} \sum_{\sigma} c_{\mathbf{q}\sigma}^{\dagger} X_{\mathbf{j}}^{0 \leftarrow \sigma} \, , \, v_{\mathbf{k}} e^{i\mathbf{kj}} \sum_{\sigma'} \left(X_{\mathbf{j}}^{\sigma' \leftarrow 0} + \eta(\sigma') X_{\mathbf{j}}^{d \leftarrow -\sigma'} \right) c_{\mathbf{k}\sigma'} \right]$$

$$= v_{\mathbf{k}} v_{\mathbf{q}}^* e^{i(\mathbf{k}-\mathbf{q})\mathbf{j}} \left(\frac{1}{\epsilon_{\mathbf{k}} - \epsilon_f} + \frac{1}{\epsilon_{\mathbf{q}} - \epsilon_f} \right) \tag{D.10}$$

$$\cdot \left\{ X_{\mathbf{j}}^{\uparrow \leftarrow \downarrow} c_{\mathbf{q}\downarrow}^{\dagger} c_{\mathbf{k}\uparrow} + X_{\mathbf{j}}^{\downarrow \leftarrow \uparrow} c_{\mathbf{q}\uparrow}^{\dagger} c_{\mathbf{k}\downarrow} + \sum_{\sigma} X_{\mathbf{j}}^{\sigma \leftarrow \sigma} c_{\mathbf{q}\sigma}^{\dagger} c_{\mathbf{k}\sigma} + X_{\mathbf{j}}^{0 \leftarrow 0} \sum_{\sigma} c_{\mathbf{q}\sigma}^{\dagger} c_{\mathbf{k}\sigma} \right\} .$$

The first two terms describe spin-flip scattering. In view of the global spin-rotational invariance, it is desirable to rewrite the other terms so that an isotropic c–f spin–spin coupling can be identified. This is achieved by introducing $2S_{\mathbf{j}}^z = X_{\mathbf{j}}^{\uparrow \leftarrow \uparrow} - X_{\mathbf{j}}^{\downarrow \leftarrow \downarrow}$ for the f-spins, and making use of identities like $X_{\mathbf{j}}^{\downarrow \leftarrow \downarrow} + X_{\mathbf{j}}^{0 \leftarrow 0} = 1 - \hat{n}_{\mathbf{j}\uparrow}^f = f_{\mathbf{j}\uparrow} f_{\mathbf{j}\uparrow}^{\dagger}$. The abovestanding expression can be rewritten as

$$v_{\mathbf{k}} v_{\mathbf{q}}^* e^{i(\mathbf{k}-\mathbf{q})\mathbf{j}} \left(\frac{1}{\epsilon_{\mathbf{k}} - \epsilon_f} + \frac{1}{\epsilon_{\mathbf{q}} - \epsilon_f} \right) \cdot$$

$$\left\{ \mathbf{S}_{\mathbf{j}} \cdot \sum_{\alpha,\beta} c_{\mathbf{q}\alpha}^{\dagger} \vec{\sigma}_{\alpha\beta} c_{\mathbf{k}\beta} + \frac{1}{2} \left(\sum_{\sigma} f_{\mathbf{j}\sigma} f_{\mathbf{j}\sigma}^{\dagger} \right) \left(\sum_{\sigma'} c_{\mathbf{q}\sigma'}^{\dagger} c_{\mathbf{k}\sigma'} \right) \right\} .$$

Similar contributions can be collected with $(\epsilon_{\mathbf{k}} - \epsilon_f - U_f)^{-1}$ from

$$- \frac{v_{\mathbf{k}} v_{\mathbf{q}}^* e^{i(\mathbf{k}-\mathbf{q})\mathbf{j}}}{\epsilon_{\mathbf{k}} - \epsilon_f - U_f} \left[\sum_{\sigma} \eta(\sigma) X_{\mathbf{j}}^{d \leftarrow -\sigma} c_{\mathbf{k}\sigma} \, , \, \sum_{\sigma'} c_{\mathbf{q}\sigma'}^{\dagger} \left(X_{\mathbf{j}}^{0 \leftarrow \sigma'} + \eta(\sigma') X_{\mathbf{j}}^{-\sigma' \leftarrow d} \right) \right]$$

$$= \frac{v_{\mathbf{k}} v_{\mathbf{q}}^* e^{i(\mathbf{k}-\mathbf{q})\mathbf{j}}}{\epsilon_{\mathbf{k}} - \epsilon_f - U_f} \left\{ -\mathbf{S_j} \cdot \sum_{\alpha,\beta} c_{\mathbf{q}\alpha}^\dagger \vec{\sigma}_{\alpha\beta} c_{\mathbf{k}\beta} + \frac{1}{2} \left(\sum_{\sigma} f_{\mathbf{j}\sigma}^\dagger f_{\mathbf{j}\sigma} \right) \left(\sum_{\sigma'} c_{\mathbf{q}\sigma'}^\dagger c_{\mathbf{k}\sigma'} \right) \right\} .$$

All in all, we find (omitting, for brevity, terms in which $\epsilon_{\mathbf{k}}$ is replaced by $\epsilon_{\mathbf{q}}$) the Kondo coupling

$$\mathcal{H}_{\text{eff}}^{(1)} = \frac{1}{2L} \sum_{\mathbf{j}} \sum_{\mathbf{k},\mathbf{q}} v_{\mathbf{q}} v_{\mathbf{k}}^* e^{i(\mathbf{q}-\mathbf{k})\mathbf{j}} \cdot \left(\frac{1}{\epsilon_{\mathbf{k}} - \epsilon_f} - \frac{1}{\epsilon_{\mathbf{k}} - \epsilon_f - U_f} \right) \mathbf{S_j} \cdot \sum_{\alpha,\beta} c_{\mathbf{k}\alpha}^\dagger \vec{\sigma}_{\alpha\beta} c_{\mathbf{q}\beta} + \text{H.c.} \,, \tag{D.11}$$

and also spin-independent ("potential") scattering

$$\mathcal{H}_{\text{eff}}^{(2)} = \frac{1}{4L} \sum_{\mathbf{j}} \sum_{\mathbf{k},\mathbf{q}} v_{\mathbf{q}} v_{\mathbf{k}}^* e^{i(\mathbf{q}-\mathbf{k})\mathbf{j}} \left\{ \frac{1}{\epsilon_{\mathbf{k}} - \epsilon_f} \cdot \sum_{\sigma} f_{\mathbf{j}\sigma} f_{\mathbf{j}\sigma}^\dagger + \frac{1}{\epsilon_{\mathbf{k}} - \epsilon_f - U_f} \cdot \sum_{\sigma} f_{\mathbf{j}\sigma}^\dagger f_{\mathbf{j}\sigma} \right\} \left(\sum_{\sigma'} c_{\mathbf{k}\sigma'}^\dagger c_{\mathbf{q}\sigma'} \right) . \tag{D.12}$$

Note that the energy denominators get added in $\mathcal{H}_{\text{eff}}^{(2)}$, while they are subtracted in $\mathcal{H}_{\text{eff}}^{(1)}$. Since the f^2 level is supposed to lie far above the Fermi level, in the relevant range $\epsilon_{\mathbf{k}} - \epsilon_f - U_f < 0$, and thus both contributions to the Kondo coupling are antiferromagnetic.

Taking terms with different site indices from the two sides of the commutators, we get a contribution which describes f-electron propagation (see, e.g., the *Appendix* of [361])

$$\mathcal{H}_{\text{eff}}^{(3)} = -\frac{1}{2L} \sum_{\mathbf{g},\mathbf{h}} \sum_{\mathbf{k}} |v_{\mathbf{k}}|^2 e^{i\mathbf{k}(\mathbf{g}-\mathbf{h})} \sum_{\sigma} \left\{ \frac{1}{\epsilon_{\mathbf{k}} - \epsilon_f} f_{\mathbf{g}\sigma}^\dagger f_{\mathbf{h}\sigma} (1 - \hat{n}_{\mathbf{h},-\sigma}^f) + \frac{1}{\epsilon_{\mathbf{k}} - \epsilon_f - U_f} f_{\mathbf{g}\sigma}^\dagger f_{\mathbf{h}\sigma} \hat{n}_{\mathbf{h},-\sigma}^f \right\} + \text{H.c.} \tag{D.13}$$

The canonical transformation generates also second-order hybridization processes whereby an f^2 site can decay into two conduction electrons, or vice versa. We trust that such processes would be eliminated at the next ($\mathcal{S}^{(2)} \propto v^2$) level of the transformation.

Finally, diagonal ($\mathbf{k} = \mathbf{q}$) contributions yield an energy shift which is always negative [361]

$$\mathcal{H}_{\text{eff}}^{(4)} = \frac{1}{4} \sum_{\mathbf{k},\sigma} |v_{\mathbf{k}}|^2 \left\{ -\frac{1 - c_{\mathbf{k}\sigma}^{\dagger} c_{\mathbf{k}\sigma}}{\epsilon_{\mathbf{k}} - \epsilon_f} + \frac{c_{\mathbf{k}\sigma}^{\dagger} c_{\mathbf{k}\sigma}}{\epsilon_{\mathbf{k}} - \epsilon_f - U_f} \right\} \propto -\frac{v^2}{W}. \tag{D.14}$$

Bibliography

[1] Abbate, M., J.C. Fuggle, A. Fujimori, L.H. Tjeng, C.T. Chen, R. Potze, G.A. Sawatzky, H. Eisaki, and S. Uchida: *Phys. Rev.* B**47**, 16124 (1993).

[2] Abragam, A., and B. Bleany: *Electron Paramagnetic Resonance in Transition Ions*, (Clarendon Press, Oxford, 1970).

[3] Abrikosov, A.A.: *Fundamentals of the Theory of Metals*, (North-Holland, Amsterdam, 1988).

[4] Abromowitz, M., and Segun, I.A.: *Handbook of Mathematical Functions*, (Dover Publications, Inc., New York, 1968).

[5] Adler, D.: *Solid State Physics* **21**, 1 (1969); gives a survey of transition metal oxides.

[6] Aeppli, G., and Z. Fisk: *Comments Cond. Mat. Phys.* **16**, 155 (1992).

[7] Affleck, I.: *J. Phys.: Condens. Matter* **1**, 3047 (1989).

[8] Affleck, I.: *J. Phys. A: Math. Gen.* **31**, 4573 (1998).

[9] Affleck, I., M.P. Gelfand, and R.R.P. Singh: *J. Phys. A: Math. Gen.* **27**, 7313 (1994).

[10] Affleck, I., T. Kennedy, E.H. Lieb, and H. Tasaki: *Phys. Rev. Lett.* **59**, 799 (1987).

[11] Affleck, I., and E.H. Lieb: *Lett. Math. Phys.* **12**, 57 (1986).

[12] Aléonard, R., and P. Morin: *Phys. Rev.* B**19**, 3868 (1979).

[13] Allen, S.J.: *Phys. Rev.* **166**, 530 (1968).

[14] Andersen, O.K.: *Phys. Rev.* B**12**, 3060 (1975).

[15] Andersen, O.K., J. Madsen, U.K. Poulsen, O. Jepsen, and J. Kollár: *Physica* **86** & **88**B, 249 (1977).

[16] Anderson, P.W.: *Phys. Rev.* **83**, 1260 (1951).

[17] Anderson, P.W.: *Phys. Rev.* **86**, 694 (1952).

[18] Anderson, P.W.: *Phys. Rev.* **102**, 1008 (1956).

[19] Anderson, P.W.: *Phys. Rev.* **124**, 41 (1961).

[20] Anderson, P.W.: *Phys. Rev.* **115**, 2 (1959); Solid State Physics **14**, 99; (1963).

[21] Anderson, P.W.: *Mater. Res. Bull.* **8**, 153 (1973).

[22] Anderson, P.W.: *Basic Notions of Condensed Matter Physics*, (The Benjamin–Cummings Publ.Co., Inc., London, 1984).

[23] Anderson, P.W.: *Science* **235**, 1196 (1987).

[24] Anderson, P.W., in: *Frontiers and Borderlines in Many-Particle Physics*, Eds. R.A. Broglia and J.R. Schrieffer, (North-Holland, Amsterdam, p. 1, 1988).

[25] Anderson, P.W.: *A Career in Theoretical Physics*, pp. 113–129, (World Scientific, Singapore, 1994).

[26] Anderson, P.W., and H. Hasegawa: *Phys. Rev.* **100**, 675 (1955).

[27] Andreani, L.C., E. Liviotti, P. Santini, and G. Amoretti: *Z. Physik* B**100**, 96 (1996).

[28] Aoki, Y., H. Sato, and H. Sugawara: *Physica* B **230&232**, 770 (1997).

[29] Arovas, D.P., and F. Guinea: cond-mat/9711145.

[30] Asamitsu, A., Y. Moritomo, R. Kumai, Y. Tomioka, and Y. Tokura: *Phys. Rev.* B**54**, 1716 (1996).

[31] Asano, T., *et al*: *J. Phys. Soc. Japan*: **66**, 460 (1997).

[32] Ashcroft, N.W.,and N.D. Mermin: *Solid State Physics* (Holt, Rinehart and Winston, New York, 1976).

[33] Auerbach, A.: *Interacting Electrons and Quantum Magnetism*, (Springer-Verlag, New York and Berlin, 1994).

[34] Baeriswyl, D., in: *Nonlinearity in Condensed Matter*, eds. A.R. Bishop *et al.*, Springer Series in Solid State Sciences **69**, p. 183, (Springer-Verlag, Berlin, 1987).

[35] Baeriswyl, D., and W. von der Linden: *Int. J. Modern Phys.* B**5**, 999 (1991).

[36] Balicas, L., G. Kriza, and F.I.B. Williams: *Phys. Rev. Lett.* **75**, 2000 (1995).

[37] Ballhausen, C.J.: *Introduction to Ligand Field Theory*, (McGraw–Hill, New York, 1962).

[38] Bao, W., C. Broholm, G. Aeppli, S. Carter, P. Dai, T.F. Rosenbaum, J.M.Honig, P. Metcalf, and S.F.Trevino: to be published.

[39] Bao, W., C. Broholm, S.A. Carter, T.F. Rosenbaum, G. Aeppli, S.F. Trevino, P. Metcalf, J.M. Honig, and J. Spalek: *Phys. Rev. Lett.* **71**, 766 (1993).

[40] Bao, W., C. Broholm, G. Aeppli, P. Dai, J.M. Honig, and P. Metcalf: *Phys. Rev. Lett.* **78**, 507 (1997).

[41] Barbiellini, B., and P.M. Platzman: cond-mat/9710216.

[42] Barentzen, H., and P. Wróbel: *Z. Phys. B – Condensed Matter* **93**, 375 (1994).

[43] Bariev, R.Z., A. Klümper, A. Schadschneider, and J. Zittartz: *J. Phys. A: Math. Gen.* **26**, 1249 (1993).

[44] Barrett, S.E., G. Dabbagh, L.N. Pfeiffer, K.W. West, and R. Tycko: *Phys. Rev. Lett.* **74**, 5112 (1995).

[45] Barzykin, V., and I. Affleck: *Phys. Rev.* B**57**, 432 (1998).

[46] Basile, A.G., and V. Elser: *Phys. Rev.* B**41**, 4842 (1990).

[47] Bednorz, J.G., and K.A. Müller: *Z. Phys. B – Condensed Matter* **64**, 189 (1986).

[48] Bethe, H.: *Z. Phys.* **71**, 205 (1931).

[49] Boebinger, G., A. Passner, and J. Bevk: *Sci. Am.* **272**, 34 (June 1995).

[50] Bonaudi, F.: *Nucl. Phys.* B *Proc. Suppl.* **33**C, 8 (1993).

[51] Brinkman, W.F., and T.M. Rice: *Phys. Rev.* B**2**, 4302 (1970).

[52] Brooks, M.S.S., B. Johansson, and H.L. Skriver, in: *Handbook on the Physics and Chemistry of the Actinides*, Vol. **1**, 153, (North-Holland, Amsterdam, 1984).

[53] Buyers, W.J.L., R.M. Morra, R.L. Armstrong, P. Gerlach, and K. Hirakawa: *Phys. Rev. Lett.* **56**, 371 (1986).

[54] Cage, M.E., B.F. Field, and R.F. Dziuba: *IEEE Trans. Instrum. Meas.* **34**, 301 (1985).

[55] Campagna, M., G.K. Wertheim, and Y. Baer: *Topics in Applied Physics* **27**: Photoemission in Solids II, Eds. C.D. Graham, Jr. et *al*, p. 217, (Springer Verlag Berlin, Heidelberg, 1979).

[56] Campbell, D.K., T.K. Gammel, and E.Y. Loh, Jr.: *Phys. Rev.* B**38**, 12043 (1988); *Phys. Rev.* B**42**, 475 (1990).

[57] Canals, B., and C. Lacroix: *Phys. Rev. Lett.* **80**, 2933 (1998).

[58] Capellmann, H. (Ed.): *Metallic Magnetism*, (Springer-Verlag, Berlin, 1987).

[59] Carlson, E.W., S.A. Kivelson, Z. Nussinov, and V.J. Emery: cond-mat/9709112.

[60] Castellani, C., C.R. Natoli, and J. Ranninger: *Phys. Rev.* B**18**, 4945; 4967; 5001 (1978).

[61] Castilla, G., S. Chakravarty, and V.J. Emery: *Phys. Rev. Lett.* **75**, 1823 (1995).

[62] Chan, S.: *J. Phys. Chem. Solids* **32**, 1111 (1971).

[63] Chao, K.A.: *Solid State Commun.* **14**, 525 (1974).

[64] Chao, K.A., J. Spałek, and A.M. Oleś: *Phys. Rev.* B**18**, 3453 (1978).

[65] Chubukov, A.V., and D.I. Golosov: *J. Phys.: Condens. Matter* **3**, 69 (1991).

[66] Cohen, R.L., M. Eibschütz, and K.W. West: *Phys. Rev. Lett.* **24**, 383 (1970).

[67] Coleman, S.: *Commun. Math. Phys.* **31**, 259 (1973).

[68] Coulson, C.A.: *Valence*, (Clarendon Press, Oxford, 1961).

[69] Coqblin, B., and J.R. Schrieffer: *Phys. Rev.* **185**, 847 (1969).

[70] Cornelius, A.L., A.K. Gangopadhyay, J.S. Schilling, and W. Assmuss: *Phys. Rev.* B**55**, 14109 (1997).

[71] Cornelius, A.L., and J.S. Schilling: *Phys. Rev.* B**49**, 3955 (1994).

[72] Cornut, B., and B. Coqblin: *Phys. Rev.* B**5**, 4541 (1972).

[73] Cox, D.L.: *Phys. Rev. Lett.* **59**, 1240 (1987).

[74] Cross, M.C., and D.D. Osheroff: *Phys. Today*, p. 35, (February 1987).

[75] Cyrot, M., and M. Lavagna: *J. Phys.* **40**, 763 (1979).

[76] Cyrot, M., and C. Lyon-Caen: *J. Phys.* **36**, 253 (1975).

[77] Dagotto, E., J. Riera, A. Sandvik, and A. Moreo: *Phys. Rev. Lett.* **76**, 1731 (1996).

[78] Dagotto, E., S. Yunoki, A.L. Malvezzi, A. Moreo, J. Hu, S. Capponi, D. Poilblanc, and N. Furukawa: cond-mat/9709029.

[79] Daul, S., and R.M. Noack: *Z. Phys. B – Condensed Matter* **103**, 293 (1997); *Phys. Rev.* B**58**, 2635 (1998).

[80] Daul, S., P. Pieri, M. Dzierzawa, D. Baeriswyl, and P. Fazekas: *Physica* B **230–232**, 1021 (1997).

[81] de Boer F.R., J.J.M. Franse, E. Louis, A. Menovsky, J.A. Mydosh, T.T.M. Palstra, U. Rauchschwalbe, W. Schlabitz, F. Steglich, and A. de Visser: *Physica* B**138**, 1 (1986).

[82] De Gennes, P.G.: *Phys. Rev.* **82** 403 (1951).

[83] de Haas, W.J., J.H. de Boer, and G.J. van der Berg: *Physica* **1**, 1115 (1934).

[84] de Jongh, L.J., and A.R. Miedema: *Advan. Phys.* **23**, 1 (1974).

[85] den Nijs, M., and K. Rommelse: *Phys. Rev.* B**40**, 4709 (1989).

[86] Des Cloizeaux, J., and J.J. Pearson: *Phys. Rev.* **128**, 2131 (1962).

[87] Dev, G., and J.K. Jain: *Phys. Rev.* B**45**, 1223 (1992).

[88] Dombre, T.: *Helv. Phys. Acta* **63**, 261 (1990).

[89] Doniach, S.: *Physica* **91** B, 231 (1977).

[90] Doniach, S., and E.H. Sondheimer: *Green's Functions for Solid State Physicists*, (Benjamin/Cummings Publ. Co., Reading, Ma., 1974).

[91] Doucot, B., and X.G. Wen: *Phys. Rev.* B**40**, 2719 (1989).

[92] Dreizler, R.M., and E.K.U. Gross: *Density Functional Theory*, (Springer-Verlag, Berlin, 1990).

[93] Du, R.R., H.L. Stormer, D.C. Tsui, L.N. Pfeiffer, and K.W. West: *Phys. Rev. Lett.* **70**, 2944 (1993).

[94] Dyson, F.J.: *Phys. Rev.* **102**, 1217 and 1230 (1956).

[95] Dyson, F.J., E.H. Lieb, and T. Spencer: *Phys. Rev. Lett.* **37**, 120 (1976).

[96] Dzierzawa, M., D. Baeriswyl, and M. Di Stasio: *Phys. Rev.* B**51**, 1993 (1995).

[97] Edwards, D.M., and E.P. Wohlfarth: *Proc. Roy. Soc. London* A**303**, 127 (1968).

[98] Eisenstein, J.P., and H.L. Stormer: *Science* **248**, 1510 (1990).

[99] Erdős, P.: *Introduction to Magnetism* (unpublished lecture notes of the International School on Magnetism, Kudowa Zdrój, Poland, 1995).

[100] Fabricius, K., A. Klümper, U. Löw, B. Büchner, T. Lorenz, G. Dhalenne, and and A. Revcholevschi: *Phys. Rev.* B**57**, 1102 (1998).

[101] Fang, Y., A.E. Ruckenstein, E. Dagotto, and S. Schmitt-Rink: *Phys. Rev.* B**40**, 7406 (1989).

[102] Fáth, G., and J. Sólyom: *Phys. Rev.* B**51**, 3620 (1995); and references therein.

[103] Fawcett, E.: *Rev. Mod. Phys.* **60**, 209 (1988) gives a review of earlier work on Cr.

[104] Fazekas P., in: *Condensed Matter Theories* **6**, Eds. S. Fantoni and S. Rosati, p. 279, (Plenum Press, New York and London, 1991).

[105] Fazekas, P.: *Phil. Mag.* B**76**, 797 (1997).

[106] Fazekas, P., and P.W. Anderson: *Phil. Mag.* **30**, 423 (1974).

[107] Fazekas, P., and B.H. Brandow: *Phys. Scripta* **36**, 809 (1987).

[108] Fazekas, P., B. Menge, and E. Müller–Hartmann: *Z. Phys. B – Condensed Matter* **78**, 69 (1990).

[109] Fazekas, P., and E. Müller–Hartmann: *Z. Phys. B – Condensed Matter* **85**, 285 (1991).

[110] Fazekas, P., and K. Penc: *Int. J. Modern Phys.* B**2** 1021 (1988).

[111] Fazekas, P., and H. Shiba: *Int. J. Modern Phys.* B **5**, 289 (1991).

[112] Feinberg, D., P. Germain, M. Grilli, and G. Seibold: *Phys. Rev.* B**57**, R5583 (1998).

[113] Feiner, L.F., and A.M. Oleś: cond-mat/9805011.

[114] Fertig, H.A., L. Brey, R. Cote, and A.H. MacDonald: *Phys. Rev.* B**50**, 11018 (1994).

[115] Föex, M.: *C.R. Hebd. Séan. Acad. Sci.* B **223**, 1126 (1946).

[116] Forgan, E.M., B.D. Rainford, S.L. Lee, J.S. Abell, Y. Bi: *J. Phys.: Cond. Matt.* **2**, 10211 (1990).

[117] Fradkin, E.: *Field Theories of Condensed Matter Systems*, (Addison-Wesley, Redwood City, 1991).

[118] Freericks, J.K., and M. Jarrell: *Phys. Rev. Lett.* **74**, 186 (1995).

[119] Fulde, P.: *Electron Correlations in Molecules and Solids*, Springer Series in Solid-State Sciences **100**, (Springer-Verlag, Berlin, 1991).

[120] Fulde, P., J. Keller, and G. Zwicknagl: *Solid State Physics* **41**, 1 ((Academic Press, New York and London, 1980).

[121] Galeczki, G., and G. Nimtz: *Solid State Commun.* **74**, 863 (1990).

[122] Garland, J.C., and R. Bowers: *Phys. Rev. Lett.* **21**, 1007 (1968).

[123] Gebhard, F.: *The Mott Metal–Insulator Transition*, Springer Tracts in Modern Physics **137**, (Springer-Verlag Berlin Heidelberg, 1997).

[124] Gebhard, F.: *Phys. Rev.* B**41**, 9452 (1990).

[125] Gebhard, F.: *Phys. Rev.* B**44**, 992 (1991).

[126] Gebhard, F., and A.E. Ruckenstein: *Phys. Rev. Lett.* **68**, 244 (1992).

[127] Gebhard, F., and D. Vollhardt: *Phys. Rev.* B**38**, 6911 (1988).

[128] Gebhard, F., and X. Zotos: *Phys. Rev.* B**43**, 1176 (1991).

[129] Gehring, G.A., and K.A. Gehring: *Rep. Prog. Phys.* **38**, 1 (1975).

[130] Georges, A., G. Kotliar, W. Krauth, and M.J. Rozenberg: *Rev. Mod. Phys.* **68**, 13 (1996).

[131] Gerhardt, W., F. Razavi, J.S. Schilling, D. Hüser, and J.A. Mydosh: *Phys. Rev.* B**24**, 6744 (1981).

[132] Goldman, V.J., J.K. Jain, and M. Shayegan: *Mod. Phys. Lett.* B **5**, 479 (1991).

[133] Gorkov, L., and L. Pitaevskii: *Sov. Phys. Dokl.* **8**, 788 (1964).

[134] Grewe, N., H.J. Leder, and B. Mühlschlegel: *Festkörperprobleme* **20**, 413 (1980).

[135] Grewe, N., and F. Steglich: *Handbook on the Physics and Chemistry of Rare Earths*, Vol. **14**, 343, (North-Holland, Amsterdam, 1991).

[136] Grimes, C.C., and G. Adams: *Phys. Rev. Lett.* **42**, 795 (1979).

[137] Gros, C., R. Joynt, and T.M. Rice: *Phys. Rev.* B**36**, 381 (1987).

[138] Grüner, G.: *Density Waves in Solids*, (Addison–Wesley Publ. Co., Reading, Mass., 1994).

[139] Grüner, G., and A. Zawadowski: *Rep. Progr. Phys.* **37**, 1497 (1974); Progr. Low Temp. Phys. B**7**, 591 (1978).

[140] Guertin, R.P., J.E. Crow, L.D. Longinotti, E. Bucher, L. Kupferberg, and S. Foner: *Phys. Rev.* B**12**, 1005 (1975).

[141] Gunnarsson, O.: *J. Phys. F: Metal Phys.* **6**, 587 (1976).

[142] Gutzwiller, M.C.: *Phys. Rev. Lett.* **10**, 159 (1963).

[143] Gutzwiller, M.C.: *Phys. Rev.* **137**, A1726 (1965).

[144] Győrffy, B.L.: in: *Electron*, ed. M. Springford, p. 60, (Cambridge University Press, Cambridge, 1997).

[145] Győrffy, B.L., Kollár, J., Pindor, A.J., Stocks, G.M., Staunton, J., and Winter, H., in: *The Electronic Structure of Complex Systems*, Eds. P. Phariseau and W.M. Temmermann, p. 593, (Plenum Press, New York, 1984).

[146] Hajdu, J., and B. Shapiro: *Europhys. Lett.* **28**, 61 (1994).

[147] Haldane, F.D.M.: *Phys. Rev.* B**25**, 4925 (1982).

[148] Haldane, F.D.M.: *Phys. Lett.* A **93**, 464 (1983); *Phys. Rev. Lett.* **50**, 1153 (1983).

[149] Haldane, F.D.M.: *Phys. Rev. Lett.* **60**, 635 (1988).

[150] Hallberg, K., X.Q.G. Wang, P. Horsch, and A. Moreo: *Phys. Rev. Lett.* **76**, 4955 (1996).

[151] Halperin, B.I.: *Helv. Phys. Acta* **56**, 75 (1983).

[152] Hanisch, T.: *Gitterabhängigkeit des Ferromagnetismus im Hubbard-Modell* (PhD Thesis, Köln, 1996, published by Cuvillier Verlag, Göttingen, 1997).

[153] Hanisch, T., B. Kleine, A. Ritzl, and E. Müller–Hartmann: *Ann. Physik* **4**, 303 (1995).

[154] Hanisch, T., and E. Müller–Hartmann: *Ann. Physik* **2**, 381 (1993).

[155] Hanisch, T., G.S. Uhrig, and E. Müller–Hartmann: *Phys. Rev.* B**56**, 13960 (1997).

[156] Harper, P.G.: *Proc. Phys. Soc. London* A **68**, 874 (1955).

[157] Harris, A.B., and R.V. Lange: *Phys. Rev.* **157**, 295 (1967).

[158] Harris, E.A., and J. Owen: *Phys. Rev. Lett.* **11**, 9 (1963).

[159] Hase, M., I. Terasaki, and K. Uchinokura: *Phys. Rev. Lett.* **70**, 3651 (1993).

[160] Held, K., M. Ulmke, N. Blümer, and D. Vollhardt: *Phys. Rev.* B**56**, 14469 (1997).

[161] Held, K., and D. Vollhardt: cond-mat/9803182; extended version in preparation.

[162] Henley, C.L.: *J. Appl. Phys.* **61**, 3962 (1987).

[163] Herring, C.: *Exchange Interactions among Itinerant Electrons*, in: *Magnetism Vol. IV*, eds. G.T. Rado and H. Suhl, Academic Press, New York (1966). – See particularly chapters VI, and IX.

[164] Hewson, A.C.: *The Kondo Problem to Heavy Fermions*, (Cambridge Univeristy Press, 1993).

[165] Hirakawa, K., I. Yamada, and Y. Kurogi: *J. Phys.* **32**, *Suppl.* C1, 890 (1970).

[166] Hirsch, J.E.: *Phys. Rev.* B**31**, 4403 (1985).

[167] Hirsch, J.E.: *Physica* C **158**, 326 (1989).

[168] Hirsch, R.: *Zum Magnetismus stark korrelierter Elektronensysteme*, (PhD Thesis, Köln, 1994, published by Verlag Shaker, Aachen, 1994).

[169] Hirst, L.L.: *Advan. Phys.* **27**, 231 (1978).

[170] Hlubina, R., S. Sorella, and F. Guinea: *Phys. Rev. Lett.* **78**, 1343 (1997).

[171] Hofstadter, D.R.: *Phys. Rev.* B**14**, 2239 (1976).

[172] Hohenberg, P., and W. Kohn: *Phys. Rev.* **136**, B864 (1964).

[173] Holstein, T., and H. Primakoff: *Phys. Rev.* **58**, 1098 (1940).

[174] Hubbard, J.: (I) Proc. Roy. Soc. London **A 276**, 238 (1963); (II) *ibid* **277**, 237 (1964); (III) *ibid* **281**. 401 (1965); (IV) *ibid* 542 (196).

[175] Hulthén, L.: *Arkiv för Matematik, Astronomy och Fysik* **26**A, No:11, 1 (1938).

[176] Hund, F.: *Z. Phys.* **33**, 345 (1925).

[177] Hund, F.: *Ann. Physik* **32**, 102 (1938); reprinted in English on the occasion of F. Hund's 100th birthday in *Ann. Physik (Leipzig)* **5**, 1 (1996).

[178] Hutchings, M.T., in *Solid State Physics* **16**, eds. F. Seitz and D. Turnbull, p. 277 (Academic Press, New York and London, 1964).

[179] Ichimura, M., M. Fujita, and K. Nakao: *J. Phys. Soc. Japan* **61**, 2027 (1992).

[180] Iordansky, S.V.: *Solid State Commun.* **48**, 1 (1982).

[181] Ishizuka, M., I. Yamada, K. Amaya, and S. Endo: *J. Phys. Soc. Japan* **65**, 1927 (1996).

[182] Jahn, H., and E. Teller: *Phys. Rev.* **49**, 874 (1936).

[183] Jain, J.K.: *Phys. Rev. Lett.* **63**, 199 (1989); *Phys. Rev.* B**41**, 7653 (1990); *J. Phys. Chem. Solids* **51**, 889 (1990).

[184] Jain, J.K.: *Advan. Phys.* **41**, 105 (1992); *Comments Cond. Mat. Phys.* **16**, 307 (1993).

[185] Jain, J.K., S.A. Kivelson, and N. Trivedi: *Phys. Rev. Lett.* **64**, 1297 (1990).

[186] Jansen, A.P.J.: *Phys. Rev.* B**33**, 6352 (1986).

[187] Janßen, M., O. Viehweger, U. Fastenrath, and J. Hajdu: *Introduction to the Theory of the Integer Quantum Hall Effect*, (Weinheim, New York, 1994).

[188] Jarrel, M., and Th. Pruschke: *Z. Phys. B – Condensed Matter* **90**, 187 (1993).

[189] Jarrel, M., and Th. Pruschke: *Phys. Rev.* B**49**, 1458 (1994).

[190] Jensen, J., M.J.M. Leask, M.R. Wells, R.C.C. Ward, S.M. Hayden, and A.P. Ramirez: *J. Magn. Magn. Mater.* **140–144**, 1191 (1995).

[191] Jensen, J., and A.R. Mackintosh: *Rare Earth Magnetism*, (Clarendon Press, Oxford, 1991).

[192] Jin, S., T.H. Tiefel, M. McCormack, R.A. Fastnacht, R. Ramesh, and L.H. Chen: *Science* **264**, 413 (1994).

[193] , Johansson, B., I.A. Abrikosov, M. Aldén, A.V. Ruban, and H.L. Skriver: *Phys. Rev. Lett.* **74**, 2335 (1995).

[194] Jolicoeur, T., and J.C. Le Gillou: *Phys. Rev.* B**40**, 2727 (1989).

[195] Jones, H.F.: *Groups, Representations and Physics*, (Adam Hilger, Bristol and New York, 1990).

[196] Jones, W., and N.H. March: *Theoretical Solid State Physics*, (Dover Publ., Inc., New York, 1973).

[197] Kadowaki, K., and S.B. Woods: *Solid State Commun.* **58**, 507 (1986).

[198] Kajueter, H., G. Kotliar, and G. Moeller: Phys. Rev. B**53**, 16214 (1996).

[199] Kalatsky, V.A., E. Müller-Hartmann, V.L. Pokrovskiĭ, and G.S. Uhrig: *Phys. Rev. Lett.* **80**, 1304 (1998).

[200] Kanamori, J.: *J. Phys. Chem. Solids* **10**, 87 (1959).

[201] Kanamori, J.: *Progr. Theor. Phys.* **30**, 275 (1963).

[202] Kaplan, T., P. Horsch, and P. Fulde: *Phys. Rev. Lett.* **49**, 889 (1982).

[203] Kasuya, T.: *Progr. Theor. Phys.* **16**, 45 (1956).

[204] Kawamura, H.: *J. Phys. Soc. Japan*, **53**, 2452 (1984).

[205] Kennedy, T., and H. Tasaki: *Phys. Rev.* B**45**, 304 (1992).

[206] Khomskii, D.I., and K.I. Kugel: *Solid State Commun.* **13**, 763 (1973).

[207] Khomskii, D.I., and G.A. Sawatzky: *Solid State Comm.* **102**, 87 (1997).

[208] Khurana, A.: *Phys. Today* , 17 (November 1989).

[209] Kittel, C.: *Introduction to Solid State Physics*, 4th ed., (Wiley, New York, 1971). See also the 7th ed. (Wiley, New York, 1996).

[210] Kittel, C.: *Quantum Theory of Solids*, Second revised printing, (John Wiley and Sons, New York, 1987).

[211] Kivelson, S., D.–H. Lee, and S.–C. Zhang: *Phys. Rev.* B**46**, 2223 (1992); Karlhede, A., S. Kivelson, and S.L. Sondhi, in: *Correlated Electron Systems* **9**, 242, Ed. V.J. Emery (World Scientific, Singapore, 1993). For a popular introduction, see [212].

[212] Kivelson, S., D.–H. Lee, and S.–C. Zhang: *Sci. Am.* **277**, 64 (1996).

[213] Kleine, B.: *Magnetische Ordnung auf frustrierten Gittern*, (PhD Thesis, Köln, 1995, published by Cuvillier Verlag, Göttingen, 1995).

[214] Kleine, B., P. Fazekas, and E. Müller–Hartmann: *Z. Phys. B – Condensed Matter*, **86**, 405 (1992).

[215] Kleine, B., E. Müller–Hartmann, K. Frahm, and P. Fazekas: *Z. Phys. B – Condensed Matter*, **87**, 103 (1992).

[216] Klümper, A.: *Europhys. Lett.* **9**, 815 (1989).

[217] Knapp, G.S., F.Y. Fradin, and H.V. Culbert: *J. Appl. Phys.* **42**, 1341 (1971).

[218] Knapp, G.S, B.W. Veal, and H.V. Culbert: *Int. J. Magn.* **1**, 93 (1971).

[219] Kohn, W., and J.M. Luttinger: *Phys. Rev. Lett.* **15**, 524 (1965).

[220] Kohn, W., and L.J. Sham: *Phys. Rev.* **140**, A1133 (1964).

[221] Kojima, K.M. *et al*: *Phys. Rev. Lett.* **78**, 1787 (1997).

[222] Kollar, M., R. Strack and D. Vollhardt: *Phys. Rev.* B**53** (1996).

[223] Kondo, J.: *Progr. Theor. Phys.* **32**, 37 (1964); *Solid State Physics* **23**, 184 (1969).

[224] Kosevich, Yu.A., and A.V. Chubukov: *Zh. Eksp. Teor. Fiz.* **91**, 1105 (1986).

[225] Kotliar, G., and H. Kajueter: *Phys. Rev.* B**54**, R14221 (1996).

[226] Kubo, K.: *J. Phys. Soc. Japan* **51**, 782 (1982): T. Momoi and K. Kubo: *Phys. Rev.* B**58**, R567 (1998).

[227] Kubo, K., and N. Ohata: *J. Phys. Soc. Japan* **33**, 21 (1972).

[228] Kubo, Y.: *J. Phys. Soc. Japan* **66**, 2236 (1997); and references therein.

[229] Kutzelnigg, W., and J.D. Morgan III: *Z. Phys. D: Atoms, Molecules and Clusters* **36**, 197 (1996).

[230] Lacroix, C.: *Solid State Comm.* **54**, 991 (1985).

[231] Laloux, L., A. Georges, and W. Krauth: *Phys. Rev.* B**50**, 3092 (1994).

[232] Landau, L.D., and E.M. Lifshitz: *Quantum Mechanics: Non-Relativistic Theory*, 2nd ed., (Pergamon Press, Oxford, 1965).

[233] Larsen, D.M.: *J. Phys. Chem. Solids* **29**, 271 (1968).

[234] Laughlin, R.: *Phys. Rev.* B**23**, 5632 (1981).

[235] Laughlin, R.: *Phys. Rev.* B**27**, 3383 (1983).

[236] Laughlin, R.: *Phys. Rev. Lett.* **50**, 1395 (1983).

[237] Lawrence, J.M., P.S. Riseborough, and R.D. Parks: *Repts. Progr. Phys.* **44**, 1 (1981).

[238] Le *et al*: *Europhys. Lett.* **15**, 547 (1991).

[239] Lea, K.R., M.J.M. Leask, and W.P. Wolf: *J. Phys. Cem. Solids* **23**, 1381 (1962).

[240] Leadley, D.R. *et al*: *Phys. Rev. Lett.* **79**, 4246 (1997).

[241] Lee, S., and K.C. Lee: *Phys. Rev.* B**57**, 8472 (1998).

[242] Liang, S., B. Doucot, and P.W. Anderson: *Phys. Rev. Lett.* **61**, 365 (1988).

[243] Lieb, E.H.: *Phys. Rev. Lett.* **62**, 1201 (1989).

[244] Lieb, E.H., in: *The Hubbard Model, its Physics and Mathematical Physics*, eds. D. Baeriswyl *et al.*, p. 1 (Plenum Press, New York, 1995).

[245] Lieb, E.H, and D.C. Mattis: *Phys. Rev.* **125**, 164 (1962).

[246] Lieb, E.H, and D.C. Mattis: *J. Math. Phys.* **3**, 749 (1962).

[247] Lieb, E.H., D. Schultz, and D.C. Mattis: *Ann. Phys. (New York)* **16**, 407 (1961).

[248] Löw, U., V.J. Emery, K. Fabricius, and S.A. Kivelson: *Phys. Rev. Lett.* **72**, 1918 (1994).

[249] Lopes, L.C., R. Jullien, A.K. Battacharjee, and B. Coqblin, in: *Valence Instabilites*, Eds. P. Wachter and H. Boppart, (North-Holland Publ. Co., Amsterdam), p. 383 (1982).

[250] Luther, A., and I. Peschel: *Phys. Rev.* B**12**, 3908 (1975).

[251] Lüthi, B., M.E. Mullen, K. Andres, E. Bucher, and J.P. Maita: *Phys. Rev.* B**8**, 2639 (1973).

[252] MacDonald, A.H., in: *Quantum Coherence in Mesoscopic Systems*, Ed. B. Kramer (Plenum Press, New York, 1991); cond-mat/9410047.

[253] MacDonald, A.H.: *Solid State Commun.* **102**, 143 (1997).

[254] MacDonald, A.H., and S.M. Girvin: *Phys. Rev.* B 34, 5639 (1986).

[255] McWhan, D.B., and J.P. Remeika: Phys. Rev. B**2**, 3734 (1970).

[256] McWhan, D.B., A. Menth, J.P. Remeika, W.F. Brinkman, and T.M. Rice: *Phys. Rev.* B**7**, 1920 (1973).

[257] Mahan, G.D.: *Many–Particle Physics*, (Plenum Press, New York and London, 1981).

[258] Majumdar, C.K.: *J. Phys. C: Solid St. Phys.* **3**, 911 (1970).

[259] Maleev, S.V.: *Zh. Eksp. Teor. Fiz.* **30**, 1010 (1957).

[260] Manousakis, E.: *Rev. Mod. Phys.* **63**, 1 (1991).

[261] Marshall, W.: *Proc. Roy. Soc. (London)* A **232**, 48 (1955).

[262] Mattheiss, L.F.: *Phys. Rev.* B**49**, 14050 (1994).

[263] Mattheiss, L.F., E.M. Gyorgy, and D.W. Johnston, Jr.: *Phys. Rev.* B**37**, 3745 (1988).

[264] Mattis, D.C.: *Theory of Magnetism I*, (Springer-Verlag, Berlin, 1981).

[265] Mattis, D.C., and R.E. Peña, *Phys. Rev.* B**10**, 1006 (1974).

[266] Mattocks, P.G., and D. Melville: *J. Phys. F: Metal Phys.* **8**, 1291 (1978).

[267] Maude, D.K. *et al*: *Phys. Rev. Lett.* **77**, 4604 (1996).

[268] Menge, B.: PhD Thesis, Universität zu Köln (1991).

[269] Mermin, N.D., and H. Wagner: *Phys. Rev. Lett.* **17**, 1133 (1966).

[270] Metzner, W.: *Z. Phys. B – Condensed Matter* **77**, 253 (1989).

[271] Metzner, W., and D. Vollhardt: *Phys. Rev. Lett.* **59**, 121 (1987); *Phys. Rev.* B**37**, 7382 (1988).

[272] Metzner, W., and D. Vollhardt: *Phys. Rev. Lett.* **62**, 324 (1989).

[273] Mielke, A.: *J. Phys. A: Math. Gen.* **25**, 4335 (1992).

[274] Mielke, A., and H. Tasaki: *Commun. Math. Phys.* **158**, 341 (1993).

[275] Miles, P.A., W.B. Westphal, and A. von Hippel: *Rev. Mod. Phys.* **29**, 279 (1957).

[276] Millis, A.J.: *Phys. Rev.* B**48**, 7183 (1993).

[277] Millis, A.J.: *Phys. Rev.* B**55**, 6405 (1997).

[278] Millis, A.J., and S.N. Coppersmith: *Phys. Rev.* B**43**, 13770 (1991).

[279] Mishra, S.K., S. Satpathy, F. Aryasetiawan, and O. Gunnarsson: *Phys. Rev.* B**55**, 2725 (1997).

[280] Miyashita, S., and H. Kawamura: *J. Phys. Soc. Japan*, **54**, 3385 (1985).

[281] Miyashita, S., and H. Shiba: *J. Phys. Soc. Japan*, **53**, 1145 (1984).

[282] Mizokawa, T., and A. Fujimori: *Phys. Rev.* B**51**, 12880 (1995).

[283] Moeller, G., Qimiao Si, G. Kotliar, M.J. Rozenberg, and D.S. Fisher: *Phys. Rev. Lett.* **74**, 2082 (1995).

[284] Moessner, R., and J.T. Chalker: *Phys. Rev. Lett.* **80**, 2929 (1998).

[285] Monachesi, P., L.C. Andreani, A. Continenza, and A.K. McMahan: *J. Appl. Phys.* **73**, 6634 (1993).

[286] Monachesi, P., Z. Domański, and M.S.S. Brooks: *Phys. Rev.* B**50**, 1013 (1994).

[287] Morgan III, J.D., and W. Kutzelnigg: *J. Phys. Chem.* **97**, 2425 (1993).

[288] Morin, P., and D. Schmitt: *Phys. Rev.* B**23**, 5936 (1981).

[289] Morita, E.: *J. Math. Phys.* **12**, 1744 (1971).

[290] Moriya, T., p. 15 in [58].

[291] Moser, M., P. Wachter, F. Hulliger, and J.R. Etorneau: *Solid State Comm.* **54**, 241 (1985).

[292] Mott, N.F.: Proc. Roy. Soc. A **62**, 416 (1949).

[293] Mott, N.F.: *Metal–Insulator Transitions*, (Taylor & Francis Ltd., London, 1974).

[294] Mott, N.F., in: *The Metallic and Non-Metallic States of Matter*, eds. P.P. Edwards and C.N.R. Rao, p. 1, (Taylor & Francis, London, 1985), gives a brief review of the early history of the theory of the electronic structure of solids.

[295] Müller–Hartmann, E.: *Z. Phys. B – Condensed Matter* **74**, 507 (1989); **76**, 211 (1989).

[296] Müller–Hartmann, E., in: *Proc. V Symp. Phys. Metals*, Eds. Ė. Talik and J. Szade, (Poland, 1991), p. 22.

[297] Müller–Hartmann, E.: *J. Low Temp. Phys.* **99**, 349 (1995).

[298] Müller–Hartmann, E.: private communication.

[299] Müller–Hartmann, E., and E. Dagotto: *Phys. Rev.* B**54**, R6819 (1996).

[300] Mulders, A.M., *et al*: *Phys. Rev.* B**56**, 8752 (1997).

[301] Murakami, Y., H. Kawada, H. Kawata, M. Tanaka, T. Arima, Y. Moritomo, and Y. Tokura: *Phys. Rev. Lett.* **80**, 1932 (1998).

[302] Nagaoka, Y.: *Solid State Commun.* **3**, 409 (1965); *Phys. Rev.* **147**, 392 (1966).

[303] Nagasima, T., and H. Ishii: *J. Phys. Soc. Japan* **67**, 1431 (1998).

[304] Negele, J.W., and H. Orland: *Quantum Many-Particle Systems*, (Addison–Wesley Publ. Co., Redwood City, 1988).

[305] Nieuwenhuys, G.J.: *Phys. Rev.* B**35**, 5260 (1987).

[306] Norman, M.R.: *Phys. Rev.* B**40**, 10632 (1989).

[307] Noziéres, P.: *Ann. Phys. Fr.* **10**, 19 (1985).

[308] Obermeier, T., T. Pruschke, and J. Keller: *Phys. Rev.* B**56**, R8479 (1997).

[309] O'Brien, M.C.M., and C.C. Chancey: *Am. J. Phys.* **61**, 688 (1993).

[310] Ogata, M., and H. Shiba: *Phys. Rev.* B**41**, 2326 (1990).

[311] Ogawa, S.: *Researches of the Electrotechnical Laboratory*, Tokyo, Japan, Nr. 735 (1972) (quoted in [442]).

[312] Ogawa, T., and K. Kanda: *Z. Physik B – Condensed Matter* **30**, 355 (1978).

[313] Okamoto, K., and K. Nomura: *Phys. Lett.* A **169**, 433 (1992); Nomura, K., and K. Okamoto: *J. Phys.* A **27**, 5773 (1994).

[314] Okuno, Y., and K. Miyake: *J. Phys. Soc. Japan* **67**, 2469 (1998).

[315] Oleś, A.M.: *J. Phys. C: Solid State Phys.* **15**, L1065 (1982).

[316] Onuki, Y., and T. Komatsubara: *J. Magn. Magn. Mater.* **63&64**, 281 (1987).

[317] Ott, H.R., in *Progr. Low Temp. Phys.* Vol. XI, Ch. 5, (Elsevier Science Publishers B.V., 1987).

[318] Palstra, T.T.M., A.A. Menovsky, J. van den Berg, A.J. Dirkmaat, P.H. Kes, G.J, Nieuwenhuys, and J.A. Mydosh: *Phys. Rev. Lett.* **55**, 2727 (1985).

[319] Park, J.-H., C.T. Chen, S-W Cheong, W. Bao, G. Meigs, V. Chakarian, and Y.U. Idzerda: *Phys. Rev. Lett.* **76**, 4215 (1996).

[320] Pauling, L.: *The Nature of the Chemical Bond* (Cornell University Press, Ithaca, N.Y., 1967).

[321] Pavarini, E., and L.C. Andreani: *unpublished.*

[322] Penc, K., and H. Shiba: *Phys. Rev.* B**52**, R715 (1995).

[323] Penc, K., H. Shiba, F. Mila, and T. Tsukagoshi: *Phys. Rev.* B**54**, 4056 (1996).

[324] Penn, D.R.: *Phys.Rev.* **142**, 350 (1966).

[325] Pfannkuche, D., and R.R. Gerhardts: *Phys. Rev.* B**46**, 12606 (1992).

[326] Pieri, P., S. Daul, D. Baeriswyl, M. Dzierzawa, and P. Fazekas: *Phys. Rev.* B**54**, 9250 (1996).

[327] Pines, D., and P. Nozières: *The Theory of Quantum Liquids I*, (W.A. Benjamin, Inc., New York, 1966).

[328] Pruschke, T., Q. Qin, T. Obermeier, and J. Keller: *J. Phys.: Condens. Matter* **8**, 3161 (1996).

[329] Putikka, W.O., M.U. Luchini, and M. Ogata: *Phys. Rev. Lett.* **69**, 2288 (1992).

[330] Ramakrishnan, T.V., in: *The Metallic and Non-Metallic States of Matter*, eds. P.P. Edwards and C.N.R. Rao, p. 23, (Taylor & Francis, London, 1985).

[331] Ramirez, A.P., P. Coleman, P. Chandra, E. Brück, A.A. Menovsky, Z. Fisk, and E. Bucher: *Phys. Rev. Lett.* **68**, 2680 (1992).

[332] Razafimandimby, H.A.: *Z. Physik B – Condensed Matter* **49**, 33 (1982).

[333] Reger, J.D., J.A. Riera, and A.P. Young: *J. Phys.: Condens. Matter* 1, 1855 (1989).

[334] Rice, T.M., and W.F. Brinkman, in: *Critical Phenomena in Alloys, Magnets, and Superconductors*, Eds. R.E. Mills, E. Ascher, and R.I. Jaffee, p. 593, (McGraw-Hill Book Co., New York, 1971).

[335] Rice, T.M., and K. Ueda: *Phys. Rev.* B**34**, 6420 (1986).

[336] Rice, T.M., K. Ueda, H.R. Ott, and H. Rudigier: *Phys. Rev.* B**31**, 594 (1985).

[337] Riera, J., K. Hallberg, and E. Dagotto: *Phys. Rev. Lett.* **79**, 713 (1997).

[338] Riseborough, P.S., and P. Hanggi: *J. Phys. Chem. Solids*, **54**, 1123 (1993).

[339] Rodbell, D.S., I.S. Jacobs, J. Owen, and E.A. Harris: *Phys. Rev. Lett.* **11**, 10 (1963).

[340] Rosenbaum, T.F., S.B. Field, D.A. Nelson, and P.B. Littlewood: *Phys. Rev. Lett.* **54**, 241 (1985).

[341] Ruderman, M.: *Phys. Rev. Lett.* **27**, 1306 (1971).

[342] Ruderman, M.A., and C. Kittel: *Phys. Rev.* **96**, 99 (1954).

[343] Rumer, G.: *Nachr. d. Ges. d. Wiss. zu Göttingen, Mat. Phys. Klasse*, **337**, Heft 4 (1932).

[344] Sakakibara, T., T. Goto, K. Yoshimura, and K. Fukamichi: *J. Phys.: Condens. Matter* **2**, 3381 (1990).

[345] Santini, P. and G. Amoretti: *Phys. Rev. Lett.* **73**, 1027 (1994); **74**, 4098 (1995); P. Santini: *Phys. Rev.* B**57**, 5191 (1998).

[346] Santoro, G., M. Airoldi, S. Sorella, and E. Tosatti: *Phys. Rev.* B**47**, 16216 (1993).

[347] Saminadayar, L., D.C. Glattli, Y. Jin, and B. Etienne: *Phys. Rev. Lett.* **79**, 2526 (1997).

[348] Sawatzky, G.A., in: *Superconductivity: From Basic Physics to the Latest Developments*, Eds. P.N. Butcher and Yu Lu, (World Scientific, Singapore, 1995).

[349] Sawatzky, G.A., W. Geertsma, and C. Haas: *J. Magn. Magn. Mater.* **3**, 37 (1976).

[350] Schlabitz, W., J. Baumann, B. Pollit, U. Rauchschwalbe, H.M. Mayer, U. Ahlheim, and C.D. Bredl: *Z. Phys. B – Condensed Matter* **62**, 171 (1986).

[351] Schoenes, J., H. Brändle, A. Weber, and F. Hulliger: *Physica* B **163**, 496 (1990).

[352] Schrieffer, J.R., and P.A. Wolff: *Phys. Rev.* **194**, 491 (1966).

[353] Schulz, H.J.: *Phys. Rev. Lett.* **64**, 1445 (1990).

[354] Schulz, H.J.: *Phys. Rev. Lett.* **77**, 2790 (1996).

[355] Schwabl, F.: *Quantenmechanik*, 4te Aufl., (Springer-Verlag, Berlin, 1993).

[356] Seiler, K., C. Grós, T.M. Rice, K. Ueda, and D. Vollhardt: *J. Low Temp. Phys.* **64**, 195 (1986).

[357] Severin, L., M.S.S. Brooks, and B. Johansson: *Phys. Rev. Lett.* **71**, 3214 (1993).

[358] Shastry, B.S.: *Phys. Rev. Lett.* **60**, 639 (1988).

[359] Shastry, B.S., H.R. Krishnamurty, and P.W. Anderson: *Phys. Rev.* B**41**, 2375 (1990).

[360] Shiba, H.: *J. Phys. Soc. Japan* **55**, 2765 (1986).

[361] Shiba, H., and P. Fazekas: *Progr. Theor. Phys. Suppl.* **101**, 403 (1990).

[362] Shiina, R., T. Nishitani, and H. Shiba: *J. Phys. Soc. Japan* **66**, 3159 (1997).

[363] Shiina, R., O. Sakai, H. Shiba, and P. Thalmeier: *J. Phys. Soc. Japan* **67**, 941 (1998).

[364] Shirahana, K, O.I. Kirichek, and K. Kono: *Phys. Rev. Lett.* **79**, 4218 (1997).

[365] Sikkema, A.E., W.J.L. Buyers, I. Affleck, and J. Gan: *Phys. Rev.* B**54**, 9322 (1996).

[366] Skriver, H.L.: *J. Phys.* F **11**, 97 (1981).

[367] Skyrme, T.H.R.: *Proc. R. Soc. London* A **262**, 237 (1961).

[368] Solovyev, I., N. Hamada, and K. Terakura: *Phys. Rev. Lett.* **76**, 4825 (1996).

[369] Sólyom, J.: *private communication.*

[370] Sondhi, S.L., A. Karlhede, S.A. Kivelson, and E.H. Rezayi: *Phys. Rev.* B**47**, 16419 (1993).

[371] Sorella, S., and E. Tosatti: *Europhys. Lett.* **19**, 699 (1992).

[372] Sorensen, E.S., and I. Affleck: *Phys. Rev.* B**53**, 9153 (1996).

[373] Spalek, J., A.M. Oleś, and J.M. Honig: *Phys. Rev.* B**28**, 6802 (1983).

[374] Staunton, J.B., and B.L. Győrffy: *Phys. Rev. Lett.* **69**, 371 (1992).

[375] Stewart, G.R.: *Rev. Mod. Phys.* **56**, 755 (1984).

[376] Stoner, E.C.: *Proc. R. Soc. London* A**165**, 372 (1938).

[377] Stormer, H.L.: *Solid State Comm.* **92**, 159 (1994).

[378] Strack, R., and D. Vollhardt: *J. Low Temp. Phys.* **99**, 385 (1995).

[379] Sugano, S., Y. Tanabe, and H.Kamimura: *Multiplets of Transition-Metal Ions in Crystals*, (Academic Press, New York and London, 1970).

[380] Sugiyama, K., H. Fuke, K. Kindo, K. Shimohata, A.A. Menovsky, J.A. Mydosh, and M. Date: *J. Phys. Soc. Japan* **59**, 3331 (1990).

[381] Sütő, A.: *Commun. Math. Phys.* **140**, 43 (1991).

[382] Sütő, A., and P. Fazekas: *Phil. Mag.* **35**, 623 (1977).

[383] Tahvildar–Zadeh, A.N., J.K. Freericks, and M. Jarrell: *Phys. Rev.* B**55**, 942 (1997).

[384] Takagi, H., R.J. Cava, M. Marezio, B. Batlogg, J.J. Krajewski, W.F. Peck, Jr., P. Bordet, D.E. Cox: *Phys. Rev. Lett.* **68**, 3777 (1992).

[385] Takahashi, M.: *J. Phys. C: Solid State Phys.* **10**, 1289 (1977).

[386] Takahashi, M.: *J. Phys. Soc. Japan* **51**, 3475 (1982).

[387] Takahashi, M.: *Progr. Theor. Phys.* **87**, 233 (1986).

[388] Takigawa, M., E.T. Ahrens, and Y. Ueda: *Phys. Rev. Lett.* **76**, 283 (1996).

[389] Taniguchi, S. *et al*: *J. Phys. Soc. Japan* **64**, 2758 (1995).

[390] Tasaki, H.: *Phys. Rev.* B**40**, 9192 (1989).

[391] Tasaki, H.: *Phys. Rev. Lett.* **64**, 2066 (1990).

[392] Tasaki, H.: *Phys. Rev. Lett.* **75**, 4678 (1995).

[393] Tasaki, H.: *Markov Processes and Related Fields* **2**, 183 (1996); cond-mat/9512169; *J. Phys.: Condens. Matter* **10**, 4353 (1998); *Progr. Theor. Phys* **99**, 489 (1998).

[394] Teller, E.: *Physica A* **114**, 14 (1982).

[395] Thessieu, C., J. Flouquet, G. Lapertot, A.N. Stepanov, and D. Jaccard: *Solid State Commun.* **95**, 707 (1995).

[396] Thompson, C.J., Y.S. Yang, A.J. Guttmann, and M.F. Sykes: *J. Phys. A: Math. Gen.* **24**, 1261 (1991).

[397] Thouless, D.J.: *Proc. Phys. Soc.* **86**, 893 (1965).

[398] Thouless, D.J., M. Kohmoto, M.P. Nightingale, and M. den Nijs: *Phys. Rev. Lett.* **49**, 405 (1982).

[399] Tinkham, M.: *Group Theory and Quantum Mechanics*, (McGraw-Hill Book Co., New York, 1964).

[400] Tokura, Y.: *J. Phys. Chem. Solids* **53**, 1619 (1992).

[401] Tokura, Y., Y. Taguchi, Y. Okada, Y. Fujishima, T. Arima, K. Kumagai, and Y. Iye: *Phys. Rev. Lett.* **70**, 2126 (1993).

[402] Tomioka, Y., A. Asamitsu, Y. Moritomo, H. Kuwahara, and Y. Tokura: *Phys. Rev. Lett.* **74**, 5108 (1995).

[403] Toulouse, G.: *Commun. Phys.* **2**, 115 (1977).

[404] Tranquada, J.M., B.J. Sternlieb, J.D. Axe, Y. Nakamura, and S. Uchida: *Nature* **375**, 561 (1995).

[405] Trivedi, N., and J.K. Jain: *Modern Phys. Lett.* B**5**, 503 (1991).

[406] Troyer, M., H. Kontani, and K. Ueda: *Phys. Rev. Lett.* **76**, 3822 (1996).

[407] Troyer, M., H. Tsunetsugu, and T.M. Rice: *Phys. Rev.* B**53**, 251 (1996).

[408] Tsui, D.C.: *Physica* B **164**, 59 (1990).

[409] Tsui, D.C., H.L. Störmer, and A.C. Gossard: *Phys. Rev. Lett.* **48**, 1559 (1982).

[410] Tsunetsugu, H., M. Sigrist, and K. Ueda: *Rev. Mod. Phys.* **69**, 809 (1997).

[411] Ueda, K.: *Physica* B **230&232**, 22 (1997).

[412] Ueda, K., H. Kontani, M. Sigrist, and P.A. Lee: *Phys. Rev. Lett.* **76**, 1932 (1996).

[413] Uhrig, G.: *Phys. Rev. Lett.* **77**, 3629 (1996).

[414] Uhrig, G., and D. Vollhardt: *Phys. Rev.* B**52**, (1995).

[415] Uimin, G.: *Phys. Rev.* B**55**, 8267 (1997).

[416] Ulmke, M.: *Eur. Phys. J.* B **1**, 301 (1998).

[417] Usher, A., R.J. Nicholas, J.J. Harris, and C.T. Foxon: *Phys. Rev.* B**41**, 1129 (1990).

[418] van Dongen, P.G.J.: *Phys. Rev. Lett.* **67**, 757 (1991).

[419] van Dongen, P.G.J.: *Phys. Rev. Lett.* **74**, 182 (1995).

[420] van Dongen, P.G.J., F. Gebhard, and D. Vollhardt: *Z. Phys. B – Condensed Matter* **76**, 199 (1989).

[421] van Elp, J., J.L. Wieland, H. Eskes, P. Kuiper, G.A. Sawatzky, F.M.F. de Groot, and T.S. Turner: *Phys. Rev.* B**44**, 6090 (1991).

[422] Van Vleck, J.H.: *Rev. Mod. Phys.* **25**, 220 (1953).

[423] Van Vleck, J.H.: *The Theory of Electric and Magnetic Susceptibilities*, (Oxford University Press, 1965).

[424] Varma, C.M.: *Rev. Mod. Phys.* **48** 219 (1976).

[425] Varma, C.M.: *Comments Solid State Phys.* **11**, 221 (1985).

[426] Varma, C.M.: *Phys. Rev. Lett.* **61**, 2713 (1988).

[427] Varma, C., W. Weber, and L.J. Randall: *Phys. Rev.* B**33**, 1015 (1986).

[428] Varma, C.M., and Y. Yafet: *Phys. Rev.* B**13**, 2950 (1976).

[429] Verschuur, G.I.: *Hidden Attraction: The Mystery and History of Magnetism*, (Oxford University Press, New York and Oxford, 1993).

[430] Visscher, P.B.: *Phys. Rev.* B**10**, 943 (1974).

[431] Vollhardt, D.: *Rev. Mod. Phys.* **56**, 99 (1984).

[432] Vollhardt, D., N. Blümer, K. Held, M. Kollar, J. Schliepf, and M. Ulmke: *Z. Phys. B – Condens. Matter* **103**, 283 (1997).

[433] von der Linden, W., and D.M. Edwards: *J. Phys.: Condens. Matter* **3**, 4917 (1991).

[434] von Klitzing, K., G. Dorda, and M. Pepper: *Phys. Rev. Lett.* **45**, 494 (1980).

[435] von Löhneysen, H. *et al*: *Physica* B **230&232**, 550 (1997).

[436] Wachter, P., in: *Handbook on the Physics and Chemistry of the Rare Earths* **19**, Ch. 132, Eds. K.A. Gschneidner Jr. *et al* (North-Holland, Amsterdam, 1993).

[437] Wahle, J., N. Blümer, J. Schlipf, K. Held, and D. Vollhardt: cond-mat 9711242.

[438] Wannier, G.H.: *Phys. Rev.* **79**, 357 (1950).

[439] White, R.M.: *Quantum Theory of Magnetism*, Springer Ser. Sol. St. Sci. **32**, (Springer-Verlag, Berlin, 1983).

[440] Wigner, E.P.: *Phys. Rev.* **46**, 1002 (1934).

[441] Wigner, E.P.: *Group Theory and Its Application to the Quantum Mechanics of Atomic Spectra*, (Academic Press, New York, 1959).

[442] Wijn, H.P.J. (ed.): *Magnetic Properties of Metals: d-Elements, Alloys and Compounds*, (Springer-Verlag, Berlin, 1991).

[443] Williamson D.L., S. Bukshpan, and R. Ingalls: *Phys. Rev.* B**6**, 4194 (1972).

[444] Wischmann, H.A., and E. Müller–Hartmann: *Phys. Rev.* B**43**, 8668 (1991).

[445] Wohlfarth, E.P., and P. Rhodes: *Philos. Mag.* **7**, 1817 (1962).

[446] Wu, X.G., G. Dev, and J.K. Jain: *Phys. Rev. Lett.* **71**, 153 (1993).

[447] Wurth, P., and E. Müller–Hartmann: unpublished.

[448] Wurth, P., G. Uhrig, and E. Müller–Hartmann: *Ann. Physik* **5**, 148 (1996).

[449] Xu, G., J.F. DiTusa, T. Ito, K. Oka, H. Takagi, C. Broholm, and G. Aeppli: *Phys. Rev.* B**54**, R6827 (1996).

[450] Yakovenko, V.M., and H.-S. Goan: *J. Phys. I France* **6**, 1917 (1996).

[451] Yamanaka, M., M. Oshikawa, and I. Affleck: *Phys. Rev. Lett.* **79**, 1110 (1997).

[452] Yang, C.N.: *Phys. Rev. Lett.* **63**, 2144 (1989).

[453] Yang, C.N., and S.C. Zhang: *Modern Phys. Lett.* B**4**, 759 (1990).

[454] Yildirim, T., A.B. Harris, O. Entin–Wohlman, and A. Aharony: *Phys. Rev. Lett.* **73**, 2919 (1994).

[455] Yokoyama, H., and Y. Saiga: *J. Phys. Soc. Japan* (1997).

[456] Yokoyama, H., and H. Shiba: *J. Phys. Soc. Japan* **56**, 3582 (1987).

[457] Yokoyama, H., and H. Shiba: *J. Phys. Soc. Japan* **59**, 3669 (1990).

[458] Yosida, K.: *Phys. Rev.* **106**, 893 (1957).

[459] Yosida, K.: *Phys. Rev.* **147**, 223 (1966).

[460] Yosida, K., and S. Inagaki: *J. Phys. Soc. Japan* **50**, 3268 (1981).

[461] Yosida, K., and A. Yoshimori, in: *Magnetism*, eds. G.T. Rado and H. Suhl, p. 253 (New York, Academic Press, 1973).

[462] Zaanen, J., G.A. Sawatzky, and J.W. Allen: *Phys. Rev. Lett.* **55**, 418 (1985).

[463] Zachar, O., S.A. Kivelson, and V.J. Emery: *Phys. Rev.* B**57**, 1422 (1998).

[464] Zawadowski, A.: *Phys. Rev. Lett.* **59**, 469 (1987).

[465] Zeiger, H.J., and G.W. Pratt: *Magnetic Interactions in Solids*, (Clarendon Press, Oxford, 1973).

[466] Zener, C.: *Phys. Rev.* **81**, 440 (1951); **82**, 403 (1951).

[467] Zhang, F.C, C. Gros, T.M. Rice and H. Shiba: *Supercond. Sci. Technol.* **1**, 36 (1988).

[468] Zhang, F.C., and T.M. Rice: *Phys. Rev.* B**37**, 3759 (1988).

[469] Zhang, S.C.: *Int. J. Mod. Phys. B* **5**, 153 (1991).

[470] Zhang, X.Y., M.J. Rozenberg, and G. Kotliar: *Phys. Rev. Lett.* **70**, 1666 (1993).

[471] Zirngiebl, E., B. Hillebrands, S. Blumenröder, G. Güntherodt, M. Loewenhaupt, J. Carpenter, K. Winzer, and Z. Fisk: *Phys. Rev.* B**30**, 4052 (1984).

Index